Milestones in Drug Therapy

For further volumes:
http://www.springer.com/series/4991

Henry M. Staines • Sanjeev Krishna
Editors

Treatment and Prevention of Malaria

Antimalarial Drug Chemistry, Action and Use

Springer

Volume Editors
Dr Henry M. Staines
Centre for Infection and Immunology
Division of Clinical Sciences
St George's, University of London
Cranmer Terrace
London SW17 0RE
United Kingdom
hstaines@sgul.ac.uk

Prof. Sanjeev Krishna
Centre for Infection and Immunology
Division of Clinical Sciences
St George's, University of London
Cranmer Terrace
London SW17 0RE
United Kingdom
skrishna@sgul.ac.uk

Series Editors
Prof. Michael J. Parnham, Ph.D.
Visiting Scientist
Research & Clinical Immunology Unit
University Hospital for Infectious Diseases "Dr. Fran Mihaljević"
Mirogojska 8
HR-10000 Zagreb
Croatia

Prof. Dr. Jacques Bruinvels
Sweelincklaan 75
NL-3723 JC Bilthoven
The Netherlands

ISBN 978-3-0346-0479-6 e-ISBN 978-3-0346-0480-2
DOI 10.1007/978-3-0346-0480-2

Printed on acid-free paper

Springer Basel AG is part of Springer Science + Business Media (www.springer.com)

HMS dedicates this book to his wife, Zoë, and children,
Talia, Luca and Oren and SK to Yasmin's memory
and Karim, and to the rest of his exceptional family.

Preface

Malaria is a devastating disease that extracts huge health and economic costs from the poorest countries in endemic regions. Malaria is caused by single celled parasites, belonging to the genus *Plasmodium* that have infected humans (and related primates) for thousands of years. In its different specific and clinical guises, malaria is one of the strongest selective forces to have shaped our recent evolution. These parasites have already evaded one attempt at eradication in the mid twentieth century. Now, there are renewed attempts to control and eventually eradicate what remains one of the world's biggest killers.

With ambitious new targets set to reduce the global burden of malaria, we must urgently develop new tools for disease control, as well as optimising and re-evaluating our current tools. An indispensable part of controlling malaria is the capability of treating the disease effectively, despite the ability of this highly mutable parasite to develop resistance sooner or later to all classes of antimalarials. Understanding of how antimalarial drugs might work, how best to use them and how to assess for resistance to them has expanded considerably in the past few years. This book aims to capture these recent advances in our understanding of all antimalarial classes, and discuss how this information is pertinent for treating patients.

The introductory chapter details the disease, its current political, financial and technical context, alongside the policies and tools required to make eradication a possibility. Subsequent chapters cover the history, chemistry, mechanisms of action and resistance, preclinical and clinical use, pharmacokinetics and safety and tolerability of our current antimalarial drug armamentarium. Each chapter reflects the unique perspectives of its expert authors, and often describes new ideas and directions for study. There is particular emphasis on artemisinins (and related next generation peroxides) that have become the frontline treatment for malaria, as part of artemisinin-based combination therapies (ACTs). The artemisinins may have become established in ACTs in the past decade, but they are now being challenged by the potential for resistance that has recently been described and is only just being defined.

Other chapters authoritatively discuss our antimalarial drug development pipeline and how this is being shaped by public/private partnerships; molecular markers of antimalarial drug resistance, their use in monitoring treatment failures and the insights they provide into the action of these drugs; malaria prevention strategies, including chemoprophylaxis, where the risk of catching malaria is balanced against the risk of side effects of drugs and the critical use of diagnostics to improve the identification of malaria and to refine treatment strategies.

The treatment and prevention of malaria is a fascinating and complex subject – made all the more interesting now that malaria eradication is back on the global agenda. We hope that readers will be stimulated by this volume and that they may find its contents useful in dealing with malaria.

London, United Kingdom

Henry M. Staines
Sanjeev Krishna

Contents

Antimalarial Drugs and the Control and Elimination of Malaria

Karen I. Barnes

Abstract Malaria remains a massive global public health problem despite being readily preventable and treatable. The past decade has seen unprecedented levels of political, technical and financial support that have facilitated the scaling-up of malaria control interventions, particularly the implementation of artemisinin-based combination therapy (ACT) policies. During this window of opportunity for reducing the burden of malaria globally and possibly eventually eliminating malaria, attention now needs to be focussed on ensuring that countries select and implement treatment policies that are not only highly effective, but will also have a prolonged useful therapeutic life, reduce malaria transmission safely and effectively and, where applicable, be active against *P. vivax*. To reduce the probability of resistance, antimalarials should be used in quality-assured fixed-dose combinations and treatment doses need to be optimised on the basis of pharmacokinetic assessments conducted within therapeutic efficacy studies in each key target population. As important is ensuring optimal targeting and adherence with these treatment policies.

1 Introduction

Malaria is a massive global public health problem. Nearly half the world's population lives at risk of malaria, which causes an estimated one million deaths and 450 million *Plasmodium falciparum* and 390 million *P. vivax* cases each year [1, 2]. Those with malaria also carry an increased burden of HIV/AIDS, measles, respiratory tract infections, diarrhoea, malnutrition and anaemia [3]. Malaria in pregnancy increases the infant risk of low birth weight, abortions and stillbirths, in addition to

K.I. Barnes (✉)
Division of Clinical Pharmacology, Department of Medicine, University of Cape Town, Anzio Road, Observatory, Cape Town 7925, South Africa
e-mail: karen.barnes@uct.ac.za

H.M. Staines and S. Krishna (eds.), *Treatment and Prevention of Malaria*,
DOI 10.1007/978-3-0346-0480-2_1, © Springer Basel AG 2012

the maternal burdens of anaemia, severe malaria and maternal mortality [4]. The indirect burden of malaria includes its adverse effects on education, worker productivity and investment. It has been estimated that malaria costs Africa $12 billion per year, with a fivefold reduction in per capita gross domestic product (GDP) after controlling for other socio-economic determinants [5].

Efforts to reduce malaria morbidity and mortality include control of the mosquito vector (using insecticide-treated bed nets and indoor residual spraying) and prompt treatment with effective antimalarials. Unprecedented levels of political, technical and financial support have facilitated the scaling-up of malaria control interventions, particularly changes in malaria treatment policy from the inexpensive yet failing monotherapies, chloroquine and sulfadoxine–pyrimethamine, to the recommended artemisinin-based combination therapies (ACTs). ACTs are generally considered as the best current treatment of uncomplicated falciparum malaria [6], as they have high cure rates, have more rapid parasite clearance times and have the potential to reduce both antimalarial resistance and malaria transmission. Over the last decade, the bar for recommending an antimalarial regimen as policy for uncomplicated falciparum malaria was raised from requiring an adequate clinical and parasitological response (ACPR) rate at 14 days of merely 75%, to at least 95% at ≥28 days [6]. Fortunately, there are now a number of ACTs in most settings that meet this stringent criterion. While most malaria endemic countries have adopted ACT policies, the implementation of these policies has been slower.

The extent to which malaria can be eradicated in the foreseeable future is a subject of active debate, but it is generally agreed that the tools are available to reduce the global burden of malaria substantially. How these tools are selected and, more importantly, deployed will be critical in determining the success achieved. Optimising the impact of ACTs on the control and eventual elimination of malaria depends on careful selection of the regimen implemented. In addition to the usual considerations of effectiveness, safety and cost, treatment policy selection should consider the likely useful therapeutic life (the time until ACPR rates at ≥28 days decrease below 90%), impact on malaria transmission and, where relevant, efficacy against non-falciparum malaria. As important are the selection of evidence-based dosage regimens that are appropriate for each key target population, especially young children and pregnant women [7], and optimising the implementation strategies deployed to ensure high coverage and adherence rates among those with malaria, while limiting use among those with non-malarial febrile illnesses [8].

2 Malaria: The Basics

All malaria is transmitted by female mosquitoes of the genus *Anopheles*. Humans are mainly infected by four species of *Plasmodium*: *P. falciparum*, *P. vivax*, *P. ovale* and *P. malariae*, although human infections with the monkey malaria parasite, *P. knowlesi* have also been reported recently in the forested regions of Southeast Asia [9]. The majority of all human malaria cases are caused by

P. falciparum and *P. vivax*, although the burden of *P. ovale* and *malariae* are poorly defined. Although almost all severe malaria is caused by *P. falciparum*, severe disease and malaria-related deaths have also been reported with *P. vivax* and *P. knowlesi*.

The sporozoite form of the parasite is inoculated into humans when bitten by an infected female *Anopholes* mosquito. Sporozoites rapidly enter the liver cells where they multiply to form thousands of merozoites. These then enter the bloodstream where they invade red blood cells and multiply to form new merozoites. Infected red blood cells burst, releasing merozoites that infect new red blood cells. This is referred to as the asexual blood stage, the stage of the plasmodial life cycle that causes the clinical signs and symptoms of malaria. Some merozoites that invade the red blood cells develop into gametocytes, the sexual stages of the parasite. Gametocytes are ingested by the mosquito when it takes a blood meal. In the mosquito gut, the gametocytes develop into gametes and fuse to form a zygote. After fertilisation, the zygote transforms into a motile ookinete, which penetrates the mosquito stomach wall and becomes an oocyst. The oocyst divides to produce sporozoites, which move into the salivary glands, from where another human can be infected when the mosquito takes a blood meal (Fig. 1).

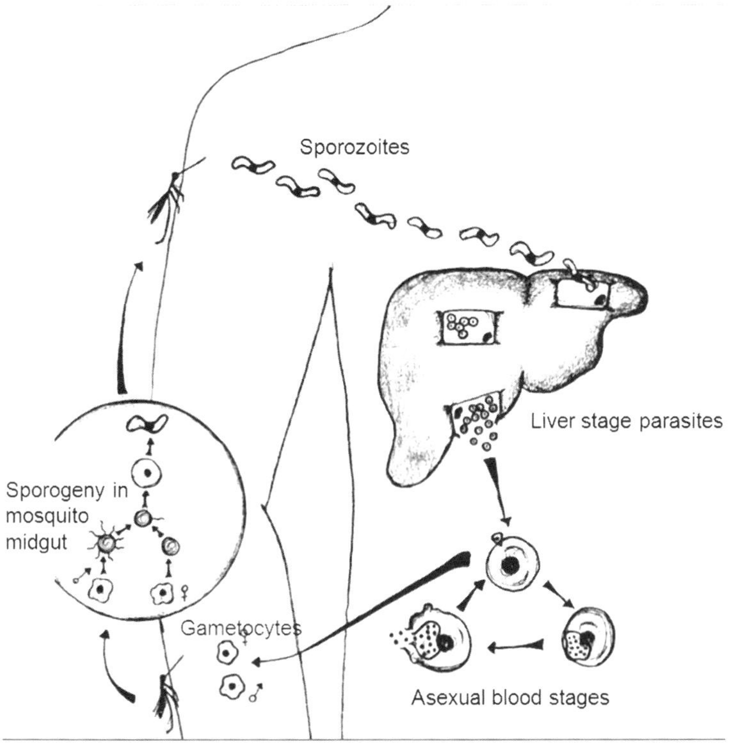

Fig. 1 The lifecycle of the malaria parasite in the human host and *Anopholene* mosquito vector

Malaria transmission rates are determined by the parasite reservoir in a community and the abundance and behaviour of the mosquito vectors [10]. The probability of a mosquito being infected depends on the prevalence, duration and density of viable gametocyte carriage in the human host, although additional immunological factors also affect transmissibility [11]. There are many factors that can lead to an increase in the duration and density of *P. falciparum* gametocyte carriage. Most of these are not well defined, but gametocyte numbers increase with the density and duration of asexual parasitaemia (emphasising the importance of prompt treatment), anaemia and drug resistance [11, 12].

3 Treating Malaria to Prevent Transmission

While eliminating the asexual stages of plasmodial infections is the focus of treatment of individual symptomatic patients, at a population level, limiting the transmission of malaria, and in particular, the transmission of resistant parasites is pivotal for decreasing the community's burden of malaria. In considering antimalarial drug effects on transmissibility, three different components need to be considered (a) activity against asexual stages and early gametocytes, (b) activity against mature infectious gametocytes and (c) sporontocidal effects in the mosquito [13].

Early access to effective treatment of the asexual blood stage can reduce the incidence and prevalence of malaria in a community, although the effects are greater in areas of low transmission where a higher proportion of the infectious reservoir is in non-immune and thus symptomatic individuals, who are more likely to seek antimalarial treatment [13]. However, achieving malaria control and eventually elimination requires a complete parasitological cure, including killing of the parasites in the sexual (gametocyte) stages that are responsible for malaria transmission [11, 14]. *P. falciparum* gametocytes are relatively insensitive to most antimalarials, other than the artemisinins and primaquine [15]. It has been suggested that artemisinins predominantly inhibit gametocyte development, whereas primaquine accelerates gametocyte clearance in *P. falciparum* malaria [16]. ACTs have the advantage of being the only antimalarials currently available that rapidly reduce both asexual and gametocyte stages of *P. falciparum*. When compared with amodiaquine plus sulfadoxine–pyrimethamine treatment, ACTs reduced the duration of gametocyte carriage (quantified using nucleic acid sequence-based amplification) fourfold [17]. The large-scale deployment of ACTs contributed to a marked reduction in the number of malaria cases seen in a number of countries, mostly in areas of low to moderate intensity transmission [18–23].

As the effect of the artemisinins on *P. falciparum* gametocytes is not complete, patients treated with artemisinins can still transmit malaria [24–26]. Mature gametocytes are resistant to almost all of the antimalarial drugs that affect the asexual stages and the only licenced drug that can ensure complete killing of *P. falciparum* gametocytes is primaquine. This 8-aminoquinoline is very effective in preventing transmission, even when administered as a single dose. The addition of primaquine to artesunate plus sulfadoxine–pyrimethamine in Tanzania, when compared with artesunate plus sulfadoxine–pyrimethamine alone, resulted in a further fourfold reduction of the duration of gametocyte carriage [17], and an even greater reduction in sub-microscopic gametocytaemia [27]. However, primaquine may cause methaemoglobinaemia and haemolysis, which can be severe and occasionally life threatening. Haemolysis occurs most frequently (but not only) in patients with certain glucose-6-phosphate dehydrogenase (G6PD)-deficiency variants, particularly when a prolonged course of treatment is used. Primaquine is contraindicated in pregnancy, lactation, infants and young children and in those with haemolytic anaemia, methaemoglobinaemia or severe G6PD deficiency. As G6PD-deficient variants protect against *P. falciparum* and *vivax* malaria, this abnormality is most prevalent in malaria-endemic areas [6, 28–32]. The key operational question now is whether the benefits of adding primaquine (probably as a single dose) to ACTs in order to further reduce transmission exceed the risks. Unfortunately, there are remarkably limited data available to inform this decision, as summarised by Baird [33]: "Despite more than 50 years of continuous use in millions of people annually as the only drug available for its therapeutic indication, it is not known how primaquine acts or how it should be taken."

Lastly, atovaquone and the antifolate antimalarials reduce transmission by decreasing the formation of sporozoites in the *Anopheline* mosquito. For antifolates, this effect is reduced by antifolate resistance, creating a further transmission advantage for resistant parasites [13]. Also, atovaquone–proguanil offers the further benefit of acting as a causal prophylactic agent, but atovaquone rapidly selects for the *cytochrome b* mutation associated with high-level resistance. The possible role of atovaquone–proguanil alone or in combination with artesunate in attempts to contain or eliminate malaria deserves further study.

4 Antimalarial Resistance: The Major Threat to Malaria Control and Elimination

Parasite resistance to antimalarial medicines is a major threat to achieving malaria control and eventual elimination. Antimalarial resistance in *P. falciparum* parasites results in an enormous public health and economic burden. The rise in malaria-related hospital admissions and malaria mortality across west, east and southern Africa during the 1990s is largely accounted for by the continued use of the cheap monotherapies, chloroquine and sulfadoxine–pyrimethamine, despite widespread high levels of resistance [34–37]. Lower levels of resistance are associated with return of illness, anaemia and increased gametocyte carriage (which fuels malaria transmission, particularly of the resistant parasites) and a higher risk of treatment failure in subsequent infections [38, 39]. Parasite resistance has been documented for all classes of antimalarials, including – in the Southeast Asian epicentre of drug resistance along the Thai–Cambodian border – the artemisinin derivatives [40, 41]. If the efficacy of the artemisinin derivatives is lost, then effective control and elimination will not be possible with currently available tools [13]. Despite these concerns, the artemisinin-resistance phenotype has been poorly characterised, and the contribution of host factors remains to be defined. The key features of the artemisinin-resistant phenotype are prolonged parasite clearance times, despite apparently adequate drug exposure, and even dose escalation [42, 43]. Although molecular markers for artemisinin resistance remain elusive, a genetic basis for this clinical phenotype has been proposed recently based on its high heritability [44]. The recent declines in the clinical effectiveness of all antimalarial drugs, including the artemisinins, have prompted suggestions to revise the definitions of antimalarial drug resistance to include a category for extensively drug-resistant (XDR) malaria, as this approach has proved useful in tuberculosis for individual patient care and for public health [45].

Antimalarial resistance spreads when parasites are exposed to the selective window of drug concentrations that are sufficient to kill sensitive but not resistant parasites [7] (Fig. 2). Drugs with longer terminal elimination half-lives have the advantage of providing a longer post-treatment prophylactic effect, which appears to be important for their action in intermittent preventive therapy (IPT) in high-risk groups such as pregnant women, infants and young children. However, these long-acting antimalarials have the disadvantage of residual concentrations inhibiting sensitive parasites far longer than resistant parasites, thus fuelling the spread of resistance. The window of selection is prolonged with an increase in resistance or in the terminal elimination half-life (unless these terminal concentrations are too low even to kill sensitive parasites) (Fig. 2).

Antimalarial resistance spreads because gametocyte carriage and infectivity to mosquitoes is consistently higher in patients infected with drug-resistant compared with drug-sensitive parasites. An increase in gametocyte numbers has been identified as the first indication that an antimalarial is beginning to fail and emphasises the need for the treatment policy implemented to include drugs that will kill the sexual stages [12, 13]. Combining antimalarials with differing modes of action is expected to reduce the probability of a resistant (mutant) parasite surviving treatment [46]. Despite their mismatched elimination half-lives, ACTs are preferred to other combination therapies given their potential to reduce malaria transmission – due to their rapid clearance of asexual parasites together with their partial gametocidal activity [6]. The gametocyte-reducing effect of widespread use of artesunate plus mefloquine therapy has resulted not only in the sustained decrease in malaria transmission described above, but also in decreased mefloquine resistance in northwest Thailand, an area of low-intensity malaria transmission notorious for multi-drug resistance [18, 47]. By contrast, the first and only effort to date in Africa documenting the routine large-scale surveillance of temporal changes in resistance after successful implementation of an ACT treatment policy, found that systematic deployment of artesunate plus sulfadoxine–pyrimethamine had not delayed the spread of sulfadoxine–pyrimethamine resistance and may in fact have contributed to the rapid increase in the proportion of parasites carrying quintuple

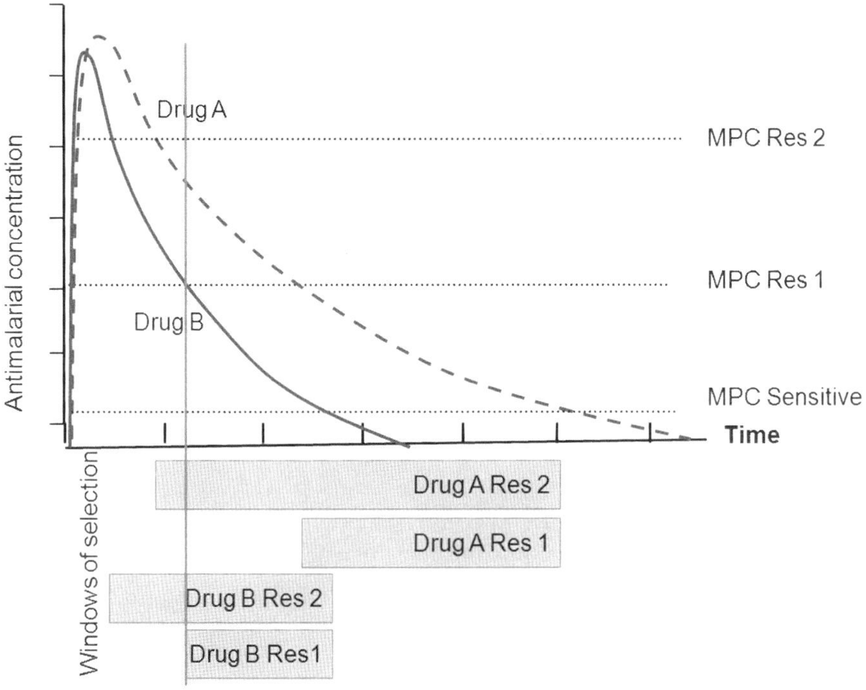

Fig. 2 Resistance selection by drugs with long elimination half-lives. The *curves* show antimalarial drug concentrations over time for Drug A (*dashed line*) and Drug B (*solid line*). The Window of Selection is the time when antimalarial concentrations are sufficient to clear sensitive but not resistant parasites. The three *dotted lines* show hypothetical minimum parasiticidal concentrations (MPCs) needed for clearing sensitive, partially resistant (Res 1) and highly resistant (Res 2) parasites. The duration of the window of selection increases with (a) increasing levels of resistance, so is longer for highly resistant than for partially resistant parasites, and (b) terminal elimination half-life, so is longer for Drug A than for Drug B

dihydrofolate reductase and *dihydropteroate synthetase* resistance markers from 11 to 75% over the 5-year study period [48]. Transmission of sulfadoxine–pyrimethamine resistance occurs intrinsically more readily than with mefloquine, probably since mefloquine resistance confers a survival disadvantage, while this does not appear to be the case with sulfadoxine–pyrimethamine.

A further challenge to limiting the rate of spread of ACT resistance is that expanding ACT access is necessary for reducing malaria morbidity and mortality. To be accessed promptly, ACTs need to be available near the home. With recent efforts to reduce the costs of ACTs dramatically, this is becoming achievable even in the poorest communities. However, such ready access creates creates the opportunity for widespread and indiscriminate use of antimalarials, which exerts a strong selective pressure towards resistant parasites towards high levels of resistance [42]. This could be addressed by limiting ACT use to those with a confirmed malaria diagnosis [6]. While 78 malaria-endemic countries (33 in Africa) have a policy that patients of all ages with suspected malaria should receive a diagnostic test before treatment, this policy is only implemented in a minority of African cases – but is used in more than 80% of suspected cases outside Africa [23]. Other challenges to the effective targeting of ACTs are that only 38 countries (16 in Africa) are deploying rapid diagnostic tests at a community level and that ACTs continue to be used by those with negative malaria tests [23, 49].

Continued use of artemisinin-based monotherapy is considered a major factor in resistance to the artemisinins emerging and spreading, emphasising the importance of oral artemisinins being used only in combination with an effective longer acting antimalarial. This makes fixed dose artemisinin-based combinations highly preferable to loose tablets or blister-packed combinations [6]. To this end, the WHO recommends the withdrawal of all oral artemisinin-based monotherapies from the market [23, 42]. Others have argued that oral artesunate monotherapies are still needed, but should be reserved for use as a 7-day treatment course for patients with uncomplicated hyperparasitaemia and pregnant women in areas of multi-drug resistance [50].

De novo antimalarial drug resistance is most likely to occur in hyperparasitaemic patients who are non-immune, particularly if their antimalarial drug exposure is inadequate [51]. In hyperparasitaemic patients, parasite populations are larger and recrudescence rates following treatment are high [52]. Drug exposure can be inadequate due to sub-standard antimalarial quality, poor adherence, vomiting, unusual pharmacokinetic behaviour or underdosing [7]. Current antimalarial dosing recommendations are generally based on the lowest effective dose seen in dose-finding studies, which are conducted early in a drug's therapeutic life and thus before parasite resistance has become apparent [7]. Recommending the lowest effective dose, while justified in terms of cost and safety/tolerability, might not be the wisest choice as this is likely to select for resistant parasites. For example, mathematical modelling suggests that, if the recommended 25-mg/kg mefloquine dose had been deployed initially, instead of 15 mg/kg, then mefloquine resistance could have been delayed [53]. Furthermore, the relationship between the drug concentrations actually achieved with the recommended antimalarial dosage regimen and the therapeutic response needs to be reassessed once resistance starts to develop because, by definition, the minimum concentrations required to clear these resistant parasites has increased [7]. Current dose recommendations also almost invariably assume that the same weight-adjusted (milligram per kilogram) dose is effective for all key target population groups [7]. This approach encourages resistance selection, particularly in patients with high parasite burdens and low drug levels [51]. These are often young children or pregnant women who lack immunity and so generally have higher parasite densities, and whose larger apparent volumes of distribution and higher apparent clearance rates result in sub-optimal drug concentrations for many antimalarials [7, 51]. Despite extensive use for four decades, it has only been recognised recently that the currently recommended doses of both sulfadoxine–pyrimethamine and chloroquine achieve substantially lower plasma drug concentrations in young children than in older patients [54, 55]. Children given the recommended dosage regimens are similarly at increased risk of inadequate exposure to both lumefantrine and piperaquine [56, 57]. Similarly, physiological changes in pregnancy result in decreased exposure to a number of key antimalarial drugs, including the artemisinins, sulfadoxine, lumefantrine and mefloquine [7]; no data on the pharmacokinetics of amodiaquine or piperaquine in pregnancy has been published yet.

To reduce the probability of resistance, quality-assured fixed-dose combination antimalarials should be used, treatment doses need to be optimised on the basis of pharmacokinetic assessments conducted within therapeutic efficacy studies in each key target population and patients with heavy parasite burdens have to be identified and receive sufficient treatment to prevent recrudescence [51].

5 *P. vivax*: A Particular Challenge to Malaria Elimination

The focus of malaria control programmes has, to date, been largely on *P. falciparum* because this is the major cause of severe malaria and malaria mortality, especially in sub-Saharan Africa. However, once elimination becomes the target, *P. vivax* needs to be given much more attention [58]. It has a more widespread distribution and infects 130–435 million people a year amongst a population at risk of approximately 2.85 billion, mostly in Central and Southeast Asia [33, 59]. *P. vivax* can undergo sporogeny in mosquitoes at lower temperatures than *P. falciparum* and forms a latent liver stage, the hypnozoite, which initiates relapses (Fig. 1) [60]. Gametocytes of *P. vivax* appear in the circulation at the same time as the asexual stages and, although killed by the antimalarial drugs that are effective against the asexual blood stages (unlike *P. falciparum*), *P. vivax* transmits well at very low parasite densities, so transmission can already have occurred before a patient has become symptomatic and sought treatment [13]. These factors, together with the low priority given by policy makers, funders and researchers to these infections that have been mislabelled "benign" [61], explain why *P. vivax* malaria is so widespread and is significantly more difficult to control or eliminate than falciparum malaria.

The asexual stages of *P. vivax* are increasingly resistant to chloroquine but remain highly sensitive to the artemisinins [6, 16]. Amodiaquine, mefloquine, piperaquine, lumefantrine, sulfadoxine-pyrimethamine and quinine are also effective in the treatment of chloroquine-resistant asexual blood stages of *P. vivax* [6, 16, 62–64]. ACTs are the preferred treatment in areas where falciparum malaria is also endemic or *P. vivax* is chloroquine resistant [6]. However, ACTs do not provide a radical cure.

Primaquine is the only radically curative drug for *P. vivax* (and *P. ovale*) malaria; it prevents relapse by clearing the hypnozoite stage when given as a 14-day course [65]. This prolonged treatment course compromises adherence and safety, with the main risk being haemolysis (as noted above). Supervision of this long course of therapy markedly reduces the risk of relapse, and almost all reports of primaquine resistant malaria are associated with lack of such supervision [66].

6 Progress Towards Malaria Control and Eventual Elimination

There has been substantial progress in reducing the burden of malaria globally over the last 60 years, with the number of countries that are malaria-free increasing from nine in 1945 to 108 today [58]. More than one-third of the 108 malaria-endemic countries documented reductions in malaria cases of >50% in 2009 compared with 2000, including 11 countries and one area in Africa and 32 countries in other regions [23]. These impressive results occurred in countries that achieved high coverage with their vector control and ACT treatment programmes. These successes have fuelled a wave of optimism that has led to renewed commitments to achieving the ambitious goal of progressively reducing the burden of malaria, leading eventually to global eradication[1], as outlined in the Roll Back Malaria Global Malaria Action Plan. This entails three components (a) effective malaria control[2] to reduce malaria morbidity in the majority of malaria-endemic countries by scaling-up and then sustaining appropriate vector and parasite control interventions, (b) progressive elimination[3] from the margins of malaria transmission, to "shrink the malaria map", and (c) research to bring forward better drugs, diagnostics, insecticides, vaccines and other tools, as well as inform policy and improve operational implementation of effective strategies [58, 67, 68]. Better drugs are needed for elimination-specific indications such as mass treatment, curing asymptomatic infections, curing relapsing liver stages of *P. vivax* and *P. ovale* and preventing transmission [69].

The ACT coverage rates (i.e. the proportion of parasitaemic patients that promptly receives an adequate dose and duration of ACT treatment) need to be high to impact on malaria transmission and the spread of resistance. One of the major deterrents to ensuring widespread access to ACTs is their cost – being tenfold more expensive than chloroquine and sulfadoxine–pyrimethamine monotherapies. The patients and governments that most need ACTs can least afford them [70]. Fortunately, international funding commitments for malaria have increased from around US$ 0.3 billion in 2003 to US$ 1.8 billion in 2010 [23], due to greater commitments by the US President's Malaria Initiative, the World Bank and primarily the emergence of the Global Fund and more recently, its innovative Affordable Medicines Facility for malaria (AMFm). This increased financial, technical and political support is resulting in dramatic scale-up of malaria control interventions in many settings and measurable reductions in malaria burden.

In general, the number of cases fell least in countries with the highest malaria incidence rates, with the notable exceptions of Zanzibar (United Republic of

---

[1] Malaria eradication is the permanent reduction to zero of the worldwide incidence of malaria infection caused by a specific agent; i.e. applies to a particular malaria parasite species.

[2] Malaria control is reducing the disease burden to a level at which it is no longer a public health problem.

[3] Malaria elimination is interrupting local mosquito-borne malaria transmission in a defined geographical area, i.e. zero incidence of locally contracted cases.

Tanzania), Zambia, Eritrea, Rwanda and Sao Tome and Principe, that illustrate that dramatic reductions in malaria morbidity and mortality can also be achieved in areas with a high malaria incidence [18–23, 71]. Similar results have also been seen in more limited geographic areas of the high malaria burden countries of Equatorial Guinea (Bioko Island), the Gambia, Kenya and Mozambique [72–75].

There is evidence from Bioko Island (Equatorial Guinea), Kenya, Sao Tome and Principe, Zanzibar and Zambia that large decreases in malaria cases and deaths have been mirrored by steep declines in all-cause deaths in children under 5 years of age [20, 71, 73, 74], suggesting that intensive malaria control in African countries could play an important role in not only achieving the Millennium Development Goal 6 of reducing malaria incidence and death rates, but also the Millennium Development Goal 4 of reducing all-cause childhood mortality by two-thirds by 2015 [76].

In 2009, however, there was evidence of an increase in malaria cases in three countries that had previously reported dramatic reductions in malaria burden (Rwanda, Sao Tome and Principe and Zambia) [23]. These resurgences highlight the fragility of malaria control and the critical importance of sustaining control interventions and surveillance rigorously – particularly in areas that have historically carried a high malaria burden.

At the other end of the malaria transmission intensity spectrum, tangible progress is being made. In 2010, both Morocco and Turkmenistan were certified as having achieved malaria elimination [23]. At least another 27 countries are working towards malaria elimination; nine countries have interrupted transmission and are in the phase of preventing re-introduction of malaria; ten countries are implementing nationwide elimination programmes and eight countries are in the pre-elimination phase [23]. In Botswana, Cape Verde, Namibia, Sao Tome and Principe, South Africa and Swaziland, large initial decreases in the number of malaria cases have been sustained but remain at 10–25% of those reported in 2000 [19, 22, 71]. However, the few remaining cases are proving more difficult to prevent, to detect and treat promptly, and additional interventions are likely to be necessary for further reductions in malaria morbidity to be achieved. Encouraging results of additive benefit are starting to be seen in studies evaluating the combination of indoor residual spraying with insecticide-treated bed-net deployment [77] and of adding primaquine to ACTs [17, 27].

Since the current levels of international financial support for malaria control fall far short of the estimated US$ 6 billion required annually to ensure maximal impact worldwide [56], it seems even less likely that international funding will be sustained for the long haul required to achieve the more expensive and ambitious yet possible goal of malaria eradication. As the risk of malaria decreases, the behaviour of patients, caregivers, healthcare providers and funders become less likely to take the steps needed to reduce the malaria burden further until it is eventually eliminated. Effective information, education and communication campaigns, strong programmes monitoring the impact of malaria control interventions on disease burden, good governance and coherent advocacy (that acknowledges the many demands on limited financial and especially human resources in malaria endemic countries) are important tools for encouraging ongoing support, once there are only a few locally transmitted malaria cases.

The goals and strategies required to achieve elimination of the parasite from low-transmission settings are very different for those needed for reducing malaria morbidity and mortality in high-transmission settings. In an elimination programme, treatment of a sufficient number of infected subjects in a community to interrupt transmission becomes the primary goal. In order to interrupt transmission, the individuals who are parasitaemic (infected) and – more importantly in terms of elimination – gametocytaemic (infectious) need to be treated even if they are asymptomatic. Two possible approaches to this objective can be adopted – mass screening and treatment of both infected and infectious individuals (regardless of whether nor not they are symptomatic), or mass drug administration (MDA) given to as large a proportion of the population as possible on the grounds that this will cover a higher proportion of those infected. The lack of a rapid diagnostic method that is suitable for field use and sensitive enough for diagnosing the lower limit of parasite and gametocyte densities able to cause and transmit malaria is currently a major obstacle to using mass screening and treatment in malaria elimination.

MDA is the administration of a complete treatment course of antimalarial medicines to every individual in a geographically defined area on a specific day. MDA is not recommended by the World Health Organization, as there is no evidence of long-term benefits in large population groups [6]. An analysis of 19 MDA projects carried out over the period 1932–1999 found only one study in the small island population ($n = 718$) of Aneityum, Vanuatu, where MDA might have contributed to the elimination of *P. falciparum* and *P. vivax* malaria [78, 79]. MDA has been highly effective in reducing parasite prevalence to a very low level, but parasitaemia soon rebounds to its previous level once MDA is stopped, as seen in Garki, Nigeria and Nicaragua [78]. Mass treatment with ACTs alone is unlikely to be sufficient for malaria elimination – and primaquine and/or atovaquone–proguanil may be worth adding. In this context, drug safety should be given priority as drugs are given to a large number of people who are not infected. Thus, more evidence is needed on the risk:benefit profile of atovaquone–proguanil and primaquine to inform mass treatment approaches in the context of malaria elimination programmes [58]. Lessons should also be learnt from the lasting legacy of MDA of chloroquine and pyrimethamine: the rapid selection of resistant parasites.

7 Conclusions

Prompt effective antimalarial treatment is and will remain pivotal in achieving malaria control and eventually elimination. The past decade has seen remarkable progress being made in the fight against malaria. Almost all countries in which *P. falciparum* malaria is endemic have adopted ACT policies. High ACT coverage, together with the scaling-up of effective vector control interventions, has resulted in documented reductions in malaria cases of >50% in 2008 compared with 2000 in

43 of the 108 malaria-endemic countries. Unprecedented levels of financial, technical and political support have made this possible. During this window of opportunity for reducing the burden of malaria globally and possibly eventually eliminating malaria, attention now needs to be focussed on ensuring that countries select treatment policies that not only achieve cure rates >95% [65], but that are also likely to have a prolonged useful therapeutic life, reduce malaria transmission safely and effectively and, where applicable, are also active against *P. vivax*. As important is ensuring optimal targeting, dosing and adherence with these policies.

References

1. Hay C, Guerra A, Tatem A, Noor R, Snow R (2005) The global distribution and population at risk of malaria: past, present, and future. Lancet Infect Dis 4:327–336
2. Hay SI, Guerra CA, Gething PW, Patil AP, Tatem AJ, Noor AM, Kabaria CW, Manh BH, Elyazar IR, Brooker S, et al (2009) A world malaria map: *Plasmodium falciparum* endemicity in 2007. PLoS Med 6:e1000048. Erratum in: PLoS Med 6
3. Malaney P, Spielman A, Sachs J (2004) The malaria gap. Am J Trop Med Hyg 71:141–146
4. Ward SA, Sevene EJ, Hastings IM, Nosten F, McGready R (2007) Antimalarial drugs and pregnancy: safety, pharmacokinetics, and pharmacovigilance. Lancet Infect Dis 7:136–144
5. Sachs J, Malaney P (2002) The economic and social burden of malaria. Nature 415:680–685
6. World Health Organization (2010). Guidelines for the treatment of malaria. Second Edition. http://www.who.int/malaria/publications/atoz/9789241547925/en/index.html. Accessed 1 July 2010
7. Barnes KI, Watkins WM, White NJ (2008) Antimalarial dosing regimens and drug resistance. Trends Parasitol 24:127–134
8. Malenga G, Palmer A, Staedke S, Kazadi W, Mutabingwa T, Ansah E, Barnes KI, Whitty CJ (2005) Antimalarial treatment with artemisinin combination therapy in Africa. BMJ 331:706–707
9. Kantele A, Jokiranta TS (2011) Review of cases with the emerging fifth human malaria parasite, *Plasmodium knowlesi*. Clin Infect Dis 52:1356–1362
10. Drakeley C, Sutherland C, Bousema JT, Sauerwein RW, Targett GA (2006) The epidemiology of *Plasmodium falciparum* gametocytes: weapons of mass dispersion. Trends Parasitol 22:424–430
11. Barnes KI, White NJ (2005) Population biology and antimalarial resistance: the transmission of antimalarial drug resistance in *Plasmodium falciparum*. Acta Trop 94:230–240
12. Barnes KI, Little F, Mabuza A, Mngomezulu N, Govere J, Durrheim D, Roper C, Watkins B, White NJ (2008) Increased gametocytemia after treatment: an early parasitological indicator of emerging sulfadoxine-pyrimethamine resistance in falciparum malaria. J Infect Dis 197:1605–1613
13. White NJ (2008) The role of anti-malarial drugs in eliminating malaria. Malar J 7:S8
14. Babiker HA, Schneider P, Reece SE (2009) Gametocytes: insights gained during a decade of molecular monitoring. Trends Parasitol 24:525–530
15. Butcher GA (1997) Antimalarial drugs and the mosquito transmission of Plasmodium. Int J Parasitol 27:975–987
16. Pukrittayakamee S, Chotivanich K, Chantra A, Clemens R, Looareesuwan S, White NJ (2004) Activities of artesunate and primaquine against asexual- and sexual-stage parasites in falciparum malaria. Antimicrob Agents Chemother 48:1329–1334
17. Bousema T, Okell L, Shekalaghe S, Griffin JT, Omar S, Sawa P, Sutherland C, Sauerwein R, Ghani AC, Drakeley C (2010) Revisiting the circulation time of *Plasmodium falciparum*

gametocytes: molecular detection methods to estimate the duration of gametocyte carriage and the effect of gametocytocidal drugs. Malar J 9:136
18. Nosten F, van Vugt M, Price R, Luxemburger C, Thway KL, Broackman A, McGready R, terKuile F, Looareesuwan S, White NJ (2000) Effects of artesunate-mefloquine combination on incidence of *Plasmodium falciparum* malaria and mefloquine resistance in western Thailand. Lancet 356:297–302
19. Barnes KI, Durrheim DN, Little F, Jackson A, Mehta U, Allen E, Dlamini SS, Tsoka J, Bredenkamp B, Mthembu DJ et al (2005) Effect of artemether-lumefantrine policy and improved vector control on malaria burden in KwaZulu-Natal, South Africa. PLoS Med 2:e330
20. Bhattarai A, Ali AS, Kachur SP, Mårtensson A, Abbas AK, Khatib R, Al-Mafazy AW, Ramsan M, Rotllant G, Gerstenmaier JF et al (2007) Impact of artemisinin-based combination therapy and insecticide-treated nets on malaria burden in Zanzibar. PLoS Med 4:e309
21. Otten M, Aregawi M, Were W, Karema C, Medin A, Bekele W, Jima D, Gausi K, Komatsu R, Korenromp E et al (2009) Initial evidence of reduction of malaria cases and deaths in Rwanda and Ethiopia due to rapid scale-up of malaria prevention and treatment. Malar J 8:14
22. Barnes KI, Chanda P, Ab Barnabas G (2009) Impact of the large-scale deployment of artemether/lumefantrine on the malaria disease burden in Africa: case studies of South Africa, Zambia and Ethiopia. Malar J 8:S8
23. World Health Organization (2010). World Malaria Report. http://www.who.int/malaria/world_malaria_report_2010/en/index.html. Accessed 12 May 2011
24. Targett G, Drakeley C, Jawara M, von Seidlein L, Coleman R, Deen J, Pinder M, Doherty T, Sutherland C, Walraven G et al (2001) Artesunate reduces but does not prevent post treatment transmission of *Plasmodium falciparum* to *Anopheles gambiae*. J Infect Dis 183:1254–1259
25. Bousema JT, Schneider P, Gouagna LC, Drakeley CJ, Tostmann A, Houben R, Githure JI, Ord R, Sutherland CJ, Omar SA et al (2006) Moderate effect of artemisinin-based combination therapy on transmission of *Plasmodium falciparum*. J Infect Dis 193:1151–1159
26. Okell LC, Drakeley CJ, Ghani AC, Bousema T, Sutherland CJ (2008) Reduction of transmission from malaria patients by artemisinin combination therapies: a pooled analysis of six randomized trials. Malar J 7:125
27. Shekalaghe S, Drakeley C, Gosling R, Ndaro A, van Meegeren M, Enevold A, Alifrangis M, Mosha F, Sauerwein R, Bousema T (2007) Primaquine clears submicroscopic *Plasmodium falciparum* gametocytes that persist after treatment with sulphadoxine-pyrimethamine and artesunate. PLoS One 2:e1023
28. Beutler E, Duparc S; G6PD Deficiency Working Group (2007) Glucose-6-phosphate dehydrogenase deficiency and antimalarial drug development. Am J Trop Med Hyg 77:779–789
29. Leslie T, Briceño M, Mayan I, Mohammed N, Klinkenberg E, Sibley CH, Whitty CJ, Rowland M (2010) The impact of phenotypic and genotypic G6PD deficiency on risk of *Plasmodium vivax* infection: A case-control study amongst Afghan refugees in pakistan. PLoS Med 7: e1000283
30. Coleman MD, Coleman NA (1996) Drug-induced methaemoglobinaemia: treatment issues. Drug Saf 14:394–405
31. Sin DD, Shafran SD (1996) Dapsone- and primaquine-induced methemoglobinemia in HIV-infected individuals. J Acquir Immune Defic Syndr Hum Retrovirol 12:477–481
32. Shekalaghe SA, terBraak R, Daou M, Kavishe R, van den Bijllaardt W, van den Bosch S, Koenderink JB, Luty AJ, Whitty CJ, Drakeley C et al (2010) In Tanzania, hemolysis after a single dose of primaquine coadministered with an artemisinin is not restricted to glucose-6-phosphate dehydrogenase-deficient (G6PD A-) individuals. Antimicrob Agents Chemother 54:1762–1768
33. Baird JK (2007) Neglect of *Plasmodium vivax* malaria. Trends Parasitol 23:533–539
34. Attaran A, Barnes KI, Curtis C, d'Alessandro U, Fanello CI, Galinski MR, Kokwaro G, Looareesuwan S, Makanga M, Mutabingwa T et al (2004) WHO, the Global Fund, and medical malpractice in malaria treatment. Lancet 363:237–240

35. Trape JF (2001) The public health impact of chloroquine resistance in Africa. Am J Trop Med Hyg 64:12–17
36. Trape JF, Pison G, Spiegel A, Enel C, Rogier C (2002) Combating malaria in Africa. Trends Parasitol 18:224–230
37. Zucker JR, Ruebush TK II, Obonyo C, Otieno J, Campbell CC (2003) The mortality consequences of the continued use of chloroquine in Africa: experience in Siaya, Western Kenya. Am J Trop Med Hyg 68:386–389
38. Price R, Nosten F, Simpson JA, Luxemburger C, Phaipun L, ter Kuile F, van Vugt M, Chongsuphajaisiddhi T, White NJ (1999) Risk factors for gametocyte carriage in uncomplicated falciparum malaria. Am J Trop Med Hyg 60:1019–1023
39. Price RN, Simpson JA, Nosten F, Luxemburger C, Hkirjaroen L, ter Kuile F, Chongsuphajaisiddhi T, White NJ (2001) Factors contributing to anaemia in uncomplicated falciparum malaria. Am J Trop Med Hyg 65:614–622
40. Dondorp AM, Nosten F, Yi P, Das D, Phyo AP, Tarning J, Lwin KM, Ariey F, Hanpithakpong W, Lee SJ et al (2009) Artemisinin resistance in *Plasmodium falciparum* malaria. N Engl J Med 361:455–67. Erratum in. N Engl J Med 361:1714
41. Noedl H, Se Y, Schaecher K, Smith BL, Socheat D, Fukuda MM; Artemisinin Resistance in Cambodia 1 (ARC1) Study Consortium (2008) Evidence of artemisinin-resistant malaria in western Cambodia. N Engl J Med 359:2619–2620
42. WHO (2010) Global Report on Antimalarial Drug Efficacy and DrugResistance: 2000–2010. WHO, Geneva. http://whqlibdoc.who.int/publications/2010/9789241500470_eng.pd. Accessed 12 May 2011
43. Bethell D, Se Y, Lon C, Tyner S, Saunders D, Sriwichai S, Darapiseth S, Teja-Isavadharm P, Khemawoot P, Schaecher K et al (2011) Artesunate dose escalation for the treatment of uncomplicated malaria in a region of reported artemisinin resistance: a randomized clinical trial. PLoS One 6:e19283
44. Anderson TJ, Nair S, Nkhoma S, Williams JT, Imwong M, Yi P, Socheat D, Das D, Chotivanich K, Day NP, White NJ, Dondorp AM (2010) High heritability of malaria parasite clearance rate indicates a genetic basis for artemisinin resistance in Cambodia. J Infect Dis 201:1326–1330
45. Wongsrichanalai C, Varma JK, Juliano JJ, Kimerling ME, MacArthur JR (2010) Extensive drug resistance in malaria and tuberculosis. Emerg Infect Dis 16:1063–1067
46. White NJ (2004) Antimalarial drug resistance. J Clin Invest 113:1084–1092
47. Brockman A, Price RN, van Vugt M, Heppner DG, Walsh D, Sookto P, Wimonwattrawatee T, Looareesuwan S, White NJ, Nosten F (2000) *Plasmodium falciparum* antimalarial drug susceptibility on the north-western border of Thailand during five years of extensive use of artesunate-mefloquine. Trans R Soc Trop Med Hyg 94:537–544
48. Raman J, Little F, Roper C, Kleinschmidt I, Cassam Y, Maharaj R, Barnes KI (2010) Five years of large-scale *dhfr* and *dhps* mutation surveillance following the phased implementation of artesunate plus sulfadoxine-pyrimethamine in Maputo Province, Southern Mozambique. Am J Trop Med Hyg 82:788–794
49. Lubell Y, Reyburn H, Mbakilwa H, Mwangi R, Chonya S, Whitty CJ, Mills A (2008) The impact of response to the results of diagnostic tests for malaria: cost-benefit analysis. BMJ 336:202–205
50. Nosten F, Ashley E, McGready R, Price R (2006) We still need artesunate monotherapy. BMJ 333:45
51. White NJ, Pongtavornpinyo W, Maude RJ, Saralamba S, Aguas R, Stepniewska K, Lee SJ, Dondorp AM, White LJ, Day NP (2009) Hyperparasitaemia and low dosing are an important source of anti-malarial drug resistance. Malar J 8:253
52. Luxemburger C, Nosten F, Raimond SD, Chongsuphajaisiddhi T, White NJ (1995) Oral artesunate in the treatment of uncomplicated hyperparasitemic falciparum malaria. Am J Trop Med Hyg 53:522–525

53. Simpson JA, Watkins ER, Price RN, Aarons L, Kyle DE, White NJ (2000) Mefloquine pharmacokinetic-pharmacodynamic models: implications for dosing and resistance. Antimicrob Agents Chemother 44:3414–3424
54. Barnes KI, Little F, Smith PJ, Evans A, Watkins WM, White NJ (2006) Sulfadoxine-pyrimethamine pharmacokinetics in malaria: paediatric dosing implications. Clin Pharmacol Ther 80:582–596
55. Ringwald P, Keundjian A, Same Ekobo A, Basco LK (2000) Chemoresistance of *P. falciparum* in urban areas of Yaounde, Cameroon. Part 1: surveillance of in vitro and in vivo resistance of Plasmodium falciparum to chloroquine from 1994 to 1999 in Yaounde, Cameroon. Trop Med Int Health 5:612–619
56. Checchi F, Piola P, Fogg C, Bajunirwe F, Biraro S, Grandesso F, Ruzagira E, Babigumira J, Kigozi I, Kiguli J et al (2006) Supervised versus unsupervised antimalarial treatment with six-dose artemether-lumefantrine: pharmacokinetic and dosage-related findings from a clinical trial in Uganda. Malar J 5:59
57. Price RN, Hasugian AR, Ratcliff A, Siswantoro H, Purba HL, Kenangalem E, Lindegardh N, Penttinen P, Laihad F, Ebsworth EP et al (2007) Clinical and pharmacological determinants of the therapeutic response to dihydroartemisinin-piperaquine for drug resistant malaria. Antimicrob Agents Chemother 51:4090–4097
58. Feachem RGA, The Malaria Elimination Group (2009) Shrinking the malaria map: a guide on malaria elimination for policy makers. The Global Health Group, Global Health Sciences, University of California, San Francisco. http://www.malariaeliminationgroup.org/publications. Accessed 1 July 2010
59. Guerra CA, Howes RE, Patil AP, Gething PW, Van Boeckel TP, Temperley WH, Kabaria CW, Tatem AJ, Manh BH, Elyazar IR et al (2010) The international limits and population at risk of *Plasmodium vivax* transmission in 2009. PLoS Negl Trop Dis 4:e774
60. Krotoski WA (1985) Discovery of the hypnozoite and a new theory of malarial relapse. Trans R Soc Trop Med Hyg 79:1–11
61. Price RN, Tjitra E, Guerra CA, Yeung S, White NJ, Anstey NM (2007) Vivax malaria: neglected and not benign. Am J Trop Med Hyg 77:79–87
62. Ratcliff A, Siswantoro H, Kenangalem E, Maristela R, Wuwung RM, Laihad F, Ebsworth EP, Anstey NM, Tjitra E, Price RN (2007) Two fixed-dose artemisinin combinations for drug-resistant falciparum and vivax malaria in Papua, Indonesia: an open-label randomized comparison. Lancet 369:757–765
63. Pukrittayakamee S, Chantra A, Simpson JA, Vanijanonta S, Clemens R, Looareesuwan S, White NJ (2000) Therapeutic responses to different antimalarial drugs in vivax malaria. Antimicrob Agents Chemother 44:1680–1685
64. Arias AE, Corredor A (1989) Low response of Colombian strains of *Plasmodium vivax* to classical antimalarial therapy. Trop Med Parasitol 40:21–23
65. Galappaththy GNL, Omari AAA, Tharyan P (2007) Primaquine for preventing relapses in people with *Plasmodium vivax* malaria. Cochrane Database Syst Rev 2007, Issue 1:CD004389. doi: 10.1002/14651858.CD004389.pub2
66. Baird JK, Hoffman SL (2004) Primaquine therapy for malaria. Clin Infect Dis 39:1336–1345
67. Mendis K, Rietveld A, Warsame M, Bosman A, Greenwood B, Wernsdorfer WH (2009) From malaria control to eradication: the WHO perspective. Trop Med Int Health 14:802–809
68. Roll Back Malaria (2008) The Global Malaria Action Plan for a malaria free world. http://www.rollbackmalaria.org/gmap/toc.pdf. Accessed 4 July 2010
69. The malERA Consultative Group on Drugs (2011) A research agenda for malaria eradication: drugs. PLoS Med 8:e1000402
70. Arrow KJ, Panosian C, and Gelband H (Eds), Institute of Medicine of the National Academies Committee on the Economics of Antimalarial Drugs. Saving lives, buying time: economics of malaria drugs in an age of resistance. The National Academies Press, Washington. http://www.nap.edu/openbook.php?isbn=0309092183. Accessed 17 May 2011

71. World Health Organization (2009) World Malaria Report. http://www.who.int/malaria/world_malaria_report_2009/en/index.html. Accessed 17 Apr 2011
72. Ceesay SJ, Casals-Pascual C, Erskine J, Anya SE, Duah NO, Fulford AJ, Sesay SS, Abubakar I, Dunyo S, Sey O et al (2008) Changes in malaria indices between 1999 and 2007 in The Gambia: a retrospective analysis. Lancet 372:1545–1554
73. Kleinschmidt I, Schwabe C, Benavente L, Torrez M, Ridl FC, Segura JL, Ehmer P, Nchama GN (2009) Marked increase in child survival after four years of intensive malaria control. Am J Trop Med Hyg 80:882–888
74. O'Meara WP, Bejon P, Mwangi TW, Okiro EA, Peshu N, Snow RW, Newton CR, Marsh K (2008) Effect of a fall in malaria transmission on morbidity and mortality in Kilifi, Kenya. Lancet 372:1555–1562
75. Sharp BL, Kleinschmidt I, Streat E, Maharaj R, Barnes KI, Durrheim DN, Ridl FC, Morris N, Seocharan I, Kunene S et al (2007) Seven years of regional malaria control collaboration–Mozambique, South Africa, and Swaziland. Am J Trop Med Hyg 76:42–47
76. United Nations Development Programme. Millennium Development Goals. http://www.undp.org/mdg/basics.shtml. Accessed 11 July 2010
77. Kleinschmidt I, Schwabe C, Shiva M, Segura JL, Sima V, Mabunda SJ, Coleman M (2009) Combining indoor residual spraying and insecticide-treated net interventions. Am J Trop Med Hyg 81:519–524
78. von Seidlein L, Greenwood BM (2003) Mass administration of antimalarial drugs. Trends Parasitol 19:790–796
79. Kaneko A, Taleo G, Kalkoa M, Yamar S, Kobayakawa T, Björkman A (2000) Malaria eradication on islands. Lancet 356:1560–1564

4-Aminoquinolines: Chloroquine, Amodiaquine and Next-Generation Analogues

Paul M. O'Neill, Victoria E. Barton, Stephen A. Ward, and James Chadwick

Abstract For several decades, the 4-aminoquinolines chloroquine (CQ) and amodiaquine (AQ) were considered the most important drugs for the control and eradication of malaria. The success of this class has been based on excellent clinical efficacy, limited host toxicity, ease of use and simple, cost-effective synthesis. Importantly, chloroquine therapy is affordable enough for use in the developing world. However, its value has seriously diminished since the emergence of widespread parasite resistance in every region where *P. falciparum* is prevalent. Recent medicinal chemistry campaigns have resulted in the development of short-chain chloroquine analogues (AQ-13), organometallic antimalarials (ferroquine) and the "fusion" antimalarial trioxaquine (SAR116242). Projects to reduce the toxicity of AQ have resulted in the development of metabolically stable AQ analogues (isoquine/*N-tert*-butyl isoquine). In addition to these developments, older 4-aminoquinolines such as piperaquine and the related aza-acridine derivative pyronaridine continue to be developed. It is the aim of this chapter to review 4-aminoquinoline structure–activity relationships and medicinal chemistry developments in the field and consider the future therapeutic value of CQ and AQ.

P.M. O'Neill (✉)
Department of Chemistry, Robert Robinson Laboratories, University of Liverpool, Liverpool L69 7ZD, UK

Department of Pharmacology, MRC Centre for Drug Safety Science, University of Liverpool, Liverpool L69 3GE, UK
e-mail: pmoneill@liverpool.ac.uk

V.E. Barton • J. Chadwick
Department of Chemistry, Robert Robinson Laboratories, University of Liverpool, Liverpool L69 7ZD, UK

S.A. Ward
Liverpool School of Tropical Medicine, Pembroke Place, Liverpool L3 5QA, UK

H.M. Staines and S. Krishna (eds.), *Treatment and Prevention of Malaria*,
DOI 10.1007/978-3-0346-0480-2_2,

1 History and Development

Quinine **1**, a member of the *cinchona* alkaloid family, is one of the oldest antimalarial agents and was first extracted from *cinchona* tree bark in the late 1600s. The *cinchona* species is native to the Andean region of South America, but when its therapeutic potential was realised, Dutch and British colonialists quickly established plantations in their south-east Asian colonies. These plantations were lost to the Japanese during World War II, stimulating research for synthetic analogues based on the quinine template, such as the 4-aminoquinoline chloroquine (CQ **2**, Fig. 1) [1].

A thorough historical review of CQ (in honour of chloroquine's 75th birthday) is available elsewhere [2]. In short, CQ was first synthesized in 1934 and became the most widely used antimalarial drug by the 1940s [3]. The success of this class has been based on excellent clinical efficacy, limited host toxicity, ease of use and simple, cost-effective synthesis. Importantly, CQ treatment has always been affordable – as little as USD 0.10 in Africa [4]. However, the value of quinoline-based antimalarials has been seriously eroded in recent years, mainly as a result of the development and spread of parasite resistance [5].

Although much of the current research effort is directed towards the identification of novel chemotherapeutic targets, we still do not fully understand the mode of action and the complete mechanism of resistance to the quinoline compounds, knowledge that would greatly assist the design of novel, potent and inexpensive alternative quinoline antimalarials. The search for novel quinoline-based antimalarials with pharmacological benefits superseding those provided by CQ has continued throughout the later part of the twentieth century and the early part of this century since the emergence of CQ resistance.

Comprehensive reviews on the pharmacology [6] and structure activity relationships [7] have been published previously, so will be only mentioned briefly. It is the aim of this chapter to review developments in the field that have led to the next-generation 4-aminoquinolines in the development "pipeline", in addition to discussion of the future therapeutic value of CQ and amodiaquine (AQ). We will begin with studies directed towards an understanding of the molecular mechanism of action of this important class of drug.

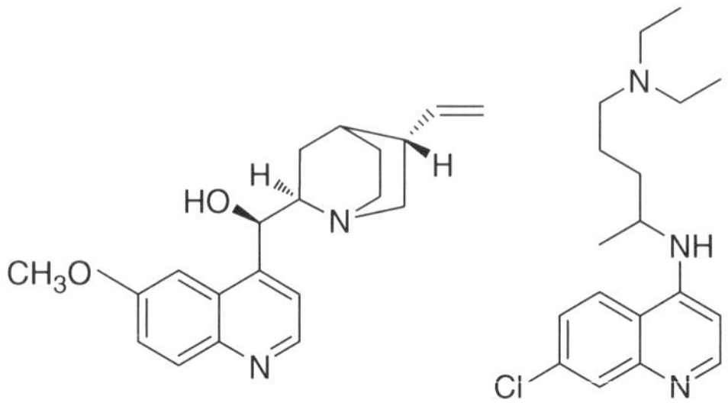

Quinine, **1**

Chloroquine, **2**

Fig. 1 Quinine **1** and related 4-aminoquinoline antimalarial chloroquine, **2**

2 Mode of Action of Quinoline Antimalarials

The precise modes of action of the quinoline antimalarials are still not completely understood, although various mechanisms have been proposed for the action of CQ and related compounds [8]. Some of the proposed mechanisms would require higher drug concentrations than those that can be achieved in vivo and, therefore, are not considered as convincing as other arguments [9]. Such mechanisms include the inhibition of protein synthesis [10], the inhibition of food vacuole phospholipases [11], the inhibition of aspartic proteinases [12] and the effects on DNA and RNA synthesis [13, 14].

CQ is active against the erythrocytic stages of malaria parasites but not against pre-erythrocytic or hypnozoite-stage parasites in the liver [15] or mature gametocytes. Since CQ acts exclusively against those stages of the intra-erythrocytic cycle during which the parasite is actively degrading haemoglobin, it was assumed that CQ somehow interferes with the parasite-feeding process. Although this is still a matter of some controversy, evidence of proposed mechanisms will be discussed in the following sections.

2.1 *Haem–CQ Drug Complexes*

To obtain essential amino acids for its growth and division, the parasite degrades haemoglobin within the host red blood cell. Digestion of its food source occurs in an acidic compartment known as the digestive vacuole (DV) (a lysosome-type structure, approximately pH 5). During feeding, the parasite generates the toxic and soluble molecule haem [ferriprotoporphyrin IX, FP Fe (II)] and biocrystallises it at, or within, the surface of lipids to form the major detoxification product haemozoin (Fig. 2) [16].

Slater et al. [17] demonstrated the ability of CQ to inhibit the in vitro FP detoxification in the high micro-molar range. The ability of CQ and a number of other quinoline antimalarial drugs to inhibit both spontaneous FP crystallisation and parasite extract catalysed crystallisation of FP has since been confirmed [18, 19].

Considerable evidence has been presented in recent years that antimalarial drugs such as CQ act by forming complexes with haem (FP Fe (II)) and/or the hydroxo- or aqua complex of haematin (ferriprotoporphyrin IX, Fe (III) FP), derived from parasite proteolysis of host haemoglobin [20–22] (Fig. 2), although the exact nature of these complexes is a matter of debate.

Dorn et al. [23, 24] confirmed that CQ forms a complex with the μ-oxo dimeric form of FP (haematin) with a stoichiometry of 1 CQ: 2 μ-oxo dimers. In other studies, CQ was found to bind to monomeric haem to form a highly toxic haem–CQ complex, which incorporates into the growing dimer chains and terminates the chain extension, blocking further sequestration of toxic haem and disrupting membrane function (Fig. 2) [25, 26].

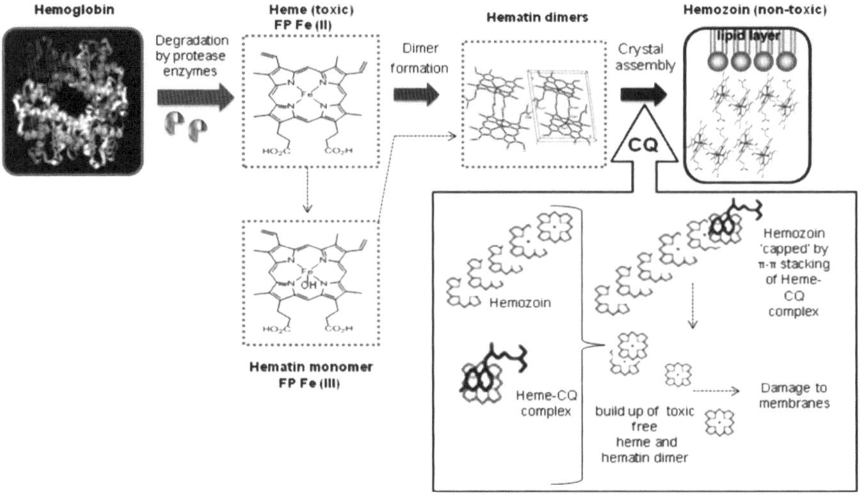

Fig. 2 Degradation of haemoglobin and detoxification mechanisms of the parasite and proposed target of CQ

2.2 Accumulation of CQ in the Acidic Food Vacuole

Due to the weak base properties of CQ and related analogues, their effectiveness has also been shown to be partly dependent upon drug accumulation in the acidic DV. A number of early studies have suggested that CQ accumulation can be explained by an ion-trapping or weak-base mechanism [27, 28]. CQ is a diprotic weak base ($pK_{a1} = 8.1$, $pK_{a2} = 10.2$) and in its unprotonated form, it diffuses through the membranes of the parasitised erythrocyte and accumulates in the acidic DV (pH 5–5.2) [27]. Once inside, the drug becomes protonated and, as a consequence, membrane impermeable and becomes trapped in the acidic compartment of the parasite (Fig. 3).

Various studies have suggested that the kinetics and saturability of CQ uptake are best explained by the involvement of a specific transporter [29, 30] or carrier-mediated mechanism for the uptake of CQ [31]. Another hypothesis by Chou et al. [32] suggests that free haematin (FP) in the DV might act as an intra-vacuolar receptor for CQ. Work by Bray et al. also strongly supports this hypothesis [33].

3 CQ Resistance Development

The first incidences of resistance to CQ were reported in 1957. The reasons for the emergence of resistance are multi-factorial: uncontrolled long-term treatment regimes, travel activity resulting in spread of resistant strains and frequent feeding of mosquitoes from several different hosts, to name but a few [34]. The mechanism by which resistance is acquired is discussed below.

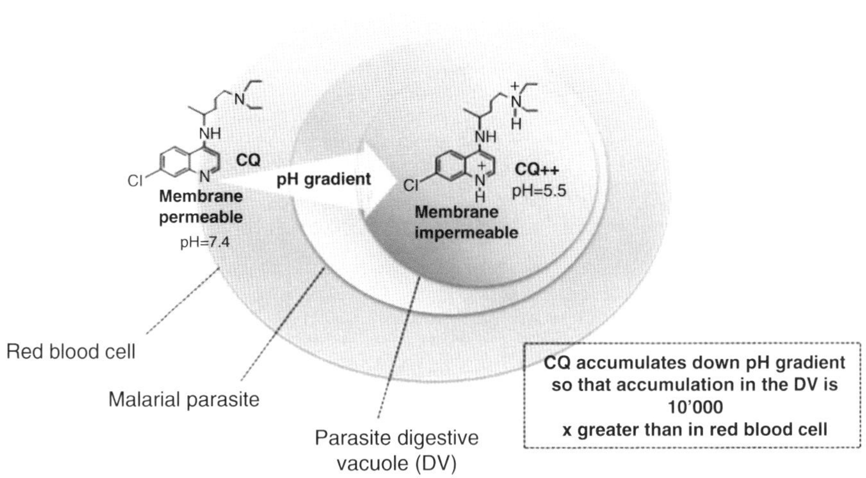

Fig. 3 Ion trapping; diffusion of CQ due to the pH gradient leads to increased concentration of CQ in the DV

3.1 Parasite-Resistance Mechanisms

It was soon proven that the concentration of CQ inside the DV was reduced in parasite-resistant strains. The powerful accumulation mechanism of CQ was therefore less effective, suggesting mutations in transporter proteins in these resistant strains. Resistant isolates also have reduced apparent affinity of CQ–FP binding in the DV, therefore CQ-resistant isolates have evolved a mechanism whereby the access of CQ to FP is reduced [35].

3.1.1 PfCRT

Another characteristic of CQ-resistant isolates is that their phenotype can be partially "reversed" by the calcium channel blocker verapamil so that the isolates become resensitised to CQ [35]. Verapamil was shown to act by increasing the access of CQ to the FP receptor and this effect is considered a phenotypic marker of CQ resistance. The characteristic effects of CQ resistance (reduced CQ sensitivity, reduced CQ uptake and the verapamil effect) have all been attributed to specific amino acid changes in an integral DV membrane protein, the *P. falciparum* chloroquine resistance transporter (PfCRT) [36, 37]. PfCRT mutated at amino acid 76 appears to be central to the chloroquine resistance phenotype. Mutant PfCRT seems to allow movement of drugs out of the DV; therefore blocking of PfCRT by verapamil restores sensitivity.

In brief, there are three proposed models for the resistance mechanism of PfCRT:

- *The partitioning model*: CQ was found to flow out of the DV of CQ-resistant strains much faster that CQ-sensitive strains, by a verapamil-blockable route [38]. Initially, this was attributed to changes in DV pH for CQ-sensitive and CQ-resistant strains. However, it was later shown that CQ-resistant parasites have a similar resting DV pH, and, therefore, must possess a CQ efflux mechanism in the DV membrane, increasing the permeability of a particular form of CQ [39].
- *The channel model*: In this model, mutated PfCRT acts as a channel, providing a leak pathway for the passive diffusion of protonated CQ, allowing it to flow freely from the DV [40, 41].
- *The carrier model*: In this alternate model, mutated PfCRT acts as a carrier, transporting protonated CQ by facilitated diffusion or active transport across the DV membrane [42, 43].

The issue of exactly how PfCRT confers this phenotype has been recently reviewed, although it remains a matter of debate [44].

3.1.2 PfMDR1

A multi-drug resistance homologue in *P. falciparum* (PfMDR1) has also been implicated in CQ resistance. PfMDR1 has been demonstrated to reside in the parasites' DV membrane with its ATP-binding domain facing the cytoplasm [45]. This suggests that PfMDR1 directs drug movement into the DV. Loss of this drug import capability could be advantageous to the parasite when the drug targets the DV. Irrespective of the specifics of MDR1-mediated chloroquine transport, the protein has been shown to contribute to chloroquine resistance. Sanchez et al. functionally expressed a number of different polymorphs of *pfmdr1* (the gene that codes for PfMDR1) in *Xenopus laevis* oocytes in order to characterize the transport properties of PfMDR1 and its interaction with antimalarial drugs. They demonstrated that PfMDR1 does indeed transport CQ and that polymorphisms within PfMDR1 affect the substrate specificity; wild-type PfMDR1 transports CQ, whereas polymorphic PfMDR1 variants from parasite lines associated with resistance apparently are not as efficient [46].

3.2 Recycling of CQ

CQ still remains the treatment of choice in a few geographical areas where it can still be relied upon, although guidelines now instruct the use of combination chemotherapy to slow the development of resistance to the partner drug [47]. In some resistance "hot spots", CQ was completely abandoned for a combination of sulfadoxine–pyrimethamine almost two decades ago. In such cases, there is evidence to suggest that CQ sensitivity can be restored [48]; 8 years after discontinuation of CQ in Malawi, the *pfcrt* T76 mutation [49] had disappeared from nearly every isolate analysed. Similar observations have been made in Tanzania, South Africa, China and parts of Thailand [50]. These results have given some hope that "drug-cycling" may be an option for the future and CQ combinations may be used effectively again in disease-endemic areas where it was once abandoned [2]. However, the concern with this strategy is that re-selection of resistance mutants is likely to be very rapid.

Ursing et al. have reported that the failure rate of CQ treatment can be decreased by giving the drug twice per day rather than as a once daily treatment regimen [51–53]. Doubling the dosing frequency in this way achieved a high cure rate despite underlying CQ resistance and without any adverse side effects [51]. This increase in efficacy can be explained by the pharmacokinetics of CQ; the second daily dose of CQ acting to raise plasma concentrations to levels where they have activity against resistant parasites [54]. It has also been shown that the use of this type of treatment regimen can stabilize the spread of CQ resistance [53, 55]. One major drawback with this type of double-dose treatment regimen is the narrow therapeutic index for CQ and, in order for such treatment to be widely used, extensive safety re-evaluation would need to be performed in large populations to ensure safety at the population level.

4 Modifications to Improve CQ

CQ, **2** contains a 7-chloroquinoline-substituted ring system with a flexible pentadiamino side chain. The haem-binding template, 7-chloro- and terminal amino group are all important for antimalarial activity, as detailed in Fig. 4.

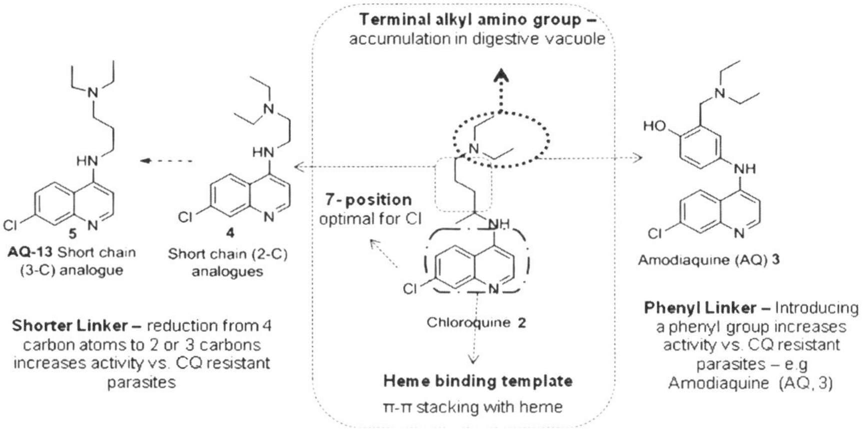

Fig. 4 Exploring the structure–activity relationship (SAR) of CQ: modifications shown led to the development of new analogues AQ (**3**), AQ-13 (**5**) and other short chain analogues (**4**) which have good activities against CQ-resistant strains

Since CQ's discovery, numerous attempts have been made to prepare a superior antimalarial quinolone-based drug. The following section briefly summarizes some of the more important recent advances in the field, with particular emphasis on 4-aminoquinolines that are in clinical and pre-clinical development. For a more in-depth discussion of 4-aminoquinoline analogue development over the last 10 years, Kaur et al. have recently published an extensive review [56].

4.1 Modifications to Overcome Resistance: Short-Chain Analogues

4.1.1 AQ-13

Studies on 4-aminoquinoline structure–activity relationships (SARs) have revealed that 2-carbon side-chain CQ analogues such as **4** retain activity against CQ-resistant *Plasmodium* parasites [57, 58]. Krogstad et al. have synthesized a series of analogues with varying diaminoalkane side chains at the 4-position [57]. Interestingly, compounds with diaminoalkyl side chains shorter than four carbon atoms or longer than seven carbon atoms were active against CQ-susceptible, CQ-resistant, and multi-drug-resistant strains of *P. falciparum* in vitro (IC_{50} values of 40–60 nM against the K1 multi-drug resistant strain) and exhibited no cross-resistance with CQ.

One of these analogues, AQ-13 **5**, a short-chain aminoquinoline antimalarial drug, underwent Phase I clinical trials. The mode of action is suggested to be the same as CQ but the presence of the short linker chain is believed to enable the molecule to circumvent the parasite-resistance mechanism (PfCRT), making **5** active against CQ-resistant parasites.

Preliminary pharmacokinetic studies indicate that AQ-13 has a similar profile to that of CQ [59] and the Phase I clinical trials were positive [60], concluding minimal difference in toxicity compared with CQ. However, since AQ-13 exhibited increased clearance compared with CQ, dose adjustment is required and an initial dose-finding Phase II (efficacy) study of AQ-13 in Mali is planned. Since clinical trials have shown that oral doses of 1,400 and 1,750 mg AQ-13 are as safe as equivalent oral doses of CQ and have similar pharmacokinetics, more recent trials were performed to determine if a 2,100 mg dose of AQ-13 (700 mg per day for 3 days) was safe to include as a third arm in Phase II studies in Mali and to investigate the effects of food (the standardised FDA fatty meal) on the bioavailability and pharmacokinetics of AQ-13. Based on the results, it is proposed to compare the 1,400, 1,700 and 2,100 mg doses of AQ-13 with each other and with Coartem in an initial dose-finding efficacy (Phase II) study of AQ-13 in Mali [61].

A possible drawback with these derivatives is the potential to undergo side-chain dealkylation (for short-chain CQ analogues such as **5** (AQ-13), deethylation is a particular problem in vivo) [62]. This metabolic transformation significantly reduces the lipid solubility of the drug and significantly increases cross-resistance up to and beyond that seen with CQ [63].

4.1.2 Ferroquine: An Organometallic Antimalarial

Metal complexes have been used as drugs in a variety of diseases [64]. Incorporation of metal fragments into CQ has generally produced an enhancement of the efficacy of CQ with no acute toxicity. Three novel CQ complexes of transition metals (Rh, Ru, Au) have been synthesized (**6**, **7** and **8**, Fig. 5) [65, 66], with the Au–CQ complex **8** in particular, displaying high in vitro activity against the asexual blood-stage of two CQ-resistant *P. falciparum* strains.

Four new ferrocene-CQ analogues were developed by Biot and co-workers, where the carbon chain of CQ was replaced by the hydrophobic ferrocenyl group [67]. Some of the compounds showed potent antimalarial activity in vivo against *P. berghei* and were 22 times more potent against schizonts than CQ in vitro against a drug-resistant strain of *P. falciparum*. The same group reported two new ferrocene-CQ compounds in 1999, one of which (**9**) showed very promising antimalarial activity in vivo against *P. berghei* and in vitro against CQ-resistant strains of *P. falciparum* [68].

Now named ferroquine (SSR-97193, FQ), **9** is the first novel organometallic antimalarial drug candidate to enter clinical trials. A multi-factorial mechanism of action is proposed including the ability to target lipids, inhibit the formation of haemozoin and generate reactive oxygen species [69]. The ferrocene group alone does not have antimalarial activity but possibly utilises the parasites' affinity for iron to increase the probability of encountering the molecule [69, 70]. In addition to its activity against CQ-resistant *P. falciparum* isolates, FQ is also highly effective against drug-resistant *P. vivax* malaria [71]. A Phase II clinical trial in combination with artesunate is to be completed by October 2011 to assess activity in reducing parasitaemia and to explore the pharmacokinetics of ferroquine and its metabolites [72].

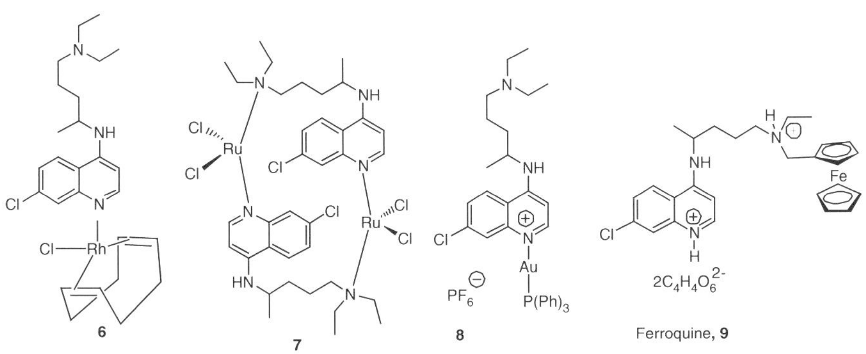

6 7 8 Ferroquine, **9**

Fig. 5 Organometallic antimalarials

4.1.3 Piperaquine

Other notable work in the chloroquine SAR field has involved the preparation of bisquinoline dimers, some of which possess excellent activity against CQ-resistant parasites. This activity against resistant parasites may be explained by their steric bulk, which prevents them from fitting into the binding site of PfCRT. Alternatively, the bisquinolines may be more efficiently trapped inside the DV because of their four positive charges.

Early examples of such agents include bis(quinolyl) piperazines such as piperaquine, **10** (Fig. 6). Piperaquine was first synthesized in the 1960s and used extensively in China for prophylaxis and treatment for the next 20 years. With the development of piperaquine-resistant strains of *P. falciparum* and the emergence of the artemisinin derivatives, its use declined during the 1980s [73].

During the next decade, piperaquine was rediscovered as one of a number of compounds suitable for combination with an artemisinin derivative. The pharmacokinetic properties of piperaquine have now been characterised [74], revealing that it is a highly lipid-soluble drug with a large volume of distribution at steady state, good bioavailability, long elimination half-life and a clearance rate that is markedly higher in children than in adults. The tolerability, efficacy, pharmacokinetic profile and low cost of piperaquine make it a promising partner drug for use as part of an artemisinin combination therapy (ACT).

Initial results were encouraging [73, 75], and Phase III clinical trials were completed in 2009 [76]. A recent report analysing individual patient data analysis of efficacy and tolerability in acute uncomplicated falciparum malaria, from seven published randomised clinical trials conducted in Africa and South East Asia concluded that dihydroartemisinin (DHA)-piperaquine is well tolerated, highly effective and safe [77]. Although not currently registered in the UK, a fixed combination called Duo-cotecxin is registered in China, Pakistan, Cambodia and Myanmar in addition to 18 African countries. Concerns with this combination lie in the fact that the calculated terminal half-life for piperaquine is around 16.5 days [78], compared with that of DHA (approximately 0.5 h) [79]; hence, the development of resistance could be a possibility due to prolonged exposure of piperaquine at sub-therapeutic levels effectively as a monotherapy.

A 1,2,4-trioxolane (RBx11160/Arterolane) has also been recently partnered with piperaquine and progressed to Phase III clinical trials. The clinical trials of RBx11160 alone identified its tendency to degrade relatively rapidly due to high levels of iron (II) in infected red blood cells, leading to a clinical efficacy of 60–70% [80]. The combination with a longer lasting drug such as piperaquine,

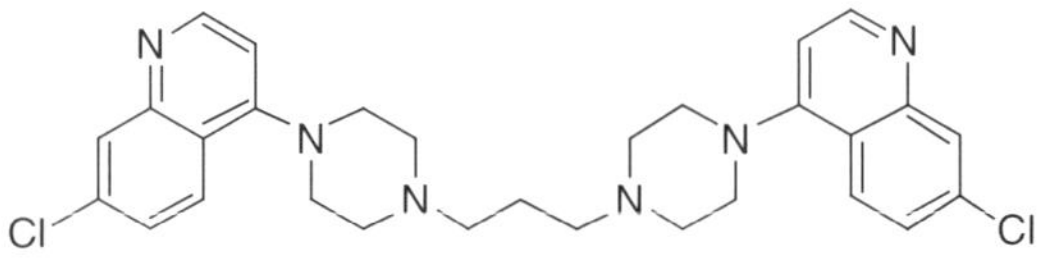

Piperaquine, **10**

Fig. 6 Structure of piperaquine **10**

with a completely different mechanism of action, may reduce the possibility of resistance and recrudescence [81]; recent results suggest the combination is highly active, with patients being free from recrudescence on day 28 after treatment [76]. This combination may also offer an advantage over DHA-piperaquine in the sense that the artemisinin-based component of the combination is a totally synthetic 1,2,4-trioxolane. This avoids over-reliance on the natural product artemisinin, whose cost and availability has been shown to fluctuate in recent years [82].

4.1.4 Trioxaquine SAR116242

Combination chemotherapy is now the mainstay of antimalarial treatment; each novel artemisinin-based antimalarial that reaches clinical trials is usually employed in an additional trial with an appropriate partner drug. However, a relatively novel approach is the concept of "covalent biotherapy" – a synthetic hybrid molecule containing two covalently linked pharmacophores [83]. The hybrid is designed to target the parasite by two distinct mechanisms thus circumventing resistance development. The hybrid also has several advantages over multi-component drugs such as:

- Expense – in principle, the risks and costs involved with a hybrid may not be any different when compared with those of a single entity.
- Safety – lower risk of drug–drug adverse interactions.
- Matched pharmacokinetics (i.e. a single entity)

A possible disadvantage, however, is that it is more difficult to adjust the ratio of activities at different targets [84]. Recent examples include trioxaquines developed by Meunier and co-workers, containing a 1,2,4-trioxane (as the artemisinin-based component) covalently bound to a 4-aminoquinoline [85]. These novel trioxaquines were found to be potent against CQ and pyrimethamine-resistant strains, and have improved antimalarial activity compared with the individual components. Several trioxaquines were developed over a number of years culminating in the selection of a drug-development candidate known as SAR116242, **11** (Fig. 7).

The superior antimalarial activity in both CQ-sensitive and CQ-resistant isolates ($IC_{50} = 10$ nM) has been attributed to its dual mechanism of haem alkylation and haemozoin inhibition. In addition, incorporation of a second cyclohexyl ring within the linker that joins the two pharmacophores increased the metabolic stability of this molecule compared with other trioxaquines containing a linear tether [86].

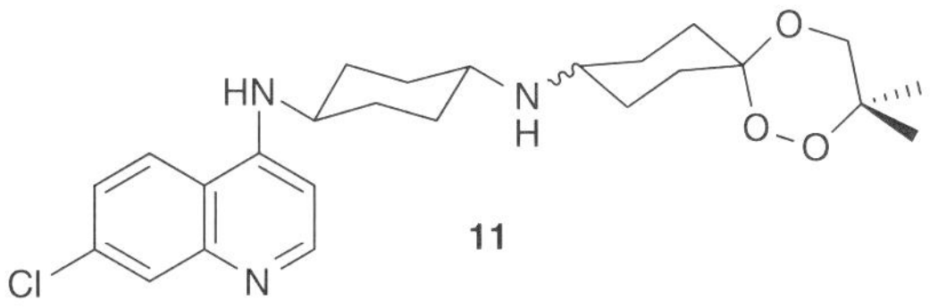

11

SAR116242

Fig. 7 Structure of SAR116242 **11**

The drug was synthesised as a mixture of diastereoisomers, but each diastereoisomer was found to be equipotent in their in vitro antiplasmodial activities and also displayed similar pharmacological profiles. However, it is not clear whether the pharmacokinetics and safety profiles of each individual diasteroisomer are the same. SAR 116242 is undergoing pre-clinical assessment by Sanofi-Aventis to determine its potential as the first "fusion" antimalarial.

4.1.5 Amodiaquine

Amodiaquine **3** (AQ), a phenyl substituted analogue of CQ, was first found to be effective against non-human malaria in 1946. Its mechanism of action is thought to be similar to CQ, but this is again a matter of some controversy [87].

Clinical use of AQ has been severely restricted because of associations with hepatotoxicity and agranulocytosis. Due to this toxicity, WHO withdrew recommendation for the drug as a monotherapy in the early 1990s. The AQ side chain contains a 4-aminophenol group; a structural alert for toxicity, because of metabolic oxidation to a quinoneimine (Fig. 8). Although cross-resistance of CQ and AQ has been documented for 20 years [88], AQ remains an important drug as it is effective against many CQ-resistant strains. Therefore, many drug design projects have since focussed on reducing this toxicity [87].

4.2 Modifications to Reduce Toxicity of AQ

4.2.1 Metabolism of CQ and AQ

CQ is highly lipophilic, as well as being a diacidic base. After oral administration, CQ is rapidly absorbed from the gastrointestinal tract, having a high bioavailability of between 80 and 90%. CQ undergoes *N*-deethylation to give the desethyl

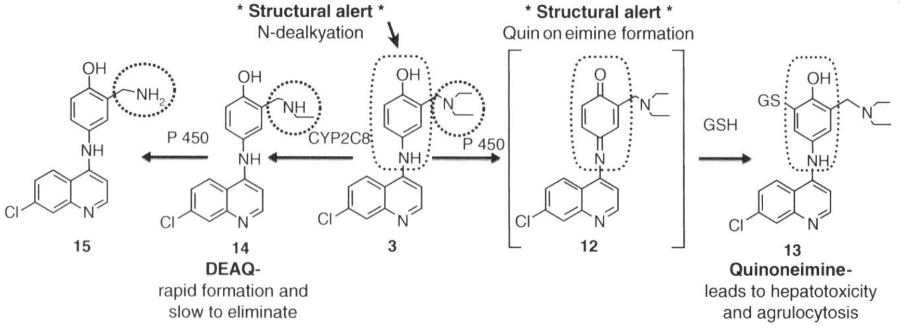

Fig. 8 Metabolism of AQ to toxic quinoneimine and DEAQ metabolites

compound as a major metabolite which has the same activity as CQ against sensitive strains, but reduced activity versus CQ-resistant strains [89].

Upon oral administration, AQ is rapidly absorbed and extensively metabolized. Although AQ has a high absorption rate from the gut due to a large first pass effect, AQ has a low bioavailability and is considered a pro-drug for desethylamodiaquine (DEAQ, **14**) [90]. In contrast to the metabolism of CQ, AQ also produces a toxic quinoneimine metabolite **12** (Fig. 8). The metabolites have been detected in vivo by the excretion of glutathione (GSH) conjugates (such as **13**) in experimental animals [91, 92]. It has been postulated that AQ toxicity involves immune-mediated mechanisms directed against the drug protein conjugates via in vivo bioactivation and covalent binding of the drug to proteins [93].

The main metabolite of AQ is DEAQ **14**, with other minor metabolites being 2-hydroxyl-DEAQ and *N*-bisdesethyl AQ (bis-DEAQ **15**) [94] (Fig. 8). The formation of DEAQ is rapid and its elimination very slow with a terminal half-life of over 100 h [95], as a result the mean plasma concentration of DEAQ is six- to sevenfold higher than the parent drug. Recent studies have established that the main P450 isoform catalysing the *N*-dealkylation of amodiaquine is CYP2C8 [96]. Mutations in PfCRT have been found in resistance isolates and correlate with high-level resistance to the AQ metabolite DEAQ in in vitro tests.

4.2.2 Modification of Metabolic Structural Alerts

Since AQ retains antimalarial activity against many CQ-resistant parasites, the next focus was to make a safer, cost-effective alternative. Initial studies involved the design and synthesis of fluoroamodiaquine (FAQ, **16**, Fig. 9) [97] since this analogue cannot form toxic metabolites by P450-mediated processes and retains substantial antimalarial activity versus CQ-resistant parasites. However, the resulting *N*-desethyl 4′-fluoro amodiaquine metabolite has significantly reduced activity against CQ-resistant parasites [97]. Concerns about cost led to the preparation of

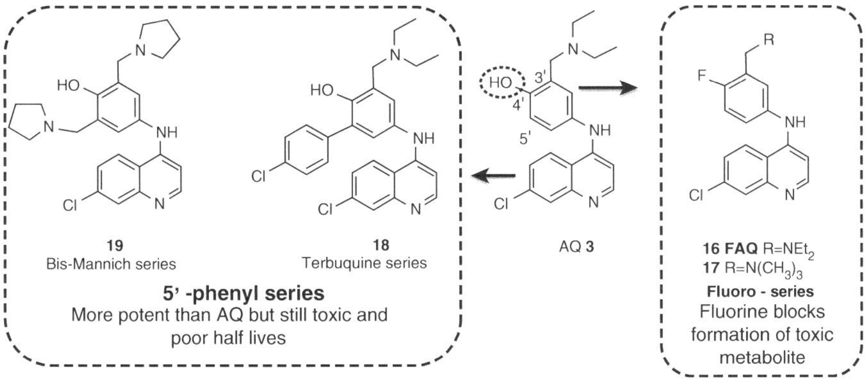

Fig. 9 Modification of structural alerts to reduce toxicity of AQ

other synthetically accessible analogues; the tebuquine series [98] and the bis-Mannich series [99] (Fig. 9).

Tebuquine (**18**), a biaryl analogue of AQ discovered by Parke-Davis, is significantly more active than AQ and CQ both in vitro and in vivo and has potent antimalarial activity and reduced cross-resistance with CQ [100, 101]. Both the bis-Mannich and terbuquine series were expected to offer advantages over AQ in the sense that they contain Mannich side chains that are more resistant to cleavage to *N*-desalkyl metabolites. A potential drawback with the bis-Mannich class of antimalarial compounds was recognized by Tingle et al. [102]. They demonstrated that such compounds have long half-lives, raising concerns over potential drug toxicity and resistance development. Compounds in the tebuquine series have also been shown to have unacceptable toxicity profiles that is exacerbated by the long half-lives [102].

Pyronaridine

Pyronaridine **20** (Fig. 10) is another member of the class of Mannich-base schizontocides; however, the usual quinoline heterocycle is replaced by an aza-acridine. Like AQ **2**, pyronaridine **20** retains the aminophenol substructure which can be oxidised to the respective quinoneimine. Since pyronaridine contains two Mannich-base side chains, it has been suggested that the second Mannich base moiety prevents the formation of the hazardous thiol addition products by sterically shielding the quinoneimine from the attack of the sulphur nucleophile [103].

Pyronaridine **20** was developed and used in China since the 1980s, but has not been registered in other countries. In a clinical study performed in Thailand, high recrudescence has been observed and in vitro assays revealed the presence of pyronaridine-resistant strains [104]. Another study in Africa showed high activity against CQ-resistant field isolates (IC_{50} values of 0.8–17.9 nM) [105]. Data suggest there may be some in vitro cross-resistance or at least cross-susceptibility between pyronaridine **20**, CQ **2** and AQ **3**. The combination of pyronaridine **20** and the artemisinin analogue artesunate (Pyramax) is in clinical development and began Phase III clinical trials in 2006. In terms of safety, pyronaridine-artesunate was well

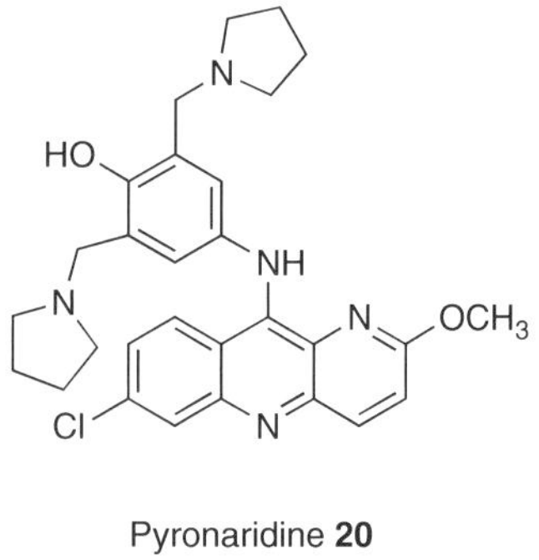

Pyronaridine **20**

Fig. 10 Structure of pyronaridine **20**

tolerated in Phase II trials. However, a few patients exhibited raised liver enzymes, therefore the risk of toxicity to the liver still needs to be closely monitored [106]. Pyramax was submitted to the European Medicines Agency (EMA) for regulatory approval at the end of March 2010 [107].

Isoquine

An approach to circumvent the facile oxidation of AQ involves the interchange of the 3′-hydroxyl and the 4′-Mannich side-chain function of AQ. This provided a new series of analogues that avoid the formation of toxic quinoneimine metabolites via cytochrome P450-mediated metabolism (Fig. 11) [108].

While several analogues displayed potent antimalarial activity against both CQ-sensitive and resistant strains, isoquine **22** (ISQ), the direct isomer of AQ, displayed potent in vitro antimalarial activity in addition to excellent oral in vivo ED_{50} and ED_{90} activity of 1.6 and 3.7 mg/kg, respectively, against the *P. yoelii* NS strain (compared with 7.9 and 7.4 mg/kg for AQ) [109]. Subsequent metabolism studies in the rat model demonstrated that **22** does not undergo in vivo bioactivation, as evidenced by the lack of glutathione metabolites in the bile. Unfortunately, pre-clinical evaluation displayed unacceptably high first pass metabolism to dealkylated metabolites, which complicated the development and compromised activity against CQ-resistant strains [110].

Since the metabolic cleavage of the *N*-diethylamino-group was an issue, the more metabolically stable *N*-*tert*-butyl analogue was developed in the hope that this

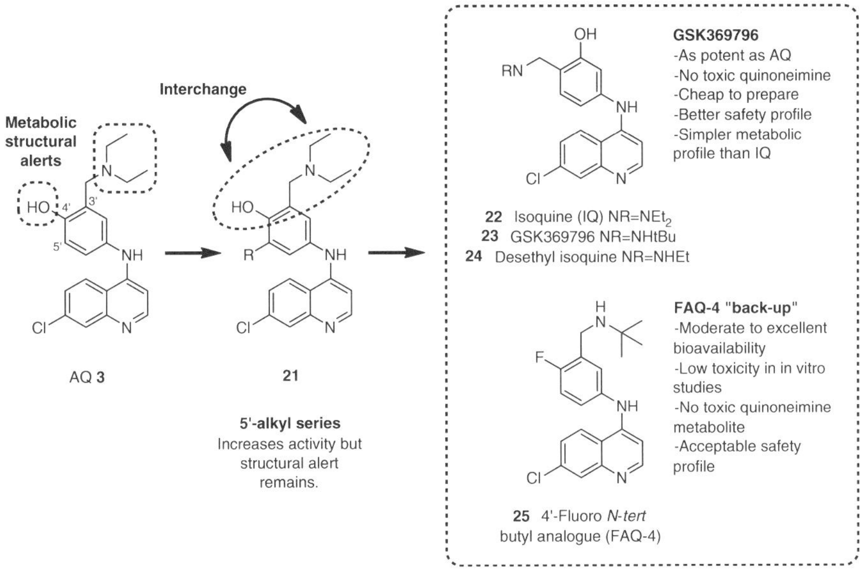

Fig. 11 Modifications of AQ to reduce toxicity of metabolic structural alerts

would lead to a much simpler metabolic profile and enhanced bioavailability. Development of the *N-tert*-butyl analogue **23** (GSK369796) followed (Fig. 11), which has superior pharmacokinetic and pharmacodynamic profiles to isoquine in pre-clinical evaluation studies performed by Glaxo SmithKline pharmaceuticals [110]. In spite of the excellent exposures and near quantitative oral bioavailabilities in animal models, development of **23** has been discontinued due to the inability to achieve exposures at doses considered to demonstrate superior drug safety compared with CQ.

4′-Fluoro-*N-tert*-butylamodiaquine FAQ-4 (**25**) was also identified as a "back-up" candidate for further development studies based on potent activity versus CQ-sensitive and resistant parasites, moderate to excellent oral bioavailability, low toxicity in in vitro studies, and an acceptable safety profile, and this molecule is undergoing formal pre-clinical evaluation [111].

5 The Future of CQ and AQ

5.1 *CQ/AQ Next-Generation Candidates in Clinical Development*

4-Aminoquinoline-based drug development projects continue to yield promising drug candidates and several molecules have entered into pre-clinical development or clinical trials over the last few years. Projects to reduce resistance development of CQ have resulted in the development of short-chain chloroquine analogues (AQ-13), organometallic antimalarials (ferroquine) and a "fusion" trioxaquine antimalarial (SAR116242). Projects to reduce the toxicity of AQ have resulted in the development of metabolically stable amodiaquine analogues (isoquine/*tert*-butyl isoquine) and aza-acridine derivatives (pyronaridine) (Table 1).

5.2 *CQ/AQ Combinations: ACTs and Non-ACTs*

The 4-aminoquinolines CQ and AQ have had a revival over the last 20 years due to the development of ACT. Artesunate-amodiaquine (Coarsucam) was approved for the WHO pre-qualification project in October 2008. It is expected to have a 25% share of the ACT market, with another ACT, Coartem (artemether/lumefantrine) taking the remaining 75% [76].

Methylene blue (MB), a specific inhibitor of *P. falciparum* glutathione reductase was the first synthetic antimalarial drug ever used in the early 1900s. Interest in its use as an antimalarial has recently been revived, due to its potential to reverse CQ resistance and its affordability [112]. It is thought that MB prevents the crystallisation of haem to haemozoin in a similar mechanism as the 4-aminoquinolines.

Table 1 Summary of 4-aminoquinolines entering or in clinical trials, modified and updated from recent reviews [4, 76]

Active ingredients (product name)	Partnership	Phase/status	Strengths	Weakness
Artesunate 50 mg Amodiaquine 135 mg (Coarsucam®)	Sanofi-Aventis, DNDi	Prequalified 2008	• Soluble tablets for paediatric use. • 1 tablet a day – 3 days • WHO prequalified • Three dose strengths • Has 25% of the ACT market	• Resistance to AQ – GI side effects • Not used as prophylactic due to toxic effect of AQ • Reports of resistant strains • No approval yet but WHO prequalified
DHA 10 mg piperaquine 80 mg (Eurartesim™), Artekin, also Duocotexin (fixed dose Holley and Cotect)	Sigma-Tau, MMV, Chongquing, Holley	III	• 1 tablet a day for 3 days • Piperaquine longest half life of all ACTs partners. • Long post-treatment prophylaxic effect • Extensive safety data	• On WHO treatment guidelines but not approved • Long half life of piperaquine could lead to resistance (16.5 days – DHA approximately 0.5 h)
Pyronaridine 60 mg artesunate 20 mg (Pyramax)	Shin Poong, MMV	III	• 1 tablet a day for 3 days • End point achieved in Phase III trials, submitted to EMEA (late 2009) • Clinical data and registration also for *P. vivax*	• Possible hepatotoxicity from pyronaridine – needs to be investigated • Long half life pyronaridine may lead to resistant strains • Paediatric formula in development (2012 release)
Azithromycin 250 mg Chloroquine 150 mg	Pfizer/MMV	III	• Fixed dose combination (four tablets) for prophylactic use during pregnancy • Long post-treatment prophylaxic effect • Extensive safety data • High efficacy in Phase III trials, even in CQ-resistant areas	• Prohibitively expensive for malaria control programmes • Regimen requires partial self-administration • Anti-CQ campaigns in some areas – may be problem with patient compliance
Rbx11160 150 mg Piperaquine 800 mg (Arterolane)	Ranbaxy	II	• No embryotoxicity concern as with artemisinin combinations • Synthetic so costs kept low • Potential activity against artemisinin-resistant strains to be established	• Efficacy concerns (poor activity of Rbx11160as a monotherapy) • As yet no studies in children, or juvenile toxicology data • Phase III India 2009 – no launch until at least 2011

(continued)

Table 1 (continued)

Active ingredients (product name)	Partnership	Phase/ status	Strengths	Weakness
			• Phase III study as a combination planned India 2009	
SSR-97193 (Ferroquine) artesunate	Sanofi-Aventis	II	• Also effective against *P. vivax* chloroquine resistant strains	• Cost of goods for metal based drugs – may be expensive
Methylene blue, chloroquine	Ruprecht-Karls-University, Heidelberg, DSM	II	• Reports of combination with AQ or artesunate planned. • MB/AQ Cost-effective	Methylene blue/chloroquine did not meet WHO criterion of 95% efficacy
AQ-13	Immtech	I	• Similar to CQ in its efficacy and PK	• Very similar structure to CQ-possible parasite could develop resistance very quickly? • AQ-13 exhibits increased clearance compared with CQ therefore higher dose required
N-tert-butyl Isoquine	GSK, MMV	I	• Excellent exposures • Near quantitative bioavailabilites • Superior PK data to ISQ	• *N-tert* discontinued due to problems with inadequate exposure levels • Phase I back-up molecule being evaluated
SAR116242 (Trioxaquine)	Sanofi, Palumed	Preclinical	• Totally synthetic, metabolically stable and cost effective	• Synthetic route produces diastereomers • Molecule has potential to express both established safety concerns of 4-aminoquinolines (narrow TI) and endoperoxides (embryotoxicity, neurotoxicity) requiring careful safety evaluation

MB was entered into clinical trials with CQ as a partner drug but this combination was not sufficiently effective, even at higher doses of MB [113]. More recent trials with AQ or artesunate as a partner drug provided more optimism; MB-artesunate achieved a more rapid clearance of *P. falciparum* parasites than MB–AQ, but MB–AQ displayed the overall highest efficacy. As MB and AQ are both available and affordable, the MB–AQ combination would be an inexpensive non-ACT antimalarial regimen. A larger multi-centre Phase III study is now planned for the near future.

Another non-ACT combination in Phase II clinical trials is azithromycin/chloroquine (AZ/CQ). Azithromycin is a newer member of the family of macrolide antibiotics. This combination has entered Phase III clinical trials and is currently the most promising non-artemisinin-based prophylactic therapy for Intermittent Preventative Treatment in Pregnant Women (IPTp) [76] and a fixed-dose combination tablet of AZ/CQ is being developed specifically for this use. The combination is synergistic against CQ-resistant strains of *P. falciparum* and has already shown efficacy in the treatment of symptomatic malaria in sub-Saharan Africa, an area of high CQ resistance [76]. Both AZ and CQ have demonstrated safety in children and pregnant women over a number of years and azithromycin provides an additional benefit in treating and preventing sexually transmitted diseases [114]. A pivotal study comparing AZ/CQ IPTp with the current adopted therapy sulfadoxine–pyrimethamine IPTp began in October 2010 and is expected to be completed by January 2013 [115].

6 Conclusions

Due to the increasing spread of malaria resistance to drugs such as CQ and AQ, current treatment regimes rely heavily on artemisinin-based therapies. This could lead to an overdependence on artemisinin availability and may influence cost, so it is extremely important that 4-aminoquinoline drug development programmes continue. Costly lessons have been learnt from the loss of sensitivity to one of the most important drugs for malaria treatment and extreme caution is now taken to ensure that with every new antimalarial developed, a partner drug is found and co-administered to reduce the spread of parasite resistance. Increased understanding of 4-aminoquinoline SARs, mechanisms of toxicity and parasite resistance has aided development of what will hopefully be the next generation of 4-aminoquinolines. The future of 4-aminoquinolines relies heavily on strong partnerships between the public health sectors, MMV (Medicines for Malaria Venture) academia and private pharmaceutical/biotechnology companies to yield a continuing pipeline of 4-aminoquinoline candidates, which not only overcome resistance development but also demonstrate increased efficacy compared with CQ. Equally important is a consideration of the safety attributes of this class since the animal toxicities observed in industry standard pre-clinical development of next-generation analogues such as NTB-isoquine (**23**) **8**, in the absence of any prior human experience, might have precluded the further development of any 4-aminoquinoline and indicates limitations of our current pre-clinical testing strategies to accurately predict human risk in malaria treatment [110].

References

1. Phillipson JD, O'Neill MJ (1986) Novel antimalarial drugs from plants? Parasitol Today 2:355–359
2. Jensen M, Mehlhorn H (2009) Seventy-five years of Resochin in the fight against malaria. Parasitol Res 105:609–627
3. Loeb LF, Clarke WM, Coatney GR, Coggeshall LT, Dieuaide FR, Dochez AR (1946) Activity of a new antimalarial agent, Chloroquine (SN 7618). JAMA 130:1069–1070
4. Wells TN, Poll EM (2010) When is enough enough? The need for a robust pipeline of high-quality antimalarials. Discov Med 9:389–398
5. Winstanley PA, Ward SA, Snow RW (2002) Clinical status and implications of antimalarial drug resistance. Microb Infect 4:157–164
6. Foley M, Tilley L (1998) Quinoline antimalarials: mechanisms of action and resistance and prospects for new agents. Pharmacol Ther 79:55–87
7. Egan TJ (2001) Quinoline antimalarials. Expert Opin Ther Patents 11:185–209
8. Tilley L, Loria P, Foley M (2001) Chloroquine and other quinoline antimalarials. In: Rosenthal PJ (ed) Antimalarial chemotherapy: mechanisms of action, resistance and new direction in drug discovery. Humana, Totowa, NJ, pp 87–121
9. Olliaro P (2001) Mode of action and mechanisms of resistance for antimalarial drugs. Pharmacol Ther 89:207–219
10. Surolia N, Padmanaban G (1991) Chloroquine inhibits heme-dependent protein synthesis in *Plasmodium falciparum*. Proc Natl Acad Sci USA 88:4786–4790
11. Ginsburg H, Geary TG (1987) Current concepts and new ideas on the mechanism of action of quinoline-containing antimalarials. Biochem Pharmacol 36:1567–1576
12. Vander Jagt DL, Hunsaker LA, Campos NM (1986) Characterization of a hemoglobin-degrading, low molecular weight protease from *Plasmodium falciparum*. Mol Biochem Parasitol 18:389–400
13. Cohen SN, Yielding KL (1965) Inhibition of DNA and RNA polymerase reactions by chloroquine. Proc Natl Acad Sci USA 54:521–527
14. Meshnick SR (1990) Chloroquine as intercalator: a hypothesis revived. Parasitol Today 6:77–79
15. Peters W (1970) Chemotherapy and drug resistance in malaria. Academic, London
16. Egan TJ (2008) Recent advances in understanding the mechanism of hemozoin (malaria pigment) formation. J Inorg Biochem 102:1288–1299
17. Slater AF, Cerami A (1992) Inhibition by chloroquine of a novel haem polymerase enzyme activity in malaria trophozoites. Nature 355:167–169
18. Egan TJ, Ross DC, Adams PA (1994) Quinoline anti-malarial drugs inhibit spontaneous formation of beta-haematin (malaria pigment). FEBS Lett 352:54–57
19. Raynes K, Foley M, Tilley L, Deady LW (1996) Novel bisquinoline antimalarials. Synthesis, antimalarial activity, and inhibition of haem polymerisation. Biochem Pharmacol 52:551–559
20. Adams PA, Berman PA, Egan TJ, Marsh PJ, Silver J (1996) The iron environment in heme and heme-antimalarial complexes of pharmacological interest. J Inorg Biochem 63:69–77
21. Egan TJ, Mavuso WW, Ross DC, Marques HM (1997) Thermodynamic factors controlling the interaction of quinoline antimalarial drugs with ferriprotoporphyrin IX. J Inorg Biochem 68:137–145

22. Egan TJ, Helder MM (1999) The role of haem in the activity of chloroquine and related antimalarial drugs. Coord Chem Rev 190–192:493–517
23. Vippagunta SR, Dorn A, Matile H, Bhattacharjee AK, Karle JM, Ellis WY, Ridley RG, Vennerstrom JL (1999) Structural specificity of chloroquine-hematin binding related to inhibition of hematin polymerization and parasite growth. J Med Chem 42:4630–4639
24. Dorn A, Vippagunta SR, Matile H, Jaquet C, Vennerstrom JL, Ridley RG (1998) An assessment of drug-haematin binding as a mechanism for inhibition of haematin polymerisation by quinoline antimalarials. Biochem Pharmacol 55:727–736
25. Sullivan DJ, Gluzman IY, Russell DG, Goldberg DE (1996) On the molecular mechanism of chloroquine's antimalarial action. Proc Natl Acad Sci USA 93:11865–11870
26. Buller R, Peterson ML, Almarsson O, Leiserowitz L (2002) Quinoline binding site on malaria pigment crystal: a rational pathway for antimalaria drug design. Cryst Growth Des 2:553–562
27. Hawley SR, Bray PG, Park BK, Ward SA (1996) Amodiaquine accumulation in *Plasmodium falciparum* as a possible explanation for its superior antimalarial activity over chloroquine. Mol Biochem Parasitol 80:15–25
28. Geary TG, Divo AD, Jensen JB, Zangwill M, Ginsburg H (1990) Kinetic modelling of the response of *Plasmodium falciparum* to chloroquine and its experimental testing *in vitro*. Implications for mechanism of action of and resistance to the drug. Biochem Pharmacol 40:685–691
29. Ferrari V, Cutler DJ (1991) Simulation of kinetic data on the influx and efflux of chloroquine by erythrocytes infected with *Plasmodium falciparum*. Evidence for a drug-importer in chloroquine-sensitive strains. Biochem Pharmacol 42(Suppl):S167–179
30. Ferrari V, Cutler DJ (1991) Kinetics and thermodynamics of chloroquine and hydroxychloroquine transport across the human erythrocyte membrane. Biochem Pharmacol 41:23–30
31. Sanchez CP, Wunsch S, Lanzer M (1997) Identification of a chloroquine importer in *Plasmodium falciparum*. Differences in import kinetics are genetically linked with the chloroquine-resistant phenotype. J Biol Chem 272:2652–2658
32. Chou AC, Chevli R, Fitch CD (1980) Ferriprotoporphyrin IX fulfills the criteria for identification as the chloroquine receptor of malaria parasites. Biochemistry 19:1543–1549
33. Bray PG, Janneh O, Raynes KJ, Mungthin M, Ginsburg H, Ward SA (1999) Cellular uptake of chloroquine is dependent on binding to ferriprotoporphyrin IX and is independent of NHE activity in *Plasmodium falciparum*. J Cell Biol 145:363–376
34. D'Alessandro U, Buttiëns H (2001) History and importance of antimalarial drug resistance. Trop Med Int Health 6:845–848
35. Bray PG, Mungthin M, Ridley RG, Ward SA (1998) Access to hematin: the basis of chloroquine resistance. Mol Pharmacol 54:170–179
36. Sidhu ABS, Verdier-Pinard D, Fidock DA (2002) Chloroquine resistance in *Plasmodium falciparum* malaria parasites conferred by *pfcrt* mutations. Science 298:210–213
37. Fidock DA, Nomura T, Talley AK, Cooper RA, Dzekunov SM, Ferdig MT, Ursos LMB, Sidhu ABS, Naude B, Deitsch KW (2000) Mutations in the *P. falciparum* digestive vacuole transmembrane protein PfCRT and evidence for their role in chloroquine resistance. Mol Cell 6:861–871
38. Krogstad DJ, Gluzman IY, Kyle DE, Oduola AMJ, Martin SK, Milhous WK, Schlesinger PH (1987) Efflux of chloroquine from *Plasmodium falciparum* – mechanism of chloroquine resistance. Science 238:1283–1285
39. Hayward R, Saliba KJ, Kirk K (2006) The pH of the digestive vacuole of *Plasmodium falciparum* is not associated with chloroquine resistance. J Cell Sci 119:1016–1025
40. Bray PG, Mungthin M, Hastings IM, Biagini GA, Saidu DK, Lakshmanan V, Johnson DJ, Hughes RH, Stocks PA, O'Neill PM (2006) PfCRT and the trans-vacuolar proton electrochemical gradient: regulating the access of chloroquine to ferriprotoporphyrin IX. Mol Microbiol 62:238–251

41. Warhurst DC, Craig JC, Adagu IS (2002) Lysosomes and drug resistance in malaria. Lancet 360:1527–1529
42. Sanchez CP, Stein WD, Lanzer M (2007) Is PfCRT a channel or a carrier? Two competing models explaining chloroquine resistance in *Plasmodium falciparum*. Trends Parasitol 23:332–339
43. Martin RE, Marchetti RV, Cowan AI, Howitt SM, Broer S, Kirk K (2009) Chloroquine transport via the malaria parasite's chloroquine resistance transporter. Science 325:1680–1682
44. Sanchez CP, Dave A, Stein WD, Lanzer M (2010) Transporters as mediators of drug resistance in *Plasmodium falciparum*. Int J Parasitol 40:1109–1118
45. van Es HH, Karcz S, Chu F, Cowman AF, Vidal S, Gros P, Schurr E (1994) Expression of the plasmodial *pfmdr1* gene in mammalian cells is associated with increased susceptibility to chloroquine. Mol Cell Biol 14:2419–2428
46. Sanchez CP, Rotmann A, Stein WD, Lanzer M (2008) Polymorphisms within PfMDR1 alter the substrate specificity for anti-malarial drugs in *Plasmodium falciparum*. Mol Microbiol 70:786–798
47. WHO (2010) Guidelines for the treatment of malaria, 2nd edn. WHO (World Health Organization), Geneva
48. Laufer MK, Thesing PC, Eddington ND, Masonga R, Dzinjalamala FK, Takala SL, Taylor TE, Plowe CV (2006) Return of chloroquine antimalarial efficacy in Malawi. New Engl J Med 355:1959–1966
49. Djimde A, Doumbo OK, Cortese JF, Kayentao K, Doumbo S, Diourte Y, Dicko A, Su XZ, Nomura T, Fidock DA et al (2001) A molecular marker for chloroquine-resistant falciparum malaria. New Engl J Med 344:257–263
50. Read AF, Huijben S (2009) Evolutionary biology and the avoidance of antimicrobial resistance. Evol Appl 2:40–51
51. Ursing J, Kofoed PE, Rodrigues A, Blessborn D, Thoft-Nielsen R, Bjorkman A, Rombo L (2011) Similar efficacy and tolerability of double-dose chloroquine and artemether-lumefantrine for treatment of *Plasmodium falciparum* infection in guinea-bissau: a randomized trial. J Infect Dis 203:109–116
52. Ursing J, Rombo L, Kofoed PE, Gil JP (2008) Carriers, channels and chloroquine efficacy in Guinea-Bissau. Trends Parasitol 24:49–51
53. Kofoed PE, Ursing J, Poulsen A, Rodrigues A, Bergquist Y, Aaby P, Rombo L (2007) Different doses of amodiaquine and chloroquine for treatment of uncomplicated malaria in children in Guinea-Bissau: implications for future treatment recommendations. Trans R Soc Trop Med Hyg 101:231–238
54. Hand CC, Meshnick SR (2011) Is chloroquine making a comeback? J Infect Dis 203:11–12
55. Ursing J, Schmidt BA, Lebbad M, Kofoed PE, Dias F, Gil JP, Rombo L (2007) Chloroquine resistant *P.falciparum* prevalence is low and unchanged between 1990 and 2005 in Guinea-Bissau: an effect of high chloroquine dosage? Infect Genet Evol 7:555–561
56. Kaur K, Jain M, Reddy RP, Jain R (2010) Quinolines and structurally related heterocycles as antimalarials. Eur J Med Chem 45:3245–3264
57. De D, Krogstad FM, Cogswell FB, Krogstad DJ (1996) Aminoquinolines that circumvent resistance in *Plasmodium falciparum in vitro*. Am J Trop Med Hyg 55:579–583
58. Ridley RG, Hofheinz W, Matile H, Jaquet C, Dorn A, Masciadri R, Jolidon S, Richter WF, Guenzi A, Girometta MA (1996) 4-aminoquinoline analogs of chloroquine with shortened side chains retain activity against chloroquine-resistant *Plasmodium falciparum*. Antimicrob Agents Chemother 40:1846–1854
59. Ramanathan-Girish S, Catz P, Creek MR, Wu B, Thomas D, Krogstad DJ, De D, Mirsalis JC, Green CE (2004) Pharmacokinetics of the antimalarial drug, AQ-13, in rats and cynomolgus Macaques. Int J Toxicol 23:179–189

60. Mzayek F, Deng H, Mather FJ, Wasilevich EC, Liu H, Hadi CM, Chansolme DH, Murphy HA, Melek BH, Tenaglia AN (2007) Randomized dose-ranging controlled trial of AQ-13, a candidate antimalarial, and chloroquine in healthy volunteers. PLoS Clin Trials 2:e6
61. Mzayek F, Deng HY, Hadi MA, Mave V, Mather FJ, Goodenough C, Mushatt DM, Lertora JJ, Krogstad D (2009) Randomized clinical trial (RCT) with a crossover study design to examine the safety and pharmacokinetics of a 2100 mg dose of AQ-13 and the effects of a standard fatty meal on its bioavailability. Am J Trop Med Hyg 81:S252
62. De D, Krogstad FM, Byers LD, Krogstad DJ (1998) Structure-activity relationships for antiplasmodial activity among 7-substituted 4-aminoquinolines. J Med Chem 41:4918–4926
63. Ward SA, Bray PG, Hawley SR, Mungthin M (1996) Physicochemical properties correlated with drug resistance and the reversal of drug resistance in *Plasmodium falciparum*. Mol Pharmacol 50:1559–1566
64. Farrel N (1989) Transition metal complexes as drugs and chemotherapeutic agents. Kluwer Academic, Dordrecht
65. Sanchez-Delgado RA, Navarro M, Perez H, Urbina JA (1996) Toward a novel metal-based chemotherapy against tropical diseases. 2. Synthesis and antimalarial activity *in vitro* and *in vivo* of new ruthenium- and rhodium-chloroquine complexes. J Med Chem 39:1095–1099
66. Sanchez-Delgado RA, Navarro M, Perez H (1997) Toward a novel metal-based chemotherapy against tropical diseases. 3. Synthesis and antimalarial activity *in vitro* and *in vivo* of the new gold-chloroquine complex [Au(PPh3)(CQ)]PF6. J Med Chem 40:1937–1939
67. Biot C, Glorian G, Maciejewski LA, Brocard JS, Domarle O, Blampain G, Millet P, Georges AJ, Abessolo H, Dive D (1997) Synthesis and antimalarial activity *in vitro* and *in vivo* of a new ferrocene-chloroquine analogue. J Med Chem 40:3715–3718
68. Biot C, Delhaes L, N'Diaye CM, Maciejewski LA, Camus D, Dive D, Brocard JS (1999) Synthesis and antimalarial activity *in vitro* of potential metabolites of ferrochloroquine and related compounds. Biorg Med Chem 7:2843–2847
69. Dubar F, Khalife J, Brocard J, Dive D, Biot C (2008) Ferroquine, an ingenious antimalarial drug – thoughts on the mechanism of action. Molecules 13:2900–2907
70. Barends M, Jaidee A, Khaohirun N, Singhasivanon P, Nosten F (2007) *In vitro* activity of ferroquine (SSR 97193) against *Plasmodium falciparum* isolates from the Thai-Burmese border. Malar J 6:81
71. Leimanis ML, Jaidee A, Sriprawat K, Kaewpongsri S, Suwanarusk R, Barends M, Phyo AP, Russell B, Renia L, Nosten F (2010) *Plasmodium vivax* susceptibility to ferroquine. Antimicrob Agents Chemother 54:2228–2230
72. Sanofi-Aventis (2000) Dose ranging study of ferroquine with artesunate in african adults and children with uncomplicated *Plasmodium falciparum* malaria (FARM). In: ClinicalTrials.gov [Internet]. National Library of Medicine (US), Bethesda (MD). http://clinicaltrials.gov/ct2/show/NCT00988507. Accessed 23 May 2011. NLM Identifier: NCT00988507
73. Davis TME, Hung TY, Sim IK, Karunajeewa HA, Ilett KF (2005) Piperaquine – a resurgent antimalarial drug. Drugs 65:75–87
74. Hung TY, Davis TME, Ilett KF, Karunajeewa H, Hewitt S, Denis MB, Lim C, Socheat D (2004) Population pharmacokinetics of piperaquine in adults and children with uncomplicated falciparum or vivax malaria. Br J Clin Pharmacol 57:253–262
75. Hien TT, Dolecek C, Mai PP, Dung NT, Truong NT, Thai LH, An DTH, Thanh TT, Stepniewska K, White NJ (2004) Dihydroartemisinin-piperaquine against multidrug-resistant *Plasmodium falciparum* malaria in Vietnam: randomised clinical trial. Lancet 363:18–22
76. Olliaro P, Wells TNC (2009) The global portfolio of new antimalarial medicines under development. Clin Pharmacol Ther 85:584–595
77. Zwang J, Ashley EA, Karema C, D'Alessandro U, Smithuis F, Dorsey G, Janssens B, Mayxay M, Newton P, Singhasivanon P (2009) Safety and efficacy of dihydroartemisinin-piperaquine in falciparum malaria: a prospective multi-centre individual patient data analysis. PLoS ONE 4:e6358

78. Price RN, Hasugian AR, Ratcliff A, Siswantoro H, Purba HLE, Kenangalem E, Lindegardh N, Penttinen P, Laihad F, Ebsworth EP (2007) Clinical and pharmacological determinants of the therapeutic response to dihydroartemisinin-piperaquine for drug-resistant malaria. Antimicrob Agents Chemother 51:4090–4097
79. Khanh NX, de Vries PJ, Ha LD, van Boxtel CJ, Koopmans R, Kager PA (1999) Declining concentrations of dihydroartemisinin in plasma during 5-day oral treatment with artesunate for falciparum malaria. Antimicrob Agents Chemother 43:690–692
80. Charman SA (2007) Synthetic peroxides: a viable alternative to artemisinins for the treatment of uncomplicated malaria? In: American Society of Tropical Medicine and Hygiene (ASTMH) 56th Annual Meeting, Philadelphia, Pennsylvania, USA, 4–8 Nov 2007
81. Snyder C, Chollet J, Santo-Tomas J, Scheurer C, Wittlin S (2007) *In vitro* and *in vivo* interaction of synthetic peroxide RBx11160 (OZ277) with piperaquine in *Plasmodium* models. Exp Parasitol 115:296–300
82. White NJ (2008) Qinghaosu (Artemisinin): the price of success. Science 320:330–334
83. Meunier B (2008) Hybrid molecules with a dual mode of action: dream or reality? Acc Chem Res 41:69–77
84. Muregi FW, Ishih A (2010) Next-generation antimalarial drugs: hybrid molecules as a new strategy in drug design. Drug Dev Res 71:20–32
85. Benoit-Vical F, Lelievre J, Berry A, Deymier C, Dechy-Cabaret O, Cazelles J, Loup C, Robert A, Magnaval JF, Meunier B (2007) Trioxaquines are new antimalarial agents active on all erythrocytic forms, including gametocytes. Antimicrob Agents Chemother 51:1463–1472
86. Cosledan F, Fraisse L, Pellet A, Guillou F, Mordmuller B, Kremsner PG, Moreno A, Mazier D, Maffrand JP, Meunier B (2008) Selection of a trioxaquine as an antimalarial drug candidate. Proc Natl Acad Sci USA 105:17579–17584
87. O'Neill PM, Bray PG, Hawley SR, Ward SA, Park BK (1998) 4-aminoquinolines – past, present, and future: a chemical perspective. Pharmacol Ther 77:29–58
88. Daily EB, Aquilante CL (2009) Cytochrome P450 2 C8 pharmacogenetics: a review of clinical studies. Pharmacogenomics 10:1489–1510
89. Fu S, Bjorkman A, Wahlin B, Ofori-Adjei D, Ericsson O, Sjoqvist F (1986) *In vitro* activity of chloroquine, the two enantiomers of chloroquine, desethylchloroquine and pyronaridine against *Plasmodium falciparum*. Br J Clin Pharmacol 22:93–96
90. White NJ, Looareesuwan S, Edwards G, Phillips RE, Karbwang J, Nicholl DD, Bunch C, Warrell DA (1987) Pharmacokinetics of intravenous amodiaquine. Br J Clin Pharmacol 23:127–135
91. Jewell H, Maggs JL, Harrison AC, O'Neill PM, Ruscoe JE, Park BK (1995) Role of hepatic metabolism in the bioactivation and detoxication of amodiaquine. Xenobiotica 25:199–217
92. Jewell H, Ruscoe JE, Maggs JL, O'Neill PM, Storr RC, Ward SA, Park BK (1995) The effect of chemical substitution on the metabolic activation, metabolic detoxication, and pharmacological activity of amodiaquine in the mouse. J Pharmacol Exp Ther 273:393–404
93. Clarke JB, Neftel K, Kitteringham NR, Park BK (1991) Detection of antidrug IgG antibodies in patients with adverse drug reactions to amodiaquine. Int Arch Allergy Appl Immunol 95:369–375
94. Churchill FC, Mount DL, Patchen LC, Bjorkman A (1986) Isolation, characterization and standardization of a major metabolite of amodiaquine by chromatographic and spectroscopic methods. J Chromatogr B 377:307–318
95. Laurent F, Saivin S, Chretien P, Magnaval JF, Peyron F, Sqalli A, Tufenkji AE, Coulais Y, Baba H, Campistron G et al (1993) Pharmacokinetic and pharmacodynamic study of amodiaquine and its two metabolites after a single oral dose in human volunteers. Arzneim-Forsch 43:612–616
96. Li XQ, Bjorkman A, Andersson TB, Ridderstrom M, Masimirembwa CM (2002) Amodiaquine clearance and its metabolism to N-desethylamodiaquine is mediated by

CYP2C8: a new high affinity and turnover enzyme-specific probe substrate. J Pharmacol Exp Ther 300:399–407

97. O'Neill PM, Harrison AC, Storr RC, Hawley SR, Ward SA, Park BK (1994) The effect of fluorine substitution on the metabolism and antimalarial activity of amodiaquine. J Med Chem 37:1362–1370
98. O'Neill PM, Willock DJ, Hawley SR, Bray PG, Storr RC, Ward SA, Park BK (1997) Synthesis, antimalarial activity, and molecular modeling of tebuquine analogues. J Med Chem 40:437–448
99. Barlin GB, Ireland SJ, Nguyen TMT, Kotecka B, Rieckmann KH (1994) Potential antimalarials. XXI. Mannich base derivatives of 4-[7-Chloro(and 7-trifluoromethyl) quinolin-4-ylamino]phenols. Aust J Chem 47:1553–1560
100. Peters W, Robinson BL (1992) The chemotherapy of rodent malaria. XLVII. Studies on pyronaridine and other Mannich base antimalarials. Ann Trop Med Parasitol 86:455–465
101. Ward SA, Hawley SR, Bray PG, O'Neill PM, Naisbitt DJ, Park BK (1996) Manipulation of the N-alkyl substituent in amodiaquine to overcome the verapamil-sensitive chloroquine resistance component. Antimicrob Agents Chemother 40:2345–2349
102. Tingle MD, Ruscoe JE, O'Neill PM, Ward SA, Park BK (1998) Effect of disposition of Mannich antimalarial agents on their pharmacology and toxicology. Antimicrob Agents Chemother 42:2410–2416
103. Biagini GA, O'Neill PM, Bray PG, Ward SA (2005) Current drug development portfolio for antimalarial therapies. Curr Opin Pharmacol 5:473–478
104. Looareesuwan S, Kyle DE, Viravan C, Vanijanonta S, Wilairatana P, Wernsdorfer WH (1996) Clinical study of pyronaridine for the treatment of acute uncomplicated falciparum malaria in Thailand. Am J Trop Med Hyg 54:205–209
105. Pradines B, Mabika Mamfoumbi M, Parzy D, Owono Medang M, Lebeau C, Mourou Mbina JR, Doury JC, Kombila M (1999) *In vitro* susceptibility of African isolates of *Plasmodium falciparum* from Gabon to pyronaridine. Am J Trop Med Hyg 60:105–108
106. Nosten FH (2010) Pyronaridine-artesunate for uncomplicated falciparum malaria. Lancet 375:1413–1414
107. Medicines for Malaria Venture. Pyramax dossier submitted to EMA. http://www.mmv.org/achievements-challenges/achievements/pyramax%C2%AE-dossier-submitted-ema?page=0. Accessed 24 May 2011
108. O'Neill PM, Mukhtar A, Stocks PA, Randle LE, Hindley S, Ward SA, Storr RC, Bickley JF, O'Neil IA, Maggs JL (2003) Isoquine and related amodiaquine analogues: a new generation of improved 4-aminoquinoline antimalarials. J Med Chem 46:4933–4945
109. Delarue S, Girault S, Maes L, Debreu-Fontaine MA, Labaeid M, Grellier P, Sergheraert C (2001) Synthesis and *in vitro* and *in vivo* antimalarial activity of new 4-anilinoquinolines. J Med Chem 44:2827–2833
110. O'Neill PM, Park BK, Shone AE, Maggs JL, Roberts P, Stocks PA, Biagini GA, Bray PG, Gibbons P, Berry N (2009) Candidate selection and preclinical evaluation of N-tert-Butyl isoquine (GSK369796), an affordable and effective 4-Aminoquinoline antimalarial for the 21st century. J Med Chem 52:1408–1415
111. O'Neill PM, Shone AE, Stanford D, Nixon G, Asadollahy E, Park BK, Maggs JL, Roberts P, Stocks PA, Biagini G (2009) Synthesis, antimalarial activity, and preclinical pharmacology of a novel series of 4-Fluoro and 4-Chloro analogues of amodiaquine. Identification of a suitable "back-up" compound for N-tert-butyl isoquine. J Med Chem 52:1828–1844
112. Schirmer RH, Coulibaly B, Stich A, Scheiwein M, Merkle H, Eubel J, Becker K, Becher H, Müller O, Zich T (2003) Methylene blue as an antimalarial agent. Redox Rep 8:272–275
113. Meissner PE, Mandi G, Coulibaly B, Witte S, Tapsoba T, Mansmann U, Rengelshausen J, Schiek W, Jahn A, Walter-Sack I (2006) Methylene blue for malaria in Africa: results from a dose-finding study in combination with chloroquine. Malar J 5:84
114. Chico RM, Pittrof R, Greenwood B, Chandramohan D (2008) Azithromycin-chloroquine and the intermittent preventive treatment of malaria in pregnancy. Malar J 7:255

115. Pfizer (2000) Evaluate azithromycin plus chloroquine and sulfadoxine plus pyrimethamine combinations for intermittent preventive treatment of falciparum malaria infection in pregnant women In Africa. In: ClinicalTrials.gov [Internet]. National Library of Medicine (US), Bethesda (MD). http://clinicaltrials.gov/ct2/show/NCT01103063. Accessed 2011 May 23. NLM Identifier: NCT01103063

Cinchona Alkaloids: Quinine and Quinidine

David J. Sullivan

Abstract For 400 years, quinine has been the effective antimalarial. From a pulverized bark, which stopped cyclic fevers, to an easily isolated crystal alkaloid, which launched many pharmaceutical companies, tons of quinine are still purified for medicinal and beverage use. The quest for quinine synthesis pioneered early medicinal dyes, antibacterials, and other drugs. In a specialized *Plasmodium* lysosome for hemoglobin degradation, quinine binds heme, which inhibits heme crystallization to kill rapidly. Although quinine drug resistance was described 100 years ago, unlike chloroquinine or the antifolates that have been rendered ineffective by the spread of resistant mutants, quinine has only a few persistent, resistant parasites worldwide. The artemisinin drugs, superior to quinine for severe malaria, have greatly reduced the use of quinine as an antimalarial. Evidence for prolonged artemisinin parasite clearance times both renews the quest for rapidly parasiticidal drugs for severe malaria and possibly holds a place for quinine.

1 Early History of Quinine

Cinchona bark extracts were identified as early as 1632 to be effective in treating fevers, particularly the tertian fever of malaria [1]. This specific symptomatic management preceded the identification of the etiologic organism of malaria by almost 250 years. Many bacterial organisms were not discovered until the use of microscopy and dyes to contrast them in the late 1800s. Laveran identified the malaria parasite in Algerian soldiers in the 1880s without dyes [2]. His first description was of an exflagellating male gametocyte, later followed by observations of

D.J. Sullivan (✉)
W. Harry Feinstone Department of Molecular Microbiology and Immunology, The Malaria Research Institute, The Johns Hopkins Bloomberg School of Public Health, 615 N. Wolfe St. Rm E5628, Baltimore, MD 21205, USA
e-mail: dsulliva@jhsph.edu

H.M. Staines and S. Krishna (eds.), *Treatment and Prevention of Malaria*,
DOI 10.1007/978-3-0346-0480-2_3,

intraerythrocytic forms containing the malaria pigment hemozoin. Later, Ehrlich in 1891 with the use of methylene blue was able to speciate and treat human *Plasmodium* and put to rest initial doubts of Laveran's descriptions [3, 4].

Despite historical accounts of the Jesuits in Peru curing the Countess of Chinchón with local bark, the countess actually died of yellow fever en route to Europe [5]. There are three other plausible historical accounts of the discovery of the curative properties of Peruvian (Cinchona) bark. One is of Peruvian Indians drinking bark teas to stop shivering while mining in the mountains. Another account relates Indians drinking the bark tea to stop shivering after crossing cold mountain rivers [6, 7]. PCC Garnham relates a third story in the region of Loxa after an earthquake felled cinchona trees into a lake, which grew brown with the bark. This water was curative of malarial fevers [8]. We do know that shipments of bark arrived in Spain in 1636. Soon cinchona bark became known as Jesuit's bark or the Cardinal's bark [5]. In contrast to some historical accounts, Oliver Cromwell may not have died from malaria after refusing to take the Catholic bark, but from septicemia related to kidney stones [9]. The Peruvian bark was much sought after and soon was worth its weight in gold. Other barks were also sampled for fevers. An English physician by the name of Stone described the curative properties of willow bark from which salicylates were later identified [10]. This was the first of many of our present drugs, which trace an origin to quinine [5, 11].

2 Quinine Extraction

There are approximately 90 varieties of Cinchona, a red bark tree, which is included in the madder family of which coffee and gardenias are also members. Different barks of cinchona contain 5–10% quinine [12]. Approximately 1–2 g of pulverized bark, seeped in boiling water for 10 min will yield about 100 mg of active ingredient. In 1820, Pelletier and Caventou identified a straightforward process for crystal purification of the active quinine salt [13]. Pulverized bark was soaked in water and crystal formed from acidified water due to the poor solubility of quinine crystals relative to the other alkaloids in the bark. Unlike others, they made their process freely available. The black residue leftover was not active but the four active alkaloids extracted were quinine and quinidine, cinchoine and cinchoinidine. Much later, Gammie reported a process from India where 100 parts of pulverized powder was mixed with 8 parts caustic soda, 500 parts water and 600 parts fuel and kerosene oil [14]. The alkaloids were absorbed by the oil; the oil was transferred to heated acid water, and then cooled for crystallization. Sulfuric acid was used to make quinine sulfate and hydrochloric acid for quinine hydrochlorate. Even today the process requires 500–1,000 metric tons of bark to produce hundreds of tons of quinine. About half of current twenty-first century production is used for beverages and the rest for medicines. Many pharmaceutical companies such as Boehringer Mannheim (which had a cinchona tree on its original logo) originated by extracting and providing quinine [15]. In addition to malaria, quinine was used in many

Table 1 List of early US Federal Drug Association approvals

Drug	Identifier	NDA number	Company	Date approved	Withdrawn	Trade name	Generic
Oxalic acid	N000001001	000001	John A Millar Co Div Chatham Pharm	9/15/38	2/5/71	Koagamin Inj	Oxalic acid
Quinine	N000027001	000027	1st Texas Pharmaceuticals Inc Sub Scherer Laboratories Inc	9/19/38	10/20/80	Private Formula	Arsenic trioxide 0.5 gr; Strychnine sulfate 0.5 gr; Iron, Reduced 15 gr; Cascara 15 gr; Quinine hydrobromide 30 gr
Quinine	N000077001	000077	Mendez Angel M	7/31/39	7/24/70	Quinarsine Inj	Quinine formate 5 g; Sodium cacodylate 2 g; Urethane 4 g; Amylocaine hydrochloride 1 g
Quinine	N000227003	000227	Modern Drugs Inc	5/2/40	7/24/70	No-Ko Tab	Quinine sulfate 0.75 gr; Belladonna extract 0.05 gr; Camphor 0.25 gr; Capsicum oleoresin 0.25 gr; Sodium salicylate 3 gr
Quinine	N000463001	000463	Portia Laboratories	2/20/39	7/24/70	Mal-Caps	Quinine sulfate 2 gr; Sodium bicarbonate 5 gr
Quinine	N000478001	000478	Joseph Personeni Inc	3/1/39	7/24/70	Treponyl Inj	Quinine bismuth iodide 0.1 gr/Ml
Quinine	N000574001	000574	Sr Seaver and Co	4/8/39	7/24/70	Ryzalen Tab	Quinine sulfate 0.25 gr; Ammonium chloride 0.5 gr; Camphor 0.5 gr; Belladonna extract 0.05 gr; Aconite 0.10 gr; Lime, Iodized 0.33 gr; Phenolphthalein 0.33 gr
Quinine	N001069001	001069	Hoysalb and Co Inc	8/30/39	7/24/70	Kenene Sol	Quinine sulfate 8 gr; Strychnine sulfate 8/64 gr; Ferric citrochloride tincture 24 min/Fl Oz
Quinine	N001180001	001180	1st Texas Pharmaceuticals Inc Sub Scherer Laboratories Inc	7/15/39	8/6/71	Private Formula Cap	Quinine sulfate 2 gr; Bismuth subnitrate 0.1 gr; Charcoal, activated 13/7 gr; Calomel 0.1 gr; Pepsin 4/7 gr
Quinine	N001431001	001431	Standard Products Co	8/25/39	7/24/70	Tonik-Tyme Elx	Quinine phosphate 2 gr/Oz; Pepsin 420 min/Oz; Ferric phosphate 16 gr/Oz; Gentian 5 gr/Oz
Quinine	N001504001	001504	Verard Co	8/30/39	7/24/70	Verard Sol	Quinine sulfate Unk; Hydrogen peroxide Unk
Quinidine	N001975001	001975	Sutliff And Case Co Inc	12/27/39	7/24/70	Quinidine Sul Tab	Quinidine sulfate 3 gr
Chloroquine	N006002002	006002	Sanofi Synthelabo			Aralen Hcl	Chloroquine hydrochloride Eq 40 mg Base/Ml
Primaquine	N008151001	008151	Eli Lilly and Co	8/10/51	8/6/71	Primaquine Tab	Primaquine phosphate 15 mg

grain (gr)= 0.06 grams (g)

remedies for headache, nausea, or as part of a mixture for a general cure-all or analgesic. In the United States, the Federal Drug Association approved quinine in combination with arsenic, amylocaine, belladonna, strychnine, bismuth and iron as part of early twentieth-century popular remedies, as seen in Table 1. The first Federal Drug Association approval for malaria was Mal-Caps in February 1939 by Portia Laboratories with the ingredients quinine sulfate and sodium bicarbonate.

3 Synthesis

The quest for synthesis was long a holy grail of medicinal chemistry, producing many useful drugs and companies along the way. Adolph Strecker in Oslo determined the chemical formula for quinine: $–C_{20}H_{24}N_2O_2$ [16]. William Henry Perkins was trying to synthesize quinine from the oxidation of allyltoluidine ($C_{10}H_{12}N$) and instead produced the color mauve, which was colorfast [17]. This was the first synthesis of a color dye rather than being a plant-derived one. The colored synthetic dye textile industry was begun [5]. Paul Ehrlich is credited with the "magic bullet" concept of specific drugs for microbes, based largely on dyes. Suramin was a dye discovered to be curative for African sleeping sickness [18] and was followed by the synthesis of methylene blue successful for both malaria and African sleeping sickness [19]. Many antipsychotic and antidepressant drugs were discovered after experiments with side-group substitutions on the methylene blue scaffold [20]. In a dye-based rational drug discovery process, the sulfur group was incorporated into Prontosil [21]. This was fortunately first tested in animals where the active sulfa group was cleaved to have an antibacterial effect. This sulfur-based scaffold was the basis for many antibacterial and anticancer therapeutics. The potent antibacterial quinolones were produced during later experiments with chloroquine synthesis [22]. Nalixic acid was not very bioavailable or potent, but addition of the fluorines greatly improved the antibacterial properties of this therapeutic class [23], which also incidentally has weak antimalarial activity first described by Krishna and colleagues with ciprofloxacin [24], which later led to more investigation of the fluoroquinolones [25].

The true synthesis of quinine is a story of deconstruction and reconstruction. Pasteur by acid catalysis and isomerization of quinine, produced *d*-quinotoxine [26]. Rabe and Kindler (1903–1918) later described the synthesis of *d*-quinotoxine back to quinine in a short communication [27]. Prelog and Prostenik first degraded quinotoxine to homomeroquinene, then described the resynthesis of quinotoxine [26, 27]. Woodward and colleagues start from 7-hydroxyisoquinoline and publish the synthesis to homomeroquinene [28]. This completed the synthesis but without any stereochemical resolution. Quinine has four stereogenic sites with a possible 16 different configurations, of which quinine is one. Woodward's student Gilbert Stork completed the full stereochemical synthesis in sufficient yield in 2001 [27]. This was one of the major challenges in medicinal chemistry [26].

4 Mechanism of Action

Quinine and quinidine act rapidly on blood-stage parasites [29]. They are not active against liver stage and gametocytes at pharmacologic relevant doses [30, 31]. Importantly, while the curative properties have saved billions of people, the lack of activity against the gametocytes infective for mosquitoes does not halt the transmission cycle. A present day hope for the artemisinin class of drugs is its relative, but not absolute, potency against the gametocytes [30]. Artemisinin lowers a day-seven gametocyte count, but its population effect on lowering transmission remains to be proven. Despite this deficiency, quinine was successful in mass quinization campaigns, which conquered malaria in temperate Italy at the start of the twentieth century [32].

Early morphologic effects of quinine show digestive vacuolar swelling and pigment clumping [33]. Krishna and colleagues delineated the effects to stages actively degrading hemoglobin from late ring to early schizont stages [29]. Paul Roepe defined high concentrations of quinolines for very short intervals as also being active [34]. Many have postulated additional mechanisms of action for quinine against blood-stage parasites [35–38]. While possible, these additional mechanisms also have to be confined to the stages actively degrading hemoglobin.

The quinolines as a class are now considered to inhibit the process of heme crystal formation in the digestive vacuole [39, 40]. Quinine binds to heme reversibly by π–π interactions with heme [36]. The quinoline nucleus is important for heme binding and the side groups are important for both heme crystal inhibition and digestive vacuole localization as weak bases [41, 42]. Roepe has also described covalent interactions of quinine and quinidine with heme [43, 44]. The abundance and activity of these covalent complexes have not been demonstrated in intraerythrocytic parasites. Stereoisomers like 9-epi-quinine are not active against the parasite which suggests specific transport might be important, as these inactive isomers can bind heme and are equally active in the inhibition of heme crystal formation [35]. Sullivan showed that quinidine has a slower off rate of heme crystals after both chloroquine and quinine were individually bound to crystals in equal amounts in a radiolabeled assay. Chloroquine may have a faster on and off rate but once bound quinidine was slower to be removed from crystals [39, 40, 45]. This differential off rate from insoluble heme crystals complicates models of drug resistance based on transport of soluble drug.

In other cell types, quinine and quinidine inhibit potassium channels and stimulate otic cells to cause tinnitus. Potassium channels have been identified in *Plasmodium* but activity of quinine or quinidine has not been tested, due to a lack of functional characterization [46–48]. In the human *ether-a-go-go* related gene (hERG) channel, quinolines bind and inhibit the channel only when it is open causing a delayed rectification of intracellular potassium [49]. Other drugs that also inhibit the channel include the tricyclic antidepressants, macrolides like erythromycin and less so azithromycin, antifungal azoles, fluoroquinolines, antihistamines and pentamidine [50, 51]. Many of these classes also exhibit moderate inhibition of *Plasmodium* signifying a decrease in parasitemia by a 10- to a 100-fold amount per 48–72 h cycle rather than the more potent quinolines which decrease by a 100- to almost a 1,000-fold every cycle [52]. Artemisinin is the most rapidly *Plasmodium*-cidal drug with a 10,000-fold drop in parasitemia with treatment. The pharmacodynamics of quinoline action has been partially described [53–56]. Some work has shown that, at low concentrations of quinine (below 500 nM) elimination of parasites is concentration dependent. Above thresholds of 500 nM to 1 μM exposure time regulates parasite clearance in vitro [57]. Most likely, some of both are demonstrated in the killing of *Plasmodium*. The quinolines do accumulate in the digestive vacuole in excess of predictions made by the Henderson–Hasselbach equilibrium based on pH gradients presumably because of either heme or heme crystal binding [35, 39, 58]. Morphologic effects on the parasite by quinine were first described in Dar es Salaam in the late 1800s by Robert Koch. Sometimes in patient blood films shrunken pynotic parasites are visualized at late ring stage with larger digestive vacuoles [33].

5 Structure Activity

None of the alterations to quinine have improved its action against the parasites. The methoxy group of the quinoline ring and the vinyl group of quinuclidine are not required for antimalarial activity. The secondary alcohol group is essential for activity. Reduction of the alcohol group increases toxicity as well as mitigating antimalarial activity [35]. The 9-epiquinine (*8S*, *9S*) and 9-epiquinidine (*8R*, *9R*) stereoisomers were 100 times less potent against sensitive parasites, with IC_{50} values of close to 3,000 nM against clone D-6, but ten times less potent against chloroquine-resistant parasite clone W-2 with IC_{50} values close to 1,000 nM [59–61]. Quinidine (*8R*, *9S*) and dihydroquinidine were slightly more potent than quinine (*8S*, *9R*) and dihydroquinine at 10 nM rather than 30 nM for D-6 and at approximately 60 nM rather than 120 nM for W-2. The dihydroquinolines have a saturated vinyl group. Replacement of the quinuclidine vinyl group with carboxylate greatly reduced activity. In mefloquine, the vinyl-quinuclidine is lost retaining the 4-quinolinemethanol. Mefloquine has equal potency of stereoisomers about the C-11 hydroxy portion [62].

6 Quinine Failure

Drug failure has to be distinguished from *Plasmodium* drug resistance in the case of quinine. The most common cause of quinine drug failure has been inadequate compliance with a full course of therapy [63]. Other causes of apparent drug failure

Table 2 Comparison of quinine and quinidine base and salt weights from formulations

Drug	Molecular weight	Base	Salt	Base/salt ratio	mg Base in 650 mg of salt dose
Quinine	324.4	100	0		
Quinine sulfate	422	100	130	0.77	500
Quinine hydrochloride	360	100	111	0.90	586
Quinine hyrobromide	405	100	125	0.80	520
Quinine ethylcarbonate	396	100	122	0.82	533
Quinine bisulfate	520	100	160	0.63	406
Quinine dihyrochloride	397	100	122	0.82	533
Quinidine gluconate	518	100	160	0.63	406
Quinidine	324	100	0		

include poor quality drug, rare variants of accelerated host metabolism of quinine in spite of adequate dosing, poor absorption from the intestinal tract, underdosing of drug and delay in initiation of treatment such that the disease progresses [63]. Because of the side effects of quinine that include tinnitus, nausea, headache, abdominal pain, and visual defects (so called cinchonism), many people do not finish a complete course, stopping when the symptoms from drug exceed symptoms from disease [64]. Counterfeit drugs, substandard drugs, or variation in drug dose has been associated with the cinchona alkaloids for centuries [65, 66]. Many of the more than 100 species of the trees had different amounts of quinine or quinidine in the bark from 1 to 15%. In the late 1800s and early 1900s, many different manufacturers had diverse quality drugs on the market. Rarely, quinine failure has been documented with inadequate levels despite similar observed dosing in cohorts treated at the same time with the same drug. William Fletcher in the 1920s documented failure in a case given adequate drugs [67]. White and colleagues also documented in a Thailand individual, suboptimum drug levels in spite of adequate dosing that resulted in prolongation of therapy [68]. If a patient is ill, poor absorption from the intestine may result in inadequate levels. The different salts of quinine also do differ substantially in molecular weight. Quinine and quinidine have a molecular weight of 324. At the extremes, quinine bisulfate salt has a molecular weight of 520 with 324 mg of quinine base, while quinine dihydrochloride salt has a molecular weight of 360 with 324 mg of quinine base. The ratio of salt to base can therefore range from 1.7 to 1.2, which can produce confusion and underdosing, if more than 500 mg of quinine is intended but 400 mg of quinine in the bisulfate salt is given. Table 2 illustrates the relative ratios.

7 Quinine Resistance

True drug resistance was first documented in 1910 in German railroad workers in Brazil. Arthur Neiva in a classic report written in German and Portuguese entitled “About the formation of a quinine resistant race of the malaria parasite” describes the setting for the requirement of increasing the dosing for prophylaxis [68]. The drug was from a reputable manufacturer (Merck). The population was 3,500 railroad workers who also brought families with them. During the preparatory phase, no work was performed during the malaria peak period. Still, 36% of the workers were sick with a high mortality. They worked in deep water and slept in open grass at night. Chagas and Neiva recommended prophylaxis of "Chinin," but the workers refused. Next, enforcement of quinine prophylaxis through controlled administration with the threat of mandatory layoff without compliance was the policy. However, the workers' families did not receive the drug and were constantly sick with malaria and required symptomatic treatment. In the first year, 1/2 of a gram was given every 3 days. Only workers not taking chinin had malaria. In the second season, however, malaria appeared in workers taking chinin. A common pattern was also that workers, on holiday in Rio, showed symptoms of malaria when not taking chinin. The dose was increased to 0.5 g every 2 days during the second season, again with 100% prevention in active workers but increasing number of cases in returning workers from Rio. Neiva called these individuals repositories of adapted parasites. During the peak malaria time of the second season, symptoms appeared in the workers who had not stopped taking chinin. The dose was increased to ½ gram every day, with prevention in the workers, except those taking lesser doses. They concluded that the parasites acquired resistance from the blood of untreated families or returning workers from Rio. The parasites had time to adjust with a subgroup that was constantly exposed to drug during all life stages. In January of 1908, the number of workers increased to 4,000, with regular chinin intake only in railroad workers. Neiva observed highly resistant parasites surviving in both workers taking daily chinin and less frequent doses. The infection rate was 8–10%. Only if there had not been a single interruption of chinin daily were patients symptom-free. Even if workers left the region and stopped taking chinin, malaria was observed 3–30 days later. Resistance was observed in setting of interruptions and where some parasites had been consistently exposed to chinin and still survived. Chinin treatment for the entire population is reasonable if everyone takes medication without interruption. A follow up paper describes the high doses of quinine necessary to treat workers returning to Germany [68]. As many as Thirty-four of 90 patients had *P. falciparum*, with some persistent parasitemias despite 32 grams in 21 days and documentation of drug in urine.

The observation that followed was of specific geographic requirements for increased dosing. Different places in Italy and Sardina required more than 8 times the doses needed for cure of Indian and African falciparum malaria in the 1930s [69]. In recent decades since culturing *P. falciparum* has become possible, a diversity of isolates have widely different concentrations needed to inhibit half of the parasites. A general agreement is an IC_{50} value of 500 nM for a cutoff of quinine-resistant parasites. Unlike chloroquine, which has a clear demarcation between chloroquine sensitive and resistant parasites around 100 nM, quinine has a continuous incremental gradient of IC_{50} values. Some studies report higher IC_{50} values of 800–1,000 nM. A group at the NIH focused on chloroquine resistance, published IC_{50} values to both chloroquine and quinine on almost 100 different laboratory isolates from around the world (Fig. 1) [70]. Only eight isolates were

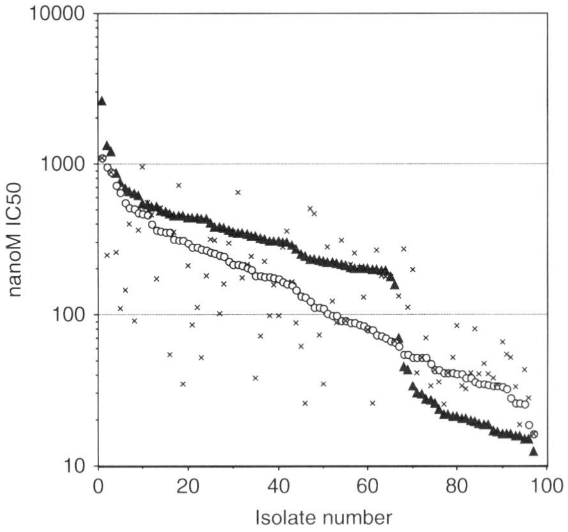

Fig. 1 Comparison of incremental IC_{50} values from more than 100 worldwide laboratory isolates shows a clear separation in chloroquine (*filled triangles*) between 50 and 100 nM. The quinine (*empty circles*) IC_{50} values when also ordered by decreasing values instead show a small step-value difference. Correlation of quinine IC_{50} value paired with the chloroquine IC_{50} value (*cross*) shows that the resistance to either is rarely correlated. The isolate number is a descending rank order number for chloroquine and quinine, except when the quinine value is paired with chloroquine

resistant to quinine, with an incremental drop in IC_{50} values amongst the isolates. Two-thirds were resistant to chloroquine, with a steep shoulder from a sensitive IC_{50} value of less than 30 nM to the resistant IC_{50} value of over 100 nM for chloroquine. There was no correlation of quinine resistance with chloroquine resistance (Fig. 1).

The spread of worldwide clinical quinine resistance has been sporadic and not sustained. A graph of chloroquine resistance as percentage of isolates tested from patient studies with more than 15 subjects depicts high chloroquine resistance in Southeast Asia, increasing chloroquine resistance in Africa but steady low quinine resistance in around 10% of isolates (Fig. 2). This has been the pattern, with quinine resistance being sporadic and staying less than 20% of isolates (Table 3). With increasing use of artemisinin in this century, documentation of quinine resistance has been decreasing. In western Cambodia over the years from 2001 to 2007, the geometric mean IC_{50} value has been 140 nM compared with 90 nM in eastern Cambodia [91]. Furthermore, looking at the trends over the same time span for quinine IC_{50}, an IC_{50} below 300 nM was noted, except for 2007 in western Cambodia. Table 3 and Fig. 2 also show that the percentage of worldwide quinine-resistant isolates in ex vivo drug-sensitivity testing is lower than half.

Quinine is active against the gametocytes of *P. vivax* and *P. malariae* but not *P. falciparum*. In general, *P. vivax* gametocytes are more sensitive to drugs than the

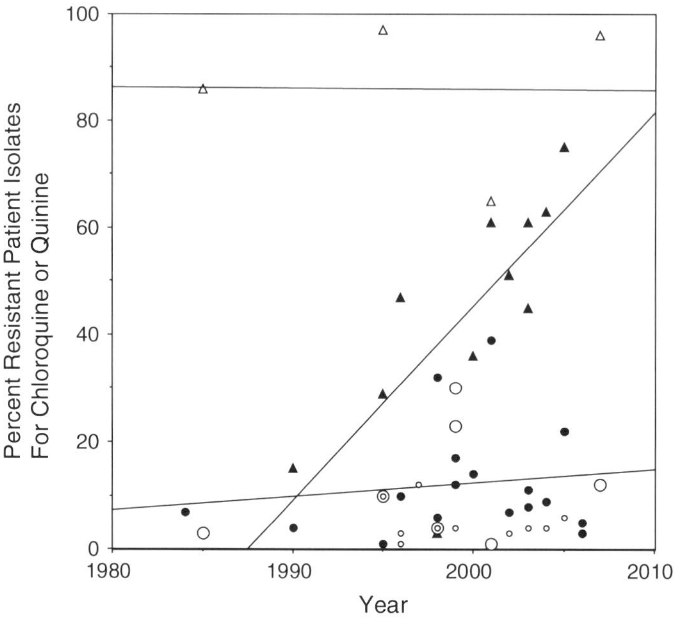

Fig. 2 Quinine-resistant IC_{50} value determined from ex vivo patient isolates not adapted to laboratory culture show that as a percent of a total number, quinine resistance from Africa (*filled circles*), from Southeast Asia (*large empty circle*) or from South America (*small empty circle*), remains approximately 10% over the past 30 years. In contrast, the percent resistant to chloroquine in Southeast Asia (*empty triangle*) was close to 90% over the past 30 years, while in Africa (*filled triangle*) the percentage rapidly climbed from 10% to more than 80%. A single value of 97% chloroquine resistance from South America was not graphed. See Table 3 for study details and references

gametocytes of *P. falciparum*. The avian malaria *P. relictum* also has somewhat drug-resistant gametocytes [67]. In separate reports quinine has a delayed parasite clearance time unique to *P. malariae*. Chloroquine or atebrine would clear parasites in 3–4 days, while quinine would take 6–9 days. Fletcher first analyzed cases in the 1920s with delayed parasite clearance for *P. malariae* (Table 4) [67]. Young and Eyles also noted the delayed parasite clearance with quinine rather than with quinacrine [92]. *P. vivax* was cleared much faster even with higher parasite densities. Delayed parasite clearance kinetics is not a phenotype unique to the artemisinin drugs and *P. falciparum*.

8 Molecular-Resistance Mechanisms

Like other quinolines, quinine resistant parasites have less intracellular drug concentrations. Recent experimental data suggest two phenotypes involved with quinine resistance: lowered drug accumulation [93] and a resistance to toxic effects

Table 3 Quinine and chloroquine resistance from patient studies with IC_{50} value determinations

Region	Country of origin	No. patients	No. with IC_{50} CQ >100 nM	No. with IC_{50} QN >500 nM	No. with IC_{50} MFQ >25 nM	Method	Year isolated	Reference
Africa	Senegal	15		1		RAD	1984	[71]
Africa	Equatorial Guinea	104	16	4		BF	1990	[72]
Africa	Senegal	85	25	1	19	RAD	1995	[73]
Africa	Senegal	161	76	16		RAD	1996	[74]
Africa	Malawi	35	1	2	1	RAD	1998	[75]
Africa	Senegal	51		6		RAD	1999	[76]
Africa	Gabon	65		11		RAD	1999	[77]
Africa	Senegal	70	25	10		LDH	2000	[78]
Africa	Senegal	70	43	27		LDH	2001	[78]
Africa	Senegal	70	36	5		LDH	2002	[78]
Africa	Senegal	70	43	8		LDH	2003	[78]
Africa	Rwanda	74	33	6		RAD	2003	[79]
Africa	Senegal	70	44	6		LDH	2004	[78]
Africa	DRCongo	110	83	24	8	RAD	2005	[80]
Africa	Uganda	196		9		HRPII	2006	[81]
Africa	Madagascar	29		1		RAD	2006	[82]
SEA	Philippine	59	51	2	4	BF	1983	[83]
SEA	Papua New Guinea	31	30	3		BF	1995	[84]
SEA	Thailand	22		5		HRPII	1999	[85]
SEA	Bangladesh	44		13		HRPII	1999	[85]
SEA	Thailand	95		4		RAD	2000	[86]
SEA	Cambodia	293	191	2		RAD	2001	[87]
SEA	Thailand–Burma	26	6	3	12	SYBR Green	2007	[88]
South America	French Guiana	22		2		RAD	1995	[89]
South America	Brazil	30	29	1	0	RAD	1996	[90]
South America	French Guiana	48		5		RAD	1996	[89]

(continued)

Table 3 (continued)

Region	Country of origin	No. patients	No. with IC_{50} CQ >100 nM	No. with IC_{50} QN >500 nM	No. with IC_{50} MFQ >25 nM	Method	Year isolated	Reference
South America	French Guiana	68		8		RAD	1997	[89]
South America	French Guiana	47		2		RAD	1998	[89]
South America	French Guiana	52		2		RAD	1999	[89]
South America	French Guiana	146		3		RAD	2002	[89]
South America	French Guiana	159		6		RAD	2003	[89]
South America	French Guiana	69		3		RAD	2004	[89]
South America	French Guiana	59		3		RAD	2005	[89]

SEA Southeast Asia, *RAD* radioactive hypoxanthine, *HRPII* histidine-rich protein II ELISA, *BF* blood film, *LDH* Lactate dehydrogenase ELISA, *SYBR Green* syber green

Table 4 Parasite clearance times for *P. malariae*, *P. vivax*, and *P. falciparum* among cinchona alkaloids adapted from Fletcher [67]

	Quinine	Quinidine	Cinchonidine
Quartan *P. malariae*			
Number of patients	6	9	1
PCT (days)	6.8	3.7	5
STD PCT (days)	1.7	2.1	
GEOMEAN Parasitemia	793.7	266	333
Par/μl per day reduction	116	70	66
Tertian *P. vivax*			
Number of patients	5	4	3
PCT (days)	2.8	2.5	2.7
STD PCT (days)	1.1	1.0	1.2
GEOMEAN Parasitemia	1,408.8	759.6	584.8
Par/μl per day reduction	503.1	303.9	219.3
Malignant tertian *P. falciparum*			
Number of patients	6	6	5
PCT (days)	2.8	2.3	5.2
STD PCT (days)	0.4	0.8	3.6
GEOMEAN Parasitemia	3,984.2	3,549.5	1,727.5
Par/μl per day reduction	1,406.2	1,521.2	332.2

of the drug on the parasite. Roepe and colleagues show, at similar intracellular drug concentrations, different toxicity on different isolates [34, 94]. Fidock and colleagues have transferred the same chloroquine-resistant transporter (PfCRT) or the sodium–proton exchanger (PfNHE) into two different isolates to show different tolerance to the same extracellular concentrations of parasiticidal quinolines [95]. While the resistance to quinolines involves different transporters and varies between *Plasmodium* species, data are emerging implicating the multidrug resistance transporter, PfMDR1, the multidrug resistance protein, PfMRP, and PfCRT and PfNHE as all being involved in quinoline resistance [93, 96]. The wild type MDR1 transports vinblastine, chloroquine, and quinine [97]. The "resistance" mutations in PfMDR1 do not transport quinine or chloroquine but can still transport vinblastine. The related halofantrine, in contrast, is not transported by wildtype mdr1 but can be effluxed out of *Xenopus* oocytes by mutant PfMDR1. Importantly, quinine can act as an inhibitor of vinblastine transport. Quinine may be acting in a dual role as both inhibitor and substrate of this transporter, compatible with the suggestion that its binding pocket is separate from its transport site. Quinine and mefloquine compete for binding of labeled chloroquine with higher affinity than chloroquine to the PfMDR1 protein [98]. Selection of PfCRT mutations leads to chloroquine resistance and increased sensitivity to quinine [99]. Clinically, the association of quinine resistance with PfHNE is not epidemiologically as strong as chloroquine with PfCRT [100].

9 Toxicity

Quinine received 2.4 million prescriptions in England in 2000–2001 [101]. Toxic overdose is associated with blindness and cardiac toxicity. The maximum daily dose for humans is 33 mg/kg or 2 g for a 60 kg human. Intake of 2–8 g can kill an adult and 900–2,400 mg can kill a child. In a patient (without malaria), levels over 15 mg/L increase risk of blindness and cardiotoxicity [102, 103]. Besides the use of oral charcoal and benzodiazepines, treatment for quinine overdose is supportive. In a case report, complete visual loss, tinnitus and bilateral hearing loss, headache, and diminished taste sensation from quinine have been managed successfully with the calcium antagonist flunarizine and stellate ganglia block [104]. However, most experts report marginal to no success with ganglion block for blindness [105–107]. In mice, the single oral lethal dose 50 is 1,100 mg/kg, rabbits 208 mg/kg, and rhesus monkeys 13 mg/kg [108].

Blackwater fever is a syndrome of hemolysis, fever, and black urine or hemoglobinuria [109]. In the first half of the nineteenth century, it was associated with Europeans residing in the tropics, having little immunity to *P. falciparum* [110]. A rate of 15% in some populations has been reported. Parasitemia at time of presentation was usually not high. While quinine was the most consistent initiator of blackwater fever, it did occur in the setting of no quinine [109]. Acute hemoglobinuria can be caused by high parasitemia in its own right, glucose-6-phosphate dehydrogenase (G6PD) deficiency-induced hemolysis, and drug-related hemolysis, most notably for sulfa drugs and quinine [110]. Largely with the implementation of chloroquine use in the 1950s, the incidence of blackwater fever decreased. Recently, anecdotal cases have been reported in African individuals with and without G6PD deficiency timed with quinine use and history of frequent recurring infections [111]. A recent larger series of 21 cases was documented in Europeans who had lived in sub-Saharan Africa from 1990 to 1999 [112]. Two-thirds had normal G6PD levels and about one-third had low parasitemias. The arylaminoalcohols, quinine, mefloquine, and halofantrine were associated in more than 80% of case presentations [113].

In the use of quinine for leg cramps in the absence of malaria parasites, a quinine-induced thrombocytopenia can occur ("cocktail purpura"). This was first noted in the literature by Dr. Vipan in 1865 in The Lancet in four cases following quinine therapy [114]. The first case was a lady of 50 years prescribed quinine for heart neuralgia. She developed purpura and a bloody discharge at the site of therapeutic chest blistering. Quinine was stopped with resolution of purpura in a few days. The purpura recurred months later with quinine retreatment for dental neuralgia. The second case was a lady with tertian malaria treated with quinine, which developed purpura on the second day. The purpura resolved by day 7 after quinine cessation. In the third case, a boy with general disability and no specific diseases developed purpura a few days following quinine therapy. Despite purpura, the physician continued quinine until bleeding gums developed. The fourth case was a patient taking quinine for 2 weeks for hay fever who developed purpura

3 days after the physician's visit. Dr. Vipan closed with the statement "If these few notes on the effects of quinine prove of any use to the profession or add one drop to the ocean of science, the purpose of the writer will be fully answered" [114].

The mechanism of purpura has been drug-induced antibodies to quinine or quinidine [115]. The precipitating drug exposure can be cocktail purpura from bitter lemon or tonic water containing about 80 mg/L or about 20 mg per cocktail. Quinine, cyclosporine, mitomycin C, and ticlopidine were the most common drugs causing thrombotic thrombocytopenic purpura and hemolytic uremic syndrome [116]. Antibodies bind tightly to drug adherent on platelets on glycoproteins GP IIb–IIIa and/or GP Ib-V–IX [117]. Hemolytic uremic syndrome can also be from antibodies binding to endothelial cells [116, 118, 119]. As much as 82% of the thrombocytopenia cases reported nocturnal leg cramps as the reason for taking quinine [120]. From 1969 to 2000, the FDA received 113 cases of quinine-induced thrombocytopenia (3.6 per year). The cases peaked in 1996 with about 20 decreasing with the FDA elimination of quinine from over-the-counter formularies [121]. After the FDA ban on quinine for leg cramps, the FDA recently reported 38 cases of quinine-associated thrombocytopenia and 2 deaths with the continued use of quinine from the period April 2005 to October 2008 (8.4 cases of thrombocytopenia each year) [122]. A recent Cochrane review for leg cramps and quinine use looked at 23 trials with 1,586 participants, who met the clinical evidence criteria [123]. Quinine at 300 mg per night reduced cramp number by almost 30% from 8.4 cramps per week to 6.3 cramps per week. They found that besides minor gastrointestinal symptoms, serious adverse effects were not greater than placebo up to 60 days of use. Equipoise exists for quinine use in leg cramps.

10 Quinine Pharmacokinetics

The usual adult dose for treating uncomplicated malaria is 10 mg/kg of salt three times a day for 7–14 days. This dosing regimen can be shortened, if quinine is combined with an antibiotic with antimalarial activity, such as clindamycin (10 mg/kg twice daily for 3–7 days) or doxycycline (100 mg once a day for 7 days, if not contraindicated). Dosing after meals decreases the gastric irritation. Totaquine is a mixture of about 75% of the total crystallizable cinchona alkaloids containing about 20% quinine [92, 124]. The cinchona alkaloids are of similar potency to quinine [125, 126]. Quinine and quinidine is usually completely absorbed from the gastrointestinal tract, reaching peak levels in 1–3 h [127]. The elimination half-time increases from healthy subjects (~10 h) to uncomplicated malaria (15 h) to severe malaria (about 20 h) [128]. In the presence of parasitemia, a greater amount of quinine is found in the red blood cell fraction, which quickly decreases. Plasma levels remain elevated because of the increase in the acidic/basic glycoprotein, which is an acute-phase serum protein. The $C_{\max}$ after a 20 mg/kg loading dose is 15–16 mg/L. Levels can sometimes reach 20 mg/L in malaria patients but free quinine remains in the nontoxic range below 2 mg/L, with the increase in the acidic/

basic glycoprotein [102]. Only 20% of quinine is removed in the urine with the rest transformed in the liver to 3- and 2-hydroxyquinine by cytochrome P450 3A4, and then to more water-soluble molecules [129]. Of the cytochrome P450s, 2D6 is inhibited by quinine or quinidine via the nitrogen association with aspartate amino acid number 301 [130]. A growing much-needed evidence base is beginning to explore the interaction of the quinolines, like quinine, with the highly active antiretrovirals [131–133]. The ingestion of a liter of tonic water with 80 mg of quinine does not inhibit cytochrome 2D6 [134].

The pharmacodynamic relationship of quinine was studied in 30 adult patients [54]. The mean parasite clearance time was 73 h with a standard deviation of 24 h. Patients with an area under the concentration time curve less than 20 mg/L from 3 to7 days had a 5.3 relative risk increase in recrudescence. The authors determined a minimum parasiticidal concentration of 3.4 mg/L in the plasma and minimal inhibitory concentration of 0.7 mg/L. Treatment for more than 6 days was required to cure patients. In contrast to a 4 log drop in parasites with artemisinins, the parasite reduction was 250-fold per 48 h cycle [54]. These data are in approximate agreement of the mean plasma concentration for the four cinchona alkaloids of 5 mg/L for quinine, 1 mg/L for quinidine, 0.1 mg/L for cinchonine, and 2.5 mg/L for cinchonidine [135].

11 Usage in Severe Malaria

If there has been no previous treatment with quinine or quinidine and a patient is severely ill, then a loading dose of quinine or quinidine is recommended [136], if artesunate is unavailable. For quinine treatment of adults, this would consist of an intravenous infusion of 20 mg/kg of dihydrochloride salt, given over 2–4 h, followed every 8 h by an infusion of 10 mg (salt)/kg given over 2–4 h. Children in Africa can be treated with 20 mg salt/kg as a loading dose, infused as for adults or given by intramuscular injection [137]. If the intramuscular route is used, then the quinine is diluted (1:1, vol:vol with normal saline), and the loading dose is split in two and given to the anterior thigh to minimize local toxicity while preserving appropriate pharmacokinetic profiles. Maintenance doses can be continued by the intramuscular route every 8–12 h. Conversion to the oral route of treatment is recommended as soon as this can be tolerated. If there is multiorgan failure, then a dose reduction is suggested after 48 h of treatment (usually applicable to adults).

Dosing regimens are different for quinidine, for which electrocardiographic monitoring is recommended also because of the greater risk of associated cardiotoxicity [136]. Both drugs require monitoring of glucose for hypoglycemia, which can be persistently severe in pregnancy. This occurs in 50% of pregnant women treated with quinine versus 10% in the nonpregnant population [136].

ferrochloroquine against 103 Gabonese isolates of *Plasmodium falciparum*. J Antimicrob Chemother 48:179–184

78. Agnamey P, Brasseur P, de Pecoulas PE, Vaillant M, Olliaro P (2006) *Plasmodium falciparum in vitro* susceptibility to antimalarial drugs in Casamance (southwestern Senegal) during the first 5 years of routine use of artesunate-amodiaquine. Antimicrob Agents Chemother 50:1531–1534
79. Tinto H, Rwagacondo C, Karema C, Mupfasoni D, Vandoren W, Rusanganwa E, Erhart A, Van Overmeir C, Van Marck E, D'Alessandro U (2006) *In-vitro* susceptibility of *Plasmodium falciparum* to monodesethylamodiaquine, dihydroartemisinin and quinine in an area of high chloroquine resistance in Rwanda. Trans R Soc Trop Med Hyg 100:509–514
80. Pradines B, Hovette P, Fusai T, Atanda HL, Baret E, Cheval P, Mosnier J, Callec A, Cren J, Amalvict R et al (2006) Prevalence of *in vitro* resistance to eleven standard or new antimalarial drugs among *Plasmodium falciparum* isolates from Pointe-Noire, Republic of the Congo. J Clin Microbiol 44:2404–2408
81. Nsobya SL, Kiggundu M, Nanyunja S, Joloba M, Greenhouse B, Rosenthal PJ (2010) *In vitro* sensitivities of *Plasmodium falciparum* to different antimalarial drugs in Uganda. Antimicrob Agents Chemother 54:1200–1206
82. Andriantsoanirina V, Menard D, Rabearimanana S, Hubert V, Bouchier C, Tichit M, Bras JL, Durand R (2010) Association of microsatellite variations of *Plasmodium falciparum* Na^+/H^+ exchanger (*Pfnhe-1*) gene with reduced *in vitro* susceptibility to quinine: lack of confirmation in clinical isolates from Africa. Am J Trop Med Hyg 82:782–787
83. Smrkovski LL, Buck RL, Alcantara AK, Rodriguez CS, Uylangco CV (1985) Studies of resistance to chloroquine, quinine, amodiaquine and mefloquine among Philippine strains of *Plasmodium falciparum*. Trans R Soc Trop Med Hyg 79:37–41
84. Hombhanje FW (1998) In vitro susceptibility of *Plasmodium falciparum* to four antimalarial drugs in the Central Province of Papua New Guinea. P N G Med J 41:51–58
85. Noedl H, Faiz MA, Yunus EB, Rahman MR, Hossain MA, Samad R, Miller RS, Pang LW, Wongsrichanalai C (2003) Drug-resistant malaria in Bangladesh: an *in vitro* assessment. Am J Trop Med Hyg 68:140–142
86. Chaijaroenkul W, Bangchang KN, Mungthin M, Ward SA (2005) *In vitro* antimalarial drug susceptibility in Thai border areas from 1998–2003. Malar J 4:37
87. Lim P, Chim P, Sem R, Nemh S, Poravuth Y, Lim C, Seila S, Tsuyuoka R, Denis MB, Socheat D et al (2005) *In vitro* monitoring of *Plasmodium falciparum* susceptibility to artesunate, mefloquine, quinine and chloroquine in Cambodia: 2001–2002. Acta Trop 93:31–40
88. Chaijaroenkul W, Wisedpanichkij R, Na-Bangchang K (2010) Monitoring of *in vitro* susceptibilities and molecular markers of resistance of *Plasmodium falciparum* isolates from Thai-Myanmar border to chloroquine, quinine, mefloquine and artesunate. Acta Trop 113:190–194
89. Legrand E, Volney B, Meynard JB, Mercereau-Puijalon O, Esterre P (2008) *In vitro* monitoring of *Plasmodium falciparum* drug resistance in French Guiana: a synopsis of continuous assessment from 1994 to 2005. Antimicrob Agents Chemother 52:288–298
90. Cerutti Junior C, Marques C, Alencar FE, Durlacher RR, Alween A, Segurado AA, Pang LW, Zalis MG (1999) Antimalarial drug susceptibility testing of *Plasmodium falciparum* in Brazil using a radioisotope method. Mem Inst Oswaldo Cruz 94:803–809
91. Lim P, Wongsrichanalai C, Chim P, Khim N, Kim S, Chy S, Sem R, Nhem S, Yi P, Duong S et al (2010) Decreased *in vitro* susceptibility of *Plasmodium falciparum* isolates to artesunate, mefloquine, chloroquine, and quinine in Cambodia from 2001 to 2007. Antimicrob Agents Chemother 54:2135–2142
92. Young MD, Eyles DE (1948) The efficacy of chloroquine, quinacrine, quinine and totaquine in the treatment of *Plasmodium malariae* infections (quartan malaria). Am J Trop Med Hyg 28:23–28

93. Sanchez CP, Stein WD, Lanzer M (2008) Dissecting the components of quinine accumulation in *Plasmodium falciparum*. Mol Microbiol 67:1081–1093
94. Cabrera M, Paguio MF, Xie C, Roepe PD (2009) Reduced digestive vacuolar accumulation of chloroquine is not linked to resistance to chloroquine toxicity. Biochemistry 48:11152–11154
95. Nkrumah LJ, Riegelhaupt PM, Moura P, Johnson DJ, Patel J, Hayton K, Ferdig MT, Wellems TE, Akabas MH, Fidock DA (2009) Probing the multifactorial basis of *Plasmodium falciparum* quinine resistance: evidence for a strain-specific contribution of the sodium-proton exchanger PfNHE. Mol Biochem Parasitol 165:122–131
96. Ursing J, Zakeri S, Gil JP, Bjorkman A (2006) Quinoline resistance associated polymorphisms in the *pfcrt*, *pfmdr1* and *pfmrp* genes of *Plasmodium falciparum* in Iran. Acta Trop 97:352–356
97. Sanchez CP, Rotmann A, Stein WD, Lanzer M (2008) Polymorphisms within PfMDR1 alter the substrate specificity for anti-malarial drugs in *Plasmodium falciparum*. Mol Microbiol 70:786–798
98. Pleeter P, Lekostaj JK, Roepe PD (2010) Purified *Plasmodium falciparum* multi-drug resistance protein (PfMDR 1) binds a high affinity chloroquine analogue. Mol Biochem Parasitol 173:158–161
99. Cooper RA, Ferdig MT, Su XZ, Ursos LM, Mu J, Nomura T, Fujioka H, Fidock DA, Roepe PD, Wellems TE (2002) Alternative mutations at position 76 of the vacuolar transmembrane protein PfCRT are associated with chloroquine resistance and unique stereospecific quinine and quinidine responses in *Plasmodium falciparum*. Mol Pharmacol 61:35–42
100. Henry M, Briolant S, Zettor A, Pelleau S, Baragatti M, Baret E, Mosnier J, Amalvict R, Fusai T, Rogier C et al (2009) *Plasmodium falciparum* Na^+/H^+ exchanger 1 transporter is involved in reduced susceptibility to quinine. Antimicrob Agents Chemother 53:1926–1930
101. Langford NJ, Good AM, Laing WJ, Bateman DN (2003) Quinine intoxications reported to the Scottish Poisons Information Bureau 1997–2002: a continuing problem. Br J Clin Pharmacol 56:576–578
102. Taylor WR, White NJ (2004) Antimalarial drug toxicity: a review. Drug Saf 27:25–61
103. Touze JE, Heno P, Fourcade L, Deharo JC, Thomas G, Bohan S, Paule P, Riviere P, Kouassi E, Buguet A (2002) The effects of antimalarial drugs on ventricular repolarization. Am J Trop Med Hyg 67:54–60
104. Gous P, Haus M (1990) Intravenous flunarizine therapy for acute toxicity in malaria. S Afr Med J 77:217
105. Bacon P, Spalton DJ, Smith SE (1988) Blindness from quinine toxicity. Br J Ophthalmol 72:219–224
106. Dyson EH, Proudfoot AT, Prescott LF, Heyworth R (1985) Death and blindness due to overdose of quinine. Br Med J (Clin Res Ed) 291:31–33
107. Dickinson P, Sabto J, West RH (1981) Management of quinine toxicity. Trans Ophthalmol Soc N Z 33:56–58
108. Hill J (1963) Part 2: The antimalarial drugs. In: Schnitzer RJ, Hawking F (eds) Experimental chemotherapy, vol 1. Academic, New York, pp 513–602
109. George CR (2009) Blackwater fever: the rise and fall of an exotic disease. J Nephrol22 (Suppl 14):120–128
110. Bruce-Chwatt LJ (1987) Quinine and the mystery of blackwater fever. Acta Leiden 55:181–196
111. Rogier C, Imbert P, Tall A, Sokhna C, Spiegel A, Trape JF (2003) Epidemiological and clinical aspects of blackwater fever among African children suffering frequent malaria attacks. Trans R Soc Trop Med Hyg 97:193–197
112. Bruneel F, Gachot B, Wolff M, Regnier B, Danis M, Vachon F (2001) Resurgence of blackwater fever in long-term European expatriates in Africa: report of 21 cases and review. Clin Infect Dis 32:1133–1140
113. Van den Ende J, Coppens G, Verstraeten T, Van Haegenborgh T, Depraetere K, Van Gompel A, Van den Enden E, Clerinx J, Colebunders R, Peetermans WE et al (1998) Recurrence of

blackwater fever: triggering of relapses by different antimalarials. Trop Med Int Health 3:632–639
114. Vipan WH (1865) Quinine as a cause of purpura. Lancet 86:37
115. Aster RH (1999) Drug-induced immune thrombocytopenia: an overview of pathogenesis. Semin Hematol 36:2–6
116. Zakarija A, Bennett C (2005) Drug-induced thrombotic microangiopathy. Semin Thromb Hemost 31:681–690
117. Burgess JK, Lopez JA, Berndt MC, Dawes I, Chesterman CN, Chong BH (1998) Quinine-dependent antibodies bind a restricted set of epitopes on the glycoprotein Ib-IX complex: characterization of the epitopes. Blood 92:2366–2373
118. Park YA, Hay SN, King KE, Matevosyan K, Poisson J, Powers A, Sarode R, Shaz B, Brecher ME (2009) Is it quinine TTP/HUS or quinine TMA? ADAMTS13 levels and implications for therapy. J Clin Apher 24:115–119
119. Boehme MW, Werle E, Kommerell B, Raeth U (1994) Serum levels of adhesion molecules and thrombomodulin as indicators of vascular injury in severe *Plasmodium falciparum* malaria. Clin Investig 72:598–603
120. Kojouri K, Vesely SK, George JN (2001) Quinine-associated thrombotic thrombocytopenic purpura-hemolytic uremic syndrome: frequency, clinical features, and long-term outcomes. Ann Intern Med 135:1047–1051
121. Brinker AD, Beitz J (2002) Spontaneous reports of thrombocytopenia in association with quinine: clinical attributes and timing related to regulatory action. Am J Hematol 70:313–317
122. (2010) Safety of quinine. Med Lett Drugs Ther, 52:88
123. El-Tawil S, Al Musa T, Valli H, Lunn MP, El-Tawil T, Weber M (2010) Quinine for muscle cramps. Cochrane Database Syst Rev 12:CD005044
124. Yeager OW, Reider RF, Mc DG (1946) Evaluation of totaquine in treatment of malaria in an endemic malarious area. J Am Pharm Assoc Am Pharm Assoc 35:337
125. Taggart JV, Earle DP Jr et al (1948) Studies on the chemotherapy of the human malarias; the physiological disposition and antimalarial activity of the cinchona alkaloids. J Clin Invest 27:80–86
126. Earle DP Jr, Welch WJ, Shannon JA et al (1948) Studies on the chemotherapy of the human malarias the metabolism of cinchonine in relation to its antimalarial activity. J Clin Invest 27:87–92
127. White NJ (1985) Clinical pharmacokinetics of antimalarial drugs. Clin Pharmacokinet 10:187–215
128. White NJ, Looareesuwan S, Warrell DA, Warrell MJ, Bunnag D, Harinasuta T (1982) Quinine pharmacokinetics and toxicity in cerebral and uncomplicated falciparum malaria. Am J Med 73:564–572
129. Zhao XJ, Yokoyama H, Chiba K, Wanwimolruk S, Ishizaki T (1996) Identification of human cytochrome P450 isoforms involved in the 3-hydroxylation of quinine by human live microsomes and nine recombinant human cytochromes P450. J Pharmacol Exp Ther 279:1327–1334
130. Hutzler JM, Walker GS, Wienkers LC (2003) Inhibition of cytochrome P450 2D6: structure-activity studies using a series of quinidine and quinine analogues. Chem Res Toxicol 16:450–459
131. Soyinka JO, Onyeji CO, Omoruyi SI, Owolabi AR, Sarma PV, Cook JM (2010) Pharmacokinetic interactions between ritonavir and quinine in healthy volunteers following concurrent administration. Br J Clin Pharmacol 69:262–270
132. Soyinka JO, Onyeji CO, Omoruyi SI, Owolabi AR, Sarma PV, Cook JM (2009) Effects of concurrent administration of nevirapine on the disposition of quinine in healthy volunteers. J Pharm Pharmacol 61:439–443
133. Nosten F, McGready R, D'Alessandro U, Bonell A, Verhoeff F, Menendez C, Mutabingwa T, Brabin B (2006) Antimalarial drugs in pregnancy: a review. Curr Drug Saf 1:1–15

134. Donovan JL, DeVane CL, Boulton D, Dodd S, Markowitz JS (2003) Dietary levels of quinine in tonic water do not inhibit CYP2D6 *in vivo*. Food Chem Toxicol 41:1199–1201
135. Taggart JV, Earle DP, Berliner RW, Zubrod CG, Welch WJ, Wise NB, Schroeder EF, London IM, Shannon JA (1948) Studies on the chemotherapy of the human malarias. III. The physiological disposition and antimalarial activity of the cinchona alkaloids. J Clin Invest 27:80–86
136. Krishna S, White NJ (1996) Pharmacokinetics of quinine, chloroquine and amodiaquine. Clinical implications. Clin Pharmacokinet 30:263–299
137. Krishna S, Nagaraja NV, Planche T, Agbenyega T, Bedo-Addo G, Ansong D, Owusu-Ofori A, Shroads AL, Henderson G, Hutson A et al (2001) Population pharmacokinetics of intramuscular quinine in children with severe malaria. Antimicrob Agents Chemother 45:1803–1809
138. Watt G, Loesuttivibool L, Shanks GD, Boudreau EF, Brown AE, Pavanand K, Webster HK, Wechgritaya S (1992) Quinine with tetracycline for the treatment of drug-resistant falciparum malaria in Thailand. Am J Trop Med Hyg 47:108–111
139. Pukrittayakamee S, Chantra A, Vanijanonta S, Clemens R, Looareesuwan S, White NJ (2000) Therapeutic responses to quinine and clindamycin in multidrug-resistant falciparum malaria. Antimicrob Agents Chemother 44:2395–2398
140. Sinclair D, Donegan S, Lalloo DG: (2011) Artesunate versus quinine for treating severe malaria. Cochrane Database Syst Rev 3:CD005967
141. Phu NH, Tuan PQ, Day N, Mai NT, Chau TT, Chuong LV, Sinh DX, White NJ, Farrar J, Hien TT (2010) Randomized controlled trial of artesunate or artemether in Vietnamese adults with severe falciparum malaria. Malar J 9:97
142. Lubell Y, Yeung S, Dondorp AM, Day NP, Nosten F, Tjitra E, Abul Faiz M, Yunus EB, Anstey NM, Mishra SK et al (2009) Cost-effectiveness of artesunate for the treatment of severe malaria. Trop Med Int Health 14:332–337
143. Jones KL, Donegan S, Lalloo DG: (2007) Artesunate versus quinine for treating severe malaria. Cochrane Database Syst Rev:CD005967
144. Dondorp AM, Nosten F, Yi P, Das D, Phyo AP, Tarning J, Lwin KM, Ariey F, Hanpithakpong W, Lee SJ et al (2009) Artemisinin resistance in *Plasmodium falciparum* malaria. N Engl J Med 361:455–467
145. Witkowski B, Lelievre J, Barragan MJ, Laurent V, Su XZ, Berry A, Benoit-Vical F (2010) Increased tolerance to artemisinin in *Plasmodium falciparum* is mediated by a quiescence mechanism. Antimicrob Agents Chemother 54:1872–1877

8-Aminoquinolines: Primaquine and Tafenoquine

Norman C. Waters and Michael D. Edstein

Abstract 8-Aminoquinolines are an important class of antimalarial drugs because they are effective against the liver stages of *Plasmodium* infections and thus are administered for radical cure and presumptive antirelapse therapy against relapsing malaria. In this chapter, we discuss two 8-aminoquinolines, primaquine and tafenoquine. Primaquine was identified in 1946 and has been used extensively to clear liver-stage parasites, especially those from *Plasmodium vivax*. These can persist in the liver for months, as a dormant form of the parasite (the hypnozoite), which re-emerges much later to cause clinical disease. Tafenoquine, a primaquine analog, is currently under advanced clinical development. Tafenoquine has a much longer elimination half-life compared with primaquine (14 days versus 6 h) and is highly effective both in treating relapses of *P. vivax* malaria and as a causal prophylactic agent against *P. falciparum* and *P. vivax* malaria. A major drawback to the 8-aminoquinolines is their toxicity in glucose-6-phosphate dehydrogenase (G6PD)-deficient individuals. We discuss clinical uses, pharmacokinetics and metabolism, safety and tolerability, mechanisms of action and drug resistance for both these drugs.

1 Introduction

The 8-aminoquinolines have a long history, being the first chemotype of synthetic antimalarials when pamaquine was used from the late 1920s [1]. In 1946, the screening of a large number of 8-aminoquinolines identified primaquine as a relatively safe and efficacious compound [2]. Today, additional 8-aminoquinolines have been synthesised as the search for safer and more efficacious compounds

N.C. Waters (✉) • M.D. Edstein
Australian Army Malaria Institute, Gallipoli Barracks, Weary Dunlop Drive, Enoggera, QLD 4051, Australia
e-mail: norman.waters@us.army.mil

H.M. Staines and S. Krishna (eds.), *Treatment and Prevention of Malaria*,
DOI 10.1007/978-3-0346-0480-2_4,

continues. From these efforts, tafenoquine is now in advanced clinical development and may become a new addition to the arsenal of antimalarial drugs.

The 8-aminoquinolines are effective against the exo-erythrocytic liver stages of the malaria parasite. This is central to preventing relapsing malaria as well as causal prophylaxis for malaria infections. Causal prophylaxis refers to the killing of parasites while they are in the liver, and thus prevents infection of erythrocytes and any signs of clinical disease. The efficacy of 8-aminoquinolines against liver-stage infection is especially valuable in the clearance of *P. vivax* and *P. ovale*, in which latent liver-stage forms known as hypnozoites can persist in the liver for months to years. Relapse infection occurs when the hypnozoites exit dormancy and differentiate into merozoites, which rupture from the hepatocyte to cause a blood-stage infection. In addition to being the only class of drugs with activity against hypnozoites, 8-aminoquinolines are active against gametocytes and thus interfere with malaria transmission.

2 Primaquine

2.1 Chemistry

The chemical name of primaquine is 6-methoxy-8-(4-amino-1-methylbutyl) aminoquinoline and its chemical formula is $C_{15}H_{12}N_3O$, with a molecular weight of 259 (Fig. 1). Primaquine is a racemic mixture composed of D- and L-enantiomers, due to the presence of an asymmetric chiral center. It is water soluble and solutions are stable when protected from light. Primaquine tablets are given in the form of the diphosphate salt containing either 13.2 mg (= ~7.5 mg base) or 26.3 mg (= ~15 mg base).

2.2 Clinical Use

There are three established indications for the use of primaquine: causal prophylaxis for all species of malaria, presumptive antirelapse therapy (terminal prophylaxis or postexposure prophylaxis) for *P. vivax* and *P. ovale* and radical cure of *P. vivax* and *P. ovale* infections. Since the use of primaquine depends on the species of parasite, an understanding of malaria endemicity is necessary for adequate

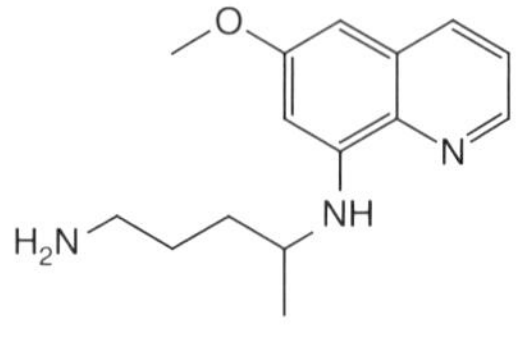

Fig. 1 Structure of Primaquine

compliance and circumvent the rise of chloroquine-resistant *P. vivax*. For antirelapse or radical cure therapy, primaquine is usually partnered with another antimalarial drug such as chloroquine (25 mg/kg over 3 days). Since primaquine lacks substantial activity against the asexual erythrocytic stages of *P. falciparum* and acts slowly against blood stages of *P. vivax*, a blood schizontocide drug should be administered with primaquine [11]. Early studies suggest that the addition of chloroquine or quinine potentiated the activity of primaquine [27]. A recent report demonstrates that synergy between primaquine and chloroquine may be attributed to the ability of primaquine to increase the accumulation of chloroquine within the parasite [28].

Evaluation of nine different trials that compared a 14-day primaquine plus chloroquine with a 5-day primaquine plus chloroquine regimen concluded that the 14-day primaquine regimen was superior to chloroquine alone or 5-day primaquine plus chloroquine [29]. This evaluation, however, did not take into account the dose of either chloroquine or primaquine. As an alternative to chloroquine plus primaquine for the treatment of vivax malaria, artesunate plus primaquine combinations have been shown to produce markedly shorter parasite and fever clearance times [24, 30]. Although artesunate has no radical curative activity, the rapid action of artemisinins on the blood stages of *P. vivax* is highly beneficial to the patient, in that infection and malaria symptoms are aborted at a much faster rate than with chloroquine. Furthermore, with increasing reports of chloroquine-resistant *P. vivax* malaria in Oceania, Southeast Asia, the Indian subcontinent and the Americas [14], artesunate may be considered a potential replacement for chloroquine for aborting an acute attack of vivax malaria.

2.6 Transmission Blocking

In a few malaria-endemic areas, the addition of a single dose of 45 mg primaquine to the treatment regimen had been advocated to reduce gametocyte burden and thus interfere with the transmission cycle of the malaria parasite. Early studies demonstrated that primaquine is a potent gametocytocidal and sporontocidal agent [31]. Several clinical studies demonstrated a primaquine-dependent reduction in gametocyte clearance times, when administered as a single 0.5-mg/kg dose to artesunate or quinine [32], a single 45-mg dose to chloroquine–sulfadoxine–pyrimethamine [33] and a single 0.75-mg/kg dose to artesunate–sulfadoxine–pyrimethamine [34] as compared with treatment groups receiving the various drug combinations without primaquine.

One study did not observe any significant advantage in adding a single 0.75-mg/kg primaquine dose to artesunate–sulfadoxine–pyrimethamine in reducing the gametocyte burden [35]. Discrepancies may be attributed to the methods and the accuracy of detecting submicroscopic levels of gametocytes, as artesunate possesses gametocytocidal activity. The absence of an additive effect of primaquine is consistent with the suggestion that the most effective way to prevent gametocytaemia is to clear asexual blood forms [36]. Since artesunate does not kill mature gametocytes [37], unlike primaquine, and treatment with particular antimalarials induces gametocytogenesis [38], additional transmission-blocking studies are required to address the benefit of adding primaquine to treatment regimens. Future studies are essential since ongoing efforts are aimed at eliminating malaria [39].

2.7 Mechanism of Action

The mechanism of action by which primaquine exerts its antimalarial activity is largely unknown but the mitochondria may be the biological target of primaquine. Specifically, primaquine accumulates within the mitochondria, resulting in swelling and structural changes within the inner membranes [40–44], thus destroying mitochondrial function [45–47]. Primaquine is quickly metabolised to several reactive intermediates that are responsible for toxicity to erythrocytes (discussed below) and also apparently for antimalarial activity [48, 49]. Several of the active metabolites are structurally similar to naphthoquinones [50]. The antimalarial activity of naphthoquinones, such as atovaquone, is due to inhibition of mitochondrial function [51, 52]. Atovaquone has been shown to collapse the mitochondrial electron membrane potential, resulting in disruption of pyrimidine biosynthesis [53, 54]. Since asexual blood stage parasites rely on glycolysis for their energy source rather than oxidative phosphorylation-generated ATP, a role in pyrimidine biosynthesis would support the essentiality of the mitochondria for asexual growth and explain the blood-stage antimalarial activity of atovaquone. Primaquine, however, is not an effective blood-stage antimalarial against *P. falciparum*. Interestingly, swelling of host cell mitochondria was not observed and hydroxynaphthoquinone and naphthoquinone are approximately 1,000-fold more potent against the plasmodial cytochrome bc_1 complex than the mammalian complex [51]. These selectivity differences are believed to be a result of structural differences within the plasmodial bc_1 complex that increases the affinity for selected antimalarials such as atovaquone and 8-aminoquinolines [55].

The metabolism of primaquine produces reactive intermediates that ultimately results in the accumulation of free radicals, hydrogen peroxides and superoxides which may be responsible for antimalarial activity [56]. Such weak activity of primaquine in vitro may be indicative of the fact that primaquine requires metabolism for antimalarial activity [57]. A similar mode of antimalarial action has been suggested for artemisinins, which are metabolized into free radicals [58]. These free radicals may disrupt oxidation–reduction systems, inactivate specific enzymes or attach to and disrupt biological membranes [59].

Although a generalised mechanism of action has been discussed for *Plasmodium*, it should be acknowledged that different mechanisms may exist depending on the species of *Plasmodium*. For example, primaquine appears to be effective against asexual blood stages of *P. berghei* [43, 60] and *P. vivax* [6–8]; however, it is a poor

NADH-dependent enzyme that converts MetHb to haemoglobin) or under extreme oxidative stress, the levels of MetHb may increase to harmful levels, resulting in cyanosis. Primaquine increases the rate of MetHb formation [79] through oxidative stress via the free-radical metabolites of primaquine. MetHb levels as high as 11% have been reported in healthy Caucasians treated with primaquine [80]. In individuals without anemia, primaquine-induced methaemoglobinaemia, however, is a well-tolerated condition that is alleviated upon the discontinuation of primaquine dosing [11].

Gastrointestinal (GI) discomfort has been associated with primaquine in a dose-dependent manner [80–82]. Symptoms include cramping, nausea, diarrhoea and vomiting. Most of these symptoms are mild and are often avoided, if primaquine is taken with food [80].

2.10 Primaquine Resistance

Experimentally induced primaquine resistance has been developed in *P. berghei* and *P. knowlesi* [83, 84]. These controlled experiments were later supported with field reports that indicated the existence of primaquine-tolerant *P. vivax* [85]. Several reports suggest resistance to standard antirelapse primaquine therapy; however, factors such as noncompliance with the 14-day treatment [12] or inadequate weight-based dose could also explain the observed failures rather than inherited resistance [86]. *P. vivax* strains from Southeast Asia and the Southwest Pacific are more tolerant to primaquine than elsewhere [19]. These tolerant strains, however, can be effectively treated with increased doses of primaquine [11]. Although little evidence exists to support primaquine-resistant exo-erythrocytic stages including hypnozoites, several reports have described multiple relapses of *P. vivax* in military personnel after primaquine treatment [87, 88]. Further well-controlled studies where treatment compliance is known and primaquine is administered in a weight-based dose would help resolve the resistance issue.

3 Tafenoquine

3.1 Historical Development

Originally labeled as WR238605 or SB-252263 and now named tafenoquine, the drug is a new 8-aminoquinoline antimalarial being codeveloped by GlaxoSmithKline Pharmaceuticals and the US Army as a replacement for primaquine for radical cure of *P. vivax* malaria and as a potential prophylactic agent [89–91]. In an effort to develop less toxic, more active and longer acting 8-aminoquinolines, tafenoquine was first synthesised by the US Army at the Walter Reed Army

Institute of Research in 1979. Although tafenoquine is a primaquine analog, it possesses different physicochemical properties, antimalarial potency and toxicological and pharmacokinetic properties compared with primaquine. In in vitro testing and in vivo preclinical animal models tafenoquine is more active than primaquine. To date, it has been evaluated in more than 2,000 human subjects in clinical studies.

On an equimolar basis, in vitro antimalarial susceptibility studies have shown tafenoquine to exhibit equivalent activity (IC_{50} of 0.7–1.5 μM) to primaquine against culture-adapted chloroquine-sensitive strains, but was considerably more active than primaquine against multidrug-resistant *P. falciparum* lines, with IC_{50} values ranging from 0.06 to 0.3 μM [92]. It is conceivable that tafenoquine's enhanced blood schizontocidal potency compared with primaquine is because it exerts greater oxidative stress on multidrug-resistant parasitised erythrocytes [93]. In contrast to culture-adapted multidrug-resistant *P. falciparum* lines, tafenoquine was only marginally more active than primaquine against wild isolates of *P. falciparum* from central, west and east Africa (mean IC_{50} values of 4.43 μM versus 6.82 μM) [94, 95]. The enantiomers of tafenoquine have similar levels of in vitro antimalarial activity against the drug-sensitive D6 and multidrug-resistant W2, TM90-C2a and TM90-C2b strains of *P. falciparum* (D. K. Kyle personal communication).

In the rodent–*P. berghei* Peters 4-day suppressive test, tafenoquine was about 9 times more active as a blood schizontocide than primaquine against the drug-sensitive *P. berghei* N strain and 4–5 times as active as primaquine against highly resistant chloroquine, mefloquine or halofantrine strains of *P. berghei* [95]. In addition to developing new schizontocidal drugs, the capacity to interrupt malaria transmission is also of great importance. Tafenoquine possesses significant sporontocidal activity against *P. berghei*, with a minimum effective dose of 25 mg/kg that prevents mosquitoes from developing sporozoites [96]. Tafenoquine also has gametocytocidal activity, with a significant reduction in the number of gametocytes in the blood of *P. berghei*-infected mice treated with 25 mg/kg, resulting in a twofold extension of mice survival time [90].

In the rhesus monkey–*P. cynomolgi* model, tafenoquine was effective as a causal prophylactic agent against pre-erythrocytic tissue stages of sporozoite-induced *P. cynomolgi* malaria [97]. The causal prophylactic ED_{50} (50% effective dose) of tafenoquine was 0.125 mg/kg/day or 0.27 μM/kg/day for 3 days, which was 14 times more effective than primaquine, with an ED_{50} of 1 mg/kg/day or 3.86 μM/kg/day for 3 days. Tafenoquine was also a highly effective agent against liver stages of *P. cynomolgi*, with an ED_{50} of 0.172 mg/kg/day or 0.371 μM/kg/day for 7 days and was 7 times more potent than primaquine, with an ED_{50} of 0.712 mg/kg/day or 2.75 μM/kg/day for 7 days [98].

Although developed primarily as an antirelapse agent, tafenoquine has also been found to possess significant blood schizontocidal activity against trophozoite-induced infections in simian–malaria models. Against *P. cynomolgi B* and *P. fragile*, which are recognised as biological counterparts of *P. vivax* and *P. falciparum* infections in humans, respectively [99], tafenoquine at a dose of

3.16 mg/kg/day for 7 days led to a cure for established trophozoite induced infections in monkeys with both these parasites [100]. In contrast, primaquine was only partially curative (25% for *P. cynomolgi B* and 67% for *P. fragile*) at a dose of 10 mg/kg/day for 7 days. Tafenoquine was also effective against blood-induced vivax malaria infections of the chloroquine-resistant AMRU1 strain in the *Aotus* monkey–*P. vivax* model. Parasite clearance of the AMRU1 strain occurred at a dose of 0.3 mg/kg tafenoquine daily for 3 days and cures were achieved at 3 mg/kg daily for 3 days [101].

In addition to tafenoquine's greater in vitro and in vivo antimalarial activities compared with primaquine in preclinical studies, it is less toxic than primaquine. In acute oral toxicity studies, tafenoquine's LD_{50} (50% lethal dose) of 0.78 and 0.64 mM/kg in rats and guinea pigs, respectively, was markedly less toxic than primaquine, with corresponding LD_{50} values of 0.46 and 0.12 mM/kg [98]. In subchronic and chronic studies of tafenoquine (WR 238605 IND #38503), the compound was also found to be less toxic than primaquine. For example, in dog toxicology studies, 3 and 9 mg/kg/day of primaquine orally for 28 days resulted in muscle necrosis, coma and death, whereas tafenoquine up to a maximum tested dose of 16 mg/kg/day for 28 days did not produce these adverse events [102].

3.2 *Chemistry*

The chemical name for the racemic tafenoquine is (±)-8-[(4-amino-1-methylbutyl) amino]-2,6-dimethoxy-4-methyl-5-(3-trifluoromethylphenoxy) quinoline succinate. The structural formula for tafenoquine is shown in Fig. 2. Its chemical formula is $C_{24}H_{28}N_3O_3 \cdot C_4H_6O_4$, with molecular weights of 463 for the base and 581 for the succinate salt. Tafenoquine is an off-white to pink/orange/brown solid powder with a strong phenolic odor. It is poorly soluble in water and stable at room temperature, when stored in amber bottles for at least 10 years. The formulated product of tafenoquine is a hard gelatin capsule containing 250 mg tafenoquine succinate equivalent to 200 mg of the free base. Tafenoquine capsules should be stored below 30°C and protected from light.

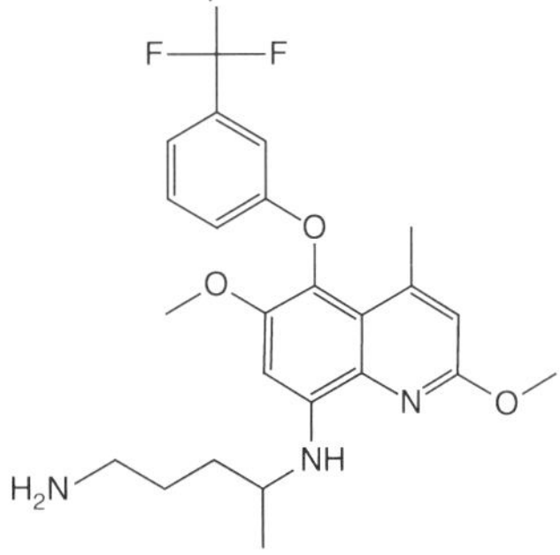

Fig. 2 Structure of Tafenoquine

3.3 Mechanism of Action and Development of Resistance

As already indicated, the exact mechanism of action of 8-aminoquinolines is not well understood. It has been proposed that the blood-stage activity of 8-aminoquinolines may be derived from an oxidative stress mechanism since it is known that primaquine stimulates the hexose monophosphate shunt, increases hydrogen peroxide and MetHb production and decreases glutathione levels in the erythrocyte [93, 103, 104]. Similar to chloroquine, the blood-stage activity of tafenoquine may be through inhibition of haematin polymerisation. In contrast to the inactive primaquine ($IC_{50} > 2{,}500$ μM), tafenoquine (IC_{50} of 16 μM) inhibited haematin polymerisation more efficiently than did chloroquine (IC_{50} of 80 μM) [92]. Other suggested modes of action of tafenoquine include drug-induced mitochondrial dysfunction or inhibition of receptor recycling by endosomes [105, 106].

In vitro studies have also shown a positive correlation between tafenoquine and primaquine ($r^2 = 0.61$) against seven *P. falciparum* lines, with different levels of susceptibility to chloroquine and mefloquine [92]. In contrast, no correlation exists between tafenoquine and either chloroquine or mefloquine, suggesting a lack of cross-resistance between tafenoquine and chloroquine or mefloquine.

3.4 Pharmacokinetics and Metabolism

The pharmacokinetics of tafenoquine has been investigated following both single and multiple oral administration of the drug in healthy subjects. Single-dose studies ranging from 4 to 600 mg tafenoquine have been carried out in 48 healthy males (Caucasian [$n = 20$], African American [$n = 12$] and Hispanic [$n = 16$]) in the fasting state [98]. The absorption half-life of tafenoquine was 1.7 h, suggesting rapid absorption of the compound. However, the t_{max} of 13.8 h implied prolonged absorption of tafenoquine from the gut. Plasma tafenoquine concentrations declined in a mono-exponential manner and the drug was slowly cleared, with an elimination half-life of 14 days. The C_{max} and area under the drug concentration curve of tafenoquine were linear over the doses studied. The tafenoquine concentration–time data were best described by a one-compartment model, with first-order absorption and elimination. Tafenoquine had a low oral clearance (CL/F, 5.7 L/h) and a large apparent volume of distribution (V/F, 2,558 L), suggesting extensive tissue binding. Whole blood concentrations of tafenoquine were 1.8-fold higher than corresponding plasma concentrations, reflecting an accumulation of the drug in erythrocytes, which may contribute to the greater potency of tafenoquine compared with primaquine, which does not concentrate in erythrocytes [61].

The population pharmacokinetics of tafenoquine has also been determined in healthy Thai and Australian soldiers after receiving tafenoquine for malaria prophylaxis. A one-compartment model with first-order absorption and elimination was found to best describe the population pharmacokinetics of tafenoquine. In the

Thai study, 104 soldiers received a loading dose of 400 mg tafenoquine daily for 3 days followed by 400 mg tafenoquine monthly for 5 consecutive months [107]. Blood samples were randomly collected from each soldier on several occasions each month. The population estimates of the first-order absorption rate constant (K_a), CL/F and V/F were 0.69/h, 3.20 L/h and 1,820 L, respectively. The absorption and elimination half-lives were 1.0 h and 16.4 days, respectively. The covariants, age and weight influenced the volume of distribution. The one subject who contracted malaria had a higher plasma clearance, but this was not considered to have sufficient impact to warrant a change in dosing.

In the Australian study, 490 soldiers received a loading dose of 200 mg tafenoquine daily for 3 days followed by a weekly dose of 200 mg tafenoquine for 6 months [108]. Blood samples were collected from each soldier after the last loading dose and then at weeks 4, 8 and 16. Typical values of K_a, CL/F and V/F were 0.24/h, 4.37 L/h and 1,901 L, respectively. The V/F was similar to that reported in the Thai soldiers, but the systemic CL/F was greater (4.37 L/h versus 3.20 L/h). The derived elimination half-life of tafenoquine in the Australian soldiers of 12.7 days was slightly shorter than the 14 and 16 days reported previously in healthy Caucasians, African-Americans and Hispanic subjects [98] and in Thai soldiers [107], respectively, which may partly reflect the fact that the last samples were drawn at only up to 1 week post dose and therefore, the presumed "terminal" phase may have included some components of the distribution phase. The mean values for CL/F and V/F obtained in the fed Australian soldiers were 30–35% lower than values derived in the fasted healthy subjects participating in the single dose escalating study of tafenoquine. A possible explanation for the disparity is that a high-fat meal can increase the oral bioavailability (F) of tafenoquine by up to 40% (A. K. Miller personal communication), which when comparing the two studies would bring the respective CL/F and V/F values into closer agreement after correcting for F.

Limited investigations have been carried out on the metabolism of tafenoquine. In vitro rat liver microsomal studies have identified tafenoquine to be metabolised to aminophenolic compounds that undergo air oxidation to a mixture of quinones and quinoneimines [109]. Similar to primaquine, the metabolism of tafenoquine is difficult to study, because its structure contains several metabolically labile constituent groups, and its intermediates are unstable and possess amphoteric properties [74]. So far, no metabolites of tafenoquine have been identified in either human plasma or urine.

3.5 Safety and Tolerability

In single dose escalating pharmacokinetic studies in healthy subjects, only a few GI side effects such as heartburn, flatulence, vomiting and diarrhoea were seen in those subjects who received the higher doses of 300–600 mg tafenoquine [98]. These side effects were few and were not unexpected, based on past experiences with primaquine. Methaemoglobinaemia, haemolytic anaemia, thrombocytopenia, or changes in white blood cell counts or electrocardiograms were not observed in the subjects. Because tafenoquine is related to primaquine, it can cause methaemoglobinaemia and haemolytic anaemia in individuals with deficiency of G6PD. Thus, all individuals who receive an 8-aminoquinoline should undergo laboratory testing for confirmation of a normal G6PD status [110]. This is potentially tafenoquine's major drawback for use worldwide as G6PD is one of the most common human genetic polymorphisms. Although malaria patients with anemia may be at greater risk, methaemoglobinaemia generally is not a serious concern when <20% of haemoglobin is in the MetHb form and only rarely will testing for methaemoglobinaemia be indicated on clinical grounds, such as the presence of bluish mucous membranes [111].

In individuals with severe G6PD deficiency, such as the Mediterranean variety, tafenoquine or primaquine should not be used. Even individuals with the low-grade deficiency (A-) variant of G6PD, which is most commonly found in Africa, can be at risk of developing haemolysis when exposed to tafenoquine. In a Kenyan field study, two women who were inadvertently given tafenoquine (400 mg daily for 3 days) experienced a haemolytic reaction when their G6PD deficiency status was incorrectly recorded during screening [112]. One woman, who was later found to be heterozygous for the (A-) G6PD variant, developed intravascular haemolysis and required a 2-unit blood transfusion. Haemolysis did not continue after the acute event, no renal compromise was seen in spite of blackwater urine, and she restored and maintained normal haematologic parameters for 6 months after the event. The other woman, who was later found to be homozygous for the (A-) G6PD variant, remained asymptomatic despite an acute 30 g/L decrease in haemoglobin, which was noticed only because of routine blood tests. She restored her haemoglobin level without intervention.

3.6 Clinical Use

3.6.1 Chemoprophylaxis against *P. falciparum* and *P. vivax* Malaria

The development and spread of multiple drug-resistant *P. falciparum* malaria in many parts of the world highlights the need to develop new, safe, well-tolerated and effective chemoprophylactic agents for travellers and in special risk groups such as military personnel. A long-acting drug that acts on all stages of the malaria parasite could be a significant addition to the limited armamentarium for protecting individuals against malaria infections. Tafenoquine is a long-acting antimalarial and, based on preclinical studies acts on all stages of the parasite, including the pre-erythrocytic stages providing causal prophylactic activity. Table 1 summarises the Phase II and III studies on the safety, tolerability and protective efficacy of tafenoquine in its clinical development.

Table 1 Studies on the safety, tolerability and protective efficacy of tafenoquine

Purpose of study	Study design	TQ Regimen	Subjects	Safety and tolerability	Efficacy
Prophylactic studies					
Prophylactic efficacy against *Pf* in a challenge model [113]	Randomised, placebo-controlled, double-blinded	600 mg	4 Adults	TQ was well tolerated, with only mild, transient headache and diarrhoea reported	3 of 4 subjects protected from developing *Pf* malaria
Minimum effective weekly dose of TQ for prevention of *Pf* malaria in Ghana [114]	Randomised, placebo-controlled, double-blinded, dose-ranging	25, 50, 100 or 200 mg ow for 12 weeks	463 Adults	All regimens were SWT. The four TQ groups demonstrated AE rates comparable to those of the placebo group and showed no evidence of a dose-related effect	Relative to placebo (86/94), the protective efficacies were 32% for 25 mg (58/93), 84% for 50 mg (13/91), 87% for 100 mg (11/94) and 86% for 200 mg (12/91)
Long-term prophylactic activity of TQ against *Pf* in Gabon [110]	Randomised, placebo-controlled, double-blinded	25, 50, 100 or 200 mg od for 3 days	410 (aged 12–20 years)	TQ were well tolerated but abdominal pain was reported more commonly in the TQ groups than in the placebo group. No other symptom such as headache, diarrhea, dizziness and was significantly associated with TQ use	Relative to placebo (14/82), the protective efficacies were 0% for 25 mg (16/79), 80% for 50 mg (3/86), 93% for 100 mg (1/79) and 100% for 200 mg (0/84)
Prophylactic efficacy of TQ against *Pf* in Kenya [112]	Randomised, placebo-controlled, double-blinded	A: LD 400 mg + placebo ow for 13 weeks; B: LD 200 mg + 200 mg ow for 13 weeks; C: LD 400 mg + 400 mg ow for 13 weeks	223 Adults	Reported AEs were similar among the subjects on the four treatment groups. The mean MetHb concentrations in subjects on 200 mg and 400 mg ow were 2.5% and 4.5%, respectively	Relative to placebo (54/59), the protective efficacies were 68% for A (16/54), 86% for B (7/53) and 89% for C (6/57)

(continued)

Table 1 (continued)

Purpose of study	Study design	TQ Regimen	Subjects	Safety and tolerability	Efficacy
Prophylactic activity of TQ against *Pf* and *Pv* malaria in Thailand [115]	Randomised, placebo-controlled, double-blinded	LD 400 mg od for 3 days + 400 mg om for 5 months	205 Thai soldiers	Monthly TQ was SWT. GI complaints (diarrhoea, nausea, or vomiting) were significantly more common in the TQ group than the placebo group	Relative to placebo (30/92), the protective efficacies were 96% against *Pv*, 97% against all species, and 100% against *Pf* (1/96)
Prophylactic trial of TQ in Timor-Leste [116]	Randomised (3:1 to TQ), double-blinded	LD 200 mg od for 3 days + 200 mg ow for 6 months or LD 250 mg od MQ for 3 days + 250 mg MQ ow for 6 months	654 AMP	Both TQ and MQ were well tolerated. In a subset of TQ individuals ($n = 98$), MetHb levels increased by 1.8% and mild vortex keratopathy (phospholipid corneal deposits) was detected in 93% (69/74) of TQ subjects	No diagnoses of malaria occurred for either treatment group in Timor-Leste, but 0.9% (4/462) and 0.7% (1/153) of recipients developed *Pv* infections in the TQ and MQ groups, respectively
Long-term safety of TQ [117]	Randomised (2:1 to TQ), placebo-controlled, double-blinded	LD 200 mg od for 3 days + 200 mg ow for 23 weeks	120 Adults	No effect on night vision or other ophthalmic indices such as colour vision and macular function. After 6 months of dosing, there was no TQ effect on renal function	

Abbreviations: *TQ* tafenoquine, *MQ* mefloquine, *LD* loading dose, *AMP* Australian military personnel, *SWT* safe and well tolerated, *AE* adverse events, *GI* gastrointestinal, *od* once daily, *om once monthly*, *ow* once weekly, *Pf P. falciparum*, *Pv P. vivax*

3.6.2 Presumptive Antirelapse Therapy and Radical Cure

Tafenoquine was also developed as a potential replacement of primaquine for presumptive antirelapse therapy and radical cure. Table 2 summarises the clinical development of tafenoquine for antirelapse therapy.

3.7 Future Potential

Tafenoquine is a unique antimalarial drug that is active against all stages of *Plasmodium* spp. Although clinical studies of tafenoquine have shown the long-acting 8-aminoquinoline to have comparable efficacy to primaquine for radical cure and presumptive antirelapse therapy, the markedly shorter regime of tafenoquine compared with primaquine (3 days versus 14 days) is more convenient and with improved compliance one could expect the number of relapses of *P. vivax* malaria to decrease markedly. For the treatment of uncomplicated *P. falciparum*, artemisinin-based combination therapies (ACTs) are now recommended for first-line treatment worldwide. Because of tafenoquine's long elimination half-life of 14 days, it could be considered as a partner drug with an artemisinin derivative such as artesunate. Today, however, we have very efficacious and well-tolerated ACTs for the treatment of falciparum malaria [125]. Thus, it may be more prudent to limit the use of tafenoquine to treating *P. vivax* and *P. ovale* infections, and for selected applications, including prophylaxis (short and long-term) for special risk groups such as military personnel.

Furthermore, since tafenoquine possesses both gametocytocidal and sporontocidal activity it is a promising candidate agent for transmission-blocking public health applications. Because of its long half-life, tafenoquine has enormous potential for malaria control and possibly the elimination of the disease. To test this latter concept will be difficult. Perhaps tafenoquine could be evaluated for transmission blocking in an area of low endemicity, with controlled geographical access such as an island. For malaria elimination, tafenoquine could be used in mass drug administration to eliminate residual parasites in an entire population [112] and, thus, would be an excellent drug for the eradication of malaria under the new initiative by the Bill and Melinda Gates Foundation [126].

Before these possible public health applications of tafenoquine can be implemented, a regimen that can safely be given to G6PD-deficient individuals needs to be developed. Alternatively, a field friendly, rapid and inexpensive G6PD test needs to be produced so that the G6PD status of the individual can be ascertained prior to tafenoquine administration. A clinical dose-escalating study in G6PD-deficient subjects is planned to better quantify and characterise the risk of tafenoquine use in this important risk group [117].

Table 2 Studies on the safety, tolerability and efficacy of tafenoquine for anti-relapse therapy

Purpose of study	Study design	Regimen	Subjects	Safety and tolerability	Efficacy – relapse frequency
Presumptive antirelapse therapy					
PNG [118]	Randomised, open-label study	A: 400 mg od TQ for 3 days; B: 7.5 mg tid PQ for 14 days	592 AMP	Increase in mild GI disturbances with TQ vs. PQ	1.9% (7/378) for A and 2.8% (6/214) for B within 12 months after leaving PNG
Timor-Leste [119]	Randomised, open-label study	A/B: 200 mg od/td TQ for 3 days; C: 400 mg od TQ for 3 days; D: 7.5 mg tid PQ for 14 days	925 AMP	GI disturbances in all groups, being twofold higher in females for both treatments [120]. Reduced AEs with reduced dose of TQ	4.9% (20/406) for A, 5.3% (4/75) for B, 11.0% (17/155) for C, and 10.0% (29/289) for D within 12 months after leaving Timor-Leste
Radical cure therapy					
TQ vs. CQ Thailand [121]	Randomised open-label study after CQ treatment (1,500 mg over 3 days)	A: 300 mg od TQ for 7 days; B: 500 mg od TQ for 3 day, repeated after 1 week; C: one dose of 500 mg TQ; D: CQ only	23 adults (completed 2–6 months of follow-up)	TQ was SWT. MetHb values peaked at 13.5%, 14.7%, and 6.4% in treatment groups A–C. Mild, transient AEs consisting of headache and GI in a minority of all patients	0% (0/7) for A, 11.1% (1/9) for B (day 120), 14.3% (1/7) for C (day 112) and 57.1% (4/7) for D (with relapse on days 40, 43, 49 and 84)
TQ vs. PQ Thailand [122]	Randomised open-label study after CQ treatment (1,500 mg over 3 days)	A: 300 mg od TQ for 7 days; B: 600 mg od TQ for 3 days; C: one dose 600 mg TQ; D: no further treatment; E: 15 mg od PQ for 14 days	46 TQ, 10 CQ and 12 CQ + PQ (completed at least 8 weeks of follow-up or had a relapse)	TQ was SWT. AEs on TQ and PQ therapy were generally mild and transient, consisting predominantly of headache, abdominal discomfort or diarrhoea and were more frequent in the TQ groups compared with the PQ group	0% (0/15) for A, 0% (0/15) for B, 6.3% (1/16) for C, 80% (8/10) for D and 25% (3/12) for E. The protective efficacy was 92.6% for CQ + TQ recipients compared with CQ + PQ recipients

TQ alone [123]	Open-label study	800 mg TQ over 3 days	2 AMP returning from PNG	TQ was well tolerated, with one patient experiencing mild diarrhoea	Parasite clearance 3 to 4 days. No recurrence after 2 years
Extended TQ regimen [124]	Open-label study after CQ treatment (1,500 mg over 3 days)	LD 200 mg od TQ for 3 days, plus 200 mg ow TQ for 8 weeks	27 AMP		Patients recruited after 2–4 clinical episodes of *P. vivax* malaria. One patient had a relapse after 6 months of observation

Abbreviations: *TQ* tafenoquine, *PQ* primaquine, *CQ* chloroquine, *LD* loading dose, PNG Papua New Guinea, *AMP* Australian military personnel, *SWT* safe and well tolerated, *AE* adverse events, *GI* gastrointestinal, *od* once daily, *ow* once weekly, *tid thrice daily*

Acknowledgments We thank Professor Dennis Shanks for review and helpful discussions with the manuscript. The opinions or assertions contained herein are the private views of the authors and are not to be construed as official, or as reflecting true views of the United States Department of the Army, the Department of Defense or the Australian Defense Force.

References

1. Peters W (1999) The evolution of tafenoquine–antimalarial for a new millennium? J R Soc Med 92:345–352
2. Alving AS, Pullman TN et al (1948) The clinical trial of 18 analogues of pamaquin (plasmochin) in vivax malaria, Chesson strain. J Clin Invest 27:34–45
3. Arnold J, Alving AS, Hockwald RS, Clayman CB, Dern RJ, Beutler E, Flannagan OL, Jeffery GM (1955) The antimalarial action of primaquine against the blood and tissue stages of falciparum malaria (Panama, P-F-6 strain). J Lab Clin Med 46:391–397
4. Baird JK, Wiady I, Sutanihardja A, Suradi P, Basri H, Sekartuti AE, Fryauff DJ, Hoffman SL (2002) Short report: therapeutic efficacy of chloroquine combined with primaquine against *Plasmodium falciparum* in northeastern Papua, Indonesia. Am J Trop Med Hyg 66:659–660
5. Basco LK, Bickii J, Ringwald P (1999) *In-vitro* activity of primaquine against the asexual blood stages of *Plasmodium falciparum*. Ann Trop Med Parasitol 93:179–182
6. Edgecomb J, Arnold J, Yount E Jr, Alving A, Eichelberger L (1950) Primaquine, SN-13272, a new curative agent in vivax malaria: a preliminary report. Nat Malar Soc 9:285–357
7. Wilairatana P, Silachamroon U, Krudsood S, Singhasivanon P, Treeprasertsuk S, Bussaratid V, Phumratanaprapin W, Srivilirit S, Looareesuwan S (1999) Efficacy of primaquine regimens for primaquine-resistant *Plasmodium vivax* malaria in Thailand. Am J Trop Med Hyg 61:973–977
8. Pukrittayakamee S, Vanijanonta S, Chantra A, Clemens R, White NJ (1994) Blood stage antimalarial efficacy of primaquine in *Plasmodium vivax* malaria. J Infect Dis 169:932–935
9. Powell RD, Brewer GJ (1967) Effects of pyrimethamine, chlorguanide, and primaquine against exoerythrocytic forms of a strain of chloroquine-resistant *Plasmodium falciparum* from Thailand. Am J Trop Med Hyg 16:693–698
10. Arnold J, Alving AS, Hockwald RS, Clayman CB, Dern RJ, Beutler E, Jeffery GM (1954) The effect of continuous and intermittent primaquine therapy on the relapse rate of Chesson strain vivax malaria. J Lab Clin Med 44:429–438
11. Hill DR, Baird JK, Parise ME, Lewis LS, Ryan ET, Magill AJ (2006) Primaquine: report from CDC expert meeting on malaria chemoprophylaxis I. Am J Trop Med Hyg 75:402–415
12. Baird JK, Hoffman SL (2004) Primaquine therapy for malaria. Clin Infect Dis 39:1336–1345
13. Fryauff D, Baird J, Basri H, Sumawinata I, Purnomo RT, Ohrt C, Mouzin E, Church C, Richards A et al (1995) Randomised placebo-controlled trial of primaquine for prophalaxis of falciparum and vivax malaria. Lancet 346:1190–1193
14. Baird JK, Lacy MD, Basri H, Barcus MJ, Maguire JD, Bangs MJ, Gramzinski R, Sismadi P, Krisin LJ et al (2001) Randomized, parallel placebo-controlled trial of primaquine for malaria prophylaxis in Papua, Indonesia. Clin Infect Dis 33:1990–1997
15. Baird JK, Fryauff DJ, Basri H, Bangs MJ, Subianto B, Wiady I, Purnomo LB, Masbar S, Richie TL et al (1995) Primaquine for prophylaxis against malaria among nonimmune transmigrants in Irian Jaya, Indonesia. Am J Trop Med Hyg 52:479–484
16. Soto J, Toledo J, Rodriquez M, Sanchez J, Herrera R, Padilla J, Berman J (1998) Primaquine prophylaxis against malaria in nonimmune Colombian soldiers: efficacy and toxicity. A randomized, double-blind, placebo-controlled trial. Ann Intern Med 129:241–244
17. Alving AS, Arnold J, Robinson DH (1952) Mass therapy of subclinical vivax malaria with primaquine. JAMA 149:1558

18. Jelinek T, Nothdurft HD, Von Sonnenburg F, Loscher T (1995) Long-term efficacy of primaquine in the treatment of vivax malaria in nonimmune travelers. Am J Trop Med Hyg 52:322–324
19. Baird JK, Rieckmann KH (2003) Can primaquine therapy for vivax malaria be improved? Trends Parasitol 19:115–120
20. Schwartz E, Regev-Yochay G, Kurnik D (2000) Short report: a consideration of primaquine dose adjustment for radical cure of *Plasmodium vivax* malaria. Am J Trop Med Hyg 62:393–395
21. Duarte EC, Pang LW, Ribeiro LC, Fontes CJ (2001) Association of subtherapeutic dosages of a standard drug regimen with failures in preventing relapses of vivax malaria. Am J Trop Med Hyg 65:471–476
22. Clyde DF, McCarthy VC (1977) Radical cure of Chesson strain vivax malaria in man by 7, not 14, days of treatment with primaquine. Am J Trop Med Hyg 26:562–563
23. Goller JL, Jolley D, Ringwald P, Biggs BA (2007) Regional differences in the response of *Plasmodium vivax* malaria to primaquine as anti-relapse therapy. Am J Trop Med Hyg 76:203–207
24. Dao NV, Cuong BT, Ngoa ND, le Thuy TT, The ND, Duy DN, Dai B, Thanh NX, Chavchich M, Rieckmann KH et al (2007) Vivax malaria: preliminary observations following a shorter course of treatment with artesunate plus primaquine. Trans R Soc Trop Med Hyg 101:534–539
25. Pukrittayakamee S, Imwong M, Chotivanich K, Singhasivanon P, Day NP, White NJ (2010) A comparison of two short-course primaquine regimens for the treatment and radical cure of *Plasmodium vivax* malaria in Thailand. Am J Trop Med Hyg 82:542–547
26. Krudsood S, Tangpukdee N, Wilairatana P, Phophak N, Baird JK, Brittenham GM, Looareesuwan S (2008) High-dose primaquine regimens against relapse of *Plasmodium vivax* malaria. Am J Trop Med Hyg 78:736–740
27. Alving AS et al (1955) Potentialtion of the curative action of primaquine in vivax malaria by quinine and chloroquine. J Lab Clin Med 46:301–306
28. Bray PG, Deed S, Fox E, Kalkanidis M, Mungthin M, Deady LW, Tilley L (2005) Primaquine synergises the activity of chloroquine against chloroquine-resistant *P. falciparum*. Biochem Pharmacol 70:1158–1166
29. Galappaththy GN, Omari AA, Tharyan P (2007) Primaquine for preventing relapses in people with *Plasmodium vivax* malaria. Cochrane Database Syst Rev CD004389
30. Silachamroon U, Krudsood S, Treeprasertsuk S, Wilairatana P, Chalearmrult K, Mint HY, Maneekan P, White NJ, Gourdeuk VR, Brittenham GM et al (2003) Clinical trial of oral artesunate with or without high-dose primaquine for the treatment of vivax malaria in Thailand. Am J Trop Med Hyg 69:14–18
31. Rieckmann KH, McNamara JV, Frischer H, Stockert TA, Carson PE, Powell RD (1968) Gametocytocidal and sporontocidal effects of primaquine and of sulfadiazine with pyrimethamine in a chloroquine-resistant strain of *Plasmodium falciparum*. Bull World Health Organ 38:625–632
32. Pukrittayakamee S, Chotivanich K, Chantra A, Clemens R, Looareesuwan S, White NJ (2004) Activities of artesunate and primaquine against asexual- and sexual-stage parasites in falciparum malaria. Antimicrob Agents Chemother 48:1329–1334
33. Lederman ER, Maguire JD, Sumawinata IW, Chand K, Elyazar I, Estiana L, Sismadi P, Bangs MJ, Baird JK (2006) Combined chloroquine, sulfadoxine/pyrimethamine and primaquine against *Plasmodium falciparum* in Central Java, Indonesia. Malar J 5:108
34. Shekalaghe S, Drakeley C, Gosling R, Ndaro A, van Meegeren M, Enevold A, Alifrangis M, Mosha F, Sauerwein R, Bousema T (2007) Primaquine clears submicroscopic *Plasmodium falciparum* gametocytes that persist after treatment with sulphadoxine-pyrimethamine and artesunate. PLoS One 2:e1023
35. El-Sayed B, El-Zaki SE, Babiker H, Gadalla N, Ageep T, Mansour F, Baraka O, Milligan P, Babiker A (2007) A randomized open-label trial of artesunate- sulfadoxine-pyrimethamine

with or without primaquine for elimination of sub-microscopic *P. falciparum* parasitaemia and gametocyte carriage in eastern Sudan. PLoS One 2:e1311

36. Suputtamongkol Y, Chindarat S, Silpasakorn S, Chaikachonpatd S, Lim K, Chanthapakajee K, Kaewkaukul N, Thamlikitkul V (2003) The efficacy of combined mefloquine-artesunate versus mefloquine-primaquine on subsequent development of *Plasmodium falciparum* gametocytemia. Am J Trop Med Hyg 68:620–623
37. Kumar N, Zheng H (1990) Stage-specific gametocytocidal effect *in vitro* of the antimalaria drug qinghaosu on *Plasmodium falciparum*. Parasitol Res 76:214–218
38. Dunyo S, Milligan P, Edwards T, Sutherland C, Targett G, Pinder M (2006) Gametocytaemia after drug treatment of asymptomatic *Plasmodium falciparum*. PLoS Clin Trials 1:e20
39. White NJ (2008) The role of anti-malarial drugs in eliminating malaria. Malar J 7(Suppl 1):S8
40. Lanners HN (1991) Effect of the 8-aminoquinoline primaquine on culture-derived gametocytes of the malaria parasite *Plasmodium falciparum*. Parasitol Res 77:478–481
41. Beaudoin RL, Aikawa M (1968) Primaquine-induced changes in morphology of exoerythrocytic stages of malaria. Science 160:1233–1234
42. Boulard Y, Landau I, Miltgen F, Ellis DS, Peters W (1983) The chemotherapy of rodent malaria. XXXIV. Causal prophylaxis Part III: Ultrastructural changes induced in exo-erythrocytic schizonts of *Plasmodium yoelii yoelii* by primaquine. Ann Trop Med Parasitol 77:555–568
43. Howells RE, Pters W, Fullard J (1970) The chemotherapy of rodent malaria. 13. Fine structural changes observed in the eryhrocytic stages of *Plasmodium berghei berghei* following exposureto primaquine and menoctone. Ann Trop Med Parasitol 64:203–207
44. Aikawa M, Beaudoin RL (1970) *Plasmodium fallax*: high-resolution autoradiography of exoerythrocytic stages treated with Primaquine *in vitro*. Exp Parasitol 27:454–463
45. Peters W, Ellis D, Boulard Y, Landau I (1984) The chemotherapy of rodent malaria XXXVI. Part IV. The activity of a new 8-aminoquinoline, WR 225,448 against exo-erythrocytic schizonts of *Plasmodium yoelii yoelii*. Ann Trop Med Parasitol 78:467–478
46. Rotman A (1975) Genetics of a primaquin-resistant yeast. J Gen Microbiol 89:1–10
47. Krungkrai J, Burat D, Kudan S, Krungkrai S, Prapunwattana P (1999) Mitochondrial oxygen consumption in asexual and sexual blood stages of the human malarial parasite, *Plasmodium falciparum*. Southeast Asian J Trop Med Public Health 30:636–642
48. Idowu OR, Peggins JO, Brewer TG (1995) Side-chain hydroxylation in the metabolism of 8-aminoquinoline antiparasitic agents. Drug Metab Dispos 23:18–27
49. Vale N, Moreira R, Gomes P (2009) Primaquine revisited six decades after its discovery. Eur J Med Chem 44:937–953
50. Grewal RS (1981) Pharmacology of 8-aminoquinolines. Bull World Health Organ 59:397–406
51. Fry M, Pudney M (1992) Site of action of the antimalarial hydroxynaphthoquinone, 2-[trans-4-(4′-chlorophenyl) cyclohexyl]-3-hydroxy-1,4-naphthoquinone (566 C80). Biochem Pharmacol 43:1545–1553
52. Fry M, Beesley JE (1991) Mitochondria of mammalian Plasmodium spp. Parasitology 102:17–26
53. Srivastava IK, Rottenberg H, Vaidya AB (1997) Atovaquone, a broad spectrum antiparasitic drug, collapses mitochondrial membrane potential in a malarial parasite. J Biol Chem 272:3961–3966
54. Painter HJ, Morrisey JM, Mather MW, Vaidya AB (2007) Specific role of mitochondrial electron transport in blood-stage *Plasmodium falciparum*. Nature 446:88–91
55. Vaidya AB, Lashgari MS, Pologe LG, Morrisey J (1993) Structural features of *Plasmodium* cytochrome b that may underlie susceptibility to 8-aminoquinolines and hydroxynaphthoquinones. Mol Biochem Parasitol 58:33–42
56. Fletcher KA, Barton PF, Kelly JA (1988) Studies on the mechanisms of oxidation in the erythrocyte by metabolites of primaquine. Biochem Pharmacol 37:2683–2690

57. Greenberg J, Taylor DJ, Josephson ES (1951) Studies on *Plasmodium gallinaceum in vitro* II. The effects of some 8-aminoquinolines against the erythrocytic parasites. J Infect Dis 88:163–167
58. Meshnick SR (1998) Artemisinin antimalarials: mechanisms of action and resistance. Med Trop 58:13–17
59. McChesney JD (1981) Considerations about the structure-activity relationships of 8- aminoquinoline antimalarial drugs. Bull World Health Organ 59:459–462
60. Peters W, Irare SG, Ellis DS, Warhurst DC, Robinson BL (1984) The chemotherapy of rodent malaria, XXXVIII. Studies on the activity of three new antimalarials (WR 194,965, WR 228,258 and WR 225,448) against rodent and human malaria parasites (*Plasmodium berghei* and *P. falciparum*). Ann Trop Med Parasitol 78:567–579
61. Mihaly GW, Ward SA, Edwards G, Nicholl DD, Orme ML, Breckenridge AM (1985) Pharmacokinetics of primaquine in man. I. Studies of the absolute bioavailability and effects of dose size. Br J Clin Pharmacol 19:745–750
62. Ward SA, Mihaly GW, Edwards G, Looareesuwan S, Phillips RE, Chanthavanich P, Warrell DA, Orme ML, Breckenridge AM (1985) Pharmacokinetics of primaquine in man. II. Comparison of acute vs chronic dosage in Thai subjects. Br J Clin Pharmacol 19:751–755
63. Bangchang KN, Songsaeng W, Thanavibul A, Choroenlarp P, Karbwang J (1994) Pharmacokinetics of primaquine in G6PD deficient and G6PD normal patients with vivax malaria. Trans R Soc Trop Med Hyg 88:220–222
64. Binh VQ, Chinh NT, Thanh NX, Cuong BT, Quang NN, Dai B, Travers T, Edstein MD (2009) Sex affects the steady-state pharmacokinetics of primaquine but not doxycycline in healthy subjects. Am J Trop Med Hyg 81:747–753
65. Mihaly GW, Ward SA, Edwards G, Orme ML, Breckenridge AM (1984) Pharmacokinetics of primaquine in man: identification of the carboxylic acid derivative as a major plasma metabolite. Br J Clin Pharmacol 17:441–446
66. Greaves J, Evans DA, Gilles HM, Fletcher KA, Bunnag D, Harinasuta T (1980) Plasma kinetics and urinary excretion of primaquine in man. Br J Clin Pharmacol 10:399–404
67. Mayorga P, Deharo E, Landau I, Couarraze G (1997) Preliminary evaluation of primaquine activity on rodent malaria model after transdermal administration. Parasite 4:87–90
68. Nishi KK, Jayakrishnan A (2007) Self-gelling primaquine-gum arabic conjugate: an injectable controlled delivery system for primaquine. Biomacromolecules 8:84–90
69. Singh KK, Vingkar SK (2008) Formulation, antimalarial activity and biodistribution of oral lipid nanoemulsion of primaquine. Int J Pharm 347:136–143
70. Stensrud G, Sande SA, Kristensen S, Smistad G (2000) Formulation and characterisation of primaquine loaded liposomes prepared by a pH gradient using experimental design. Int J Pharm 198:213–228
71. Vangapandu S, Sachdeva S, Jain M, Singh S, Singh PP, Kaul CL, Jain R (2004) 8-Quinolinamines conjugated with amino acids are exhibiting potent blood-schizontocidal antimalarial activities. Bioorg Med Chem 12:239–247
72. Rajic Z, Kos G, Zorc B, Singh PP, Singh S (2009) Macromolecular prodrugs. XII. Primaquine conjugates: synthesis and preliminary antimalarial evaluation. Acta Pharm 59:107–115
73. Borissova R, Lammek B, Stjarnkvist P, Sjoholm I (1995) Biodegradable microspheres. 16. Synthesis of primaquine-peptide spacers for lysosomal release from starch microparticles. J Pharm Sci 84:249–255
74. Brueckner RP, Ohrt C, Baird JK, Milhous WK (2001) 8-aminoquinolines. In: RP J (eds) Antimalarial chemotherapy: mechanisms of action, resistance and new directions. Humana, Totowa, NJ, pp 123–151
75. Cappellini MD, Fiorelli G (2008) Glucose-6-phosphate dehydrogenase deficiency. Lancet 371:64–74
76. Beutler E, Duparc S (2007) Glucose-6-phosphate dehydrogenase deficiency and antimalarial drug development. Am J Trop Med Hyg 77:779–789

77. Beutler E, Dern RJ, Alving AS (1955) The hemolytic effect of primaquine. VI. An *in vitro* test for sensitivity of erythrocytes to primaquine. J Lab Clin Med 45:40–50
78. Beutler E, Dern RJ, Flanagan CL, Alving AS (1955) The hemolytic effect of primaquine. VII. Biochemical studies of drug-sensitive erythrocytes. J Lab Clin Med 45:286–295
79. Srivastava P, Singh S, Jain GK, Puri SK, Pandey VC (2000) A simple and rapid evaluation of methemoglobin toxicity of 8-aminoquinolines and related compounds. Ecotoxicol Environ Saf 45:236–239
80. Clayman C, Arnold J, Hockwald R, Yount E Jr, Edgecomb J, Alving A (1952) Toxicity of primaquine in Caucasians. JAMA 149:1563–1568
81. Clyde DF (1981) Clinical problems associated with the use of primaquine as a tissue schizontocidal and gametocytocidal drug. Bull World Health Organ 59:391–395
82. Taylor WR, White NJ (2004) Antimalarial drug toxicity: a review. Drug Saf 27:25–61
83. Prakash S, Chakrabarti AK, Choudhury DS (1961) Studies on *Plasmodium berghei* Vincke and Lips, 1948. XXXI. Selection of a primaquine resistant strain. Indian J Malariol 15:115–122
84. Peters W (1966) Drug responses of mepacrine- and primaquine-resistant strains of *Plasmodium berghei* Vincke and Lips, 1948. Ann Trop Med Parasitol 60:25–30
85. Arnold J, Alvinig AS, Clayman CB (1961) Induced primaquine resistance in vivax malaria. Trans R Soc Trop Med Hyg 55:345–350
86. Baird JK (2007) A rare glimpse at the efficacy of primaquine. Am J Trop Med Hyg 76:201–202
87. Kitchener S (2002) Malaria in the Australian Defence Force associated with the InterFET peacekeeping operation in East Timor. Mil Med 167:iii–iv
88. Spudick JM, Garcia LS, Graham DM, Haake DA (2005) Diagnostic and therapeutic pitfalls associated with primaquine-tolerant *Plasmodium vivax*. J Clin Microbiol 43:978–981
89. Crockett M, Kain KC (2007) Tafenoquine: a promising new antimalarial agent. Expert Opin Investig Drugs 16:705–715
90. The GlaxoSmithKline (GSK) Clinical Study Register. http://www.gsk-clinicalstudyregister.com. Accessed 23 Apr 2010
91. Medicines for Malaria Venture Project Portfolio. http://www.mmv.org/research-development/project-portfolio/tafenoquine. Accessed 23 Apr 2010
92. Vennerstrom JL, Nuzum EO, Miller RE, Dorn A, Gerena L, Dande PA, Ellis WY, Ridley RG, Milhous WK (1999) 8-aminoquinolines active against blood stage *Plasmodium falciparum in vitro* inhibit hematin polymerization [In Process Citation]. Antimicrob Agents Chemother 43:598–602
93. Vennerstrom JL, Eaton JW (1988) Oxidants, oxidant drugs, and malaria. J Med Chem 31:1269–1277
94. Pradines B, Mamfoumbi MM, Tall A, Sokhna C, Koeck JL, Fusai T, Mosnier J, Czarnecki E, Spiegel A, Trape JF et al (2006) *In vitro* activity of tafenoquine against the asexual blood stages of *Plasmodium falciparum* isolates from Gabon, Senegal, and Djibouti. Antimicrob Agents Chemother 50:3225–3226
95. Peters W, Robinson BL, Milhous WK (1993) The chemotherapy of rodent malaria. LI. Studies on a new 8-aminoquinoline, WR 238,605. Ann Trop Med Parasitol 87:547–552
96. Coleman RE, Clavin AM, Milhous WK (1992) Gametocytocidal and sporontocidal activity of antimalarials against *Plasmodium berghei* ANKA in ICR Mice and *Anopheles stephensi* mosquitoes. Am J Trop Med Hyg 46:169–182
97. Heisy GE, Milhous WK, Hansuklarita P, Theoharides AD, Schuster BG, Davidson DE (1988) Radical curative properties of Tafenoquine (WR 238605, SB252263), Abstr. No. 323. The American Society of Tropical Medicine and Hygiene, Washington DC, p 217
98. Brueckner RP, Lasseter KC, Lin ET, Schuster BG (1998) First-time-in-humans safety and pharmacokinetics of WR 238605, a new antimalarial. Am J Trop Med Hyg 58:645–649
99. Coatney GE, Collins WE, Warren M, Contacos PG (1971) Primate malarias. Department of Health and Welfare, US Govt Publication, Washington, DC

100. Puri SK, Dutta GP (2003) Blood schizontocidal activity of WR 238605 (Tafenoquine) against *Plasmodium cynomolgi* and *Plasmodium fragile* infections in rhesus monkeys. Acta Trop 86:35–40
101. Obaldia N 3rd, Rossan RN, Cooper RD, Kyle DE, Nuzum EO, Rieckmann KH, Shanks GD (1997) WR 238605, chloroquine, and their combinations as blood schizonticides against a chloroquine-resistant strain of *Plasmodium vivax* in *Aotus* monkeys. Am J Trop Med Hyg 56:508–510
102. Lee CC, Kinter LD, Heiffer MH (1981) Subacute toxicity of primaquine in dogs, monkeys, and rats. Bull World Health Organ 59:439–448
103. Augusto O, Weingrill CL, Schreier S, Amemiya H (1986) Hydroxyl radical formation as a result of the interaction between primaquine and reduced pyridine nucleotides. Catalysis by hemoglobin and microsomes. Arch Biochem Biophys 244:147–155
104. Baird JK, Davidson DJ, Decker JJ (1986) Oxidative activity of hydroxylated primaquine analogs. Non-toxicity to glucose-6-phosphate dehydrogenase-deficient human red blood cells *in vitro*. Biochem Pharmacol 35:1091–1098
105. Hiebsch RR, Raub TJ, Wattenberg BW (1991) Primaquine blocks transport by inhibiting the formation of functional transport vesicles. Studies in a cell-free assay of protein transport through the Golgi apparatus. J Biol Chem 266:20323–20328
106. van Weert AW, Geuze HJ, Groothuis B, Stoorvogel W (2000) Primaquine interferes with membrane recycling from endosomes to the plasma membrane through a direct interaction with endosomes which does not involve neutralisation of endosomal pH nor osmotic swelling of endosomes. Eur J Cell Biol 79:394–399
107. Edstein MD, Kocisko DA, Brewer TG, Walsh DS, Eamsila C, Charles BG (2001) Population pharmacokinetics of the new antimalarial agent tafenoquine in Thai soldiers. Br J Clin Pharmacol 52:663–670
108. Charles BG, Miller AK, Nasveld PE, Reid MG, Harris IE, Edstein MD (2007) Population pharmacokinetics of tafenoquine during malaria prophylaxis in healthy subjects. Antimicrob Agents Chemother 51:2709–2715
109. Idowu OR, Peggins JO, Brewer TG, Kelley C (1995) Metabolism of a candidate 8-aminoquinoline antimalarial agent, WR 238605, by rat liver microsomes. Drug Metab Dispos 23:1–17
110. Lell B, Faucher JF, Missinou MA, Borrmann S, Dangelmaier O, Horton J, Kremsner PG (2000) Malaria chemoprophylaxis with tafenoquine: a randomised study. Lancet 355:2041–2045
111. Shanks GD, Kain KC, Keystone JS (2001) Malaria chemoprophylaxis in the age of drug resistance. II. Drugs that may be available in the future. Clin Infect Dis 33:381–385
112. Shanks GD, Oloo AJ, Aleman GM, Ohrt C, Klotz FW, Braitman D, Horton J, Brueckner R (2001) A new primaquine analogue, tafenoquine (WR 238605), for prophylaxis against *Plasmodium falciparum* malaria. Clin Infect Dis 33:1968–1974
113. Brueckner RP, Coster T, Wesche DL, Shmuklarsky M, Schuster BG (1998) Prophylaxis of *Plasmodium falciparum* infection in a human challenge model with WR 238605, a new 8-aminoquinoline antimalarial. Antimicrob Agents Chemother 42:1293–1294
114. Hale BR, Owusu-Agyei S, Fryauff DJ, Koram KA, Adjuik M, Oduro AR, Prescott WR, Baird JK, Nkrumah F, Ritchie TL et al (2003) A randomized, double-blind, placebo-controlled, dose-ranging trial of tafenoquine for weekly prophylaxis against *Plasmodium falciparum*. Clin Infect Dis 36:541–549
115. Walsh SD, Eamsila C, Sasiprapha T, Sangkharomya S, Khaewsathien P, Supakalin P, Tang DB, Jarasrumgsichol P, Chuenchitra SC, Edstein MD, et al (2004) Randomized, double-blind, placebo controlled evaluation of monthly tafenoquine (WR 238605) for *Plasmodium falciparum* and *P. vivax* malaria in Royal Thai Army soldiers. J Inf Dis 190:1456-1463
116. Nasveld PE, Edstein MD, Reid M, Brennan L, Harris IE, Kitchener SJ, Leggat PA, Pickford P, Kerr C, Ohrt C et al (2010) Randomized, double-blind study of the safety, tolerability, and efficacy of tafenoquine versus mefloquine for malaria prophylaxis in nonimmune subjects. Antimicrob Agents Chemother 54:792–798

117. Leary KJ, Riel MA, Roy MJ, Cantilena LR, Bi D, Brater DC, van de Pol C, Pruett K, Kerr C, Veazey JM Jr et al (2009) A randomized, double-blind, safety and tolerability study to assess the ophthalmic and renal effects of tafenoquine 200 mg weekly versus placebo for 6 months in healthy volunteers. Am J Trop Med Hyg 81:356–362
118. Nasveld P, Kitchener S, Edstein M, Rieckmann K (2002) Comparison of tafenoquine (WR238605) and primaquine in the post-exposure (terminal) prophylaxis of vivax malaria in Australian Defence Force personnel. Trans R Soc Trop Med Hyg 96:683–684
119. Elmes NJ, Nasveld PE, Kitchener SJ, Kocisko DA, Edstein MD (2008) Comparison of three different dose regimens of tafenoquine versus primaquine for post exposure prophylaxis of vivax malaria in the South West Pacific. Trans Roy Soc Trop Med Hyg 102:1095–1101
120. Edstein MD, Nasveld PE, Kocisko DA, Kitchener SJ, Gatton ML, Rieckmann KH (2007) Gender differences in gastrointestinal disturbances and plasma concentrations of tafenoquine in healthy volunteers after tafenoquine administration for post-exposure vivax malaria prophylaxis. Trans R Soc Trop Med Hyg 101:226–230
121. Walsh DS, Looareesuwan S, Wilairatana P, Heppner DG Jr, Tang DB, Brewer TG, Chokejindachai W, Viriyavejakul P, Kyle DE, Milhous WK et al (1999) Randomized dose-ranging study of the safety and efficacy of WR 238605 (Tafenoquine) in the prevention of relapse of *Plasmodium vivax* malaria in Thailand. J Infect Dis 180:1282–1287
122. Walsh DS, Wilairatana P, Tang DB, Heppner DG Jr, Brewer TG, Krudsood S, Silachamroon U, Phumratanaprapin W, Siriyanonda D, Looareesuwan S (2004) Randomized trial of 3-dose regimens of tafenoquine (WR238605) versus low-dose primaquine for preventing *Plasmodium vivax* malaria relapse. Clin Infect Dis 39:1095–1103
123. Nasveld P, Kitchener S (2005) Treatment of acute vivax malaria with tafenoquine. Trans R Soc Trop Med Hyg 99:2–5
124. Kitchener S, Nasveld P, Edstein MD (2007) Tafenoquine for the treatment of recurrent *Plasmodium vivax* malaria. Am J Trop Med Hyg 76:494–496
125. Nosten F, White NJ (2007) Artemisinin-based combination treatment of falciparum malaria. Am J Trop Med Hyg 77:181–192
126. Roberts L, Enserink M (2007) Malaria. Did they really say ... eradication? Science 318:1544–1545

Other 4-Methanolquinolines, Amyl Alcohols and Phentathrenes: Mefloquine, Lumefantrine and Halofantrine

Francois Nosten, Penelope A. Phillips-Howard, and Feiko O. ter Kuile

Abstract This chapter describes mefloquine, pyronaridine, halofantrine, piperaquine and lumefantrine under the broader title of the 4-methanolquinolines, amyl alcohols and phentathrenes. We provide a brief resume of each drug, in terms of their chemical properties, formulae, pharmacokinetics, clinical indications for use, and their efficacy and safety. Recognizing the limited number of antimalarials available, and in the developmental pipeline, attention is focussed on describing the history of each drug and how their indications have evolved as data on safety in human populations accumulates over time, and how patterns of use have changed with growing multiple drug resistance. Their combined use with the artemisinin derivatives is briefly described and readers are recommended to consult other chapters in this book which more fully describe such combinations.

F. Nosten (✉)
Nuffield Department of Clinical Medicine, Centre for Tropical Medicine, Churchill Hospital, University of Oxford, Oxford OX3 7LJ, UK

Shoklo Malaria Research Unit, PO Box 46, 68/30 Baan Tung Road, Mae Sot, Tak Province 63110, Thailand

Mahidol University, Bangkok, Thailand
e-mail: SMRU@tropmedres.ac

P.A. Phillips-Howard
Centre for Public Health, Liverpool John Moores University, Liverpool, Henry Cotton Campus, 15–21 Webster Street, L3 2ET Liverpool, UK
e-mail: p.phillips-howard@ljmu.ac.uk

F.O. ter Kuile
Malaria Epidemiology Unit Child and Reproductive Health Group, Liverpool School of Tropical Medicine, Pembroke Place, Liverpool L3 5QA, UK
e-mail: terkuile@liverpool.ac.uk

H.M. Staines and S. Krishna (eds.), *Treatment and Prevention of Malaria*,
DOI 10.1007/978-3-0346-0480-2_5,

1 Mefloquine

1.1 Structure and Action

Mefloquine hydrochloride is a 4-quinolinemethol derivative synthesised as a structural analogue (2-aryl substituted chemical) of quinine. Its full chemical name is (R*, S*)-(±)-α-2-piperidinyl-2,8-bis (trifluoromethyl)-4-quinolinemethanol hydrochloride. Its formula is $C_{17}H_{16}F_6N_2O$ (Fig. 1). Mefloquine was discovered by the Experimental Therapeutics Division of the Walter Reed Army Institute of Research (WRAIR) in the 1970s for chemoprophylaxis (250 mg weekly) and therapy (15–25 mg/kg) and was approved by the U.S. Food and Drug Administration in 1989.

Mefloquine is a blood schizonticide, active against the erythrocytic stages of *Plasmodium falciparum* and *P. vivax*, with no effect on the exoerythrocytic (hepatic) stages of the parasite, and with limited information of its effect on *P. ovale, P. malariae* and *P. knowlesi*. Studies indicate mefloquine interferes with the transport of haemoglobin and other substances from the host erythrocyte to the food vacuole of the malaria parasite, causing swelling and cytotoxicity [1]. Mefloquine strongly inhibits endocytosis in the D10 strain of *P. falciparum* using several lines of evidence: a reduction in haemoglobin levels in the parasite as assessed by Western blotting, decreased levels of accumulation of biotinylated dextran by the parasite in preloaded erythrocytes, significantly lower concentrations of fluorescent dextran in the food vacuole, and a reduced percentage of parasites with multiple transport vesicles [2].

Mefloquine is a chiral molecule; it has two asymmetric carbon atoms and exists in two racemic forms (erythro and threo), each of which is composed of a pair of optical isomers, i.e. (±)-erythro-enantiomers and the (±)-threo-epimers. Clinically, the racemic mixture of the erythro-enantiomers is used [3]. Unlike other antimalarial drugs such as chloroquine, halofantrine, and lumefantrine, there is stereoselectivity

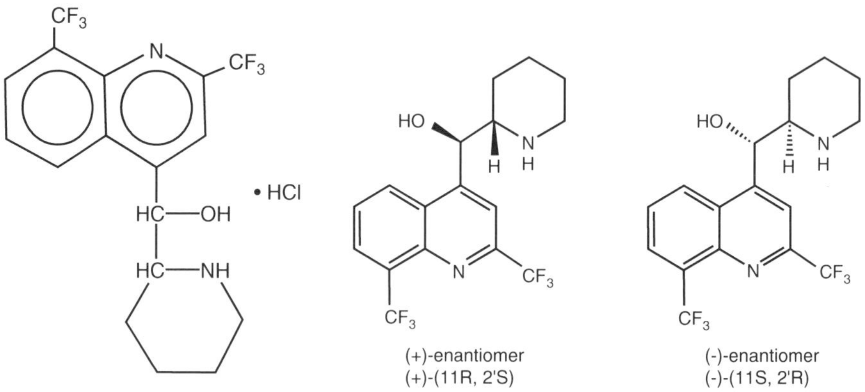

Fig. 1 Structure of mefloquine hydrochloride and its two enantiomers

in its antimalarial activity, with the (+)-isomer ~1.7 times more potent than the (−)-isomer in vitro [4, 5].

1.2 Pharmacokinetics

Mefloquine is moderately well absorbed orally and extensively distributed and is 98% bound to plasma proteins. Splitting 25 mg/kg mefloquine into 2 or 3 doses given 16–24 h apart reduces vomiting, improves oral bioavailability and the therapeutic response in the treatment of acute falciparum malaria [6]. Food increases its bioavailability by up to 40%. The parent compound is metabolized by the cytochrome P450 enzyme CYP3A4 to two major metabolites: carboxy- and hydroxyl-mefloquine, which are inactive against *P. falciparum*. Mefloquine is eliminated slowly and has a terminal elimination half-life of ~3 weeks in volunteers and 2 weeks in patients. Total clearance, which is essentially hepatic, is 30 ml/min in volunteers. A steady-state plasma concentration of 1,000–2,000 μg/l is reached after 7–10 weeks following weekly 250 mg prophylaxis and it is therefore recommended to start medication at least 2 weeks before travel. There are stereoselective differences in their pharmacokinetics and the ability of the mefloquine enantiomers to cause certain adverse effects [5]. In humans, the plasma concentration of the (−) enantiomer is approximately threefold higher than the (+) enantiomer, reflecting the stereoselectivity in the clearance and volume distribution [5, 7–11]. Co-administration with artemisinin does not appear to influence mefloquine enantiomer pharmacokinetics [12].

1.3 Clinical Use

One tablet of 250 mg mefloquine hydrochloride per week (adult dose; equivalent to 228 mg of the free base) has been used for prophylaxis in travellers, including for young children and pregnant women. The limited data available on the use of mefloquine in human pregnancy are reassuring and do not indicate an increased teratogenic risk [13]. A retrospective study of 208 women on the Thai–Burmese border treated with mefloquine found an increased risk of stillbirths [14], however, this finding was not confirmed in a large prospective trial of mefloquine prophylaxis in Malawian pregnant women [15] and remains unexplained. As a treatment, it is now mainly used in combination with artesunate, a water-soluble artemisinin derivative. It is available as a loose combination (as part of a blister pack containing both drugs), and as a new fixed-dose combination developed with support from the Drugs for Neglected Diseases Initiative (DNDi) and produced by Farmanguinhos/Fiocruz, Brazil and Cipla, India. The treatment dose is 25 mg/kg of mefloquine and 12 mg/kg of artesunate given as 8.3 and 4 mg/kg/day over 3 days, respectively. Tablets of the new fixed-dose combination come in adult and child "strengths", with several co-blistered formulations of the loose combinations made by different manufacturers. Mefloquine is also being explored for a new indication as intermittent preventive therapy (IPT) against malaria. Two trials evaluating the role of mefloquine as IPT in infants (IPTi) [16] and in pregnant women (IPTp) [17] found mefloquine to be very effective, but the low tolerability limited its acceptance for use as IPTi. Further IPTp studies are ongoing in five countries in Africa with the lower 15 mg/kg dose in pregnancy.

1.4 Resistance

Most experience with mefloquine as monotherapy and later in combination with the artemisinin derivatives has been gained from areas of multiple drug resistance in Southeast Asia, such as on the Thai–Burmese (Myanmar) border. Thailand was the first country to use mefloquine for first-line treatment of acute malaria. From 1985 to 1990, it was recommended in combination with sulfadoxine and pyrimethamine, as "MSP" in a fixed-dose combination, at a single dose of ~15/30/1.5 mg/kg, providing a 98% cure rate after its introduction in 1985, however, this dropped to <50% in children by 1990 [18]. Because of high levels of existing parasite resistance to SP, and lack of additional therapeutic efficacy over mefloquine alone, the SP component was dropped [18, 19] and replaced by mefloquine monotherapy; initially at a single dose of 15 mg/kg. High levels of treatment failure with this dose [18, 20] prompted 25 mg/kg dosing [21–23], split (750 and 500 mg, 16–24 h apart) to reduce vomiting [24]. Within 8 years, mefloquine monotherapy failure rates on the Thai–Burmese border reached 60% and, following extensive testing, the combination of mefloquine 25 mg/kg with artesunate 12 mg/kg given over 3 days (MAS_3) became the new standard therapy [25]. This therapy, combined with early diagnosis and use of insecticide treated nets, reduced *P. falciparum* malaria incidence, and halted, and later reversed the progression of mefloquine resistance [26–28]. The combination offered a potential public health solution for multiple drug-resistant *P. falciparum*, and allowed time for the development of other new drugs [1].

In vivo resistance to mefloquine, mediated mainly by an increase in gene copy number and expression of the *P. falciparum* multi-drug resistance (MDR) gene-1 (*pfmdr1*), a gene encoding a parasite-transport protein [29, 30], has been confirmed. This has been reported on the borders of Thailand with Burma (Myanmar) and Cambodia, in the western provinces of Cambodia, the eastern states of Burma (Myanmar) and its border with China, along the Laos and Burma borders, the adjacent Thai–Cambodian border and in southern Vietnam. It is likely the initial deployment of low dose of mefloquine may have encouraged resistance. Theoretical evidence suggests that initial use of higher doses, preferably in combination with an artemisinin derivative, is less likely to lead to resistance [31].

1.5 Tolerability

Preclinical studies demonstrate mefloquine to be safe and effective [32], and extensive clinical experience to date supports this. Nevertheless, widespread deployment of mefloquine for treatment and prophylaxis has been hampered by concerns about its tolerability. Side effects following treatment are common; they are usually mild and restricted to dizziness/vertigo and gastro-intestinal disturbances [24]. Vomiting after mefloquine is a problem in young children, but can be mitigated by splitting the dose over 2 or 3 days, and by fever reduction [24, 33]. In older children and adults, mild neuro-psychiatric events (headache, dizziness, insomnia and vivid dreams) are reported in ~25% of patients treated with 25 mg/kg mefloquine.

Mefloquine is also associated with a self-limiting acute neuropsychiatric syndrome manifest by encephalopathy, convulsions or psychosis [34–36], apparent in international travellers taking 250 mg mefloquine each week for prophylaxis [35, 37]. Mefloquine is thus contraindicated for prophylaxis in patients with active depression, a recent history of depression, generalised anxiety disorder, psychosis, or schizophrenia or other major psychiatric disorders, or with a history of convulsions [38]. These contraindications are prevalent in 9–10% of the military [39] and civilian [40] populations presenting for malaria chemoprophylaxis [41], but are not documented in endemic populations. While the mechanism is not yet fully understood, neuropsychiatric events have been demonstrated to be associated with dose in humans [42, 43]. The rates in travellers are estimated to be 1:10,000 persons, equally frequent with chloroquine prophylaxis, but higher than in similar populations that used other forms of prophylaxis [35, 44]. The incidence following treatment doses is 1:1,000 in Asian patients [45], 1:200 in Caucasian or African patients with uncomplicated malaria and 1:20 in patients recovering from severe malaria [46]. Previous history of psychiatric illness or epilepsy is a risk factor. Females and individuals of low body mass index are also at apparent greater risk. Neuropsychiatric reactions are more common if mefloquine was used in the previous 2 months, and thus should not be used to treat recrudescent infections within 2 months of treatment.

Total (racemic) concentrations of mefloquine are ~30-fold higher in brain than in plasma [47]. In man, approximately threefold higher concentrations of the (−)-enantiomer is observed in plasma, and 1.5-fold higher in brain, but post-mortem studies demonstrated stereoselective brain penetration, greater for the (+)-enantiomer, with (−) and (+) concentrations at ~23- versus 56-fold higher in the brain's white matter compared with plasma (the reverse is found in rat models where the penetration of the (−)-enantiomer is greater than that of its antipode) [7, 47–49]. There is a growing body of evidence on the mechanisms of possible neurotoxicity (see reviews [50] and [41]). The high level of accumulation of mefloquine in brain tissue may be associated with direct neurotoxic damage and cell death, with binding to neuroreceptors and cholinesterases, inhibition of sarco (endo)plasmic reticulum Ca^{2+} ATPase (SERCA) activity and interference with cellular Ca^{2+} homeostasis and reductions in central nervous system efflux in individuals possessing certain (human) MDR1 polymorphisms [50].

1.6 The Future

Using mefloquine as a scaffold, WRAIR has constructed a library of 200 potential next generation quinoline methanol compounds to identify leads that possess biological properties consistent with the target product profile for malaria chemoprophylaxis but less susceptible to passage across the blood–brain barrier (to reduce adverse neurological effects) [51, 52]. During a programme to examine the biochemical basis of side effects, investigators discovered that the (−)-(R,S)-enantiomer is a potent adenosine A2A receptor antagonist, resulting in a programme to develop novel adenosine A2A antagonists for the management of Parkinson's disease [53, 54]. Mefloquine is effective against JC virus and is reported to have successfully treated progressive multifocal leuko-encephalopathy (a progressive, usually fatal, demyelinating disease caused by the JC virus) [55, 56]. Mefloquine is undergoing in vitro and in vivo studies to evaluate its effectiveness for the treatment of helminth infections [57], including those caused by *Schistosoma* [58], *Clonorchis* and *Paragonimus* [59].

2 Pyronaridine

2.1 Structure and Action

Pyronaridine is a Mannich base with a pyronaridine nucleus synthesised from mepacrine (9 amino acridine). Its formula is $C_{29}H_{32}ClN_5O_2$ (Fig. 2). It was synthesised in 1970 in China and is available as a free base and as a tetraphosphate, the salt used in current formulations. It was used in China as a monotherapy in the 1980s and 1990s but has now been developed as a combination therapy with

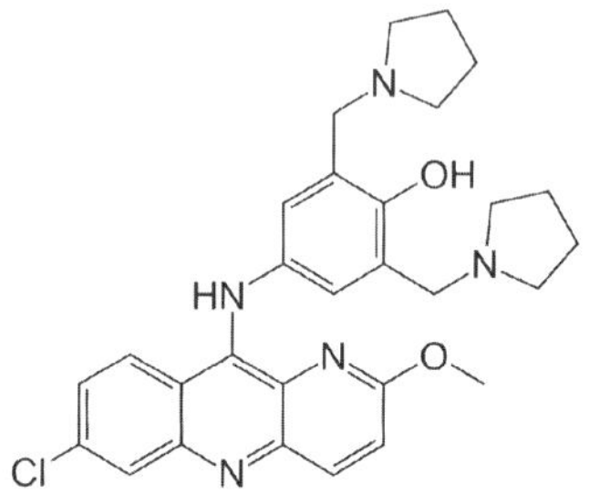

Fig. 2 Structure of pyronaridine

artesunate by Shin Poong Pharmaceutical (Korea) and Medicines for Malaria Venture (MMV, Switzerland).

Pyronaridine is active against asexual forms of *Plasmodium* by forming complexes with ferriprotoporphyrin IX. Growth studies of *P. falciparum* K1 in culture demonstrate the ability of pyronaridine to inhibit in vitro β-haematin formation, to form a complex with a stoichiometry of 1:2, to enhance haematin-induced red blood cell lysis, and to inhibit glutathione-dependent degradation of haematin [60]. However, observations that pyronaridine exerted this mechanism of action in situ, based on showing antagonism of pyronaridine in combination with antimalarials (chloroquine, mefloquine, and quinine) that inhibit β-haematin formation, were equivocal. Interestingly pyronaridine is also active against young gametocytes (stage II and III) in vitro [61], although recent clinical studies did not detect a difference in gametocyte carriage following treatment with the fixed combination artesunate–pyronaridine when compared with artemether– lumefantrine [62]. The compound is also a poor substrate and an inhibitor of the Permeability glycoprotein (P-gp) ATP-dependant transporter, a product of the human multidrug resistance-1 (MDR1;ABCB1) gene that influences the passage of many drugs across epithelial barriers. The P-gp-mediated efflux could attenuate oral absorption of the drug when the luminal concentration falls; however, this is likely to play a minimal role in the initial absorption of the drug.

2.2 Pharmacokinetics

The pharmacokinetic properties of pyronaridine are not well characterised. The drug is readily absorbed from the small intestine following oral administration and is widely distributed in most tissues [63]. The peak value of the drug in the blood is reached at around 8 h post-administration, and it shows a poor permeability across the blood–brain barrier. Pyronaridine concentrates preferentially in infected red blood cells and its distribution and elimination are influenced by age and disease status. It is eliminated slowly, with the half-life currently estimated at 18 days in patients with malaria [64]. Improvements to the assay that measures blood concentrations are likely to reveal a longer half-life.

2.3 Resistance

Resistance is known to have developed in *P. falciparum* when the drug was used in China but the molecular mechanism is unknown. High recurrence rates were noted in early clinical trials in Thailand [65] and cross-resistance with chloroquine was suggested in vitro [66]. Pyronaridine is not used as a single agent anymore but only in combination with artesunate to prevent the emergence of de novo resistance.

3 Halofantrine

3.1 Structure, Action and Resistance

Halofantrine belongs to the phenanthrene methanol group and was developed by WRAIR in the 1960s and then by Smith Klein, now GlaxoSmithKlein. It is a small chiral molecule and the chemical formula is $C_{26}H_{30}Cl_2F_3NO$ (Fig. 3). The oral formulation is in tablets containing 250 mg of hydrochloride salt. Like other quinoline derivatives, the mode of action of halofantrine appears to be in the inhibition of the formation of β-haematin crystals but the precise mechanism of action is unclear. Recently it was shown that halofantrine forms complexes with ferriprotoporphyrin IX and that the inhibition of the haemozoin formation occurs principally at the lipid–aqueous interface, an environment more compatible with the crystal structure of halofantrine–ferriprotoporphyrin IX [67]. Halofantrine is poorly and erratically absorbed due to its low solubility. Absorption is enhanced by fat co-administration. The principal active metabolite is *N*-desbutyl-halofantrine. The terminal half-life in patients with malaria is ~4.7 days, making it a relatively rapidly eliminated antimalarial when compared with other drugs of the same group. As halofantrine was withdrawn from use due to the discovery of halofantrine cardiotoxicity (see below), there are few studies on the development of resistance to this drug. Molecular studies have demonstrated that mutations in *pfmdr1* result in altered halofantrine transport, suggesting a role for this efflux transport mechanism in resistance to this drug [68]. Cross-resistance with mefloquine was shown in clinical studies in South East Asia [69] and it is likely that the drug shares common resistance mechanisms with mefloquine and lumefantrine due to their chemical similarities.

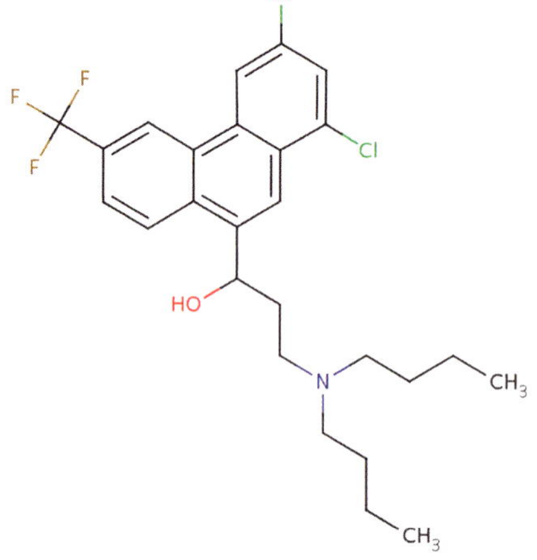

Fig. 3 Structure of halofantrine

3.2 Tolerability

Initially, halofantrine looked like a promising drug for the treatment of uncomplicated falciparum infections caused by chloroquine-resistant parasites. It rapidly cleared parasites and was well tolerated. The first report of the cardiotoxicity of halofantrine in 1993 [70], came as a surprise since the drug had been developed in full compliance with GCP standards. Halofantrine and its principal quinidine-like metabolite have a Class III effect on cardiac repolarization [71]. It causes a dose-dependent blockade of the I_{kr} channel (through hERG) by binding to the open or inactivated state. This translates on an ECG to a marked prolongation of the QT interval and is more marked when halofantrine is given after mefloquine, probably as a result of the inhibition of the slow delayed rectifier potassium channel I_{ks} [72]. This QT prolongation, seen at therapeutic doses, increases the risk of the potentially fatal Torsade de Pointes and since the first report, several sudden deaths have been related to the drug. Because of this, halofantrine has been withdrawn in many countries and from international guidelines on the treatment of malaria.

4 Piperaquine

4.1 Structure and Action

Piperaquine is a bis-amino 4 quinoline synthesised more than 50 years ago by Rhone-Poulenc (France). It was abandoned and rediscovered in China by the Shanghai Research Institute of Pharmaceutical Industry. The drug was used on a large scale (140,000,000 doses) for prophylaxis and treatment of chloroquine resistant *P. falciparum* between 1978 and 1994, but resistance developed in the 1990s. Piperaquine is a lipophilic compound and its chemical formula is $C_{29}H_{32}Cl_2N_6$ (Fig. 4). The full mechanism of action is unknown but piperaquine concentrates in the parasite food vacuole and inhibits the dimerisation of haematin by binding. Recent work by Warhurst and colleagues has shed more light on the possible mode of action [73]. The high activity of piperaquine against chloroquine-resistant falciparum could be explained by its high lipid accumulation ratio (LAR) leading to an increase in β-haematin inhibition in vacuolar lipids where the crystals of haemozoin are produced. The drug may also act by blocking efflux from the food

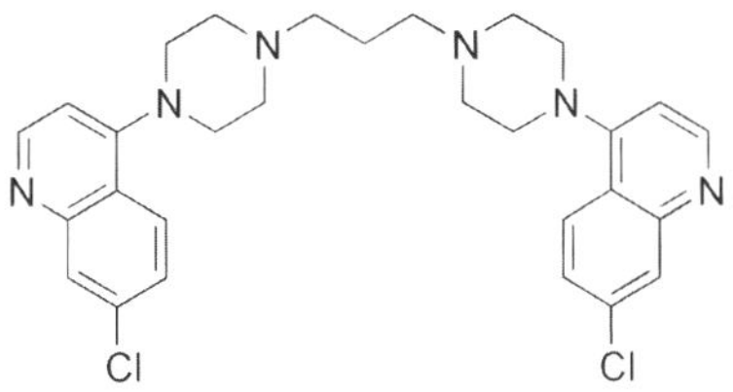

Fig. 4 Structure of piperaquine

vacuole by hydrophobic interaction with the parasite chloroqine-resistance transporter, *pfcrt* [73].

4.2 Pharmacokinetics

Piperaquine is lipophilic and exhibits considerable inter-individual variability in pharmacokinetics. It accumulates preferentially in infected red blood cells and this affects the plasma/blood concentration ratio. Like chloroquine, it has a large apparent volume of distribution and a slow elimination. The terminal elimination half-life is probably longer than previously thought and could exceed 30 days [74]. This is valuable for ensuring a prolonged post-treatment prophylactic effect and when the drug is used for IPT. In children, studies have shown that, dose adjustment may be needed [75]. Likewise, pregnant women may have lower piperaquine exposure than non-pregnant women, and the dosage in pregnancy may also need to be adjusted. Piperaquine has two major metabolites: a carboxylic acid and a mono-*N*-oxidated piperaquine product.

4.3 Resistance

Resistance to piperaquine is known to have developed in vivo in China when it was used as monotherapy but there is no indication that it has spread elsewhere. There is no specific molecular marker of resistance to piperaquine and the role of the *P. falciparum* transport proteins *pfmdr1* and *pfcrt* remains unclear [76, 77]. In clinical use, piperaquine is now used in a fixed combination with dihydroartemisinin (DHA) developed by Holleypharm (China) and Sigma-Tau (Italy) in partnership with MMV (Switzerland). DHA-piperaquine is one of the most promising artemisinin-based combination therapies (ACTs) in the antimalarial armamentarium.

5 Lumefantrine

5.1 Structure and Action

Lumefantrine is a racemic 2,4,7,9-substituted fluorine derivative belonging to the arylamino-alcohol group of antimalarials with a molecular structure reminiscent of halofantrine. It was originally synthesised by the Academy of Military Sciences in Beijing (PRC) in the 1980s. Its chemical formula is $C_{30}H_{32}Cl_3NO$ (Fig. 5). It is insoluble in water and the two enantiomers have equal antimalarial activities.

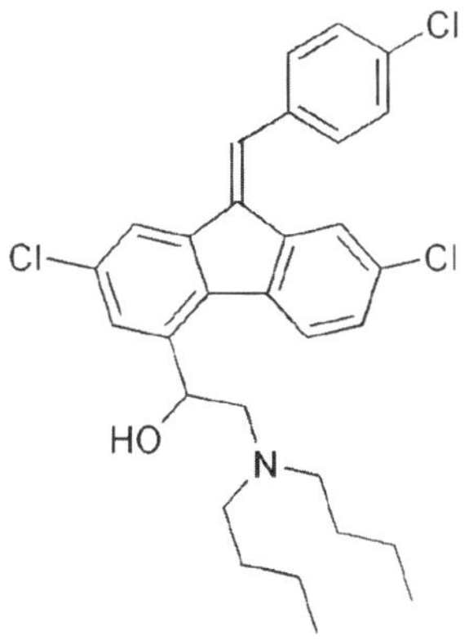

Fig. 5 Structure of lumefantrine

Lumefantrine is active against *P. falciparum* and *P. vivax* asexual stages but not against pre-erythrocytic liver stages, including hynozoites, or against gametocytes. The mode of action of lumefantrine is not known precisely but by similarity of structure with the antimalarials of the same group, it is assumed that lumefantrine kills parasites by inhibiting the polymerisation of heme.

5.2 *Pharmacokinetics*

The pharmacokinetics of lumefantrine have been extensively described in various populations (European, Asian and African) in adults, children and also in pregnant women. The drug is slowly absorbed (time to peak concentrations is approximately 6 h) and metabolised to desbutyl-lumefantrine via CYP3A4 but largely eliminated as parent compound via the liver in faeces and urine. The absorption is dose limited, so the total daily dose must be given on two separate occasions in order to be absorbed, the first serious impediment to observance [78]. The terminal elimination half-life is approximately 4.5 days, reminiscent of halofantrine and much shorter than mefloquine or piperaquine. Lumefantrine is highly lipophilic and has a low and variable bioavailability. This is a major contributor to the observed inter-individual variability in its kinetics. The relative fraction of the dose absorbed is also highly variable between patients and between doses. This is probably explained by the combined effects of illness on intestinal mobility, increased food intake with recovery and decreasing parasitaemia. Co-administration of food (or some fat) has a marked effect on the absorption of lumefantrine and this is the second major impediment to observance. Recent studies have shown that as little as 1.2 g of fat was needed for optimum lumefantrine absorption [71]. Unfortunately, this was not taken into account when the paediatric formulation of artemether–lumefantrine (Coartem) was developed. Initially, a 4-dose regimen of Coartem was recommended based on the initial Chinese trials. However this resulted in low cure rates in Thailand [79] but it helped in defining the lumefantrine exposure–cure rate relationship. The most determinant factors of cure were found to be the initial parasite load and the Area Under the plasma concentration Curve (AUC) of lumefantrine. A useful surrogate of the AUC is the day 7 lumefantrine concentration. A plasma lumefantrine concentration of 280 ng/ml was found to be a useful discriminating cut-off to determine subsequent risk of recrudescence [80] and, in the absence of resistance, a day 7 concentration of 500 ng/ml would be expected to cure >90% of patients [81]. After too many years of delay, the 6-dose regimen became universally recommended. For pregnancy, however, this standard regimen is associated with lower plasma concentrations because of the increased volume of distribution and faster elimination, which will lead to treatment failures [82]. Modelling suggests that longer courses are needed to achieve lumefantrine concentrations comparable to that in non-pregnant patients [75].

5.3 Resistance

Resistance to lumefantrine can be readily obtained in vitro and in animal models. In vitro, single-nucleotide polymorphisms in the *pfmdr1* gene have been associated with increased IC_{50} values for lumefantrine [83]. An increase in *pfmdr1* copy number also resulted in decreased in vitro susceptibility and increased risk of failure in patients receiving the 4-dose regimen [80]. Interestingly, resistance to lumefantrine in *P. falciparum* could be associated with the loss of chloroquine resistance (i.e. the loss of the *pfcrt* K76T mutation) [84]. Lumefantrine is not used as a single drug but only in combination with artemether (see Coartem). Its future depends on controlling the emergence of resistance to the artemisinin derivatives and/or to the emergence of resistance to lumefantrine itself or through cross resistance with mefloquine.

References

1. Olliaro P (2001) Mode of action and mechanisms of resistance for antimalarial drugs. Pharmacol Ther 89:207–219
2. Hoppe HC, van Schalkwyk DA, Wiehart UI, Meredith SA, Egan J, Weber BW (2004) Antimalarial quinolines and artemisinin inhibit endocytosis in *Plasmodium falciparum*. Antimicrob Agents Chemother 48:2370–2378
3. Sweeney TR (1981) The present status of malaria chemotherapy: mefloquine, a novel antimalarial. Med Res Rev 1:281–301
4. Karle JM, Olmeda R, Gerena L, Milhous WK (1993) *Plasmodium falciparum*: role of absolute stereochemistry in the antimalarial activity of synthetic amino alcohol antimalarial agents. Exp Parasitol 76:345–351
5. Brocks DR, Mehvar R (2003) Stereoselectivity in the pharmacodynamics and pharmacokinetics of the chiral antimalarial drugs. Clin Pharmacokinet 42:1359–1382

6. Simpson JA, Price R, ter Kuile F, Teja-Isavatharm P, Nosten F, Chongsuphajaisiddhi T, Looareesuwan S, Aarons L, White NJ (1999) Population pharmacokinetics of mefloquine in patients with acute falciparum malaria. Clin Pharmacol Ther 66:472–484
7. Gimenez F, Pennie RA, Koren G, Crevoisier C, Wainer IW, Farinotti R (1994) Stereoselective pharmacokinetics of mefloquine in healthy Caucasians after multiple doses. J Pharm Sci 83:824–827
8. Martin C, Gimenez F, Bangchang KN, Karbwang J, Wainer IW, Farinotti R (1994) Whole blood concentrations of mefloquine enantiomers in healthy Thai volunteers. Eur J Clin Pharmacol 47:85–87
9. Hellgren U, Jastrebova J, Jerling M, Krysen B, Bergqvist Y (1996) Comparison between concentrations of racemic mefloquine, its separate enantiomers and the carboxylic acid metabolite in whole blood serum and plasma. Eur J Clin Pharmacol 51:171–173
10. Bourahla A, Martin C, Gimenez F, Singhasivanon V, Attanath P, Sabchearon A, Chongsuphajaisiddhi T, Farinotti R (1996) Stereoselective pharmacokinetics of mefloquine in young children. Eur J Clin Pharmacol 50:241–244
11. Kerb R, Fux R, Morike K, Kremsner PG, Gil JP, Gleiter CH, Schwab M (2009) Pharmacogenetics of antimalarial drugs: effect on metabolism and transport. Lancet Infect Dis 9:760–774
12. Svensson US, Alin H, Karlsson MO, Bergqvist Y, Ashton M (2002) Population pharmacokinetic and pharmacodynamic modelling of artemisinin and mefloquine enantiomers in patients with falciparum malaria. Eur J Clin Pharmacol 58:339–351
13. Phillips-Howard PA, Steffen R, Kerr L, Vanhauwere B, Schildknecht J, Fuchs E, Edwards R (1998) Safety of mefloquine and other antimalarial agents in the first trimester of pregnancy. J Travel Med 5:121–126
14. Nosten F, Vincenti M, Simpson J, Yei P, Thwai KL, de Vries A, Chongsuphajaisiddhi T, White NJ (1999) The effects of mefloquine treatment in pregnancy. Clin Infect Dis 28:808–815
15. Steketee RW, Wirima JJ, Slutsker L, Roberts JM, Khoromana CO, Heymann DL, Breman JG (1996) Malaria parasite infection during pregnancy and at delivery in mother, placenta, and newborn: efficacy of chloroquine and mefloquine in rural Malawi. Am J Trop Med Hyg 55(1 Suppl):24–32
16. Gosling RD, Gesase S, Mosha JF, Carneiro I, Hashim R, Lemnge M, Mosha FW, Greenwood B, Chandramohan D (2009) Protective efficacy and safety of three antimalarial regimens for intermittent preventive treatment for malaria in infants: a randomised, double-blind, placebo-controlled trial. Lancet 374:1521–1532
17. Briand V, Bottero J, Noel H, Masse V, Cordel H, Guerra J, Kossou H, Fayomi B, Ayemonna P, Fievet N et al (2009) Intermittent treatment for the prevention of malaria during pregnancy in Benin: a randomized, open-label equivalence trial comparing sulfadoxine-pyrimethamine with mefloquine. J Infect Dis 200:991–1001
18. Nosten F, ter Kuile F, Chongsuphajaisiddhi T, Luxemburger C, Webster HK, Edstein M, Phaipun L, Thew KL, White NJ (1991) Mefloquine-resistant falciparum malaria on the Thai-Burmese border. Lancet 337:1140–1143
19. Nosten F, Price RN (1995) New antimalarials. A risk-benefit analysis. Drug Saf 12:264–273
20. White NJ (1992) Antimalarial drug resistance: the pace quickens. J Antimicrob Chemother 30:571–585
21. ter Kuile FO, Nosten F, Thieren M, Luxemburger C, Edstein MD, Chongsuphajaisiddhi T, Phaipun L, Webster HK, White NJ (1992) High-dose mefloquine in the treatment of multidrug-resistant falciparum malaria. J Infect Dis 166:1393–1400
22. Fontanet AL, Johnston DB, Walker AM, Rooney W, Thimasarn K, Sturchler D, Macdonald M, Hours M, Wirth DF (1993) High prevalence of mefloquine-resistant falciparum malaria in eastern Thailand. Bull World Health Organ 71:377–383
23. Fontanet AL, Johnston BD, Walker AM, Bergqvist Y, Hellgren U, Rooney W (1994) Falciparum malaria in eastern Thailand: a randomized trial of the efficacy of a single dose of mefloquine. Bull World Health Organ 72:73–78

24. ter Kuile FO, Nosten F, Luxemburger C, Kyle D, Teja-Isavatharm P, Phaipun L, Price R, Chongsuphajaisiddhi T, White NJ (1995) Mefloquine treatment of acute falciparum malaria: a prospective study of non-serious adverse effects in 3673 patients. Bull World Health Organ 73:631–642
25. Ter Kuile FO, Teja-Isavatharm P, Edstein MD, Keeratithakul D, Dolan G, Nosten F, Phaipun L, Webster HK, White NJ (1994) Comparison of capillary whole blood, venous whole blood, and plasma concentrations of mefloquine, halofantrine, and desbutyl-halofantrine measured by high-performance liquid chromatography. Am J Trop Med Hyg 51:778–784
26. Nosten F, van Vugt M, Price R, Luxemburger C, Thway KL, Brockman A, McGready R, ter Kuile F, Looareesuwan S, White NJ (2000) Effects of artesunate-mefloquine combination on incidence of *Plasmodium falciparum* malaria and mefloquine resistance in western Thailand: a prospective study. Lancet 356:297–302
27. Carrara VI, Sirilak S, Thonglairuam J, Rojanawatsirivet C, Proux S, Gilbos V, Brockman A, Ashley EA, McGready R, Krudsood S et al (2006) Deployment of early diagnosis and mefloquine-artesunate treatment of falciparum malaria in Thailand: the Tak Malaria Initiative. PLoS Med 3:e183
28. Carrara VI, Zwang J, Ashley EA, Price RN, Stepniewska K, Barends M, Brockman A, Anderson T, McGready R, Phaiphun L et al (2009) Changes in the treatment responses to artesunate-mefloquine on the northwestern border of Thailand during 13 years of continuous deployment. PLoS One 4:e4551
29. Price R, Robinson G, Brockman A, Cowman A, Krishna S (1997) Assessment of *pfmdr 1* gene copy number by tandem competitive polymerase chain reaction. Mol Biochem Parasitol 85:161–169
30. Price RN, Uhlemann AC, Brockman A, McGready R, Ashley E, Phaipun L, Patel R, Laing K, Looareesuwan S, White NJ et al (2004) Mefloquine resistance in *Plasmodium falciparum* and increased *pfmdr1* gene copy number. Lancet 364:438–447
31. Simpson JA, Watkins ER, Price RN, Aarons L, Kyle DE, White NJ (2000) Mefloquine pharmacokinetic-pharmacodynamic models: implications for dosing and resistance. Antimicrob Agents Chemother 44:3414–3424
32. Palmer KJ, Holliday SM, Brogden RN (1993) Mefloquine: a review of its antimalarial activity, pharmacokinetic properties and therapeutic efficacy. Drugs 45:430–475
33. Ashley EA, McGready R, Hutagalung R, Phaiphun L, Slight T, Proux S, Thwai KL, Barends M, Looareesuwan S, White NJ et al (2005) A randomized, controlled study of a simple, once-daily regimen of dihydroartemisinin-piperaquine for the treatment of uncomplicated, multi-drug-resistant falciparum malaria. Clin Infect Dis 41:425–432
34. Bem JL, Kerr L, Stuerchler D (1992) Mefloquine prophylaxis: an overview of spontaneous reports of severe psychiatric reactions and convulsions. J Trop Med Hyg 95:167–179
35. Phillips-Howard PA, ter Kuile FO (1995) CNS adverse events associated with antimalarial agents. Fact or fiction? Drug Saf 12:370–383
36. WHO (1991) Review of central nervous system adverse events related to the antimalarial drug, mefloquine (1985–1990). World Health Organization, Geneva
37. Schlagenhauf P, Steffen R, Lobel H, Johnson R, Letz R, Tschopp A, Vranjes N, Bergqvist Y, Ericsson O, Hellgren U et al (1996) Mefloquine tolerability during chemoprophylaxis: focus on adverse event assessments, stereochemistry and compliance. Trop Med Int Health 1:485–494
38. Roche Laboratories Inc (2008) Lariam® (mefloquine hydrochloride) Complete Product Information. US Package Insert, Nutley, NJ
39. Nevin RL, Pietrusiak PP, Caci JB (2008) Prevalence of contraindications to mefloquine use among USA military personnel deployed to Afghanistan. Malar J 7:30
40. Hill DR (1991) Pre-travel health, immunization status, and demographics of travel to the developing world for individuals visiting a travel medicine service. Am J Trop Med Hyg 45:263–270

41. Nevin RL (2009) Epileptogenic potential of mefloquine chemoprophylaxis: a pathogenic hypothesis. Malar J 8:188
42. Taylor WR, White NJ (2004) Antimalarial drug toxicity: a review. Drug Saf 27:25–61
43. Baird JK (2005) Effectiveness of antimalarial drugs. N Engl J Med 352:1565–1577
44. Weinke T, Trautmann M, Held T, Weber G, Eichenlaub D, Fleischer K, Kern W, Pohle HD (1991) Neuropsychiatric side effects after the use of mefloquine. Am J Trop Med Hyg 45:86–91
45. Luxemburger C, Nosten F, ter Kuiile F, Frejacques L, Chongsuphajaisiddhi T, White NJ (1991) Mefloquine for multidrug-resistant malaria. Lancet 338:1268
46. Nguyen TH, Day NP, Ly VC, Waller D, Mai NT, Bethell DB, Tran TH, White NJ (1996) Post-malaria neurological syndrome. Lancet 348:917–921
47. Pham YT, Nosten F, Farinotti R, White NJ, Gimenez F (1999) Cerebral uptake of mefloquine enantiomers in fatal cerebral malaria. Int J Clin Pharmacol Ther 37:58–61
48. Gimenez F, Gillotin C, Basco LK, Bouchaud O, Aubry AF, Wainer IW, Le Bras J, Farinotti R (1994) Plasma concentrations of the enantiomers of halofantrine and its main metabolite in malaria patients. Eur J Clin Pharmacol 46:561–562
49. Pham YT, Regina A, Farinotti R, Couraud P, Wainer IW, Roux F, Gimenez F (2000) Interactions of racemic mefloquine and its enantiomers with P-glycoprotein in an immortalised rat brain capillary endothelial cell line, GPNT. Biochim Biophys Acta 1524:212–219
50. Toovey S (2009) Mefloquine neurotoxicity: a literature review. Travel Med Infect Dis 7:2–6
51. Milner E, McCalmont W, Bhonsle J, Caridha D, Carroll D, Gardner S, Gerena L, Gettayacamin M, Lanteri C, Luong T et al (2010) Structure-activity relationships amongst 4-position quinoline methanol antimalarials that inhibit the growth of drug sensitive and resistant strains of *Plasmodium falciparum*. Bioorg Med Chem Lett 20:1347–1351
52. Milner E, McCalmont W, Bhonsle J, Caridha D, Cobar J, Gardner S, Gerena L, Goodine D, Lanteri C, Melendez V et al (2010) Anti-malarial activity of a non-piperidine library of next-generation quinoline methanols. Malar J 9:51
53. Weiss SM, Benwell K, Cliffe IA, Gillespie RJ, Knight AR, Lerpiniere J, Misra A, Pratt RM, Revell D, Upton R et al (2003) Discovery of nonxanthine adenosine A2A receptor antagonists for the treatment of Parkinson's disease. Neurology 61:S101–S106
54. Gillespie RJ, Adams DR, Bebbington D, Benwell K, Cliffe IA, Dawson CE, Dourish CT, Fletcher A, Gaur S, Giles PR et al (2008) Antagonists of the human adenosine A2A receptor. Part 1: Discovery and synthesis of thieno[3,2-d]pyrimidine-4-methanone derivatives. Bioorg Med Chem Lett 18:2916–2919
55. Gofton TE, Al-Khotani A, O'Farrell B, Ang LC, McLachlan RS (2011) Mefloquine in the treatment of progressive multifocal leukoencephalopathy. J Neurol Neurosurg Psychiatry 82:452–455
56. Brickelmaier M, Lugovskoy A, Kartikeyan R, Reviriego-Mendoza MM, Allaire N, Simon K, Frisque RJ, Gorelik L (2009) Identification and characterization of mefloquine efficacy against JC virus *in vitro*. Antimicrob Agents Chemother 53:1840–1849
57. Keiser J, Utzinger J (2010) The drugs we have and the drugs we need against major helminth infections. Adv Parasitol 73:197–230
58. Holtfreter MC, Loebermann M, Klammt S, Sombetzki M, Bodammer P, Riebold D, Kinzelbach R, Reisinger EC (2011) *Schistosoma mansoni*: schistosomicidal effect of mefloquine and primaquine *in vitro*. Exp Parasitol 127:270–276
59. Xiao SH, Xue J, Li-Li X, Zhang YN, Qiang HQ (2010) Effectiveness of mefloquine against *Clonorchis sinensis* in rats and *Paragonimus westermani* in dogs. Parasitol Res 107:1391–1397
60. Auparakkitanon S, Chapoomram S, Kuaha K, Chirachariyavej T, Wilairat P (2006) Targeting of hematin by the antimalarial pyronaridine. Antimicrob Agents Chemother 50:2197–2200
61. Chavalitshewinkoon-Petmitr P, Pongvilairat G, Auparakkitanon S, Wilairat P (2000) Gametocytocidal activity of pyronaridine and DNA topoisomerase II inhibitors against multi-drug-resistant *Plasmodium falciparum in vitro*. Parasitol Int 48:275–280

62. Tshefu AK, Gaye O, Kayentao K, Thompson R, Bhatt KM, Sesay SS, Bustos DG, Tjitra E, Bedu-Addo G, Borghini-Fuhrer I et al (2010) Efficacy and safety of a fixed-dose oral combination of pyronaridine-artesunate compared with artemether-lumefantrine in children and adults with uncomplicated *Plasmodium falciparum* malaria: a randomised non-inferiority trial. Lancet 375:1457–1467
63. Park SH, Pradeep K (2010) Absorption, distribution, excretion, and pharmacokinetics of 14 C-pyronaridine tetraphosphate in male and female Sprague-Dawley rats. J Biomed Biotechnol 2010:590707
64. Wattanavijitkul T. (2010) Population pharmacokinetics of pyronaridine in the treatment of malaria. Ph.D. dissertation, University of Iowa. http://ir.uiowa.edu/etd/622. Accessed 15 Apr 2011
65. Looareesuwan S, Kyle DE, Viravan C, Vanijanonta S, Wilairatana P, Wernsdorfer WH (1996) Clinical study of pyronaridine for the treatment of acute uncomplicated falciparum malaria in Thailand. Am J Trop Med Hyg 54:205–209
66. Elueze EI, Croft SL, Warhurst DC (1996) Activity of pyronaridine and mepacrine against twelve strains of *Plasmodium falciparum in vitro*. J Antimicrob Chemother 37:511–518
67. de Villiers KA, Marques HM, Egan TJ (2008) The crystal structure of halofantrine-ferriprotoporphyrin IX and the mechanism of action of arylmethanol antimalarials. J Inorg Biochem 102:1660–1667
68. Sanchez CP, Rotmann A, Stein WD, Lanzer M (2008) Polymorphisms within PfMDR1 alter the substrate specificity for anti-malarial drugs in *Plasmodium falciparum*. Mol Microbiol 70:786–798
69. ter Kuile FO, Dolan G, Nosten F, Edstein MD, Luxemburger C, Phaipun L, Chongsuphajaisiddhi T, Webster HK, White NJ (1993) Halofantrine versus mefloquine in treatment of multidrug-resistant falciparum malaria [see comments]. Lancet 341:1044–1049
70. Nosten F, ter Kuile F, Luxemburger C, Woodrow C, Kyle DE, Chongsuphajaisiddhi T, White NJ (1993) Cardiac effects of antimalarial treatment with halofantrine. Lancet 341:1054–1056
71. Ashley EA, Stepniewska K, Lindegardh N, Annerberg A, Kham A, Brockman A, Singhasivanon P, White NJ, Nosten F (2007) How much fat is necessary to optimize lumefantrine oral bioavailability? Trop Med Int Health 12:195–200
72. Tie H, Walker BD, Singleton CB, Valenzuela SM, Bursill JA, Wyse KR, Breit SN, Campbell TJ (2000) Inhibition of HERG potassium channels by the antimalarial agent halofantrine. Br J Pharmacol 130:1967–1975
73. Warhurst DC, Craig JC, Adagu IS, Guy RK, Madrid PB, Fivelman QL (2007) Activity of piperaquine and other 4-aminoquinoline antiplasmodial drugs against chloroquine-sensitive and resistant blood-stages of *Plasmodium falciparum*. Role of beta-haematin inhibition and drug concentration in vacuolar water- and lipid-phases. Biochem Pharmacol 73:1910–1926
74. Tarning J, Lindegardh N, Annerberg A, Singtoroj T, Day NP, Ashton M, White NJ (2005) Pitfalls in estimating piperaquine elimination. Antimicrob Agents Chemother 49:5127–5128
75. Tarning J, Ashley EA, Lindegardh N, Stepniewska K, Phaiphun L, Day NP, McGready R, Ashton M, Nosten F, White NJ (2008) Population pharmacokinetics of piperaquine after two different treatment regimens with dihydroartemisinin-piperaquine in patients with *Plasmodium falciparum* malaria in Thailand. Antimicrob Agents Chemother 52:1052–1061
76. Briolant S, Henry M, Oeuvray C, Amalvict R, Baret E, Didillon E, Rogier C, Pradines B (2010) Absence of association between piperaquine *in vitro* responses and polymorphisms in the *pfcrt*, *pfmdr1*, *pfmrp*, and *pfnhe* genes in *Plasmodium falciparum*. Antimicrob Agents Chemother 54:3537–3544
77. Muangnoicharoen S, Johnson DJ, Looareesuwan S, Krudsood S, Ward SA (2009) Role of known molecular markers of resistance in the antimalarial potency of piperaquine and dihydroartemisinin *in vitro*. Antimicrob Agents Chemother 53:1362–1366
78. Ashley EA, Stepniewska K, Lindegardh N, McGready R, Annerberg A, Hutagalung R, Singtoroj T, Hla G, Brockman A, Proux S et al (2007) Pharmacokinetic study of artemether-lumefantrine given once daily for the treatment of uncomplicated multidrug-resistant falciparum malaria. Trop Med Int Health 12:201–208

79. van Vugt M, Wilairatana P, Gemperli B, Gathmann I, Phaipun L, Brockman A, Luxemburger C, White NJ, Nosten F, Looareesuwan S (1999) Efficacy of six doses of artemether-lumefantrine (benflumetol) in multidrug-resistant *Plasmodium falciparum* malaria. Am J Trop Med Hyg 60:936–942
80. Price RN, Uhlemann AC, van Vugt M, Brockman A, Hutagalung R, Nair S, Nash D, Singhasivanon P, Anderson TJ, Krishna S et al (2006) Molecular and pharmacological determinants of the therapeutic response to artemether-lumefantrine in multidrug-resistant *Plasmodium falciparum* malaria. Clin Infect Dis 42:1570–1577
81. White NJ, van Vugt M, Ezzet F (1999) Clinical pharmacokinetics and pharmacodynamics and pharmacodynamics of artemether-lumefantrine. Clin Pharmacokinet 37:105–125
82. McGready R, Tan SO, Ashley EA, Pimanpanarak M, Viladpai-Nguen J, Phaiphun L, Wustefeld K, Barends M, Laochan N, Keereecharoen L et al (2008) A randomised controlled trial of artemether-lumefantrine versus artesunate for uncomplicated *Plasmodium falciparum* treatment in pregnancy. PLoS Med 5:e253
83. Anderson TJ, Nair S, Qin H, Singlam S, Brockman A, Paiphun L, Nosten F (2005) Are transporter genes other than the chloroquine resistance locus (*pfcrt*) and multidrug resistance gene (*pfmdr*) associated with antimalarial drug resistance? Antimicrob Agents Chemother 49:2180–2188
84. Johnson DJ, Fidock DA, Mungthin M, Lakshmanan V, Sidhu AB, Bray PG, Ward SA (2004) Evidence for a central role for PfCRT in conferring *Plasmodium falciparum* resistance to diverse antimalarial agents. Mol Cell 15:867–877

Antifolates: Pyrimethamine, Proguanil, Sulphadoxine and Dapsone

Alexis Nzila

Abstract The inhibition or disruption of folate metabolism remains an attractive target for the discovery of new antimalarial drugs. The importance of this pathway was proved in the 1940s with the discovery of the triazine proguanil. Proguanil is converted in vivo to the active metabolite, cycloguanil, an inhibitor of the dihydrofolate reductase enzyme. Proguanil has mainly been used for prophylaxis and currently is used in combination with atovaquone (Malarone®) for this purpose. Pyrimethamine was discovered based on its similarity to cycloguanil, and has been combined with the sulpha drug sulphadoxine. This combination of pyrimethamine/sulphadoxine has been the drug of choice to replace chloroquine in the treatment of uncomplicated malaria. However, resistance to pyrimethamine/sulphadoxine is now common, and its use is now restricted to the treatment of malaria in pregnancy, and "intermittent preventive treatment." Efforts are under way to discover and develop new antifolates. In this chapter, I summarize our knowledge of folate metabolism in the malarial parasite, and discuss the role and place of antifolates in the treatment of malaria and new strategies of folate disruption as a drug target.

1 Folate Biochemistry

Antifolates block the synthesis or conversion of folate derivatives. Folate derivatives are important cellular cofactors for the production of deoxythymidylate (dTMP) and, thus, synthesis of DNA. In mammals and plants, the folate pathway also generates the amino acid methionine, mediates the metabolism of histidine, glutamic acid and serine and controls the initiation of protein synthesis in mitochondria through formylation of methionine [1, 2]. Thus, rapidly dividing cells, such as cancer cells

A. Nzila (✉)
Biological studies Group, Department of Chemistry King Fayhd University of Petroleum and Minerals, PO Box 468, Dharan 31261, Saudi Arabia
e-mail: alexisnzila@kfupm.edu.sa

H.M. Staines and S. Krishna (eds.), *Treatment and Prevention of Malaria*,
DOI 10.1007/978-3-0346-0480-2_6,

and malaria parasites, rely on the availability of folates. The antifolate methotrexate has been in use for over 60 years as an anticancer drug, and currently, disruption of folate metabolism is central in anticancer chemotherapy [2].

Plasmodium falciparum relies completely on the de novo synthesis pathway of dTMP and is unable to salvage pyrimidine from the exogenous medium. Thus, the folate pathway is critical to the parasite's survival [3]. Figure 1 summarizes the folate biochemical pathway in *P. falciparum* [4]. The important enzymes, with regard to antifolate activity, are dihydrofolate reductase (DHFR) and dihydropteroate synthase (DHPS). Indeed, all antifolates currently in use target these two enzymes. Pyrimethamine and proguanil are the archetypal DHFR inhibitors (Fig. 2a). Drugs that target DHPS are sulpha-based and they include sulphonamide (sulphadoxine) and sulfone (dapsone) (Fig. 2b). In the past, attempts were made to use the DHPS inhibitors alone as antimalarial agents [5]. However, this approach was abandoned because of their low efficacy and unacceptable toxicity. The interest in this class of antifolates was fostered when it was demonstrated that they synergized with DHFR inhibitors [6], leading to their use in combination.

2 Combination with Proguanil

2.1 Proguanil with the Antifolate Dapsone

Proguanil was the first antifolate against malaria to be discovered [7]. This drug is metabolized to its triazine form, cycloguanil, an inhibitor of DHFR [8] (Fig. 2a). It has been used alone as a prophylactic agent against malaria [9]. However, selection of mutations in DHFR, as the result of pyrimethamine use, is associated with decreased cycloguanil activity [10]. Thus, proguanil efficacy is reduced in areas of pyrimethamine resistance (see Sect. 6.5.1). An attempt was made to combine proguanil with dapsone for malaria treatment [11], and an artemisinin-based combination of proguanil/dapsone/artesunate was also evaluated [12]. However, these combinations have been abandoned because of the reduced activity of cycloguanil (the active metabolite of proguanil). Even if these drugs were developed, dapsone toxicity would have been a limitation (see Sect. 4).

2.2 Proguanil with a Non-Sulpha-Based Drug, Atovaquone

Proguanil has also been combined with atovaquone, an inhibitor of electron-transport to the cytochrome bc_1 complex (coenzyme Q); this combination, known as Malarone®, is primarily used as a prophylactic agent against malaria [13]. Proguanil is converted to cycloguanil, an inhibitor of DHFR, and at the same time, proguanil synergizes with atovaquone (as an inhibitor of electron transport).

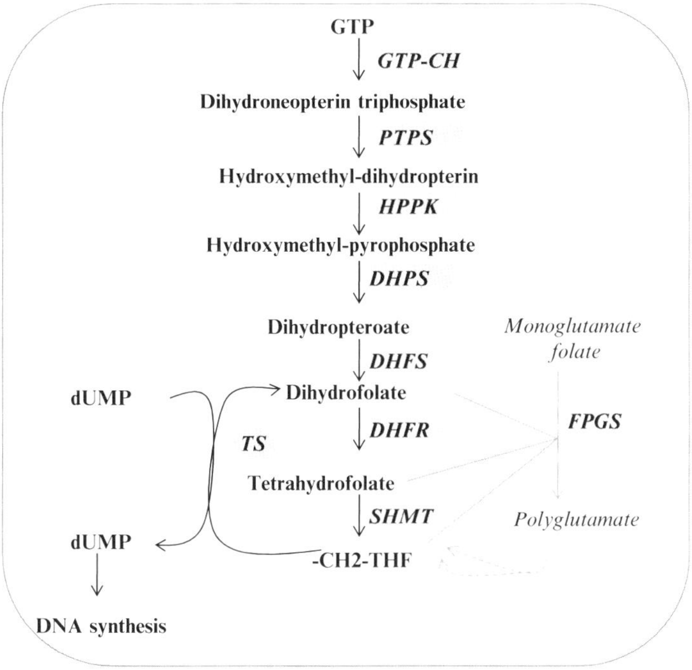

Fig. 1 Folate biochemical pathway in *P. falciparum*. The following abbreviations are used: *GTP-CH* GTP-cyclohydrolase I, *PTPS* pyruvyl tetrahydropterin synthase III, *HPPK* hydroxymethyl dihydropterin pyrophosphokinase, *DHPS* dihydropteroate synthase, *DFHS* dihydrofolate synthase, *FPGS* folylpoly-gamma-glutamate synthase, *DHFR* dihydrofolate reductase, *SHMT* serine hydroxymethyltransferase, *TS* thymidylate synthase. PTPS has recently been characterized in *P. falciparum* [4]

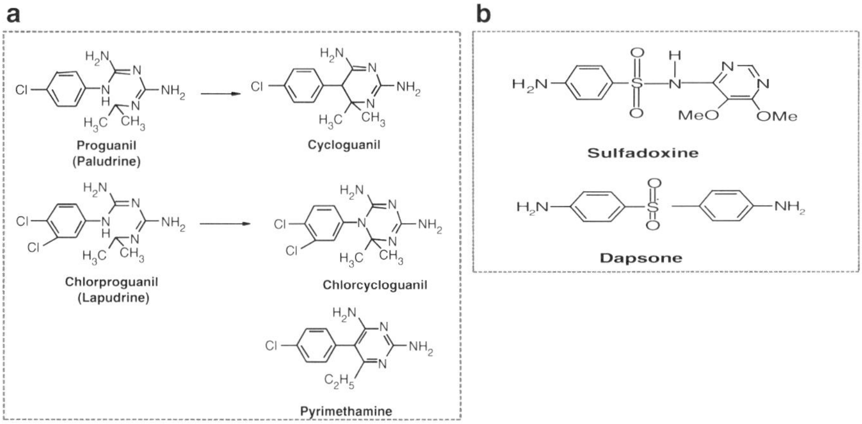

Fig. 2 Chemical structures of inhibitors of dihydrofolate reductase (**a**) and dihydropteroate synthase (**b**) enzymes used in the treatment of malaria

As a result, the sum of both effects account for the in vivo efficacy of Malarone® but in areas of high pyrimethamine resistance (where cycloguanil would be inactive), the potency of Malarone® would primarily be borne by the synergistic effect of proguanil/atovaquone on electron transport (see Sect. 5).

3 Combination with Pyrimethamine

Pyrimethamine, an inhibitor of DFHR, is a derivative of 2,4-diaminopyrimidine that was initially synthesized as an anticancer drug and was identified as an antimalarial based on its structural similarity to proguanil [14]. This drug has been the most widely used antimalarial antifolate agent thus far. It was first used as a monotherapy (known as Daraprim®) [15] and then was combined with sulphadoxine or sulphalene (see below).

3.1 *Uncomplicated Malaria Treatment*

Pyrimethamine was combined with sulphadoxine (this drug is known as Fansidar®) or sulphalene, under the name of Metakelfin®, for the treatment of uncomplicated malaria. These two sulpha drugs have a long elimination profile, like pyrimethamine, with half-lives >80 h [16]. On the other hand, pyrimethamine has been combined with the short-acting dapsone, under the name of Maloprim®. The half-life of dapsone is around 24 h [17] and, as a result, from the second day of treatment, the synergistic property of the combination is substantially reduced, decreasing the drug efficacy, and explaining the relatively low efficacy of this drug combination [18].

Though the efficacy of Fansidar® and Metakelfin® is comparable [19], Fansidar® has been more widely used than Metakelfin® (partly due to dapsone related toxicity) [19]. However, the efficacy of Fansidar® has been compromised as a result of the emergence of pyrimethamine- and sulphadoxine-resistant parasites, leading to its discontinuation as a drug for mass treatment of malaria in Africa [20, 21]. Combinations of pyrimethamine/sulphadoxine/artesunate and pyrimethamine/sulphadoxine/amodiaquine have been evaluated in several malaria-endemic areas [22]; however, a combination including a failing drug would be of limited benefit.

3.2 *Intermittent Preventive Treatment*

A body of evidence has shown that the administration of one dose of pyrimethamine/sulphadoxine in the second and third trimester in asymptomatic pregnant women is associated with a reduction of placental parasitaemia, maternal anemia and low birth weight [23, 24]. This led the World Health Organization to recommend Intermittent Preventive Treatment (IPT) in pregnant women living in malaria-endemic areas, and this concept has now been extended to infants and children [23]. Pyrimethamine/sulphadoxine has been the drug of choice for IPT, even in areas with moderate levels of pyrimethamine/sulphadoxine resistance [23, 25]; however, in the context of high pyrimethamine/sulphadoxine resistance, its efficacy is greatly compromised [26]. Alternative drugs are being evaluated, including the combination of pyrimethamine/sulphadoxine with other antimalarials such as piperaquine and mefloquine [27, 28].

4 Combination with Chlorproguanil

The potency of proguanil led to the search for analogs with higher activity. These studies resulted in the discovery of two important molecules, among others, chlorproguanil and BRL 6331 (WR 99210). Only the information on chlorproguanil is summarized here since BRL 6331 has not reached clinical stages in human trials; however, information on this compound can be found elsewhere [10, 29].

Chlorproguanil is generated by chlorination of proguanil, and is metabolized in vivo to chlorcycloguanil, the active metabolite (Fig. 2a), which is more active than cycloguanil and pyrimethamine [30, 31]. Chlorproguanil was recommended for prophylaxis but was not used as much as proguanil [9]. This antifolate has been combined with dapsone for the treatment of uncomplicated malaria, and a combination of chlorproguanil/dapsone/artesunate was also investigated. However, these drugs have been discontinued because of dapsone toxicity in glucose-6-phosphate-dehydrogenase (G6PD)-deficient patients [32, 33].

5 Antifolate Resistance

5.1 Resistance to the Combination Pyrimethamine/Sulphadoxine

Predominantly, pyrimethamine/sulphadoxine resistance is attributable to parasites that carry point mutations at codons 108 (Ser to Asn), 51 (Asn to Ile) and 59 (Cys to Arg) of *dhfr*. These are triple-mutant parasites. Resistance is augmented by point mutations at codons 437 (Ala to Gly) and/or 540 (Lys to Glu) or 437 and/or 581 (Ala to Lys) of the *dhps* gene [20, 21]. High levels of pyrimethamine/sulphadoxine resistance are associated with the selection of an additional mutation at codon 164 (Ile to Leu) of *dhfr*. Interestingly, the existence of this mutation in Africa has been a matter of debate [34–37] but there is now compelling evidence that it does occur in African isolates [30, 38–40]. While 164-Leu was not found to be associated with in vivo pyrimethamine/sulphadoxine resistance in Kenya [41], recently, Karema et al. have demonstrated substantially reduced pyrimethamine/sulphadoxine efficacy in an area of Rwanda with high 164-Leu prevalence [38].

Resistance to cycloguanil (the active metabolite of proguanil) is associated with mutation at codons 108 (Ser to Thr) and 16 (Ala to Val). At present, these mutations have been described in South America only, an area where proguanil has been widely as a prophylactic agent [20, 21]. The selection of pyrimethamine resistance has a bearing on the activity of cycloguanil. Indeed, parasites resistant to pyrimethamine are also resistant to cycloguanil, and this cross resistance is the result of the selection of point mutations in DHFR. This explains why the efficacy of proguanil is compromised in areas where resistance to Fansidar is high [20, 21].

The GTP-cyclohydrolase I enzyme (GTP-CH) catalyzes the first step of folate biosynthesis (Fig. 1). Interestingly, investigations have demonstrated that parasites highly resistant to the antifolate pyrimethamine, mainly those with the 164-Leu mutation, have an increased copy number (up to 11) of the *gtp-ch* gene [42]. The presence of mutations in DHFR could be associated with reduced enzyme kinetic properties, leading to a fitness cost [35, 42], though in vitro, these reduced enzyme kinetic properties have so far yielded contradictory results [43, 44]. Thus, the increase in copy number of *gtp-ch*, which is an adaptive phenomenon, could reflect compensatory mechanisms to maintain sufficient folate product (tetrahydrofolate). This increase in copy number in association with the presence of the 164-Leu mutation is a clear indication that GTP-CH modulates antifolate activity [42], along with changes of amino acids in key positions in DHFR and DHPS enzymes.

5.2 Origin of Antifolate-Resistant Parasites

Several studies have been dedicated to defining the origin of *dhfr*-mutant parasites. Similar work was undertaken to study chloroquine resistance, and the results indicate that chloroquine-resistant strains originated in South-East Asia – an area known to be the focus for the emergence of the multidrug-resistant parasites – and spread out globally, and into Africa [45]. The origin of new alleles is studied using polymerase chain reaction (PCR)-based assays of microsatellite markers, which consist of many repeated short sequences spread throughout the genome [46]. Reduced microsatellite diversity around the drug-resistant gene indicates selection of resistant strains as a result of drug pressure, and the evolutionary history of this drug resistance can be established by comparing microsatellite haplotypes in regions around the drug-resistant gene. This approach has revealed that African *dhfr* triple-mutant parasites originated mainly from South-East Asia [37, 47–50], though an indigenous origin of some triple mutants, and a few parasites also carrying 164-Leu was suggested [48, 50, 51]. Interestingly, all studies carried out so far indicate an African indigenous origin for *dhfr* double mutants (mutations at codon 108 and 51 or 108 and 59) [40, 48–52]. These observations indicate that highly antifolate-resistant parasites may have originated from outside Africa. Recently, a similar study was carried out to map the origin of *dhps* mutants. The data show that African *dhps* mutants arose from five geographical foci within Africa, an indication that mutations in this gene may not have been imported from outside Africa [53].

6 Inhibition of Folate Salvage as a Strategy to Increase Antifolate Activity

The malaria parasite can both salvage folate and synthesize it de novo. Both pathways increase the availability of folate in cells. The addition of folate derivatives decreases the activity of antifolate drugs in vitro and in vivo [54, 55]. This clearly shows that folate uptake makes a significant contribution to antifolate drug efficacy. Therefore, inhibition of this salvage pathway could provide a rationale for the development of agents that could potentiate the activity of antifolate drugs acting downstream of the folate uptake processes.

We have demonstrated that probenecid, an inhibitor of anion transporters, at a concentration of $<100\ \mu M$, which is readily achievable in vivo in humans, substantially increases the activity of inhibitors of DHFR and DHPS [55, 56]. This "chemosensitization" (or increase in antifolate activity) was also associated with a decrease in folate uptake into the parasite. This provides an explanation for how probenecid increases antifolate activity [56, 57]. Interestingly, this "probenecid effect" has been tested in vivo and the results indicate that probenecid significantly increases the efficacy of pyrimethamine/sulphadoxine in African children suffering from *P. falciparum* infection [58–60], although in one in vitro study, probenecid did not potentiate the activity of pyrimethamine [61]. This indicates that inhibitors of the anion transporters (responsible for the uptake of folate derivatives) could be of clinical importance in the treatment of malaria with antifolate drugs, as it has now been proven in cancer treatment [62].

7 Antifolate Anticancer Drugs for the Treatment of Malaria

Our group and others have demonstrated that the anticancer antifolate methotrexate is active against *P. falciparum* pyrimethamine-sensitive and pyrimethamine-resistant laboratory strains and field isolates, including those carrying the 164-Leu *dhfr* mutation [30, 55, 63–65]. In addition, we have shown that the anticancer antifolates aminopterin and trimetrexate are potent inhibitors of *P. falciparum* growth in vitro [30, 55]. $IC_{90/99}$ values (inhibitory concentration that kills 90 and 99% of parasites, respectively) of all these folate-antagonist agents fall between 150 and 350 nM. Thus, if such concentrations can be achieved in vivo with an acceptable toxicity profile, these compounds could potentially be used as antimalarial drugs. However, anticancer agents in general, and methotrexate in particular, are perceived to be toxic and thus not suitable for malaria treatment.

A law in toxicology, known as "Paracelsus' law" states that "all substances are poisons, and there are none which are not." The right dose differentiates a poison from a remedy; this principle is also known as the "dose–response effect" [66, 67]. Thus, a molecule becomes a drug if the dose required to treat a complication is pharmacologically active with minimal toxicity. This principle has been exploited in the use of anticancer drugs (thus, known to be toxic at high dose) for the treatment of nonneoplastic diseases, at low and relatively safe doses [68]. Methotrexate is one of the examples that vindicate Paracelsus' law.

Methotrexate is used at high dose, up to 5,000–12,000 mg/m^2/week (130–300 mg kg^{-1}) for several weeks for the treatment of cancer, and this dose can yield serum concentrations >1,000 μM, i.e. within the range of concentrations that are associated with methotrexate life-threatening toxicity [69]. On the other hand, a 1,000-fold lower dose of methotrexate (LD-methotrexate) [0.1–0.4 mg kg^{-1}] is used once weekly in the treatment of rheumatoid arthritis, juvenile idiopathic arthritis in children (including infants <1 year old), and psoriasis [70, 71]. This use is associated with a relatively low toxicity, and the drug has become one of the important drugs in the treatment for rheumatoid arthritis.

Interestingly, the proof of concept that methotrexate could be used to treat malaria was established 40 years ago. Two relatively small clinical trials have demonstrated that doses as low as 2.5 mg per day for 3–5 days were safe and effective in treating malaria infection in humans (caused by *P. falciparum* and/or *P. vivax*) [72, 73]. This in vivo efficacy of LD-methotrexate is also supported by its pharmacokinetic behavior. Indeed, a daily dose of 5 mg in adults (0.035–0.1 mg kg^{-1}) could yield serum methotrexate concentrations between 250 and 500 nM [74, 75], concentrations that exceed the $IC_{90/99}$ concentrations required to kill malaria parasites in vitro [30]. Taken together, this information warrants further investigation of the drug as an antimalarial. A Phase I evaluation of LD-methotrexate in 25 volunteers has been carried out (clinicaltrial.gov; NCT 00791531), a step towards its development as an antimalarial. Low dose of the anticancer trimetrexate also has the potential to become an antimalarial [55, 68].

8 Concluding Remarks

In the face of the burgeoning problem of antimalarial drug resistance, new drugs are needed. The disruption of folate metabolism as a strategy to identify new antimalarials, has been exploited for almost 50 years, beginning with the discovery of proguanil and pyrimethamine. Understanding of both the mode of action of, and mechanisms of resistance to, these drugs has permitted the search for and design of new antifolate agents with greater potency, using structure-based drug design, quantitative structure–activity relationships (QSAR), and X-ray crystallography among others [10, 76–80]. This effort has led to the discovery of new antifolates and some of them have reached advanced preclinical stages [81].

However, while this effort should be pursued, the folate pathway offers other opportunities for drug discovery. The experience gained from using antifolates in cancer has generated useful information that can be exploited to discover new antimalarial strategies. For example, we have provided evidence that the activity of antifolates can be enhanced despite the development of resistance. This concept has been vindicated in vivo. We have also demonstrated that some anticancer antifolates can be used at a low and safe dose to treat malaria. Thus, the opportunity exists to discover new antifolate drugs and extend the therapeutic life of existing ones.

Acknowledgments The author thanks the director of Kenya Medical Research Institute for permission to publish these data. This work was supported by the European Developing Countries Clinical Trials Partnership (EDTCP).

Transparency Declarations None to declare.

References

1. Nzila A, Ward SA, Marsh K, Sims PF, Hyde JE (2005) Comparative folate metabolism in humans and malaria parasites (part II): activities as yet untargeted or specific to *Plasmodium*. Trends Parasitol 21:334–339
2. Nzila A, Ward SA, Marsh K, Sims PF, Hyde JE (2005) Comparative folate metabolism in humans and malaria parasites (part I): pointers for malaria treatment from cancer chemotherapy. Trends Parasitol 21:292–298
3. Hyde JE (2007) Targeting purine and pyrimidine metabolism in human apicomplexan parasites. Curr Drug Targets 8:31–47
4. Dittrich S, Mitchell SL, Blagborough AM, Wang Q, Wang P, Sims PF, Hyde JE (2008) An atypical orthologue of 6-pyruvoyltetrahydropterin synthase can provide the missing link in the folate biosynthesis pathway of malaria parasites. Mol Microbiol 67:609–618
5. Michel R (1968) Comparative study of the association of sulfalene and pyrimethamine and of sulfalene alone in mass chemoprophylaxis of malaria. Med Trop (Mars) 28:488–494
6. Bushby SR (1969) Combined antibacterial action *in vitro* of trimethoprim and sulphonamides. The *in vitro* nature of synergy. Postgrad Med J 45(Suppl):10–18
7. Curd FHS, Davey DG, Rose FL (1945) Study on synthetic antimalarial drugs. X. Some biguanide derivatives as new types of antimalarial substances with both therapeutic and causal prophylactic activity. Ann Trop Med Parasitol 39:208–216
8. Carrington HC, Crowther AF, Davey DG, Levi AA, Rose FL (1951) A metabolite of paludrine with high antimalarial activity. Nature 168:1080
9. Wernsdorfer WH (1990) Chemoprophylaxis of malaria: underlying principles and their realization. Med Trop (Mars) 50:119–124
10. Nzila A (2006) The past, present and future of antifolates in the treatment of *Plasmodium falciparum* infection. J Antimicrob Chemother 57:1043–1054
11. Mutabingwa TK, Maxwell CA, Sia IG, Msuya FH, Mkongewa S, Vannithone S, Curtis J, Curtis CF (2001) A trial of proguanil-dapsone in comparison with sulfadoxine-pyrimethamine for the clearance of *Plasmodium falciparum* infections in Tanzania. Trans R Soc Trop Med Hyg 95:433–438

12. Krudsood S, Imwong M, Wilairatana P, Pukrittayakamee S, Nonprasert A, Snounou G, White NJ, Looareesuwan S (2005) Artesunate-dapsone-proguanil treatment of falciparum malaria: genotypic determinants of therapeutic response. Trans R Soc Trop Med Hyg 99:142–149
13. Nakato H, Vivancos R, Hunter PR (2007) A systematic review and meta-analysis of the effectiveness and safety of atovaquone proguanil (Malarone) for chemoprophylaxis against malaria. J Antimicrob Chemother 60:929–936
14. Falco EA, Goodwin LG, Hitchings GH, Rollo IM, Russell PB (1951) 2:4-diaminopyrimidines – a new series of antimalarials. Br J Pharmacol Chemother 6:185–200
15. Peters W (1987) Experimental resistance IV: dihydrofolate reductase inhibitors and drugs with related activities. In: Peters W (ed) Chemotherapy and drug resistance in malaria. Academic, London, pp 481–521
16. Watkins WM, Mosobo M (1993) Treatment of *Plasmodium falciparum* malaria with pyrimethamine- sulfadoxine: selective pressure for resistance is a function of long elimination half-life. Trans R Soc Trop Med Hyg 87:75–78
17. Winstanley P, Watkins W, Muhia D, Szwandt S, Amukoye E, Marsh K (1997) Chlorproguanil/dapsone for uncomplicated *Plasmodium falciparum* malaria in young children: pharmacokinetics and therapeutic range. Trans R Soc Trop Med Hyg 91:322–327
18. Segal HE, Chinvanthananond P, Laixuthai B, Pearlman EJ, Hall AP, Phintuyothin P, Na-Nakorn A, Castaneda BF (1975) Comparison of diaminodiphenylsulphonepyrimethamine and sulfadoxine-pyrimethamine combinations in the treatment of falciparum malaria in Thailand. Trans R Soc Trop Med Hyg 69:139–142
19. Peters W (1987) Resistance in human malaria III: dihydrofolate reductase inhibitors. In: Peters W (ed) Chemotherapy and drug resistance in malaria. Academic, London, pp 593–658
20. Gregson A, Plowe CV (2005) Mechanisms of resistance of malaria parasites to antifolates. Pharmacol Rev 57:117–145
21. Sibley CH, Hyde JE, Sims PF, Plowe CV, Kublin JG, Mberu EK, Cowman AF, Winstanley PA, Watkins WM, Nzila AM (2001) Pyrimethamine-sulfadoxine resistance in *Plasmodium falciparum*: what next? Trends Parasitol 17:582–588
22. Nosten F, White NJ (2007) Artemisinin-based combination treatment of falciparum malaria. Am J Trop Med Hyg 77:181–192
23. Grobusch MP, Egan A, Gosling RD, Newman RD (2007) Intermittent preventive therapy for malaria: progress and future directions. Curr Opin Infect Dis 20:613–620
24. Grobusch MP, Gabor JJ, Aponte JJ, Schwarz NG, Poetschke M, Doernemann J, Schuster K, Koester KB, Profanter K, Borchert LB et al (2009) No rebound of morbidity following intermittent preventive sulfadoxine-pyrimethamine treatment of malaria in infants in Gabon. J Infect Dis 200:1658–1661
25. Aponte JJ, Schellenberg D, Egan A, Breckenridge A, Carneiro I, Critchley J, Danquah I, Dodoo A, Kobbe R, Lell B et al (2009) Efficacy and safety of intermittent preventive treatment with sulfadoxine-pyrimethamine for malaria in African infants: a pooled analysis of six randomised, placebo-controlled trials. Lancet 374:1533–1542
26. Gosling RD, Gesase S, Mosha JF, Carneiro I, Hashim R, Lemnge M, Mosha FW, Greenwood B, Chandramohan D (2009) Protective efficacy and safety of three antimalarial regimens for intermittent preventive treatment for malaria in infants: a randomised, double-blind, placebo-controlled trial. Lancet 374:1521–1532
27. Briand V, Bottero J, Noel H, Masse V, Cordel H, Guerra J, Kossou H, Fayomi B, Ayemonna P, Fievet N et al (2009) Intermittent treatment for the prevention of malaria during pregnancy in Benin: a randomized, open-label equivalence trial comparing sulfadoxine-pyrimethamine with mefloquine. J Infect Dis 200:991–1001
28. Cisse B, Cairns M, Faye E, O ND, Faye B, Cames C, Cheng Y, M ND, Lo AC, Simondon K et al (2009) Randomized trial of piperaquine with sulfadoxine-pyrimethamine or dihydroartemisinin for malaria intermittent preventive treatment in children. PLoS One 4:e7164
29. Diaz DS, Kozar MP, Smith KS, Asher CO, Sousa JC, Schiehser GA, Jacobus DP, Milhous WK, Skillman DR, Shearer TW (2008) Role of specific cytochrome P450 isoforms in the

conversion of phenoxypropoxybiguanide analogs in human liver microsomes to potent antimalarial dihydrotriazines. Drug Metab Dispos 36:380–385
30. Kiara SM, Okombo J, Masseno V, Mwai L, Ochola I, Borrmann S, Nzila A (2009) *In vitro* activity of antifolate and polymorphism in dihydrofolate reductase of *Plasmodium falciparum* isolates from the Kenyan coast: emergence of parasites with Ile-164-Leu mutation. Antimicrob Agents Chemother 53:3793–3798
31. Nzila-Mounda A, Mberu EK, Sibley CH, Plowe CV, Winstanley PA, Watkins WM (1998) Kenyan *Plasmodium falciparum* field isolates: correlation between pyrimethamine and chlorcycloguanil activity *in vitro* and point mutations in the dihydrofolate reductase domain. Antimicrob Agents Chemother 42:164–169
32. Tiono AB, Dicko A, Ndububa DA, Agbenyega T, Pitmang S, Awobusuyi J, Pamba A, Duparc S, Goh LE, Harrell E et al (2009) Chlorproguanil-dapsone-artesunate versus chlorproguanil-dapsone: a randomized, double-blind, phase III trial in African children, adolescents, and adults with uncomplicated *Plasmodium falciparum* malaria. Am J Trop Med Hyg 81:969–978
33. Premji Z, Umeh RE, Owusu-Agyei S, Esamai F, Ezedinachi EU, Oguche S, Borrmann S, Sowunmi A, Duparc S, Kirby PL et al (2009) Chlorproguanil-dapsone-artesunate versus artemether-lumefantrine: a randomized, double-blind phase III trial in African children and adolescents with uncomplicated *Plasmodium falciparum* malaria. PLoS One 4:e6682
34. Hastings MD, Bates SJ, Blackstone EA, Monks SM, Mutabingwa TK, Sibley CH (2002) Highly pyrimethamine-resistant alleles of dihydrofolate reductase in isolates of *Plasmodium falciparum* from Tanzania. Trans R Soc Trop Med Hyg 96:674–676
35. Nzila A, Ochong E, Nduati E, Gilbert K, Winstanley P, Ward S, Marsh K (2005) Why has the dihydrofolate reductase 164 mutation not consistently been found in Africa yet? Trans R Soc Trop Med Hyg 99:341–346
36. Ochong E, Nzila A, Kimani S, Kokwaro G, Mutabingwa T, Watkins W, Marsh K (2003) Molecular monitoring of the Leu-164 mutation of dihydrofolate reductase in a highly sulfadoxine/pyrimethamine-resistant area in Africa. Malar J 2:46
37. Roper C, Pearce R, Nair S, Sharp B, Nosten F, Anderson T (2004) Intercontinental spread of pyrimethamine-resistant malaria. Science 305:1124
38. Karema C, Imwong M, Fanello CI, Stepniewska K, Uwimana A, Nakeesathit S, Dondorp A, Day NP, White NJ (2010) Molecular correlates of high-level antifolate resistance in Rwandan children with *Plasmodium falciparum* malaria. Antimicrob Agents Chemother 54:477–483
39. Lynch C, Pearce R, Pota H, Cox J, Abeku TA, Rwakimari J, Naidoo I, Tibenderana J, Roper C (2008) Emergence of a *dhfr* mutation conferring high-level drug resistance in *Plasmodium falciparum* populations from southwest Uganda. J Infect Dis 197:1598–1604
40. McCollum AM, Poe AC, Hamel M, Huber C, Zhou Z, Shi YP, Ouma P, Vulule J, Bloland P, Slutsker L et al (2006) Antifolate resistance in *Plasmodium falciparum*: multiple origins and identification of novel *dhfr* alleles. J Infect Dis 194:189–197
41. Hamel MJ, Poe A, Bloland P, McCollum A, Zhou Z, Shi YP, Ouma P, Otieno K, Vulule J, Escalante A et al (2008) Dihydrofolate reductase I164L mutations in *Plasmodium falciparum* isolates: clinical outcome of 14 Kenyan adults infected with parasites harbouring the I164L mutation. Trans R Soc Trop Med Hyg 102:338–345
42. Nair S, Miller B, Barends M, Jaidee A, Patel J, Mayxay M, Newton P, Nosten F, Ferdig MT, Anderson TJ (2008) Adaptive copy number evolution in malaria parasites. PLoS Genet 4: e1000243
43. Sandefur CI, Wooden JM, Quaye IK, Sirawaraporn W, Sibley CH (2007) Pyrimethamine-resistant dihydrofolate reductase enzymes of *Plasmodium falciparum* are not enzymatically compromised *in vitro*. Mol Biochem Parasitol 154:1–5
44. Sirawaraporn W, Sathitkul T, Sirawaraporn R, Yuthavong Y, Santi DV (1997) Antifolate-resistant mutants of *Plasmodium falciparum* dihydrofolate reductase. Proc Natl Acad Sci USA 94:1124–1129

45. Wootton JC, Feng X, Ferdig MT, Cooper RA, Mu J, Baruch DI, Magill AJ, Su XZ (2002) Genetic diversity and chloroquine selective sweeps in *Plasmodium falciparum*. Nature 418:320–323
46. Anderson TJ, Su XZ, Bockarie M, Lagog M, Day KP (1999) Twelve microsatellite markers for characterization of *Plasmodium falciparum* from finger-prick blood samples. Parasitology 119:113–125
47. Maiga O, Djimde AA, Hubert V, Renard E, Aubouy A, Kironde F, Nsimba B, Koram K, Doumbo OK, Le Bras J et al (2007) A shared Asian origin of the triple-mutant *dhfr* allele in *Plasmodium falciparum* from sites across Africa. J Infect Dis 196:165–172
48. McCollum AM, Basco LK, Tahar R, Udhayakumar V, Escalante AA (2008) Hitchhiking and selective sweeps of *Plasmodium falciparum* sulfadoxine and pyrimethamine resistance alleles in a population from central Africa. Antimicrob Agents Chemother 52:4089–4097
49. Mita T (2009) Origins and spread of *pfdhfr* mutant alleles in *Plasmodium falciparum*. Acta Trop 58:201–209
50. Mita T, Tanabe K, Takahashi N, Culleton R, Ndounga M, Dzodzomenyo M, Akhwale WS, Kaneko A, Kobayakawa T (2009) Indigenous evolution of *Plasmodium falciparum* pyrimethamine resistance multiple times in Africa. J Antimicrob Chemother 63:252–255
51. Certain LK, Briceno M, Kiara SM, Nzila AM, Watkins WM, Sibley CH (2008) Characteristics of *Plasmodium falciparum dhfr* haplotypes that confer pyrimethamine resistance, Kilifi, Kenya, 1987–2006. J Infect Dis 197:1743–1751
52. Roper C, Pearce R, Bredenkamp B, Gumede J, Drakeley C, Mosha F, Chandramohan D, Sharp B (2003) Antifolate antimalarial resistance in southeast Africa: a population-based analysis. Lancet 361:1174–1181
53. Pearce RJ, Pota H, Evehe MS, Ba el H, Mombo-Ngoma G, Malisa AL, Ord R, Inojosa W, Matondo A, Diallo DA et al (2009) Multiple origins and regional dispersal of resistant *dhps* in African *Plasmodium falciparum* malaria. PLoS Med 6:e1000055
54. Carter JY, Loolpapit MP, Lema OE, Tome JL, Nagelkerke NJ, Watkins WM (2005) Reduction of the efficacy of antifolate antimalarial therapy by folic acid supplementation. Am J Trop Med Hyg 73:166–170
55. Nduati E, Diriye A, Ommeh S, Mwai L, Kiara S, Masseno V, Kokwaro G, Nzila A (2008) Effect of folate derivatives on the activity of antifolate drugs used against malaria and cancer. Parasitol Res 102:1227–1234
56. Nzila A, Mberu E, Bray P, Kokwaro G, Winstanley P, Marsh K, Ward S (2003) Chemosensitization of *Plasmodium falciparum* by probenecid *in vitro*. Antimicrob Agents Chemother 47:2108–2112
57. Wang P, Wang Q, Sims PF, Hyde JE (2007) Characterisation of exogenous folate transport in *Plasmodium falciparum*. Mol Biochem Parasitol 154:40–51
58. Sowunmi A, Adedeji AA, Fateye BA, Babalola CP (2004) *Plasmodium falciparum* hyperparasitaemia in children. Risk factors, treatment outcomes, and gametocytaemia following treatment. Parasite 11:317–323
59. Sowunmi A, Fehintola FA, Adedeji AA, Gbotosho GO, Falade CO, Tambo E, Fateye BA, Happi TC, Oduola AM (2004) Open randomized study of pyrimethamine-sulphadoxine vs. pyrimethamine-sulphadoxine plus probenecid for the treatment of uncomplicated *Plasmodium falciparum* malaria in children. Trop Med Int Health 9:606–614
60. Sowunmi A, Adedeji AA, Fateye BA, Fehintola FA (2004) Comparative effects of pyrimethamine-sulfadoxine, with and without probenecid, on *Plasmodium falciparum* gametocytes in children with acute, uncomplicated malaria. Ann Trop Med Parasitol 98:873–878
61. Tahar R, Basco LK (2007) Molecular epidemiology of malaria in Cameroon. XXVI. Twelve-year *in vitro* and molecular surveillance of pyrimethamine resistance and experimental studies to modulate pyrimethamine resistance. Am J Trop Med Hyg 77:221–227
62. Fury MG, Krug LM, Azzoli CG, Sharma S, Kemeny N, Wu N, Kris MG, Rizvi NA (2005) A phase I clinical pharmacologic study of pralatrexate in combination with probenecid in adults with advanced solid tumors. Cancer Chemother Pharmacol 57:671–677

63. Dar O, Khan MS, Adagu I (2008) The potential use of methotrexate in the treatment of falciparum malaria: *in vitro* assays against sensitive and multidrug-resistant falciparum strains. Jpn J Infect Dis 61:210–211
64. Fidock DA, Nomura T, Wellems TE (1998) Cycloguanil and its parent compound proguanil demonstrate distinct activities against *Plasmodium falciparum* malaria parasites transformed with human dihydrofolate reductase. Mol Pharmacol 54:1140–1147
65. Walter RD, Bergmann B, Kansy M, Wiese M, Seydel JK (1991) Pyrimethamin-resistant *Plasmodium falciparum* lack cross-resistance to methotrexate and 2,4-diamino-5-(substituted benzyl) pyrimidines. Parasitol Res 77:346–350
66. Langman LJ, Kapur BM (2006) Toxicology: then and now. Clin Biochem 39:498–510
67. Rozman KK, Doull J (2001) Paracelsus, Haber and Arndt. Toxicology 160:191–196
68. Nzila A, Okombo J, Becker RP, Chilengi R, Lang T, Niehues T (2010) Anticancer agents against malaria: time to revisit? Trends Parasitol 26:125–129
69. Chabner BA, Amrein P, Drucker B, Michealson M, Mitsiades C, Goss P, Ryan D, Ramachandra S, Richardson P, Supko J et al (2006) Antineoplastic agents. In: Brunton L (ed) The pharmacological basis of therapeutics 9/e. McGrwa-Hill, New York, pp 1315–1465
70. Niehues T, Lankisch P (2006) Recommendations for the use of methotrexate in juvenile idiopathic arthritis. Paediatr Drugs 8:347–356
71. Swierkot J, Szechinski J (2006) Methotrexate in rheumatoid arthritis. Pharmacol Rep 58:473–492
72. Sheehy TW, Dempsey H (1970) Methotrexate therapy for *Plasmodium vivax* malaria. JAMA 214:109–114
73. Wildbolz A (1973) Methotrexate in the therapy of malaria. Ther Umsch 30:218–222
74. Chladek J, Grim J, Martinkova J, Simkova M, Vaneckova J (2005) Low-dose methotrexate pharmacokinetics and pharmacodynamics in the therapy of severe psoriasis. Basic Clin Pharmacol Toxicol 96:247–248
75. Chladek J, Grim J, Martinkova J, Simkova M, Vaniekova J, Koudelkova V, Noiekova M (2002) Pharmacokinetics and pharmacodynamics of low-dose methotrexate in the treatment of psoriasis. Br J Clin Pharmacol 54:147–156
76. Chiu TL, So SS (2004) Development of neural network QSPR models for Hansch substituent constants. 2. Applications in QSAR studies of HIV-1 reverse transcriptase and dihydrofolate reductase inhibitors. J Chem Inf Comput Sci 44:154–160
77. Dasgupta T, Chitnumsub P, Kamchonwongpaisan S, Maneeruttanarungroj C, Nichols SE, Lyons TM, Tirado-Rives J, Jorgensen WL, Yuthavong Y, Anderson KS (2009) Exploiting structural analysis, *in silico* screening, and serendipity to identify novel inhibitors of drug-resistant falciparum malaria. ACS Chem Biol 4:29–40
78. Hecht D, Fogel GB (2009) A novel *in silico* approach to drug discovery via computational intelligence. J Chem Inf Model 49:1105–1121
79. Maitarad P, Kamchonwongpaisan S, Vanichtanankul J, Vilaivan T, Yuthavong Y, Hannongbua S (2009) Interactions between cycloguanil derivatives and wild type and resistance-associated mutant *Plasmodium falciparum* dihydrofolate reductases. J Comput Aided Mol Des 23:241–252
80. Thongpanchang C, Taweechai S, Kamchonwongpaisan S, Yuthavong Y, Thebtaranonth Y (2007) Immobilization of malarial (*Plasmodium falciparum*) dihydrofolate reductase for the selection of tight-binding inhibitors from combinatorial library. Anal Chem 79:5006–5012
81. MMV. http://www.mmv.org/IMG/pdf/Global_Malaria_FINALq42009.pdf. Accessed 11 Jan 2011

Naphthoquinones: Atovaquone, and Other Antimalarials Targeting Mitochondrial Functions

Akhil B. Vaidya

Abstract Mitochondria in malaria parasites are highly divergent from their counterparts in mammalian hosts. This degree of divergence underlies the validity of mitochondrial functions as targets for antimalarial drugs. The mitochondrial electron transport chain (mtETC) at the cytochrome bc_1 complex is selectively inhibited in malaria parasites by atovaquone. Proguanil, the synergistic partner of atovaquone, appears to target an alternative pathway that generates electropotential across the inner membrane of parasite mitochondria. However, the rapid emergence of atovaquone-resistance mutations effectively negates the synergistic effect of proguanil. New antimalarials targeting the mtETC with reduced propensity for resistance development could overcome this challenge. A critical function of the mtETC is to serve mitochondrially located dihydroorotate dehydrogensae (DHODH), an enzyme of the pyrimidine biosynthesis pathway. Compounds with selective activity against parasite DHODH are under development as potential new antimalarials. Recent studies on unusual tricarboxylic acid metabolism and ATP synthase structure point to additional opportunities for investigations aimed to identify other selective inhibitors.

1 Introduction

Some of the most potent poisons such as cyanide affect mitochondrial respiration, blocking electron transport and resulting in bioenergetic crisis leading to death. Thus, the idea of using antimalarial drugs that affect the mitochondrial electron transport chain (mtETC) might at first appear to be risky. However, this is precisely what a currently used antimalarial drug, atovaquone, does by selectively inhibiting

A.B. Vaidya (✉)
Department of Microbiology and Immunology Drexel University College of Medicine, 2900 Queen Lane, Philadelphia, PA 19129, USA
e-mail: Akhil.Vaidya@DrexelMed.edu

H.M. Staines and S. Krishna (eds.), *Treatment and Prevention of Malaria*,
DOI 10.1007/978-3-0346-0480-2_7,

the parasite mtETC without affecting host mitochondria. The reason for this selectivity lies in structural differences in the cytochrome *b* encoded by the parasite mitochondrial DNA (mtDNA) that distinguishes it from that encoded by the host mtDNA [1]. Clearly, the great evolutionary distance between the parasites and their hosts is reflected in the significant differences observed for their essential physiological processes [2]. Such differences provide opportunities for devising means for selective interference as the basis for developing new antiparasitic drugs. This chapter will focus on some of the recent findings on the unusual mitochondrion of malaria parasites, describing its essential functions and the mechanisms by which some of the antimalarial compounds exert their effects. Some of the insights into mechanisms of action and synergy between atovaquone and proguanil – components of the drug Malarone™ – are important in decisions relating to partner drugs to be used with compounds that target the mtETC.

2 Selective Inhibition of Parasite the mtETC

As with all mitochondria, the main mobile electron carrier for the mtETC in malaria parasites is ubiquinone (also called coenzyme Q, CoQ; see Fig. 1 for the structures of CoQ and other compounds discussed here). Unlike the host mitochondria, however, pyruvate oxidation through the citric acid cycle does not appear to be the source of reducing equivalents in the parasite mitochondria [2, 3]. Also, the parasites do not possess a proton-pumping multisubunit NADH dehydrogenase, but encode a single subunit type II NADH dehydrogenase (NDH2) that reduces CoQ without contributing to the proton gradient [4, 5]. In addition to the NDH2, four other dehydrogenases – dihydroorotate dehydrogenase (DHODH), glycerol-3-phosphate dehydrogenase (GPDH), succinate dehydrogenase (SDH), and malate–quinone oxidoreductase (MQO) – require CoQ as the electron acceptor. The reduced ubihydroquinone ($CoQH_2$) is oxidized by the dimeric cytochrome bc_1 complex (Complex III), which serves an essential step for the continued provision of CoQ to mitochondrial dehydrogenases. As depicted in Fig. 2, $CoQH_2$ oxidation by Complex III is achieved through a process called the Q cycle (see [6, 7] for reviews) in which electrons from $CoQH_2$ are bifurcated at the quinone oxidation (Q_O) site with one electron passed on to the iron–sulfur (2Fe2S) cluster of the Rieske protein on the way to cytochrome c_1 and cytochrome *c*, while the other passes on to the quinone reduction (Q_i) site through the low- and high-potential heme b_L and b_H, reducing a resident quinone to semiquinone. Another round of $CoQH_2$ oxidation results in the reduction of another molecule of cytochrome *c* and full reduction of the semiquinone to $CoQH_2$ at the Q_i site. This process results in the translocation of four protons across the inner mitochondrial membrane and the reduction of two cytochrome *c* molecules. The transfer of electrons by the Rieske iron–sulfur protein requires a large-scale domain movement in which the 2Fe2S cluster moves from a Q_O proximal position to the cytochrome c_1 proximal position. Analysis of limited proteolysis and electron paramagnetic resonance measurements

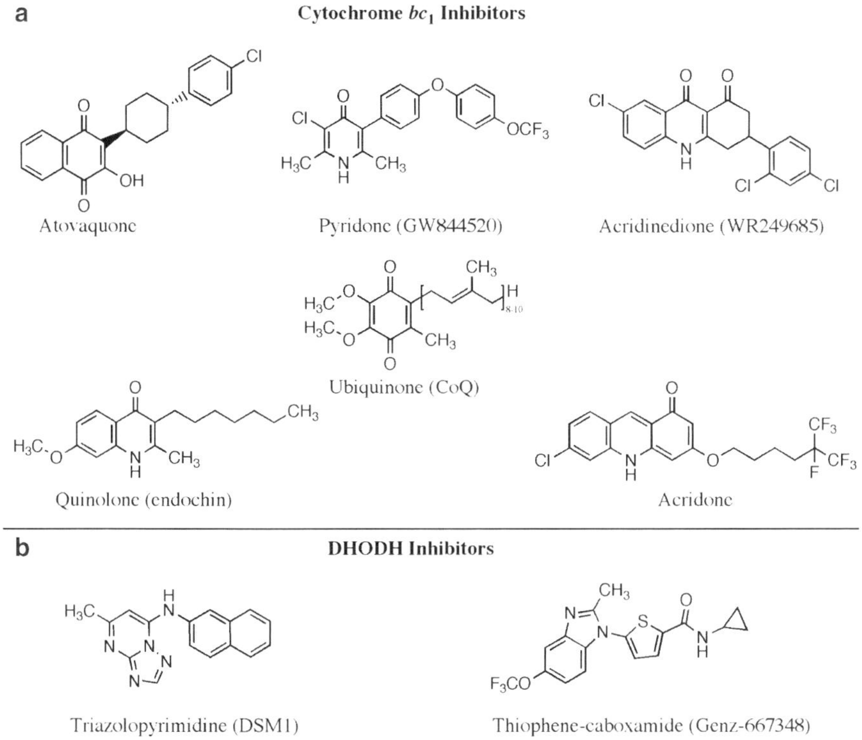

Fig. 1 Structures of compounds discussed in the text. (**a**) Structures of CoQ and five chemical classes of cytochrome bc_1 complex inhibitors. (**b**) Structures of two chemical classes that exhibit selective inhibition of parasite DHODH

of engineered bacterial cytochrome bc_1 complexes showed that atovaquone inhibited the movement of the Rieske protein [8]. Subtle differences in the highly conserved Q_O site residues appear to be responsible for the selectively greater inhibition of the parasite complex compared to the mammalian complex [9]. Molecular modeling studies based on crystal structures of the yeast and bovine Complex III in the presence of known Q_O site inhibitors have postulated an atovaquone binding site, and have proposed the molecular basis for preferential binding of atovaquone to the parasite Complex III [10, 11]. Although there are sufficient caveats in using this surrogate approach, in the absence of authentic crystal structures of the parasite cytochrome bc_1 complex, these studies could provide guidelines for understanding drug action and approaches for further drug design. A critical point is the necessity for identifying compounds that would give the largest therapeutic window so as to minimize the risk of toxicity arising from inhibition of the host Complex III.

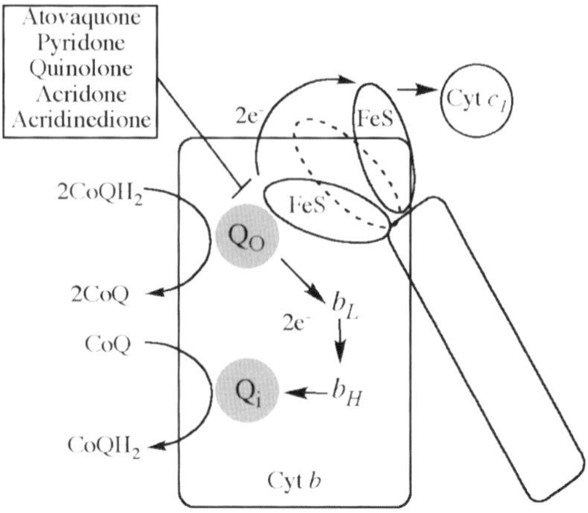

Fig. 2 A schematic representation of the Q cycle within the Complex III for $CoQH_2$ oxidation. With two molecules of $CoQH_2$ being oxidized at the Q_O site of the complex, two electrons are passed on to cytochrome *c*, two electrons to a CoQ molecule resident at the Q_i site, and four protons are translocated across the membrane. Transfer of electrons to cytochrome c_1 involves a large-scale movement of the 2Fe2S domain of the Rieske protein. The five chemical classes of antimalarials shown in the box inhibit this step of electron transport at the Q_O site

3 Atovaquone Resistance

In early clinical trials with atovaquone as a single agent, recrudescent parasites emerged in about 30% of the patients [12]. The recrudescent parasites were >1,000-fold resistant to atovaquone. Even in cultures of *P. falciparum*, atovaquone-resistant parasites could be observed at the frequency as high as 10^{-7} in some strains [13]. Under conditions of suboptimal treatment, several independent lines of the rodent malaria parasite, *P. yoelii*, resistant to atovaquone were readily derived [9]. DNA sequencing of the mitochondrial genome of these resistant parasites revealed mutations centered around the Q_O site of cytochrome *b*, delineating the putative atovaquone-binding region of the parasite Complex III [9]. The subtle differences in the structural features of the Q_O site between the parasite and the host Complex III that form the basis for selective toxicity of atovaquone could explain the propensity of resistance development: mutations could arise in the parasite cytochrome *b* that would make it more like the host protein, thereby gaining resistance to atovaquone.

Could this high frequency of resistance development be due to polymorphisms within the mitochondrial DNA (mtDNA) of malaria parasites in the population? On the contrary, extensive sequence analyses of mtDNA from geographically distant *P. falciparum* have revealed extraordinary sequence conservation [14, 15]. Furthermore, the ease with which atovaquone-resistant parasites could be derived in cultures of cloned parasites argues against the idea that atovaquone-resistant mutants are prevalent in the parasite populations. It is worth noting that no wild-type mtDNA can be detected once atovaquone resistance has arisen in cultured *P. falciparum*. Given that the parasite mitochondria contain multiple copies of mtDNA [16], this observation is consistent with the idea that under selection, the drug-resistant mutations quickly sweep through the mtDNA population. The suggested mode of parasite mtDNA replication involves extensive gene conversion events [17], which could account for the rapid selective sweeps of atovaquone resistant mutations.

4 Atovaquone–Proguanil Combination

The high frequency of treatment failures in the clinic clearly precluded the use of atovaquone as a single antimalarial agent. A search for a potential partner showed proguanil to have synergistic effect with atovaquone in inhibiting *P. falciparum* growth in vitro [18]. As an antimalarial known since the 1950s [19, 20], proguanil already had an established record of efficacy and safety, favoring its choice as a partner for atovaquone compared to other compounds. A fixed combination of atovaquone and proguanil, registered as Malarone, was extensively tested in clinical trials and found to be highly effective for the prevention and treatment of malaria [21]. No treatment failures due to resistance development were reported in clinical trials of Malarone; however, sporadic cases of prophylaxis and treatment failures due to cytochrome *b* mutations have been reported since its wider use in the field [22–24]. These worrisome reports as well as the high cost of Malarone have restricted its use primarily to a prophylactic drug prescribed for visitors to malaria endemic regions.

By inhibiting electron transport through Complex III, atovaquone also prevents proton translocation, thus decreasing the electropotential across the inner mitochondrial membrane of the parasite. Dose-dependent collapse of mitochondrial membrane potential was demonstrated in a flow-cytometric assay with a cationic fluorescent probe in live intact *P. yoelii* [25]. The same assay also revealed that proguanil by itself did not inhibit mtETC nor did it affect mitochondrial membrane potential [26]. However, when used in combination with atovaquone, proguanil reduced the concentration at which atovaquone collapsed the mitochondrial membrane potential by about six- to eightfold; it also enhanced the magnitude of the collapse [26]. This observation provided an initial indication as to the underlying mechanism for the synergy between atovaquone and proguanil. Importantly, the active metabolite of proguanil, cycloguanil (a selective inhibitor of the parasite dihydrofolate reductase, DHFR), did not demonstrate synergy with atovaquone for neither parasite growth inhibition [18] nor the collapse of the mitochondrial membrane potential [26]. Thus, the synergistic action of proguanil is the property of the prodrug itself; its metabolic activation is not required. More recent studies described later have confirmed and extended these findings.

5 Inhibition of mtETC and the Viability of *P. falciparum*

In metazoan cells, mitochondria play a major role in programmed cell death with mtETC inhibition as well as other insults to the mitochondrial physiology leading to apoptosis. However, apoptotic dismantling of atovaquone-treated parasites has not been observed [27]. Monitoring the viability of parasites exposed to atovaquone and the atovaquone–proguanil combination revealed that ring and schizont-stage *P. falciparum* remained viable for 48 h, and the trophozoite stages remained viable for 24 h under treatment [28]. Thus, an apoptotic program is not initiated by the inhibition of mitochondrial respiration and the collapse of mitochondrial membrane potential. Instead, the parasites seem to enter a "static" state and can survive for up to 48 h in culture. The atovaquone–proguanil combination appears to be a static rather than a cidal drug in vitro. Conversion of proguanil to cycloguanil in vivo, however, could affect the pharmacodynamics of the combination when administered to patients. Nonetheless, it would be prudent to assess the killing rates of antimalarial compounds under development (e.g. pyridones and quinolones; discussed later) that target mitochondrial physiology, so as to optimize their pharmacokinetic and pharmacodynamic properties.

6 Specific Role of mtETC in Blood Stage *P. falciparum*

Among the five mitochondrial dehydrogenases that require CoQ as the electron acceptor, DHODH has been considered the most important as it carries out the fourth step in the pyrimidine biosynthesis pathway, an essential step in parasites unable to salvage pyrimidine. In most eukaryotic organisms, DHODH is localized to mitochondria, but in the yeast *Saccharomyces cerevisiae* the enzyme is localized to the cytosol and uses fumarate as the electron acceptor [29]. Painter et al. [30] generated transgenic *P. falciparum* lines that expressed the yeast DHODH (yDHODH) under the control of a constitutive promoter. Interestingly, the yDHODH-transgenic parasites were completely resistant to atovaquone as well as other mtETC inhibitors. The resistance was not due to mutations affecting mtETC components since mitochondria isolated from the transgenic parasites were fully susceptible to atovaquone. Thus, acquisition of a single metabolic bypass within the pyrimidine biosynthesis pathway was sufficient to render the blood-stage *P. falciparum* independent of the mitochondrial electron transport chain. A critical function of the mtETC in blood-stage *P. falciparum*, therefore, is to provide oxidized CoQ to serve as the electron acceptor for DHODH [30].

These findings further validated the parasite DHODH as a target for antimalarial drug discovery and development, and also raised questions as to the role played by mitochondrial dehydrogenases other than DHODH in blood stage *P. falciparum*. Recent studies have revealed that *P. falciparum* strains of different genetic background have varying reliance on the mtETC. Whereas some of the yDHODH-transgenic strains could be propagated for long-term growth in the presence of atovaquone, other yDHODH-transgenic strains stopped their growth after ~96 h of exposure to atovaquone. Interestingly, this growth inhibition could be reversed by the inclusion of oxidized decyl-ubiquinone in the culture medium (Ke et al. unpublished results). This suggests that there are strain-specific variations in demands for CoQ under conditions where the mtETC is uncoupled from pyrimidine biosynthesis.

The robust drug resistance imparted by transgenic yDHODH has permitted the construction of transfection vectors with yDHODH as the selectable marker [31]. While atovaquone can be used as the selective agent, strain-specific variations limit its use. Instead, parasite-specific DHODH inhibitors such as triazolopyrimidines [32] (e.g. DSM1; see Fig. 1 for structure) work as better selective agents applicable to all *P. falciparum* strains [31]. Because of the paucity of useful selectable markers for genetic manipulation of *P. falciparum*, the advent of yDHODH as a new selectable marker will likely be of great utility in the field, permitting genetic complementation as well as double- and triple-knockout experiments.

7 The Reversal of mtETC Inhibitor Resistance in yDHODH-Transgenic Parasites by Proguanil

The antimalarial activity of proguanil requires its metabolic conversion to the triazine compound cycloguanil, which is a potent inhibitor of parasite DHFR [19, 20]. However, as mentioned above, synergy with atovaquone is mediated by proguanil, not cycloguanil. Proguanil by itself has minimal intrinsic activity against *P. falciparum* with an EC_{50} of about 50 μM [33], which was also the value observed for the effect of proguanil against the yDHODH-transgenic parasites [30]. Surprisingly, when proguanil was tested in the presence of 100-nM atovaquone, the yDHODH-transgenic parasites had an EC_{50} of about 60 nM for the effect of proguanil, an almost 1,000-fold drop in EC_{50}. Conversely, the inclusion of 1-μM proguanil reduced the atovaquone EC_{50} from >2 μM to 0.7 nM in the transgenic parasites; proguanil completely reversed atovaquone resistance in yDHODH-transgenic *P. falciparum*. Resistance to all mtETC inhibitors in yDHODH-transgenic parasites was similarly reversed by proguanil. The mitochondrial membrane potential was fully collapsed when atovaquone and proguanil were combined [30].

These intriguing observations could be explained by the proposition that maintenance of mitochondrial membrane potential is essential for survival even in yDHODH-transgenic parasites, and that there are two paths operational for maintaining the membrane potential. The dominant path for generating mitochondrial membrane potential is through the mtETC, which, when targeted by atovaquone in the wild-type parasites, results in parasite inhibition due to the block of pyrimidine biosynthesis. Although the yDHODH-transgenic parasites can overcome atovaquone inhibition through the metabolic bypass afforded by the transgene, for survival they still require mitochondrial membrane potential, which is established through a secondary pathway independent of the mtETC. It is this secondary pathway that is proposed to be inhibited by proguanil, and thus proguanil in the presence of atovaquone results in reversal of the atovaquone resistance of the yDHODH-transgenic parasite. The intrinsic EC_{50} of proguanil appears to be about 60 nM, which becomes apparent only in yDHODH-transgenic parasites in the presence of atovaquone [30]. The dominant mtETC-dependent path for mitochondrial membrane potential would be fully functional in the transgenic parasites in the absence of atovaquone, and thus proguanil by itself would not affect the parasites unless at a 1,000-fold higher concentration, most likely due to its off-target activity. The molecular nature of the proposed secondary pathway targeted by proguanil is unclear at this point.

8 Problems with Atovaquone–Proguanil Combination Therapy

On the face of it, proguanil would appear to be an ideal partner drug for atovaquone as well as other anti-mtETC antimalarials under development. However, a close examination in light of the observations described above raises significant doubts as to the use of the atovaquone–proguanil combination as antimalarials in the field. The atovaquone–proguanil combination could actually be considered as a triple-drug combination (1) atovaquone as the mtETC inhibitor, (2) proguanil with its intrinsic activity against a secondary target that can generate mitochondrial membrane potential, and (3) proguanil as a prodrug for cycloguanil inhibiting the parasite DHFR. Unfortunately, single point mutations within the parasite cytochrome *b* that give rise to atovaquone resistance would render both atovaquone and the intrinsic proguanil activity ineffective. As described in Fig. 3, metabolic conversion of proguanil to cycloguanil requires the host cytochrome P_{450} enzyme CYP2C19 [34]. Genetic polymorphisms resulting in poor metabolism through CYP2C19 are quite prevalent, especially among the populations residing in malaria-endemic areas [35–38]; thus, such individuals will not convert proguanil to cycloguanil in an efficient manner. Furthermore, cycloguanil resistance due to

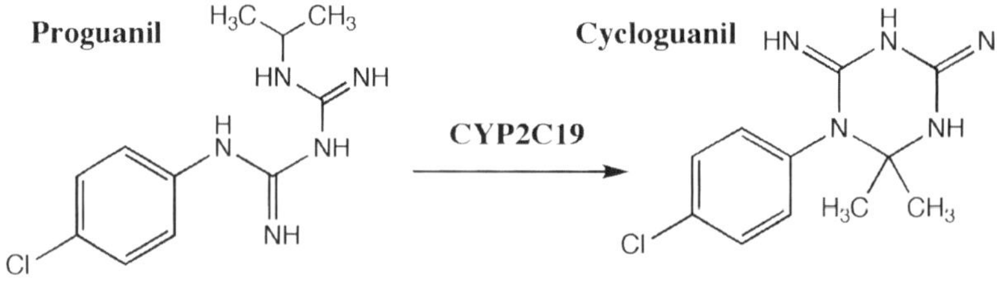

Fig. 3 Structures of proguanil and its metabolite cycloguanil. This conversion is carried out by the cytochrome P_{450} isoform CYP2C19. Genetic polymorphisms involving CYP2C19 with high frequency among certain populations inhibit this metabolic conversion

mutations in the parasite DHFR is quite widespread among the field isolates of *P. falciparum*. Thus, the efficacy of cycloguanil as a stand-alone antimalarial is highly diminished. In light of these points, the atovaquone–proguanil combination in effect turns out to be a monotherapy in two groups of individuals. First, in poor metabolizing individuals, proguanil to cycloguanil conversion would be highly diminished, leaving the parasites unexposed to a DHFR inhibitor. Second, individuals who do metabolize proguanil to cycloguanil, but get infected with cycloguanil-resistant parasites, would not benefit from cycloguanil. Atovaquone resistance could arise very rapidly in such patients due to cytochrome *b* mutations, and, once such parasites arise, proguanil as a prodrug will not provide its synergistic effect, since this effect is apparent only in parasites with atovaquone-sensitive mtETC. Indeed, reports of Malarone prophylaxis and treatment failures [22–24] provide considerable support for this proposition.

The choice of a partner drug for new anti-mtETC inhibitors under development will need to be guided by these insights. A key feature of such compounds has to be a much lower frequency of resistance development compared to atovaquone. Also, the use of Malarone in artemisinin-resistance containment campaigns could become problematic because of emergent resistance to the containment drug combination.

9 Use of yDHODH-Transgenic Parasites to Decipher Mode of Drug Action

The yDHODH-transgenic parasites are resistant to all mtETC inhibitors as well as to the parasite DHODH inhibitors, such as the triazolopyrimidines [32]. Importantly, proguanil is able to reverse the resistance to mtETC inhibitors but not so for the parasite DHODH inhibitors (Morrisey et al. unpublished data). These observations could be used to assess the mode of action for antimalarial compounds identified through high-throughput empirical screens based on parasite growth inhibition. Indeed, recent studies have employed the yDHODH-transgenic parasites in this manner to deconvolute the mode of action for thousands of compounds identified through their antimalarial activities in whole cell-based assays [39, 40]. Several additional mtETC and DHODH inhibitors were identified through this approach.

At least four additional chemical classes of compounds under investigation appear to target the parasite mtETC in a selective manner: pyridones [41], quinolones [42, 43], acridones [44], and acridinediones [45]. Testing representative compounds from each of these classes in yDHODH-transgenic parasites in the presence and the absence of proguanil confirmed that they act through inhibition of the mtETC (Morrisey et al. unpublished data). Furthermore, these assays also revealed that these compounds did not have any off-target activity in *P. falciparum*.

The mode of action for artemisinin and related compounds has been unclear for many years. A few studies have suggested that artemisinin may work through inhibition of the parasite mitochondrial respiration [46, 47]. However, examination of a series of artemisinins for their inhibitory activity against the yDHODH-transgenic parasites failed to demonstrate any reduction in potency (Morrisey et al. unpublished data). Therefore, it appears that artemisinin and derivatives are unlikely to work through parasite mtETC inhibition.

10 DHODH Inhibitors as Antimalarials

Because malaria parasites are completely dependent on de novo pyrimidine biosynthesis, enzymes involved in this pathway are potentially attractive targets for antimalarial drug discovery and development. DHODH carries out the only redox step in this pathway and is a "druggable" enzyme as shown by successful deployment of leflunomide, a human DHODH inhibitor approved for the treatment of rheumatoid arthritis. Two groups have conducted high-throughput screening of large libraries of compounds for their ability to inhibit the parasite DHODH enzymatic activity, and have identified several chemical scaffolds with selective inhibitory activities [32, 48, 49]. Among these compounds, the triazolopyrimidine [32] and substituted thiophene–carboxamide [49] series (see Fig. 1 for structures) have good in vivo activity and are under further development. These are encouraging findings that hold promise for new antimalarials targeting a mitochondrial function.

11 Other Potential Targets in Mitochondrial Functions

Mitochondria in malaria parasites are highly divergent from their hosts' organelles [2]. Indeed, the organelle appears to have much diminished functions compared to the mammalian mitochondria. Many of these functions, however, are critical for parasite survival, and their divergence from their host counterparts presents opportunities for selective antiparasitic agents [50]. Recent studies have revealed that the architecture of the mitochondrial tricarboxylic acid (TCA) metabolism in *P. falciparum* is dramatically different from that of the human mitochondria [51]. Some of the TCA metabolism enzymes, such as malate–quinone oxidoreductase and fumarate hydratase, are significantly different in malaria parasites [50]. Similarly, the ATP synthase complex in *Plasmodium* is also highly divergent in a manner similar to recent observations on a ciliate ATP synthase [52]. These and several other features of *Plasmodium* mitochondria call for detailed studies with the hope of identifying and exploiting their druggable properties.

Acknowledgments I thank colleagues and students in the Center for Molecular Parasitology at Drexel University College of Medicine for discussions and dynamism. I am also grateful for funding from the US National Institutes of Health (Grant number: AI028398) and Medicines for Malaria Venture.

References

1. Vaidya AB, Lashgari MS, Pologe LG, Morrisey J (1993) Structural features of *Plasmodium* cytochrome b that may underlie susceptibility to 8-aminoquinolines and hydroxynaphthoquinones. Mol Biochem Parasitol 58:33–42
2. Vaidya AB, Mather MW (2009) Mitochondrial evolution and functions in malaria parasites. Annu Rev Microbiol 63:249–267
3. Foth BJ, Stimmler LM, Handman E, Crabb BS, Hodder AN, McFadden GI (2005) The malaria parasite *Plasmodium falciparum* has only one pyruvate dehydrogenase complex, which is located in the apicoplast. Mol Microbiol 55:39–53
4. Gardner MJ, Hall N, Fung E, White O, Berriman M, Hyman RW, Carlton JM, Pain A, Nelson KE, Bowman S et al (2002) Genome sequence of the human malaria parasite *Plasmodium falciparum*. Nature 419:498–511
5. Fisher N, Bray PG, Ward SA, Biagini GA (2007) The malaria parasite type II NADH:quinone oxidoreductase: an alternative enzyme for an alternative lifestyle. Trends Parasitol 23:305–310
6. Crofts AR (2004) The cytochrome bc1 complex: function in the context of structure. Annu Rev Physiol 66:689–733
7. Cooley JW (2010) A structural model for across membrane coupling between the Qo and Qi active sites of cytochrome bc1. Biochim Biophys Acta 1797:1842–1848
8. Mather MW, Darrouzet E, Valkova-Valchanova M, Cooley JW, McIntosh MT, Daldal F, Vaidya AB (2005) Uncovering the molecular mode of action of the antimalarial drug atovaquone using a bacterial system. J Biol Chem 280:27458–27465
9. Srivastava IK, Morrisey JM, Darrouzet E, Daldal F, Vaidya AB (1999) Resistance mutations reveal the atovaquone-binding domain of cytochrome b in malaria parasites. Mol Microbiol 33:704–711
10. Kessl JJ, Lange BB, Merbitz-Zahradnik T, Zwicker K, Hill P, Meunier B, Palsdottir H, Hunte C, Meshnick S, Trumpower BL (2003) Molecular basis for atovaquone binding to the cytochrome bc1 complex. J Biol Chem 278:31312–31318
11. Kessl JJ, Meshnick SR, Trumpower BL (2007) Modeling the molecular basis of atovaquone resistance in parasites and pathogenic fungi. Trends Parasitol 23:494–501
12. Looareesuwan S, Viravan C, Webster HK, Kyle DE, Hutchinson DB, Canfield CJ (1996) Clinical studies of atovaquone, alone or in combination with other antimalarial drugs, for treatment of acute uncomplicated malaria in Thailand. Am J Trop Med Hyg 54:62–66
13. Rathod PK, McErlean T, Lee PC (1997) Variations in frequencies of drug resistance in *Plasmodium falciparum*. Proc Natl Acad Sci USA 94:9389–9393
14. McIntosh MT, Srivastava R, Vaidya AB (1998) Divergent evolutionary constraints on mitochondrial and nuclear genomes of malaria parasites. Mol Biochem Parasitol 95:69–80
15. Joy DA, Feng X, Mu J, Furuya T, Chotivanich K, Krettli AU, Ho M, Wang A, White NJ, Suh E et al (2003) Early origin and recent expansion of *Plasmodium falciparum*. Science 300:318–321
16. Vaidya AB, Arasu P (1987) Tandemly arranged gene clusters of malarial parasites that are highly conserved and transcribed. Mol Biochem Parasitol 22:249–257
17. Preiser PR, Wilson RJ, Moore PW, McCready S, Hajibagheri MA, Blight KJ, Strath M, Williamson DH (1996) Recombination associated with replication of malarial mitochondrial DNA. EMBO J 15:684–693

18. Canfield CJ, Pudney M, Gutteridge WE (1995) Interactions of atovaquone with other antimalarial drugs against *Plasmodium falciparum in vitro*. Exp Parasitol 80:373–381
19. Carrington HC, Crowther AF, Davey DG, Levi AA, Rose FL (1951) A metabolite of "Paludrine" with high antimalarial activity. Nature 168:1080
20. Crowther AF, Levi AA (1953) Proguanil – the isolation of a metabolite with high antimalarial activity. Br J Pharmacol 8:93–97
21. Looareesuwan S, Chulay JD, Canfield CJ, Hutchinson DB (1999) Malarone (atovaquone and proguanil hydrochloride): a review of its clinical development for treatment of malaria. Malarone Clinical Trials Study Group. Am J Trop Med Hyg 60:533–541
22. Fivelman QL, Butcher GA, Adagu IS, Warhurst DC, Pasvol G (2002) Malarone treatment failure and *in vitro* confirmation of resistance of *Plasmodium falciparum* isolate from Lagos, Nigeria. Malar J 1:1
23. Kuhn S, Gill MJ, Kain KC (2005) Emergence of atovaquone-proguanil resistance during treatment of *Plasmodium falciparum* malaria acquired by a non-immune north American traveller to west Africa. Am J Trop Med Hyg 72:407–409
24. Krudsood S, Patel SN, Tangpukdee N, Thanachartwet W, Leowattana W, Pornpininworakij K, Boggild AK, Looareesuwan S, Kain KC (2007) Efficacy of atovaquone-proguanil for treatment of acute multidrug-resistant *Plasmodium falciparum* malaria in Thailand. Am J Trop Med Hyg 76:655–658
25. Srivastava IK, Rottenberg H, Vaidya AB (1997) Atovaquone, a broad spectrum antiparasitic drug, collapses mitochondrial membrane potential in a malarial parasite. J Biol Chem 272:3961–3966
26. Srivastava IK, Vaidya AB (1999) A mechanism for the synergistic antimalarial action of atovaquone and proguanil. Antimicrob Agents Chemother 43:1334–1339
27. Nyakeriga AM, Perlmann H, Hagstedt M, Berzins K, Troye-Blomberg M, Zhivotovsky B, Perlmann P, Grandien A (2006) Drug-induced death of the asexual blood stages of *Plasmodium falciparum* occurs without typical signs of apoptosis. Microbes Infect 8:1560–1568
28. Painter HJ, Morrisey JM, Vaidya AB (2010) Mitochondrial electron transport inhibition and viability of intraerythrocytic *Plasmodium falciparum*. Antimicrob Agents Chemother 54:5281–5287
29. Nagy M, Lacroute F, Thomas D (1992) Divergent evolution of pyrimidine biosynthesis between anaerobic and aerobic yeasts. Proc Natl Acad Sci USA 89:8966–8970
30. Painter HJ, Morrisey JM, Mather MW, Vaidya AB (2007) Specific role of mitochondrial electron transport in blood-stage *Plasmodium falciparum*. Nature 446:88–91
31. Ganesan SM, Morrisey JM, Ke H, Painter HJ, Laroiya K, Phillips MA, Rathod PK, Mather MW, Vaidya AB (2011) Yeast dihydroorotate dehydrogenase as a new selectable marker for *Plasmodium falciparum* transfection. Mol Biochem Parasitol 177:29–34
32. Phillips MA, Gujjar R, Malmquist NA, White J, El Mazouni F, Baldwin J, Rathod PK (2008) Triazolopyrimidine-based dihydroorotate dehydrogenase inhibitors with potent and selective activity against the malaria parasite *Plasmodium falciparum*. J Med Chem 51:3649–3653
33. Fidock DA, Nomura T, Wellems TE (1998) Cycloguanil and its parent compound proguanil demonstrate distinct activities against *Plasmodium falciparum* malaria parasites transformed with human dihydrofolate reductase. Mol Pharmacol 54:1140–1147
34. Helsby NA, Ward SA, Howells RE, Breckenridge AM (1990) *In vitro* metabolism of the biguanide antimalarials in human liver microsomes: evidence for a role of the mephenytoin hydroxylase (P450 MP) enzyme. Br J Clin Pharmacol 30:287–291
35. Herrlin K, Massele AY, Jande M, Alm C, Tybring G, Abdi YA, Wennerholm A, Johansson I, Dahl ML, Bertilsson L et al (1998) Bantu Tanzanians have a decreased capacity to metabolize omeprazole and mephenytoin in relation to their CYP2C19 genotype. Clin Pharmacol Ther 64:391–401
36. Wanwimolruk S, Bhawan S, Coville PF, Chalcroft SC (1998) Genetic polymorphism of debrisoquine (CYP2D6) and proguanil (CYP2C19) in South Pacific Polynesian populations. Eur J Clin Pharmacol 54:431–435

37. Xie HG, Kim RB, Stein CM, Wilkinson GR, Wood AJ (1999) Genetic polymorphism of (S)-mephenytoin 4′-hydroxylation in populations of African descent. Br J Clin Pharmacol 48:402–408
38. Desta Z, Zhao X, Shin JG, Flockhart DA (2002) Clinical significance of the cytochrome P450 2 C19 genetic polymorphism. Clin Pharmacokinet 41:913–958
39. Guiguemde WA, Shelat AA, Bouck D, Duffy S, Crowther GJ, Davis PH, Smithson DC, Connelly M, Clark J, Zhu F et al (2010) Chemical genetics of *Plasmodium falciparum*. Nature 465:311–315
40. Gamo FJ, Sanz LM, Vidal J, de Cozar C, Alvarez E, Lavandera JL, Vanderwall DE, Green DV, Kumar V, Hasan S et al (2010) Thousands of chemical starting points for antimalarial lead identification. Nature 465:305–310
41. Yeates CL, Batchelor JF, Capon EC, Cheesman NJ, Fry M, Hudson AT, Pudney M, Trimming H, Woolven J, Bueno JM et al (2008) Synthesis and structure-activity relationships of 4-pyridones as potential antimalarials. J Med Chem 51:2845–2852
42. Winter RW, Kelly JX, Smilkstein MJ, Dodean R, Hinrichs D, Riscoe MK (2008) Antimalarial quinolones: synthesis, potency, and mechanistic studies. Exp Parasitol 118:487–497
43. Winter R, Kelly JX, Smilkstein MJ, Hinrichs D, Koop DR, Riscoe MK (2011) Optimization of endochin-like quinolones for antimalarial activity. Exp Parasitol 127(2):545–51
44. Winter RW, Kelly JX, Smilkstein MJ, Dodean R, Bagby GC, Rathbun RK, Levin JI, Hinrichs D, Riscoe MK (2006) Evaluation and lead optimization of anti-malarial acridones. Exp Parasitol 114:47–56
45. Biagini GA, Fisher N, Berry N, Stocks PA, Meunier B, Williams DP, Bonar-Law R, Bray PG, Owen A, O'Neill PM et al (2008) Acridinediones: selective and potent inhibitors of the malaria parasite mitochondrial bc1 complex. Mol Pharmacol 73:1347–1355
46. Li W, Mo W, Shen D, Sun L, Wang J, Lu S, Gitschier JM, Zhou B (2005) Yeast model uncovers dual roles of mitochondria in action of artemisinin. PLoS Genet 1:e36
47. Wang J, Huang L, Li J, Fan Q, Long Y, Li Y, Zhou B (2010) Artemisinin directly targets malarial mitochondria through its specific mitochondrial activation. PLoS One 5:e9582
48. Phillips MA, Rathod PK (2010) *Plasmodium* dihydroorotate dehydrogenase: a promising target for novel anti-malarial chemotherapy. Infect Disord Drug Targets 10:226–239
49. Booker ML, Bastos CM, Kramer ML, Barker RH Jr, Skerlj R, Sidhu AB, Deng X, Celatka C, Cortese JF, Guerrero Bravo JE et al (2010) Novel inhibitors of *Plasmodium falciparum* dihydroorotate dehydrogenase with anti-malarial activity in the mouse model. J Biol Chem 285:33054–33064
50. Mather MW, Henry KW, Vaidya AB (2007) Mitochondrial drug targets in apicomplexan parasites. Curr Drug Targets 8:49–60
51. Olszewski KL, Mather MW, Morrisey JM, Garcia BA, Vaidya AB, Rabinowitz JD, Llinas M (2010) Branched tricarboxylic acid metabolism in *Plasmodium falciparum*. Nature 466:774–778
52. Balabaskaran Nina P, Dudkina NV, Kane LA, van Eyk JE, Boekema EJ, Mather MW, Vaidya AB (2010) Highly divergent mitochondrial ATP synthase complexes in Tetrahymena thermophila. PLoS Biol 8:e1000418

Non-Antifolate Antibiotics: Clindamycin, Doxycycline, Azithromycin and Fosmidomycin

Sanjeev Krishna and Henry M. Staines

Abstract A range of antibiotics, in addition to those that target folate metabolism, have demonstrated antimalarial activity. They include those belonging to the lincosamides, tetracyclines and macrolides classes and fosmidomycin, a derivative of phosphonic acid. Predominantly, they target pathways within the apicoplast, a relict plastid found in most apicomplexan parasites including *Plasmodium*. In general, they are not highly active against malarial parasites and are slow acting but are clinically useful when used in combination with other antimalarial drug classes. In addition, some are safe to use in pregnancy and for the treatment of small children. Here, we review the current understanding of their mechanisms of action and clinical use.

1 Introduction

There are many non-antifolate antibiotics that have antimalarial activity including members belonging to the following classes: fluoroquinolones, lincosamides, tetracyclines and macrolides. The fluoroquinolones will not be discussed further here, as they were moderately active against *Plasmodium falciparum* in vitro [1, 2], but this property did not extend to useful antimalarial efficacy when tested in vivo [3, 4]. Fosmidomycin is a phosphonic acid derivative that is also included in discussion, as a potentially interesting new class of antimalarial. The lincosamides, tetracyclines and macrolides have established antimalarial properties and are clinically useful in particular circumstances. None of these antibiotics is highly active and rapidly acting; so, their usage requires combination with other rapidly effective classes of antimalarial, where they contribute to increasing cure rates even with

S. Krishna (✉) • H.M. Staines
Division of Clinical Sciences, Centre for Infection and Immunology, St. George's University of London, Cranmer Terrace, London SW17 0RE, UK
e-mail: sgjf100@sgul.ac.uk; hstaines@sgul.ac.uk

H.M. Staines and S. Krishna (eds.), *Treatment and Prevention of Malaria*,

DOI 10.1007/978-3-0346-0480-2_8,

shorter courses of treatment. Some antibiotics have the additional advantages of being safe to use in pregnancy or for small children (discussed below).

2 Clindamycin

The lincosamides are named after lincomycin that was extracted from *Streptomyces lincolnensis* in a soil sample in 1962. The structure of lincosamides is unusual, as are their antimicrobial properties. Clindamycin was the only semi-synthetic lincosamide to be developed for clinical use (Fig. 1). Here, it has activity against Gram-positive bacteria and many anaerobes, but not against Gram-negative aerobes. More relevant to its antimalarial properties, clindamycin is active against several apicomplexans including *Plasmodium* spp., *Toxoplasma gondii* and *Babesia microti*.

2.1 Mechanism of Action

Clindamycin acts at the same site on ribosomes as erythromycin and chloramphenicol. It inhibits protein synthesis in bacteria by binding to the 50S ribosomal unit and it can exert concentration-dependent bactericidal activity beyond its accepted bacteriostatic effects [5]. In its actions against apicomplexans, clindamycin targets their apicoplast organelles and interferes with function and survival after one or two rounds of parasite replication have taken place, resulting in a "delayed-death" phenotype. This is well described for *Toxoplasma* [6] as well as for *P. falciparum* with correlations between in vitro and in vivo pharmacodynamics being observed [7–9]. Further evidence for this proposed mechanism of action for clindamycin comes from observations of mutations in the apicoplast genome of parasites obtained from the Peruvian Amazon. A point-mutation in the apicoplast-encoded 23S rRNA gene that confers resistance to lincosamides in other organisms was identified in parasites that were 100-fold more resistant to clindamycin than wild-type parasites [10].

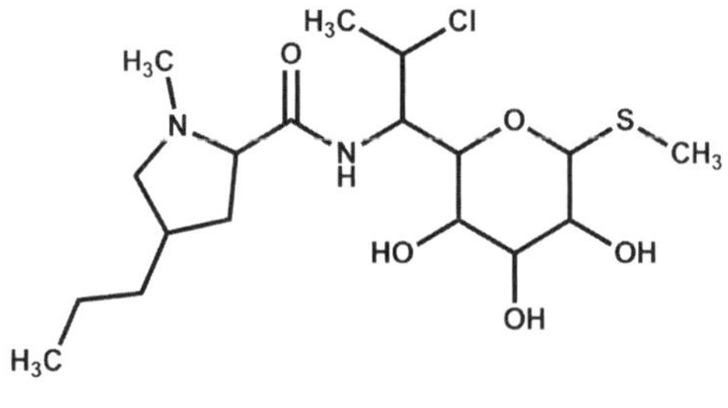

Fig. 1 Structure of clindamycin

2.2 Pharmacokinetics

Clindamycin has good (>90%) bioavailability after oral dosing, is protein bound (>90%) and widely distributed in tissues and has an elimination half-time of approximately 2–3 h. It is eliminated mainly by biliary excretion (with 20% excreted by kidneys) after metabolism to three major derivatives [11].

2.3 Usage in Malaria

The antimalarial properties of clindamycin have been confirmed in many studies, including earlier studies after the first one in 1975 that used clindamycin as a monotherapy before its more successful use in combination treatment regimens [11]. The interesting mechanism of action of clindamycin and its pharmacokinetic properties suggest that a minimum of 5 days treatment with at least twice-daily dosing is needed to achieve adequate cure rates in uncomplicated malaria [11], a regimen that is not currently recommended.

Table 1 summarises the antimalarial properties of clindamycin combined with quinine. Clindamycin dosages ranging from 5 mg/kg twice daily given with quinine, to 8 mg/kg three times a day or 12 mg/kg twice daily in general give acceptable cure rates. One small study [21] found that a 3-day treatment of the combination was as effective as a 7-day monotherapy regimen with quinine alone in travellers returning with uncomplicated malaria. Clindamycin can be used in children and pregnant women and may be combined with artesunate as a partner antimalarial, although short-course regimens may need further study to confirm efficacy [24].

Clindamycin may be less effective in treating *P. vivax* infections [11] even when prolonged courses of monotherapy are used, although a later review suggested that clindamycin is more effective than tetracyclines [25]. It also has no apparent synergism with quinine for *P. vivax* and nor does it have anti-relapse properties [11].

2.4 Tolerance and Safety

Non-specific and relatively common gastro-intestinal side effects of nausea, vomiting, abdominal pain and diarrhoea are associated with clindamycin use. Toxin-producing *Clostridium difficile* colitis can also complicate clindamycin therapy, most commonly in hospitalised patients treated with prolonged courses of clindamycin, and not with the antimalarial regimes that have been most studied in recent years. Rashes may be relatively common side effects, but severe eruptions are not. Transient reversible neutropenia and thrombocytopenia have also occurred.

Table 1 Clinical trials of clindamycin plus quinine against *P. falciparum* malaria

Study details					Regimen							Efficacy (%)	Ref.
Year	Place	Design[a]	Pop.[b]	*N*[c]	Clindamycin		Quinine		Route[d]	Days	Dosing[e]		
					Dosage, form[f]	No. of doses/day	Dosage	No. of doses/day					
1974	United States	WHO	A	5	450 mg, salt	4	540 mg	3	p.o.	3	Yes	100	[12]
1975	United States	WHO	A	5	450 mg, salt	3	560 mg	3	p.o.	3	Yes	60	[13]
1975	United States	WHO	A	2	600 mg, salt	1	560 mg	3	p.o.	3	No	50	[13]
1975	Thailand	WHO	A	4	450 mg, salt	3	540 mg	3	p.o.	3	Yes	100	[14]
1975	Thailand	WHO	A	5	150 mg, salt	3	270 mg	3	p.o.	3	No	60	[14]
1988	Brazil	WHO, RCT	A	40	10 mg/kg, base	2	12 mg/kg	2	p.o.	3	Yes	90	[15]
1994	Gabon	WHO, RCT	C	34	5 mg/kg, base	2	12 mg/kg	2	p.o.	3	Yes	88	[16]
1995	Gabon	WHO, RCT	C[g]	50	5 mg/kg, base	3	8 mg/kg	3	i.v.	4	Yes	96	[17]
1995	Gabon	WHO, RCT	A	40	5 mg/kg, base	2	12 mg/kg	2	p.o.	3[h]	Yes	92	[18]
1997	Gabon	WHO[i]	C	161	8 mg/kg, salt	2	8 mg/kg	2	p.o.	3	Yes	97	[19]
2000	Thailand	WHO, RCT	A	68	5 mg/kg, base	4	8 mg/kg	3	p.o.	7	Yes	100	[20]
2001	France	WHO, RCT	A	53	5 mg/kg, salt	3	8 mg/kg	3	i.v.	3	Yes	100	[21]
2001	Thailand	WHO, RCT	P	65	5 mg/kg, NS[j]	3	8 mg/kg	3	p.o.	7	Yes	100	[22]

[a]WHO, study conducted according to World Health Organisation guidelines [23]; *RCT* randomised controlled trial
[b]Pop., study population; A, adults; C, children; P, pregnant women
[c]*N*, number of subjects
[d]i.v., intravenous; p.o., oral
[e]Adequate dosing (i.e., clindamycin given at least twice daily and more than 3 days)
[f]8 mg of clindamycin hydrochloride salt is equivalent to 5 mg of base
[g]Severe malaria
[h]Quinine was administered for only 1.5 days
[i]Short follow-up (3 weeks)
[j]*NS* not specified. Taken from [11] with kind permission from the the authors and the American Society for Microbiology

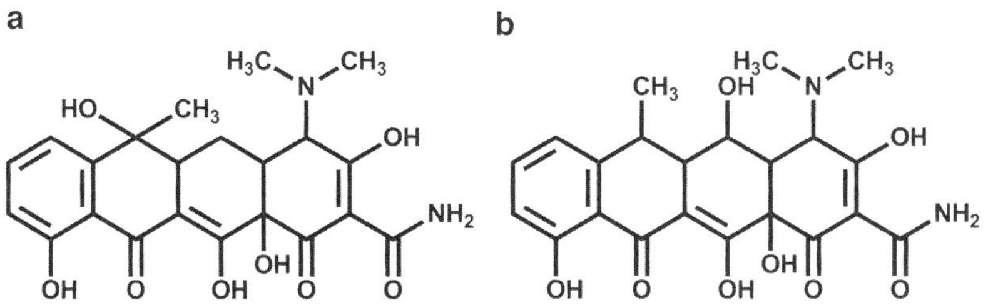

Fig. 2 Structures of tetracycline (**a**) and doxycycline (**b**)

3 Tetracyclines Including Doxycycline

The discovery of chlortetracycline in 1948 preceded that of clindamycin, although it was also from another soil organism *Streptomyces aureofasciens*. Catalytic dehydrogenation of chlortetracycline gave tetracycline in 1953. This class of antibiotic has wide antimicrobial properties, ranging from Gram-positive and Gram-negative bacteria, rickettsiae (where its use can be diagnostic), mycoplasmas, chlamydia and protozoa. Doxycycline is a semi-synthetic derivative of tetracycline first developed in 1967. The structures of tetracycline and doxycycline are given in Fig. 2.

3.1 Mechanism of Action

The tetracyclines including doxycycline act by inhibiting protein synthesis by binding to the 30S ribosomal subunit. They are relatively slow acting as antimalarial agents. Doxycycline is useful as a prophylactic agent on its own, as well as being used in combination treatment regimens to increase cure rates of conventional antimalarial agents [26]. It has limited casual prophylactic activity [26] and does not kill hypnozoites.

3.2 Pharmacokinetics

Doxycyline has ~90% oral bioavailability, which is reduced by food and delayed by alcohol, with peaks without delay occurring at ~2 h after administration [26]. Co-administration with cations such as iron supplements reduces absorption by chelation of the antibiotic, and milk may reduce absorption [26]. Plasma protein binding of doxycycline is 90% and elimination half-time is 18 h, making this one of the most convenient tetracyclines to use in practise. Most elimination of unchanged drug is through the gastrointestinal tract, with about a third being renally excreted.

3.3 Usage in Malaria

Tetracycline requires dosing 4 times as day and is therefore cumbersome to use as an antimalarial. Doxycyline (for adults up to 100 mg base twice daily, with another antimalarial) should only be used in combination with rapidly acting antimalarials such as quinine to increase cure rates, and not on its own. Table 2 summarises antimalarial treatment properties of doxycycline [26]. It can be used as a prophylactic antimalarial (for adults up to 100 mg base once daily starting before travel and continuing for 4 weeks after return), even in areas of multidrug-resistant parasites [26].

3.4 Safety and Toxicity

The tetracyclines are not recommended in pregnancy, and in children whose teeth can be stained (8 years or less). Doxycycline is a photosensitising antibiotic and can also cause oesophageal erosions and heartburn if taken incorrectly. Nausea and abdominal pain are relatively common, but frequency can be reduced by taking doxycycline with food [26].

4 Azithromycin

Azithromycin is a semi-synthetic derivative of erythromycin, whereby a methyl-substituted nitrogen atom is incorporated into the lactone ring system to improve acid stability (Fig. 3). In contrast to doxycycline, azithromycin can be used in younger children and in pregnancy, but its much increased cost compared with doxycycline may limit its applicability for management of malaria. Azithromycin is useful in the management of respiratory tract infections and has activity against *Toxoplasma* and can be used to treat uncomplicated babesiosis when combined with atovaquone in immunocompetent individuals [32].

4.1 Mechanism of Action

Azithromycin acts by binding to the 50S ribosomal sub-unit of susceptible microorganisms and inhibiting protein synthesis. Consistent with the activities of other antimalarial antibiotics, azithromycin is relatively slow acting and should be used in combination therapies for malaria. There may be some cross-resistance between antibiotics as observed for clindamycin-resistant *T. gondii*.

Table 2 Efficacy of doxycycline with or without quinine for treatment of *P. falciparum* malaria

Country (population)	Study type	Sample size/doxycycline dose	Duration of treatment	Cure rate (95% CI)	Ref.
United States (mixture of immune and non-immune)	Challenge, open-label	*P. vivax:*			[27]
		1/100 mg twice a day	4 days	0	
		1/100 mg twice a day	6 days	0	
		P. falciparum:			
		4/100 mg twice a day	5 days	0	
		9/100 mg twice a day	7 days	100%	
West Malaysia (varied, children aged 2 months–8 years)	Challenge, open-label	9/4 mg/kg daily	4 days	44.4%	[28]
		26/4 mg/kg daily	7 days	84.6%	
Irian Jaya, Indonesia (non-immune)	Randomised, open-label, comparator = CQ ($N = 30$) and doxycycline + CQ ($N = 39$)	20/100 mg twice a day	7 days	*P. falciparum*: 64.7% (42.0–87.4%) *P. vivax*: 33.3% (6.6–59.5%)	[29]
Brazil (semi-immune)	Randomised, open-label, comparator = AL ($N = 28$)	31/100 mg twice a day + 500 mg quinine every 8 h	5 days doxycycline 3 days quinine	100%	[30]
Pakistan (unknown immune status)	Challenge, open-label	100/100 mg twice a day + 10 mg/kg quinine every 8 h	7 days doxycycline 3 days quinine	100%	[31]

CQ chloroquine, *AL* artemether-lumefantrine. Taken from [26] with kind permission from the authors and the American Journal of Tropical Medicine and Hygiene

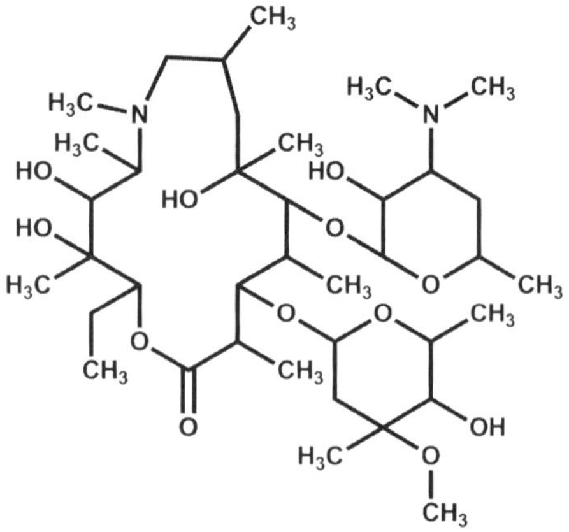

Fig. 3 Structure of azithromycin

4.2 Pharmacokinetics

Azithromycin has moderate oral bioavailability (37% after 500 mg) and this is reduced significantly (by 50%) if given with food; hence, it should be used either 1 h before food or 2 h after. Antacids also reduce availability. The degree of protein binding of azithromycin is dose-dependent, but not usually >50%. The drug is extensively distributed in tissues, which act as a depot with an elimination half-time of ~70 h. Most drugs are probably eliminated unchanged by biliary and gastrointestinal excretion.

4.3 Clinical Studies

Azithromycin by itself is not useful to treat uncomplicated *P. falciparum* or *P. vivax* infections [33] as failure rates can exceed 50% when conventional antibacterial doses for adults of 0.5 g per day for 3 days are used. Cure rates may be increased if the dose of azithromycin is also increased (to for example, 1 g per day) in combination therapies, but comparisons with other antimalarial combinations are not so encouraging (Fig. 4). Current evidence suggests that other combinations should be used to treat malaria unless there is no choice.

4.4 Safety and Toxicity

Azithromycin is relatively well tolerated, with mild gastrointestinal side effects being most commonly reported. Occasional abnormalities in liver function are observed and allergic reactions are rare. Nausea may be more common with higher dose regimens [33].

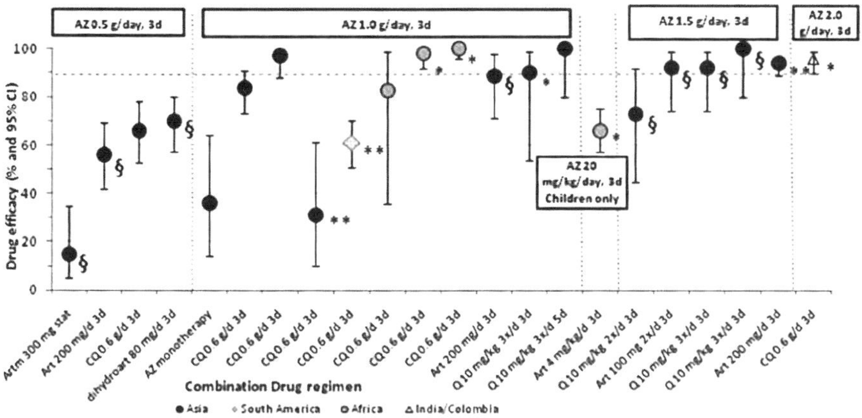

Fig. 4 Efficacy of azithromycin containing treatment regimens for *P. falciparum* (28 day follow-up). *Abbr*: *AZ* azithromycin, *CQ* chloroquine, *Artm* artemether, *Art* artesunate, *dihydroart* dihydroartemisinin, *Q* quinine, *CI* confidence interval, *d* days. *Symbols*: *Asterisk* PCR-corrected, *double asterisk* partially PCR-corrected, *section sign* study conducted in an area without malaria transmission (Bangkok). The AZ dose in the combination with artemisinin (300 mg) was 500 mg at start, followed by 250 mg after 24 h and 48 h. An interrupted line has been drawn at the 90% efficacy level, the minimum level for the 95% confidence interval for a potentially useful drug regimen as recommended by WHO [34]. Taken from [33] with kind permission from the authors and John Wiley & Sons, Ltd on behalf of The Cochrane Collaboration. Copyright Cochrane Collaboration.

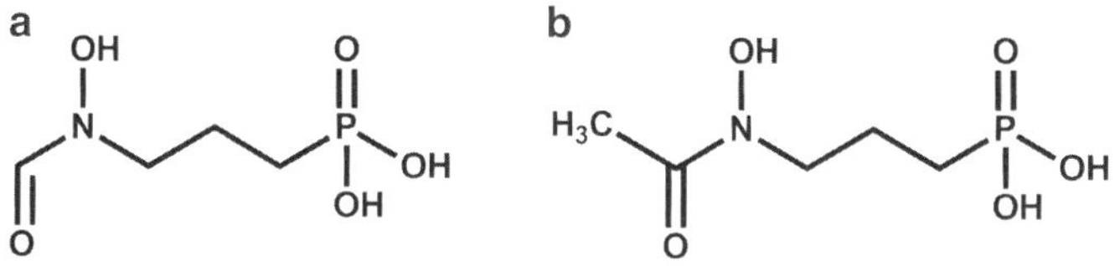

Fig. 5 Structure of fosmidomycin (**a**) and FR-900098 (**b**)

5 Fosmidomycin

Fosmidomycin is a natural antibiotic, originally derived from *Streptomyces lavendulae*. It is currently being investigated as a combination partner in antimalarial chemotherapy regimens, with the idea that it represents a non-artemisinin class of antimalarial with an unusual mode of action. Fosmidomycin's relatively simple chemical structure (Fig. 5) makes it amenable to complete synthesis, which is how it is currently made for investigational studies. Fujisawa Pharmaceutical Company in Osaka, Japan [35], originally developed fosmidomycin as an antibacterial agent [36–39] for treating urinary tract infections approximately three decades ago. It is most effective against enterobacteria and not against Gram-positive organisms or anaerobes. Since its discovery, cephalosporins have emerged as being more effective for recurrent infections and the development of fosmidomycin as an antibacterial agent has not been taken further forward, until it has been repurposed as an antimalarial.

6 Mechanism of Action

The repurposing of fosmidomycin as an antimalarial depended on several advances in understanding the biology of the malarial parasite. One advance lay in the recognition of the plastid organelle as an excellent target for new drugs, as no similar structure exists in animal cells. Another advance was the identification of an alternative [nonmevalonate or 1-deoxy-D-xylulose 5-phosphate (DOXP)] pathway for isoprenoid synthesis in parasites that was hitherto described in plants and eubacteria. Key enzymes forming part of this synthesis pathway include DOXP reductoisomerase and 2*C*-methyl-D-erythritol-4-(cytidine-5-diphospho) transferase (Fig. 6) and are located in the parasite's apicoplast. This pathway contributes to many functions such as prenylation of membrane-bound proteins and synthesis of carotenoids and terpenoids.

Hassan Jomaa and colleagues [40] tested the idea that inhibiting this pathway would prove lethal to parasites by expressing a recombinant version of DOXP reductoisomerase and showing that it was inhibited by fosmidomycin (with an inhibitory constant of ~30 nM). They also showed that fosmidomycin and compound FR-900098 (Fig. 5), a compound that acts as a pro-drug for fosmidomycin, killed cultured *P. falciparum*, including highly chloroquine-resistant strains (IC_{50} values ranging from 300 to 1,200 nM), as well as being able to cure mice infected with *P. vinckei*. Recently, it has been reported that fosmidomycin also targets a second enzyme in the DOXP pathway, 2*C*-methyl-D-erythritol-4-(cytidine-5-diphospho) transferase (Fig. 6) [41]. Other recent studies have suggested that the uptake of fosmidomycin, which is highly charged, requires specific transport mechanisms [42, 43]. This includes the new permeability pathways (NPP) that are responsible for altered plasma membrane permeability of the host erythrocyte as intracellular plasmodial parasites mature [42, 44]. These new data not only explain the selectively of fosmidomycin for plasmodial parasites over closely related

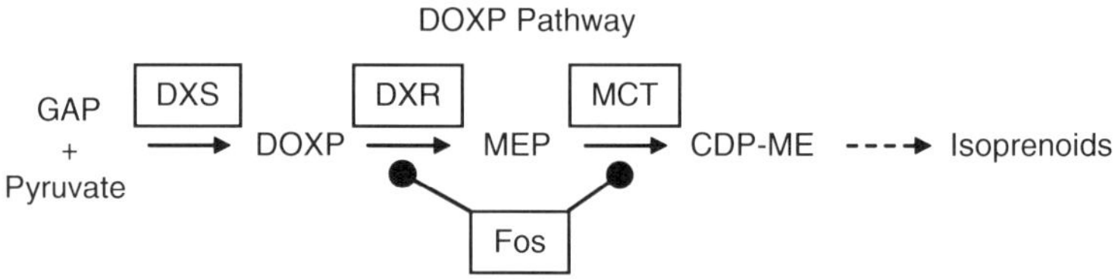

Fig. 6 The 1-deoxy-D-xylulose 5-phosphate (DOXP) pathway for isoprenoid synthesis and the sites of action of fosmidomycin (Fos). *GAP* glyceraldehyde-3-phosphate, *DXS* DOXP synthase, *DXR* DOXP reductoisomerase, *MEP* 2*C*-methyl-D-erythritol-4-phosphate, *MCT* 2*C*-methyl-D-erythritol-4-(cytidine-5-diphospho) transferase, *CDP-ME* 4-(cytidine-5-diphospho)-2*C*-methyl-D-erythritol

Table 3 Fosmidomycin clinical trials data

Region of study	Study design	Regimen	Subjects	Safety and tolerability	Efficacy
Monotherapy					
Gabon/Thailand [48]	Uncontrolled, open-label study	1,200 mg every 8 h for 7 days	26 adults	Well tolerated with mild GI disturbances in 5 patients	At day 28, cure rates were 78% (7/9) in Gabon and 22% (2/9) in Thailand
Gabon [49]	Uncontrolled, open-label study	1,200 mg every 8 h for 5, 4 or 3 days	32 adults	Well tolerated with most frequent AEs being headache, weakness, myalgia, abdominal pain, and loose stools	At day 14, cure rates were 89% (8/9), 88% (7/8) and 60% (6/10) for 5, 4 and 3 day regimens
Combination therapy					
Gabon [50]	Randomised, controlled open-label study	30/5 mg/kg FC, 30 mg/kg F or 5 mg/kg C every 12 h for 5 days	36 children (7–14 years)	Well tolerated with no serious AEs (mostly GI disturbances)	At day 28, cure rates were 100% (12/12), 42% (5/12) and 100% (12/12) for FC, F and C regimens
Gabon [51]	Uncontrolled, dose-reduction study	30/10 mg/kg FC every 12 h for 5, 4, 3, 2, 1 days	52 children (7–14 years)	Well tolerated with most frequent AEs being GI disturbances [mostly loose stools (13 events) and abdominal pain (9 events)]	At day 28, cure rates were 100% (10/10), 100% (10/10), 90% (9/10), 70% (7/10) and 10% (1/10) for 5, 4, 3, 2 and 1 day regimens
Gabon [52]	Uncontrolled, dose-reduction study	30/2 mg/kg F/AS every 12 h for 5, 4, 3, 2, 1 days	50 children (6–12 years)	Well tolerated with most frequent AEs being GI events	At day 28, cure rates were 100% (9/9), 90% (9/10), 100% (10/10), 60% (6/10) and 40% (4/10) for 5, 4, 3, 2 and 1 day regimens
Gabon [53]	Uncontrolled, Phase IIb, single-arm study	30/10 mg/kg FC every 12 h for 3 days	51 children (1–14 years)	Well tolerated with GI disturbances but relatively high rates of treatment-associated	At day 28, cure rates were 89% (42/47) and 62% (5/8) for children 3–14 years and 1–2 years

(continued)

Table 3 (continued)

Region of study	Study design	Regimen	Subjects	Safety and tolerability	Efficacy
				neutropenia (8/51) and falls of haemoglobin concentrations of ≥2 g/dl (7/51)	
Gabon [54]	Randomised study	30/10 mg/kg FC every 12 h for 3 days or 25/1.25 mg/kg SP single dose	105 children (3–14 years)	Both treatments were well tolerated with slightly more AEs occurring in the SP group	At day 28, cure rates were 94% (46/49) and 94% (46/49) for FC and SP regimens
Thailand [46]	Uncontrolled, open-label study	A: 1,200 mg F every 12 h for 7 days or B: 900/600 mg FC every 12 h for 7 days	33 adults	Both treatments were well tolerated with no serious AEs. Mild events included headache and GI disturbance	At day 28, 22% (2/9) and 100% (12/12) for A and B regimens
Thailand [47]	Randomised, uncontrolled open-label study	A: 900/300 mg FC every 6 h for 3 days or B: 1,800/600 mg FC every 12 h for 3 days	114 Adults	Both regimens well tolerated with no serious AEs (GI events were most with abdominal pain developing in 10% of patients, diarrhoea in 5% and vomiting in 4%)	At day 28, 91% (21/23) and 90% (70/78) for A and B regimens

F fosmidomycin, *C* clindamycin, *SP* sulfadoxine-pyrimethamine, *AS* artesunate, *AE* adverse event, *GI* gastro-intestinal species (e.g. *Toxoplasma*) but can also be used to test potential resistance mechanisms in *Plasmodium* spp.

6.1 Pharmacokinetics

Fosmidomycin has relatively poor oral bioavailability (~25%) and is poorly protein bound (<5%) with an elimination half-time of 1.6–1.8 h [45]. This elimination half time is somewhat prolonged in subjects with malaria to 3.4 h (range 1.4–11.8 h) and not altered importantly after co-administration with clindamycin in one study [46], although particular dosing regimens of this combination (fosmidomycin and clindamycin) may influence pharmacokinetic behaviour of each drug [47].

6.2 Fosmidomycin in Malaria

Table 3 summarises studies of fosmidomycin in children and adults with malaria. Monotherapy studies with fosmidomycin have confirmed its antimalarial activity in patients, and also that, unlike clindamycin, it has relatively rapid antimalarial actions. However, recrudescence rates are rather high, compelling the choice of an appropriate combination partner to achieve adequate clinical and parasitological cure rates. Clindamycin has been the most thoroughly studied partner drug and results of various studies are also included in Table 3. In general, a 3-day treatment regimen with 30 mg/kg fosmidomycin and 10 mg/kg clindamycin given in a 8 or 12 h interval and in the populations reported provided adequate responses in adults and children except for one study in children aged <3 years [53]. This combination was also well tolerated, with mild gastrointestinal side effects being some of the most frequent (Table 3).

Acknowledgments We thank Peter Kremsner and David Hutchinson for useful discussion and the EDTCP (project code: IP.2008.31060.003) for financial support.

References

1. Krishna S, Davis TM, Chan PC, Wells RA, Robson KJ (1988) Ciprofloxacin and malaria. Lancet 1:1231–1232
2. Tripathi KD, Sharma AK, Valecha N, Biswas S (1993) *In vitro* activity of fluoroquinolones against chloroquine-sensitive and chloroquine-resistant *Plasmodium falciparum*. Indian J Malariol 30:67–73
3. Havemann K, Bhibi P, Hellgren U, Rombo L (1992) Norfloxacin is not effective for treatment of *Plasmodium falciparum* infection in Kenya. Trans R Soc Trop Med Hyg 86:586

4. Watt G, Shanks GD, Edstein MD, Pavanand K, Webster HK, Wechgritaya S (1991) Ciprofloxacin treatment of drug-resistant falciparum malaria. J Infect Dis 164:602–604
5. Reusser F (1975) Effect of lincomycin and clindamycin on peptide chain initiation. Antimicrob Agents Chemother 7:32–37
6. Fichera ME, Bhopale MK, Roos DS (1995) *In vitro* assays elucidate peculiar kinetics of clindamycin action against *Toxoplasma gondii*. Antimicrob Agents Chemother 39:1530–1537
7. Burkhardt D, Wiesner J, Stoesser N, Ramharter M, Uhlemann AC, Issifou S, Jomaa H, Krishna S, Kremsner PG, Borrmann S (2007) Delayed parasite elimination in human infections treated with clindamycin parallels 'delayed death' of *Plasmodium falciparum in vitro*. Int J Parasitol 37:777–785
8. Fichera ME, Roos DS (1997) A plastid organelle as a drug target in apicomplexan parasites. Nature 390:407–409
9. Kohler S, Delwiche CF, Denny PW, Tilney LG, Webster P, Wilson RJ, Palmer JD, Roos DS (1997) A plastid of probable green algal origin in Apicomplexan parasites. Science 275:1485–1489
10. Dharia NV, Plouffe D, Bopp SE, Gonzalez-Paez GE, Lucas C, Salas C, Soberon V, Bursulaya B, Kochel TJ, Bacon DJ et al (2010) Genome scanning of Amazonian *Plasmodium falciparum* shows subtelomeric instability and clindamycin-resistant parasites. Genome Res 20:1534–1544
11. Lell B, Kremsner PG (2002) Clindamycin as an antimalarial drug: review of clinical trials. Antimicrob Agents Chemother 46:2315–2320
12. Miller LH, Glew RH, Wyler DJ, Howard WA, Collins WE, Contacos PG, Neva FA (1974) Evaluation of clindamycin in combination with quinine against multidrug-resistant strains of *Plasmodium falciparum*. Am J Trop Med Hyg 23:565–569
13. Clyde DF, Gilman RH, McCarthy VC (1975) Antimalarial effects of clindamycin in man. Am J Trop Med Hyg 24:369–370
14. Hall AP, Doberstyn EB, Nanokorn A, Sonkom P (1975) Falciparum malaria semi-resistant to clindamycin. Br Med J 2:12–14
15. Kremsner PG, Zotter GM, Feldmeier H, Graninger W, Rocha RM, Wiedermann G (1988) A comparative trial of three regimens for treating uncomplicated falciparum malaria in Acre, Brazil. J Infect Dis 158:1368–1371
16. Kremsner PG, Winkler S, Brandts C, Neifer S, Bienzle U, Graninger W (1994) Clindamycin in combination with chloroquine or quinine is an effective therapy for uncomplicated *Plasmodium falciparum* malaria in children from Gabon. J Infect Dis 169:467–470
17. Kremsner PG, Radloff P, Metzger W, Wildling E, Mordmuller B, Philipps J, Jenne L, Nkeyi M, Prada J, Bienzle U et al (1995) Quinine plus clindamycin improves chemotherapy of severe malaria in children. Antimicrob Agents Chemother 39:1603–1605
18. Metzger W, Mordmuller B, Graninger W, Bienzle U, Kremsner PG (1995) High efficacy of short-term quinine-antibiotic combinations for treating adult malaria patients in an area in which malaria is hyperendemic. Antimicrob Agents Chemother 39:245–246
19. Vaillant M, Millet P, Luty A, Tshopamba P, Lekoulou F, Mayombo J, Georges AJ, Deloron P (1997) Therapeutic efficacy of clindamycin in combination with quinine for treating uncomplicated malaria in a village dispensary in Gabon. Trop Med Int Health 2:917–919
20. Pukrittayakamee S, Chantra A, Vanijanonta S, Clemens R, Looareesuwan S, White NJ (2000) Therapeutic responses to quinine and clindamycin in multidrug-resistant falciparum malaria. Antimicrob Agents Chemother 44:2395–2398
21. Parola P, Ranque S, Badiaga S, Niang M, Blin O, Charbit JJ, Delmont J, Brouqui P (2001) Controlled trial of 3-day quinine-clindamycin treatment versus 7-day quinine treatment for adult travelers with uncomplicated falciparum malaria imported from the tropics. Antimicrob Agents Chemother 45:932–935
22. McGready R, Cho T, Samuel VL, Brockman A, van Vugt M, Looareesuwan S, White NJ, Nosten F (2001) Randomized comparison of quinine-clindamycin versus artesunate in the treatment of falciparum malaria in pregnancy. Trans R Soc Trop Med Hyg 95:651–656

23. WHO (1996) Assessment of therapeutic efficacy of antimalarial drugs for uncomplicated falciparum malaria in areas with intense trassmission. World Health Organisation, Geneva
24. Ramharter M, Oyakhirome S, Klein Klouwenberg P, Adegnika AA, Agnandji ST, Missinou MA, Matsiegui PB, Mordmuller B, Borrmann S, Kun JF et al (2005) Artesunate-clindamycin versus quinine-clindamycin in the treatment of *Plasmodium falciparum* malaria: a randomized controlled trial. Clin Infect Dis 40:1777–1784
25. Pukrittayakamee S, Imwong M, Looareesuwan S, White NJ (2004) Therapeutic responses to antimalarial and antibacterial drugs in vivax malaria. Acta Trop 89:351–356
26. Tan KR, Magill AJ, Parise ME, Arguin PM (2011) Doxycycline for malaria chemoprophylaxis and treatment: report from the CDC expert meeting on malaria chemoprophylaxis. Am J Trop Med Hyg 84:517–531
27. Clyde DF, Miller RM, DuPont HL, Hornick RB (1971) Antimalarial effects of tetracyclines in man. J Trop Med Hyg 74:238–242
28. Ponnampalam JT (1981) Doxycycline in the treatment of falciparum malaria among aborigine children in West Malaysia. Trans R Soc Trop Med Hyg 75:372–377
29. Taylor WR, Widjaja H, Richie TL, Basri H, Ohrt C, Tjitra TE, Jones TR, Kain KC, Hoffman SL (2001) Chloroquine/doxycycline combination versus chloroquine alone, and doxycycline alone for the treatment of *Plasmodium falciparum* and *Plasmodium vivax* malaria in northeastern Irian Jaya, Indonesia. Am J Trop Med Hyg 64:223–228
30. Alecrim MG, Lacerda MV, Mourao MP, Alecrim WD, Padilha A, Cardoso BS, Boulos M (2006) Successful treatment of *Plasmodium falciparum* malaria with a six-dose regimen of artemether-lumefantrine versus quinine-doxycycline in the Western Amazon region of Brazil. Am J Trop Med Hyg 74:20–25
31. Ejaz A, Haqnawaz K, Hussain Z, Butt R, Awan ZI, Bux H (2007) Treatment of uncomplicated *Plasmodium falciparum* malaria with quinine-doxycycline combination therapy. J Pak Med Assoc 57:502–505
32. Krause PJ, Lepore T, Sikand VK, Gadbaw J Jr, Burke G, Telford SR 3rd, Brassard P, Pearl D, Azlanzadeh J, Christianson D et al (2000) Atovaquone and azithromycin for the treatment of babesiosis. N Engl J Med 343:1454–1458
33. van Eijk AM, Terlouw DJ (2011) Azithromycin for treating uncomplicated malaria. Cochrane Database Syst Rev CD006688
34. WHO (2006) Guidelines for the treatment of malaria. World Health Organisation, Geneva
35. Kamiya T, Hashimoto M, Hemmi K, Takeno H (1980) Hydroxyaminohydrocarbonphosphonic acids. US Patent 4206156
36. Kanimoto Y, Greenwood D (1987) Activity of fosmidomycin in an *in vitro* model of the treatment of bacterial cystitis. Infection 15:465–468
37. Kuroda Y, Okuhara M, Goto T, Okamoto M, Terano H, Kohsaka M, Aoki H, Imanaka H (1980) Studies on new phosphonic acid antibiotics. IV. Structure determination of FR-33289, FR-31564 and FR-32863. J Antibiot (Tokyo) 33:29–35
38. Mine Y, Kamimura T, Nonoyama S, Nishida M, Goto S, Kuwahara S (1980) *In vitro* and *in vivo* antibacterial activities of FR-31564, a new phosphonic acid antibiotic. J Antibiot (Tokyo) 33:36–43
39. Okuhara M, Kuroda Y, Goto T, Okamoto M, Terano H, Kohsaka M, Aoki H, Imanaka H (1980) Studies on new phosphonic acid antibiotics. III. Isolation and characterization of FR-31564, FR-32863 and FR-33289. J Antibiot (Tokyo) 33:24–28
40. Jomaa H, Wiesner J, Sanderbrand S, Altincicek B, Weidemeyer C, Hintz M, Turbachova I, Eberl M, Zeidler J, Lichtenthaler HK et al (1999) Inhibitors of the nonmevalonate pathway of isoprenoid biosynthesis as antimalarial drugs. Science 285:1573–1576
41. Zhang B, Watts KM, Hodge D, Kemp LM, Hunstad DA, Hicks LM, Odom AR (2011) A second target of the antimalarial and antibacterial agent fosmidomycin revealed by cellular metabolic profiling. Biochemistry 50:3570–3577
42. Baumeister S, Wiesner J, Reichenberg A, Hintz M, Bietz S, Harb OS, Roos DS, Kordes M, Friesen J, Matuschewski K et al (2011) Fosmidomycin uptake into *Plasmodium* and *Babesia*-

infected erythrocytes is facilitated by parasite-induced new permeability pathways. PLoS One 6:e19334
43. Nair SC, Brooks CF, Goodman CD, Strurm A, McFadden GI, Sundriyal S, Anglin JL, Song Y, Moreno SN, Striepen B (2011) Apicoplast isoprenoid precursor synthesis and the molecular basis of fosmidomycin resistance in *Toxoplasma gondii*. J Exp Med 208:1547–1559
44. Staines HM, Ellory JC, Chibale K (2005) The new permeability pathways: targets and selective routes for the development of new antimalarial agents. Comb Chem High Throughput Screen 8:81–88
45. Kuemmerle HP, Murakawa T, De Santis F (1987) Pharmacokinetic evaluation of fosmidomycin, a new phosphonic acid antibiotic. Chemioterapia 6:113–119
46. Na-Bangchang K, Ruengweerayut R, Karbwang J, Chauemung A, Hutchinson D (2007) Pharmacokinetics and pharmacodynamics of fosmidomycin monotherapy and combination therapy with clindamycin in the treatment of multidrug resistant falciparum malaria. Malar J 6:70
47. Ruangweerayut R, Looareesuwan S, Hutchinson D, Chauemung A, Banmairuroi V, Na-Bangchang K (2008) Assessment of the pharmacokinetics and dynamics of two combination regimens of fosmidomycin-clindamycin in patients with acute uncomplicated falciparum malaria. Malar J 7:225
48. Lell B, Ruangweerayut R, Wiesner J, Missinou MA, Schindler A, Baranek T, Hintz M, Hutchinson D, Jomaa H, Kremsner PG (2003) Fosmidomycin, a novel chemotherapeutic agent for malaria. Antimicrob Agents Chemother 47:735–738
49. Missinou MA, Borrmann S, Schindler A, Issifou S, Adegnika AA, Matsiegui PB, Binder R, Lell B, Wiesner J, Baranek T et al (2002) Fosmidomycin for malaria. Lancet 360:1941–1942
50. Borrmann S, Adegnika AA, Matsiegui PB, Issifou S, Schindler A, Mawili-Mboumba DP, Baranek T, Wiesner J, Jomaa H, Kremsner PG (2004) Fosmidomycin-clindamycin for *Plasmodium falciparum* infections in African children. J Infect Dis 189:901–908
51. Borrmann S, Issifou S, Esser G, Adegnika AA, Ramharter M, Matsiegui PB, Oyakhirome S, Mawili-Mboumba DP, Missinou MA, Kun JF et al (2004) Fosmidomycin-clindamycin for the treatment of *Plasmodium falciparum* malaria. J Infect Dis 190:1534–1540
52. Borrmann S, Adegnika AA, Moussavou F, Oyakhirome S, Esser G, Matsiegui PB, Ramharter M, Lundgren I, Kombila M, Issifou S et al (2005) Short-course regimens of artesunate-fosmidomycin in treatment of uncomplicated *Plasmodium falciparum* malaria. Antimicrob Agents Chemother 49:3749–3754
53. Borrmann S, Lundgren I, Oyakhirome S, Impouma B, Matsiegui PB, Adegnika AA, Issifou S, Kun JF, Hutchinson D, Wiesner J et al (2006) Fosmidomycin plus clindamycin for treatment of pediatric patients aged 1 to 14 years with *Plasmodium falciparum* malaria. Antimicrob Agents Chemother 50:2713–2718
54. Oyakhirome S, Issifou S, Pongratz P, Barondi F, Ramharter M, Kun JF, Missinou MA, Lell B, Kremsner PG (2007) Randomized controlled trial of fosmidomycin-clindamycin versus sulfadoxine-pyrimethamine in the treatment of *Plasmodium falciparum* malaria. Antimicrob Agents Chemother 51:1869–1871

Artemisinins: Artemisinin, Dihydroartemisinin, Artemether and Artesunate

Harin A. Karunajeewa

Abstract Artemisinin comes from the plant, *Artemisia annua*, an ancient Chinese herbal remedy for relapsing fever. Rediscovery of its antimalarial action in China in the 1970s has seen it and its semisynthetic derivatives become the most useful drugs for most malarial illness. Artemisinins have a sesquiterpene lactone structure. Their anti-microbial action relates to a characteristic endoperoxide moiety. The precise mechanism of action remains controversial. Experimental induction of parasite resistance both in vitro and in vivo has been followed by recent initial clinical reports of resistance. Artemisinins are currently preferred as parenteral treatment of severe malaria, pre-referral rectal treatment and, as part of artemisinin combination therapy (ACT) oral treatment of uncomplicated falciparum malaria. Currently, amongst the most widely used drugs in the world, their future will be determined by the rate and extent of development of resistance. Better understanding of mechanisms of action and resistance and policy initiatives to prevent or delay resistance will be crucial.

1 History

The rediscovery of the antimalarial properties of artemisinin by Chinese scientists in the early 1970s has revolutionised the treatment of malaria. Almost all of the world's countries with endemic *P. falciparum* have now enacted policies favouring use of an artemisinin derivative-based combination therapy (ACT) as first-line treatment for uncomplicated malaria and an artemisinin derivative is now the preferred drug for treating severe malaria [1, 2]. When we consider the estimated 400 million annual clinical presentations with malaria [3], in terms of global health

H.A. Karunajeewa (✉)
Harin Karunajeewa, 170 Gladstone Avenue, Northcote, Vic 3070, Australia
e-mail: Harin.Karunajeewa@wh.org.au

H.M. Staines and S. Krishna (eds.), *Treatment and Prevention of Malaria*,
DOI 10.1007/978-3-0346-0480-2_9,

impact, the development of the artemisinin derivatives represents one of the most important achievements in recent medical history.

Like quinine, artemisinin is one of many plant-derived traditional medicines with antimalarial properties [4]. However, written records of artemisinin's therapeutic use predate that of Cinchona bark in South America by more than 1,000 years. The first known descriptions of its medicinal use are by the revered Chinese physician Ge Hong (284–363 AD) and appear in his text "Emergency prescriptions kept up one's sleeve," dated approximately 340 AD [5]. This recommends use of an extract of the herb, *qinghao* for the treatment of a condition designated by a Chinese medical term best translated as "intermittent fevers." This term, no doubt, included (though by no means exclusively) relapsing fever due to malaria infection. The name, *Qinghao*, means literally "the blue-green hao," referring to the plant's leaves that maintain their green colour into autumn. This description may have been used to distinguish it from a similar plant whose leaves turned yellow, and that has been subsequently described in Chinese medical literature as *huanghuahao* (the "yellow-blossom hao"). The distinction seems to align *Qinghao* with the modern botanical classification of *Artemisia annua* and *huanghuahao* with the species, *Artemisia apiacea*. This is probably important, because *A. annua* yields much higher quantities of artemisinin than other *Artemisia* species. *Artemisia annua*, known in English variously as "sweet wormwood," "sweet Annie," "sweet sagewort" and "annual wormwood" and other species belonging to the genus *Artemisia* are distributed widely throughout the world. In Europe and elsewhere, they have been used as herbal medicine for a myriad of indications, some of which have now been validated by modern medical research (use as a "vermifuge" for the treatment of parasitic worm infections) and some which have not (use as an aphrodisiac) [6]. Perhaps the most famous use of *Artemisia* plants in the Western world has been as a constituent of the drinks, absinthe and vermouth.

The medicinal extract of *qinghao* is known as *quinhaosu*, meaning "essence" of qinghao, and Ge Hong's original description of its preparation is instructive. Rather than the more conventional method of combining dried herbs with hot water to prepare a herbal tea, Ge Hong takes the unusual step of specifically instructing the reader to soak the entire fresh plant in water, to then "wring it out" and ingest the juice. With the benefit of current knowledge of artemisinin's physicochemical properties [7], this unconventional approach may have been crucial in yielding an efficacious preparation, given the poor water solubility of the active constituent, artemisinin. Ge Hong's method is likely to have produced an emulsion with plant oils containing the water-insoluble artemisinin [4] and recent testing of his original technique has yielded concentrations of artemisinin 20-times higher than those of a tea prepared from dried herbs [8].

The modern-day rediscovery of the antimalarial properties of artemisinin was made possible by a unique nexus between traditional Chinese, and conventional Western medicine, that occurred in the unprecedented geo-political context of what is referred to in the West as the Chinese Cultural Revolution (1966–1976). During this time, nationalistic pride saw a rejection of perceived Western influences in all fields, including medicine and science. "Classes for Western medical doctors learning Chinese medicine" had become mandatory [5]. Mao Tse Tung described China's national heritage as a "treasure house" of knowledge and wisdom that could be used for "serving the people" – regarded as particularly pertinent to the "task of combating malaria" (enumerated as Mao's "task number five hundred and twenty-three") [5]. The importance of malaria to Chinese national interests during this time also, no doubt, lent much to its opposition to American interests in the wars of Indo-China where the burden of malarial disease on combatants had attained crucial strategic importance. It therefore mirrored a similarly massive up-scaling of research into novel antimalarial compounds conducted by the US military that concomitantly led to the development of mefloquine.

The institution credited with the discovery of artemisinin's antimalarial properties is the Academy of Traditional Chinese Medicine. In association with the People's Liberation Army Research Unit, in the late 1960s, they began screening traditional medicines for antimalarial activity. *Qinghao* was included amongst the first ten products screened but initially showed no activity. In retrospect, this was probably because an extraction method suitable to the water insoluble active component was not used. By 1972, however, a group at the Academy led by Professor Tu Youyou had successfully prepared an ether extract described as having 95–100% efficacy when used for treatment in a rodent malaria model. Crucially, Professor Youyou and her team claim to have heeded the original advice of Ge Hong when developing their extraction method, with special attention to avoiding the use of excessive heat. The use of the aprotic solvent, ether, probably also aided the extraction of the oil and water-insoluble active compound and the use of neutral pH would have been important, now that artemisinin is known to be unstable in the presence of both acid and alkali.

Within the same year as the first report of its activity in rodent malaria in 1972, a trial was reported in 21 humans with malaria (both *P. vivax* and *P. falciparum*) in Beijing with a reported efficacy of >90%. The principal investigator in the early human trials of artemisinin derivatives was Professor Li Guoqiao of the Guangzhou University of Traditional Chinese Medicine. The professor and his group subsequently described the efficacy of artemisinin derivatives in severe (including cerebral) malaria as well as demonstrating their extraordinarily rapid in vivo activity on trophozoite blood stages and their potent gametocidal activity [9].

In 1973, Professor Youyou synthesised dihydroartemisinin, the first semi-synthetic artemisinin "derivative." The new compound had higher water solubility, a characteristic later to prove pharmacologically advantageous. Subsequently, further semi-synthetic derivatisation of dihydroartemisinin was employed to produce ethers and esters from dihydroartemisinin. Several hundred of these semi-synthetic derivatives were produced and subjected to preliminary investigation by Chinese scientists in the 1970s. Together with dihydroartemisinin, they are now collectively referred to as the "1st generation endoperoxides" or "artemisinin derivatives" to distinguish them from the subsequent development of other compounds (such as artemisone) and wholly-synthetic endoperoxides. Only a small percentage of these many hundred compounds have been evaluated in human subjects and a smaller number still have achieved widespread therapeutic use. The most-used therapeutic compounds include artemisinin itself, dihdyroartemisinin, artemether, arteether, and artesunate. These were developed for mass marketing by nationalised pharmaceutical companies in other parts of China (Guilin and Kunming), in regions where China's highest malaria prevalence occurred [10]. The first marketed pharmaceutical preparations were officially available from 1986 onwards. However, it is likely that very widespread use of artemisinin derivatives was employed much earlier in China and also in Cambodia (formerly Kampuchea), as part of Chinese military assistance to the Khmer Rouge regime in the late 1970s.

The first English-language publications regarding clinical efficacy appeared in 1982 [11, 12]. These came at a time of great pessimism and alarm in the face of burgeoning parasite resistance to existing agents, coupled with a paucity of attractive alternative drugs in the existing antimalarial drug development "pipeline," whether in the private (pharmaceutical companies) or public (military programmes) sectors of Western countries [13, 14]. At this time, Thailand and its border areas seemed to face the world's greatest problem of drug-resistant malaria, with the loss of mefloquine efficacy here being viewed as a harbinger of the catastrophic scenario of untreatable *P. falciparum* malaria [14]. In 1991, the first clinical trials demonstrating efficacy against multi-drug resistant strains of *P. falciparum* in Thailand were published [15]. Subsequently, the efficacy of the combination of artesunate and mefloquine for uncomplicated malaria appeared to be excellent, with the introduction of this combination coinciding with a reversal in the trend towards worsening mefloquine resistance in the area [16]. Since this time, susceptibility to mefloquine has not declined further in this area despite use of the artesunate–mefloquine combination for many years [16].

From the early 1990s until the current time, an explosion in interest in the artemisinin derivatives led to development of assays to measure concentrations of drug in biological fluids [17], documentation of the pharmacokinetic properties of the individual derivatives [7], theories regarding their mode of action. [18, 19], development of rectally administered formulations [20] and an extraordinarily large body of clinical trials exploring their efficacy in both uncomplicated and severe malaria [21–24]. At the time of writing well over 12,000 subjects had been enrolled in clinical studies evaluating artemisinin derivatives [23]. At least 50 randomised controlled trials have now been published evaluating the efficacy of ACTs for treatment of uncomplicated malaria [24]. In 2005, the landmark SEAQAMAT study was published [25]. This well-powered multi-centre randomised controlled trial established parenteral artesunate in preference to quinine as the therapy of choice for severe malaria in adults. In doing so, it represented the first therapeutic intervention in recent history to have a demonstrated impact on mortality in severe malaria. More recently, a mortality benefit of artesunate over quinine was also confirmed in a large study of African children with severe malaria [26], effectively ending any controversy as to its preferred status as first-line treatment for severe malaria in most contexts [2]. Until recently, the widespread use of parenteral artesunate had been impeded by its lack of approval by pharmaceutical regulatory bodies. Fortunately, however, the major manufacturer of parenteral artesunate (Guilin Pharmaceuticals, China) has recently achieved prequalification status for its product by the World Health Organisation (WHO), meaning it is likely to be widely available, including in developed countries.

By the end of the millennium, in the face of burgeoning evidence supporting the use of artemisinin derivatives, opinion leaders, led by Professor Nick White of Oxford and Mahidol Universities, were already calling for a radical re-thinking of malaria treatment policies and guidelines [27]. In particular, the strategy of ACT use was advocated as an alternative to conventional therapies compromised by parasite resistance. In 2006, the WHO officially endorsed ACT as the treatment of choice of uncomplicated *P. falciparum* malaria [28], with emphasis on the theoretical benefits of the strategy of combination therapy to limit the potential for development of parasite resistance [29].

The prototypical ACT was the "loose" combination of separate tablets of artesunate and mefloquine first used in Thailand [30]. Other "loose" ACTs have included sulphadoxine–pyrimethamine, amodiaquine and atovaquone–proguanil as partner drugs. However, "fixed" co-formulations containing both the artemisinin derivative and a non-artemisinin partner drug have been the focus of ACT development in recent years (Table 2). The first co-formulated ACT to achieve regulatory approval was artemether–lumefantrine, endorsed by the WHO in 2006 for use in areas of multi-drug resistance and in Africa, and was shortly followed by endorsement of co-formulated artesunate with amodiaquine [28]. Crucially, manufacturers of these products have provided drug at substantially lower cost for developing countries. Other promising combinations in various stages of development have utilised piperaquine, mefloquine, amodiaquine, fosmidomycin, chloroproguanil–dapsone, pyronaridine and naphthoquine as partner drugs [31].

The effective use of ACTs has been credited as a major factor in very significant reductions in malaria prevalence in many areas of South-East Asia and Sub-Saharan Africa [32, 33]. This has helped reignite interest in the potential for attempts at malaria eradication [34, 35].

The experimental induction of artemisinin resistance in rodent malaria was first reported in 1999 [36]. In 2006, given mounting concerns regarding the potentially disastrous consequences of artemisinin resistance [27, 28, 37–41], the WHO recommended the phasing-out of oral mono-preparations of artemisinin, in order to prevent their use as mono-, rather than combination therapy [42]. Soon after, the first reports indicating impaired responses to artemininin derivatives in vivo were reported from Cambodia in 2008 [43, 44]. At the time of writing, a massive effort is underway to monitor and prevent the spread of parasite resistance in this region [43, 45].

2 Chemistry

Artemisinin ($C_{15}H_{22}O_5$, IUPAC name: (3*R*, 5aS,6*R*,8a*S*,9*R*, 12*S*,12a*R*)-octahydro-3,6,9-trimethyl-3,12-epoxy-12H-pyranol[4,3-*j*]-1,2-benzodioxepin-10(3*H*)-one) is a sesquiterpene lactone, with a distinctive 1,2,4 triaxone ring structure (Fig. 1).

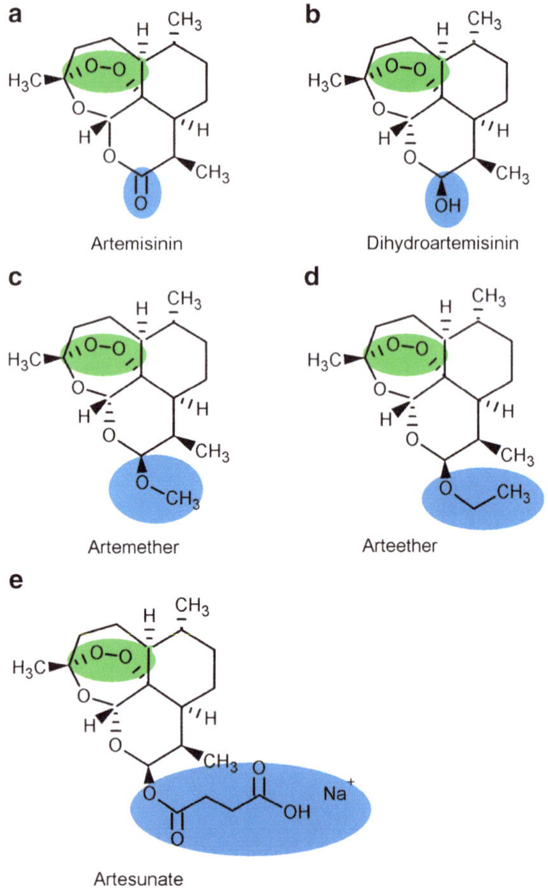

Fig. 1 Chemical structure of artemisinin (**a**) and its derivatives: dihydroartemisinin (**b**), artemether (**c**), arteether (**d**) and artesunate (**e**). The essential endoperoxide pharmacophore common to artemisinin and all its derivatives is highlighted in *green*. The C_{10} position radical that is unique to each individual derivative and that determines its water and lipid solubility (and therefore some of its pharmacokinetic properties) is highlighted in *blue*

Whilst the structure of this "backbone" varies considerably amongst artemisinin's various derivatives and synthetic analogues, all active compounds retain a distinctive endoperoxide moiety that is primarily responsible for the antimalarial activity of this drug class (Fig. 1). Artemisinin itself is poorly soluble in both water and oil, but soluble in many aprotic organic solvents. It has a melting point of 156–157°C and the molecule breaks down at temperatures above 190°C. It's molecular mass is 282.3 Da. It is unstable in the presence of both alkali and acid [46].

In the presence of sodium borohydride, a hydroreduction reaction transforms the lactone group of artemisinin to a lactol, whilst preserving the crucial endoperoxide moiety [46]. This produces dihydroartemisinin. Further semi-synthetic derivatisation produces ethers and esters of the lactol. In addition to artemisinin itself, and dihdyroartemisinin, the most-used derivatives include artemisinin's methyl ether, artemether, β-ethyl ether, arteether, and hemisuccinate ester, artesunate. These all share the same basic chemical structure of artemisinin (Fig. 1) with different chemical groups at the C_{10} (or 2-keto) position [47]. The C_{10} position determines the solubility of each artemisinin derivative and influences its diffusion across mucosal membranes as well as other pharmacokinetic properties. Artesunate has significantly greater solubility in water than either artemisinin, dihydroartemisinin or artemether [48]. Second generation artemisinins (such as artemisone) are also under investigation.

3 Mechanism of Action

The antimalarial mechanism of action of artemisinin and its derivatives remains unclear. However a number of theories have been proposed that are currently the subject of intense debate. The complex biochemistry underpinning these theories is outlined extensively in a number of recent reviews [46, 49, 50]. Any understanding of the mechanisms of antimalarial activity of the artemisinins must reconcile each of the following considerations.

Firstly, like any good antimicrobial agent that is both safe and effective, the agent produces differential toxicity on pathogen in preference to host cells. Therefore, activity must exploit metabolic pathways or processes, which are either unique to, or preferentially expressed in the parasite. In the case of other antimalarials, this has often related to the food vacuole of the parasite and its role in detoxifying digested haem arising from breakdown of host-cell haemoglobin. The haem pathway has also been extensively investigated as a mode of action of artemisinin and is discussed below. *PfATP6* is an enzyme essential for oxidative metabolism in the parasite, which is absent from mammalian cells. Specific inhibition of this enzyme has been postulated as a mechanism of action and is also discussed below [19]. Much interest has centred on the relationship between artemisinins and free haem and iron that exist in concentrations within the parasite not seen in normal host cells. If these are responsible for "activating" artemisinin, they could explain the differential toxicity of drug in parasitic versus host mammalian cells.

Secondly, the structure–efficacy relationships of artemisinin and other endoperoxides now clearly implicate the endoperoxide moiety as the major determinant of efficacy. Any proposed pathway for the toxicity of artemisinins on plasmodia must incorporate the direct or indirect action of the endoperoxide moiety on a putative molecular target.

Thirdly, mechanistic hypotheses must also reconcile the now well-documented activity of artemisinins on a broad range of organisms and tumour cells [49]. These now include a variety of parasites including helminths (a wide range of trematodes including *Schistosoma*, *Clonorchis*, *Opisthorchis* and *Fasciola* spp.), non-plasmodial *Apicomplexa* (*Toxoplasma gondii*) and protozoa and metazoa (*Trypanosoma* and *Leishmania* spp.). In vitro, and in some cases, in vivo anti-viral activity has been demonstrated against HIV1, Influenza A, hepatitis B and C and human herpes viruses, including CMV where artemisinin therapy has been used with some success in human patients [51]. Some anti-fungal activity has been demonstrated in vitro for *Pneumocystis jirovecii* [49]. A very broad range of tumour cell lines have been shown to be susceptible in vivo and artemisinin is now being examined as a clinical therapy in a variety of malignancies [49]. It has even shown activity in autoimmune conditions including rheumatoid arthritis and lupus nephritis [49]. All this suggests either multiple or broad and relatively non-specific mechanisms of action. However, it must be said that the sensitivity of other organisms to artemisinin is generally orders of magnitude less than that seen in plasmodia, where IC_{50} values are in the low nanomolar range.

Current concepts under debate include, firstly, the means by which artemisinin becomes biologically activated within the parasite, and secondly, the proposed molecular targets of activated artemisinin [52]. Bioactivation of artemisinins within parasitised erythrocytes is important in explaining the selective toxicity of the artemisinins on the parasite rather than on uninfected host cells and was first proposed by the Meshnick group to be triggered by iron (II) to generate toxic activated oxygen [53]. Subsequently, two opposing models, a "reductive scission model" [54, 55] and an "open peroxide model" [56] have been proposed to further explain this process. These two models differ by their dependency on iron and involvement of carbon-centred radicals [52].

As to the proposed molecular targets of artemisinin, a number of theories have been put forward, four of which are summarised below. These are by no means mutually exclusive and it seems likely that the anti-plasmodial activity of the artemisinins is a multi-faceted process [52].

3.1 Haem Pathway

This theory proposes that the free radicals generated following activation of artemisinin's endoperoxide then alkylate intracellular haem. The alkylated haem is then unable to undergo its usual detoxification by the parasite that ordinarily leads to the formation of the non-toxic crystalline haemozoin pigment. Supporting this theory, haem-drug adducts have clearly been demonstrated by mass spectrometry following exposure of parasites to both artemisinin and synthetic endoperoxides in vitro [57, 58]. However, others [59, 60] have demonstrated a lack of inhibition of haem polymerisation (and therefore of haemozoin formation) under carefully controlled in vitro conditions using stable artemisinin derivatives with good antimalarial activity. They have suggested that previously demonstrated haem polymerisation is an artefact of the instability of artemisinin and dihydroartemisinin in aqueous experimental conditions previously used. They argue that ring-opened products of artemisinin or dihydroartemisinin had been responsible for the haem binding seen previously. Adding to this argument are studies using rodent malaria models that have demonstrated a relative paucity of haem–artemisinin adducts in vitro [60]. In addition, the haem hypothesis fails to explain the toxicity of artemisinins for on other parasite species (such as *Toxoplasma* and *Babesia* spp.) that are not exposed to significant concentrations of haem in vivo.

3.2 *Protein Alkylation*

Free-radical generation within the parasite may also, through the formation of covalent bonds, alter the function of key parasite proteins involved in a variety of functions [18]. Suggested targets may include membrane transporters, proteases involved in haemoglobin degradation and a variety of other cellular enzymes.

3.3 *Inhibition of PfATP6*

Recently, specific inhibition of an enzyme, PfATP6, essential for oxidative metabolism in the parasite, was postulated as a mechanism of action [19]. This theory developed following interest in a compound called thapsigargin, which shares artemisinin's sesquiterpnene lactone structure. Thapsigargin is known to be a selective inhibitor of the mammalian sarco/endoplasmic reticulum membrane calcium ATPase (SERCA) that actively concentrates calcium within membrane-bound stores, thereby maintaining low cytosolic free calcium concentrations that are necessary to ensure cell survival. Artemisinin showed similar activity on mammalian SERCA as thapsigargin [19]. Because the PfATP6 of *P. falciparum* is the only enzyme orthologous to that of mammalian SERCA, it was hypothesised that this could be artemisinin's molecular target in the parasite. An elegant series of experiments showed firstly that both artemisinin and thapsigargin did, indeed inhibit PfATP6. Secondly, the mechanism appeared to be highly specific, without effects on non-SERCA Ca^{2+} ATPase or other malarial transporters. Thirdly, incubation of parasitised erythrocytes with the iron chelator, desferrioxamine, abolished the effect of artemisinin (but not thapsigargin), thus supporting the theory that bioactivation of artemisinin by iron was necessary for inhibition of the enzyme, and providing a rationale for the selective toxicity of artemisinin to parasitised cells. Studies of *T. gondii* added weight to the theory by demonstrating a similar effect on this organism's SERCA homologue [61]. A subsequent study suggested that a single amino acid residue could determine the sensitivity of SERCAs to artemisinin, thereby raising the possibility of a single mutation in genes encoding PfATP6 producing artemisinin resistance [62]. This mutation has also been examined by transfection of parasites with interesting results [63].

However, induction of stable parasite artemisinin-resistance in rodent malaria did not show mutations in the parasite's PfATP6 orthologue [39] or suggested alternative resistance pathways [64]. Studies of the synthetic endoperoxide OZ277 have shown that despite its highly potent antimalarial activity, and its competitive inhibition of artemisinin in vitro, it is a very weak inhibitor of PfATP6 [52]. Taken together, these all suggest that if PfATP6 is a molecular target of artemisinin, then it may be one of many through which endoperoxides act and through which parasite resistance could develop [65, 66]. Nevertheless, much more study is required to examine PfATP6 as a target and its mutations in contributing to resistance.

3.4 Mitochondrial Function

Studies of yeast have demonstrated an effect of artemisinin on mitochondrial membrane potential through effects on the electron transport chain [67]. Further studies have shown artemisinins to be distributed to malarial mitochondria, to directly impair mitochondrial function and to induce a strikingly rapid and dramatic production of reactive oxygen species within the mitochondria of both yeast and malaria but not mammalian cells, thereby suggesting a mechanism for selective toxicity [68]. Mitochondrial toxicity demonstrated in these studies appears to be dependent on the peroxide moiety, with derivatives such as deoxyartemisinin (which lacks an endoperoxide bridge) failing to demonstrate production of reactive oxygen species or functional toxicity in mitochondria. Further evidence of the importance of this pathway has been shown by the amelioration of artemisinin's antimalarial toxicity through interference of the mitochondrial electron transport chain, including by use of the iron chelator, desferrioxamine [68]. However, studies in transgenic parasites (see AB Vaiya, Antimalarials targeting mitochondrial functions) do not support the idea that artemisinins kill parasites by a mitochondrial-based mechanism.

4 Resistance

In theory, the artemisinins should be relatively protected from the development of resistance by their extremely short half-lives [7, 69, 70]. This reduces the overall exposure of parasite populations to low or intermediate residual drug concentrations following administration of a conventional treatment course. Historically, this seems to have been important in providing the selection pressure that has driven rapid development of resistance to longer-acting drugs such as mefloquine and sulphadoxine–pyrimethamine [69, 71]. Rapid development of high-level resistance can also occur when a single mutation affects a single enzymatic pathway critical to the drug's activation or mechanism of action. This is the case with the drug atovaquone, which is rendered ineffective by a single base-pair mutation of the gene encoding the *cyt b* enzyme [72]. Therefore, artemisinin resistance has not been ruled out and the implications of widespread artemisinin resistance have been seen as so potentially disastrous that it has long been at the forefront of debate and policy considerations by the WHO and others [28, 37, 38]. It has underpinned calls for the use of artemisinin to be restricted to combination therapies [27], leading to the WHO's recommendation that artemisinin monopreparations for oral treatment be phased out altogether in favour of co-formulated ACTs [42]. Part of the rationale for ACT relies on the supposition that parasite mutants that spontaneously arise with resistance to one drug will still be susceptible to the second, and therefore will not survive to pass on their resistance characteristics to the next generation [71, 73]. A similar approach underlies combination treatment of tuberculosis, HIV, Hepatitis B infections and most cancers. However, in the case of malaria, the rationale for combination therapy is subject to a number of caveats that mean that ACT should not be regarded as a panacea [29]. In particular, the relatively complex genome of plasmodia (*P. falciparum*'s genome is 2.3×10^7 bp) gives it the advantage of greater number of possible genetic loci that could mediate resistance, when compared with simpler organisms such as *M. tuberculosis* (genome 4.4×10^6 bp) or viruses (3.2×10^3 bp for Hepatitis B). This has also led to the development of more complex drug resistance mechanisms than those seen in bacteria and viruses. For example, mutations affecting drug transport mechanisms can simultaneously lead to resistance to a range of drugs with different molecular targets and modes of action. In *P. falciparum*, the best example is the *P. falciparum* multidrug resistance gene (*Pfmdr1*) that encodes the P-glycoprotein homologue (*Pgh*), involved in drug efflux into the parasite's food vacuole [74, 75]. Resistance can be mediated either by single nucleotide polymorphisms or by increases in gene copy number and *Pfmdr1* has been implicated in resistance to a range of diverse antimalarials including chloroquine, amodiaquine, mefloquine, halofantrine and lumefantrine [76, 77]. The rationale for ACT may also be undermined by the possibility of parasites being "protected" from the activity of one or other drug in a combination either due to sequestration in sites exposed to lower drug concentrations or to drug exposure occurring during non-susceptible phases of the parasite's life cycle [29]. Also, whilst the use of a (conventionally, more longer-acting) partner drug may "protect" the shorter-acting artemisinin component from the development of resistance, it seems less likely that the converse will be true. One study has already demonstrated the unabated spread of sulphadoxine and pyrimethamine resistance despite the introduction of an artesunate plus sulphadoxine–pyrimethamine ACT [78].

A number of groups have been able to select for parasite resistance in rodent malaria in vivo and more recently in *P. falciparum* in vitro [36, 39, 40, 64, 79] adding to concerns regarding the potential for development of resistance. In a study by Peters and Robinson reported in 1999 moderate artemisinin resistance (a 4- to 27-fold difference in sensitivity compared with sensitive strains) was selected in *P. berghei* and *P. yoelli* following sustained selection pressure in a rodent model. However this rapidly resolved with the removal of drug-selection pressure [36]. No specific mechanism for resistance could be identified based on evaluation of putative molecular targets being considered around this time, however resistant strains were shown to accumulate 43% less (radiolabeled) drug in vitro than sensitive strains [40]. The induction of stable artemisinin resistance was not reported until 2006 in a study by Afonso et al. of a *P. chabaudi* rodent model [39]. Strains with 6- to 15-fold increased resistance to artemisinin and artesunate were induced through exposure to increasing drug concentrations during several consecutive passages in mice. Resistance remained stable after cloning, freeze thawing, transmission to mosquitoes and following removal of drug pressure. Nucleotide sequences and gene copy number of putative genetic determinants of resistance (including orthologues of the *P. falciparum* genes, *pfmdr1, pfcg10, pftctp and pfatp6*) were compared between resistant and sensitive strains but failed to demonstrate any differences. Subsequent genetic linkage analysis identified two genetic mutations in a locus encoding a deubiquinating enzyme (UBP-1) [80]. In addition to artesunate, the same mutations could also be induced independently under drug selection using chloroquine, thereby demonstrating a mechanism for cross-resistance to structurally unrelated antimalarials arising as a single event. However, examination of *P. falciparum* isolates with reduced artemisinin sensitivity failed to identify mutations in the orthologous *P. falciparum* locus. More recently, an artemisinin-tolerant strain of *P. falciparum* has been developed through sustained in vitro drug pressure [64]. Morphologic evaluation revealed a sub-population of ring-form parasites surviving in maturation arrest. Observed transcriptomic modifications included down-regulation of a cell-cycle regulator and DNA biosynthesis protein. Taken together, the authors argued that artemisinin tolerance might be mediated by a quiescence mechanism in *P. falciparum*. A further study by Chavchich et al. induced stable resistance to artemisinin and artelinic acid in three laboratory reference strains of *P. falciparum* [79]. Importantly, this was also associated with decreased susceptibility to mefloquine, quinine, halofantrine and lumefantrine. Again, no mutations were detected in putative genetic determinants of artemisinin resistance (including *pfmdr1*, *pfcrt*, *pfatp6*, *pfctp*, and *pfubcth* genes) or in expression of *pfatp6* or *pftctp*. However, in contrast to the rodent model data from Afonso et al., this study clearly demonstrated increases in *pfmdr1* gene copy number as well as its mRNA and protein expression, suggesting a mechanism for cross-resistance to mefloquine and other drugs. By demonstrating a possible resistance mechanism common to artemisinin and other drug classes, it raises some concerns regarding the underlying rationale for ACT, particularly when drugs such as mefloquine and lumefantrine are used as the partner drug [29]. Together, these studies have provided a proof of principle that induction of stable artemisinin resistance is possible and, even more importantly, that this could arise through more than one mechanism, including some that may simultaneously mediate increased resistance to other structurally unrelated drugs (including ACT partner drugs).

Monitoring of clinical isolates for artemisinin sensitivity using in vitro drug susceptibility assays has methodological challenges due to difficulties in standardisation and, until very recently, the lack of confirmed in vivo resistance to enable clinical correlation. However, a study by Jambou et al. reported in 2005, using a WHO microtest, demonstrated a very wide variability in artemether and artesunate IC_{50} values from 530 clinical isolates from Senegal, French Guiana and Cambodia [41]. Whereas IC_{50} values had previously been reported in a range <30 nmol/l, values as high as 117 nmol/l were reported from French Guiana. Parasites from French Guiana also showed correlation in artemether and mefloquine IC_{50} values. Importantly, polymorphisms of the coding sequences of *pfatp6* were identified and one of these, a S769N mutation, was highly significantly associated with higher IC_{50} values from isolates from French Guiana. Again, analysis of *pftctp*, *pfmdr1* and *pfcrt* failed to suggest a role for these, although copy number was not assessed. Originally, the S769N mutation was only found in French Guiana (an area of South America with little preceding exposure to artemisinin derivatives) and was not found in a broad range of isolates from Africa, South America and Southeast Asia [81]. Recently, travellers returning to Canada from many African countries with malaria have provided parasites with mutations in *pfatp6* (A623E; S769N). Some of these have significant increases in IC_{50} values for the effect of artesunate and other derivatives on in vitro parasite growth, and interestingly there may be an interaction between these mutations and mutations in *pfmdr1* [82]. Overall, there was a high level of polymorphic diversity of the *pfatp6* gene throughout these regions though the nature of this could not be explained by genetic drift alone [83]. Given the lack of preceding drug pressure in French Guiana, alternative evolutionary mechanisms are being considered, including host factors, such as endemic haemoglobinopathies that alter intra-erythrocytic calcium homeostasis in a manner that higher cytosolic calcium concentrations might select for *pfatp6* mutants better able to regulate parasite calcium concentrations [81].

The most significant event in the recent history of the artemisinins is the first descriptions of clinical artemisinin resistance, from Western Cambodia from 2008 onwards [43, 44]. An initial report by Noedl et al. prospectively evaluated a 7-day course of artesunate monotherapy in adults with *P. falciparum* in Battambang province of Cambodia. A definition of "resistance" was used that included persistence of parasites 7 days after treatment commencement or re-emergence within 28 days in the presence of acceptable concentrations of dihdyroartemisinin, prolonged time to parasite clearance and reduced in vitro susceptibility to dihydroartemisinin. Of 60 subjects enrolled in the artesunate arm of the study, two met the criteria for artemisinin resistance. IC_{50} values for these two patients were 4 times the mean value for cured patients and 10 times that of a reference strain.

A subsequent study by Dondorp et al. prospectively evaluated the in vivo efficacy of 7-day artesunate monotherapy and of combined artesunate and mefloquine, this time comparing response in 40 patients from each of two locations: a site in Western Cambodia where concerns regarding resistance had emerged and a "control" site in Northeastern Thailand [43]. Significantly longer parasite clearance times were seen at the Cambodian site (84 h vs. 46 h). Although a higher rate of treatment failure (recrudescence) was also seen in the Cambodian patients (total 7/40), this was not statistically significantly higher than at the other site (3/40). When interpreting these data, it is important to consider that parasite clearance rates may reflect not only intrinsic parasite susceptibility to the drug, but numerous host factors, particularly those related to the degree of pre-existing malaria-specific immunity. Because of the study design used, the two groups were not necessarily matched for important factors (such as age, sex, parasite density, disease severity and the malaria transmission intensities that each population was subjected to). These factors could easily have become confounding variables in the analysis. However, the degree of difference was striking and the parasite clearance times documented from the Cambodian site were longer than any described previously worldwide. Moreover, this area has historically seen the early emergence of drug resistance to a number of agents [84]. The emergence of artemisinin resistance here becomes especially plausible when it is considered that artemisinin treatments have probably been used here longer than anywhere else in the world, having been employed since the time of the Khmer Rouge in the late 1970s. Much of this has been poorly regulated monotherapy with, quite possibly, sub-standard drugs [43, 85]. Also, subsequent studies of parasites from the patients studied by Dondorp et al. in Western Cambodia have now suggested that much of the variability in parasite clearance rates can be attributed to parasite genetic co-lineage, suggesting heritable (and therefore transmissible) parasites, rather than host factors in explaining this phenomenon [86]. However, specific genetic loci associated with delayed parasite clearance are yet to be determined and host genotypes may also need assessment.

Given concerns regarding the apparent development of early artemisinin resistance in Cambodia, a massive effort by the WHO with financial assistance from the Bill & Melinda Gates Foundation is now underway to monitor, control and limit its spread [43, 45]. Strategies will include improved surveillance, enhanced case management (early diagnosis and appropriate treatment), reducing drug pressure (by preventing inappropriate empirical therapy), targeting of mobile populations and optimising vector control and bed-net coverage, with the ultimate aim of effectively eradicating falciparum malaria from this region [43, 87]. A further development of concern in the region has been the development of resistance to commonly used ACT partner drugs, with unacceptably high treatment failure rates emerging for the artesunate–mefloquine combination [88].

5 Basic Clinical Pharmacology of the Artemisinin Derivatives

Artemisinin and its derivatives now exist in a myriad of pharmaceutical products that can be administered either orally (either as mono- or co-formulations), rectally or parenterally. The most widely used in clinical practise include products containing artemisinin itself, dihydroartemisinin, artemether and artesunate. The pharmacology of these four drugs are summarised in Table 1. All derivatives are likely to share a common mechanism of action due to the ubiquitous endoperoxide moiety. There is probably little to separate them in terms of intrinsic efficacy, as IC_{50} values exist in a similar range for each drug [91]. What is more, artesunate and, to a lesser extent artemether, are essentially pro-drugs of dihydroartemisinin. However, each derivative has different physicochemical properties due to the nature of their unique C_{10} (or 2-keto) position radical (Fig. 1). This influences the drug's solubility in oil and water and therefore the molecule's movement across biological membranes. This has important implications for pharmacokinetic properties that may impact on therapeutic response. These are discussed in more detail below.

5.1 Pharmacokinetics

5.1.1 Absorption and Bioavailability

Artemisinin, dihydroartemisinin, artemether and artesunate can all be administered orally or rectally. Artemisinin, artemether and artesunate can also be given by intramuscular injection. However artesunate is the only drug that can be

Table 1 Pharmacokinetic properties of the commonly used artemisinin derivatives

	Elimination half-life	T_{max} – oral administration	T_{max} – intramuscular injection	T_{max} – rectal (suppository) administration
Artemisinin (MW = 282.35)	2.9 h [7]	1–3 h [7]	3.4 h [7]	5.6 h [20]
Dihydroartemisinin (MW = 284.4)	40 min [7]	0.9–1.6 h [7]	NA	4 h [20]
Artemether (MW = 298.4)	3.6 h [7]	1.7–6 h [7]	1.3–8.7 h [7]	3.1 h [20]
Arteether (MW = 312.4)	12.4–30.2 h [89, 90]	NA	4.8–7.0 h [89, 90]	NA
Artesunate (MW = 384.4)	2–5 min [91–93]	15–39 min [93]	12 min [91]	1.6–3 h [20]

T_{max} time to maximal concentration, *MW* molecular weight

administered intravenously [7]. The lack of an intravenous preparation for the other drugs makes it difficult to calculate absolute bioavailability when they are administered by other routes. However oral bioavailability of artesunate has been calculated at between 61 and 88% [94, 95] and is likely to be greater than the less water soluble drugs [91, 96, 97]. Artesunate is clearly the most rapidly absorbed with time to maximal concentration (T_{max}) considerably shorter than those of the other derivatives (Table 1) [7].

Solubility of the different drugs is also likely to influence both bioavailability and absorption profile after rectal administration. All products show a wide degree of inter-individual variability in the rapidity and extent of absorption [20]. However, artesunate is the most rapidly absorbed and achieves the highest maximal drug concentrations following conventional dosing (Table 1) [20].

There are fewer options for parenteral administration of artemisinins. Because of stability issues, artesunate is available as a powder for reconstitution in sodium bicarbonate immediately prior to intravenous injection. However, intravenous administration is more costly and requires equipment and skilled staff often not available in resource-poor settings. Intramuscular injection is likely to be both more practical and have a comparable pharmacokinetic profile, achieving maximal concentrations within 12 min [91]. Artemether has been used widely as a preparation for intramuscular injection [98]. Because of its very poor water solubility it is available in ampoules containing the drug dissolved in peanut oil. It suffers from poor and erratic absorption from the intramuscular injection site. T_{max} values range from 1.3 to 8.7 h [7] and studies from children with severe malaria indicate a wide inter-individual variability in the extent of absorption, such that many children fail to demonstrate detectable concentrations of drug in plasma following intramuscular injection [99]. Its pharmacokinetic profile therefore seems to be inferior to that of artesunate when administered as an intramuscular injection for severe malaria. Arteether and artemisinin have also been administered by intramuscular injection, though much less extensively than artemether. Although data are limited, pharmacokinetic properties of these preparations are similar to those of artemether (Table 1) [7, 89, 90].

5.1.2 Metabolism and Elimination

All artemisinin derivatives are characterised by their extremely rapid elimination from plasma, with elimination half-lives that are mostly less than a few hours (Table 1) [7]. Both artemether and artesunate are metabolised to dihydroartemisinin by rapid esteratic hydrolysis of artesunate or slower cytochrome P450-mediated demethylation of artemether. Dihydroartemisinin itself and artemisinin are probably metabolised in the liver to inactive metabolites (Table 1) [7].

The pharmacokinetics of artemisinin derivatives do not appear to be influenced significantly by severity of infection [92] and dose adjustment in patients with hepatic or renal impairment is not required. However auto-induction of metabolism of artemisinin has been observed that results in a 5.5-fold increase in oral clearance following 10 days of treatment [100]. This effect appears to be largely attributable to induction of CYP2B6 (a member of the cytochrome P450 enzyme family) and is specific to artemisinin itself rather than to its derivatives.

5.1.3 Pharmacokinetic–Pharmacodynamic Relationships

A clear understanding of the manner by which pharmacokinetic parameters impact on clinical response is lacking for this drug class. Clearly, the rate of absorption (measured by T_{max} or absorption half-life) is important, particularly in severe malaria, where it is important rapidly to achieve parasiticidal drug concentrations. For other antimicrobials, including other classes of antimalarial agents, time-above mean inhibitory concentration (MIC) and area under curve (AUC) relative to pathogen MIC may be the important determinants of efficacy [101, 102]. However, given the excellent efficacy of a very rapidly eliminated drug like artesunate, it seems unlikely that either of these are important determinants of efficacy of artemisinin derivatives. It seems more likely that maximal concentration (C_{max}) is important. Indeed there is some evidence that for artemisinin, C_{max} correlates with parasite clearance time, which is probably the best available surrogate marker for clinical response [103]. Studies of other derivatives, including artesunate have also suggested relationships between absolute dose and parasite clearance time [104]. Whilst maximal concentration may be the important determinant of efficacy, it seems that AUC is likely to be the principal determinant of neurological toxicity (see Sect. 5.3) [105–107]. Therefore, in terms of optimising both efficacy and safety, the ideal pharmacokinetic profile would be one that rapidly achieves high drug concentrations but also sees the rapid elimination of the drug (thereby minimising the AUC). Artesunate appears, therefore, to have a more favourable profile than the other derivatives in this regard, whether administered orally, rectally or by injection.

5.2 Antimalarial Activity

Each of the artemisinin derivatives is active at nanomolar concentrations against asexual blood forms of all the plasmodia that infect humans. The in vitro *P. falciparum* IC_{50} value is similar for all four main artemisinin drugs [7], and is a number of orders of magnitude less than plasma drug concentrations achieved through routine therapeutic administration. The initial reduction in parasitaemia is the most rapid of all available antimalarial drugs, and seems to be a consistent feature, regardless of the derivative or dosing regimen used (Fig. 2) [21, 22, 109]. The parasite reduction ratio (PRR) has been estimated at 4 $\log_{10}$ per 48 h erythrocytic cycle [110], meaning that parasite density characteristically falls by approximately 90% of baseline within the first 12 h (a 1 $\log_{10}$ reduction), by 99% in the first 24 h (2 $\log_{10}$ reduction) and by 99.99% by 48 h [20, 108, 111, 112]. Most clinical infections with *P. falciparum* are thought to represent an initial total body parasite burden of the order of 10^{12} total parasites prior to treatment being commenced [110]. Therefore following artemisinin treatment, a 4 $\log_{10}$ reduction over 48 h will result in a total parasite burden in the region of 10^{8} parasites. This is often below the pyrogenic threshold and below the level of detection of peripheral blood parasitaemia using conventional microscopy. Therefore, total fever and parasite clearance times are often less than 48 h following treatment with an artemisinin or derivative-containing regimen [20, 108, 111, 112]. Although the patient may be feeling well and no longer have detectable parasitaemia by this time, if the residual parasite burden is 10^{8}, assuming an ongoing 4 $\log^{10}$ PRR, another 96 h (at least two 48-h erythrocytic parasite life cycles) will be required before the entire parasite burden has been eradicated and definitive cure ensured. Presumably this is why that, given the very short half-lives of artemisinin and its derivatives in blood, short courses (3–5 days) are associated with recrudescence rates that are typically greater than 25% [22, 109, 113].

Given that artemisinins have very short half-lives, and that parasite clearance persists for many hours after drug is no longer detectable in the circulation, it is likely that the equivalent of a "post-antibiotic effect" exists, where there is persistent suppression of microbial growth following limited exposure to an antimicrobial agent. Artemisinins are also active against some stages of the sexual blood form of the parasites (stage I and II gametocytes) and therefore may have the potential to reduce transmission rates (Fig. 2b) [114]. They do not have significant activity against exoerythrocytic (hepatic) stages, such as the dormant hypnozoite forms in *P. vivax*, and therefore have no ability to prevent *P. vivax* relapse.

5.3 Safety and Tolerability

The artemisinin derivatives appear to have a wide therapeutic margin, being generally well-tolerated at standard therapeutic doses in humans [21, 22, 109, 115].

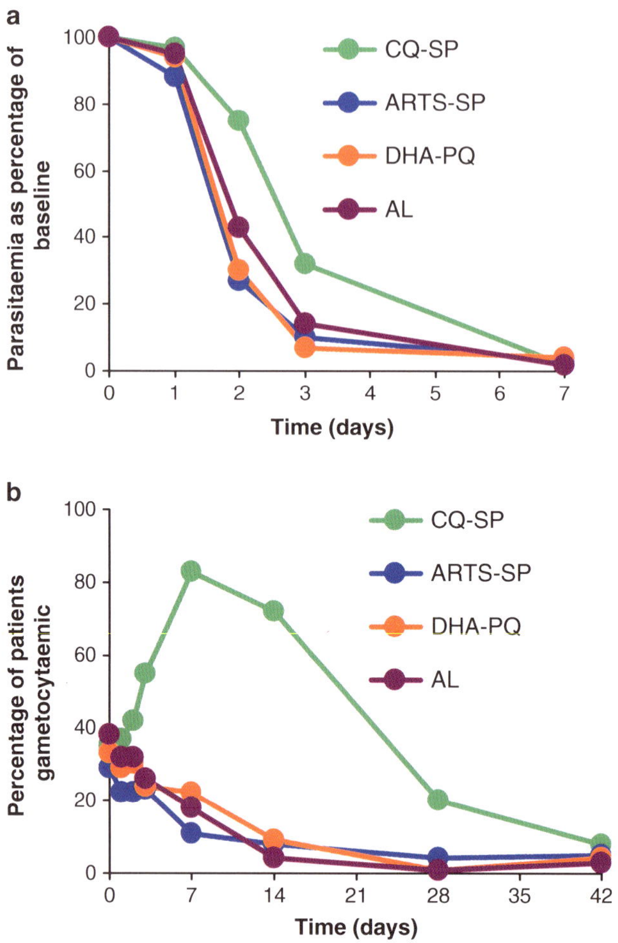

Fig. 2 Comparison of *Plasmodium falciparum* parasite clearance rates in children treated with a non-artemisinin derivative-containing regimen, chloroquine + sulphadoxine/pyrimethamine (CQ: SP) versus three artemisinin-based regimens: artesunate + sulphadoxine–pyrimethamine (ARTS–SP), dihydroartemisinin–piperaquine (DHA–PQ) and artemether–lumefantrine (AL). (**a**) trophozoite (ring forms) and (**b**) gametocytes. Data are from the study by Karunajeewa et al. of Papua New Guinean children with uncomplicated malaria [108]

Reported adverse effects have included nausea, vomiting, bowel disturbance, abdominal pain, headache and dizziness, symptoms that can result from malaria infection itself and are usually mild and self-limiting [115]. Mild and reversible haematological abnormalities, are observed infrequently with routine use at conventional dosages (e. g., artesunate 2–4 mg/kg/day) [115]. However, higher than usual therapeutic doses administered in long courses (artesunate 6 mg/kg/day for 7 days) demonstrated a high (19%) rate of neutropenia, suggesting there is an upper limit to safe dosing [116]. Minor electrocardiographic changes described [115] may reflect malarial disease itself rather than drug action and are almost certainly not clinically relevant. Animal toxicity studies using these compounds at supra-therapeutic doses in rats and dogs demonstrated a characteristic fatal neurotoxicity affecting predominantly the brainstem of animals receiving high doses over long periods [117, 118]. This toxicity now appears to be dose-related and to be restricted to compounds and dosing regimens that result in sustained high plasma concentrations of the primary drug or its active metabolite dihydroartemisinin [105]. Therefore, it is likely that the depot-like slow release of lipophilic artemisinin derivatives when administered by intramuscular injection facilitates this toxicity [106]. This theory is supported by data showing that mice fed comparable oral doses of more rapidly cleared water soluble artemisinin derivatives do not develop neurotoxicity [107]. Although somewhat controversial, there is little convincing evidence to suggest neurotoxicity has ever manifested in human subjects. Neurological symptoms of ataxia, slurred speech and hearing loss have been reported in a small numbers of humans treated with artemisinin derivatives [119, 120]. Reports that have attempted to ascribe these to side-effects of artemisinin derivatives (rather than as manifestations of malarial disease itself) have been re-examined in subsequent prospective studies that have failed to substantiate any causal associations [121, 122]. The limited penetration of artemisinin derivatives into the cerebrospinal fluid also makes the possibility of neurotoxicity at therapeutic doses less likely [123]. Nonetheless, some lingering concerns still exist that neurotoxicity may manifest in certain vulnerable individuals, including children, for whom the developing central nervous system may be at greater risk [124]. Ongoing safety monitoring and attention to pharmacokinetic disposition are warranted in order that optimally safe and effective regimens can be developed, particularly if higher dose regimens are to be considered in the face of emerging artemisinin resistance.

A large body of data from animal models (including rats, rabbits and monkeys) now raises significant concerns about the reproductive toxicity of artemisinins, demonstrating embryonic death, foetal resorption, limb dysgenesis and cardiac defects following treatment with therapeutic doses in the first trimester [125]. By contrast, birth outcomes have now been documented in >500 women exposed to artemisinins during their pregnancy, and adverse events have not been reported [126]. Nonetheless, given the strength of the animal data, artemisinin derivatives are considered contraindicated in the first trimester of pregnancy [125].

6 Clinical Uses

The great clinical utility of the artemisinin derivatives owes much to their advantages of extremely rapid parasiticidal activity, overall excellent safety and tolerability, and at the current time, the lack of widespread parasite resistance. However given the reproductive toxicity data from animal studies, their use in pregnancy is limited (see above), at least for the time-being, pending further data. Also, because of their extremely short half-lives, they are not well-suited to use as prophylaxis (although they have had some use in intermittent presumptive treatment strategies that have utilised ACTs). Their major clinical applications, therefore, are for community/outpatient-based treatment of uncomplicated malaria, parenteral treatment of severe malaria and pre-referral rectal administration of suppositories in presumptive malaria or after confirmation of diagnosis and before definitive treatments can be used. Their use as they apply to each of these three major clinical scenarios is summarised below.

6.1 Treatment of Uncomplicated Malaria

Artemisinin drugs have quickly attained popularity over conventional antimalarials for the treatment of uncomplicated malaria for a number of reasons. Patients and prescribers recognise that they are relatively free from the unpleasant side effects that compromise drugs like quinine, chloroquine and mefloquine. They also appreciate the more rapid relief of symptoms of malaria when compared with other drugs [127]. This is supported by clinical data that consistently demonstrate more rapid fever clearance and parasite clearance rates when compared with conventional therapy (Fig. 2a), with little evidence to separate one derivative from another [22, 108]. However, from very early on in their clinical development it was recognised that although initial cure rates were almost invariably close to 100%, courses of artemisinin-derivative monotherapy $\leq$5 days were associated with recrudescence rates typically greater than 25% [22, 109]. This likely reflects the very short plasma half-lives of artemisinin derivatives coupled with the need for parasiticidal drug concentrations throughout a number of parasite erythrocytic life cycles (each lasting 48 h) to eliminate the entire body parasite burden and thereby produce definitive cure. For long half-life drugs such as chloroquine and mefloquine, this is feasible with short-course therapy because a long "tail" of persisting drug is present for many days after the last dose, and this is sufficient to kill any remaining parasites. This issue can be dealt with by employing courses of monotherapy of at least 5–7 days duration [128]. Prior to the recent development of resistance in Cambodia, 7 day regimens had been described as having cure rates as high as >95% [22]. However, in the developing world, such long courses are likely to be compromised by poor compliance, particularly as defervescence usually occurs within 48 h of artemisinin treatment commencement, diminishing the incentive for the patient to complete the full course.

In addition to concerns regarding the potential for development of parasite resistance, the logistical problems associated with long-course artemisinin monotherapy have added to the rationale for ACT. The artemisinin component provides rapid early reduction in parasite burden and symptom relief for the patient. The longer half-life partner drug achieves parasite killing many days after the last dose has been taken, thereby eliminating the parasite residuum and enabling the feasibility of short-course therapy with high definitive cure rates. Given recent WHO policy directives, it is hoped that in the near future, artemisinin monotherapy will no longer be used widely and that most artemisinin-based therapy will be in the form of short-course ACT.

A large number of clinical trials have been conducted in recent years evaluating the clinical efficacy of various ACTs [24]. It is likely, however, that rather than the artemisinin component, it is the choice of partner drug that is most crucial in determining the combination's effectiveness [108, 129]. Pre-existing resistance to partner drugs in the area where the combination is used may be especially important [108, 130]. Issues of cost, tolerability/toxicity and dosing convenience must also be considered [31]. The prototypical ACT is the "loose" combination of separate tablets of artesunate and mefloquine first used in Thailand and subsequently adopted as first-line treatment policy for uncomplicated *P. falciparum* infection in much of Southeast Asia [30]. Other "loose" ACTs have included sulphadoxine-–pyrimethamine, amodiaquine and atovaquone–proguanil as partner drugs [31]. However, "fixed" co-formulations containing both the artemisinin derivative and the partner drug have overwhelming operational advantages by preventing inadvertent monotherapy. They have been the focus of ACT development in recent years (Table 2). The most widely used co-formulation, and the first to be endorsed by the WHO, is the combination artemether–lumefantrine. This has consistently demonstrated 28–42 day cure rates >95% when administered as a twice daily regimen for 3 days [131]. However, it has some potential problems that may affect its operational deployment, including the need to be administered with food or drink containing some fat in order to ensure adequate absorption [132] and a relatively complex twice-daily regimen that may compromise compliance. It has now been adopted as first-line treatment policy for laboratory-confirmed uncomplicated *P. falciparum* malaria by at least 46 countries, mostly in Africa and Asia [1]. Importantly, a suspension preparation suitable for administration to infants is also now available [133].

At the time of writing, artesunate–amodiaquine had achieved widespread use in Africa, having been adopted as treatment policy in at least 20 African countries [1]. It is now available as a co-formulation, conveniently administered as three daily doses. In many parts of the world, particularly where amodiaquine has been widely used, the efficacy of this combination is likely to be compromised by pre-existing resistance to amodiaquine. For this reason, it has not achieved widespread use outside Africa [1].

Other promising combinations in various stages of development have utilised piperaquine, mefloquine, amodiaquine, fosmidomycin, chloroproguanil–dapsone, pyronaridine and naphthoquine as partner drugs [31]. A large body of clinical trial data has been accrued to support the dihydroartemisinin–piperaquine combination that is available at an attractive price and has mostly shown excellent efficacy and tolerability, when given as three daily doses. [24, 134] It is used as first-line treatment in Vietnam, China, Myanmar and parts of Indonesia. [1] However, the hurdles to be overcome to achieve regulatory approval have delayed WHO endorsement of this and other preparations manufactured in China or elsewhere in the developing world.

Table 2 Artemisinin derivatives available as co-formulations

	Partner drugs used	Manufacturer(s) and product proprietary-name
Artemisinin	Piperaquine	Artepharm (Artequick®)
	Naphthoquine	Kunming pharmaceuticals (Arco®)
Dihydroartemisinin	Piperaquine	Beijing Holley-Cotec (Duocotecxin®) Sigma-Tau (Eurartekin®) GPC ShenZhen (Combimal®) GVS labs (P-Alaxin®)
Artemether	Lumefantrine	Novartis Pharma (Coartem® and Riamet®)[a] Addis (Artemine®) Ajanta (Artefan®) Medinomics (Fantem®) Cipla (Lumartem® and Lumet®)
Artesunate	Amodiaquine	Sanofi Aventis (Coarsucam®)[a] Guilin Pharmaceuticals (Co-Artusan®)[a] Medinomics (MalmedFD®)
	Mefloquine	Mepha (Artequin®)
	Pyronaridine	Sin Phoong (Pyramax®)

[a]At the time of writing, only these three products (Coartem®, Coarsucam® and Co-artusan®) had achieved WHO pre-qualification

6.2 *Parenteral Treatment of Severe Malaria*

Prior to the availability of the artemisinin derivatives, the only options for parenteral treatment of severe malaria in most of the world had been quinine (or quinidine), administered either by intravenous infusion or intramuscular injection [135], chloroquine (administered by intramuscular or subcutaneous injection) [136] or amodiaquine (given intravenously) [137]. However, whilst clearly highly efficacious in the treatment of chloroquine-sensitive *P. falciparum* malaria, parenteral chloroquine and amodiaquine have been rendered obsolete in most malaria-endemic regions of the world due to the development of widespread 4-aminoquinoline resistance. Therefore, for practical purposes, this leaves quinine as the only viable therapy for severe malaria in most parts of the world.

The possibility of using artemisinin derivatives to treat severe malaria has obvious theoretical advantages over conventional quinine therapy, including faster clearance of trophozoites [21, 22] and a lack of serious toxicity issues (including hypoglycaemia, "blackwater fever," haemodynamic instability, cardiac arrhythmias and tinnitus) that may complicate quinine therapy [115, 138–142]. Artemisinin was first used to treat severe malaria in China as early as 1972, but widespread use for severe malaria would require manufacture of a derivative in a form that was both stable and suitable for injection into patients who were either unconscious or too ill to take medicine orally. Initial interest centred on the oil-based preparations of arteether and artemether that could be administered by intramuscular injection. Numerous trials evaluated their use in severe malaria when compared with parenteral quinine therapy [21]. These consistently demonstrated superiority with regard to the surrogate efficacy endpoints of parasite clearance and fever clearance with artemether.

However, all studies were inadequately powered to demonstrate improvements in the clinically important endpoints of mortality or serious disability. A meta-analysis of data from studies comparing artemether with quinine did show an overall lower mortality in artemether-treated patients but the difference did not reach statistical significance [98]. A number of countries have adopted intramuscular artemether as the standard recommended treatment for severe malaria[1]. However, most early studies investigating the use of artemether for treatment of severe malaria [98] were conducted at a time before the development of robust drug assays and therefore, pharmacokinetic data were limited [97, 99, 143]. It has since become clear that both artemether and arteether have poor and highly variable absorption into the systemic circulation from the intramuscular injection site, raising concerns that a proportion of patients treated might be at risk of sub-optimal therapeutic response [144]. The pharmacokinetic profile of the water-soluble artemisinins appears to be more favourable for treatments of severe malaria. In contrast to artemether, both intravenous and intramuscular injections of artesunate result in consistent and rapid absorption, with peak levels occurring within minutes of administration [92]. The depot effect seen with artemether does not occur so that the drug's apparent plasma elimination half-lives are not distorted by absorption from the injection site. All of these suggested artesunate as a more promising replacement for parenteral quinine.

In 2005, a large multi-centre comparison of parenteral quinine with intravenous artesunate, the South-East Asian Quinine Artesunate (SEAQUAMAT) trial, was published [25]. This showed a clear mortality benefit (relative risk reduction of 35%, absolute risk reduction of 7%, number needed to treat 14) for artesunate over quinine. Because the majority of participants in this study were Southeast Asian non-immune adults, it was not immediately clear whether this mortality benefit could be extrapolated to children living in malaria-endemic areas of the world such as Africa. However, a more recent large multi-centre study of African children with severe malaria, [26] has also demonstrated a clear, although somewhat smaller mortality benefit (relative risk reduction 22.5%, absolute risk reduction 1.4%, number needed to treat 71) Parenteral artesunate is now, therefore, widely considered as the treatment of choice for severe malaria, in both children and adults regardless of context, with the only possible exception being women in the first trimester of pregnancy [2].

At the present time, it is not clear whether or to what extent early reports of delayed parasite clearance following artemisinin treatment in Cambodia will affect the efficacy of artemisinins for treatment of severe malaria. However, for the time being parenteral administration of artesunate should be considered the gold standard by which other treatments for severe malaria will be judged.

6.3 Rectal Administration

Because the majority of deaths from *P. falciparum* infection occur in children living in remote rural areas with limited access to medical care, a rapidly acting antimalarial drug in suppository form has great potential value [145, 146]. In contrast to oral therapy, rectal administration is not precluded by vomiting, prostration or impaired consciousness, all common features of severe malaria. Although parenteral therapy can also be used when oral dosing is problematic, equipment and trained personnel necessary to safely administer injections are often unavailable in these areas. An effective rectally administered treatment could mean that those too sick to take medication orally could be treated in the community, either as emergency "pre-referral" treatment (before transportation to a health facility for higher-level treatment) or, if referral is not possible, as part of a complete treatment course given at home [28, 38]. It may be feasible for suppository administration and subsequent monitoring for extrusion to be performed by a village health worker, or even by a sick child's mother, with minimal training. Because the artemisinin drugs have an excellent safety profile and a reasonably wide therapeutic margin, rectal administration could be safely and effectively performed by individuals with little or no medical training.

Artemisinin derivatives that have been formulated for rectal administration have included artemether, artemisinin, dihydroartemisinin and artesunate (Table 1). At present, most data accrued regarding rectally administered artemisinin derivatives comes from studies of artesunate suppositories, which of all the rectally administered artemisinins, appear to have the most favourable pharmacokinetic properties based on existing data [20]. These have been produced by two manufacturers and are usually formulated as 50 mg or 200 mg suppositories (Plasmotrim Rectocaps®, Mepha Pharmaceuticals, Aesch-Basel, Switzerland; Scanpharma A/E Denmark).

Rectally administered artemisinins are likely to have their greatest value in critically ill patients residing in geographically remote settings in whom definitive therapy may be many hours or days away. Therefore, the primary imperative is for early and rapid parasite clearance. This should prevent or delay the progression of microvascular parasite sequestration and the consequent metabolic derangements that can lead to death,[147] and buy time while the patient is being transported to a health facility. For this reason, the most useful indicators of clinical efficacy are those that describe the rapidity of parasite clearance, particularly in the first 12–24 h [20, 147]. These have been shown to be equivalent to or superior to those of both artemisinin derivative or quinine-based parenteral treatments for severe malaria [20, 112, 148]. Nevertheless, the high degree of inter-individual variability in absorption raises the possibility of sub-optimal response in a subset of children [20, 149]. Therefore, parenteral administration (intramuscular or intravenous artesunate) should be considered more reliable when facilities for safe injection are available. Definitive clinical and parasitological cure is determined largely by either the duration of therapy or the efficacy of the longer acting partner drug used as part of an ACT once the patient is able to swallow tablets and is therefore of secondary interest in evaluation of rectal artemisinin drugs.

A recent large multi-centre placebo-controlled trial has shown that rectal artesunate given as pre-referral therapy to children with suspected severe malaria results in a significant reduction in overall death and disability, especially when arrival at the health clinic is delayed by >6 h from the time of administration [145]. In common with oral and parenteral formulations of artemisinin derivatives, their introduction into clinical use has not followed classical pathways of rational drug development, and dosage regimens have largely been derived empirically. With the exception of artesunate, pharmacokinetic data describing the adequacy and consistency of absorption of other rectally administered derivatives is lacking. The situation is complicated by the number of rectal artemisinin derivatives on the market with the potential for multiple manufacturers of single formulations. Because pharmaceutical factors may also influence pharmaceutical disposition, preparations from different manufacturers cannot necessarily be assumed to be bioequivalent.

7 Future Potential

The future of the artemisinin-derivative drug class will be dependent on the extent to which significant parasite resistance can be avoided. The principles underlying ACT as a means of preventing artemisinin resistance are now widely accepted and implemented as health policy throughout most of the malaria-endemic world. However, the theory that ACT will prevent resistance developing is one that is difficult to test and to prove. The real effects of this strategy on the population genetics of antimalarial drug resistance remain to be seen. Its success will rely on availability of a suite of suitably effective partner drugs and on operational factors that limit the use of inappropriate artemisinin monotherapy. The development of apparent artemisinin resistance in Cambodia is of particular concern. However, this phenomenon will become doubly alarming if resistance mechanisms evolve that affect a pathway common to both drugs within ACTs, through, for example, drug efflux mechanisms mediated by *pfmdr1* copy number [76, 77, 79].

Improved understanding is required of the mechanisms of artemisinin resistance, including their underlying parasite genetics. This may enable molecular tools and PCR-based diagnostics for monitoring the spread and the development of resistance. Elucidation of the metabolic pathways involved may also shed much-needed light on the mechanisms of action of the artemisinin derivative drug class. This will be useful for the development of subsequent generations of endoperoxide drugs that must aim to subvert these resistance mechanisms.

For the most part, the introduction of the artemisinin derivatives into clinical use has not followed classical pathways of rational drug development, and dosage regimens have largely been derived empirically. However, when dealing with sensitive parasites, the therapeutic margin appears to have been wide enough that therapeutic outcomes have been universally good, regardless of the dosages administered up until now. However, this may change with the advent of reduced susceptibility to artemisinin derivatives. As the therapeutic margin becomes narrower, improved understanding of pharmacokinetic–pharmacodynamic relationships will become more important for optimally effective and safe dosing regimens to be devised.

With recent evidence that community-based pre-referral rectal administration of artesunate suppositories may have a population mortality benefit [145], artemisinin derivatives formulated for rectal administration have the potential to be amongst the most widely used of all drugs in tropical countries. Because currently available suppository formulations are all mono-preparations of an artemisinin, and because use in this context will be difficult to regulate, this raises concerns about the potential for this strategy to contribute to driving selection pressure for artemisinin resistance, and suggests a need for development of co-formulated (ACT) suppositories [38]. A wide range of preparations of different artemisinin derivatives formulated for rectal administration is now available. However, robust pharmacokinetic data are lacking for many of these and it is not clear if these can be considered therapeutically equivalent to rectal preparations of artesunate that have been the subject of most clinical studies [20]. In addition, optimal strategies for deploying intrarectal artemisinin drugs at community level are yet to be determined. In particular, it is not clear to what extent this strategy is socially or culturally acceptable and feasible across the range of societies in which this intervention could be deployed. Information on existing attitudes to rectal administration in specific cultural contexts should help to inform the development of the most appropriate deployment strategies [127].

Because *A. annua* can be easily grown in a wide range of climates and environments, it has been suggested that it could be grown as a crop at the village-level to enable preparation of a "home-grown" local malaria remedy, similar to its original use in ancient China [5, 150]. This could be useful in remote areas with poor access to conventional health-care and in some societies, the mode of delivery might be more culturally acceptable than modern pharmaceutical administration. However it grows less well and yields less active drug in tropical climates, where the world's greatest malaria burden occurs. In addtion, administration in this way would effectively constitute monotherapy with raw artemisinin, a compound with lesser activity than its pharmaceutical derivatives (such as artesunate). In particular, because it would be difficult to regulate dosage, it seems unlikely that such a strategy would receive official endorsement from the WHO or others, given current existing concerns about the development of resistance.

Although the recent advent of resistance in a small focus in Southeast Asia is concerning, it should be remembered this has emerged up to 30 years after artemisinin drugs were first used in this area. Artemisinin derivatives have now been used heavily for many years in many other parts of the world, without such problems emerging. They represent the most important class of antimalarial drug at the current time and are likely to remain so for the foreseeable future. Following recent WHO policy directives, in the near future, a substantial proportion (if not most) of the estimated 400 million annual clinical presentations with suspected malarial illness will be treated with an artemisinin-containing regimen [1, 28]. Artemisinin derivatives could therefore become one of the most-used classes of drug in the world, raising significant issues regarding global production capacity.

An estimated 10,000 ha of *A. annua* will be required to produce the equivalent of 100 million adult treatment courses [151]. To address these massive global demands, higher yielding varieties of *A. annua* are being investigated [152] and alternatives to agricultural manufacture of raw artemisinin, including recombinant microbial production and wholly synthetic methods of producing artemisinin and other synthetic endoperoxides, may also emerge as feasible means of ensuring a consistent supply of affordable drug .

Acknowledgements Tim Davis and Ken Ilett have been an invaluable source of information and ideas used in the writing of this chapter. Figure 2 is reproduced from Karunajeewa et al. New England Journal of Medicine, 2008, 359 (24) pp. 2545–57 [108] with the permission of the authors.

References

1. WHO. Malaria Treatment Policies (by region). http://www.who.int/malaria/publications/treatment-policies/en/. Accessed 24 Apr 2010
2. WHO (2010) Guidelines for the treatment of malaria, 2nd edn. World Health Organization, Geneva
3. Snow RW, Guerra CA, Noor AM, Myint HY, Hay SI (2005) The global distribution of clinical episodes of *Plasmodium falciparum* malaria. Nature 434:214–217
4. Willcox M, Bodeker G, Bourdy G, Dhingra V, Falquet J, Ferreira JFS, Graz B, Hirt H-M (2004) *Artemisia annua* as a traditional herbal antimalarial. In: Willcox M, Bodeker G, Raoanaivo P (eds) Traditional medicinal plants and malaria. CRC, Boca Raton, pp 43–59
5. Hsu E (2006) Reflections on the 'discovery' of the antimalarial qinghao. Br J Clin Pharmacol 61:666–670
6. Keiser J, Utzinger J (2007) Artemisinins and synthetic trioxolanes in the treatment of helminth infections. Curr Opin Infect Dis 20:605–612
7. Ilett K, Batty K (2005) Artemisinin and its derivatives. In: Yu VL, Edwards G, McKinnon PS, Peloquin CA (eds) Antimicrobial therapy and vaccines. ESun Technologies, Pitsburg, PA, pp 981–1002
8. Wright CW, Linley PA, Brun R, Wittlin S, Hsu E (2010) Ancient Chinese methods are remarkably effective for the preparation of artemisinin-rich extracts of Qing Hao with potent antimalarial activity. Molecules 15:804–812
9. China Cooperative Research Group on qinghaosu and its derivatives as antimalarials (1982) Clinical studies on the treatment of malaria with qinghaosu and its derivatives. J Tradit Chin Med 2:45–50
10. Schlitzer M (2007) Malaria chemotherapeutics. Part I: History of antimalarial drug development, currently used therapeutics, and drugs in clinical development. ChemMedChem 2:944–986
11. Bruce-Chwatt LJ (1982) Qinghaosu: a new antimalarial. Br Med J (Clin Res Ed) 284:767–768
12. Jiang JB, Li GQ, Guo XB, Kong YC, Arnold K (1982) Antimalarial activity of mefloquine and qinghaosu. Lancet 2:285–288
13. Bruce-Chwatt LJ (1985) Recent trends of chemotherapy and vaccination against malaria: new lamps for old. Br Med J (Clin Res Ed) 291:1072–1076
14. Nosten F, ter Kuile F, Chongsuphajaisiddhi T, Luxemburger C, Webster HK, Edstein M, Phaipun L, Thew KL, White NJ (1991) Mefloquine-resistant falciparum malaria on the Thai-Burmese border. Lancet 337:1140–1143

15. Bunnag D, Viravan C, Looareesuwan S, Karbwang J, Harinasuta T (1991) Clinical trial of artesunate and artemether on multidrug resistant falciparum malaria in Thailand. A preliminary report. SE Asian J Trop Med Publ Health 22:380–385
16. Nosten F, van Vugt M, Price R, Luxemburger C, Thway KL, Brockman A, McGready R, ter Kuile F, Looareesuwan S, White NJ (2000) Effects of artesunate-mefloquine combination on incidence of *Plasmodium falciparum* malaria and mefloquine resistance in western Thailand: a prospective study. Lancet 356:297–302
17. Edwards G (1994) Measurement of artemisinin and its derivatives in biological fluids. Trans R Soc Trop Med Hyg 88:37–39
18. Meshnick SR (2002) Artemisinin: mechanisms of action, resistance and toxicity. Int J Parasitol 32:1655–1660
19. Eckstein-Ludwig U, Webb RJ, Van Goethem ID, East JM, Lee AG, Kimura M, O'Neill PM, Bray PG, Ward SA, Krishna S (2003) Artemisinins target the SERCA of *Plasmodium falciparum*. Nature 424:957–961
20. Karunajeewa HA, Manning L, Mueller I, Ilett KF, Davis TM (2007) Rectal administration of artemisinin derivatives for the treatment of malaria. JAMA 297:2381–2390
21. McIntosh HM, Olliaro P (2000) Artemisinin derivatives for treating severe malaria. Cochrane Database Syst Rev CD000527
22. McIntosh HM, Olliaro P (2000) Artemisinin derivatives for treating uncomplicated malaria. Cochrane Database Syst Rev CD000256
23. Myint HY, Tipmanee P, Nosten F, Day NPJ, Pukrittayakamee S, Looareesuwan S, White NJ (2004) A systematic overview of published antimalarial drug trials. Trans R Soc Trop Med Hyg 98:73–81
24. Sinclair D, Zani B, Donegan S, Olliaro P, Garner P (2009) Artemisinin-based combination therapy for treating uncomplicated malaria. Cochrane Database Syst Rev CD007483
25. Dondorp A, Nosten F, Stepniewska K, Day N, White N (2005) Artesunate versus quinine for treatment of severe falciparum malaria: a randomised trial. Lancet 366:717–725
26. Dondorp AM, Fanello CI, Hendriksen IC, Gomes E, Seni A, Chhaganlal KD, Bojang K, Olaosebikan R, Anunobi N, Maitland K et al (2010) Artesunate versus quinine in the treatment of severe falciparum malaria in African children (AQUAMAT): an open-label, randomised trial. Lancet 376:1647–1657
27. White NJ, Nosten F, Looareesuwan S, Watkins WM, Marsh K, Snow RW, Kokwaro G, Ouma J, Hien TT, Molyneux ME et al (1999) Averting a malaria disaster. Lancet 353:1965–1967
28. WHO (2006) Guidelines for the treatment of malaria. WHO, Geneva
29. Kremsner PG, Krishna S (2004) Antimalarial combinations. Lancet 364:285–294
30. Nosten F, Luxemburger C, ter Kuile FO, Woodrow C, Eh JP, Chongsuphajaisiddhi T, White NJ (1994) Treatment of multidrug-resistant *Plasmodium falciparum* malaria with 3-day artesunate-mefloquine combination. J Infect Dis 170:971–977
31. Davis A (2004) Clinical trials in parasitic diseases. Trans R Soc Trop Med Hyg 98:139–141
32. Barnes KI, Chanda P, Ab Barnabas G (2009) Impact of the large-scale deployment of artemether/lumefantrine on the malaria disease burden in Africa: case studies of South Africa, Zambia and Ethiopia. Malar J 8:S8
33. Barnes KI, Durrheim DN, Little F, Jackson A, Mehta U, Allen E, Dlamini SS, Tsoka J, Bredenkamp B, Mthembu DJ et al (2005) Effect of artemether-lumefantrine policy and improved vector control on malaria burden in KwaZulu-Natal, South Africa. PLoS Med 2: e330
34. Greenwood B (2009) Can malaria be eliminated? Trans R Soc Trop Med Hyg 103:S2–S5
35. White NJ (2008) The role of anti-malarial drugs in eliminating malaria. Malar J 7:S8
36. Peters W, Robinson BL (1999) The chemotherapy of rodent malaria. LVI. Studies on the development of resistance to natural and synthetic endoperoxides. Ann Trop Med Parasitol 93:325–329
37. WHO (2001) Antimalarial drug combination therapy. Report of a technical consultation. WHO, Geneva

38. WHO (2007) Use of rectal artesmisinin based suppositories in the management of severe malaria. Report of a WHO Informal Consultation, 27–28 Mar 2006
39. Afonso A, Hunt P, Cheesman S, Alves AC, Cunha CV, do Rosario V, Cravo P (2006) Malaria parasites can develop stable resistance to artemisinin but lack mutations in candidate genes *atp6* (encoding the sarcoplasmic and endoplasmic reticulum Ca^{2+} ATPase), *tctp*, *mdr1*, and *cg10*. Antimicrob Agents Chemother 50:480–489
40. Walker DJ, Pitsch JL, Peng MM, Robinson BL, Peters W, Bhisutthibhan J, Meshnick SR (2000) Mechanisms of artemisinin resistance in the rodent malaria pathogen *Plasmodium yoelii*. Antimicrob Agents Chemother 44:344–347
41. Jambou R, Legrand E, Niang M, Khim N, Lim P, Volney B, Ekala MT, Bouchier C, Esterre P, Fandeur T et al (2005) Resistance of *Plasmodium falciparum* field isolates to *in-vitro* artemether and point mutations of the SERCA-type PfATPase6. Lancet 366:1960–1963
42. Rehwagen C (2006) WHO ultimatum on artemisinin monotherapy is showing results. BMJ 332:1176
43. Dondorp AM, Yeung S, White L, Nguon C, Day NP, Socheat D, von Seidlein L (2010) Artemisinin resistance: current status and scenarios for containment. Nat Rev Microbiol 8:272–280
44. Noedl H, Se Y, Schaecher K, Smith BL, Socheat D, Fukuda MM (2008) Evidence of artemisinin-resistant malaria in western Cambodia. N Engl J Med 359:2619–2620
45. Samarasekera U (2009) Countries race to contain resistance to key antimalarial. Lancet 374:277–280
46. Li J, Zhou B (2010) Biological actions of artemisinin: insights from medicinal chemistry studies. Molecules 15:1378–1397
47. Schlitzer M (2008) Antimalarial drugs – what is in use and what is in the pipeline. Arch Pharm (Weinheim) 341:149–163
48. Chemical Abstracts Services SciFinder Scholar database. American Chemical Society, Washington DC, USA
49. Krishna S, Bustamante L, Haynes RK, Staines HM (2008) Artemisinins: their growing importance in medicine. Trends Pharmacol Sci 29:520–527
50. O'Neill PM, Barton VE, Ward SA (2010) The molecular mechanism of action of artemisinin–the debate continues. Molecules 15:1705–1721
51. Efferth T, Romero MR, Wolf DG, Stamminger T, Marin JJ, Marschall M (2008) The antiviral activities of artemisinin and artesunate. Clin Infect Dis 47:804–811
52. O'Neil MT, Korsinczky ML, Gresty KJ, Auliff A, Cheng Q (2007) A novel *Plasmodium falciparum* expression system for assessing antifolate resistance caused by mutant *P. vivax* dihydrofolate reductase-thymidylate synthase. J Infect Dis 196:467–474
53. Meshnick SR (1994) The mode of action of antimalarial endoperoxides. Trans R Soc Trop Med Hyg 88:31–32
54. Jefford CW (2001) Why artemisinin and certain synthetic peroxides are potent antimalarials. Implications for the mode of action. Curr Med Chem 8:1803–1826
55. Posner GH, Wang D, Cumming JN, Oh CH, French AN, Bodley AL, Shapiro TA (1995) Further evidence supporting the importance of and the restrictions on a carbon-centered radical for high antimalarial activity of 1,2,4-trioxanes like artemisinin. J Med Chem 38:2273–2275
56. Haynes RK, Chan WC, Lung CM, Uhlemann AC, Eckstein U, Taramelli D, Parapini S, Monti D, Krishna S (2007) The Fe^{2+}-mediated decomposition, PfATP6 binding, and antimalarial activities of artemisone and other artemisinins: the unlikelihood of C-centred radicals as bioactive intermediates. ChemMedChem 2:1480–1497
57. Creek DJ, Charman WN, Chiu FC, Prankerd RJ, Dong Y, Vennerstrom JL, Charman SA (2008) Relationship between antimalarial activity and heme alkylation for spiro- and dispiro-1,2,4-trioxolane antimalarials. Antimicrob Agents Chemother 52:1291–1296
58. Meshnick SR, Haynes RK, Monti D, Taramelli D, Basilico N, Parapini S, Olliaro P (2003) Artemisinin and heme. Antimicrob Agents Chemother 47:2712–2713

59. Haynes RK, Monti D, Taramelli D, Basilico N, Parapini S, Olliaro P (2003) Artemisinin antimalarials do not inhibit hemozoin formation. Antimicrob Agents Chemother 47:1175
60. Krishna S, Woodrow CJ, Staines HM, Haynes RK, Mercereau-Puijalon O (2006) Re-evaluation of how artemisinins work in light of emerging evidence of in vitro resistance. Trends Mol Med 12:200–205
61. Nagamune K, Beatty WL, Sibley LD (2007) Artemisinin induces calcium-dependent protein secretion in the protozoan parasite *Toxoplasma gondii*. Eukaryot Cell 6:2147–2156
62. Uhlemann AC, Cameron A, Eckstein-Ludwig U, Fischbarg J, Iserovich P, Zuniga FA, East M, Lee A, Brady L, Haynes RK et al (2005) A single amino acid residue can determine the sensitivity of SERCAs to artemisinins. Nat Struct Mol Biol 12:628–629
63. Valderramos SG, Scanfeld D, Uhlemann AC, Fidock DA, Krishna S (2010) Investigations into the role of the *Plasmodium falciparum* SERCA (PfATP6) L263E mutation in artemisinin action and resistance. Antimicrob Agents Chemother 54:3842–3852
64. Witkowski B, Lelievre J, Barragan MJ, Laurent V, Su XZ, Berry A, Benoit-Vical F (2010) Increased tolerance to artemisinin in *Plasmodium falciparum* is mediated by a quiescence mechanism. Antimicrob Agents Chemother 54:1872–1877
65. Krishna S, Pulcini S, Fatih F, Staines H (2010) Artemisinins and the biological basis for the PfATP6/SERCA hypothesis. Trends Parasitol 26:517–523
66. Woodrow CJ, Bustamante LY (2011) Mechanisms of artemisinin action and resistance: wider focus is needed. Trends Parasitol 27:2–3; author reply 3–4
67. Li W, Mo W, Shen D, Sun L, Wang J, Lu S, Gitschier JM, Zhou B (2005) Yeast model uncovers dual roles of mitochondria in action of artemisinin. PLoS Genet 1:e36
68. Wang J, Huang L, Li J, Fan Q, Long Y, Li Y, Zhou B (2010) Artemisinin directly targets malarial mitochondria through its specific mitochondrial activation. PLoS One 5:e9582
69. Hastings IM, Watkins WM, White NJ (2002) The evolution of drug-resistant malaria: the role of drug elimination half-life. Philos Trans R Soc Lond B Biol Sci 357:505–519
70. Stepniewska K, White NJ (2008) The pharmacokinetic determinants of the window of selection for antimalarial drug resistance. Antimicrob Agents Chemother 52(5):1589–96
71. White NJ (2004) Antimalarial drug resistance. J Clin Invest 113:1084–1092
72. Korsinczky M, Chen N, Kotecka B, Saul A, Rieckmann K, Cheng Q (2000) Mutations in *Plasmodium falciparum* cytochrome b that are associated with atovaquone resistance are located at a putative drug-binding site. Antimicrob Agents Chemother 44:2100–2108
73. White N (1999) Antimalarial drug resistance and combination chemotherapy. Philos Trans R Soc Lond B Biol Sci 354:739–749
74. Cowman AF, Galatis D, Thompson JK (1994) Selection for mefloquine resistance in *Plasmodium falciparum* is linked to amplification of the *pfmdr1* gene and cross-resistance to halofantrine and quinine. Proc Natl Acad Sci USA 91:1143–1147
75. Reed MB, Saliba KJ, Caruana SR, Kirk K, Cowman AF (2000) Pgh1 modulates sensitivity and resistance to multiple antimalarials in *Plasmodium falciparum*. Nature 403:906–909
76. Price RN, Uhlemann AC, Brockman A, McGready R, Ashley E, Phaipun L, Patel R, Laing K, Looareesuwan S, White NJ et al (2004) Mefloquine resistance in *Plasmodium falciparum* and increased *pfmdr1* gene copy number. Lancet 364:438–447
77. Price RN, Uhlemann AC, van Vugt M, Brockman A, Hutagalung R, Nair S, Nash D, Singhasivanon P, Anderson TJ, Krishna S et al (2006) Molecular and pharmacological determinants of the therapeutic response to artemether-lumefantrine in multidrug-resistant Plasmodium falciparum malaria. Clin Infect Dis 42:1570–1577
78. Raman J, Little F, Roper C, Kleinschmidt I, Cassam Y, Maharaj R, Barnes KI (2010) Five years of large-scale *dhfr* and *dhps* mutation surveillance following the phased implementation of artesunate plus sulfadoxine-pyrimethamine in Maputo Province, Southern Mozambique. Am J Trop Med Hyg 82:788–794
79. Chavchich M, Gerena L, Peters J, Chen N, Cheng Q, Kyle DE (2010) Induction of resistance to artemisinin derivatives in *Plasmodium falciparum*: role of *Pfmdr1* amplification and expression. Antimicrob Agents Chemother 54(6):2455–64

80. Hunt P, Afonso A, Creasey A, Culleton R, Sidhu AB, Logan J, Valderramos SG, McNae I, Cheesman S, do Rosario V et al (2007) Gene encoding a deubiquitinating enzyme is mutated in artesunate- and chloroquine-resistant rodent malaria parasites. Mol Microbiol 65:27–40
81. Jambou R, Martinelli A, Pinto J, Gribaldo S, Legrand E, Niang M, Kim N, Pharath L, Volnay B, Ekala MT et al (2010) Geographic structuring of the *Plasmodium falciparum* sarco(endo) plasmic reticulum Ca^{2+} ATPase (PfSERCA) gene diversity. PLoS One 5:e9424
82. Shahinas D, Lau R, Khairnar K, Hancock D, Pillai DR (2011) Artesunate misuse and *Plasmodium falciparum* malaria in traveler returning from Africa. Emerg Infect Dis 16:1608–1610
83. Tanabe K, Zakeri S, Palacpac NM, Afsharpad M, Randrianarivelojosia M, Kaneko A, Marma AS, Horii T, Mita T (2011) Spontaneous mutations in the *Plasmodium falciparum* sarcoplasmic/ endoplasmic reticulum Ca^{2+}-ATPase (*PfATP6*) gene among geographically widespread parasite populations unexposed to artemisinin-based combination therapies. Antimicrob Agents Chemother 55:94–100
84. Vinayak S, Alam MT, Mixson-Hayden T, McCollum AM, Sem R, Shah NK, Lim P, Muth S, Rogers WO, Fandeur T et al (2010) Origin and evolution of sulfadoxine resistant *Plasmodium falciparum*. PLoS Pathog 6:e1000830
85. Newton P, Proux S, Green M, Smithuis F, Rozendaal J, Prakongpan S, Chotivanich K, Mayxay M, Looareesuwan S, Farrar J et al (2001) Fake artesunate in southeast Asia. Lancet 357:1948–1950
86. Anderson TJ, Nair S, Nkhoma S, Williams JT, Imwong M, Yi P, Socheat D, Das D, Chotivanich K, Day NP et al (2010) High heritability of malaria parasite clearance rate indicates a genetic basis for artemisinin resistance in western Cambodia. J Infect Dis 201:1326–1330
87. Maude RJ, Pontavornpinyo W, Saralamba S, Aguas R, Yeung S, Dondorp AM, Day NP, White NJ, White LJ (2009) The last man standing is the most resistant: eliminating artemisinin-resistant malaria in Cambodia. Malar J 8:31
88. Rogers WO, Sem R, Tero T, Chim P, Lim P, Muth S, Socheat D, Ariey F, Wongsrichanalai C (2009) Failure of artesunate-mefloquine combination therapy for uncomplicated *Plasmodium falciparum* malaria in southern Cambodia. Malar J 8:10
89. Li Q, Lugt CB, Looareesuwan S, Krudsood S, Wilairatana P, Vannaphan S, Chalearmrult K, Milhous WK (2004) Pharmacokinetic investigation on the therapeutic potential of artemotil (beta-arteether) in Thai patients with severe *Plasmodium falciparum* malaria. Am J Trop Med Hyg 71:723–731
90. Sabarinath SN, Asthana OP, Puri SK, Srivastava K, Madhusudanan KP, Gupta RC (2005) Clinical pharmacokinetics of the diastereomers of arteether in healthy volunteers. Clin Pharmacokinet 44:1191–1203
91. Ilett KF, Batty KT, Powell SM, Binh TQ, le Thu TA, Phuong HL, Hung NC, Davis TM (2002) The pharmacokinetic properties of intramuscular artesunate and rectal dihydroartemisinin in uncomplicated falciparum malaria. Br J Clin Pharmacol 53:23–30
92. Davis TM, Phuong HL, Ilett KF, Hung NC, Batty KT, Phuong VD, Powell SM, Thien HV, Binh TQ (2001) Pharmacokinetics and pharmacodynamics of intravenous artesunate in severe falciparum malaria. Antimicrob Agents Chemother 45:181–186
93. Batty KT, Thu LT, Davis TM, Ilett KF, Mai TX, Hung NC, Tien NP, Powell SM, Thien HV, Binh TQ et al (1998) A pharmacokinetic and pharmacodynamic study of intravenous vs oral artesunate in uncomplicated falciparum malaria. Br J Clin Pharmacol 45:123–129
94. Binh TQ, Ilett KF, Batty KT, Davis TM, Hung NC, Powell SM, Thu LT, Thien HV, Phuong HL, Phuong VD (2001) Oral bioavailability of dihydroartemisinin in Vietnamese volunteers and in patients with falciparum malaria. Br J Clin Pharmacol 51:541–546
95. Newton P, Suputtamongkol Y, Teja-Isavadharm P, Pukrittayakamee S, Navaratnam V, Bates I, White N (2000) Antimalarial bioavailability and disposition of artesunate in acute falciparum malaria. Antimicrob Agents Chemother 44:972–977

96. Titulaer HA, Zuidema J, Kager PA, Wetsteyn JC, Lugt CB, Merkus FW (1990) The pharmacokinetics of artemisinin after oral, intramuscular and rectal administration to volunteers. J Pharm Pharmacol 42:810–813
97. Karbwang J, Na-Bangchang K, Congpuong K, Molunto P, Thanavibul A (1997) Pharmacokinetics and bioavailability of oral and intramuscular artemether. Eur J Clin Pharmacol 52:307–310
98. Artemether-Quinine Meta-analysis Study Group (2001) A meta-analysis using individual patient data of trials comparing artemether with quinine in the treatment of severe falciparum malaria. Trans R Soc Trop Med Hyg 95:637–650
99. Murphy SA, Mberu E, Muhia D, English M, Crawley J, Waruiru C, Lowe B, Newton CR, Winstanley P, Marsh K et al (1997) The disposition of intramuscular artemether in children with cerebral malaria; a preliminary study. Trans R Soc Trop Med Hyg 91:331–334
100. Simonsson US, Jansson B, Hai TN, Huong DX, Tybring G, Ashton M (2003) Artemisinin autoinduction is caused by involvement of cytochrome P450 2B6 but not 2 C9. Clin Pharmacol Ther 74:32–43
101. Czock D, Markert C, Hartman B, Keller F (2009) Pharmacokinetics and pharmacodynamics of antimicrobial drugs. Expert Opin Drug Metab Toxicol 5:475–487
102. Gautam A, Ahmed T, Batra V, Paliwal J (2009) Pharmacokinetics and pharmacodynamics of endoperoxide antimalarials. Curr Drug Metab 10:289–306
103. Alin MH, Ashton M, Kihamia CM, Mtey GJ, Bjorkman A (1996) Clinical efficacy and pharmacokinetics of artemisinin monotherapy and in combination with mefloquine in patients with falciparum malaria. Br J Clin Pharmacol 41:587–592
104. Angus BJ, Thaiaporn I, Chanthapadith K, Suputtamongkol Y, White NJ (2002) Oral artesunate dose-response relationship in acute falciparum malaria. Antimicrob Agents Chemother 46:778–782
105. Li QG, Mog SR, Si YZ, Kyle DE, Gettayacamin M, Milhous WK (2002) Neurotoxicity and efficacy of arteether related to its exposure times and exposure levels in rodents. Am J Trop Med Hyg 66:516–525
106. Gordi T, Lepist EI (2004) Artemisinin derivatives: toxic for laboratory animals, safe for humans? Toxicol Lett 147:99–107
107. Genovese RF, Newman DB, Brewer TG (2000) Behavioral and neural toxicity of the artemisinin antimalarial, arteether, but not artesunate and artelinate, in rats. Pharmacol Biochem Behav 67:37–44
108. Karunajeewa HA, Mueller I, Senn M, Lin E, Law I, Gomorrai PS, Oa O, Griffin S, Kotab K, Suano P et al (2008) A trial of combination antimalarial therapies in children from Papua New Guinea. N Engl J Med 359:2545–2557
109. Hien TT, White NJ (1993) Qinghaosu. Lancet 341:603–608
110. White NJ (2003) Malaria. In: Cook G, Zumla A (eds) Manson's tropical diseases. Elsevier, Amsterdam, pp 1205–1296
111. Karunajeewa HA, Kemiki A, Alpers MP, Lorry K, Batty KT, Ilett KF, Davis TM (2003) Safety and therapeutic efficacy of artesunate suppositories for treatment of malaria in children in Papua New Guinea. Pediatr Infect Dis J 22:251–256
112. Karunajeewa HA, Reeder J, Lorry K, Dabod E, Hamzah J, Page-Sharp M, Chiswell GM, Ilett KF, Davis TM (2006) Artesunate suppositories versus intramuscular artemether for treatment of severe malaria in children in Papua New Guinea. Antimicrob Agents Chemother 50:968–974
113. Borrmann S, Adegnika AA, Moussavou F, Oyakhirome S, Esser G, Matsiegui PB, Ramharter M, Lundgren I, Kombila M, Issifou S et al (2005) Short-course regimens of artesunate-fosmidomycin in treatment of uncomplicated *Plasmodium falciparum* malaria. Antimicrob Agents Chemother 49:3749–3754
114. Price RN, Nosten F, Luxemburger C, ter Kuile FO, Paiphun L, Chongsuphajaisiddhi T, White NJ (1996) Effects of artemisinin derivatives on malaria transmissibility. Lancet 347:1654–1658

115. Price R, van Vugt M, Phaipun L, Luxemburger C, Simpson J, McGready R, ter Kuile F, Kham A, Chongsuphajaisiddhi T, White NJ et al (1999) Adverse effects in patients with acute falciparum malaria treated with artemisinin derivatives. Am J Trop Med Hyg 60:547–555
116. Bethell D, Se Y, Lon C, Socheat D, Saunders D, Teja-Isavadharm P, Khemawoot P, Darapiseth S, Lin J, Sriwichai S et al (2010) Dose-dependent risk of neutropenia after 7-day courses of artesunate monotherapy in Cambodian patients with acute *Plasmodium falciparum* malaria. Clin Infect Dis 51:e105–e114
117. Brewer TG, Peggins JO, Grate SJ, Petras JM, Levine BS, Weina PJ, Swearengen J, Heiffer MH, Schuster BG (1994) Neurotoxicity in animals due to arteether and artemether. Trans R Soc Trop Med Hyg 88:33–36
118. Brewer TG, Grate SJ, Peggins JO, Weina PJ, Petras JM, Levine BS, Heiffer MH, Schuster BG (1994) Fatal neurotoxicity of arteether and artemether. Am J Trop Med Hyg 51:251–259
119. Miller LG, Panosian CB (1997) Ataxia and slurred speech after artesunate treatment for falciparum malaria. N Engl J Med 336:1328
120. Toovey S, Jamieson A (2004) Audiometric changes associated with the treatment of uncomplicated falciparum malaria with co-artemether. Trans R Soc Trop Med Hyg 98:261–267
121. Kissinger E, Hien TT, Hung NT, Nam ND, Tuyen NL, Dinh BV, Mann C, Phu NH, Loc PP, Simpson JA et al (2000) Clinical and neurophysiological study of the effects of multiple doses of artemisinin on brain-stem function in Vietnamese patients. Am J Trop Med Hyg 63:48–55
122. Davis TM, Edwards GO, McCarthy JS (1997) Artesunate and cerebellar dysfunction in falciparum malaria. N Engl J Med 337:792; author reply 793
123. Davis TM, Binh TQ, Ilett KF, Batty KT, Phuong HL, Chiswell GM, Phuong VD, Agus C (2003) Penetration of dihydroartemisinin into cerebrospinal fluid after administration of intravenous artesunate in severe falciparum malaria. Antimicrob Agents Chemother 47:368–370
124. Johann-Liang R, Albrecht R (2003) Safety evaluations of drugs containing artemisinin derivatives for the treatment of malaria. Clin Infect Dis 36:1626–1627; author reply 1627–1628
125. Clark RL (2009) Embryotoxicity of the artemisinin antimalarials and potential consequences for use in women in the first trimester. Reprod Toxicol 28(3):285–96
126. McGready R, Cho T, Keo NK, Thwai KL, Villegas L, Looareesuwan S, White NJ, Nosten F (2001) Artemisinin antimalarials in pregnancy: a prospective treatment study of 539 episodes of multidrug-resistant *Plasmodium falciparum*. Clin Infect Dis 33:2009–2016
127. Hinton RL, Auwun A, Pongua G, Oa O, Davis TM, Karunajeewa HA, Reeder JC (2007) Caregivers' acceptance of using artesunate suppositories for treating childhood malaria in papua new Guinea. Am J Trop Med Hyg 76:634–640
128. Giao PT, Binh TQ, Kager PA, Long HP, Van Thang N, Van Nam N, de Vries PJ (2001) Artemisinin for treatment of uncomplicated falciparum malaria: is there a place for monotherapy? Am J Trop Med Hyg 65:690–695
129. Davis TM, Karunajeewa HA, Ilett KF (2005) Artemisinin-based combination therapies for uncomplicated malaria. Med J Aust 182:181–185
130. Adjuik M, Babiker A, Garner P, Olliaro P, Taylor W, White N (2004) Artesunate combinations for treatment of malaria: meta-analysis. Lancet 363:9–17
131. Omari AAA, Preston C, Garner PA (2005) Artemether-lumefantrine (six-dose regimen) for treating uncomplicated falciparum malaria. Cochrane Database Syst Rev CD005564
132. Ashley EA, Stepniewska K, Lindegardh N, Annerberg A, Kham A, Brockman A, Singhasivanon P, White NJ, Nosten F (2007) How much fat is necessary to optimize lumefantrine oral bioavailability? Trop Med Int Health 12:195–200
133. Juma EA, Obonyo CO, Akhwale WS, Ogutu BR (2008) A randomized, open-label, comparative efficacy trial of artemether-lumefantrine suspension versus artemether-lumefantrine tablets for treatment of uncomplicated *Plasmodium falciparum* malaria in children in western Kenya. Malar J 7:262

134. Karunajeewa H (2010) Piperaquine. In: Grayson ML (eds): Kucers' the use of antibiotics. Hodder Education, London
135. Winstanley P, Newton C, Watkins W, Mberu E, Ward S, Warn P, Mwangi I, Waruiru C, Pasvol G, Warrell D et al (1993) Towards optimal regimens of parenteral quinine for young African children with cerebral malaria: the importance of unbound quinine concentration. Trans R Soc Trop Med Hyg 87:201–206
136. White NJ, Miller KD, Churchill FC, Berry C, Brown J, Williams SB, Greenwood BM (1988) Chloroquine treatment of severe malaria in children. Pharmacokinetics, toxicity, and new dosage recommendations. N Engl J Med 319:1493–1500
137. Looareesuwan S, Phillips RE, White NJ, Karbwang J, Benjasurat Y, Attanath P, Warrell DA (1985) Intravenous amodiaquine and oral amodiaquine/erythromycin in the treatment of chloroquine-resistant falciparum malaria. Lancet 2:805–808
138. Roche RJ, Silamut K, Pukrittayakamee S, Looareesuwan S, Molunto P, Boonamrung S, White NJ (1990) Quinine induces reversible high-tone hearing loss. Br J Clin Pharmacol 29:780–782
139. Krishna S, White NJ (1996) Pharmacokinetics of quinine, chloroquine and amodiaquine. Clinical implications. Clin Pharmacokinet 30:263–299
140. Touze JE, Heno P, Fourcade L, Deharo JC, Thomas G, Bohan S, Paule P, Riviere P, Kouassi E, Buguet A (2002) The effects of antimalarial drugs on ventricular repolarization. Am J Trop Med Hyg 67:54–60
141. Taylor WR, White NJ (2004) Antimalarial drug toxicity: a review. Drug Saf 27:25–61
142. White NJ (1994) Clinical pharmacokinetics and pharmacodynamics of artemisinin and derivatives. Trans R Soc Trop Med Hyg 88:S41–S43
143. Karbwang J, Na-Bangchang K, Tin T, Sukontason K, Rimchala W, Harinasuta T (1998) Pharmacokinetics of intramuscular artemether in patients with severe falciparum malaria with or without acute renal failure. Br J Clin Pharmacol 45:597–600
144. Mithwani S, Aarons L, Kokwaro GO, Majid O, Muchohi S, Edwards G, Mohamed S, Marsh K, Watkins W (2004) Population pharmacokinetics of artemether and dihydroartemisinin following single intramuscular dosing of artemether in African children with severe falciparum malaria. Br J Clin Pharmacol 57:146–152
145. Gomes MF, Faiz MA, Gyapong JO, Warsame M, Agbenyega T, Babiker A, Baiden F, Yunus EB, Binka F, Clerk C et al (2009) Pre-referral rectal artesunate to prevent death and disability in severe malaria: a placebo-controlled trial. Lancet 373:557–566
146. TDR/GEN/01.5 Special Programme for Research and Training in Tropical Diseases. 15th Programme Report. Progress 1999–2000
147. White NJ (1997) Assessment of the pharmacodynamic properties of antimalarial drugs *in vivo*. Antimicrob Agents Chemother 41:1413–1422
148. Gomes M, Ribeiro I, Warsame M, Karunajeewa H, Petzold M (2008) Rectal artemisinins for malaria: a review of efficacy and safety from individual patient data in clinical studies. BMC Infect Dis 8:39
149. Karunajeewa HA, Ilett KF, Dufall K, Kemiki A, Bockarie M, Alpers MP, Barrett PH, Vicini P, Davis TM (2004) Disposition of artesunate and dihydroartemisinin after administration of artesunate suppositories in children from Papua New Guinea with uncomplicated malaria. Antimicrob Agents Chemother 48:2966–2972
150. de Ridder S, van der Kooy F, Verpoorte R (2008) Artemisia annua as a self-reliant treatment for malaria in developing countries. J Ethnopharmacol 120:302–314
151. Hommel M (2008) The future of artemisinins: natural, synthetic or recombinant? J Biol 7:38
152. Centre for Novel Agricultural Products DoB, University of York. The CNAP Artemisia Research Project: Project Update Number 9, May 2011. http://www.york.ac.uk/cnap/artemisaproject/pdfs/update-009.pdf. Accessed 15 May 2011

Second-Generation Peroxides: The OZs and Artemisone

Dejan M. Opsenica and Bogdan A. Šolaja

Abstract The emergence of multi-drug resistant strains of *Plasmodium falciparum* has rendered many affordable antimalarials, such as chloroquine, much less effective in addressing the severe health issues in sub-Saharan Africa, Southeast Asia and the Amazon region. In order to overcome the neurotoxicity of an initial series of artemisinin-derived drugs and their relatively high production costs, an intensive and all-inclusive research programme to develop new derivatives has been undertaken. Two efficient antimalarial drug candidates of different chemotype have been devised, the artemisinin derivative artemisone and 1,2,4-troxolane **OZ277**. Both are nontoxic, more potent than artemisinin and should be affordable to people of endemic regions. The same may hold for the backup candidates artemiside and **OZ439**.

1 Introduction

Great efforts have been expended over the last 30 years to discover new and efficacious endoperoxide antimalarials based on a natural product isolated from sweet wormwood – artemisinin, and related derivatives (collectively known as artemisinins) - primarily because, until very recently [1, 2], resistance to this type of drug had not been observed (Fig. 1) [3].

The main drawback of the initial series of artemisinin derivatives (Fig. 1) is their relative metabolic instability, i.e., susceptibility to hydrolysis (esters), or

D.M. Opsenica
Institute of Chemistry, Technology and Metallurgy, University of Belgrade, Studentski trg 16, Belgrade 11158, Serbia
e-mail: dopsen@chem.bg.ac.rs

B.A. Šolaja (✉)
Faculty of Chemistry, University of Belgrade, Studentski trg 16, P.O. Box 51, Belgrade 11158, Serbia
e-mail: bsolaja@chem.bg.ac.rs

H.M. Staines and S. Krishna (eds.), *Treatment and Prevention of Malaria*,
DOI 10.1007/978-3-0346-0480-2_10,

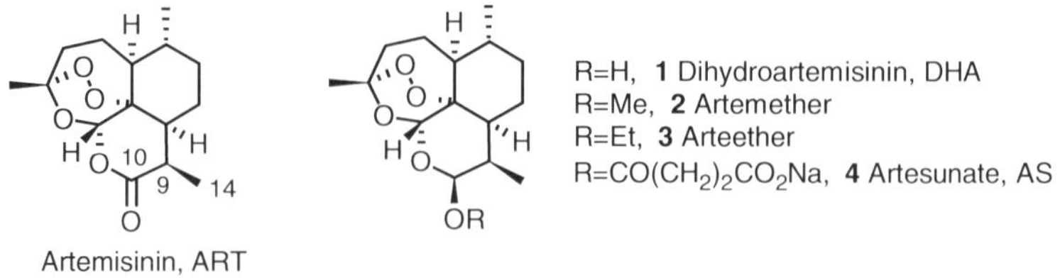

Fig. 1 Artemisinin and its early derivatives

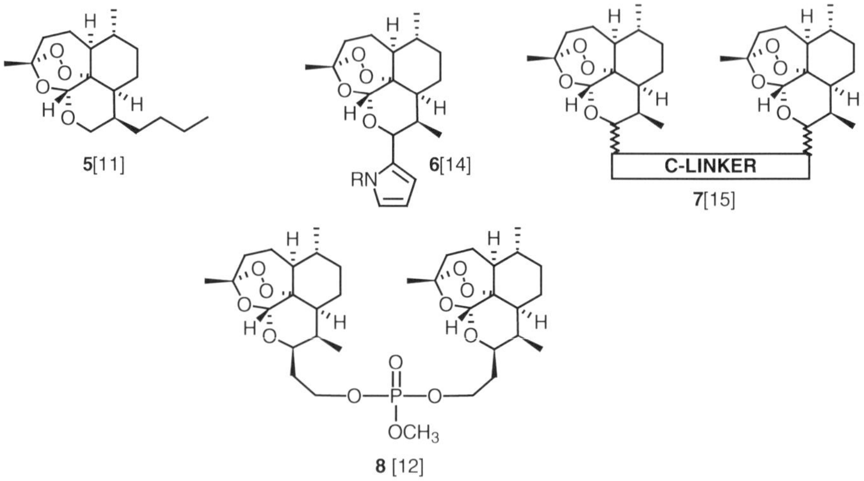

Fig. 2 Semi-synthetic artemisinins: C(14) and C(10) carba derivatives

oxidative-dealkylation (ethers) to dihydroartemisinin [4]. Dihydroartemisinin, apart from being a more potent antimalarial than artemisinin [5], was found to be neurotoxic in vitro [6, 7] and in mice [8]. However, similar neurotoxicity has not been observed in humans [9]. Although embryotoxicity has been detected within a narrow window of embryogenesis, this has not been convincingly observed in clinical trials from 1,837 pregnant women [10].

These potential toxicity issues provoked an extensive search for metabolically stable artemisinin derivatives and two major semi-synthetic artemisinin sub-classes were developed: C(10) deoxyartemisinin derivatives usually functionalized at C(9) or at C(14) (**5**) [11], and derivatives possessing a new C(10)-NR, or C(10)-C bond (**6**–**8**) [12–15]. The second sub-class is more diversified with artemisinin-derived dimers and artemisone (**9e**) as the most prominent members (Figs. 2 and 3).

The quest for efficacious and cheap antimalarial drugs that can be administered as a single dose, preferably orally, is not limited to natural product-derived semi-synthetic artemisinins, like artemisone, but also to other classes of peroxide antimalarials. Of these, the most prominent peroxide classes are the fully synthetic 1,2,4-trioxanes[16, 17], 1,2,4-trioxolanes (ozonides, OZs) and mixed 1,2,4,5-

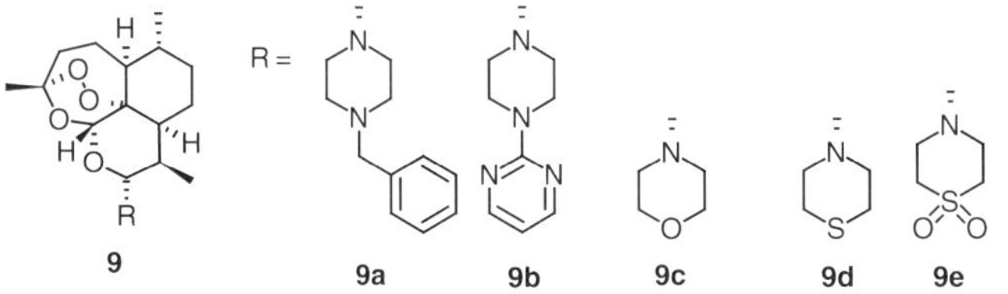

	ED_{90} (mg/kg) s.c.	p.o.
9a	0.78[a]	2.40[a]
9b	0.46[a]	1.50[a]
9c	0.18[a]	1.30[a]
9d	0.51[b]	1.90[b]
9e	1.50[b]	3.10[b]
4	4.60[a]	9.30[a]

[a] Data taken from ref. 22; [b] Data taken from ref. 21.

Fig. 3 Structures and antimalarial activities of derivatives **9**

tetraoxanes [18–20]. Two representatives of this class that have entered studies in humans are discussed below in detail.

2 Artemisone (Artemifon)

With the aim of preventing metabolic transformation of artemisinins to dihydroartemisinin (**1**, Fig. 1) and the drive to improve pharmacokinetic characteristics, several 10-(alkylamino)artemisinins **9** were designed (Fig. 3) [21, 22]. All the tested compounds showed excellent in vivo activities against *P. berghei* (**9c** was the most active derivative, almost 25 times (given *s.c.*) and 7 times (given *p.o.*) more active than artesunate **4**). Unfortunately, **9c** suffers from serious neurotoxicity issues even at low doses, thus indicating, that more lipophilic artemisinins are more toxic [21, 23].

A detailed antimalarial efficacy and drug–drug interaction study of artemisone (a drug examined recently in clinical Phase II trials [24]) was performed (**9e**, Fig. 3) [25]. It was shown that the in vitro antimalarial activities of artemisone against 12 different *P. falciparum* strains were comparable (exhibiting a mean IC_{50} value of 0.83 nM), independent of their drug-susceptibility profile to other antimalarial drug classes [21, 25]. During examination of the in vitro drug–drug interaction against drug sensitive 3D7 and multi-drug resistant K1 strains, it was noticed that artemisone showed slight antagonistic effects with chloroquine, amodiaquine, tafenoquine, atovaquone or pyrimethamine, and slight synergism with mefloquine. In vivo screening using the 4-day Peters test against drug-susceptible (NY), primaquine-resistant (P) and sulphadoxine/pyrimethamine-resistant (KFY) lines of *P. berghei*, chloroquine-resistant (NS) and artemisinin-resistant lines of *P. yoelii* NS and drug-susceptible *P. chabaudi* (AS) showed artemisone has superior ED_{50} and ED_{90} activity in comparison with artesunate (**4**, Fig. 1). Artemisone exhibited a 7-times greater activity (lower dosage) than artesunate (artemisone $ED_{90} = 12.13$ mg/kg vs. artesunate $ED_{90} = 87.50$ mg/kg) against the *P. yoelii* artemisinin-resistant line. The above results appear quite important in light of the recent in vitro isolation of artemether-resistant *P. falciparum* strains from humans [26] and

9e*

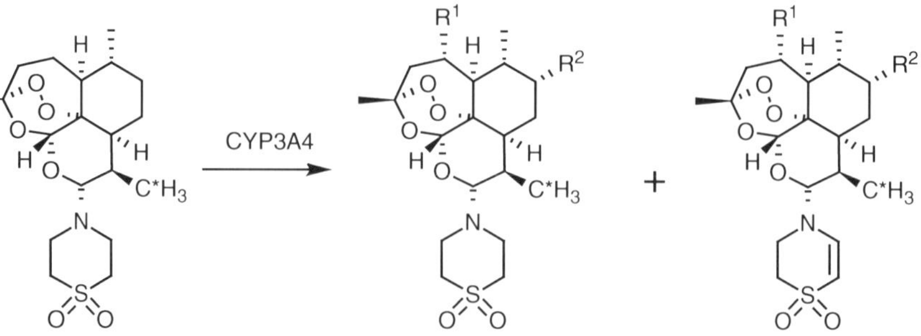

10: R^1 = OH, R^2 = H
11: R^1 = H, R^2 = OH

12: R^1 = R^2 = H
13: R^1 = OH, R^2 = H
14: R^1 = H, R^2 = OH

Scheme 1 Metabolic transformations of artemisone

emerging evidence for resistance in vivo [1, 2], suggesting more potent derivatives may be efficacious where first generation compounds are failing.

During in vivo drug–drug interaction examinations against the drug-susceptible *P. berghei* NY and the mefloquine-resistant *P. berghei* N1100 lines, artemisone showed synergism with mefloquine against both parasite lines. In combination with chloroquine, no interaction against drug-susceptible *P. berghei* NY parasites was detected; however, a synergistic effect against the chloroquine-resistant line, *P. yoelii* NS, was observed.

No dihydroartemisinin had been produced after 30 min, when isotopically labelled artemisone **9e*** (Scheme 1) was incubated with human liver microsomes. Only dehydrogenated **12** and mono-hydroxylated metabolites **10** and **11** and **13** and **14**, with *syn*-hydroxyl and peroxide groups were observed (Scheme 1) [21], clearly distinguishing artemisone and similar compounds from the first-generation artemisinins.[1] Incubation with microsomes and 14 recombinant CYP isoforms, together with selective inhibitors of CYP, showed that artemisone was primarily metabolised by recombinant CYP3A4. Interestingly, artemisinin induces its own metabolism [27] and is metabolised principally by CYP2B6 [28]. These results indicate that artemisone and artemisinin, in spite of being structurally similar, have a different metabolic profile in *P. falciparum;* therefore, it is possible that they can exert their antimalarial activity through similar but not identical mechanisms (vide infra). Isolated artemisone metabolites were tested against the *P. falciparum* K1 strain and were also found to be potent antimalarials, with **11** and **12** being the most active (**11**: $IC_{50} = 5.51$ nM and **12**: $IC_{50} = 4.26$ nM, artemisone: $IC_{50} = 1.99$ nM) [21, 29].

[1] However, dihydroartemisinin was detected in plasma during assessment of the safety of artemisone [27]. The concentrations were low, with geometric mean C_{max} values of 10 ng/ml after an 80 mg dose.

In preclinical studies, artemisone showed enhanced efficacy and improved pharmacokinetics in comparison with artesunate and did not demonstrate neurotoxicity in vitro and in vivo [21, 29], which is a characteristic of the current artemisinin derivatives used in clinical treatment [30]. Administration of artemisone as a single dose (10–80 mg) or multiple doses (40 mg or 80 mg given once daily for 3 days) was well tolerated. It appears that artemisone is devoid of time-dependent pharmacokinetics (unlike artemisinin and artemether), with comparable C_{max}, AUC and $t_{½}$ values after the first and third doses following the 3-day courses of artemisone. Although not being so active in vitro as the parent drug, the relatively high concentrations of metabolites obtained after artemisone administration probably add to the overall parasiticidal effect of artemisone [29]. In vivo testing on non-immune *Aotus* monkeys infected with *P. falciparum* showed that a single dose of artemisone (10 mg/kg) in combination with single doses of mefloquine (12.5 mg/kg or 5 mg/kg) cleared parasitaemia by day 1, with complete cure for all four monkeys tested [31]. With a single dose of mefloquine (2.5 mg/kg) parasitaemia was cleared by day 1 but without cure. For 3 days of treatment with a combination of artemisone (10 mg/kg/day) and amodiaquine (20 mg/kg/day), all three monkeys tested were cured, in contrast to those that were administered with the individual drugs for 3 days. From this study, it is clear that various total dosages of artemisone (20–90 mg/kg) alone, administered over 1–3 days, were unable to cure non-immune *Aotus* monkeys infected with *P. falciparum*. However, cure can be achieved when artemisone is combined with a single, sub-curative dose of mefloquine, or with a 3-day treatment course of amodiaquine (or clindamycin).

Efficacy of several artemisinin derivatives was examined for the treatment of murine cerebral malaria, CM, ECM [32]. It was shown that artemisone and artemiside (**9d**, Fig. 3) were more effective in the treatment of ICR and C57BL mice in comparison with dihydroartemisinin or artesunate. In all experiments performed on *P. berghei*-infected mice, treatment with artemisone led to complete cure (at least parasitaemia was reduced to an undetectable level). The observed recrudescent parasites were successfully eradicated by repeating the treatment with artemisone without selecting for parasite resistance. In addition, it was shown that complete cure of infected mice could be achieved even when treatment with artemisone was commenced at late stages of pathogenesis, 6 days post-infection. These results opened a considerable time window for adequate treatment, since human malaria is diagnosed after clinical symptoms are apparent. Following WHO's instructions for artemisinin combination therapy (ACT), the efficacy of artemisone–CQ combination was examined and it was shown that this combination was more successful than single therapy of both drugs individually. Combination of these two drugs prevented recrudescence and cured all mice (2×5 mg/kg/day artemisone + 15 mg/kg CQ). In this study, artemiside appeared to be even more successful then artemisone, but this derivative has to be submitted to more detailed preclinical toxicological evaluation. However, the preliminary results suggest that artemiside may represent a new option for antimalarial therapy based on artemisinin derivatives [32].

2.1 *Possible Mechanisms: Fe(II) Interaction and the Peroxide Bond*

In an effort to define more clearly a reasonable mechanism for the action of artemisone and related derivatives, experiments with various Fe(II) salts were performed [33]. It was found that artemisone, unlike other examined aminoartemisinin derivatives **9a**, **9b** (Fig. 3) and **15–17** (Fig. 4), reacts with Fe (II) salts under aqueous conditions to afford the products **18–20**, which are structurally similar to the ones obtained from artemisinin and dihydroartemisinin. The relative ratio of products **18–20** (Scheme 2) depends on the employed salts and solvent mixtures. A reasonable mechanism was proposed for Fe(II)-induced generation of the products, including artemisone → **18**, **19** formation via iminium cations. Evidence that a primary C-radical is generated during Fe(II) peroxide scission came from the isolation of adduct **22** (Scheme 2), arising from reaction in the presence of the radical trapper 4-oxo-TEMPO. However, the yield of this

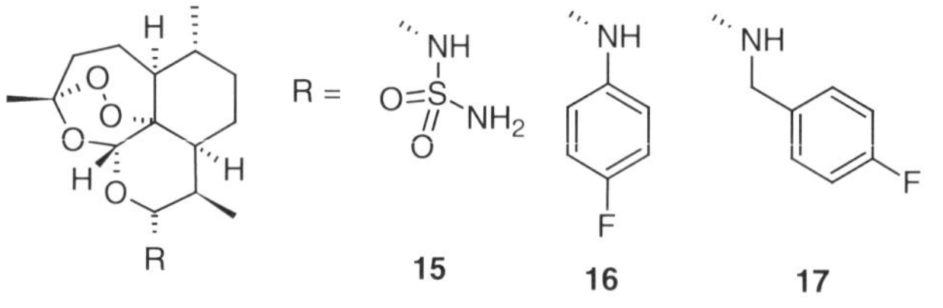

	IC_{50} (nM)[a] W2[b]	D10[b]
9a	0.66	1.17
9b	1.53	2.04
9e	0.60	1.17
15	3.03	7.89
16	0.82	2.07
ART	4.82	7.33

[a] Data taken from ref. 33; [b] W2 is *P. f.* CQR strain, D10 is *P. f.* CQS strain.

Fig. 4 Semi-synthetic artemisinins: C(10) aza derivatives

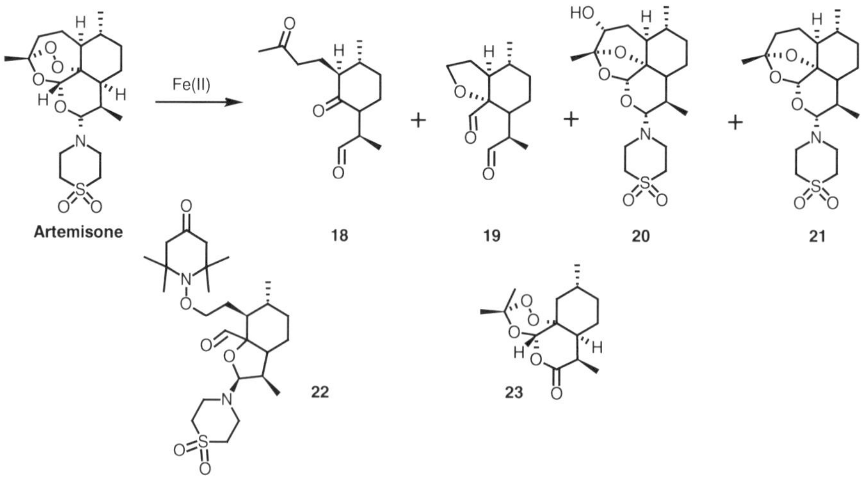

Scheme 2 Products **18–21** obtained after reaction of artemisone with various Fe(II) salts

product is low (10%) and the authors argued that the role of Fe(II) is during decomposition rather than activation of artemisone [33].

Others have shown that artemisone readily reacts with haem in vitro producing alkylated products [34] very similar to artemisinins [35]. The ability of artemisinin itself to alkylate haem has been confirmed in vitro [36, 37] and in vivo [38]. These data indicate that Fe(II) species are able to activate artemisinins, including artemisone, in their antimalarial action. However, it should be noted that the conditions used to determine alkylation by artemisinins have been questioned, in particular the use of dimethyl sulphoxide that produces a reducing environment [39, 40], and, currently, there is no evidence that artemisone alkylates haem in vivo. In addition, the theory of Fe(II)-initiated activation of artemisinins could not explain the pronounced antimalarial activity of 5-nor-4,5-seco-artemisinin **23** [41–43], which is clearly less capable of forming C radicals upon peroxide bridge cleavage by Fe(II). Another inconsistency in applying the general Fe(II) mechanism (vide supra) to artemisinin and artemisone, and their derivatives, arises from their activities in the presence of desferrioxamine B (DFO). DFO acts as a free radical scavenger (for hydroxyl and peroxyl radicals) and as a Fe(III) chelator and reduces artemisinin inhibition of PfATP6 (a proposed site of action; see below) [44]. In contrast, artemisone, which exhibits approximately the same reactivity with aqueous Fe(II) as artemisinin, retains inhibition potency towards PfATP6 in the presence of DFO [33]. This detail strongly indicates that although the compounds are of similar structure, they might not exert their antimalarial activity by sharing the same mechanism of action.

In order to understand better the manner in which artemisinins exert their antimalarial activity, extensive research was launched bearing in mind that cleavage of the peroxide bridge via an iron-dependent mechanism (as the sole mechanism) does not provide complete answers to observed peroxide drug behaviour. Based on the finding that the in vitro antimalarial activity of artemisinin was significantly increased under a 2% carbon monoxide atmosphere (by 40–50%) and under a 20% oxygen atmosphere (by 20–30%) [45], a detailed study of the interaction of artemisone and related derivatives with various forms of haemoglobin (Hb), haem, as well as an analysis of their antimalarial activity was performed [39]. In contrast to artemisinin, artemisone does not react with Hb–Fe(II) and oxyHb–Fe(II) to produce metHb–Fe(III); in addition, it was shown that both artemisinin and artemisone (and **9a**, **9b**, **16**) are inactive towards Hb–Fe(II) in a 2% carbon monoxide atmosphere. On the other hand, both compounds induced the oxidation of the haemoglobin catabolic product, haem–Fe(II), to produce haem–Fe (III). In addition, on exposure of haem–Fe(II) and Hb–Fe(II) to carbon monoxide, stable complexes were formed, which are also inactive for reactions with peroxide antimalarials. However, in a biologically relevant experiment, both compounds showed increased antimalarial activity against *P. falciparum* W2 strain under a 2% carbon monoxide atmosphere, while the activity of chloroquine remained unchanged. The authors suggested [39] that peroxide antimalarials behave as reactive oxygen species (ROS), or that they produce ROS via Haber–Weiss chemistry. Furthermore, it was concluded that passivation of haem–Fe(II) by its conversion into the CO complex results in decreased decomposition of artemisinins (artemisone included), thus making them more available for reaction with their actual target. The authors proposed that Fe(II), regardless of its origin (Hb–Fe(II) or haem–Fe(II)), is not the sole initiator of O–O scission, and suggested that another mechanism be sought [39].

2.2 *Possible Mechanisms: Primary Targets, Accumulation and Co-Factors*

The discovery that artemisinins target SERCA orthologues in *P. falciparum* (PfATP6) and *P. vivax* (PvSERCA) [44] shed new light and encouraged research towards the same and other novel targets for peroxide antimalarials. Thus, it was found that artemisone is more potent against plasmodial SERCA orthologues [$K_i = 1.7 \pm 0.6$ nM (PfATP6) and $K_i = 0.072 \pm 0.012$ nM (PvSERCA)], as compared with values for artemisinin [$K_i = 169 \pm 31$ nM (PfATP6) and $K_i = 7.7 \pm 4.9$ nM (PvSERCA)] [46].

In order to better understand the mechanism of the accumulation into infected and uninfected erythrocytes, radiolabelled artemisone **9e*** was submitted to uptake assays [47]. It was found that artemisone was accumulated in infected erythrocytes at significantly higher levels than in uninfected ones, by a saturable, competitive, time- and temperature-dependent mechanism. Most radioactivity was detected in pellet fractions which predominantly contain proteins suggesting that artemisone is probably associated with proteins [48]. The distribution of artemisone is dependent on the stage of parasite maturation: after incubation with immature parasites, which are highly susceptible to action of artemisinins in vitro, as well as in vivo, most artemisone was found in a Triton-soluble fraction, which contains proteins, DNA and RNA. In mature parasites most artemisone was found in the Triton-insoluble fraction, which predominantly contains haemozoin. This finding suggests that mature infected erythrocytes can act by removing the artemisinins. This was further supported by the observation that mature infected erythrocytes take up artemisone much faster than do the more sensitive ring-stage-infected erythrocytes.

Methylene blue (MB) was the first synthetic drug ever used in humans, and it was Paul Erlich who cured two patients from malaria using MB in 1891 [49]. The discovery that artemisinin and artemether exhibit synergic effects with methylene blue [50], unlike chloroaminoquinolines, initiated systematic examination of reactivity relationships between the drugs with the aim to possibly correlate their mechanism of antimalarial action. Subsequent research by the same group [51] revealed MB as a redox-cycling agent that produces H_2O_2 at the expense of O_2 and of NADPH in each cycle (Fig. 5). Thus, MB consumes NADPH and O_2 needed for the pathogen's metabolism, with probably serious consequences to the NADPH/NADP ratio.

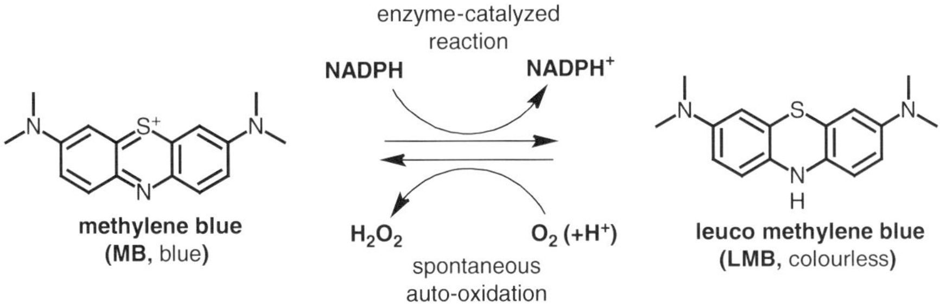

Fig. 5 Redox-cycling interconversion between methylene blue (MB) and leuco methylene blue (LMB)

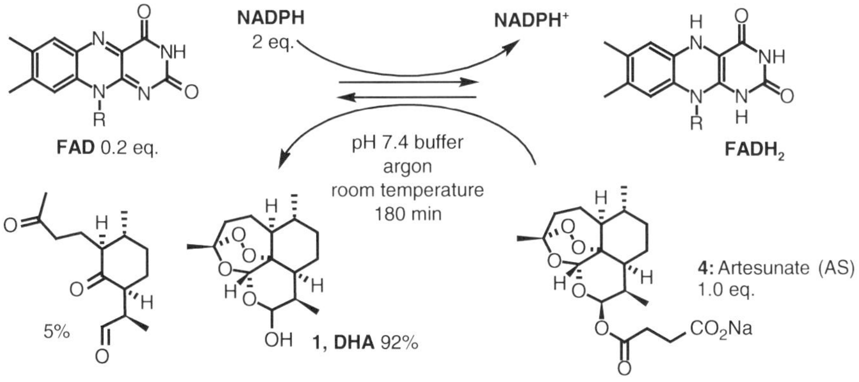

Scheme 3 Biomimetic catalytic system under physiological conditions: works with other flavins, all examined artemisinins, tetraoxane and trioxolane antimalarials

The results of subsequent research revealed that artemisone, other artemisinins [52] and other peroxide antimalarials like ozonides and tetraoxanes [53] are active in the presence of MB-ascorbic acid, MB–*N*-benzyl-1,4-dihydronicotinamide (BNAH), riboflavin–BNAH and riboflavin–NADPH systems and yield identical products to those that were isolated from the reaction of the same antimalarials with Fe(II). According to this proposal, antimalarial peroxides act as oxidants re-oxidising LMB and $FADH_2$, so contributing to depletion of NADPH (Scheme 3).

The observed results suggest that peroxide antimalarials disrupt a highly sensitive redox balance established in the parasite and thereby cause its death. The results also correlate well with the observed SAR of artemisinins, i.e., artemisinins that exhibited low antimalarial activity, like 9-epiartemisinin [54, 55], also show low reactivity under the applied reaction conditions. In addition, high antimalarial activity of 5-nor-4,5-seco-artemisinin **23** can easily be comprehended using the new mechanism proposal. The proposed explanation involves iron-free reactions; however, Fe(III) rapidly oxidises $FADH_2$, and thus contributes to redox cycling without interfering with artemisinins.

3 1,2,4-Trioxolanes (Ozonides)

1,2,4-Trioxolanes are a very well-known class of organic compounds. They are intermediates in the transformation of olefins into carbonyls during ozonolysis. It was an unexpected and surprising discovery [56] that ozonides are relatively stable and that many of them express excellent activity against malaria parasites, as do the structurally similar 1,2,4-trioxanes.

3.1 Development of 1,2,4-Trioxolanes: OZ209, OZ277 and OZ339

Preparation of 1,2,4-trioxolanes is relatively simple and relies on the Griesbaum co-ozonolysis of suitable methyl oximes and ketones (Scheme 4) [57–59]. This method provides a highly applicable synthetic approach to tetrasubstituted 1,2,4-trioxolanes (ozonides), otherwise accessible only with great difficulty by ozonolysis of the corresponding alkenes, or by other means.

The synthesis of the final OZ antimalarial drug significantly benefits from the stability of the ozonide peroxide bridge to reduction and alkylation conditions, as well as to other standard reaction conditions. This enables transformation of the initial co-ozonolysis products into a vast array of unsymmetrically substituted ozonides ($n > 500$!) [60]. Usually, the final 1,2,4-trioxolane antimalarials are prepared in 1–4 steps, depending on the modifications required [57–59, 61–63], affording the final products in yields up to 75%. The vast majority of these antimalarials are achiral, which greatly facilitates production in the developmental step. A substantial improvement in drug design in the antimalarial field was the development of the 4-substituted cyclohexyl-adamantyl ozonide (adamantane-2-spiro-3′-1′,2′,4′-trioxaspiro[4.5]decane) chemotype. The advantage of OZ antimalarials is the use of the adamantane moiety that has lipophilic functionality, allowing the opposite part of the molecule to be fine-tuned using a number of polar functional groups, preferably basic in nature [56, 58–62]. Many of the OZ compounds obtained in this way were active in all stages of development of the malaria parasite and are more active than artemether and artesunate, both in vivo and in vitro. All these findings, gathered during much experimentation, contributed to the discovery of amines **OZ209** (**24** as mesylate, Fig. 6) and **OZ277** (**25** as tosylate, Fig. 6) as the best drug candidates.

In comparison with other OZ compounds, trioxolanes **OZ209** and **OZ277** showed superior pharmacokinetic results, such as prolonged half-life and enhanced

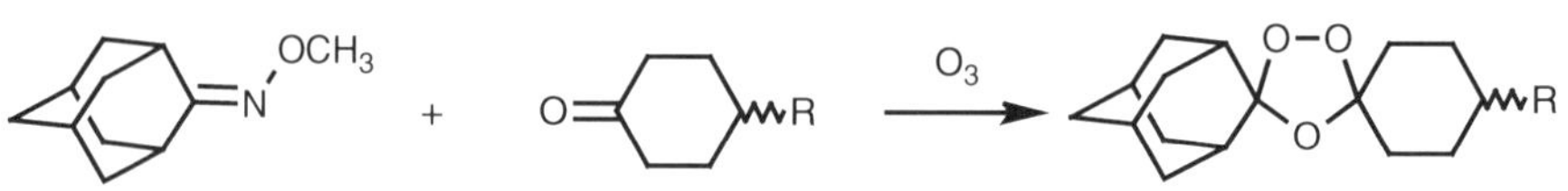

Scheme 4 Griesbaum co-ozonolysis: Ozonides

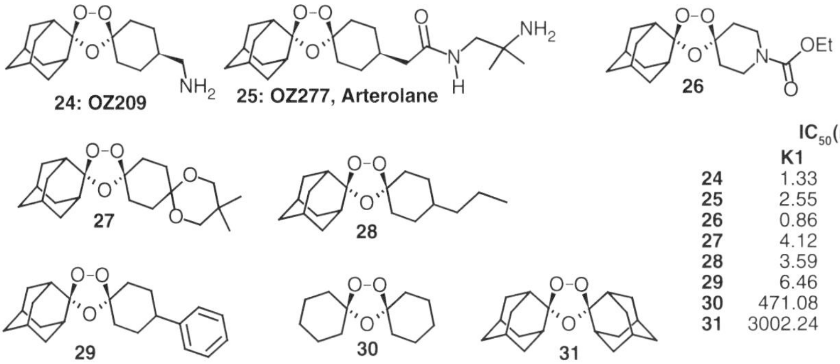

	IC$_{50}$(nM) K1	NF54
24	1.33	1.43
25	2.55	2.32
26	0.86	1.69
27	4.12	7.41
28	3.59	6.85
29	6.46	14.10
30	471.08	~2000
31	3002.24	2461.84

Fig. 6 1,2,4-Troxolanes **OZ209** and **OZ277**, and selected congeners. Compound **OZ277** has been advanced to Phase III clinical trials (vide infra)

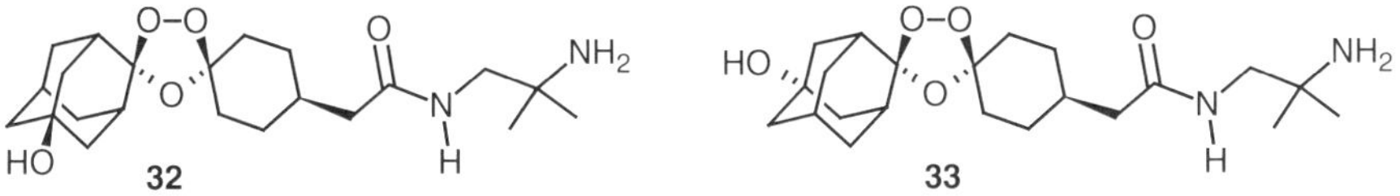

Fig. 7 Microsomal metabolites of ozonide **OZ277**

bioavailability after a single oral dose. Compound **OZ209** had somewhat better antimalarial results and a lower recrudescence level. However, **OZ277** was chosen as the development candidate, primarily because of its improved toxicological profile and reduced concentrations in brain tissue after oral dosing [56]. For example, 2 h after dosing, both **OZ209** and **OZ277** were distributed throughout the liver, kidney, lung and heart, while after 18 h, **OZ277** was detected only in the lungs and in several-fold lower concentrations than **OZ209**.

Unlike **OZ209**, which was quantified in brain tissue after both 2 h and 18 h, trioxolane **OZ277** could not be quantified in this organ at all. In view of potential neurotoxicity issues, these findings were taken as a considerable advantage of **OZ277** over **OZ209**. Trioxolane **OZ277** appeared quite stable to metabolic transformation ($t_{½} = 17$ h, *p.o.* in healthy rats) [56]. The metabolic profile of **OZ277** was studied with human liver microsomes and only two, monohydroxylated derivatives at the adamantane angular positions (**32** and **33**, Fig. 7), were identified as major metabolites, thus confirming the stability of the trioxolane moiety to metabolic transformation. Interestingly, both metabolites were inactive against the *P. falciparum* K1 strain (IC$_{50}$ > 100 ng/ml), thus demonstrating the indispensability of the unsubstituted spiro-adamantane moiety to the antimalarial activity of **OZ277** (IC$_{50}$ (K1) = 1.0 ng/ml) [64]. Unlike the artemisone products (Scheme 1), OZ metabolic derivatives **32** and **33** very probably lower the overall **OZ277** antimalarial activity. The other derivatives **26**–**29** (Fig. 5) afforded further insight into SAR in the context of the physico-chemical, biopharmaceutical and toxicological profiles of trioxolanes [61].

Recently, the same authors revealed data for a series of OZ compounds with weak base functional groups, which were responsible for a high antimalarial efficacy in *P. berghei*-infected mice [65]. Their antimalarial efficacy and ADME profiles are equal or superior to **OZ277**. One of the most promising is **OZ339** (as tosylate salt). The two trioxolanes, **OZ339** and **OZ277** are evaluated in Table 1, with artesunate added for comparison. Despite the obvious difference in in vitro activity, both ozonides eradicate parasitaemia below the detectable level 1 day after administration (99.9%, 1 $\times$10 mg/kg, and 3 $\times$ 3 mg/kg). The drug candidate **OZ277** is a powerful fast-acting antimalarial with a 67% cure record at a 3 $\times$ 10 mg/kg dosage (mice) [56]. However, at a 3 $\times$ 3 mg/kg dosage, the same compound cured no mice, while trioxolane **OZ339** cured 3/5 mice with an excellent survival time of 27 days (**OZ277** had a 2.4 times lower survival time). These good pharmacokinetic characteristics are additionally enhanced by the favourable bioavailability data for **OZ339** (78%, Table 1). In all experiments, artesunate showed inferior activity. Inhibition assays revealed that **OZ339**, like **OZ277**, did not inhibit CYP3A4, CYP2C9 and 2D6CYP450 at concentrations up to 50 μM. Finally, preliminary toxicological experiments indicated that **OZ339** was minimally toxic (liver) and, similar to **OZ277**, demonstrated no detectable signs of neurotoxicity.

As mentioned above, the tolerance of the 1,2,4-trioxolane moiety to diverse reaction conditions [57] and resistance to metabolic transformation [64] enabled the synthesis of a significant number of derivatives and many of them showed very good antimalarial activity, e.g., derivatives **34**–**38** (Fig. 8) [66], and derivatives which contain aliphatic and aromatic amino functional groups or azole heterocycles as substituents (**39**–**45**) (Fig. 8) [62].

The lack of activity of trioxolane **46** [62], and the isolation of inactive hydroxylated **OZ277** metabolites [64], point to the essential contribution of an unsubstituted spiro-adamantane system to the antimalarial properties of this class of compounds.

Many of the examined derivatives exhibited excellent in vitro results, but failed during in vivo tests, toxicity trials or metabolic stability and bioavailability tests. More lipophilic trioxolanes tend to have better oral activities and are metabolically less stable than their more polar counterparts. Such behaviour is consistent with results obtained for other classes of synthetic peroxides. Trioxolanes with a wide range of neutral and basic groups had good antimalarial profiles, unlike derivatives with acidic groups. Based on the collected extensive screening results, the authors concluded that in vitro activities of 1,2,4-trioxolanes are not (always) a reliable predictor of in vivo potency [66]. Rather, their experiments in *P. berghei*-infected mice confirmed that in vivo results were essential for compound differentiation and selection for further metabolic and pharmacokinetic profiling [65].

Trioxolane **OZ277** alkylates haem (Fig. 9) [67], and its in vitro activity against *P. falciparum* is antagonised by DFO [68]. In vitro, artesunate and **OZ277** act antagonistically against *P. falciparum*. These findings, together with only weak interaction with the proposed artemisinin target PfATP6 [44], unlike artemisone [33], suggest that interaction with food vacuole-generated haem is probably how trioxolanes are activated. Further support can be found in the fact that the

Table 1 Comparative data for fast-acting antimalarials: **33** (**OZ339**), **24** (**OZ277**) and sodium **4** (**AS**)

Compound	IC_{50} (ng/mL) K1	IC_{50} (ng/mL) NF54	Activity (1 × 10 mg/kg, *p.o.* %)[b]	ER[c]	In vivo activity (3 × 3 mg/kg, *p.o.*)[a] Activity (%)	In vivo activity Survival (days)	In vivo activity Cure[d] (%)	$t½$ (*i.v.*, min)	V_d (*i.v.*, L/kg)	Bioavailability (*p.o.*, %)
34 (**OZ339**)[e,f]	0.35	0.39	>99.9	0.45	>99.9	27.0	60	83	19	78
25 (**OZ277**)[e,f]	1.0	0.91	99.9	0.32	>99.9	11.4	0	76	16	26
4 (**AS**)[f]	1.3	1.6	67	0.43	70	9.2	0	40 (DHA)[g]	3.0 (DHA)	N.D.

N.D. not dosed

[a]Groups of five *P. berghei*-infected NMRI mice were treated orally on days +1, +2, and + 3 (3 × 3 mg/kg). Activity measured on day +4

[b]Groups of five *P. berghei*-infected NMRI mice were treated orally on day +1 (1 × 10 mg/kg). Activity measured on day +2

[c]Predicted hepatic extraction ratios (ERs) using human microsomes

[d]Percentage of mice alive on day 30 with no evidence of blood parasites

[e]Tosylate salt

[f]Taken from [61]

[g]Taken from [58, 59]

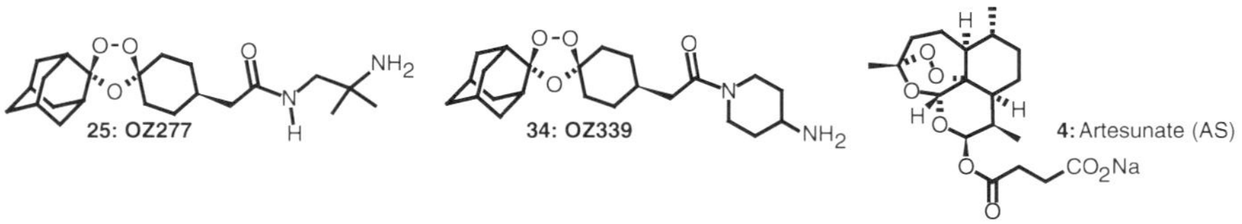

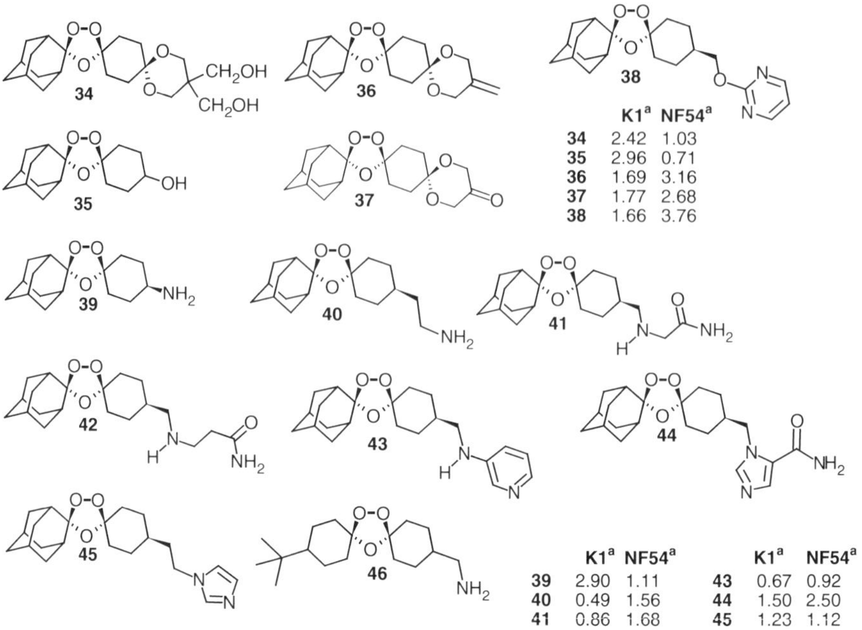

	K1[a]	NF54[a]
34	2.42	1.03
35	2.96	0.71
36	1.69	3.16
37	1.77	2.68
38	1.66	3.76

	K1[a]	NF54[a]
39	2.90	1.11
40	0.49	1.56
41	0.86	1.68
42	1.12	2.50

	K1[a]	NF54[a]
43	0.67	0.92
44	1.50	2.50
45	1.23	1.12
46	inactive	

[a] Activities are represented as IC_{50} values, expressed in nM

Fig. 8 Structures of ozonides **33–45** with their in vitro antimalarial activity

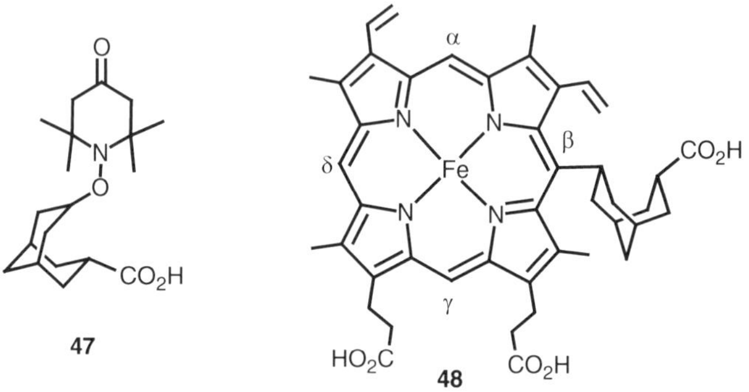

Fig. 9 Structures of adducts of the secondary radical derived from **OZ277** with 4-oxo-TEMPO and haem

stereochemistry of a given compound has little effect on the in vitro potency of trioxolane antimalarials, thereby strongly pointing to the interaction of an antimalarial peroxide (chiral or achiral) with an achiral target (haem). The selectivity of trioxolanes towards infected and not-healthy erythrocytes may be explained based on their reactivity towards free haem and stability in the presence of oxy- and deoxyHb [69].

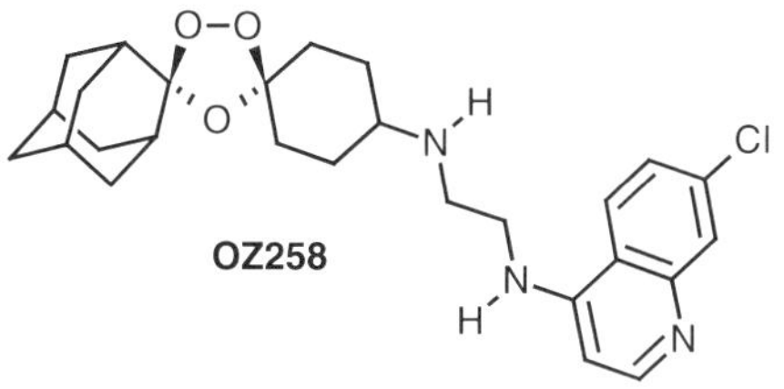

IC_{50} (nM)		% Activity (10 mg/kg)	
K1	NF54	p.o.	s.c.
11.98	10.74	75	99.95

Fig. 10 Chimaeric trioxolane **OZ258**

Based on the concept that compounds with two integrated pharmacophores might have enhanced activity [70], the chimaeric trioxolane **OZ258** was prepared (Fig. 10) [58, 59]. Although it is very active in vitro against the K1 and NF54 strains of *P. falciparum* and in vivo against the ANKA strain of *P. berghei*, **OZ258** did not achieve the synergic effect of two pharmacophores, especially when compared with trioxolanes **39** and **43**. The same holds for other chloroaminoquinoline and acridine chimaeras [71].

Very often, promising peroxide drugs eradicate parasitaemia quickly, which is crucial for rapid treatment of life-threatening cerebral malaria, and this property is inherently protective against the development of resistance. Since the drugs are typically administered for only a few days and they have short half-lives, the recrudescence of malaria parasites occurs frequently (artemisone cf. [24, 32]; **OZ277** cf. [58]). In an attempt to overcome this problem, artemisinin-based combination therapies (ACTs) are recommended (as indicated for artemisone, see above) by the WHO. The WHO currently distributes, under a no-loss and no-profit agreement with Novartis, the fixed-dose ACT drug Coartem® (artemether 20 mg/ lumefantrine 120 mg) [72]. The drug has been recently approved by the FDA for the treatment of acute, uncomplicated malaria infections [73]. Although each ACT is specific [73, 74], the following concept can be applied to all: antimalarial peroxides eliminate most of the infection and the remaining parasites are then exposed to high concentrations of the slow-acting partner drugs; because of the rapid reduction in parasites, the selective pressure for the emergence of mutant parasites is greatly reduced. In accord, **OZ277** (RBX-11160) entered Phase III clinical trials in combination with piperaquine (arterolane maleate + piperaquine phosphate) [75].

3.2 The Second Generation of 1,2,4-Trioxolane Drug Candidates: OZ439

In Phase I clinical trials, the half-life of **OZ277** in healthy volunteers was only about two- to threefold longer than that of dihydroartemisinin. OZ277's possible first-generation ozonide alternative, **OZ339**, only has a slightly higher *t*½ value

Table 2 Comparative data for first and second generation of ozonide antimalarials: **25** (**OZ277**) vs. **49** (**OZ439**)

Compound	IC_{50} (ng/mL)		Postinfection in vivo activity[a]		In vivo prophylactic activity (1 × 30 mg/kg, *p.o.*)[b]					
	K1	NF54	1 × 30 mg/kg, *p.o.* %	Cure[c]	Activity (%)	Survival (days)	Cure[c] (%)	$t_{1/2}$ (*p.o.*, min)	V_d (*i.v.*, L/kg)	Bioavailability (*p.o.*, %)
25 (**OZ277**)[d,e]	0.71	0.63	99.9	0	0	7	0	55	4	13
49 (**OZ439**)[e]	1.6	1.9	>99.9	100	99.8	>30	100	1380	15	76
4 (**AS**)[e]	1.2	1.5	92	0	21	7	0	40 (DHA)[f]	3.0 (DHA)	N.D.

N.D. not dosed
[a]Groups of *P. berghei*-infected ANKA mice ($n = 5$) were treated orally on day +1 (1 × 30 mg/kg). Activity measured on day +3
[b]Groups of *P. berghei*-infected ANKA mice ($n = 5$) were treated orally on day −1 (1 × 30 mg/kg). Activity measured on day +3
[c]Percentage of mice alive on day 30 with no evidence of blood parasites
[d]Tosylate salt
[e]Taken from [76]
[f]Intravenous, taken from [58, 59]

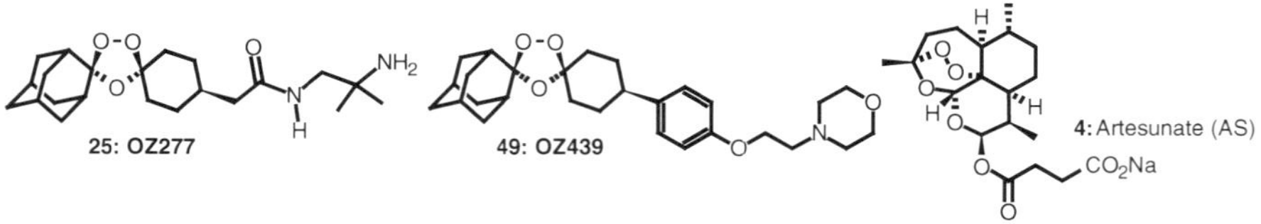

(Table 1); so the search for an ozonide with significantly increased half-life continued.

As a result, screening of the second generation of ozonide antimalarials has been completed, recently [76]. Of the several very active OZ compounds of undisclosed structure it appears that the most promising antimalarial candidate is **OZ439** (Table 2) [76]. Initial results indicate that this compound provides single-dose oral cure in a murine malaria model at 20 mg/kg, a situation not known for any of the peroxide antimalarials except for artelininc acid at >7 times higher dose [76]. The second-generation ozonide **OZ439** completed Phase I studies and is currently undergoing Phase IIa clinical trials. In accord with other ozonide antimalarials, **OZ439** is considered to be an Fe(II)-initiated pro-drug. However, it is >50-fold more stable to Fe(II)-mediated degradation compared with **OZ277** [76]. Consistent with proposed Fe(II) degradation of ozonides [67] is the significantly enhanced stability (15–20 times) of **OZ439** over **OZ277** in healthy and infected human and rat blood. This prolonged blood stability and improved pharmacokinetic characteristics (Table 2) led to the positioning of **OZ439** as the current major OZ drug candidate – with respect to post-infection cure (3 × 5 mg/kg/day and 20 mg/kg single dose), and exclusive prophylactic characteristics (Table 2). The absence of metabolic products significantly contributes to overall activity (prophylactic and post-infection) of **OZ439** relative to other peroxide antimalarial drugs [76].

To conclude, as a consequence of intensive and comprehensive research, efficient antimalarial drug candidates of different chemotype have been devised: artemisone and 1,2,4-trioxolane **OZ277**. They are nontoxic, effective at small doses and very probably inexpensive to produce[2] [77]. The same may hold for a prospective backup candidate artemiside and the newest breakthrough drug candidate **OZ439**. It would be of benefit if their combination partner would cure malaria through different mechanisms, since resistance is then less likely to occur.

References

1. Dondorp AM, Nosten F, Yi P, Das D, Phyo AP, Tarning J, Lwin KM, Ariey F, Hanpithakpong W, Lee SJ et al (2009) Artemisinin resistance in *Plasmodium falciparum* malaria. N Engl J Med 361:455–467
2. Noedl H, Se Y, Schaecher K, Smith BL, Socheat D, Fukuda MM (2008) Evidence of artemisinin-resistant malaria in western Cambodia. N Engl J Med 359:2619–2620
3. Jefford CW (2007) New developments in synthetic peroxidic drugs as artemisinin mimics. Drug Discov Today 12:487–495

[2] The annual demand for artemisinin as a starting material for transformation into semi-synthetic products amounts to ca. 114 tonnes for Coartem® production only. Since synthesis of artemisinin is uneconomical currently, Novartis initiated an increase of the agricultural cultivation of *Artemisia annua* in Kenya, Tanzania and Uganda and extraction of artemisinin therefrom, in addition to Chinese supplies of artemisinin.

4. Lin AJ, Miller RE (1995) Antimalarial activity of new dihydroartemisinin derivatives. 6. alpha-Alkylbenzylic ethers. J Med Chem 38:764–770
5. Lin AJ, Lee M, Klayman DL (1989) Antimalarial activity of new water-soluble dihydroartemisinin derivatives. 2. Stereospecificity of the ether side chain. J Med Chem 32:1249–1252
6. Fishwick J, McLean WG, Edwards G, Ward SA (1995) The toxicity of artemisinin and related compounds on neuronal and glial cells in culture. Chem Biol Interact 96:263–271
7. Wesche DL, DeCoster MA, Tortella FC, Brewer TG (1994) Neurotoxicity of artemisinin analogs *in vitro*. Antimicrob Agents Chemother 38:1813–1819
8. Nontprasert A, Pukrittayakamee S, Prakongpan S, Supanaranond W, Looareesuwan S, White NJ (2002) Assessment of the neurotoxicity of oral dihydroartemisinin in mice. Trans R Soc Trop Med Hyg 96:99–101
9. Davis TM, Binh TQ, Ilett KF, Batty KT, Phuong HL, Chiswell GM, Phuong VD, Agus C (2003) Penetration of dihydroartemisinin into cerebrospinal fluid after administration of intravenous artesunate in severe falciparum malaria. Antimicrob Agents Chemother 47:368–370
10. Li Q, Weina PJ (2010) Severe embryotoxicity of artemisinin derivatives in experimental animals, but possibly safe in pregnant women. Molecules 15:40–57
11. Avery MA, Alvim-Gaston M, Vroman JA, Wu B, Ager A, Peters W, Robinson BL, Charman W (2002) Structure-activity relationships of the antimalarial agent artemisinin. 7. Direct modification of (+)-artemisinin and *in vivo* antimalarial screening of new, potential preclinical antimalarial candidates. J Med Chem 45:4321–4335
12. Chadwick J, Mercer AE, Park BK, Cosstick R, O'Neill PM (2009) Synthesis and biological evaluation of extraordinarily potent C-10 carba artemisinin dimers against *P. falciparum* malaria parasites and HL-60 cancer cells. Bioorg Med Chem 17:1325–1338
13. Pacorel B, Leung SC, Stachulski AV, Davies J, Vivas L, Lander H, Ward SA, Kaiser M, Brun R, O'Neill PM (2010) Modular synthesis and *in vitro* and *in vivo* antimalarial assessment of C-10 pyrrole mannich base derivatives of artemisinin. J Med Chem 53:633–640
14. Posner GH, Parker MH, Northrop J, Elias JS, Ploypradith P, Xie S, Shapiro TA (1999) Orally active, hydrolytically stable, semisynthetic, antimalarial trioxanes in the artemisinin family. J Med Chem 42:300–304
15. Posner GH, Ploypradith P, Parker MH, O'Dowd H, Woo SH, Northrop J, Krasavin M, Dolan P, Kensler TW, Xie S et al (1999) Antimalarial, antiproliferative, and antitumor activities of artemisinin-derived, chemically robust, trioxane dimers. J Med Chem 42:4275–4280
16. Jefford CW (2001) Why artemisinin and certain synthetic peroxides are potent antimalarials. Implications for the mode of action. Curr Med Chem 8:1803–1826
17. McCullough KJ, Nojima M (2001) Recent advances in the chemistry of cyclic peroxides. Curr Org Chem 5:601–636
18. O'Neill PM, Amewu RK, Nixon GL, Bousejra ElGarah F, Mungthin M, Chadwick J, Shone AE, Vivas L, Lander H, Barton V et al (2010) Identification of a 1,2,4,5-tetraoxane antimalarial drug-development candidate (RKA 182) with superior properties to the semisynthetic artemisinins. Angew Chem Int Ed Engl 49:5693–5697
19. Opsenica I, Opsenica D, Smith KS, Milhous WK, Šolaja BA (2008) Chemical stability of the peroxide bond enables diversified synthesis of potent tetraoxane antimalarials. J Med Chem 51:2261–2266
20. Šolaja B, Opsenica DM, Pocsfalvi G, Milhous WK, Kyle DE (2005) Mixed steroidal 1,2,4,5-tetraoxane compounds and methods of making and using thereof. US Patent 6906098 B2
21. Haynes RK, Fugmann B, Stetter J, Rieckmann K, Heilmann HD, Chan HW, Cheung MK, Lam WL, Wong HN, Croft SL et al (2006) Artemisone–a highly active antimalarial drug of the artemisinin class. Angew Chem Int Ed Engl 45:2082–2088
22. Haynes RK, Ho WY, Chan HW, Fugmann B, Stetter J, Croft SL, Vivas L, Peters W, Robinson BL (2004) Highly antimalaria-active artemisinin derivatives: biological activity does not correlate with chemical reactivity. Angew Chem Int Ed Engl 43:1381–1385

23. Bhattacharjee AK, Karle JM (1999) Stereoelectronic properties of antimalarial artemisinin analogues in relation to neurotoxicity. Chem Res Toxicol 12:422–428
24. http://www.public-health.tu-dresden.de/dotnetnuke3/Portals/0/VKliPhaDresden%202005.pdf. Accessed 04 Jan 2011
25. Vivas L, Rattray L, Stewart LB, Robinson BL, Fugmann B, Haynes RK, Peters W, Croft SL (2007) Antimalarial efficacy and drug interactions of the novel semi-synthetic endoperoxide artemisone *in vitro* and *in vivo*. J Antimicrob Chemother 59:658–665
26. Jambou R, Legrand E, Niang M, Khim N, Lim P, Volney B, Ekala MT, Bouchier C, Esterre P, Fandeur T et al (2005) Resistance of *Plasmodium falciparum* field isolates to *in-vitro* artemether and point mutations of the SERCA-type PfATPase6. Lancet 366:1960–1963
27. Ashton M, Hai TN, Sy ND, Huong DX, Van Huong N, Nieu NT, Cong LD (1998) Artemisinin pharmacokinetics is time-dependent during repeated oral administration in healthy male adults. Drug Metab Dispos 26:25–27
28. Svensson US, Ashton M (1999) Identification of the human cytochrome P450 enzymes involved in the *in vitro* metabolism of artemisinin. Br J Clin Pharmacol 48:528–535
29. Nagelschmitz J, Voith B, Wensing G, Roemer A, Fugmann B, Haynes RK, Kotecka BM, Rieckmann KH, Edstein MD (2008) First assessment in humans of the safety, tolerability, pharmacokinetics, and *ex vivo* pharmacodynamic antimalarial activity of the new artemisinin derivative artemisone. Antimicrob Agents Chemother 52:3085–3091
30. Brewer TG, Grate SJ, Peggins JO, Weina PJ, Petras JM, Levine BS, Heiffer MH, Schuster BG (1994) Fatal neurotoxicity of arteether and artemether. Am J Trop Med Hyg 51:251–259
31. Obaldia N 3rd, Kotecka BM, Edstein MD, Haynes RK, Fugmann B, Kyle DE, Rieckmann KH (2009) Evaluation of artemisone combinations in *Aotus* monkeys infected with *Plasmodium falciparum*. Antimicrob Agents Chemother 53:3592–3594
32. Waknine-Grinberg JH, Hunt N, Bentura-Marciano A, McQuillan JA, Chan HW, Chan WC, Barenholz Y, Haynes RK, Golenser J (2010) Artemisone effective against murine cerebral malaria. Malar J 9:227
33. Haynes RK, Chan WC, Lung CM, Uhlemann AC, Eckstein U, Taramelli D, Parapini S, Monti D, Krishna S (2007) The Fe^{2+}-mediated decomposition, PfATP6 binding, and antimalarial activities of artemisone and other artemisinins: the unlikelihood of C-centered radicals as bioactive intermediates. ChemMedChem 2:1480–1497
34. Bousejra-El Garah F, Meunier B, Robert A (2008) The antimalarial artemisone is an efficient heme alkylating agent. Eur J Inorg Chem 2008:2133–2135
35. Laurent SA, Loup C, Mourgues S, Robert A, Meunier B (2005) Heme alkylation by artesunic acid and trioxaquine DU1301, two antimalarial trioxanes. ChemBioChem 6:653–658
36. Robert A, Cazelles J, Meunier B (2001) Characterization of the alkylation product of heme by the antimalarial drug artemisinin. Angew Chem Int Ed Engl 40:1954–1957
37. Robert A, Coppel Y, Meunier B (2002) Alkylation of heme by the antimalarial drug artemisinin. Chem Commun 2002:414–415
38. Robert A, Benoit-Vical F, Claparols C, Meunier B (2005) The antimalarial drug artemisinin alkylates heme in infected mice. Proc Natl Acad Sci USA 102:13676–13680
39. Coghi P, Basilico N, Taramelli D, Chan WC, Haynes RK, Monti D (2009) Interaction of artemisinins with oxyhemoglobin Hb-FeII, Hb-FeII, carboxyHb-FeII, heme-FeII, and carboxyheme FeII: significance for mode of action and implications for therapy of cerebral malaria. ChemMedChem 4:2045–2053
40. Haynes RK (2005) Reply to comments on "Highly antimalaria-active artemisinin derivatives: biological activity does not correlate with chemical reactivity". Angew Chem Int Ed Engl 44:2064–2065
41. Avery MA, Chong WKM, Bupp JE (1990) Tricyclic analogues of artemisinin: synthesis and antimalarial activity of (+)-4,5=secoartemisinin and (−)-5-nor-4,5-secoartemisinin. J Chem Soc Chem Commun 1487–1489

42. Avery MA, Gao F, Chong WKM, Hendrickson TF, Inman WD, Crews P (1994) Synthesis, conformational analysis, and antimalarial activity of tricyclic analogs of artemisinin. Tetrahedron 50:952–957
43. Haynes RK, Vonwiller SC (1996) The behaviour of qinghaosu (artemisinin) in the presence of non-heme iron(II) and (III). Tetrahedron Lett 37:257–260
44. Eckstein-Ludwig U, Webb RJ, Van Goethem ID, East JM, Lee AG, Kimura M, O'Neill PM, Bray PG, Ward SA, Krishna S (2003) Artemisinins target the SERCA of *Plasmodium falciparum*. Nature 424:957–961
45. Parapini S, Basilico N, Mondani M, Olliaro P, Taramelli D, Monti D (2004) Evidence that haem iron in the malaria parasite is not needed for the antimalarial effects of artemisinin. FEBS Lett 575:91–94
46. Uhlemann AC, Cameron A, Eckstein-Ludwig U, Fischbarg J, Iserovich P, Zuniga FA, East M, Lee A, Brady L, Haynes RK et al (2005) A single amino acid residue can determine the sensitivity of SERCAs to artemisinins. Nat Struct Mol Biol 12:628–629
47. Pooley S, Fatih FA, Krishna S, Gerisch M, Haynes RK, Wong HN, Staines HM (2011) Artemisone uptake in *Plasmodium falciparum*-infected erythrocytes. Antimicrob Agents Chemother 55:550–556
48. Yang YZ, Little B, Meshnick SR (1994) Alkylation of proteins by artemisinin. Effects of heme, pH, and drug structure. Biochem Pharmacol 48:569–573
49. Rosenthal PJ (2001) Antimalarial chemotherapy: mechanisms of action, resistance, and new directions in drug discovery. Humana, Totowa, NJ
50. Akoachere M, Buchholz K, Fischer E, Burhenne J, Haefeli WE, Schirmer RH, Becker K (2005) *In vitro* assessment of methylene blue on chloroquine-sensitive and -resistant *Plasmodium falciparum* strains reveals synergistic action with artemisinins. Antimicrob Agents Chemother 49:4592–4597
51. Buchholz K, Schirmer RH, Eubel JK, Akoachere MB, Dandekar T, Becker K, Gromer S (2008) Interactions of methylene blue with human disulfide reductases and their orthologues from *Plasmodium falciparum*. Antimicrob Agents Chemother 52:183–191
52. Haynes RK, Chan WC, Wong HN, Li KY, Wu WK, Fan KM, Sung HH, Williams ID, Prosperi D, Melato S et al (2010) Facile oxidation of leucomethylene blue and dihydroflavins by artemisinins: relationship with flavoenzyme function and antimalarial mechanism of action. ChemMedChem 5:1282–1299
53. Haynes RK, Cheu KW, Tang MM, Chen MJ, Guo ZF, Guo ZH, Coghi P, Monti D (2011) Reactions of antimalarial peroxides with each of leucomethylene blue and dihydroflavins: flavin reductase and the cofactor model exemplified. ChemMedChem 6:279–291
54. Avery MA, Gao F, Chong WK, Mehrotra S, Milhous WK (1993) Structure-activity relationships of the antimalarial agent artemisinin. 1. Synthesis and comparative molecular field analysis of C-9 analogs of artemisinin and 10-deoxoartemisinin. J Med Chem 36:4264–4275
55. Jefford CW, Burger U, Millasson-Schmidt P, Bernardinelli G, Robinson BL, Peters W (2000) Epiartemisinin, a remarkably poor antimalarial: implications for the mode of action. Helv Chim Acta 83:1239–1246
56. Vennerstrom JL, Arbe-Barnes S, Brun R, Charman SA, Chiu FC, Chollet J, Dong Y, Dorn A, Hunziker D, Matile H et al (2004) Identification of an antimalarial synthetic trioxolane drug development candidate. Nature 430:900–904
57. Tang Y, Dong Y, Karle JM, DiTusa CA, Vennerstrom JL (2004) Synthesis of tetrasubstituted ozonides by the Griesbaum coozonolysis reaction: diastereoselectivity and functional group transformations by post-ozonolysis reactions. J Org Chem 69:6470–6473
58. Vennerstrom JL, Dong Y, Chollet J, Matile H (2002) Spiro and dispiro 1,2,4-trioxolane antimalarials. US Patent 6486199
59. Vennerstrom JL, Tang Y, Dong Y, Chollet J, Matile H, Padmanilayam M, Charman WN (2004) Spiro and dispiro 1,2,4-trioxolane antimalarials. US Patent 6825230

60. Vennerstrom JL, Dong Y, Charman SA, Wittlin S, Chollet J, Wang X, Sriraghavan K, Zhou L, Matile H, Charman WN (2008) Spiro and dispiro 1,2,4-trioxolane antimalarials. US Patent 2008/0125441 A1
61. Dong Y, Chollet J, Matile H, Charman SA, Chiu FC, Charman WN, Scorneaux B, Urwyler H, Santo Tomas J, Scheurer C et al (2005) Spiro and dispiro-1,2,4-trioxolanes as antimalarial peroxides: charting a workable structure-activity relationship using simple prototypes. J Med Chem 48:4953–4961
62. Tang Y, Dong Y, Wittlin S, Charman SA, Chollet J, Chiu FC, Charman WN, Matile H, Urwyler H, Dorn A et al (2007) Weak base dispiro-1,2,4-trioxolanes: potent antimalarial ozonides. Bioorg Med Chem Lett 17:1260–1265
63. Vennerstrom JL, Dong Y, Chollet J, Matile H, Wang X, Sriraghavan K, Charman WN (2008) Spiro and dispiro 1,2,4-trioxolane antimalarials. US Patent 7371778
64. Zhou L, Alker A, Ruf A, Wang X, Chiu FC, Morizzi J, Charman SA, Charman WN, Scheurer C, Wittlin S et al (2008) Characterization of the two major CYP450 metabolites of ozonide (1,2,4-trioxolane) OZ277. Bioorg Med Chem Lett 18:1555–1558
65. Dong Y, Wittlin S, Sriraghavan K, Chollet J, Charman SA, Charman WN, Scheurer C, Urwyler H, Santo Tomas J, Snyder C et al (2010) The structure-activity relationship of the antimalarial ozonide arterolane (OZ277). J Med Chem 53:481–491
66. Dong Y, Tang Y, Chollet J, Matile H, Wittlin S, Charman SA, Charman WN, Tomas JS, Scheurer C, Snyder C et al (2006) Effect of functional group polarity on the antimalarial activity of spiro and dispiro-1,2,4-trioxolanes. Bioorg Med Chem 14:6368–6382
67. Creek DJ, Charman WN, Chiu FC, Prankerd RJ, Dong Y, Vennerstrom JL, Charman SA (2008) Relationship between antimalarial activity and heme alkylation for spiro- and dispiro-1,2,4-trioxolane antimalarials. Antimicrob Agents Chemother 52:1291–1296
68. Uhlemann AC, Wittlin S, Matile H, Bustamante LY, Krishna S (2007) Mechanism of antimalarial action of the synthetic trioxolane RBX11160 (OZ277). Antimicrob Agents Chemother 51:667–672
69. Creek DJ, Ryan E, Charman WN, Chiu FC, Prankerd RJ, Vennerstrom JL, Charman SA (2009) Stability of peroxide antimalarials in the presence of human hemoglobin. Antimicrob Agents Chemother 53:3496–3500
70. O'Neill PM, Posner GH (2004) A medicinal chemistry perspective on artemisinin and related endoperoxides. J Med Chem 47:2945–2964
71. Araujo NC, Barton V, Jones M, Stocks PA, Ward SA, Davies J, Bray PG, Shone AE, Cristiano ML, O'Neill PM (2009) Semi-synthetic and synthetic 1,2,4-trioxaquines and 1,2,4-trioxolaquines: synthesis, preliminary SAR and comparison with acridine endoperoxide conjugates. Bioorg Med Chem Lett 19:2038–2043
72. WHO. http://www.who.int/malaria/publications/atoz/coa_website5.pdf. Accessed 19 Dec 2010
73. FDA. http://www.fda.gov/NewsEvents/Newsroom/PressAnnouncements/ucm149559.htm. Accessed 19 Dec 2010
74. MMV. http://www.mmv.org/research-development/project-portfolio/phase-iia. Accessed 20 Dec 2010
75. Ranbaxy. http://www.ranbaxy.com/socialresposbility/mm.aspx. Accessed 20 Dec 2010
76. Charman SA, Arbe-Barnes S, Bathurst IC, Brun R, Campbell M, Charman WN, Chiu FC, Chollet J, Craft JC, Creek DJ et al (2011) Synthetic ozonide drug candidate OZ439 offers new hope for a single-dose cure of uncomplicated malaria. Proc Natl Acad Sci USA 108:4400–4405
77. http://www.corporatecitizenship.novartis.com/patients/access-medicines/intellectual-property/biodiversity.shtml. Accessed 05 Jan 2011

Combination Therapy in Light of Emerging Artemisinin Resistance

Harald Noedl

Abstract Within less than a decade virtually all malaria-endemic countries have adopted one of the WHO-recommended artemisinin-based combination therapies (ACTs) for the treatment of falciparum malaria. In 2006, the first cases of clinical artemisinin resistance were reported from the Thai–Cambodian border. A number of factors are likely to have contributed to the development of artemisinin resistance in Southeast Asia. However, current evidence suggests that artemisinin resistance is simply a natural consequence of the massive deployment of ACTs in the region. The potentially devastating implications of resistance to a drug class to which there is currently no real alternative call for cost-effective strategies to extend the useful life spans of currently available antimalarial drugs. At the same time, major efforts to develop novel combination therapies not based on artemisinins are required.

1 Introduction

"The history of malaria contains a great lesson for humanity – that we should be more scientific in our habit of thought, and more practical in our habits of government. The neglect of this lesson has already cost many countries an immense loss in life and prosperity" [1].

With almost 800,000 deaths and hundreds of millions of clinical cases every year, much of what Sir Ronald Ross expressed almost exactly 100 years ago still holds true today [2]. In spite of major advances in the development of new artemisinin-based combination therapies (ACTs), the fact that malaria control is almost entirely reliant on a single class of antimalarials makes malaria control more

H. Noedl (✉)
Institute of Specific Prophylaxis and Tropical Medicine, Medical University of Vienna, Kinderspitalgasse 15, Vienna 1090, Austria
e-mail: harald.noedl@meduniwien.ac.at

H.M. Staines and S. Krishna (eds.), *Treatment and Prevention of Malaria*,
DOI 10.1007/978-3-0346-0480-2_11,

vulnerable than ever before. Sir Ronald Ross was a British–Indian physician and entomologist, primarily noted for identifying the link between mosquitoes and malaria in the late nineteenth century for which he was awarded the Nobel Prize in Medicine in 1902. By that time, quinine was already firmly established in western medicine as the treatment of choice for malaria and the detection of the first cases of antimalarial drug resistance to quinine in South America was only a few years away. The discovery of the antimalarial properties of the bark of *Arbor febrifuga* (*Cinchona* spp.), a tree native to tropical South America, in the early seventeenth century had revolutionised malaria therapy. With the extraction of the main *Cinchona* alkaloids by Pelletier and Caventou in the early nineteenth century, the era of the "Peruvian bark" came to an end and the medicinal use of the bark was largely abandoned for the use of one of its main alkaloids, quinine [3]. Quinine was also the first antimalarial drug to which resistance was reported. In fact, the first reports of resistance (a series of treatment failures) emerged as early as in 1910 from South America [4, 5]. Surprisingly, throughout the twentieth century quinine resistance proved to have relatively little impact on the therapeutic use of the drug in most parts of the world and up till now it has never reached a level comparable to that seen with some of the synthetic antimalarials. Quinine is still widely used in malaria therapy and remains one of the most important partner drugs in antimalarial combination therapy. However, in recent years, the class of drugs that has drawn most of the attention and which is the basis for the majority of currently available combination therapies is the artemisinins.

Artemisinin is a sesquiterpene lactone extracted from sweet wormwood (*Artemisia annua* or Chinese: *qinghao*), a common plant native to temperate Asia, but naturalised and recently cultivated throughout the world. The first recorded use of the plant *qinghao* for the treatment of febrile illnesses dates back to the fourth century AD in China. Artemisinin was finally extracted and its antimalarial properties characterised in the early 1970s by Chinese scientists. Since then the use of the parent compound has largely been replaced by the use of its semisynthetic derivatives. Artesunate and artemether, the most commonly used artemisinin derivatives, are hydrolysed to dihydroartemisinin, which has a very short plasma half-life. This also means that virtually all artemisinin derivatives are likely to share an identical mode of action. Artemisinins are active against all asexual stages of malaria parasites and seem to exert some activity also against gametocytes [6]. Although the endoperoxide bridge seems to be vital for their antimalarial activity, the mechanism of action of the artemisinin compounds is still not fully understood [7].

More recently, fully synthetic peroxides have been developed as a promising alternative to currently used artemisinin derivatives. They contain the same peroxide bond that confers the antimalarial activity of artemisinins. One such peroxide, the ozonide OZ277 or arterolane, has recently entered Phase III clinical trials in the form of an arterolane maleate–piperaquine phosphate combination [8]. Originally, these compounds were developed as an alternative to circumvent the dependency on agricultural production of artemisinin. In the light of emerging artemisinin resistance, their performance against artemisinin-resistant parasites may now decide their future more than anything else.

Unfortunately, the poor pharmacokinetic properties of artemisinins, particularly their short half-lives and unpredictable drug levels in individual patients, translate into substantial treatment failure rates when used as monotherapy, thereby suggesting their combination with longer half-life partner drugs. In the past decade, artemisinin and its semisynthetic derivatives have therefore become the most important basis for antimalarial combination therapies.

2 Combination Therapy

Combination therapy has a long history of use in the treatment of chronic and infectious diseases such as tuberculosis, leprosy, and HIV infections. More recently, it has also been applied to malaria treatment [9–11]. The theory underlying antimalarial combination therapy is that if two drugs are used with different modes of action, and ideally, also different resistance mechanisms, then the per-parasite probability of developing resistance to both drugs is the product of their individual per-parasite probabilities [12]. This is based on the assumption that throughout its history (e.g., chloroquine resistance has independently arisen only on a very limited number of occasions) this would make selection for resistance to a treatment combining two drugs with different modes of action extremely unlikely [13].

The WHO has recently defined antimalarial combination therapy as "the simultaneous use of two or more blood schizontocidal drugs with independent modes of action and thus unrelated biochemical targets in the parasite. The concept is based on the potential of two or more simultaneously administered schizontocidal drugs with independent modes of action to improve therapeutic efficacy and also to delay the development of resistance to the individual components of the combination" [6]. This definition specifically excludes a number of combinations commonly used in malaria therapy, such as atovaquone–proguanil or sulphadoxine–pyrimethamine, based on the assumption that the respective partners share similar modes of action and further reduces the number of currently available non-ACT combinations [14]. The WHO currently recommends five different ACTs (Table 1).

Compared to chloroquine, the cost of modern combination therapies is almost prohibitive. During the first years of deployment, the high cost of the new combination treatments therefore remained a major limiting factor. However, the past years have seen a major increase in donor funding. The Global Fund to Fight AIDS, Tuberculosis and Malaria alone has committed almost US $20 billion to support large-scale prevention, treatment and care programmes, including the massive deployment of combination therapies.

Table 1 List of ACTs recommended for the treatment of uncomplicated falciparum malaria by the World Health Organization [15]

Artemisinin derivative	Partner drug(s)	Formulation[a]	Resistance
Artemether	Lumefantrine	Coformulated	MDR
Artesunate	Amodiaquine	Coformulated	–
Artesunate	Mefloquine	Coblistered or codispensed	MDR
Artesunate	Sulfadoxine–pyrimethamine	Coblistered or codispensed	–
Dihydroartemisinin	Piperaquine	Coformulated	MDR

[a]The WHO recommends fixed-dose combinations over coblistered or codispensed formulations
MDR: recommended in areas of multidrug resistance (East Asia), artesunate plus mefloquine, or artemether plus lumefantrine or dihydroartemisinin plus piperaquine

3 Pharmacokinetic Mismatch and Compliance

In essence, the main concept behind combination therapy in malaria is to delay the development of resistance, to improve therapeutic efficacy, and to reduce malaria transmission. However, the optimal pharmacokinetic properties for an antimalarial drug (whether used in combination or as a single agent) have been a matter of debate. Ideally, antimalarial drugs should be present in the blood stream just long enough to cover the approximately three parasite life cycles (i.e., 6 days for *P. falciparum*) needed to eliminate all asexual parasites. In reality, this is difficult to achieve and a key to many limitations associated with ACTs seems to be the pharmacokinetic mismatch of the partner drugs [16]. A pharmacokinetic mismatch can also be a major factor contributing to resistance of the long-acting partner drug, which in the later stages of its presence in the blood stream is not protected by the short-acting artemisinins. This is not a problem as long as both drugs are fully efficacious on their own and as long as the drug levels of both drugs remain above the minimum inhibitory concentrations until all asexual parasites have been cleared. However, with a reasonable duration of drug administration, short half-life drugs will not be able to cover the minimum duration of drug exposure. At the same time, long half-life drugs will result inevitably in a long tail, during which the drug levels of the partner drugs will be below the minimum inhibitory concentrations and without protection from the artemisinin compound. This particularly applies to the use of ACTs in high transmission areas [17].

Compliance also remains a key factor in the rational use of antimalarial drugs. In many settings, directly observed therapy is not an option. While rapid elimination reduces the selective pressure by avoiding a long tail of subtherapeutic concentrations, antimalarial drugs with a half-life of less than 24 h (such as artemisinins or quinine) need to be administered for at least 7 days to be fully efficacious. Although compliance with malaria treatment is difficult to assess in study settings and shows significant variations across different studies, there is a general consensus that antimalarial treatment regimens lasting up to 3 days are likely to give good compliance [18].

4 Coformulation

Another potential problem of treating patients with more than one antimalarial drug is the fact that many currently available combination therapies are not coformulated, greatly increasing complexity of treatment and the chances of misuse. Currently, the only widely used coformulated combination is artemether–lumefantrine. More recently, coformulated combinations of artesunate–amodiaquine and dihydroartemisinin–piperaquine have become available but coformulated combinations of many other ACTs are still unavailable. Even though coformulated, artemether–lumefantrine remains a relatively complex regimen (with an adult dose of four tablets twice daily for 3 days) and compliance, and therefore programmatic effectiveness, is not optimal [19].

5 ACT and Antimalarial Drug Resistance

Throughout the past 100 years, drug resistance has emerged as one of the biggest challenges for malaria control. The extensive deployment of antimalarial drugs since the introduction of chloroquine in the 1940s has provided a remarkable selection pressure on malaria parasites to evolve resistance mechanisms to virtually all available antimalarial drugs. In essence, it is continuous drug pressure that results in the selection of parasite populations with genetically reduced drug sensitivity. The widespread and indiscriminate use of antimalarial drugs places a strong selective pressure on malaria parasites to develop resistance. Malaria parasites can acquire high levels of resistance, both in the individual parasite as well as on a population basis. The high degree of resistance expressed by malaria parasites is at least in part attributable to their high diversity and genetic complexity, resulting in a variety of potential mechanisms to evade drug activity. This way *P. falciparum* has developed resistance to virtually all antimalarials in current use, drugs that once were considered the front line against the disease. However, the geographical distribution and extent of resistance to any single antimalarial drug show major variations. Originally believed to be limited to *P. falciparum*, antimalarial drug resistance is now also known to affect other species [20]. *P. vivax* has rapidly developed resistance to sulfadoxine–pyrimethamine in many parts of the world, whereas high-level resistance to chloroquine remains confined largely to Indonesia, East Timor, Papua New Guinea and other parts of Oceania [6].

In the late 1990s, combination therapy was introduced to overcome the quickening pace of drug resistance development in Southeast Asia. However, by that time drug resistance had reached many of the potential partner drugs or at least drugs structurally related to those used in combination with artemisinins. This particularly applies to mefloquine, an arylaminoalcohol, which had previously been used extensively as monotherapy in Southeast Asia. Based on large-scale field trials starting in 1983 mefloquine was introduced as standard therapy by the

Thai Ministry of Public Health as early as 1985 to overcome increasing chloroquine and sulfadoxine–pyrimethamine resistance [21]. Interestingly, mefloquine was used in combination with sulfadoxine–pyrimethamine initially before being deployed as monotherapy. Rising numbers of failures with the standard-dose mefloquine (15 mg/kg) resulted in an increase of the dose to 25 mg/kg and in 1995 mefloquine monotherapy had to be replaced by the combination of mefloquine with artesunate, making this combination the first ACT to be deployed on a large scale in Southeast Asia.

In an attempt to limit the impact of increasing levels of resistance to traditional antimalarial drugs, in 2001 the World Health Organization recommended that all countries experiencing resistance to conventional monotherapies, such as chloroquine, sulfadoxine–pyrimethamine, or mefloquine should use combination therapies, preferably based on artemisinin derivatives for the treatment of uncomplicated falciparum malaria [21]. However, although ACTs have shown high efficacy in the treatment of malaria in Southeast Asia, where transmission is typically low, concerns remain about their long-term implementation as first-line therapy in high-transmission areas in Africa [19, 22].

6 Cross-resistance

The problem of antimalarial drug resistance is even further aggravated by the existence of cross-resistance among drugs with related chemical structures. This particularly applies to the 4-aminoquinolines (e.g. chloroquine–amodiaquine) and the arylaminoalcohols (e.g. mefloquine–lumefantrine) but also to artemisinin and its semisynthetic derivatives (e.g. artesunate–artemether). Malaria control has largely been relying on a small number of structurally related drugs essentially belonging to just a very few different classes. Once malaria parasites develop resistance to a single member of any of these classes, their sensitivity to most other antimalarials sharing a similar mode of action (i.e., typically belonging to the same class) is also compromised. This means that (e.g., in an area like Southeast Asia where the combination of artesunate and mefloquine is loosing its clinical efficacy) the introduction of ACTs using chemically related compounds, such as artemether–lumefantrine, may not be an option.

Interestingly, activity correlations derived from in vitro studies indicate that in spite of their different chemical structure there may be a certain level of cross sensitivity between artemisinin derivatives and certain arylaminoalcohols, currently the most commonly used partner drugs in ACTs [23]. The existence of this link is also supported by the potential role that the *P. falciparum* multidrug resistance 1 (*pfmdr1*) gene may be playing in simultaneously mediating the sensitivity to arylaminoalcohols and artemisinins [24].

7 Resistance to Partner Drugs

Preventing resistance to the partner drugs is obviously crucial. If the partner drug is not 100% successful in eliminating the parasites surviving the initial impact from the (subcurative) 3-day artemisinin treatment, ACTs are likely to select for artemisinin resistance. However, resistance has been reported against most of the commonly used partner drugs or at least to structurally closely related drugs resulting in activities considerably below 100%. Currently the most important partner drugs used in ACTs are lumefantrine, mefloquine, amodiaquine, and more recently piperaquine [22]. Mefloquine and lumefantrine are structurally related and belong to the class of the arylaminoalcohol antimalarials, whereas amodiaquine and piperaquine are both closely related to chloroquine.

Mefloquine is a synthetic antimalarial widely used throughout Southeast Asia as a combination partner for artesunate. It was introduced in Thailand in the mid 1980s. By the mid 1990s, resistance had reached a level that necessitated its combination with artesunate to reach adequate cure rates [21]. In spite of high levels of mefloquine resistance, particularly in Thailand, Cambodia, and Myanmar, mefloquine remains the most important ACT partner drug in the region.

Lumefantrine is commonly used coformulated with artemether and has never been used in monotherapy on any significant scale. Although clinical resistance to lumefantrine has not explicitly been reported, there is a strong indication of cross-resistance with mefloquine [25]. The use of lumefantrine is therefore not advisable in areas where high levels of mefloquine resistance have been reported.

Amodiaquine is a 4-aminoquinoline antimalarial originally developed in the 1940s. It is structurally closely related to chloroquine but due to its higher potency shows considerable activity also against chloroquine-resistant parasites. Resistance to both drugs also seems to be mediated by the same genetic mechanism [26]. Resistance to amodiaquine was reported soon after the advent of chloroquine resistance but has never reached its magnitude [27].

Although piperaquine, a bisquinolone antimalarial, is still considered to be a relatively new antimalarial drug throughout most of the malaria-endemic world, it has a long history of use in malaria treatment and prophylaxis in China [28]. Consequently, high levels of resistance have been reported from parts of southern China and resistance can relatively easily be induced in a *P. berghei* model [29].

8 Artemisinin Resistance

The statement "Resistance has arisen to all classes of antimalarials except, as yet, to the artemisinin derivatives" [6] in the WHO treatment guidelines from 2006 unfortunately does not hold true any longer. Clinical artemisinin resistance was first identified in 2006 in Ta Sanh, a small town in close proximity to the Thai–Cambodian border, a known hotspot of antimalarial drug resistance [30].

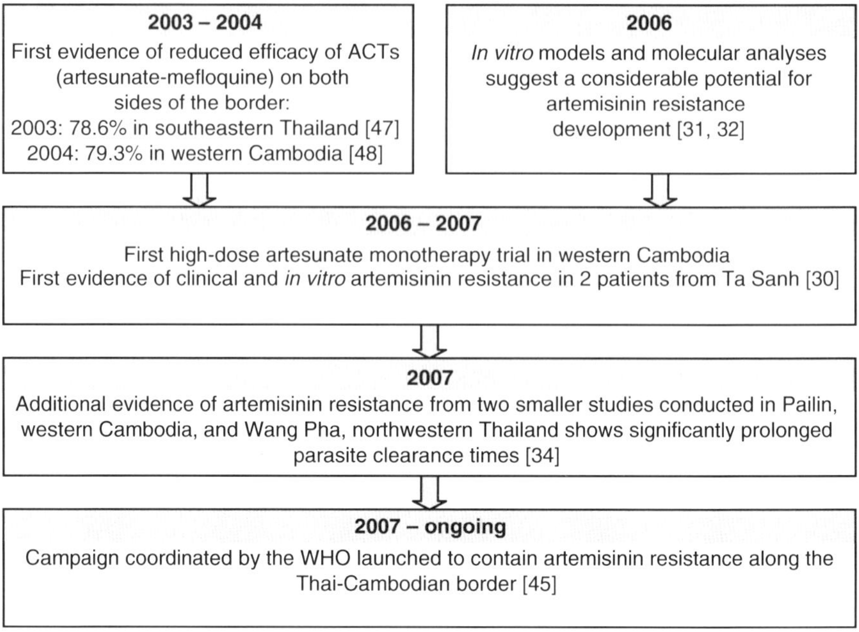
2003 – 2004
First evidence of reduced efficacy of ACTs (artesunate-mefloquine) on both sides of the border:
2003: 78.6% in southeastern Thailand [47]
2004: 79.3% in western Cambodia [48]
2006
In vitro models and molecular analyses suggest a considerable potential for artemisinin resistance development [31, 32]
2006 – 2007
First high-dose artesunate monotherapy trial in western Cambodia
First evidence of clinical and in vitro artemisinin resistance in 2 patients from Ta Sanh [30]
2007
Additional evidence of artemisinin resistance from two smaller studies conducted in Pailin, western Cambodia, and Wang Pha, northwestern Thailand shows significantly prolonged parasite clearance times [34]
2007 – ongoing
Campaign coordinated by the WHO launched to contain artemisinin resistance along the Thai-Cambodian border [45]

Fig. 1 Evidence of emerging artemisinin resistance along the Thai–Cambodian border

Even before the discovery of clinical resistance, in vitro models and molecular analysis had suggested a considerable potential for resistance development [7, 31, 32]. Moreover, ex vivo data and clinical treatment response seem to indicate that artemisinin sensitivity is also compromised in western Thailand [33, 34] (Fig. 1).

Not since the discovery of chloroquine resistance in the 1950s has malaria control been reliant upon the efficacy of a single class of drugs as much as it does currently. In the past 10 years, virtually all falciparum malaria-endemic countries have adopted some kind of ACT as first- or second-line therapy for uncomplicated falciparum malaria. In the absence of a defined artemisinin resistance mechanism and mechanism of action, as well as data from clinical trials using the new synthetic peroxides, it is hard to tell what impact the recent developments in Southeast Asia will have. However, in the current situation losing a single drug to resistance could potentially endanger virtually all malaria control efforts worldwide.

9 Defining Artemisinin Resistance

Defining artemisinin resistance remains a major challenge. The most commonly used definition of antimalarial drug resistance dates back to 1973 and defines resistance as “The ability of a parasite strain to survive and/or multiply despite the administration and absorption of a drug given in doses equal to or higher than those usually recommended but within tolerance of the subject" [38]. Although, – with minor modifications, – this definition remains valid as of today, it was developed for traditional monotherapies and in its original version has therefore only limited applicability to modern combination treatments. The most obvious problem in defining resistance to combination therapies is the fact that, with two or more combination partners, resistance might arise to a single component without ever becoming clinically evident. Since in ACTs artemisinins are generally responsible for the initial reduction in the parasite burden, the most obvious clinical parameter indicating reduced susceptibility to artemisinins is prolonged parasite clearance [39, 40]. Although data from ACT trials can provide important initial data on compromised drug sensitivity, a detailed assessment of treatment response to artemisinins requires extensive controlled monotherapy studies with artemisinin derivatives (including basic pharmacokinetics and a reliable way of excluding reinfections). These studies are currently being conducted in Asia and Africa. Ex vivo drug sensitivity data can be extremely helpful in interpreting geographical or temporal trends but need to be interpreted in a broader context with clinical data [33]. In spite of a number of interesting leads, as yet reliable molecular markers of artemisinin resistance have not been identified [31]. Artemisinin resistance can have major implications for malaria control programs in the affected countries and should therefore only be considered after careful analysis of treatment response parameters and treatment success in relation to ex vivo drug sensitivity.

10 The Causes of Artemisinin Resistance

The general view has been that several factors protect artemisinins from the development of resistance: the short plasma half-life of both the parent compound and its active metabolite, the rapidity with which the drugs clear asexual parasites, and the presence of an effective partner drug from a different class of antimalarials, which is expected to protect the artemisinin component [41]. It has been hypothesised that the de novo emergence of antimalarial drug resistance can be prevented by use of combination therapies [12, 42]. The assumption is that because of the logarithmic distribution of parasite numbers in human malaria infections, inadequately treated high biomass infections are a major source of de novo emergence of resistance. However, the recent events suggest that the hope that resistance can actually be prevented through combinations may have been overly optimistic.

Much of the blame for emerging artemisinin resistance in Southeast Asia has been assigned to the extensive use of artemisinin monotherapy. Compliance issues and counterfeited or substandard tablets that contain smaller amounts of or less active ingredients are considered to be additional sources of drug pressure [43]. Specific epidemiological, pharmacokinetic, and parasite factors in Southeast Asia have also been implicated in the development of artemisinin resistance [44]. Consequently, even before artemisinin resistance had been reported for the first time the WHO banned artemisinin monotherapy in an attempt to protect the artemisinins and to slow down the development of resistance. Interestingly Vietnam, a country with one of the longest histories of artemisinin monotherapy use in Southeast Asia, was by far not he first to report artemisinin resistance. In Vietnam, artemisinins have been used for malaria control since 1989 and although the current national guidelines recommend the use of ACTs, artemisinin and artesunate are still widely available as monotherapy through the private sector [45]. Vietnam is also one of the few countries where artemisinin monotherapy can be linked directly to a highly successful malaria control program. Between 1991 and 2006, malaria cases in the country have diminished from 1,672,000 clinical cases with 4,650 deaths, originally, to 91,635 with 43 deaths [46].

Although the development of artemisinin resistance is likely to have been a complex multifactorial event, the actual explanation for artemisinin resistance is likely to be rather simple. ACTs have simply gone down the same road that all previous antimalarials have. Artemisinin resistance is probably a natural consequence of the extensive deployment of ACTs and was bound to happen sooner or later. "Preventing" artemisinin resistance was never a real option.

11 Measures to Limit the Spread of Resistance

When the first evidence of artemisinin resistance became available in 2006, the World Health Organization launched an ambitious campaign to contain artemisinin resistance along the Thai–Cambodian border. These efforts involve the early detection and rapid treatment of all malaria infections on both sides of the border, preferably with a non-ACT combination therapy. In addition, the deployment of insecticide-treated nets to decrease malaria transmission and the screening and treatment of migrants were intensified, together with a more thorough mapping of the geographic boundaries of artemisinin resistance [15, 35]. Recently, Maude et al. concluded that containment of artemisinin-resistant malaria could also be achieved by eliminating malaria using ACT [47]. However, as ACTs are more effective against infections with artemisinin-sensitive parasites resulting in a relative increase in the proportion of artemisinin-resistant parasite isolates, this approach would require malaria elimination down to the last parasite. Unfortunately, this is unlikely to ever happen in a landlocked environment surrounded by malaria-endemic countries.

Although the history of malaria control teaches us that the Thai–Cambodian border has always been a hotspot of antimalarial drug resistance development, it also teaches us that sooner or later resistance is likely to emerge independently in other parts of the world. With the unprecedented deployment of ACTs throughout the malaria-endemic world, artemisinin resistance is likely to eventually emerge in other parts of the world.

The potentially devastating implications of resistance to a drug to which there is currently no real alternative calls for cost-effective strategies to extend the useful life spans of currently available antimalarial drugs while at the same time investing into major efforts to develop novel compounds as a replacement for the artemisinins.

References

1. Ross R (1911) The prevention of malaria. John Murray, London
2. World Health Organization (2010) World Malaria Report 2009. WHO, Geneva
3. Bruce-Chwatt L (1988) History of malaria from prehistory to eradication. In: Wernsdorfer WH, McGregor IA (eds) Malaria. Principles and Practice of malariology. Churchill Livingstone, Edinburgh, pp 1–59
4. Nocht B, Werner H (1910) Beobachtungen ueber eine relative Chininresistenz bei Malaria aus Brasilien. Dtsch Med Wschr 36:1557–1560
5. Neiva A (1910) Ueber die Bildung einer chininresistenten Rasse des Malariaparasiten. Mem Inst Oswaldo Cruz 2:131–140
6. Chen PQ, Li GQ, Guo XB, He KR, Fu YX, Fu LC, Song Y (1994) The infectivity of gametocytes of *Plasmodium falciparum* from patients treated with artemisinin. Chin Med J 107:709–711
7. Krishna S, Woodrow CJ, Staines HM, Haynes RK, Mercereau-Puijalon O (2006) Re-evaluation of how artemisinins work in light of emerging evidence of *in vitro* resistance. Trends Mol Med 12:200–205
8. Olliaro P, Wells TN (2009) The global portfolio of new antimalarial medicines under development. Clin Pharmacol Ther 85(6):584–595
9. Peters W (1969) Drug resistance in malaria–a perspective. Trans R Soc Trop Med Hyg 63:25–45
10. Li GQ, Arnold K, Guo XB, Jian HX, Fu LC (1984) Randomised comparative study of mefloquine, qinghaosu and pyrimethamine-sulfadoxine in patients with falciparum malaria. Lancet 2:1360–1361
11. Peters W (1990) The prevention of antimalarial drug resistance. Pharmacol Ther 47:499–508
12. White NJ, Pongtavornpinyo W (2003) The *de novo* selection of drug-resistant malaria parasites. Proc Biol Sci 270:545–554
13. Su X, Kirkman LA, Fujioka H, Wellems TE (1997) Complex polymorphisms in an approximately 330 kDa protein are linked to chloroquine-resistant *P. falciparum* in Southeast Asia and Africa. Cell 91:593–603
14. Kremsner PG, Krishna S (2004) Antimalarial combinations. Lancet 364:285–294
15. Campbell CC (2009) Malaria control – addressing challenges to ambitious goals. N Engl J Med 361:522–523
16. Giao PT, de Vries PJ (2001) Pharmacokinetic interactions of antimalarial agents. Clin Pharmacokinet 40:343–373
17. Bloland PB, Ettling M, Meek S (2004) Combination therapy for malaria in Africa: hype or hope? Bull World Health Organ 78:1378–1388
18. Congpuong K, Bualombai P, Banmairuroi V, Na-Bangchang K (2010) Compliance with a three-day course of artesunate-mefloquine combination and baseline anti-malarial treatment in an area of Thailand with highly multidrug resistant falciparum malaria. Malar J 9:43
19. Bloland PB (2003) A contrarian view of malaria therapy policy in Africa. Am J Trop Med Hyg 68:125–126
20. Baird JK, Basri H, Purnomo BMJ, Subianto B, Patchen LC, Hoffman SL (1991) Resistance to chloroquine by *Plasmodium vivax* in Irian Jaya, Indonesia. Am J Trop Med Hyg 44:547–552
21. Wongsrichanalai C, Pickard AL, Wernsdorfer WH, Meshnick SR (2002) Epidemiology of drug-resistant malaria. Lancet Infect Dis 2:209–218

22. Eastman RT, Fidock DA (2009) Artemisinin-based combination therapies: a vital tool in efforts to eliminate malaria. Nat Rev Microbiol 7:864–874
23. Noedl H, Wernsdorfer WH, Krudsood S, Wilairatana P, Viriyavejakul P, Kollaritsch H, Wiedermann G, Looareesuwan S (2001) *In vivo-in vitro* model for the assessment of clinically relevant antimalarial cross-resistance. Am J Trop Med Hyg 65:696–699
24. Sidhu AB, Uhlemann AC, Valderramos SG, Valderramos JC, Krishna S, Fidock DA (2006) Decreasing *pfmdr1* copy number in *Plasmodium falciparum* malaria heightens susceptibility to mefloquine, lumefantrine, halofantrine, quinine, and artemisinin. J Infect Dis 194:528–535
25. Basco LK, Bickii J, Ringwald P (1998) *In vitro* activity of lumefantrine (benflumetol) against clinical isolates of *Plasmodium falciparum* in Yaoundé, Cameroon. Antimicrob Agents Chemother 42:2347–2351
26. Holmgren G, Gil JP, Ferreira PM, Veiga MI, Obonyo CO, Björkman A (2006) Amodiaquine resistant *Plasmodium falciparum* malaria *in vivo* is associated with selection of *pfcrt* 76 T and *pfmdr1* 86Y. Infect Genet Evol 6(4):309–314
27. Olliaro P, Mussano P (2003) Amodiaquine for treating malaria. Cochrane Database Syst Rev CD000016
28. Tran TH, Dolecek C, Pham PM, Nguyen TD, Nguyen TT, Le HT, Dong TH, Tran TT, Stepniewska K, White NJ, Farrar J (2004) Dihydroartemisinin-piperaquine against multi-drug-resistant *Plasmodium falciparum* malaria in Vietnam: randomised clinical trial. Lancet 363:18–22
29. Giao PT, de Vries PJ, le Hung Q, Binh TQ, Nam NV, Kager PA (2004) CV8, a new combination of dihydroartemisinin, piperaquine, trimethoprim and primaquine, compared with atovaquone-proguanil against falciparum malaria in Vietnam. Trop Med Int Health 9:209–216
30. Noedl H, Se Y, Schaecher K, Smith BL, Socheat D, Fukuda MM; Artemisinin Resistance in Cambodia 1 (ARC1) Study Consortium (2008) Evidence of artemisinin-resistant malaria in western Cambodia. N Engl J Med 359:2619–2620
31. Jambou R, Legrand E, Niang M, Khim N, Lim P, Volney B, Ekala MT, Bouchier C, Esterre P, Fandeur T, Mercereau-Puijalon O (2006) Resistance of *Plasmodium falciparum* field isolates to *in-vitro* artemether and point mutations of the SERCA-type PfATPase6. Lancet 366:1960–1963
32. Krishna S, Pulcini S, Fatih F, Staines H (2010) Artemisinins and the biological basis for the PfATP6/SERCA hypothesis. Trends Parasitol 26(11):517–23
33. Noedl H, Socheat D, Satimai W (2009) Artemisinin-resistant malaria in Asia. N Engl J Med 361:540–541
34. Dondorp AM, Nosten F, Yi P, Das D, Phyo AP, Tarning J, Lwin KM, Ariey F, Hanpithakpong W, Lee SJ et al (2009) Artemisinin resistance in *Plasmodium falciparum* malaria. N Engl J Med 361:455–467
35. Enserink M (2008) Signs of drug resistance rattle experts, Trigger Bold Plan. Science 322:1776
36. Vijaykadga S, Rojanawatsirivej C, Cholpol S, Phoungmanee D, Nakavej A, Wongsrichanalai C (2006) *In vivo* sensitivity monitoring of mefloquine monotherapy and artesunate-mefloquine combinations for the treatment of uncomplicated falciparum malaria in Thailand in 2003. Trop Med Int Health 11:211–219
37. Denis MB, Tsuyuoka R, Poravuth Y, Narann TS, Seila S, Lim C, Incardona S, Lim P, Sem R, Socheat D et al (2006) Surveillance of the efficacy of artesunate and mefloquine combination for the treatment of uncomplicated falciparum malaria in Cambodia. Trop Med Int Health 11:1360–1366
38. World Health Organization (1973) Chemotherapy of malaria and resistance to antimalarials. Report of a WHO Scientific Group. WHO Technical Report Series, Geneva, 529
39. Noedl H (2005) Artemisinin resistance: how can we find it? Trends Parasitol 21:404–405
40. Stepniewska K, Ashley E, Lee SJ, Anstey N, Barnes KI, Binh TQ, D'Alessandro U, Day NP, de Vries PJ, Dorsey G et al (2010) *In vivo* parasitological measures of artemisinin susceptibility. J Infect Dis 201:570–579

41. White NJ, Nosten F, Looareesuwan S, Watkins WM, Marsh K, Snow RW, Kokwaro G, Ouma J, Hien TT, Molyneux ME et al (1999) Averting a malaria disaster. Lancet 353:1965–1967
42. White NJ (1998) Preventing antimalarial drug resistance through combinations. Drug Resist Updat 1:3–9
43. Newton PN, Dondorp A, Green M, Mayxay M, White NJ (2003) Counterfeit artesunate antimalarials in southeast Asia. Lancet 362:169
44. Dondorp AM, Yeung S, White L, Nguon C, Day NP, Socheat D, von Seidlein L (2010) Artemisinin resistance: current status and scenarios for containment. Nat Rev Microbiol 8:272–280
45. Thanh NV, Toan TQ, Cowman AF, Casey GJ, Phuc BQ, Tien NT, Hung NM, Biggs BA (2010) Monitoring for *Plasmodium falciparum* drug resistance to artemisinin and artesunate in Binh Phuoc Province, Vietnam: 1998–2009. Malar J 9:181
46. Thang NDEA, le Hung X, le Thuan K, Xa NX, Thanh NN, Ky PV, Coosemans M, Speybroeck N, D'Alessandro U (2009) Rapid decrease of malaria morbidity following the introduction of community-based monitoring in a rural area of central Vietnam. Malar J 8:3
47. Maude RJ, Pontavornpinyo W, Saralamba S, Aguas R, Yeung S, Dondorp AM, Day NP, White NJ, White LJ (2009) The last man standing is the most resistant: eliminating artemisinin-resistant malaria in Cambodia. Malar J 8:31
48. World Health Organization (2006) Guidelines for the treatment of malaria. WHO, Geneva

New Medicines to Combat Malaria: An Overview of the Global Pipeline of Therapeutics

Timothy N.C. Wells

Abstract Over the last 5 years, there has been an increased investment in malaria medicines, with the emergence of a range of new fixed dose artemisinin combination therapies (FACTs). Of the six FACTS, two are now approved and prequalified by the WHO, with another two submitted for stringent regulator approval. Malaria treatments are therefore available to more than 160 million patients, which cure the disease in more than 98% of cases, with 3 days of therapy, and cost as little as $0.30 per treatment for infants. Molecules currently in the pipeline offer the possibility that this could be replaced by a single dose cure in the next few years. Many new compounds have entered the pipeline as a result of advances in phenotypic screening and new targets identified from the parasite genomes. Artemisinin resistance has been confirmed in Cambodia, and new classes of medicines will be needed in case artemisinin resistance spreads. The recent call for the eradication of malaria has set new objectives for the drug discovery and development field. There is an additional focus on having compounds, which can block transmission. In addition, safe medicines to kill liver stages, and block the relapse of *P. vivax* and *P. ovale* are needed, which are more convenient than the 14-day primaquine treatment, and better tolerated in G6PD-deficient patients. The next decade will be a defining one, as we implement new generations of therapies in the field, and bring forward the ones needed for the next stages of malaria eradication.

1 Introduction

The challenges of malaria drug discovery and development have never been greater. On the positive side, there is a wealth of activity all the way from early discovery through registration of new medicines through to optimizing their use in

T.N.C. Wells (✉)
Medicines for Malaria Venture, 20 route de Pré-Bois, Geneva CH-1215, Switzerland
e-mail: wellst@mmv.org

H.M. Staines and S. Krishna (eds.), *Treatment and Prevention of Malaria*,
DOI 10.1007/978-3-0346-0480-2_12,

Phase IV clinical trials. In the last 2 years, two classes of fixed dose artemisinin combination therapies have been approved by the World Health Authority, and treat more than 160 million patients. Two more are currently being reviewed by the European Medicines Agency (EMA), in the first step toward World Health Organization (WHO) prequalification. This brings the total number of fixed dose artemisinin combination therapies to six; each has its particular strengths (Table 1). There is such a wide choice for the "customer" – in most cases the national malaria control programs, is positive. These medicines are affordable – costing as little as $0.30 for a child dose and $1.20 for an adult dose. The approval by WHO of two of them means they can be supplied at subsidized prices with the help of international agencies, or for free in Africa. There is a continual need to build on this success by optimizing this therapeutics for their use in the clinic, especially with pediatric patients.

In the midst of this success, there is always the need to keep on improving the medicines we have. Resistance to malaria drugs is always a possibility. With the ongoing campaign to eradicate malaria [1], the urgency to have new classes of drugs is compounded since the more we use therapies, and the closer we get to eradication, the more likely resistance is to become evident [1].

Over the last 12 months, since the pipeline was last reviewed [2], there has been a lot of progress (Fig. 1 shows the current state of the global antimalarial drug portfolio, not simply the projects which are part of collaborations with Medicines for Malaria Venture). There are many challenges that any medicines must address [3]. First, to simplify treatment regimens, with the hope of bringing forward a single dose cure. Second, to have medicines that will block transmission of the parasite from one infected individual to another via the mosquito and thus break the cycle of infection. Third, to target dormant *P. vivax* and *P. ovale* forms in the liver, known as hypnozoites, which have the potential to relapse after a few weeks or even months, causing new outbreaks of disease, in the absence of an infectious bite. Finally, new medicines are urgently needed to overcome the potential threat of artemisinin resistance.

Malaria offers challenges beyond those seen in other therapeutic areas; most of the patients are children under 5 years of age, half of them are less than 2 years, and the disease is especially prevalent in expectant mothers. This places additional stringency on the safety profiles of the potential new medicines. The portfolio of projects, which we hope will address these challenges, comes from a variety of sources. Some are from rationally designed approaches using clinically validated targets, and as much as possible based on the technologies of rational drug design. Others are coming from new, innovative high-throughput screens where millions of compounds have been tested for their activity on the whole parasite. This has radically changed the face of discovery. We now have tens of thousands of starting points for medicinal chemistry active at submicromolar concentrations, where only 5 years ago there was just a handful. Taken together, there are big challenges ahead, but the next decade promises to be a very exciting one. There will be new medicines launched to help the short-term fight against the parasite, alongside bednets, insecticides and the potential contribution of vaccines; and new classes of

Table 1 Characteristics of fixed dose artemisinin combination therapies available or soon to be launched

Fixed dose ACT	Coartem-D	Coarsucam	Eurartesim	Pyramax	AS/MQ	Arco
Partner	Novartis/MMV	Sanofi/DNDi	Sigma-Tau/Pfizer/MMV	ShinPoong/MMV	Farmanguinhos DNDi; Mepha	Kunming Pharmaceutical
	Artemether Lumefantrine	Artesunate Amodiaquine	Dihydro artemesinin Piperaquine	Artesunate Pyronaridine	Artesunate Mefloquine	Artemisinin Naphthoquine
Launch	1Q'09	4Q'08	4Q'11	1Q'12	–	
Key selling point	Market leader safety: >300 million treatments Anti-Gametocyte?	First-line therapy in Francophone Africa; Once per day	Long half-life Once per day; Good posttreatment prophylaxis at 42 days	*P. vivax* clinical data (schizonticide) Potential combination with Primaquine?	First-line treatment in Thailand *P. vivax*	Single dose cure
Key weakness	Twice per day, strong food effect	resistance	Stability – interaction between DHA and PQP No pediatric formulation	Limited field use single supplier	Psychiatric, GI adverse events. No SRA or WHO approval. Expensive	Limited data available. No SRA or WHO approval. Expensive
Stability	24 months	36 months	18–24 months	24 months	No data	No data
Pediatric	formulation Dispersible taste masked	Dissolves no taste mask	Pediatric version under development for 2013	Sachet taste masked for 2012		
Comments			DHA-piperaquine marketed by Holley-Cotec but no SRA or WHO approval			

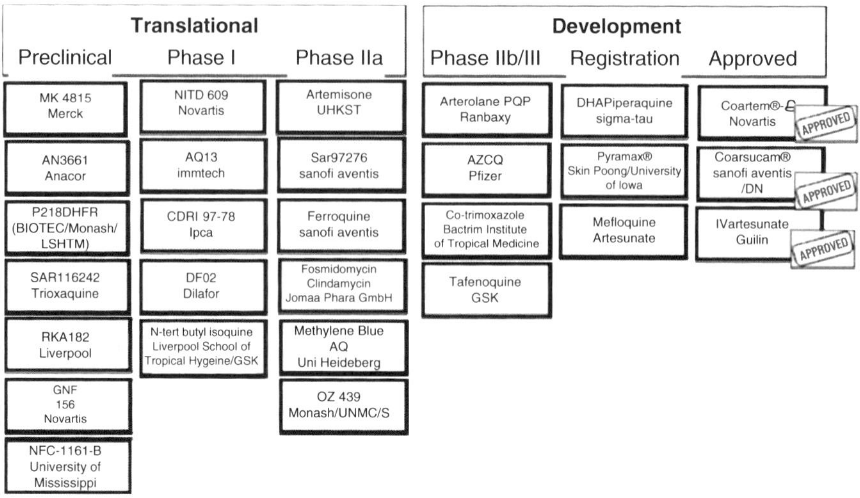

Fig. 1 Global portfolio of medicines under development for malaria March 2011. Projects are only reported which have already entered lead optimization. Preclinical refers to molecules in the process of obtaining GLP safety, toxicology and pharmacokinetic data

medicines being developed, which will help drive the long-term agenda leading to eradication.

2 The Next Generation of Artemisinin Combination Therapies: New Medicines About to Be Registered

Artemisinin was brought forward as a new treatment for malaria over 40 years ago [4] and has long been known to be an effective therapy against fevers. It acts rapidly to kill parasites, clearing fever and bringing parasitemia below the level of detection within 24 h. On its own, 7 days of treatment is needed for a cure in the majority of patients, and so a second medicine is needed to sustain the antiparasitic activity in a 3-day regimen. Artemisinin combination therapies are now extremely effective, surpassing the WHO's guidelines that 95% of patients should not have recrudesced after 28 days – many clinical trials showing efficacies of 98–99%. The combination partner also reduces the possibility of artemisinin resistance emerging, and to reinforce this value the WHO in January 2006 withdrew its recommendation for the use of artemisinin-based monotherapy for the treatment of uncomplicated malaria, and since has been campaigning for its complete withdrawal. Fixed dose combinations therefore offer two advantages, first a much simpler regimen, and second that artemisinin monotherapy is no longer an option for the patient (older combination therapies were coblistered, and field workers reported that often only the artemisinin tablet was taken). The last 2 years have marked a turning point for these fixed dose combinations. Two have been prequalified by the WHO, or by Stringent Regulatory Authorities (Tables 1 and 2). Coartem and the pediatric taste-masked dispersible form Coartem-D (artemether–lumefantrine, by Novartis in collaboration with MMV), and Coarsucam (artesunate–amodiaquine, by sanofi-aventis in collaboration with Drugs for Neglected Diseases Initiative, DND*i*), as has been discussed in chapter "Artemisinins: Artemisinin, Dihydroartemisinin, Artemether and Artesunate". Enough of these WHO prequalified medicines, and their generic versions were purchased by disease endemic countries to treat more than 160 million patients in 2010. This represents almost two thirds of the people who need treatment. In the past, cost has often been cited as a barrier to health care. Current medicines are produced with the cost of an adult course of treatment being generally around US $1.20, and a pediatric dose being a quarter of this price, and this represents a significant drop from the cost of manufacture of around $2.50 10 years ago. Several generic versions of artemether–lumefantrine are now being produced, which brings additional price competition into the market. WHO prequalification is a necessary control step here – generic medicines which have not been approved by a stringent regulatory agency must show evidence that they are manufactured to Good Manufacturing Practice standards, and also show clinical data to confirm bioequivalency to the innovator medicine (see http://www.ema.europa.eu/pdfs/human/qwp/140198enrev1.pdf for more details).

A second wave of fixed dose artemisinin combination therapies has progressed through pivotal clinical trials in the last 2 years, with two more being submitted to regulatory agencies in 2010. The first is dihydroartemisinin–piperaquine, (provisionally planned to be marketed as Eurartesim). This has been developed by sigma-tau in collaboration with MMV, which was submitted for review by the EMA, in July 2009 with approval currently anticipated in the summer of 2011. This combination is now included in the revised Malaria Treatment Guidelines [5], and would be submitted for WHO prequalification as soon as it is registered. This project was originally a collaboration with Holley-Cotec, who have their own version of the medicine launched in China and many other countries, which is being used to treat around two million patients a year. So far this medicine has not been approved by a stringent regulatory authority or prequalified by the WHO, but this situation may now change with the inclusion of DHA–piperaquine in the WHO's treatment guidelines for malaria in 2010. The second fixed dose ACT currently under regulatory review is pyronaridine–artesunate, which is being developed by Shin Poong in collaboration with MMV, and has an adult form and a pediatric taste-masked granule form for children less than 15 kg. This was submitted to the EMA in March 2010, this time under the new article 58 legislation. This allows the EMA to give scientific advice that can be used to support the use of the medicine in countries outside of Europe. The issue is that if a medicine is approved by the EMA under the normal orphan legislation, it must actually be marketed in Europe or else the authorization will be withdrawn. Article 58 allows the stringent review by the EMA, but without the need to actually market and sell the drug in Europe. A decision is anticipated at the end of in 2012. Pyronaridine–artesunate is not included in the current WHO treatment guidelines, but will be discussed over the

Table 2 The pipeline of Medicines in Development for treating malaria – April 2010

Active Ingredients	Partnership	Comments
Products marketed, registered, in submission or in pivotal clinical trials (with product names under the active ingredient)		
Artemether 20 mg/Lumefantrine 120 mg	Novartis/ MMV	Four Dose strengths are registered in 83 countries. Coartem-D is a special dispersible formulation with taste masking, developed for children. Over 350 million Coartem treatments have been used, of which in the first year of launch 53 million were Coartem-D. Generic versions, prequalified by WHO as meeting acceptable quality are available from Ajanta Pharma Ltd (India), Ipca Laboratories Ltd (India) and Cipla , Ltd (India)
Artesunate 50 mg/ Amodiaquine 135mg (Coarsucam, Artesunate-Amodiaquine Winthrop®)	Sanofi-aventis/DNDi MMV	Developed by sanofi-aventis and DNDi. Registered in Morocco in March 2007, and in 24 other Afrian countries prequalified by WHO in October 2008. Manufactured by sanofi-aventis (MAPHAR Laboratories, Morocco), with 45 million treatments at no-profit price in 2009 Other prequalified versions of the medicine, are loose combinations – from Arsuamoon® (Guilin, China), Larimal (Ipca, India) and Falcimon (Cipla, India)
Dihydroartemisinin 10 mg/ Piperaquine 80 mg	Sigma-tau/MMV Holley-Cotec	Submitted registration dossier to the EMEA in July 2009, with an expected approval date over the summer of 2011 Holley-Cotec has registered Duo-Cotexin (40 mg Dihydroartemisinin, 320 mg Piperaquine phosphate) in China, 18 African countries, plus Pakistan Cambodia and Myanmar. Other generic versions are now being produced following the inclusion into the WHO treatment guidelines in 2010
Artesunate 60 mg/ Pyronaridine 180 mg (Pyramax)	Shinpoong Pharmaceutical Co., Ltd/ MMV	Developed in two formulations a tablet for patients over 15 kg, and a sachet of granules (20 mg artesunate/60 mg pyronaridne) for patients > 5 kg. Submitted to the EMA in March 2010 with approval for adult form anticipated mid-2012

Artesunate 100 mg/ Mefloquine 220 mg	Farmanguinhos /DNDi	Product registered in Brazil in June 2008. Not prequalified by WHO, or approved by stringent regulatory authorities Cipla agreed to supply the drug in South East Asia in April 2008 Mepha has a version which is available in the premium markets in Africa, treating around 300 000 patients per year
Naphthoquine phosphate 78.3 mg/ Artemisinin 125 mg (ARCO)	Kunming Pharmaceutical Corp.(KPC)	Marketed as a single dose cure of 8 tablets: representing a dose of 20 mg/kg artemisinin. Naphthoquine phosphate represents 50 mg of free base. No ICH-GCP studies have been published to date, although clinical studies are starting now to be publishd
Azithromycin 250mg/ Chloroquine 150 mg	Pfizer/ MMV	Fixed dose combinations developed for prevention of infection in pregnancy. The treatment dose was established by an adult Phase III study, completed in September 2007. The study on intermittent preventive treatment in Pregnancy (IPTp) is currently being designed by Pfizer/MMV and the London School of Hygeine and Tropical Medicine and is now actively recruiting patients
Arterolane malate 150 mg/ Piperaquine phosphate 750 mg	Ranbaxy/ MMV	Arterolane (Rbx11160, OZ277): fully synthetic endoperoxide, originally discovered by an MMV consortium. Phase IIa showed a low exposure in patients, but this was reversed by combination partner, piperaquine. Ranbaxy started a Phase III targeting Indian registration in 2011
Sulphamethoxazole 800 mg/ Trimethoprim 160 mg (Co-trimoxazole Bactrim)	Institute of Tropical Medicine, Belgium	Testing as a preventative medicine in pregnancy (IPTp) as an alternative to SP. Trial targeted both HIV infected and non-infected pregnant women with CD4$\geq$ 200/µL. Study to complete April 2011
Products in Phase II – defining dose for pivotal trials		
Tafenoquine 200 mg	GlaxoSmith Kline	8-aminoquinoline (WR-238605; SB-252263) for the radical cure (relapse prevention) of *P. vivax*. Tafenoquine induces hemolysis in patients with a G6PD deficiency, and a study is ongoing to determine the safe dose in G6PD A- patients. A combination Phase II/III study has been discussed with US FDA and WHO and will start in 2011

(continued)

Table 2 (continued)

Active Ingredients	Partnership	Comments
Ferroquine	Sanofi-aventis	Ferroquine (SSR-97193 - originally from the University of Lille) completed Phase IIa in asymptomatic adult malaria patients in combination with artesunate in October 2008. The Phase IIb study of ferroquine in adolescents and children with symptomatic malaria started in October 2009. Efficacy was deemed insufficient, and so additional Phase II trials with Ferroquine have been discussed
Albitiazolium bromide	Sanofi-aventis	Albitiazolium bromide (SAR-97276, TE3) from Henri Vial at the University of Montpellier, is a choline uptake inhibitor. A multicentre open lablel efficacy and safety Phase II study with 180 patients started in August 2008. and was terminated due to insufficient level of efficacy. Further clinical studies are planned increasing the dose, with a view to providing a treatment for severe malaria in the case of artemisinin resistance
Fosmidomycin/ Clindamycin (Fosclin)	Jomaa Pharma GmbH	Fosclin is an oral fixed-dose combination of fosmidomycin (a 1-deoxy-D-xylulose 5-phosphate reductoisomerase inhibitor) and clindamycin (a lincosamide ribosomal protein synthesis inhibitor). Clinically it has a rapid parasite clearance time (median 18 h) primarily driven by fosmidomycin. A Phase II study in 40 patients with uncomplicated *P falciparum* malaria was initiated in November 2009. Alternative partners for Fosmidomycin are also in discussion
Artemisone	Hong Kong University of Science and Technology	Artemisone (BAY 44–9585) is a semi-synthetic artemsinin derivative, developed originally by Bayer, the University of Hong Kong University of Science and Technology and MMV. It has been completed Phase II studies in adult malaria patients, and is efficacious. It is being discussed for treatment for Artemisinin resistant patients in the Thai-Cambodia border region, although the current formulation is not sufficiently stable to allow these studies to take place

Methylene Blue/ Amodiaquine	Ruprects-Karls- University Heidelberg SFB544	Shown to be active in a Phase II study of 180 patients in Burkina Faso (Zoungrana et al., 2008). Methylene blue 10mg/kg/day and Amodiaquine 4 mg/kg/day for 3 days resulted in 95% efficacy at day 28 (PCR corrected), Vomitting and dysurea were significantly higher than with Amodiaquine-Artesunate
OZ439	MMV (Nebraska Monash Swiss Tropical Institute)	Synthetic 1,2,4 trioxalane - completed Phase I, with 800 mg dose giving plasma exposure over 60 hours, supporting its use as part of the long sought after single dose cure. Phase IIa studies started in late 2010 to confirm this result in patients, and the drug-drug interaction studies with potential partners are currently being planned
Phase I		
AQ-13	Immtech/ Tulane	Immtech is developing AQ-13 for the treatment of malaria and malaria prophylaxis in travelers (Form 10Q 2009). Additional genotoxicology studies are required. Tulane has independently shown that 700 mg AQ13 gives the same exposure as 600 mg Cholorquine in human volunteers
CDRI 97/78	Ipca	1,2,4 trioxane originally developed by India's Central Drug Research Institute. Started human volunteer studies in 2008 for safety evaluation. Human pharmacokinetic measurements are planned for a later study
GSK 932121	GlaxoSmith Kline/ MMV	GSK-932121A is 4(1H)-pyridone, mitochondrial electron transport blocker. It entered a Phase I trial in December 2008, and this trial was suspended in April 2009 because of safety concerns from studies of a pro-drug
N-tert butyl isoquine (GSK-369796)	Liverpool School of Tropical Hygeine/ GSK / MMV	N-tert butyl isoquine started Phase I in 2008, and showed a maximum tolerated dose of 1000 mg. Currently the clinical project is on hold. More supporting data is being obtained by the Liverpool group as to whether this represents an acceptable therapeutic window in man are obtained

(continued)

Table 2 (continued)

Active Ingredients	Partnership	Comments
Late stage Preclinical		
BCX 4208	Biocryst/ MMV/ Albert Einstein College of Medicine	Purine nucleoside phosphorylase inhibitor, in preclinical evaluation for treatment of P falciparum infection.. BCX 4208 has already been tested in man for other diseases (gout), and is being evaluated
MK 4815	MMV/ Merck	MK4815 has been shown to be active against P falciparum in vivo. Preclinical studies in rodents and dogs suggested additional safety studies in primates prior to commencing Phase I. Concerns about the therapeutic window have slowed down the progression of this molecule
Trioxaquine SAR116242	Sanofi-aventis/ Palumed	Trioxaquine SAR-116242 (PA-1103) is a fusion compound containing a cholorquine moiety and a trioxane endoperoxide – currently in GLP preclinical evaluation
RKA-182	Liverpool School of Tropical Medicine/ University of Liverpool	Synthetic endoperoxide containing tetroxane active group. Currently undergoing preclinical safety evaluation. Phase I studies could be expected to start in late 2011
NITD-609	Novartis led consortium/ MMV	A Novartis led consortium (with Swiss Tropical Insititute and the Biomedical Primate Research Institute) identified a spiroindolone, KAE609, optimized from a high throughput screening hit. This entered Phase I in late 2010
P218 DHFR inhibitor	MMV	P218 is a Dihydrofolate reductase inhibitor, active the resistant quadruple mutant of plasmodium DHFR. Identified by a consortium led by Thailand's BIOTEC with Monash University and London School of Hygeine and Tropical Medicine. It entered preclinical development in the spring of 2010
NPC-1161-B	University of Mississippi	Next Generation 8-aminoquinoline for the radical cure of P vivax. NPC1161B is the (−) enantiomer of the 8-aminoquinoline racemic mixture of NPC1161C retaining activity, but with potentially less hematological side effects. Partnership with Cumberland Pharmaceuticals (Nashville, TN)

DHODH inhibitors	Three consortia in the MMV portfolio	Three ongoing collaborations – one led by University of Texas, one by the Genzyme, and the other by GlaxoSmithKline. All three are being optimized using high resolution three dimensional structures. The Texas project is funded through an NIH grant. A preclinical candidate was defined in June 2011
Aminoindoles	Genzmye, Broad Institute/ MMV	Series of aminoindole inhibitors based on a whole cell screening hit, Genz 644442. Current compounds are active *in vivo* and are in late lead optimization
Imidazolidinediones	Walter Reid Army Institute of Research	WR182393 has radical curative and causal prophylactic activity in Rhesus infected with *P. cynomolgi* given intramuscularly, but not orally
dUTP'ase inhibitors	Medivir AB, University of York	Medivir is leading a project to identify dUTP'ase inhibitors, active against Plasmodium and other protozoal parasites. Current compounds are active in vivo, and expected to enter preclinical safety assessment within the next 18 months
Cell based lead	MerckSerono, WHO/TDR	MerckSerono has been collaborating with WHO TDR to develop inhibitors originally identified from cell based screening. A lead series was approved in late 2000 aiming for a preclinical candidate within 24 months
Cell based lead	Pfizer, WHO/TDR	In 2006 Pfizer started a program with WHO/TDR to screen 12 000 selected compounds against a variety of protozoan parasites. The best compounds are in lead optimization and a preclinical candidate is expected in the next 12 months
GNF-156 series	Novartis-led consortium/ MMV	This is the second compound optimized a whole cell active from HTS of the Novartis corporate collection, being developed by the consortium that includes the Swiss Tropical and Public Health Insitute and the Biomedical Primate Research Centre. It hasentered into preclinical development by the end of 2010, and could enter Phase I early in 2012
AN3661	Anacor/ MMV	AN3661 is the first of a novel class of boron containing compounds to be moved to preclinical development for the treatment of malaria. Phase I trials are anticipated to start in early 2012

(continued)

Table 2 (continued)

Active Ingredients	Partnership	Comments
Stopped in the last 12 months		
Tinidazole	Walter Reed Army Institute of Research	Based on a previously published Indian report, this compound was tested directly in patients. However no benefit on preventing relapse was observed
AD452	Treague/MMV	AD452 is the (+)erythro isomer of mefloquine. Based on rodent data it was proposed that the psychiatric and gastric adverse events experienced by patients receiving mefloquine may be caused by the (−) isomer. Pure (+)erythro-mefloquine was tested in a Phase I study with detailed psychiatric profiling, in comparison with the racemate and no significant difference was seen

Details of the prequalified medicines were downloaded from the WHO web-site at http://apps.who.int/prequal/ on January 14th 2010 next year based on recent clinical data and publications. A fixed dose combination of mefloquine–artesunate was developed by DND*i* and Farminguinhos of Brazil, and launched in 2008. This is currently being used only in the Brazilian market, although there are plans to license the technology to Ipca in India and expand its use in south-east Asia, where it is often the first-line therapy. Another version of mefloquine–artesunate is produced by Mepha, and this has been successfully marketed in the private market in Africa. One issue here is the relative price of artesunate–mefloqine, which at $2.50 is over twice as expensive as the other treatments. For the newer ACTs, the major cost component comes from the partner drug, since a larger dose is required. The cost per kilogram of the partner drugs is often higher than artesunate (currently costing $300/kg). Mefloquine is a good illustration: the dose of drug needed is three times that of artesunate, and the bulk price for mefloquine can be as high as $1,200/kg. New approaches to synthesizing mefloquine have been developed by MMV and our partners as a result of the (+)-mefloquine isomer project (see below), and there is no reason why in the future the cost of mefloquine–artesunate fixed dose combinations should not be at the same price as other ACTs. One remaining ACT is naphthoquine–artemisinin, marketed as ARCO by Kunming Pharmaceuticals. The combination tablet was initially developed by the Academy of Military Medical Sciences (AMMS), Beijing, China and is currently administered as a single-dose regimen. Although clinical experiences and clinical data are very limited, available preclinical data suggest that naphthoquine is relatively potent [6, 7]. The major concerns are those of safety (most 4-aminoquiniolines have safety issues if the entire dose is given at once), and the efficacy of artemisinin given as a single dose. Some clinical studies have been published, but they are lacking validation according to ICH Good Clinical Practice, and the safety database does not include the 2,000 patients normally required for registration.

Beyond ACTs, there are two other combination treatments in development. Artemisinin derivatives cause embryotoxicity in rats, and therefore have been contraindicated in the first trimester of pregnancy [8]. Pregnant women are especially at risk of malaria, since the parasite adheres to the placenta. Azithromycin and chloroquine are both known to be safe in pregnancy, and the combination has two other advantages. First, there appears to be clinical synergy, in that azithromycin overcomes chloroquine resistance. Second, the use of an antibacterial in pregnant women will reduce the incidence of sexually transmitted infections, with a concomitant effect on survival of the baby [9]. The second combination is sulphamethoxazole 800 mg with trimethoprim (known as Co-trimoxazole), which is being tested in HIV-infected pregnant women, and also has an impact on the malaria infection.

3 Next Generation Therapies: Overcoming Artemisinin Resistance

The major challenge of any anti-infective program is that sooner or later the infectious agent will develop resistance to therapy. Presently, the mechanism of action of and possible mechanisms of resistance to artemisinins are not fully understood. Thus far, the development of artemisinin resistance has been relatively slow considering the number of treatments that have been used, and the fact that monotherapies are still available in many disease endemic countries. The first reports of such resistance are now emerging [10, 11], where there is an increase in the parasite clearance time from patients along the Thai–Cambodia border. This is also reflected in some cases in a small change in sensitivity of parasite isolates to drugs, but so far no stable isolates of resistant parasites have been identified.

Unless the mechanism of resistance is an enzyme which specifically degrades endoperoxides, it is likely that the new semi- and fully synthetic endoperoxides: Arterolane [12] – formerly known as OZ277 or Rbx11160, OZ439 [13], Artemisone [14], RKA182 (a 1,2,4,5 tetraoxane) [15], CDRI 97/78 [16] and Trioxaquine [17] (Table 2) will have clinical activity against these resistant parasites. Since cell biology is unable to give an unambiguous answer to the question of the mechanism of resistance, the only route forward at this present time is to test these medicines in artemisinin resistance areas, and measure parasite clearance times. Such studies are now planned, and would be possible for compounds that have already been tested in uncomplicated malaria patients, such as Arterolane, Artemisone, and OZ439. Despite repeated attempts to start clinical studies in artemisinin insensitive patients, we have so far been unable to commence such studies. The relative difficulty in obtaining patients should be seen as a positive sign about the relative numbers of resistant patients, and the focal nature of the resistance to date. In addition, these studies are always difficult to interpret since parasite clearance times are also dependent on the role of the immune response in parasite infection. If these new generations of endoperoxide therapy are shown to be clinically active in artemisinin-resistant patients, then a 3-day combination therapy could be developed and registered within 5–7 years, perhaps faster depending on local need and local regulatory requirements. It is most unlikely that the entire class is compromised by an artemisinin resistance mechanism – amodiaquine is active against chloroquine resistance strains, for example. The next generation endoperoxide OZ439 offers an additional advantage over artemisinin in that it has a long half life, and is predicted to be able to maintain a therapeutically useful concentration in patients for as long as 96 h from a single dose.

In the case that artemisinin resistance destroyed the entire endoperoxide class of medicines, we will have to fall back on the antibiotic classes. Fosmidomycin–clindamycin, a mixture of an inhibitor of 1-deoxy-D-xylulose 5-phosphate reductoisomerase and an antibiotic, is currently in Phase IIa clinical trials and seems to have a rapid parasite clearance time, driven by the parasite reduction ratio of fosmidomycin [18]. If the speed of parasite killing can be confirmed in other studies, then it would be possible to imagine a combination with other partner medicines, including the 4-aminoquinoline family members such as piperaquine.

Other new approaches are still progressing through preclinical development. NITD609 is the first antimalarial to come from the phenotypic screen of more than two million compounds, carried out by Novartis. It moved from the initial high-throughput screen to a first administration in human volunteers in less than 4 years, which is extremely fast. Recently, AN3661 entered preclinical development. This is an extremely exciting new compound with unknown mechanism of action discovered by Anacor based on its activity in whole parasite assays. Its active moiety contains an oxaborole ring, and it has excellent drug-like properties. However, both of these are still at least 7 years away from being launched, and the complications of combination therapy and funding issues make this more likely to be closer to 10 years. Also, it is worth remembering that historically the chances that a molecule entering Phase I for anti-infective disease will make it all the way through to registration are at best two out of seven, and that this ratio falls as more and more innovative classes of drugs are tested (data from CMR International).

The second generation endoperoxide OZ439 is also exciting, since it has a half life of 12–24 h in man – and this means that currently a single dose in volunteers is enough to keep the plasma exposure above that required to kill the parasite for up to 60 h. It thus represents the first step toward having a single dose cure for malaria. This would be a dramatic step forward from the current 3 days of treatment – not only in terms of patient compliance, but also the ability for the health worker to observe treatment directly. The challenge over the next 2 years will be to find a partner for OZ439 in the combination. Most of the 4-aminoquinolines, aminoalcohols and related molecules have very long half lives, and so would be ideal; however, they cannot be given as a single dose because of the side effects. The hunt is therefore on for a long-acting partner where all the treatment can be given as a single administration. This also has to be a more potent medicine on a milligrams per patient basis, since the current dose of most partner drugs is as mush as 2 g. Some of the newer drugs such as naphthoquine or ferroquine [19] (a 4-aminoquinoline developed by sanofi-aventis) may have potential here.

4 Finding New Starting Points: Genomes and Screens

Two events over the last few years have helped the community find new starting points for antimalarial medicines. The first is technological. The development of high-throughput screening platforms, and their associated robotics and image processing reached such a state that by 2007 it was possible to consider testing compounds for their actions against the erythrocytic stages of *Plasmodium falciparum* whole parasites in 384 and 1,536 well formats [20, 21] by Novartis and GlaxoSmithKline. The high density of these formats has the advantage that the wells are smaller and use less compound and less cells, making the assays significantly cheaper. The cost of assays has come down by almost 100-fold in the last few years, and the throughput has gone up by two orders of magnitude – to date MMV and our partners have screened almost five million compounds. The availability of a high-throughput assay in a university context (at the Eskitis Institute, in Queensland, Australia) has meant that we have been able to screen the collections of companies who for various reasons do not want to have parasites or human blood in their HTS facility. This approach has also been adopted by other academic collaborators [22, 23]. Overall, the results have been impressive, and we are seeing a hit rate of around 0.5% of compounds which reproducibly have shown the ability to inhibit or kill the parasite at a concentration of less than 1 μM. This offers a major repository of structures of new potential antimalarials, which can be the start of medicinal chemistry programs. The recent decision to publish the structures in the public domain from their high-throughput screens now gives a map of "whole cell screening space" for the world to work from. In addition, Medicines for Malaria Venture has a program for making representative compounds available. The 25,000 currently identified hits have been clustered into around 1,000 families, and plates of these compounds will be available to malaria researchers in the second half of 2011. The availability of all these information should help groups to focus on areas where they have a unique contribution, and reduce duplication in hits to leads programs. The speed at which these molecules can be brought forward is illustrated by NITD609, which came from the Novartis screen, and moved from the original identification to preclinical evaluation in less than 3 years, a speed that is considered state of the art for commercial product development. However, the challenge will be to do this as efficiently as possible, and for this we will have to have a unique blend of using industry experience, academic insight and working with partners in countries such as India, South Africa, Brazil, and China to reduce costs.

The second major step forward has been the availability of the plasmodial genome sequences, allowing us to know a great many of the potential targets in the parasite [24]. Based on the structural knowledge of what makes good drug targets, these potential targets can be prioritized [25]. It also allows an "ortholog" approach to be followed by some groups, and this work has been pioneered by sanofi-aventis. The idea here is that among the targets identified in the parasite, many will have orthologs in the human genome, and these will already have been the subject of a human inhibitor search (more than 1,000 targets from the human genome have been worked on by the major pharmaceutical companies). By selecting compounds known to be active against the human target, and a representative set of structurally related inactives, then the hit-rate can be further improved to an astonishing 20–25%. In addition, much is already known about these compounds, such as their pharmacokinetics, metabolism and potential safety issues, meaning that the subsequent optimization is much faster. Often this work is further aided by the availability of high-resolution three-dimensional structures, so that compounds can be optimized by structure-based design [26]. These techniques have been used to great advantage on two of the projects listed in Table 2 – the identification of dihydrofolate reductase inhibitors, which overcome trimethoprim resistance [27], and the rational design of dihydro orotate dehydrogenase inhibitors [28].

5 Challenges of the Eradication Era of Gametocytes and Hypnozoites

The announcement in 2007 of a plan to eradicate malaria by the WHO and the Bill and Melinda Gates Foundation [29], raised the need for new classes of medicines. Two areas are most clearly highlighted in the search for new medicines. First, there is a need for a medicine that would destroy the dormant forms of *P. vivax* and *P. ovale,* known as hypnozoites [30]. When an infected mosquito bites, the first major parasite target is the liver, and in the case of these two parasite species, some of the parasites remain in the liver as small dormant forms – for anywhere between a few days and several years. The only registered therapy against hypnozoites is primaquine, a medicine developed during the Korean war. This has two major issues, first it requires 14 days of therapy, and so compliance is virtually nonexistent in the field, to the extent that some countries do not even have 14 days therapy on their treatment guidelines. Second, it causes hemolysis in patients who have a deficiency in the enzyme glucose 6-phosphate 1-dehydrogenase [31] (G6PD) – a condition that is as common as 15–20% in some malaria endemic regions. Tafenoquine was discovered originally by the Walter Reed Army Institute of Research, and has been tested in several thousand soldiers by the US army, and has the advantage of a longer half life, meaning that the course of treatment could be reduced to a much more manageable 3 days [32]. Currently, GlaxoSmithKline is undergoing studies to determine the safe dose in G6PD deficient individuals, as a prelude to a pivotal development program, which is planned to start in early 2011. This study will include women, and children, and should allow a registration of the product with the US FDA within 5 years. Beyond Tafenoquine, there is one other 8-aminoquinoline, NPC-1161C [33], but it is not clear that this offers an advantage in terms of potential safety in G6PD deficient patients, and this will require further *in vivo* assessment. There are currently no other chemical series in clinical development for antirelapse therapy. This is because up to now, there has been no cell biological model of the hypnozoite, and all compounds had to be tested in primates, which is extremely laborious and time consuming. The hypnozoites themselves are difficult to detect, and although primary hepatocyte systems have been available for 25 years [34], it has not been easy to produce cell lines, which support infection and produce hypnozoites. To date, the most robust system is from the related parasite *P. cynomolgi*, which is able to produce small stable forms when used to infect primary rhesus hepatocytes [35]. This gives a starting point for the evaluation of new compounds, although currently only perhaps hundreds can be tested per year.

The other challenge of the eradication era is to have molecules that will stop the transmission of the parasite to the mosquito vector. It is known that only the gametocytes (the sexual forms) of the parasite are transmitted, and that *P. falciparum* produces gametocytes in response to stress, including drug stress. The assays for gametocyte maturation and for the development of the ookinete inside the insect gut are well described [36]. Already we have been able to test the known malaria portfolio (all of the compounds listed in Table 2, plus all the compounds which are currently marketed antimalarials, and all the leads, compounds which are currently too early to be described in this review). Putting together a complete data set of the activity of compounds against the erythrocytic stages, the gametocyte and insect stages, and indeed the liver stages is part of what MMV has termed "Malaria Lifecycle Fingerprinting." The advantage of having all the data collected from identical assays carried out in identical conditions should not be underestimated. For the transmission blocking targets, a number of genes have already been identified as being involved from gene disruption (knock out) studies either in *P. falciparum* or *P. berghei* [37–39]. By selecting targets which inhibit gametocytes and transmission, it may be possible to identify molecules only active on these stages. The advantage would be that the transmission stages are relatively rare (there are less than five ookinetes, compared with as many as 10^{12} parasites at the erythrocytic stages). This means that development of resistance should be much less likely for the parasite.

The ultimate medicine would therefore be one which combined all the best attributes of a single dose cure of malaria, with something that was able to block transmission and also prevent relapses of disease caused by the hypnozoites. Such a medicine has been termed the "single exposure, radical cure" by the Malaria Eradication research working party (malERA) [40].

6 Conclusions

The last decade has seen considerable progress in the pipeline of antimalarial medicines. The prioritization of fixed dose artemisinin combination therapies, and their registration with stringent regulatory authorities and the WHO means now that the National Malaria Control Programs in disease endemic countries have a choice of treatment, and more importantly an evidence base on which to make their choice of treatment. Beyond this, the technologies which have come from high-throughput screening and from the genomic revolution have led to an acceleration of discovery efforts, and a wide array of new and exciting starting points for medicinal chemistry. The focus of molecules of the future cannot simply be about treating the patient – and coming up with the next generation of therapies to overcome resistance. The key to the future is the development of medicines, which are better suited to the needs of the patient – a single dose cure for *P. falciparum* malaria, and a shorter and safer course of treatment for destroying the hypnozoite reservoirs of *P. vivax*. Finally by having the tools that will prevent the infection of the insect vector, the community will be in a position to break the cycle of transmission, leading to the potential of the ultimate eradication of this devastating disease.

Acknowledgments I would like to thank the Medicines for Malaria Research and Development team for all their insights into this article, and for the many advisors, especially our External Scientific Advisory Committee, who have done so much to transform the pipeline of antimalarial therapeutics.

References

1. Maude RJ, Pontavornpinyo W, Saralamba S, Aguas R, Yeung S, Dondorp AM, Day NP, White NJ, White LJ (2009) The last man standing is the most resistant: eliminating artemisinin-resistant malaria in Cambodia. Malar J 8:31
2. Oliaro P, Wells TN (2009) The global portfolio of new antimalarial medicines under development. Clin Pharm Ther 85:584–595
3. Wells TN, Alonso PL, Gutteridge WE (2009) New medicines to improve control and contribute to the eradication of malaria. Nat Rev Drug Discov 8:879–891
4. White NJ (2008) Qinghaosu (artemisinin): the price of success. Science 320:330–334
5. World Health Organization. Malaria treatment guidelines, June 2010. http://www.who.int/malaria/publications/atoz/9789241547925/en/index.html. Accessed 22 May 2011
6. Tun T, Tint HS, Lin K, Kyaw TT, Myint MK, Khaing W, Tun ZW (2009) Efficacy of oral single dose therapy with artemisinin-naphthoquine phosphate in uncomplicated falciparum malaria. Acta Trop 111:275–278
7. Qu HY, Gao HZ, Hao GT, Li YY, Li HY, Hu JC, Wang XF, Liu WL, Liu ZY (2010) Single-dose safety, pharmacokinetics, and food effects studies of compound naphthoquine phosphate tablets in healthy volunteers. J Clin Pharmacol 50:1310–1318
8. Clark RL (2009) Embryotoxicity of the artemisinin antimalarials and potential consequences for use in women in the first trimester. Reprod Toxicol 28:285–296
9. Chico RM, Pittrof R, Greenwood B, Chandramohan D (2008) Azithromycin-chloroquine and the intermittent preventive treatment of malaria in pregnancy. Malar J 7:255
10. Noedl H, Se Y, Schaecher K, Smith BL, Socheat D, Fukuda MM; Artemisinin Resistance in Cambodia 1 (ARC1) Study Consortium (2008) Evidence of artemisinin-resistant malaria in western Cambodia. N Engl J Med 359:2619–2620
11. Dondorp AM, Nosten F, Yi P, Das D, Phyo PA, Tarning J, Lwin KM, Ariey F, Hanpithakpong W, Lee SJ et al (2009) Artemisinin resistance in *Plasmodium falciparum* malaria. N Engl J Med 361:455–467
12. Vennerstrom JL, Arbe-Barnes S, Brun R, Charman SA, Chiu FC, Chollet J, Dong Y, Dorn A, Hunziker D, Matile H et al (2004) Identification of an antimalarial synthetic trioxolane drug development candidate. Nature 430:900–904
13. Charman SA, Arbe-Barnes S, Bathurst IC, Brun R, Campbell M, Charman WN, Chiu FC, Chollet J, Craft JC, Creek DJ et al (2011) Synthetic ozonide drug candidate OZ439 offers new hope for a single-dose cure of uncomplicated malaria. Proc Natl Acad Sci USA 108:4400–4405
14. Nagelschmitz J, Voith B, Wensing G, Roemer A, Fugmann B, Haynes RK, Kotecka BM, Rieckmann KH, Edstein MD (2008) First assessment in humans of the safety, tolerability, pharmacokinetics, and *ex vivo* pharmacodynamic antimalarial activity of the new artemisinin derivative artemisone. Antimicrob Agents Chemother 52:3085–3091
15. O'Neill PM et al. (2010) http://www.a-star.edu.sg/Portals/0/media/Press%20Release/PressRelease_UKSIN_FINAL_media.pdf
16. Singh RP, Sabarinath S, Gautam N, Gupta RC, Singh SK (2009) Liquid chromatographic tandem mass spectrometric assay for quantification of 97/78 and its metabolite 97/63: A promising trioxane antimalarial in monkey plasma. J Chromatogr B 877:2074–2080
17. Coslédan F, Fraisse L, Pellet A, Guillou F, Mordmüller B, Kremsner PG, Moreno A, Mazier D, Maffrand JP, Meunier B (2008) Selection of a trioxaquine as an antimalarial drug candidate. Proc Natl Acad Sci USA 105:17579–17584
18. Na-Bangchang K, Ruengweerayut R, Karbwang J, Chauemung A, Hutchinson D (2007) Pharmacokinetics and pharmacodynamics of fosmidomycin monotherapy and combination therapy with clindamycin in the treatment of multidrug resistant falciparum malaria. Malar J 6:70
19. Dive D, Biot C (2008) Ferrocene conjugates of chloroquine and other antimalarials: the development of ferroquine, a new antimalarial. ChemMedChem 3:383–391

20. Plouffe D, Brinker A, McNamara C, Henson K, Kato N, Kuhen K, Nagle A, Adrián F, Matzen JT, Anderson P et al (2008) *In silico* activity profiling reveals the mechanism of action of antimalarials discovered in a high-throughput screen. Proc Natl Acad Sci USA 105:9059–9064
21. Gamo FJ, Sanz LM, Vidal J, Cozar C, Alarez E, Lavandera JL, Vanderwall DE, Green DVS, Kumar V, Hasan S et al (2010) Thousands of chemical starting points for antimalarial lead identification. Nature 465:305–310
22. Baniecki ML, Wirth DF, Clardy J (2007) High-throughput *Plasmodium falciparum* growth assay for malaria drug discovery. Antimicrob Agents Chemother 51:716–723
23. Guiguemde WA, Shelat AA, Bouck D, Duffy S, Crowther GJ, Davis PH, Smithson DC, Connelly M, Clark J, Zhu F et al (2010) Chemical genetics of *Plasmodium falciparum*. Nature 465:311–315
24. Gardner MJ, Hall N, Fung E, White O, Berriman M, Hyman RW, Carlton JM, Pain A, Nelson KE, Bowman S et al (2002) Genome sequence of the human malaria parasite *Plasmodium falciparum*. Nature 419:498–511
25. Aguero F, Al-Lazikani B, Aslett M, Berriman M, Buckner FS, Campbell RK, Carmna S, Carruthers IM, Chan AW, Chen F et al (2008) Genomic-scale prioritization of drug targets: the TDR Targets database. Nat Rev Drug Discov 7:900–907
26. Grundner C, Perrin D, Hooft van Huijsduijnen R, Swinnen D, Gonzalez J, Gee CL, Wells TN, Alber T (2007) Structural basis for selective inhibition of *Mycobacterium tuberculosis* protein tyrosine phosphatase PtpB. Structure 15:499–509
27. Yuthavong Y, Yuvaniyama J, Chitnumsub P, Vanichtanankul J, Chusacultanachai S, Tarnchompoo B, Vilaivan T, Kamchonwongpaisan S (2005) Malarial (*Plasmodium falciparum*) dihydrofolate reductase-thymidylate synthase: structural basis for antifolate resistance and development of effective inhibitors. Parasitology 130:249–259
28. Deng X, Gujjar R, El Mazouni F, Kaminsky W, Malmquist NA, Goldsmith EJ, Rathod PK, Phillips MA (2009) Structural plasticity of malaria dihydroorotate dehydrogenase allows selective binding of diverse chemical scaffolds. J Biol Chem 284:26999–27009
29. Roberts L, Enserink M (2007) Malaria. Did they really say ... eradication? Science 318:1544–1545
30. Wells TN, Burrows J, Baird JK (2010) Targeting the hypnozoite reservoir of *Plasmodium vivax*: the hidden obstacle to malaria elimination. Trend Parasitol 26:1471–1477
31. Beutler E, Duparc S; G6PD Deficiency Working Group (2007) Glucose-6-phosphate dehydrogenase deficiency and antimalarial drug development. Am J Trop Med Hyg 77:779–789
32. Kitchener S, Nasveld P, Edstein MD (2007) Tafenoquine for the treatment of recurrent *Plasmodium vivax* malaria. Am J Trop Med Hyg 76:494–496
33. Nanayakkara NP, Ager AL Jr, Bartlett MS, Yardley V, Croft SL, Khan IA, McChesney JD, Walker LA (2008) Antiparasitic activities and toxicities of individual enantiomers of the 8-aminoquinoline 8-[(4-amino-1-methylbutyl)amino]-6-methoxy-4-methyl-5-[3,4-dichlorophenoxy]quinoline succinate. Antimicrob Agents Chemother 52:2130–2137
34. Mazier D, Landau I, Druilhe P, Miltgen F, Guguen-Guillouzo C, Baccam D, Baxter J, Chigot JP, Gentilini M (1984) Cultivation of the liver forms of *Plasmodium vivax* in human hepatocytes. Nature 307:367–369
35. Dembele L, Gego A, Zeeman AM, Franetich JF, Silvi O, Rametti A, LeGrand R, Dereuddre-Bosquet N, Sauerwein R, van Gemert GJ (2011) Towards an *in vitro* model of *Plasmodium* hypnozoites suitable for drug discovery. PLoS One 6:e18162
36. Sinden RE (1998) Gametocytes and sexual development. In: Sherman IW (ed) Malaria: parasite biology, pathogenesis and protection. ASM, Washington, DC, pp 25–48
37. Dorin-Semblat D, Sicard A, Doerig C, Ranford-Cartwright L, Doerig C (2008) Disruption of the PfPK7 gene impairs schizogony and sporogony in the human malaria parasite *Plasmodium falciparum*. Eukaryot Cell 7:279–285
38. Kumar S, Molina-Cruz A, Gupta L, Rodrigues J, Barillas-Mury C (2010) A peroxidase/dual oxidase system modulates midgut epithelial immunity in *Anopheles gambiae*. Science 327:1644–1648

39. McRobert L, Taylor CJ, Deng W, Fivelman QL, Cummings RM, Polley SD, Billker O, Baker DA (2008) Gametogenesis in malaria parasites is mediated by the cGMP-dependent protein kinase. PLoS Biol 6:e139
40. Alonso PL, Djimde A, Kremsner P, Magill A, Milman J, Nájera J, Plowe CV, Rabinovich R, Wells T, Yeung S (2011) The malERA Consultative Group on Drugs, A research agenda for malaria eradication: drugs. PLoS Med 8:e1000402

Molecular Markers of *Plasmodium* Resistance to Antimalarials

Andrea Ecker, Adele M. Lehane, and David A. Fidock

Abstract Investigations into the molecular basis of *Plasmodium* parasite resistance to antimalarial drugs have made strong progress in defining key determinants. Mutations in the digestive vacuole transmembrane proteins *P. falciparum* chloroquine resistance transporter (PfCRT) and *P. falciparum* multidrug resistance protein 1 (PfMDR1) are important drivers of parasite resistance to several quinoline-based drugs including chloroquine, amodiaquine, and to a lesser extent quinine. Amplification of *pfmdr1* can also mediate resistance to mefloquine and impact lumefantrine efficacy. Parasite resistance to antifolates has been mapped to point mutations in the target enzymes dihydrofolate reductase and dihydropteroate synthase, and mutations in cytochrome *b* have been found to ablate atovaquone efficacy. Antibiotic resistance has been associated with mutations that preclude drug inhibition of protein translation in the parasite apicoplast. The study of resistance to artemisinin derivatives and several partner drugs used in artemisinin-based combination therapies is an area of active research that has yet to define clearly how in vitro resistance can translate into predictions of treatment failures. Research in this area is important not only for its ability to generate molecular markers of

A. Ecker
Department of Microbiology and Immunology, Columbia University Medical Center, Room 1502, Hammer Health Sciences Center, 701 West 168th Street, New York, NY 10032, USA

A.M. Lehane
Department of Microbiology and Immunology, Columbia University Medical Center, Room 1502, Hammer Health Sciences Center, 701 West 168th Street, New York, NY 10032, USA

Research School of Biology, The Australian National University, Canberra, ACT 0200, Australia

D.A. Fidock (✉)
Department of Microbiology and Immunology, Columbia University Medical Center, Room 1502, Hammer Health Sciences Center, 701 West 168th Street, New York, NY 10032, USA

Division of Infectious Diseases, Department of Medicine, Columbia University Medical Center, New York, NY 10032, USA
e-mail: df2260@columbia.edu

H.M. Staines and S. Krishna (eds.), *Treatment and Prevention of Malaria*,
DOI 10.1007/978-3-0346-0480-2_13,

treatment failure but also for the insights it can provide into drug mode of action and the development of chemical and pharmacological strategies to overcome resistance mechanisms.

1 Introduction

The ability to monitor and predict parasite drug resistance is crucial to the rational implementation of antimalarial treatment policies. The determination of resistance in vivo is costly and confounded by the effects of host immunity, pharmacokinetic variables, and the possibility of reinfection, among other factors. Similarly, determination of drug resistance of field isolates in vitro presents many technical caveats associated with performing drug susceptibility assays on non-culture-adapted and frequently polyclonal isolates from patients who may have an unknown treatment history. Not surprisingly, results from these in vitro assays may not always correlate with clinical outcome. Alternatively, molecular markers can serve as sentinels in the surveillance of drug resistance. However, the clinical relevance of a putative molecular marker must first be established.

While a wealth of studies attempting to correlate molecular markers and treatment outcome have been published, the lack of standardization of trial design and data reporting (as recommended in [1, 2]) can render comparison between trials difficult, and trials may fail to find an existing correlation due to lack of statistical power, if too few patients are included or the prevalence of a mutation is near saturation. Ideally, the contribution of a molecular marker to drug responses should further be determined in defined genetic backgrounds by allelic exchange. Importantly, with few exceptions, resistance phenotypes are multifactorial, and conflicting results in different genetic backgrounds, or experimental or geographical settings are therefore to be expected. The reader is referred to recent reviews for a general introduction to genetic mapping of drug resistance [3, 4].

In the following sections we will discuss molecular markers for drug resistance as they relate to *P. falciparum* (Table 1). Molecular markers for drug resistance in *P. vivax* are summarized at the end of this chapter.

2 Molecular Markers for Resistance to Quinoline-Based Drugs

2.1 P. falciparum *Chloroquine Resistance Transporter*

The primary determinant of chloroquine (CQ) resistance (CQR), *P. falciparum* chloroquine resistance transporter (*pfcrt*), was identified through a genetic cross between the CQ-sensitive and CQ-resistant clones HB3 (Honduras) and Dd2 (Indochina), respectively. In the progeny of this cross, inheritance of verapamil-reversible CQR segregated with a single genetic locus on chromosome 7, and was

Table 1 Summary of the major molecular markers for clinical resistance that have been identified for antimalarial drugs

Drug	Molecular marker for resistance	Comments
Amodiaquine	K76T mutation in PfCRT and N86Y mutation in PfMDR1	Other mutations in *pfcrt* and *pfmdr1* also affect amodiaquine response; the South American 7G8 *pfcrt* allele confers greater resistance to monodesethylamodiaquine (the primary metabolite) as compared to other *pfcrt* alleles investigated to date
Atovaquone	Y268S/C/N mutation in cytochrome *b*	This mutation mediates atovaquone resistance; it has been shown in the murine parasite *P. yoelii* that mutations in cytochrome *b* that cause atovaquone resistance also ablate the synergistic effects of the atovaquone–proguanil combination
Azithromycin	G76V mutation in the ribosomal protein L4	Based on an in vitro study; azithromycin is under investigation with partner drugs in malaria treatment trials
Chloroquine	K76T mutation in PfCRT	This mutation is accompanied by other mutations in PfCRT; polymorphisms in *pfmdr1* and perhaps *pfmrp* may augment or otherwise play a role in *pfcrt*-mediated chloroquine resistance
Clindamycin	A1875C mutation in apicoplast 23S rRNA	A2058C mutation at equivalent position in *E. coli* 23S rRNA is known to mediate resistance to clindamycin
Cycloguanil	Mutations in DHFR: A16V and S108T	Certain combinations of pyrimethamine-resistance-conferring mutations in DHFR also reduce parasite susceptibility to cycloguanil
Fosmidomycin	*pfdxr* amplification	Based on an in vitro study; fosmidomycin has not been registered for malaria treatment
Lumefantrine	*pfmdr1* amplification	The wild-type *pfcrt* allele (K76), and possibly also *pfmdr1* N86, confer a reduction in parasite susceptibility to lumefantrine as compared to other *pfcrt* and *pfmdr1* alleles
Mefloquine	*pfmdr1* amplification	Most important predictor of mefloquine resistance discovered to date; however mefloquine resistance has also been observed in the absence of *pfmdr1* amplification
Pyrimethamine	Mutations in DHFR: S108N, N51I, C59R and I164L	Not all of these mutations need be present; the extremely resistant quadruple mutant is widespread in Southeast Asia, whereas the triple mutant lacking I164L is prevalent in Africa
Quinine	No reliable marker identified	Quinine resistance is multifactorial; mutations in *pfcrt*, *pfmdr1* and *pfnhe-1*, and copy number variations in *pfmdr1*, may play a role

(continued)

Table 1 (continued)

Drug	Molecular marker for resistance	Comments
Sulfadoxine	Mutations in DHPS: S436A/F, A437G, K540E, A581G and A613S/T	Not all mutations need be present; the A437G mutation is the most commonly observed

ultimately mapped to mutations in the highly interrupted 13-exon *pfcrt* gene [5]. *pfcrt* accounts for >95% of the CQ response variation among the progeny, as revealed by quantitative trait loci (QTL) analysis [6] (Fig. 1). A strong association of *pfcrt* with parasite response to CQ has also been confirmed in a recent genome-wide association study [7]. *pfcrt* encodes a 49 kDa putative transporter of unknown function that localizes to the membrane of the parasite's digestive vacuole (DV) [5]. Based on bioinformatic analyses, PfCRT is a member of the drug/metabolite transporter superfamily [8].

Allelic exchange experiments have demonstrated unequivocally that mutant *pfcrt* can confer verapamil-reversible CQR to sensitive parasites [9]. This may to some extent depend on the genetic background as in the D10 (Papua New Guinea) strain the introduction of mutant *pfcrt* only resulted in a moderate CQ tolerance phenotype [10]. Importantly, a single mutation, resulting in the amino acid change K76T, has been shown to be essential for CQR, as removal of this mutation from CQ-resistant strains (Dd2 and 7G8, representative of resistant parasites from Southeast Asia and South America, respectively) resulted in complete loss of the resistance phenotype [11]. Replacement of the positively charged lysine (K) residue with a neutral threonine (T) may enable PfCRT to transport diprotonated CQ out of the DV, either through active transport or via facilitated diffusion, thereby decreasing access to its heme target in the DV (reviewed in [12]). This transport activity of mutant PfCRT was also studied in heterologous expression systems such as yeast [13], *Dictyostelium discoideum* [14] and *Xenopus laevis* oocytes [15]. In the latter study, mutant PfCRT was shown directly to mediate CQ transport, while no transport activity was seen for the wild type form.

Notably, K76T is always accompanied by at least three and up to eight additional polymorphisms. These are mainly found in or immediately adjacent to one of the ten transmembrane domains and may compensate for altered PfCRT function, due to the K76T polymorphism, and/or modulate drug susceptibility [5]. Amino acid sequences surrounding K76T (position 72–76) distinguish African and Southeast Asian (predominantly CVIET) and South American, Philippino, and Papua New Guinean strains (predominantly SVMNT) from the invariant wild type, CQ-sensitive haplotype (CVMNK) [16]. All together, 16 variant residues have been identified, rendering PfCRT an extraordinarily polymorphic protein (Table 2) [5, 17–28].

An association of *pfcrt* K76T and in vitro CQR has also been confirmed in many, although not all, studies using field isolates (reviewed in [29]). This might reflect the requirements for additional genes in drug resistance, an interplay with other undetected *pfcrt* polymorphisms, or technical caveats associated with determining the susceptibility of non-culture-adapted, frequently polyclonal patient isolates.

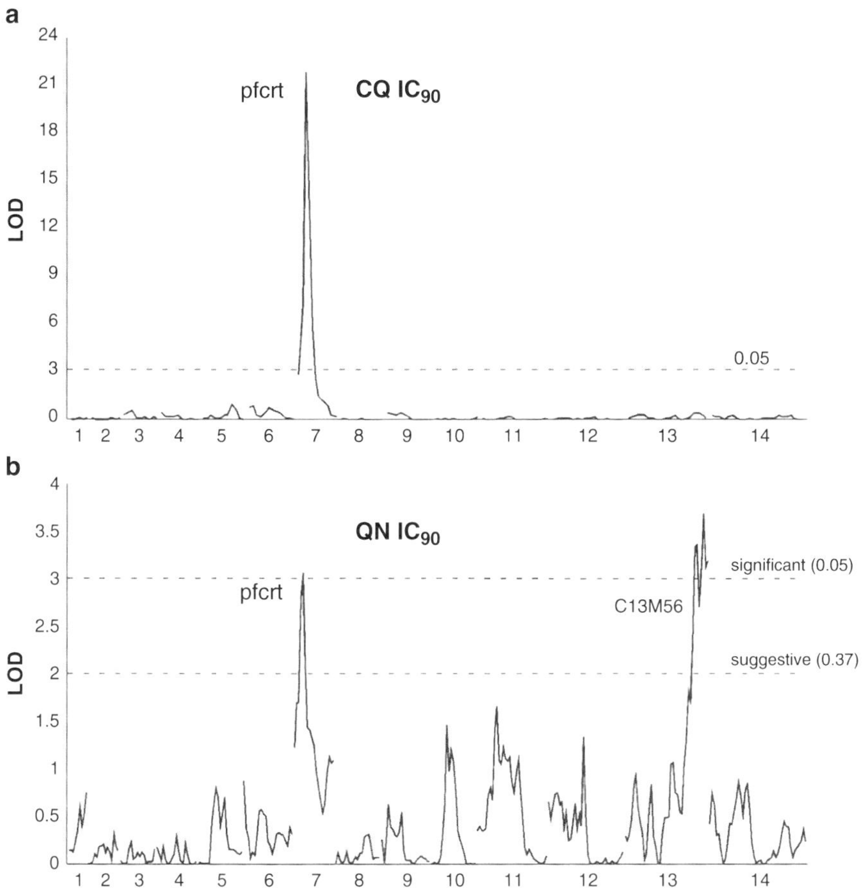

Fig. 1 Quantitative trait loci (QTL) analysis of a cross between the chloroquine (CQ)- and quinine (QN)-sensitive HB3 strain and the CQ-resistant and low-level QN-resistant Dd2 strain. *pfcrt* accounts for >95% of the variation in CQ responses amongst the progeny, evident as a sharp peak on chromosome 7 with a log of differences (LOD) score of >20 (**a**), however this gene has a more modest effect on QN responses (**b**). Most of the QN response variation can be attributed to two main additive QTL: 35% to a chromosome 13 peak centered on the C13M56 marker (thought to be a result of mutations in *pfnhe-1* or a neighboring gene [6, 90]), and another 30% to a chromosome 7 peak centered on *pfcrt*. This figure was reproduced in modified form with permission from [6]

pfcrt K76T has also been associated with CQ treatment failure in clinical settings (e.g. [30–32]). A recent review and meta-analysis of 25 studies confirmed that *pfcrt* K76T increased the risk of therapeutic failure after CQ treatment [1]. This validates the usefulness of *pfcrt* K76T as a sensitive molecular marker for CQR in the field. Nevertheless, *pfcrt* K76T has proven less predictive of clinical failure in patient populations with high pre-existing partial immunity; this "premunition" allows

Table 2 PfCRT polymorphisms

Region of origin[a]	Isolates (examples)	CQ response[b]	PfCRT position and encoded amino acid[c,d]																
			72	74	75	76	97	144	148	160	194	220	271	326	333	334	356	371	Ref.
Wild type haplotype																			
All regions	HB3	S	C	M	N	K	H	A	L	L	I	A	Q	N	T	S	I	R	[5]
Mutant haplotypes																			
A	106/1[e]	S	C	I	E	K	H	A	L	L	I	S	E	S	T	S	I	I	[5]
A, SEA	PAR, FCB	R	C	I	E	T	H	A	L	L	I	S	E	S	T	S	I	I	[5]
A, SEA	102/1, Dd2	R	C	I	E	T	H	A	L	L	I	S	E	S	T	S	T	I	[5]
SEA	field isolate	n.d.	C	I	E	T	H	A	–	L	–	S	E	–	–	–	–	R	[28]
SEA	Cam742	R	C	I	E	T	H	A	L	L	I	S	E	N	T	S	I	I	[19]
SEA	Cam783	R	C	I	E	T	H	A	L	L	I	S	E	N	T	S	T	I	[19]
SEA	TM93-C1088	R	C	I	E	T	L	A	–	L	–	S	E	S	–	–	T	I	[18]
SEA	field isolate	n.d.	C	I	E	T	H	Y	–	L	–	A	E	–	–	–	–	R	[28]
SEA	field isolate	n.d.	C	I	D	T	H	Y	–	L	–	A	E	–	–	–	–	R	[28]
SEA	field isolate	n.d.	C	I	D	T	H	Y	–	L	–	A	E	–	–	–	–	I	[28]
SEA	Cam738	R	C	I	D	T	H	A	I	L	T	S	E	N	S	S	I	R	[19]
SEA	Cam734	R	C	I	D	T	H	F	I	L	T	S	E	N	S	S	I	R	[19]
SEA	PH1	R	C	M	N	T	H	T	L	Y	I	A	Q	D	T	S	I	R	[18]
SEA	PH2	n.d.	S	M	N	T	H	T	–	Y	–	A	Q	D	–	–	I	R	[18]
SEA	PNG4	R	S	M	N	T	H	–	–	–	–	A	Q	D	–	–	L	R	[27]
SA	7 G8	R	S	M	N	T	H	A	L	L	I	S	Q	D	T	S	L	R	[5]
SA	Ecu1110	R	C	M	N	T	H	–	–	–	–	S	Q	D	–	–	L	R	[5]
SA	TU741	R	C	M	N	T	H	A	L	L	I	S	Q	D	T	N	L	R	[20]
SA	Jav	R	C	M	E	T	Q	A	L	L	I	S	Q	N	T	S	I	T	[5]
SA	TA4641	R	C	M	E	T	Q	A	L	L	I	S	E	N	T	S	I	–	[20]
SA	TA4640	R	C	M	E	T	Q	A	L	L	I	S	Q	N	S	S	I	I	[20]

[a] *A* Africa, *SEA* Southeast Asia, *SA* South America

[b] *S* CQ-sensitive, *R* CQ-resistant, *n.d.* not determined

[c] Gray shading indicates residues that differ from the wild type allele. Residues that were not reported are indicated by a dash

[d] Additional partial haplotypes for position 72–76 have been reported in [17, 21–26]

[e] Revertant?

some patients with infections carrying *pfcrt* T76 to respond adequately to CQ treatment [31, 33].

pfcrt K76T has also been used as an epidemiological tool to monitor parasite populations in response to changes in drug policy, as exemplified in a study that surveyed *pfcrt* genotypes after discontinuation of CQ use in Malawi. Strikingly, CQ-resistant strains disappeared within 10 years of CQ withdrawal, which suggests that the CVIET-type mutant *pfcrt* allele imparts a substantial fitness cost in this high transmission setting [34]. Nevertheless, minority *pfcrt* K76T genotypes may still be present in Malawi, as standard PCRs may fail to correctly identify these in polyclonal infections. They can be detected with greater sensitivity using a multiple site-specific heteroduplex tracking assay [35].

In addition to their central role in parasite response to CQ, *pfcrt* mutations can significantly alter the degree of susceptibility to a number of structurally unrelated, clinically important drugs, including several of the partner drugs used in artemisinin-based combination therapy. Introduction of mutant *pfcrt* into the CQ-sensitive GC03 strain conferred a slight increase in resistance to amodiaquine (AQ) and to its primary and active metabolite desethylamodiaquine (DEAQ) [9]. Field studies have also demonstrated selection of *pfcrt* K76T in recrudescences and reinfections following AQ treatment, although the presence or absence of this mutation before treatment was not predictive of treatment outcome [36]. *pfcrt* mutations, particularly in the 3′ region of the gene, were enriched in Colombian field isolates with decreased susceptibility to AQ and DEAQ [20]. Indeed, the analysis of two genetic crosses revealed that South American *pfcrt* alleles, in combination with South American *pfmdr1* alleles (see below for a discussion of *pfmdr1*), mediate high DEAQ resistance but only moderate CQR [37]. AQ has been used in South America since the late 1940s [37]. Thus, historically DEAQ may have been the main driving force in selection of these PfCRT haplotypes.

In contrast, the *pfcrt* K76T mutation was reported to increase parasite susceptibility to lumefantrine, and was selected against in reinfections following artemether–lumefantrine treatment [38] (note, it is generally assumed that these reinfections represent reinoculations and that selective pressure is exerted by subtherapeutic levels of the long half-life artemisinin partner drug, in this case lumefantrine). *pfcrt* mutations, particularly at position 76, can also increase parasite susceptibility to quinine, mefloquine, halofantrine, and artemisinin and its derivatives [9, 39]. In the case of quinine, the effects of *pfcrt* mutations depend on the parasite genetic background [11], and the pattern of cross-resistance with its diastereomer quinidine is complex [39, 40]. This suggests that *pfcrt* is but one component of a multifactorial basis of quinine resistance [6].

2.2 P. falciparum *Multidrug Resistance Protein 1*

P. falciparum multidrug resistance protein 1 (*pfmdr1*) encodes an ATP-binding cassette (ABC) transporter orthologous to mammalian P-glycoproteins that mediate verapamil-reversible multidrug resistance in mammalian cancer cells. PfMDR1, also known as P-glycoprotein homologue 1 (Pgh-1), is composed of two homologous halves, each with six predicted transmembrane domains and a conserved nucleotide-binding domain [12]. PfMDR1 localizes to the DV membrane and may function in importing solutes into the DV [41]. Together with PfCRT, PfMDR1 is likely to be an important regulator of drug accumulation in this organelle.

pfmdr1 was identified based on the similarity of CQR and tumor multidrug resistance and the observation that CQR can be reversed by the calcium channel blocker verapamil [42], which suggested the involvement of an *mdr* ortholog [43, 44]. However, the analysis of the HB3 × Dd2 genetic cross revealed that CQR was not linked to *pfmdr1* [45], and led to the identification of *pfcrt* as the major determinant of resistance [5].

Nevertheless, it was subsequently shown that *pfmdr1* plays a central role in parasite responses to many drugs. Amplification of *pfmdr1* was selected in vitro by mefloquine pressure [44, 46, 47], and was shown to correlate with increased transcript [43, 46, 47] and protein [46] abundance. The mefloquine-selected lines displayed cross-resistance to halofantrine and quinine [46, 47] and vice-versa [48]. This cross-resistance has also been observed in the field [49] and may explain why *pfmdr1* amplifications were observed in field isolates obtained before mefloquine was in clinical use [46].

Increased *pfmdr1* copy number was also associated with reduced in vitro susceptibility to mefloquine, halofantrine, quinine, dihydroartemisinin, and artesunate in Thai isolates [49–51]. Importantly, *pfmdr1* gene amplification was shown to be the most important predictor for treatment failure following mefloquine monotherapy [49, 52] or artesunate–mefloquine combination therapy [49, 53]. Increased *pfmdr1* copy number was also associated with a four-fold increase in the risk of treatment failure of a 4-dose artemether–lumefantrine regimen, while the 6-dose regimen remained effective in the presence of *pfmdr1* amplification [54]. Nevertheless, some mefloquine-resistant parasites lack *pfmdr1* amplification, illustrating that mefloquine (and halofantrine) cross-resistance can also be mediated by other mechanisms [48, 49, 55]. It is worth noting that *pfmdr1* transcript levels are reportedly rapidly upregulated in vitro following parasite exposure to CQ, mefloquine or quinine, but not to pyrimethamine [56]. While *pfmdr1* amplification was originally reported in CQ-resistant lines [43], several subsequent studies have demonstrated an inverse relationship between *pfmdr1* copy number and the level of CQR, suggesting that *pfmdr1* overexpression tends to reduce the level of CQR [46, 47, 57].

The genetic disruption of one of the two *pfmdr1* copies in the drug resistant FCB line increased susceptibility to mefloquine, lumefantrine, halofantrine, quinine, and artemisinin [58], further illustrating the central role that *pfmdr1* copy number plays in parasite responses to multiple antimalarials. However, no change in IC_{50} values for CQ was observed, suggesting that the effect of *pfmdr1* copy number on CQR may be strain-specific or depend on specific combinations of *pfcrt* and *pfmdr1* polymorphisms [58].

In addition to amplification, *pfmdr1* polymorphisms have also been associated with drug susceptibility. These occur in several different haplotypes (Table 3) [18, 20, 37, 59–67]. The mutations likely alter the substrate specificity of PfMDR1, as illustrated by the observation that PfMDR1 S1034C/N1042D failed to complement yeast deficient for a mating factor export molecule, while wild type PfMDR1 could [68]. Furthermore, it was shown that expression of wild type PfMDR1 in mammalian cells increased CQ uptake, while PfMDR1 S1034C/N1042D had no effect [69]. Similarly, in heterologous expression studies in *Xenopus* oocytes wild type PfMDR1 transported CQ and quinine but not halofantrine, whereas PfMDR1 variants transported halofantrine but not the other two drugs [70]. Of note, with the exception of Y184F, mutations are generally absent from *pfmdr1* alleles that have also undergone amplification [49, 54].

An association of *pfmdr1* mutations and response to CQ has been reported in some, but not other studies using field isolates (reviewed in [29, 71, 72]). A recent review and meta-analysis of 29 studies supported a role for *pfmdr1* N86Y in CQ and AQ treatment failure, although the association with CQ was weak [1]. Other than N86Y, D1246Y may also be involved in AQ/DEAQ resistance. In East Africa, D1246Y (together with N86Y and Y184) was selected in recrudescent infections after AQ therapy [73], and in a set of Colombian field isolates the parasites with

Table 3 PfMDR1 polymorphisms

Region of origin	Isolates (examples)	PfMDR1 position and encoded amino acid[a,b]					Reference
		86	184	1034	1042	1246	
Wild type							
All regions	D10	N	Y	S	N	D	
Major mutant haplotypes							
K1 haplotype (predominant in Asia and Africa)	Dd2	Y	Y	S	N	D	[37, 62]
7G8 haplotype (predominant in South America)	7G8	N	F	C	D	Y	[37, 62]
Minor mutant haplotypes							
Africa, Southeast Asia	Ghana, Nigeria 2	N	F	S	N	D	[62, 65, 67]
Southeast Asia	Field isolates	N	F	C	N	D	[67]
South America, Southeast Asia	Jav, Ecu1110, HB3, BT3	N	F	S	D	D	[20, 37, 59, 62, 67]
South America, Southeast Asia	PH4	N	F	C	D	D	[18, 59, 67]
South America	Field isolates	N	F	S	D	Y	[20, 64]
Africa	Field isolates	N	F	–	N	Y	[60]
Africa	Field isolates	N	Y	–	N	Y	[60]
Africa	GB4, Nigeria 32	Y	F	S	N	D	[37, 62]
Africa	Field isolates	Y	Y	S	N	Y	[63, 66]
Africa	Field isolates	Y	F	–	N	Y	[60]
Africa	Field isolates	F	Y	S	N	D	[61]

[a]Gray shading indicates residues that differ from the wild type allele. – not determined
[b]Additional polymorphisms may be present at other positions

pfmdr1 D1246Y tended to have the highest AQ IC_{50} values [20]. In South America, the 7G8-type *pfcrt* and *pfmdr1* alleles appear to interact to confer high-level AQ/DEAQ resistance [37]. *pfmdr1* N1042D was strongly associated with lumefantrine IC_{50} values in isolates from the Thai–Myanmar border [74]. Furthermore, in two studies in Africa the wild type N86 and the Y184F and 1246D mutations had significantly higher allelic frequencies in reinfections following artemether–lumefantrine treatment [75, 76], suggesting that *pfmdr1* polymorphisms may prove useful as a molecular marker for lumefantrine resistance. In contrast, while certain *pfmdr1* point mutations did alter susceptibility of field isolates to mefloquine and artesunate in vitro [49, 50, 74], they were not predictive for treatment outcome following mefloquine monotherapy or mefloquine–artesunate combination therapy [49, 52, 77].

Allelic exchange studies have confirmed the influence of 3′ *pfmdr1* point mutations on parasite responses to numerous drugs. In the CQ-sensitive D10 and CQ-resistant 7G8 parasite strains, the S1034C/N1042D/D1246Y mutations were sufficient to confer a significant decrease in quinine susceptibility [78]. A strong influence of position 1042 on quinine response was confirmed in the GC03 and 3BA6 genetic backgrounds, representing progeny of the HB3 × Dd2 genetic cross, although the three mutations were insufficient to confer quinine resistance to GC03 when compared with the wild type allele [72]. These findings corroborate a QTL analysis of the same genetic cross that attributed 10% of the total variation in quinine response to a region on chromosome 5 containing *pfmdr1* [6], and a recent genome-wide association study that reported an association of *pfmdr1* with quinine and, more weakly, CQ responses [7]. In contrast, the presence of the S1034C/N1042D/D1246Y mutations was correlated with increased susceptibility to artemisinin, mefloquine, and halofantrine in all genetic backgrounds used in these allelic exchange studies (D10, 7G8, GC03, and 3BA6) [72, 78]. These findings corroborate the linkage between *pfmdr1* mutations and increased susceptibility to mefloquine, halofantrine, lumefantrine, artemisinin, artemether, and arteflene (Ro 42–1611) that had previously been observed in the genetic cross between HB3 and the CQ-sensitive 3D7 strain [79].

The role of 3′ *pfmdr1* polymorphisms in parasite responses to CQ has been less clear. While reversal of S1034C/N1042D/D1246Y in a CQ-resistant clone (7G8) halved the level of CQR, introduction of the same mutations into another CQ-resistant clone (3BA6) and two sensitive clones (D10 and GC03) had no effect on parasite susceptibility to CQ [72, 78]. These findings suggested that *pfmdr1* polymorphisms can enhance CQR in a strain-dependent manner, but are insufficient to confer CQR to a CQ-sensitive strain. *pfmdr1* may contribute to CQR directly by transporting drug or by influencing physiological parameters that influence drug accumulation or parasite susceptibility to the toxic consequences of drug interference with heme detoxification. Alternatively, *pfmdr1* polymorphisms might be beneficial indirectly by compensating for physiological perturbations induced by the expression of mutant *pfcrt* [29]. This epistatic interaction may in part explain the observed linkage disequilibrium between *pfcrt* and *pfmdr1* (reviewed in [29]).

2.3 P. falciparum *Multidrug Resistance-Associated Protein and Other Transporters*

In addition to *pfcrt* and *pfmdr1*, many other transporters have earlier been reported to associate with parasite responses to quinine and CQ in a study that genotyped 97 culture-adapted isolates for SNPs in 49 genes encoding predicted or known transporters and transport regulatory proteins [80]. The identified SNPs included two point mutations in the plasma membrane ABC transporter *P. falciparum* multidrug resistance-associated protein (PfMRP) (PFA0590w), which were weakly associated with CQR in Asia (Y191H) and the Americas (A437S), and quinine resistance in the Americas (Y191H, A437S) [80]. However, studies on *P. falciparum* isolates from French travelers returning from Africa [81], and on clinical isolates from the Thailand–Myanmar border [74], failed to find significant associations between SNPs in these transporter genes and drug responses, with the single exception of a borderline significant association of a 3 bp insertion in the ABC transporter G7 (PF13_0371) and increased artesunate IC_{50} values [74], the relevance of which remains to be determined. Of note, PfMRP Y191H could not be analyzed in the Thai–Myanmar isolates due to insufficient variation [74]. Recent studies have also reported evidence for the possible selection of certain PfMRP alleles in reinfections following artemether–lumefantrine treatment (those with 876I) [82], and in recrudescence following sulfadoxine–pyrimethamine treatment (those with 1466K) [83].

Disruption of PfMRP in the CQ-resistant strain W2 rendered the parasite more susceptible to CQ and quinine, possibly as a result of reduced efflux of these drugs by the knockout parasite [84]. PfMRP may transport multiple chemically unrelated drugs, as illustrated by the reduced IC_{50} values that were also reported for artemisinin, piperaquine, and primaquine. PfMRP-deficient parasites also accumulated more glutathione, indicating that PfMRP may transport glutathione out of the cell. Overall the observed reductions in IC_{50} values were modest (38–57%), suggesting that PfMRP might serve as a secondary determinant to modulate parasite resistance to these antimalarials.

2.4 P. falciparum *Na^+/H^+ Exchanger 1*

Earlier QTL mapping of 35 progeny of a cross between the low-level quinine-resistant Dd2 clone and the quinine-sensitive HB3 clone identified five genomic regions with additive or pairwise effects on quinine response [6]. Most of the quinine response was attributed to two additive loci, a chromosome 7 peak that was centered around *pfcrt* (30% of the variation) and a peak on chromosome 13 (35% of the variation) that interacted with a locus on chromosome 9 (Fig. 1). Within the chromosome 13 region, the authors identified the *P. falciparum* Na^+/H^+ exchanger 1 (*pfnhe-1*) gene that encodes a putative Na^+/H^+ exchanger as a potential candidate gene and demonstrated an association of a D- and N-rich repeat polymorphism (microsatellite ms4760-1) with quinine response in 71 isolates from Africa, South east Asia, and Central and South America [6]. Of note, this locus is highly polymorphic, with >30 different genotypes reported thus far [85, 86]. However, subsequent studies have reported conflicting results on the association of this locus with quinine responses, with increasing repeat number associated with increased quinine IC_{50} values in one study [86], but decreased IC_{50} values in another study [87].

Some strains with higher levels of quinine resistance were reported to possess elevated Na^+/H^+ exchange activity and either a higher cytosolic pH or vacuolar-to-cytosolic pH gradient [88], although the experimental basis of these observations has been called into question in other studies [89, 90]. Importantly, a genetic knockdown of *pfnhe-1* expression through truncation of the 3′ untranslated region resulted in a statistically significant ~30% decrease in quinine mean IC_{50} values in the CQ- and (moderately) quinine-resistant 1BB5 and 3BA6 parasite lines, but not in the CQ-sensitive GC03 line [90]. These data support a role for *pfnhe-1* in contributing to quinine susceptibility in a strain-specific manner but also imply that other parasite factors are required for the parasite to be resistant to quinine.

3 Molecular Markers for Resistance to Antifolates

Antifolates that have been used for malaria treatment include pyrimethamine, sulfadoxine, and proguanil. In the case of proguanil, it is the cycloguanil metabolite that interferes with folate metabolism. Resistance of *P. falciparum* parasites to all of these agents is widespread [91]. Pyrimethamine and cycloguanil inhibit the parasite dihydrofolate reductase (DHFR) and sulfadoxine inhibits a downstream enzyme in the folate biosynthetic pathway, dihydropteroate synthase (DHPS). The sulfadoxine–pyrimethamine combination was used extensively as a first-line treatment against *P. falciparum* in many malaria-endemic regions when CQ lost its efficacy [92]. The efficacy of this combination is highly dependent on synergy between the components; when resistance to either component exists, the efficacy of the combination is severely reduced [93]. Readers are referred to Gregson and Plowe [94] for a comprehensive review on the mechanisms of antifolate action and resistance in *P. falciparum*.

Resistance of *P. falciparum* parasites to antifolates is mediated by mutations in the target proteins – DHFR in the case of pyrimethamine and cycloguanil and DHPS in the case of sulfadoxine. Evidence for an association between mutations in DHFR/DHPS and resistance to the antifolates came from sequencing of the *dhfr* and *dhps* genes in strains with varying susceptibilities to antifolate drugs [95–100]. A causal effect for mutations in these proteins was subsequently established by transfecting parasites with mutant forms of the genes [101, 102]. The effects of mutations in DHFR and DHPS on enzymatic activity and sensitivity to antifolate-mediated inhibition have also been ascertained with recombinant proteins from *P. falciparum* or by characterizing corresponding mutations in the DHFR ortholog from *Toxoplasma gondii* [103–105].

A number of different mutant forms of DHFR have been identified in the field. Pyrimethamine-resistant isolates contain a S108N mutation, and more highly resistant isolates possess additional mutations at codons 51 (N51I), 59 (C59R), and sometimes 164 (I164L). An A16V/S108T form has also been discovered and confers greater resistance to cycloguanil than to pyrimethamine [97]. The other forms confer a greater degree of resistance to pyrimethamine than to cycloguanil with the exception of the "quadruple mutant" (the S108N/N51I/C59R/I164L combination), which confers high-level resistance to both drugs [97, 98]. The quadruple mutant is of particular concern as it has been associated with very high levels of sulfadoxine–pyrimethamine treatment failure [106]. It is widespread in Southeast Asia but remains rare in Africa [107]. Studies in which *P. falciparum* DHFR variants were expressed in yeast or *E. coli* revealed that the addition of the I164L mutation to the triple mutant form of DHFR augmented resistance to pyrimethamine but also had a negative impact on enzyme activity (as inferred from the rate of yeast/bacterial growth) [108, 109]. Thus, quadruple mutant parasites may be at a fitness disadvantage, and more fit, less resistant parasites may have the upper hand in Africa where asymptomatic untreated infections are common [94]. Five mutations associated with sulfadoxine resistance have been identified in DHPS: S436A/F, A437G, K540E, A581G, and A613S/T. Among these mutations, the A437G change (either alone or in combination with other mutations) is observed most often in field isolates [93].

A recent study analyzed the literature (consisting predominantly of studies performed in Africa within the past decade) on the effect of mutations in DHFR on treatment outcome with sulfadoxine–pyrimethamine [1]. When considered in isolation and without knowledge of whether other DHFR mutations were present, the mutations at positions 108, 51, and 59 gave odds ratios (ORs; the proportion of therapeutic failures with mutation-bearing parasites divided by the proportion of therapeutic failures with wild type parasites) of 2.1, 1.7, and 1.9, respectively. For the 51 + 59 + 108 triple mutant, the OR was 4.3 [1]. No data were provided for the quadruple mutant, presumably because fewer studies on its association with treatment failure have been performed.

From studies investigating the influence of DHPS mutations on sulfadoxine–pyrimethamine treatment outcome, Picot et al. determined an OR of 1.5 for the A437G mutation and an OR of 3.9 for the 437 + 540 double mutant. For "quintuple mutants" with mutations at positions 51, 59, and 108 in DHFR and 437 and 540 in DHPS, the OR was 5.2 [1]. Two studies reported that the presence of two mutations, DHFR C59R and DHPS K540E, was highly predictive of the presence of the quintuple mutation set [110, 111]. Thus, monitoring just these two mutations could be sufficient to enable predictions of sulfadoxine–pyrimethamine efficacy in a given area [94]. As with all antimalarial treatment regimens, parasite genotype is not the only determinant of treatment outcome. Host factors including immunity, nutritional status (including blood folate concentrations), and drug metabolism rates will also impact on patient response.

Although the move towards artemisinin-based combination therapies has lessened the use of sulfadoxine–pyrimethamine for malaria treatment, the latter continues to occupy an important place in the "intermittent preventative treatment" of pregnant women. A recent study investigated whether this choice of treatment of pregnant women was of benefit in an area of Tanzania in which mutant *dhfr* and *dhps* alleles are prevalent. Worryingly, the authors found that women who received treatment in this area were more likely to have a higher parasitemia and increased level of inflammation in the placenta than those who did not [112]. This would imply that the monitoring of sulfadoxine–pyrimethamine resistance in areas in which it is being considered for use in intermittent preventative treatment is of utmost importance for informing policy choices.

There is no evidence for an amplification of the genes encoding DHFR or DHPS in antifolate-resistant *P. falciparum* field isolates [94]. However, the amplification of a gene upstream in the folate biosynthesis pathway, *gtp-cyclohydrolase I*, has been observed in certain strains [113] and has been reported to associate with the I164L mutation in DHFR [114]. An increase in the level of GTP-cyclohydrolase I (the rate-limiting enzyme in the folate biosynthesis pathway) might be an adaptation to compensate for a reduced efficiency of 164L-containing DHFR [114].

4 Molecular Markers for Resistance to Artemisinins

Clinical resistance to artemisinin derivatives has not yet emerged, although data from artemisinin monotherapy treatment studies in west Cambodia provide evidence of delayed parasite clearance times, which might presage the emergence of bona fide resistance leading to treatment failures [115, 116]. Thus there is a clear imperative to identify molecular markers of artemisinin resistance. The major candidate that has been proposed is the ATP-consuming calcium-dependent *P. falciparum* SERCA ortholog, PfATP6. Initial studies in transfected *Xenopus laevis* oocytes provided evidence that PfATP6 could be specifically inhibited by artemisinin, an effect that could be antagonized by the mammalian SERCA inhibitor thapsigargin [117]. When expressed and purified from *Saccharomyces cerevisiae*, PfATP6 incorporated into lipid vesicles could not be inhibited by artemisinin and was only partially sensitive to high concentrations of the SERCA inhibitor thapsigargin [118]. The latter suggests that caution should be taken when interpreting these results, which do not rule out the PfATP6 hypothesis (reviewed in [119]). Modeling studies have predicted that introducing an L263E mutation into PfATP6 could result in a loss of artemisinin binding, which was subsequently confirmed in oocyte assays [120]. This mutation has now been engineered into the *pfatp6* locus in *P. falciparum* [121]. Assays with cultured asexual blood stage parasites found no difference between L263E and wild type lines in their IC_{50} values with artemisinin, dihydroartemisinin, and artesunate. However, L263E mutant lines showed a clear difference in the distribution of their IC_{50} values, and IC_{50} values normalized to wild type controls showed a significant increase to artemisinin and dihydroartemisinin that was particularly evident in the D10 (Papua New Guinea) genetic background.

Studies have not observed the L263E mutation in field isolates. However, multiple other mutations have been observed, including the S769N mutation that was detected in a few isolates from French Guiana that yielded high IC_{50} values when tested against artemether [119, 122, 123]. Studies to define their impact in *P. falciparum* should shed further light on the relationship between PfATP6 and parasite susceptibility to artemisinins.

The search for molecular markers for artemisinin resistance via a genetic approach of selecting for resistance and identifying mutations that might be causal for resistance has been severely hampered by the difficulties in selecting stably resistant lines in vitro. In genetically modified parasite lines, *pfcrt* and *pfmdr1* mutations were observed to increase parasite susceptibility to artemisinin and its derivatives [9, 39, 72, 78], but the observed changes were small and neither gene therefore appears to be a major resistance determinant. *pymdr1* amplification was reported in unstably artemisinin-resistant *P. yoelii* [124], and some – but not all – artemisinin- or artelinic acid-selected *P. falciparum* lines had also increased their *pfmdr1* copy number [125, 126]. However, when the artelinic-acid resistant *P. falciparum* line was cultured in the absence of drug, parasites rapidly deamplified *pfmdr1*, concomitant with only a partial reversal of drug resistance [127]. Stable resistance to artemisinin and its derivatives has to date only been obtained in *P. chabaudi*; this was achieved in the absence of mutations or copy number changes in the *pfcrt* and *pfmdr1* orthologs and other genes previously associated with the mechanism of action of artemisinins [128].

Linkage group selection performed with the uncloned progeny of a genetic cross between the artemisinin-resistant *P. chabaudi* line and a sensitive clone subsequently identified a region on chromosome 2 that appeared to be selected by artemisinin [129]. Within this region, the authors detected two nonsynonymous mutations within a gene encoding a putative deubiquitinating protease termed ubiquitin-specific protease-1 (UBP-1) or ubiquitin carboxyl terminal hydrolase (UBCTH) (the ortholog of *P. falciparum* MAL1P1.34b). These mutations arose during selection with CQ and with CQ and mefloquine (V2728F), or with CQ and artesunate (V2697F), respectively [130]. This puzzling observation of similar mutations in the same gene occurring under selection with two different drugs could be explained if (1) mutant *pcubp1* compensated for a fitness cost in the last common progenitor instead of directly contributing to resistance to either drug; or (2) if both drugs compromised a common cellular function, such as the regulation of oxidative stress; or (3) if UBP-1 mediated the posttranslational modification of a target that modulated parasite responses to both drugs [129]. Of note, no mutations in *pfubp1* were found in unstably artemisinin- or artelinic acid-resistant *P. falciparum* [125, 129]. Without allelic exchange studies, one cannot rule out that artemisinin and artesunate resistance in *P. chabaudi* is instead caused by a different gene, either physically linked or epistatic to *pcubp1*.

Unstably artemisinin-resistant *P. yoelii* parasites were also shown to overexpress translationally controlled tumor protein homolog (TCTP) [131], a protein previously identified as potentially interacting with dihydroartemisinin [132]. However, no change in PfTCTP expression level was observed in artemisinin- or artelinic acid-resistant *P. falciparum* [125]. There is to date no evidence that TCTP can contribute to reduced parasite susceptibility to artemisinin.

Parasite responses to the artemisinin combination therapy partner drugs lumefantrine, AQ, and mefloquine can be impacted by *pfcrt* and *pfmdr1*, and the reader is referred to the respective section of this chapter. Resistance to piperaquine has been reported following its use as monotherapy in China, but the genetic resistance determinant has not been identified [133]. Resistance to pyronaridine has not yet been observed.

5 Molecular Markers for Resistance to Antibiotics

Doxycycline is a tetracycline antibiotic that is used widely for antimalarial chemoprophylaxis. Doxycycline and tetracycline can also be used in combination with quinine, quinidine, or artesunate [91]. In bacteria, tetracyclines exert their toxic effects by binding to the 30S subunit of the 70S ribosome (comprising the 16S rRNA and ribosomal proteins) and thereby inhibiting protein translation [134]. It is likely that in *P. falciparum* they target the bacterial-like protein translation machinery in the apicoplast [135]. There have been reports of prophylactic failures with doxycycline, although this could relate to inadequate doses or to poor compliance. Clinical resistance has not been demonstrated.

In a recent study, the doxycycline susceptibilities of 747 *P. falciparum* isolates from different African countries were tested in vitro with the [^{3}H]-hypoxanthine incorporation method [136]. The strains were classified into three different phenotypic groups, with IC_{50} values for doxycycline stratified as 4.9 ± 2.1, 7.7 ± 1.2 or 17.9 ± 1.4 μM. The drug assays were performed over 42 h (<1 complete asexual cycle), despite previous evidence showing that doxycycline becomes markedly more potent in the second cycle of parasite proliferation [137]. A follow-up study reported that increases in the copy number of *pfmdt* (PFE0825w; a putative drug/metabolite transporter) and *pftetQ* (PFL1710c; a putative tetQ family GTPase) were more frequent in parasites in the "higher doxycycline IC_{50} group" than in parasites from the other two groups [138]. Furthermore, the number of KYNNNN amino acid motif repeats in TetQ was on average slightly lower in the higher doxycycline IC_{50} group [138]. It is likely that the second-cycle potency of doxycycline is more relevant to its clinical activity than its first-cycle potency, thus it would be premature to consider *pftetQ* or *pfmdt* as potential molecular markers for reduced parasite susceptibility to doxycycline without first knowing whether a higher first-cycle IC_{50} value is predictive of a higher second-cycle IC_{50} value.

Clindamycin is a lincosamine antibiotic that, like the tetracyclines, inhibits prokaryotic protein translation, but in this case by binding to the 50S ribosomal subunit, which comprises the 23S rRNA, the 5S rRNA, and ribosomal proteins [134]. Clindamycin is also recommended for malaria treatment in combination with quinine, quinidine, or artesunate, and unlike the tetracyclines is deemed safe for pregnant women and small children [139]. The existence of clindamycin-resistant *P. falciparum* parasites was first uncovered in 2010, when genomic analyses of isolates from the Peruvian Amazon led to the discovery of two point mutations in the apicoplast 23S rRNA, one of which (A1875C) had a similar location to mutations known to confer clindamycin resistance in various bacteria [140]. In vitro drug susceptibility testing revealed that the A1875C mutation was associated with a marked decrease in parasite sensitivity to clindamycin (a >100-fold increase in IC_{50}).

The erythromycin derivative azithromycin, which is being investigated as a potential component of an antimalarial combination therapy, also acts against bacteria by binding to the 50S ribosomal subunit [134]. It binds in the polypeptide exit tunnel adjacent to the peptidyl transferase center and hence interferes with the translocation of nascent polypeptides through this tunnel [141]. The binding site includes the 23S rRNA and the proteins L4 and L22 [142–144].

In a recent study, *P. falciparum* parasites resistant to azithromycin were generated through in vitro drug pressure [145]. The drug-selected lines had IC_{50} values for azithromycin that were 16- to 17-fold higher than those of the parental lines (7G8 and Dd2). Studies on the mechanisms of azithromycin action and resistance in bacteria allowed Sidhu and colleagues to take a candidate gene approach to determining the molecular basis of azithromycin resistance. Genes encoding the *P. falciparum* orthologs of the 23S rRNA and the proteins L4 and L22 were PCR-amplified from the azithromycin-resistant lines and sequenced. A single point mutation was found in the *Pfrpl4* gene (encoding L4) in the azithromycin-resistant lines selected from the 7G8 and Dd2 lines, and not in the parent lines themselves. This mutation was identical in both azithromycin-resistant lines and would result in a glycine to valine substitution at position 76 in the L4 protein. Similarly-placed L4 mutations have been reported in macrolide-resistant bacteria, and the region of the protein in which they reside appears to be key in determining the access of drug to the binding site in the ribosome (reviewed in [146]) (Fig. 2). Thus, this study strongly suggests that azithromycin acts against *P. falciparum* by inhibiting protein translation in the apicoplast and that, like in bacteria, resistance can be mediated by alterations in the components that are predicted to form the drug-binding site.

Another antibiotic that has shown some promise in clinical studies is fosmidomycin [147–149]. Fosmidomycin inhibits 1-deoxy-D-xylulose 5-phosphate (DOXP) reductoisomerase (DXR), the second enzyme in the DOXP pathway for isoprenoid biosynthesis [139]. A recent study investigated whether *P. falciparum* parasites resistant to this antibiotic could be generated through in vitro drug

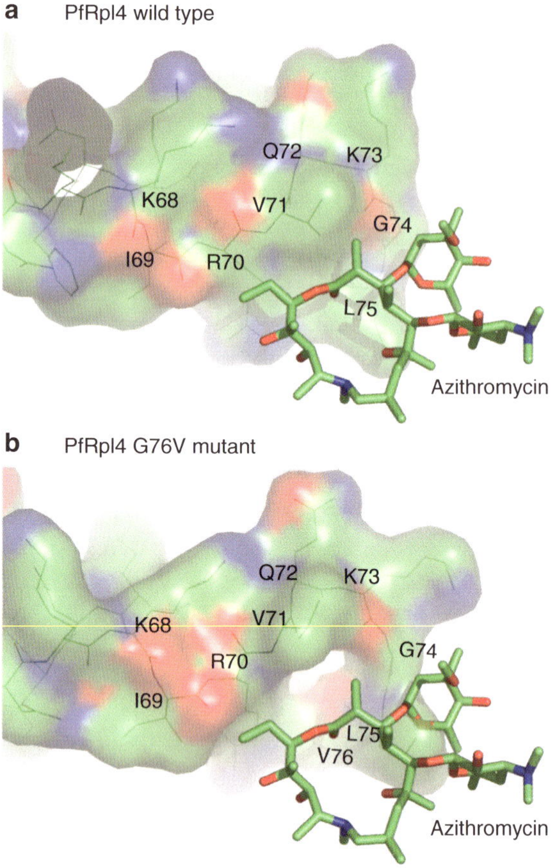

Fig. 2 Structural models of (**a**) wild type and (**b**) the G76V mutant Pfrpl4 with azithromycin. These models, derived using the *Deinococcus radiodurans* crystal structure of protein L4 as a template, revealed a steric clash between the side chain of 76V in the G76V mutant and azithromycin, consistent with this mutation conferring azithromycin resistance. Modeling was performed by Qingan Sun and James C. Sacchettini (Texas A&M University, College Station, TX). This figure was originally published in [145]. © The American Society for Biochemistry and Molecular Biology

pressure, and if so, whether the resistance mechanism could be elucidated using a high-density tiling microarray [150]. Fosmidomycin-resistant parasites (with an ~8-fold elevated IC_{50} value) were generated by sequentially increasing the drug concentration applied to Dd2 parasites. DNA from fosmidomycin-resistant clones and the parental line was then hybridized onto the microarrays, and a single large amplification event on chromosome 4 of the fosmidomycin-resistant clones was observed. The amplified region was ~100 kb in size and contained 23 genes, the first of which was *pfdxr*, the gene encoding the putative target of fosmidomycin. Subsequent analyses revealed a ~3.8-fold increase in the level of *pfdxr* transcript and a ~3-fold increase in the gene copy number in a fosmidomycin-resistant clone compared with the Dd2 parental line, with no changes in the gene sequence [150].

Thus, potential molecular markers for resistance of *P. falciparum* parasites to azithromycin and fosmidomycin have been identified before these compounds have been approved for use as antimalarial drugs. The identification of these markers was aided greatly by previous studies on antibiotic action and resistance in bacteria.

6 Molecular Markers for Resistance to Atovaquone

Atovaquone was developed in the 1980s and found to possess activity against a number of parasites including *Plasmodium* [151]. When tested as a monotherapy for *P. falciparum* malaria, atovaquone treatment was associated with a high risk of parasite recrudescence (~30%) [152–154]. Recrudescent parasites isolated from patients treated with atovaquone were found to be highly resistant to the drug, with a >1,000-fold elevated IC_{50} value when tested in vitro [153]. Despite atovaquone's shortcomings when used as a single agent, the synergistic combination of atovaquone and proguanil (Malarone) has proven very effective against malaria, and is used for treatment and prophylaxis [155]. This synergy is attributable to proguanil, not its DHFR-inhibiting metabolite cycloguanil [156]. A study with the murine malaria parasite *P. yoelii* revealed that once parasites are resistant to atovaquone, synergy with proguanil is lost [157].

Studies on the mechanism of action of atovaquone have revealed that the drug inhibits mitochondrial electron transport by interacting with the cytochrome *bc*1 complex [158]. It was recently shown, by generating transgenic parasites expressing the *S. cerevisiae* dihydroorotate dehydrogenase enzyme in the cytosol, that the parasite's electron transport chain is essential only for the regeneration of ubiquinone, which is the required electron acceptor for the parasite (but not the *S. cerevisiae*) dihydroorotate dehydrogenase [159]. The transgenic parasites were found to be highly resistant to atovaquone and other electron transport inhibitors, yet remained highly susceptible to the atovaquone–proguanil combination. The authors hypothesized that the maintenance of a membrane potential across the mitochondrial membrane is essential for parasite viability, and that in addition to the electron transport chain, a proguanil-sensitive pathway for the generation of a mitochondrial membrane potential exists in the parasite. Atovaquone alone did not abrogate the membrane potential in either the parental or transgenic parasites; however, the atovaquone–proguanil combination was highly effective in dissipating the membrane potential in both cases [159].

There is evidence to support the use of the cytochrome *b* gene (encoded on the mitochondrial genome) as a molecular marker to monitor atovaquone resistance. Mutations in this gene were found in each of nine independent atovaquone-resistant *P. yoelii* lines selected using suboptimal treatment of *P. yoelii*-infected mice [157]. The mutations in cytochrome *b* observed in the atovaquone-resistant *P. yoelii* lines all fell within a 15 amino acid region that forms part of the catalytic domain, or Q_o site, where ubiquinol oxidation occurs [160]. Mutations in the Q_o region of the cytochrome *b* gene have subsequently been reported in atovaquone-resistant *P. falciparum* [161], *P. berghei* [162], and *T. gondii* [163] parasites.

Atovaquone–proguanil treatment failures have been associated with the mutation of Y268 in *P. falciparum* cytochrome *b* to S, C or N (Y268S/C/N; [164, 165], and references therein). It has been shown that parasites bearing a mutation at position 268 can arise and spread within individual patients receiving atovaquone–proguanil treatment [166]. Introducing a mutation in the bacterial cytochrome *b* at the position corresponding to residue 268 in *P. falciparum* (Y302C in the bacterial protein) was found to render the bacterial cytochrome *bc*1 complex much less sensitive to inhibition by atovaquone [167], thereby establishing a causal effect for this mutation in the phenomenon of atovaquone resistance.

7 Molecular Markers for Drug Resistance in *P. vivax*

The lack of a long term in vitro culture system for *P. vivax* remains a major obstacle to the study of drug resistance in this important human pathogen. Insights into the genetic basis of resistance therefore remain limited and have been gained mainly by comparison with *P. falciparum*.

Orthologs of *pfcrt* and *pfmdr1* have been identified in the *P. vivax* genome. Multiple polymorphisms have been reported in these genes, *pvcrt-o* (also known as *pvcg10*) and *pvmdr1*, albeit not at positions homologous to the mutations found in the *P. falciparum* genes [168, 169]. Moreover, *pvcrt-o* and *pvmdr1* mutations did not associate with CQR in patient isolates or monkey-adapted lines [168, 170–172]. One exception may be *pvmdr1* Y976F, which was found to associate with higher CQ IC_{50} values in Thai isolates [173]. However, this mutation had reached near fixation in the studied parasite isolates from Papua, Indonesia [173], Madagascar [172], and Brazil [174], despite low rates of CQ treatment failure reported in the latter two studies, and may thus be of limited value as a molecular marker. These findings suggest that CQR may have a different genetic basis in *P. vivax*. It is worth noting, however, that expression of wild-type *pvcrt-o* in *P. falciparum* increased the CQ IC_{50} value by 2.2-fold, and in *D. discoideum* resulted in reduced CQ accumulation in acid endosomes [175], suggesting that *pvcrt-o* could nevertheless play some role in CQR in *P. vivax*.

A mechanism of drug resistance that may more readily translate from *P. falciparum* is the role of *mdr1* amplification in mefloquine resistance. Two studies have reported increased *pvmdr1* copy numbers in *P. vivax* isolates from Tak province in Thailand, where mefloquine has been heavily used, but not in isolates from regions where parasites have not been exposed to this drug (other Thai provinces, as well as Laos, Myanmar, Papua) [169, 173].

There is also emerging evidence that mutations in *dhfr* and *dhps* are associated with a reduced parasite sensitivity and a higher risk of treatment failure to sulfadoxine–pyrimethamine in *P. vivax* (reviewed in [176]). More than 20 *dhfr* alleles have been identified in *P. vivax* [176]. This includes the pyrimethamine-resistance-conferring S58R/S117N form, whose crystal structure has recently been elucidated, along with wild type *P. vivax* DHFR ([177]; Fig. 3). The S58R/S117N mutations are homologous to the C59R/S108N mutations in PfDHFR that cause moderate resistance to pyrimethamine. The residue at position 58 in PvDHFR was found not to interact with pyrimethamine, and it was suggested that it may be involved in binding the natural substrate. The mutation at codon 117 was found to perturb the

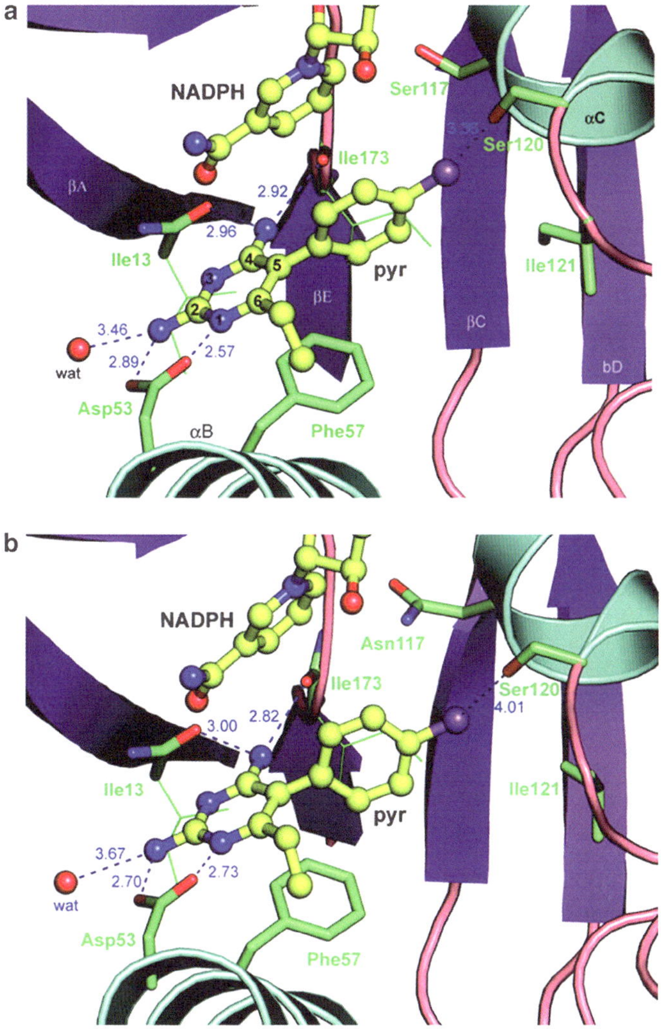

Fig. 3 Binding of pyrimethamine in the active site of *Plasmodium vivax* DHFR. The pyrimethamine (pyr) and NADPH cofactor are shown as *balls* and *sticks* with carbon, nitrogen, and chlorine colored *yellow*, *blue*, and *magenta*, respectively. (**a**) Pyr binding with the wild type *P. vivax* DHFR. Interactions between the enzyme and the pyrimidine ring of the inhibitor include electrostatic interactions and H-bonds indicated by dotted lines. *Numbers* next to the lines indicate distances in Å. (**b**) Pyr binding with the S58R/S117N double-mutant enzyme. X-ray crystallography revealed that the interactions around the pyrimidine ring were similar to the wild type enzyme. The mutation at amino acid 117 from S to N increases a steric factor in the active site. As a result, the positions of both NADPH and pyr were perturbed from their optimum binding, reducing the efficiency of pyr by as much as 300-fold. The mutant R residue at position 58 did not directly interact with the inhibitor and was proposed to affect substrate binding. Data were published by [177]. These crystal structure images were reproduced with kind permission from Yongyuth Yuthavong and PNAS

binding of pyrimethamine [177], as shown previously for the S108N mutation in the *P. falciparum* enzyme [178].

Acknowledgments Funding for this work was provided in part by the NIH (R01 AI50234 to D.A.F.) and an Investigator in Pathogenesis of Infectious Diseases Award from the Burroughs Wellcome Fund (to D.A.F.). A. M. L. is supported by an Australian NHMRC Overseas Biomedical Fellowship (585519) and A.E. is supported by a Human Frontier Science Program Long Term Postdoctoral Fellowship.

References

1. Picot S, Olliaro P, de Monbrison F, Bienvenu AL, Price RN, Ringwald P (2009) A systematic review and meta-analysis of evidence for correlation between molecular markers of parasite resistance and treatment outcome in falciparum malaria. Malar J 8:89
2. Sibley CH, Barnes KI, Watkins WM, Plowe CV (2008) A network to monitor antimalarial drug resistance: a plan for moving forward. Trends Parasitol 24:43–48
3. Hayton K, Su XZ (2008) Drug resistance and genetic mapping in *Plasmodium falciparum*. Curr Genet 54:223–239
4. Ekland EH, Fidock DA (2007) Advances in understanding the genetic basis of antimalarial drug resistance. Curr Opin Microbiol 10:363–370
5. Fidock DA, Nomura T, Talley AK, Cooper RA, Dzekunov SM, Ferdig MT, Ursos LM, Sidhu AB, Naude B, Deitsch KW et al (2000) Mutations in the *P. falciparum* digestive vacuole transmembrane protein PfCRT and evidence for their role in chloroquine resistance. Mol Cell 6:861–871
6. Ferdig MT, Cooper RA, Mu J, Deng B, Joy DA, Su XZ, Wellems TE (2004) Dissecting the loci of low-level quinine resistance in malaria parasites. Mol Microbiol 52:985–997
7. Mu J, Myers RA, Jiang H, Liu S, Ricklefs S, Waisberg M, Chotivanich K, Wilairatana P, Krudsood S, White NJ et al (2010) *Plasmodium falciparum* genome-wide scans for positive selection, recombination hot spots and resistance to antimalarial drugs. Nat Genet 42:268–271
8. Martin RE, Kirk K (2004) The malaria parasite's chloroquine resistance transporter is a member of the drug/metabolite transporter superfamily. Mol Biol Evol 21:1938–1949
9. Sidhu AB, Verdier-Pinard D, Fidock DA (2002) Chloroquine resistance in *Plasmodium falciparum* malaria parasites conferred by *pfcrt* mutations. Science 298:210–213
10. Valderramos SG, Valderramos JC, Musset L, Purcell LA, Mercereau-Puijalon O, Legrand E, Fidock DA (2010) Identification of a mutant PfCRT-mediated chloroquine tolerance phenotype in *Plasmodium falciparum*. PLoS Pathog 6:e1000887
11. Lakshmanan V, Bray PG, Verdier-Pinard D, Johnson DJ, Horrocks P, Muhle RA, Alakpa GE, Hughes RH, Ward SA, Krogstad DJ et al (2005) A critical role for PfCRT K76T in *Plasmodium falciparum* verapamil-reversible chloroquine resistance. EMBO J 24:2294–2305
12. Valderramos SG, Fidock DA (2006) Transporters involved in resistance to antimalarial drugs. Trends Pharmacol Sci 27:594–601
13. Zhang H, Paguio M, Roepe PD (2004) The antimalarial drug resistance protein *Plasmodium falciparum* chloroquine resistance transporter binds chloroquine. Biochemistry 43:8290–8296
14. Naude B, Brzostowski JA, Kimmel AR, Wellems TE (2005) *Dictyostelium discoideum* expresses a malaria chloroquine resistance mechanism upon transfection with mutant, but not wild-type, *Plasmodium falciparum* transporter PfCRT. J Biol Chem 280:25596–25603

15. Martin RE, Marchetti RV, Cowan AI, Howitt SM, Broer S, Kirk K (2009) Chloroquine transport via the malaria parasite's chloroquine resistance transporter. Science 325:1680–1682
16. Bray PG, Martin RE, Tilley L, Ward SA, Kirk K, Fidock DA (2005) Defining the role of PfCRT in *Plasmodium falciparum* chloroquine resistance. Mol Microbiol 56:323–333
17. Best Plummer W, Pinto Pereira LM, Carrington CV (2004) *Pfcrt* and *pfmdr1* alleles associated with chloroquine resistance in *Plasmodium falciparum* from Guyana, South America. Mem Inst Oswaldo Cruz 99:389–392
18. Chen N, Kyle DE, Pasay C, Fowler EV, Baker J, Peters JM, Cheng Q (2003) *pfcrt* allelic types with two novel amino acid mutations in chloroquine-resistant *Plasmodium falciparum* isolates from the Philippines. Antimicrob Agents Chemother 47:3500–3505
19. Durrand V, Berry A, Sem R, Glaziou P, Beaudou J, Fandeur T (2004) Variations in the sequence and expression of the *Plasmodium falciparum* chloroquine resistance transporter (*pfcrt*) and their relationship to chloroquine resistance *in vitro*. Mol Biochem Parasitol 136:273–285
20. Echeverry DF, Holmgren G, Murillo C, Higuita JC, Bjorkman A, Gil JP, Osorio L (2007) Polymorphisms in the *pfcrt* and *pfmdr1* genes of *Plasmodium falciparum* and *in vitro* susceptibility to amodiaquine and desethylamodiaquine. Am J Trop Med Hyg 77:1034–1038
21. Hatabu T, Iwagami M, Kawazu S, Taguchi N, Escueta AD, Villacorte EA, Rivera PT, Kano S (2009) Association of molecular markers in *Plasmodium falciparum crt* and *mdr1* with *in vitro* chloroquine resistance: a Philippine study. Parasitol Int 58:166–170
22. Huaman MC, Yoshinaga K, Suryanatha A, Suarsana N, Kanbara H (2004) Polymorphisms in the chloroquine resistance transporter gene in *Plasmodium falciparum* isolates from Lombok, Indonesia. Am J Trop Med Hyg 71:40–42
23. Lim P, Chy S, Ariey F, Incardona S, Chim P, Sem R, Denis MB, Hewitt S, Hoyer S, Socheat D et al (2003) *pfcrt* polymorphism and chloroquine resistance in *Plasmodium falciparum* strains isolated in Cambodia. Antimicrob Agents Chemother 47:87–94
24. Menard D, Yapou F, Manirakiza A, Djalle D, Matsika-Claquin MD, Talarmin A (2006) Polymorphisms in *pfcrt*, *pfmdr1*, *dhfr* genes and *in vitro* responses to antimalarials in *Plasmodium falciparum* isolates from Bangui, Central African Republic. Am J Trop Med Hyg 75:381–387
25. Nagesha HS, Casey GJ, Rieckmann KH, Fryauff DJ, Laksana BS, Reeder JC, Maguire JD, Baird JK (2003) New haplotypes of the *Plasmodium falciparum* chloroquine resistance transporter (*pfcrt*) gene among chloroquine-resistant parasite isolates. Am J Trop Med Hyg 68:398–402
26. Restrepo E, Carmona-Fonseca J, Maestre A (2008) *Plasmodium falciparum*: high frequency of *pfcrt* point mutations and emergence of new mutant haplotypes in Colombia. Biomedica 28:523–530
27. Wootton JC, Feng X, Ferdig MT, Cooper RA, Mu J, Baruch DI, Magill AJ, Su XZ (2002) Genetic diversity and chloroquine selective sweeps in *Plasmodium falciparum*. Nature 418:320–323
28. Yang Z, Zhang Z, Sun X, Wan W, Cui L, Zhang X, Zhong D, Yan G (2007) Molecular analysis of chloroquine resistance in *Plasmodium falciparum* in Yunnan Province, China. Trop Med Int Health 12:1051–1060
29. Waller KL, Lee S, Fidock DA (2004) Molecular and cellular biology of chloroquine resistance in *Plasmodium falciparum*. In: Waters AP, Janse CJ (eds) Malaria parasites: genomes and molecular biology. Caister Academic, Wymondham, pp 501–540
30. Pillai DR, Labbe AC, Vanisaveth V, Hongvangthong B, Pomphida S, Inkathone S, Zhong K, Kain KC (2001) *Plasmodium falciparum* malaria in Laos: chloroquine treatment outcome and predictive value of molecular markers. J Infect Dis 183:789–795
31. Djimde A, Doumbo OK, Cortese JF, Kayentao K, Doumbo S, Diourte Y, Dicko A, Su XZ, Nomura T, Fidock DA et al (2001) A molecular marker for chloroquine-resistant falciparum malaria. N Engl J Med 344:257–263

32. Babiker HA, Pringle SJ, Abdel-Muhsin A, Mackinnon M, Hunt P, Walliker D (2001) High-level chloroquine resistance in Sudanese isolates of *Plasmodium falciparum* is associated with mutations in the chloroquine resistance transporter gene *pfcrt* and the multidrug resistance gene *pfmdr1*. J Infect Dis 183:1535–1538
33. Wellems TE, Plowe CV (2001) Chloroquine-resistant malaria. J Infect Dis 184:770–776
34. Kublin JG, Cortese JF, Njunju EM, Mukadam RA, Wirima JJ, Kazembe PN, Djimde AA, Kouriba B, Taylor TE, Plowe CV (2003) Reemergence of chloroquine-sensitive *Plasmodium falciparum* malaria after cessation of chloroquine use in Malawi. J Infect Dis 187:1870–1875
35. Juliano JJ, Kwiek JJ, Cappell K, Mwapasa V, Meshnick SR (2007) Minority-variant *pfcrt* K76T mutations and chloroquine resistance, Malawi. Emerg Infect Dis 13:872–877
36. Holmgren G, Gil JP, Ferreira PM, Veiga MI, Obonyo CO, Bjorkman A (2006) Amodiaquine resistant *Plasmodium falciparum* malaria *in vivo* is associated with selection of *pfcrt* 76T and *pfmdr1* 86Y. Infect Genet Evol 6:309–314
37. Sa JM, Twu O, Hayton K, Reyes S, Fay MP, Ringwald P, Wellems TE (2009) Geographic patterns of *Plasmodium falciparum* drug resistance distinguished by differential responses to amodiaquine and chloroquine. Proc Natl Acad Sci USA 106:18883–18889
38. Sisowath C, Petersen I, Veiga MI, Martensson A, Premji Z, Bjorkman A, Fidock DA, Gil JP (2009) *In vivo* selection of *Plasmodium falciparum* parasites carrying the chloroquine-susceptible *pfcrt* K76 allele after treatment with artemether-lumefantrine in Africa. J Infect Dis 199:750–757
39. Cooper RA, Ferdig MT, Su XZ, Ursos LM, Mu J, Nomura T, Fujioka H, Fidock DA, Roepe PD, Wellems TE (2002) Alternative mutations at position 76 of the vacuolar transmembrane protein PfCRT are associated with chloroquine resistance and unique stereospecific quinine and quinidine responses in *Plasmodium falciparum*. Mol Pharmacol 61:35–42
40. Cooper RA, Lane KD, Deng B, Mu J, Patel JJ, Wellems TE, Su X, Ferdig MT (2007) Mutations in transmembrane domains 1, 4 and 9 of the *Plasmodium falciparum* chloroquine resistance transporter alter susceptibility to chloroquine, quinine and quinidine. Mol Microbiol 63:270–282
41. Rohrbach P, Sanchez CP, Hayton K, Friedrich O, Patel J, Sidhu AB, Ferdig MT, Fidock DA, Lanzer M (2006) Genetic linkage of *pfmdr1* with food vacuolar solute import in *Plasmodium falciparum*. EMBO J 25:3000–3011
42. Martin SK, Oduola AM, Milhous WK (1987) Reversal of chloroquine resistance in *Plasmodium falciparum* by verapamil. Science 235:899–901
43. Foote SJ, Thompson JK, Cowman AF, Kemp DJ (1989) Amplification of the multidrug resistance gene in some chloroquine-resistant isolates of *P. falciparum*. Cell 57:921–930
44. Wilson CM, Serrano AE, Wasley A, Bogenschutz MP, Shankar AH, Wirth DF (1989) Amplification of a gene related to mammalian *mdr* genes in drug-resistant *Plasmodium falciparum*. Science 244:1184–1186
45. Wellems TE, Panton LJ, Gluzman IY, do Rosario VE, Gwadz RW, Walker-Jonah A, Krogstad DJ (1990) Chloroquine resistance not linked to *mdr*-like genes in a *Plasmodium falciparum* cross. Nature 345:253–255
46. Cowman AF, Galatis D, Thompson JK (1994) Selection for mefloquine resistance in *Plasmodium falciparum* is linked to amplification of the *pfmdr1* gene and cross-resistance to halofantrine and quinine. Proc Natl Acad Sci USA 91:1143–1147
47. Peel SA, Bright P, Yount B, Handy J, Baric RS (1994) A strong association between mefloquine and halofantrine resistance and amplification, overexpression, and mutation in the P-glycoprotein gene homolog (*pfmdr*) of *Plasmodium falciparum in vitro*. Am J Trop Med Hyg 51:648–658
48. Ritchie GY, Mungthin M, Green JE, Bray PG, Hawley SR, Ward SA (1996) *In vitro* selection of halofantrine resistance in *Plasmodium falciparum* is not associated with increased expression of Pgh1. Mol Biochem Parasitol 83:35–46
49. Price RN, Uhlemann AC, Brockman A, McGready R, Ashley E, Phaipun L, Patel R, Laing K, Looareesuwan S, White NJ et al (2004) Mefloquine resistance in *Plasmodium falciparum* and increased *pfmdr1* gene copy number. Lancet 364:438–447

50. Price RN, Cassar C, Brockman A, Duraisingh M, van Vugt M, White NJ, Nosten F, Krishna S (1999) The *pfmdr1* gene is associated with a multidrug-resistant phenotype in *Plasmodium falciparum* from the western border of Thailand. Antimicrob Agents Chemother 43:2943–2949
51. Wilson CM, Volkman SK, Thaithong S, Martin RK, Kyle DE, Milhous WK, Wirth DF (1993) Amplification of *pfmdr1* associated with mefloquine and halofantrine resistance in *Plasmodium falciparum* from Thailand. Mol Biochem Parasitol 57:151–160
52. Nelson AL, Purfield A, McDaniel P, Uthaimongkol N, Buathong N, Sriwichai S, Miller RS, Wongsrichanalai C, Meshnick SR (2005) *pfmdr1* genotyping and *in vivo* mefloquine resistance on the Thai-Myanmar border. Am J Trop Med Hyg 72:586–592
53. Alker AP, Lim P, Sem R, Shah NK, Yi P, Bouth DM, Tsuyuoka R, Maguire JD, Fandeur T, Ariey F et al (2007) *Pfmdr1* and *in vivo* resistance to artesunate-mefloquine in falciparum malaria on the Cambodian-Thai border. Am J Trop Med Hyg 76:641–647
54. Price RN, Uhlemann AC, van Vugt M, Brockman A, Hutagalung R, Nair S, Nash D, Singhasivanon P, Anderson TJ, Krishna S et al (2006) Molecular and pharmacological determinants of the therapeutic response to artemether-lumefantrine in multidrug-resistant *Plasmodium falciparum* malaria. Clin Infect Dis 42:1570–1577
55. Chaiyaroj SC, Buranakiti A, Angkasekwinai P, Looressuwan S, Cowman AF (1999) Analysis of mefloquine resistance and amplification of *pfmdr1* in multidrug-resistant *Plasmodium falciparum* isolates from Thailand. Am J Trop Med Hyg 61:780–783
56. Myrick A, Munasinghe A, Patankar S, Wirth DF (2003) Mapping of the *Plasmodium falciparum* multidrug resistance gene 5′-upstream region, and evidence of induction of transcript levels by antimalarial drugs in chloroquine sensitive parasites. Mol Microbiol 49:671–683
57. Barnes DA, Foote SJ, Galatis D, Kemp DJ, Cowman AF (1992) Selection for high-level chloroquine resistance results in deamplification of the *pfmdr1* gene and increased sensitivity to mefloquine in *Plasmodium falciparum*. EMBO J 11:3067–3075
58. Sidhu AB, Uhlemann AC, Valderramos SG, Valderramos JC, Krishna S, Fidock DA (2006) Decreasing *pfmdr1* copy number in *Plasmodium falciparum* malaria heightens susceptibility to mefloquine, lumefantrine, halofantrine, quinine, and artemisinin. J Infect Dis 194:528–535
59. Bacon DJ, McCollum AM, Griffing SM, Salas C, Soberon V, Santolalla M, Haley R, Tsukayama P, Lucas C, Escalante AA et al (2009) Dynamics of malaria drug resistance patterns in the Amazon basin region following changes in Peruvian national treatment policy for uncomplicated malaria. Antimicrob Agents Chemother 53:2042–2051
60. Bonizzoni M, Afrane Y, Baliraine FN, Amenya DA, Githeko AK, Yan G (2009) Genetic structure of *Plasmodium falciparum* populations between lowland and highland sites and antimalarial drug resistance in Western Kenya. Infect Genet Evol 9:806–812
61. Dlamini SV, Beshir K, Sutherland CJ (2010) Markers of anti-malarial drug resistance in *Plasmodium falciparum* isolates from Swaziland: identification of *pfmdr1*-86 F in natural parasite isolates. Malar J 9:68
62. Foote SJ, Kyle DE, Martin RK, Oduola AM, Forsyth K, Kemp DJ, Cowman AF (1990) Several alleles of the multidrug-resistance gene are closely linked to chloroquine resistance in *Plasmodium falciparum*. Nature 345:255–258
63. Humphreys GS, Merinopoulos I, Ahmed J, Whitty CJ, Mutabingwa TK, Sutherland CJ, Hallett RL (2007) Amodiaquine and artemether-lumefantrine select distinct alleles of the *Plasmodium falciparum mdr1* gene in Tanzanian children treated for uncomplicated malaria. Antimicrob Agents Chemother 51:991–997
64. Mehlotra RK, Mattera G, Bockarie MJ, Maguire JD, Baird JK, Sharma YD, Alifrangis M, Dorsey G, Rosenthal PJ, Fryauff DJ et al (2008) Discordant patterns of genetic variation at two chloroquine resistance loci in worldwide populations of the malaria parasite *Plasmodium falciparum*. Antimicrob Agents Chemother 52:2212–2222
65. Mita T, Kaneko A, Hombhanje F, Hwaihwanje I, Takahashi N, Osawa H, Tsukahara T, Masta A, Lum JK, Kobayakawa T et al (2006) Role of *pfmdr1* mutations on chloroquine resistance

in *Plasmodium falciparum* isolates with *pfcrt* K76T from Papua New Guinea. Acta Trop 98:137–144

66. Nsobya SL, Dokomajilar C, Joloba M, Dorsey G, Rosenthal PJ (2007) Resistance-mediating *Plasmodium falciparum pfcrt* and *pfmdr1* alleles after treatment with artesunate-amodiaquine in Uganda. Antimicrob Agents Chemother 51:3023–3025
67. Pickard AL, Wongsrichanalai C, Purfield A, Kamwendo D, Emery K, Zalewski C, Kawamoto F, Miller RS, Meshnick SR (2003) Resistance to antimalarials in Southeast Asia and genetic polymorphisms in *pfmdr1*. Antimicrob Agents Chemother 47:2418–2423
68. Volkman SK, Cowman AF, Wirth DF (1995) Functional complementation of the *ste6* gene of *Saccharomyces cerevisiae* with the *pfmdr1* gene of *Plasmodium falciparum*. Proc Natl Acad Sci USA 92:8921–8925
69. van Es HH, Karcz S, Chu F, Cowman AF, Vidal S, Gros P, Schurr E (1994) Expression of the plasmodial *pfmdr1* gene in mammalian cells is associated with increased susceptibility to chloroquine. Mol Cell Biol 14:2419–2428
70. Sanchez CP, Rotmann A, Stein WD, Lanzer M (2008) Polymorphisms within PfMDR1 alter the substrate specificity for anti-malarial drugs in *Plasmodium falciparum*. Mol Microbiol 70:786–798
71. Duraisingh MT, Cowman AF (2005) Contribution of the *pfmdr1* gene to antimalarial drug-resistance. Acta Trop 94:181–190
72. Sidhu AB, Valderramos SG, Fidock DA (2005) *pfmdr1* mutations contribute to quinine resistance and enhance mefloquine and artemisinin sensitivity in *Plasmodium falciparum*. Mol Microbiol 57:913–926
73. Holmgren G, Hamrin J, Svard J, Martensson A, Gil JP, Bjorkman A (2007) Selection of *pfmdr1* mutations after amodiaquine monotherapy and amodiaquine plus artemisinin combination therapy in East Africa. Infect Genet Evol 7:562–569
74. Anderson TJ, Nair S, Qin H, Singlam S, Brockman A, Paiphun L, Nosten F (2005) Are transporter genes other than the chloroquine resistance locus (*pfcrt*) and multidrug resistance gene (*pfmdr*) associated with antimalarial drug resistance? Antimicrob Agents Chemother 49:2180–2188
75. Dokomajilar C, Nsobya SL, Greenhouse B, Rosenthal PJ, Dorsey G (2006) Selection of *Plasmodium falciparum pfmdr1* alleles following therapy with artemether-lumefantrine in an area of Uganda where malaria is highly endemic. Antimicrob Agents Chemother 50:1893–1895
76. Sisowath C, Ferreira PE, Bustamante LY, Dahlstrom S, Martensson A, Bjorkman A, Krishna S, Gil JP (2007) The role of *pfmdr1* in *Plasmodium falciparum* tolerance to artemether-lumefantrine in Africa. Trop Med Int Health 12:736–742
77. Pillai DR, Hijar G, Montoya Y, Marouino W, Ruebush TK 2nd, Wongsrichanalai C, Kain KC (2003) Lack of prediction of mefloquine and mefloquine-artesunate treatment outcome by mutations in the *Plasmodium falciparum* multidrug resistance 1 (*pfmdr1*) gene for *P. falciparum* malaria in Peru. Am J Trop Med Hyg 68:107–110
78. Reed MB, Saliba KJ, Caruana SR, Kirk K, Cowman AF (2000) Pgh1 modulates sensitivity and resistance to multiple antimalarials in *Plasmodium falciparum*. Nature 403:906–909
79. Duraisingh MT, Roper C, Walliker D, Warhurst DC (2000) Increased sensitivity to the antimalarials mefloquine and artemisinin is conferred by mutations in the *pfmdr1* gene of *Plasmodium falciparum*. Mol Microbiol 36:955–961
80. Mu J, Ferdig MT, Feng X, Joy DA, Duan J, Furuya T, Subramanian G, Aravind L, Cooper RA, Wootton JC et al (2003) Multiple transporters associated with malaria parasite responses to chloroquine and quinine. Mol Microbiol 49:977–989
81. Cojean S, Noel A, Garnier D, Hubert V, Le Bras J, Durand R (2006) Lack of association between putative transporter gene polymorphisms in *Plasmodium falciparum* and chloroquine resistance in imported malaria isolates from Africa. Malar J 5:24
82. Dahlstrom S, Ferreira PE, Veiga MI, Sedighi N, Wiklund L, Martensson A, Farnert A, Sisowath C, Osorio L, Darban H et al (2009) *Plasmodium falciparum* multidrug resistance protein 1 and artemisinin-based combination therapy in Africa. J Infect Dis 200:1456–1464

83. Dahlstrom S, Veiga MI, Martensson A, Bjorkman A, Gil JP (2009) Polymorphism in PfMRP1 (*Plasmodium falciparum* multidrug resistance protein 1) amino acid 1466 associated with resistance to sulfadoxine-pyrimethamine treatment. Antimicrob Agents Chemother 53:2553–2556
84. Raj DK, Mu J, Jiang H, Kabat J, Singh S, Sullivan M, Fay MP, McCutchan TF, Su XZ (2009) Disruption of a *Plasmodium falciparum* multidrug resistance-associated protein (PfMRP) alters its fitness and transport of antimalarial drugs and glutathione. J Biol Chem 284:7687–7696
85. Vinayak S, Alam MT, Upadhyay M, Das MK, Dev V, Singh N, Dash AP, Sharma YD (2007) Extensive genetic diversity in the *Plasmodium falciparum* Na^+/H^+ exchanger 1 transporter protein implicated in quinine resistance. Antimicrob Agents Chemother 51:4508–4511
86. Andriantsoanirina V, Menard D, Rabearimanana S, Hubert V, Bouchier C, Tichit M, Bras JL, Durand R (2010) Association of microsatellite variations of *Plasmodium falciparum* Na^+/H^+ exchanger (Pfnhe-1) gene with reduced *in vitro* susceptibility to quinine: lack of confirmation in clinical isolates from Africa. Am J Trop Med Hyg 82:782–787
87. Henry M, Briolant S, Zettor A, Pelleau S, Baragatti M, Baret E, Mosnier J, Amalvict R, Fusai T, Rogier C et al (2009) *Plasmodium falciparum* Na^+/H^+ exchanger 1 transporter is involved in reduced susceptibility to quinine. Antimicrob Agents Chemother 53:1926–1930
88. Bennett TN, Patel J, Ferdig MT, Roepe PD (2007) *Plasmodium falciparum* Na^+/H^+ exchanger activity and quinine resistance. Mol Biochem Parasitol 153:48–58
89. Spillman NJ, Allen RJ, Kirk K (2008) Acid extrusion from the intraerythrocytic malaria parasite is not via a $Na(^+)/H(^+)$ exchanger. Mol Biochem Parasitol 162:96–99
90. Nkrumah LJ, Riegelhaupt PM, Moura P, Johnson DJ, Patel J, Hayton K, Ferdig MT, Wellems TE, Akabas MH, Fidock DA (2009) Probing the multifactorial basis of *Plasmodium falciparum* quinine resistance: evidence for a strain-specific contribution of the sodium-proton exchanger PfNHE. Mol Biochem Parasitol 165:122–131
91. Vinetz JM, Clain J, Bounkeua V, Eastman RT, Fidock DA (2011) Chemotherapy of malaria. In: Brunton LL, Chabner B, Parker KL (eds) Goodman & Gilman's the pharmacological basis of therapeutics. McGraw-Hill Medical, New York, pp 1383–1418
92. Wongsrichanalai C, Pickard AL, Wernsdorfer WH, Meshnick SR (2002) Epidemiology of drug-resistant malaria. Lancet Infect Dis 2:209–218
93. Sibley CH, Hyde JE, Sims PF, Plowe CV, Kublin JG, Mberu EK, Cowman AF, Winstanley PA, Watkins WM, Nzila AM (2001) Pyrimethamine-sulfadoxine resistance in *Plasmodium falciparum*: what next? Trends Parasitol 17:582–588
94. Gregson A, Plowe CV (2005) Mechanisms of resistance of malaria parasites to antifolates. Pharmacol Rev 57:117–145
95. Peterson DS, Walliker D, Wellems TE (1988) Evidence that a point mutation in dihydrofolate reductase-thymidylate synthase confers resistance to pyrimethamine in falciparum malaria. Proc Natl Acad Sci USA 85:9114–9118
96. Cowman AF, Morry MJ, Biggs BA, Cross GA, Foote SJ (1988) Amino acid changes linked to pyrimethamine resistance in the dihydrofolate reductase-thymidylate synthase gene of *Plasmodium falciparum*. Proc Natl Acad Sci USA 85:9109–9113
97. Foote SJ, Galatis D, Cowman AF (1990) Amino acids in the dihydrofolate reductase-thymidylate synthase gene of *Plasmodium falciparum* involved in cycloguanil resistance differ from those involved in pyrimethamine resistance. Proc Natl Acad Sci USA 87:3014–3017
98. Peterson DS, Milhous WK, Wellems TE (1990) Molecular basis of differential resistance to cycloguanil and pyrimethamine in *Plasmodium falciparum* malaria. Proc Natl Acad Sci USA 87:3018–3022
99. Triglia T, Cowman AF (1994) Primary structure and expression of the dihydropteroate synthetase gene of *Plasmodium falciparum*. Proc Natl Acad Sci USA 91:7149–7153
100. Brooks DR, Wang P, Read M, Watkins WM, Sims PF, Hyde JE (1994) Sequence variation of the hydroxymethyldihydropterin pyrophosphokinase: dihydropteroate synthase gene in lines

of the human malaria parasite, *Plasmodium falciparum*, with differing resistance to sulfadoxine. Eur J Biochem 224:397–405
101. Wu Y, Kirkman LA, Wellems TE (1996) Transformation of *Plasmodium falciparum* malaria parasites by homologous integration of plasmids that confer resistance to pyrimethamine. Proc Natl Acad Sci USA 93:1130–1134
102. Triglia T, Wang P, Sims PF, Hyde JE, Cowman AF (1998) Allelic exchange at the endogenous genomic locus in *Plasmodium falciparum* proves the role of dihydropteroate synthase in sulfadoxine-resistant malaria. EMBO J 17:3807–3815
103. Sirawaraporn W, Sathitkul T, Sirawaraporn R, Yuthavong Y, Santi DV (1997) Antifolate-resistant mutants of *Plasmodium falciparum* dihydrofolate reductase. Proc Natl Acad Sci USA 94:1124–1129
104. Triglia T, Menting JG, Wilson C, Cowman AF (1997) Mutations in dihydropteroate synthase are responsible for sulfone and sulfonamide resistance in *Plasmodium falciparum*. Proc Natl Acad Sci USA 94:13944–13949
105. Reynolds MG, Roos DS (1998) A biochemical and genetic model for parasite resistance to antifolates. *Toxoplasma gondii* provides insights into pyrimethamine and cycloguanil resistance in *Plasmodium falciparum*. J Biol Chem 273:3461–3469
106. Kublin JG, Witzig RS, Shankar AH, Zurita JQ, Gilman RH, Guarda JA, Cortese JF, Plowe CV (1998) Molecular assays for surveillance of antifolate-resistant malaria. Lancet 351:1629–1630
107. Kiara SM, Okombo J, Masseno V, Mwai L, Ochola I, Borrmann S, Nzila A (2009) *In vitro* activity of antifolate and polymorphism in dihydrofolate reductase of *Plasmodium falciparum* isolates from the Kenyan coast: emergence of parasites with Ile-164-Leu mutation. Antimicrob Agents Chemother 53:3793–3798
108. Cortese JF, Plowe CV (1998) Antifolate resistance due to new and known *Plasmodium falciparum* dihydrofolate reductase mutations expressed in yeast. Mol Biochem Parasitol 94:205–214
109. Lozovsky ER, Chookajorn T, Brown KM, Imwong M, Shaw PJ, Kamchonwongpaisan S, Neafsey DE, Weinreich DM, Hartl DL (2009) Stepwise acquisition of pyrimethamine resistance in the malaria parasite. Proc Natl Acad Sci USA 106:12025–12030
110. Kublin JG, Dzinjalamala FK, Kamwendo DD, Malkin EM, Cortese JF, Martino LM, Mukadam RA, Rogerson SJ, Lescano AG, Molyneux ME et al (2002) Molecular markers for failure of sulfadoxine-pyrimethamine and chlorproguanil-dapsone treatment of *Plasmodium falciparum* malaria. J Infect Dis 185:380–388
111. Kyabayinze D, Cattamanchi A, Kamya MR, Rosenthal PJ, Dorsey G (2003) Validation of a simplified method for using molecular markers to predict sulfadoxine-pyrimethamine treatment failure in African children with falciparum malaria. Am J Trop Med Hyg 69:247–252
112. Harrington WE, Mutabingwa TK, Muehlenbachs A, Sorensen B, Bolla MC, Fried M, Duffy PE (2009) Competitive facilitation of drug-resistant *Plasmodium falciparum* malaria parasites in pregnant women who receive preventive treatment. Proc Natl Acad Sci USA 106:9027–9032
113. Kidgell C, Volkman SK, Daily J, Borevitz JO, Plouffe D, Zhou Y, Johnson JR, Le Roch K, Sarr O, Ndir O et al (2006) A systematic map of genetic variation in *Plasmodium falciparum*. PLoS Pathog 2:e57
114. Nair S, Miller B, Barends M, Jaidee A, Patel J, Mayxay M, Newton P, Nosten F, Ferdig MT, Anderson TJ (2008) Adaptive copy number evolution in malaria parasites. PLoS Genet 4: e1000243
115. Dondorp AM, Nosten F, Yi P, Das D, Phyo AP, Tarning J, Lwin KM, Ariey F, Hanpithakpong W, Lee SJ et al (2009) Artemisinin resistance in *Plasmodium falciparum* malaria. N Engl J Med 361:455–467
116. Noedl H, Se Y, Schaecher K, Smith BL, Socheat D, Fukuda MM (2008) Evidence of artemisinin-resistant malaria in western Cambodia. N Engl J Med 359:2619–2620

117. Eckstein-Ludwig U, Webb RJ, Van Goethem ID, East JM, Lee AG, Kimura M, O'Neill PM, Bray PG, Ward SA, Krishna S (2003) Artemisinins target the SERCA of *Plasmodium falciparum*. Nature 424:957–961
118. Cardi D, Pozza A, Arnou B, Marchal E, Clausen JD, Andersen JP, Krishna S, Moller JV, le Maire M, Jaxel C (2010) Purified E255L mutant SERCA1a and purified PfATP6 are sensitive to SERCA-type inhibitors but insensitive to artemisinins. J Biol Chem 285: 26406–26416
119. Krishna S, Pulcini S, Fatih F, Staines H (2010) Artemisinins and the biological basis for the PfATP6/SERCA hypothesis. Trends Parasitol 26:517–523
120. Uhlemann AC, Cameron A, Eckstein-Ludwig U, Fischbarg J, Iserovich P, Zuniga FA, East M, Lee A, Brady L, Haynes RK et al (2005) A single amino acid residue can determine the sensitivity of SERCAs to artemisinins. Nat Struct Mol Biol 12:628–629
121. Valderramos SG, Scanfeld D, Uhlemann AC, Fidock DA, Krishna S (2010) Investigations into the role of the *Plasmodium falciparum* SERCA (PfATP6) L263E mutation in artemisinin action and resistance. Antimicrob Agents Chemother 54:3842–3852
122. Jambou R, Legrand E, Niang M, Khim N, Lim P, Volney B, Ekala MT, Bouchier C, Esterre P, Fandeur T et al (2005) Resistance of *Plasmodium falciparum* field isolates to *in vitro* artemether and point mutations of the SERCA-type PfATPase6. Lancet 366:1960–1963
123. Jambou R, Martinelli A, Pinto J, Gribaldo S, Legrand E, Niang M, Kim N, Pharath L, Volnay B, Ekala MT et al (2010) Geographic structuring of the *Plasmodium falciparum* sarco(endo) plasmic reticulum Ca^{2+} ATPase (PfSERCA) gene diversity. PLoS One 5:e9424
124. Ferrer-Rodriguez I, Perez-Rosado J, Gervais GW, Peters W, Robinson BL, Serrano AE (2004) *Plasmodium yoelii*: identification and partial characterization of an MDR1 gene in an artemisinin-resistant line. J Parasitol 90:152–160
125. Chavchich M, Gerena L, Peters J, Chen N, Cheng Q, Kyle DE (2010) Induction of resistance to artemisinin derivatives in *Plasmodium falciparum*: role of *Pfmdr1* amplification and expression. Antimicrob Agents Chemother 54:2455–2464
126. Witkowski B, Lelievre J, Barragan MJ, Laurent V, Su XZ, Berry A, Benoit-Vical F (2010) Increased tolerance to artemisinin in *Plasmodium falciparum* is mediated by a quiescence mechanism. Antimicrob Agents Chemother 54:1872–1877
127. Chen N, Chavchich M, Peters JM, Kyle DE, Gatton ML, Cheng Q (2010) De-amplification of *pfmdr1*-containing amplicon on chromosome 5 in *Plasmodium falciparum* is associated with reduced resistance to artelinic acid *in vitro*. Antimicrob Agents Chemother 54:3395–3401
128. Afonso A, Hunt P, Cheesman S, Alves AC, Cunha CV, do Rosario V, Cravo P (2006) Malaria parasites can develop stable resistance to artemisinin but lack mutations in candidate genes *atp6* (encoding the sarcoplasmic and endoplasmic reticulum Ca^{2+} ATPase), *tctp*, *mdr1*, and *cg10*. Antimicrob Agents Chemother 50:480–489
129. Hunt P, Afonso A, Creasey A, Culleton R, Sidhu AB, Logan J, Valderramos SG, McNae I, Cheesman S, do Rosario V et al (2007) Gene encoding a deubiquitinating enzyme is mutated in artesunate- and chloroquine-resistant rodent malaria parasites. Mol Microbiol 65:27–40
130. Hunt P, Martinelli A, Modrzynska K, Borges S, Creasey A, Rodrigues L, Beraldi D, Loewe L, Fawcett R, Kumar S et al (2010) Experimental evolution, genetic analysis and genome re-sequencing reveal the mutation conferring artemisinin resistance in an isogenic lineage of malaria parasites. BMC Genomics 11:499
131. Walker DJ, Pitsch JL, Peng MM, Robinson BL, Peters W, Bhisutthibhan J, Meshnick SR (2000) Mechanisms of artemisinin resistance in the rodent malaria pathogen *Plasmodium yoelii*. Antimicrob Agents Chemother 44:344–347
132. Bhisutthibhan J, Pan XQ, Hossler PA, Walker DJ, Yowell CA, Carlton J, Dame JB, Meshnick SR (1998) The *Plasmodium falciparum* translationally controlled tumor protein homolog and its reaction with the antimalarial drug artemisinin. J Biol Chem 273:16192–16198
133. Eastman RT, Fidock DA (2009) Artemisinin-based combination therapies: a vital tool in efforts to eliminate malaria. Nat Rev Microbiol 7:864–874

134. Poehlsgaard J, Douthwaite S (2005) The bacterial ribosome as a target for antibiotics. Nat Rev Microbiol 3:870–881
135. Dahl EL, Rosenthal PJ (2008) Apicoplast translation, transcription and genome replication: targets for antimalarial antibiotics. Trends Parasitol 24:279–284
136. Briolant S, Baragatti M, Parola P, Simon F, Tall A, Sokhna C, Hovette P, Mamfoumbi MM, Koeck JL, Delmont J et al (2009) Multinormal *in vitro* distribution model suitable for the distribution of *Plasmodium falciparum* chemosusceptibility to doxycycline. Antimicrob Agents Chemother 53:688–695
137. Dahl EL, Shock JL, Shenai BR, Gut J, DeRisi JL, Rosenthal PJ (2006) Tetracyclines specifically target the apicoplast of the malaria parasite *Plasmodium falciparum*. Antimicrob Agents Chemother 50:3124–3131
138. Briolant S, Wurtz N, Zettor A, Rogier C, Pradines B (2010) Susceptibility of *Plasmodium falciparum* isolates to doxycycline is associated with *pftetQ* sequence polymorphisms and *pftetQ* and *pfmdt* copy numbers. J Infect Dis 201:153–159
139. Wiesner J, Reichenberg A, Heinrich S, Schlitzer M, Jomaa H (2008) The plastid-like organelle of apicomplexan parasites as drug target. Curr Pharm Des 14:855–871
140. Dharia NV, Plouffe D, Bopp SE, Gonzalez-Paez GE, Lucas C, Salas C, Soberon V, Bursulaya B, Kochel TJ, Bacon DJ et al (2010) Genome scanning of Amazonian *Plasmodium falciparum* shows subtelomeric instability and clindamycin-resistant parasites. Genome Res 20:1534–1544
141. Hansen JL, Ippolito JA, Ban N, Nissen P, Moore PB, Steitz TA (2002) The structures of four macrolide antibiotics bound to the large ribosomal subunit. Mol Cell 10:117–128
142. Douthwaite S, Hansen LH, Mauvais P (2000) Macrolide-ketolide inhibition of MLS-resistant ribosomes is improved by alternative drug interaction with domain II of 23 S rRNA. Mol Microbiol 36:183–193
143. Gabashvili IS, Gregory ST, Valle M, Grassucci R, Worbs M, Wahl MC, Dahlberg AE, Frank J (2001) The polypeptide tunnel system in the ribosome and its gating in erythromycin resistance mutants of L4 and L22. Mol Cell 8:181–188
144. Hansen LH, Mauvais P, Douthwaite S (1999) The macrolide-ketolide antibiotic binding site is formed by structures in domains II and V of 23 S ribosomal RNA. Mol Microbiol 31:623–631
145. Sidhu AB, Sun Q, Nkrumah LJ, Dunne MW, Sacchettini JC, Fidock DA (2007) *In vitro* efficacy, resistance selection, and structural modeling studies implicate the malarial parasite apicoplast as the target of azithromycin. J Biol Chem 282:2494–2504
146. Franceschi F, Kanyo Z, Sherer EC, Sutcliffe J (2004) Macrolide resistance from the ribosome perspective. Curr Drug Targets Infect Disord 4:177–191
147. Borrmann S, Adegnika AA, Matsiegui PB, Issifou S, Schindler A, Mawili-Mboumba DP, Baranek T, Wiesner J, Jomaa H, Kremsner PG (2004) Fosmidomycin-clindamycin for *Plasmodium falciparum* infections in African children. J Infect Dis 189:901–908
148. Borrmann S, Adegnika AA, Moussavou F, Oyakhirome S, Esser G, Matsiegui PB, Ramharter M, Lundgren I, Kombila M, Issifou S et al (2005) Short-course regimens of artesunate-fosmidomycin in treatment of uncomplicated *Plasmodium falciparum* malaria. Antimicrob Agents Chemother 49:3749–3754
149. Borrmann S, Issifou S, Esser G, Adegnika AA, Ramharter M, Matsiegui PB, Oyakhirome S, Mawili-Mboumba DP, Missinou MA, Kun JF et al (2004) Fosmidomycin-clindamycin for the treatment of *Plasmodium falciparum* malaria. J Infect Dis 190:1534–1540
150. Dharia NV, Sidhu AB, Cassera MB, Westenberger SJ, Bopp SE, Eastman RT, Plouffe D, Batalov S, Park DJ, Volkman SK et al (2009) Use of high-density tiling microarrays to identify mutations globally and elucidate mechanisms of drug resistance in *Plasmodium falciparum*. Genome Biol 10:R21
151. Hudson AT, Dickins M, Ginger CD, Gutteridge WE, Holdich T, Hutchinson DB, Pudney M, Randall AW, Latter VS (1991) 566 C80: a potent broad spectrum anti-infective agent with

activity against malaria and opportunistic infections in AIDS patients. Drugs Exp Clin Res 17:427–435

152. Chiodini PL, Conlon CP, Hutchinson DB, Farquhar JA, Hall AP, Peto TE, Birley H, Warrell DA (1995) Evaluation of atovaquone in the treatment of patients with uncomplicated *Plasmodium falciparum* malaria. J Antimicrob Chemother 36:1073–1078
153. Looareesuwan S, Chulay JD, Canfield CJ, Hutchinson DB (1999) Malarone (atovaquone and proguanil hydrochloride): a review of its clinical development for treatment of malaria. Am J Trop Med Hyg 60:533–541
154. Looareesuwan S, Viravan C, Webster HK, Kyle DE, Hutchinson DB, Canfield CJ (1996) Clinical studies of atovaquone, alone or in combination with other antimalarial drugs, for treatment of acute uncomplicated malaria in Thailand. Am J Trop Med Hyg 54:62–66
155. Nakato H, Vivancos R, Hunter PR (2007) A systematic review and meta-analysis of the effectiveness and safety of atovaquone proguanil (Malarone) for chemoprophylaxis against malaria. J Antimicrob Chemother 60:929–936
156. Srivastava IK, Vaidya AB (1999) A mechanism for the synergistic antimalarial action of atovaquone and proguanil. Antimicrob Agents Chemother 43:1334–1339
157. Srivastava IK, Morrisey JM, Darrouzet E, Daldal F, Vaidya AB (1999) Resistance mutations reveal the atovaquone-binding domain of cytochrome *b* in malaria parasites. Mol Microbiol 33:704–711
158. Fry M, Pudney M (1992) Site of action of the antimalarial hydroxynaphthoquinone, 2-[trans-4-(4′-chlorophenyl) cyclohexyl]-3-hydroxy-1,4-naphthoquinone (566 C80). Biochem Pharmacol 43:1545–1553
159. Painter HJ, Morrisey JM, Mather MW, Vaidya AB (2007) Specific role of mitochondrial electron transport in blood-stage *Plasmodium falciparum*. Nature 446:88–91
160. Vaidya AB, Mather MW (2000) Atovaquone resistance in malaria parasites. Drug Resist Updat 3:283–287
161. Korsinczky M, Chen N, Kotecka B, Saul A, Rieckmann K, Cheng Q (2000) Mutations in *Plasmodium falciparum* cytochrome *b* that are associated with atovaquone resistance are located at a putative drug-binding site. Antimicrob Agents Chemother 44:2100–2108
162. Syafruddin D, Siregar JE, Marzuki S (1999) Mutations in the cytochrome *b* gene of *Plasmodium berghei* conferring resistance to atovaquone. Mol Biochem Parasitol 104:185–194
163. McFadden DC, Tomavo S, Berry EA, Boothroyd JC (2000) Characterization of cytochrome *b* from *Toxoplasma gondii* and Q(o) domain mutations as a mechanism of atovaquone-resistance. Mol Biochem Parasitol 108:1–12
164. Sutherland CJ, Laundy M, Price N, Burke M, Fivelman QL, Pasvol G, Klein JL, Chiodini PL (2008) Mutations in the *Plasmodium falciparum* cytochrome *b* gene are associated with delayed parasite recrudescence in malaria patients treated with atovaquone-proguanil. Malar J 7:240
165. Perry TL, Pandey P, Grant JM, Kain KC (2009) Severe atovaquone-resistant *Plasmodium falciparum* malaria in a Canadian traveller returned from the Indian subcontinent. Open Med 3:e10–e16
166. Musset L, Le Bras J, Clain J (2007) Parallel evolution of adaptive mutations in *Plasmodium falciparum* mitochondrial DNA during atovaquone-proguanil treatment. Mol Biol Evol 24:1582–1585
167. Mather MW, Darrouzet E, Valkova-Valchanova M, Cooley JW, McIntosh MT, Daldal F, Vaidya AB (2005) Uncovering the molecular mode of action of the antimalarial drug atovaquone using a bacterial system. J Biol Chem 280:27458–27465
168. Orjuela-Sanchez P, de Santana Filho FS, Machado-Lima A, Chehuan YF, Costa MR, Alecrim MG, del Portillo HA (2009) Analysis of single-nucleotide polymorphisms in the *crt-o* and *mdr1* genes of *Plasmodium vivax* among chloroquine-resistant isolates from the Brazilian Amazon region. Antimicrob Agents Chemother 53:3561–3564
169. Imwong M, Pukrittayakamee S, Pongtavornpinyo W, Nakeesathit S, Nair S, Newton P, Nosten F, Anderson TJ, Dondorp A, Day NP et al (2008) Gene amplification of the multidrug

resistance 1 gene of *Plasmodium vivax* isolates from Thailand, Laos, and Myanmar. Antimicrob Agents Chemother 52:2657–2659

170. Nomura T, Carlton JM, Baird JK, del Portillo HA, Fryauff DJ, Rathore D, Fidock DA, Su X, Collins WE, McCutchan TF et al (2001) Evidence for different mechanisms of chloroquine resistance in 2 *Plasmodium* species that cause human malaria. J Infect Dis 183:1653–1661
171. Sa JM, Nomura T, Neves J, Baird JK, Wellems TE, del Portillo HA (2005) *Plasmodium vivax*: allele variants of the *mdr1* gene do not associate with chloroquine resistance among isolates from Brazil, Papua, and monkey-adapted strains. Exp Parasitol 109:256–259
172. Barnadas C, Ratsimbasoa A, Tichit M, Bouchier C, Jahevitra M, Picot S, Menard D (2008) *Plasmodium vivax* resistance to chloroquine in Madagascar: clinical efficacy and polymorphisms in *pvmdr1* and *pvcrt-o* genes. Antimicrob Agents Chemother 52:4233–4240
173. Suwanarusk R, Russell B, Chavchich M, Chalfein F, Kenangalem E, Kosaisavee V, Prasetyorini B, Piera KA, Barends M, Brockman A et al (2007) Chloroquine resistant *Plasmodium vivax*: *in vitro* characterisation and association with molecular polymorphisms. PLoS One 2:e1089
174. Gama BE, Oliveira NK, Souza JM, Daniel-Ribeiro CT, Ferreira-da-Cruz Mde F (2009) Characterisation of *pvmdr1* and *pvdhfr* genes associated with chemoresistance in Brazilian *Plasmodium vivax* isolates. Mem Inst Oswaldo Cruz 104:1009–1011
175. Sa JM, Yamamoto MM, Fernandez-Becerra C, de Azevedo MF, Papakrivos J, Naude B, Wellems TE, Del Portillo HA (2006) Expression and function of *pvcrt-o*, a *Plasmodium vivax* ortholog of *pfcrt*, in *Plasmodium falciparum* and *Dictyostelium discoideum*. Mol Biochem Parasitol 150:219–228
176. Hawkins VN, Joshi H, Rungsihirunrat K, Na-Bangchang K, Sibley CH (2007) Antifolates can have a role in the treatment of *Plasmodium vivax*. Trends Parasitol 23:213–222
177. Kongsaeree P, Khongsuk P, Leartsakulpanich U, Chitnumsub P, Tarnchompoo B, Walkinshaw MD, Yuthavong Y (2005) Crystal structure of dihydrofolate reductase from *Plasmodium vivax*: pyrimethamine displacement linked with mutation-induced resistance. Proc Natl Acad Sci USA 102:13046–13051
178. Yuvaniyama J, Chitnumsub P, Kamchonwongpaisan S, Vanichtanankul J, Sirawaraporn W, Taylor P, Walkinshaw MD, Yuthavong Y (2003) Insights into antifolate resistance from malarial DHFR-TS structures. Nat Struct Biol 10:357–365

Prevention of Malaria

Patricia Schlagenhauf and Eskild Petersen

Motto/Epigraph. Travellers to malaria-endemic areas need:

- Information on malaria risk, mode of transmission of the disease, incubation period and symptoms
- Advice on measures against mosquito bites
- Chemoprophylaxis for high-risk areas such as sub-Saharan Africa
- Advice regarding prompt diagnosis and self-treatment of malaria if appropriate

Abstract An estimated 80–90 million travellers visit malaria-endemic areas annually. Not all travellers have a similar risk. The risk of acquiring malaria will depend on many factors including the type and intensity of malaria transmission at the destination, the duration and style of travel, prevention measures used and individual characteristics. Primary prevention strategies are mosquito bite prevention measures (such as insecticide impregnated bednets, repellents and insecticides) and chemoprophylaxis. The three priority antimalaria chemoprophylactic regimens for travellers to areas with a high risk of *Plasmodium falciparum* malaria are: atovaquone/proguanil, doxycycline and mefloquine. In some countries, the strategy of stand-by emergency self-treatment is recommended for travellers to areas with a low risk of malaria. Malaria prevention advice and strategies should as far as possible be evidence based using sound epidemiological data when these are available. This chapter focuses on the chemoprevention and self-treatment of malaria in travellers.

P. Schlagenhauf
University of Zurich Centre for Travel Medicine, WHO Collaborating Centre for Travellers' Health, Institute for Social and Preventive Medicine, University of Zürich, Zurich, Switzerland

E. Petersen (✉)
Department of Infectious Diseases, Aarhus University Hospital Skejby, Brendstrupgardsvej 100 DK-8200 Aarhus N. Aarhus, Denmark
e-mail: eskildp@dadlnet.dk

H.M. Staines and S. Krishna (eds.), *Treatment and Prevention of Malaria*,
DOI 10.1007/978-3-0346-0480-2_14,

1 Chemoprophylaxis

The decision to use chemoprophylaxis will depend on a risk benefit analysis weighting the risk of malaria against the risk of possible adverse drug reactions. The risk of acquiring malaria will depend on many factors including the type and intensity of malaria transmission at the destination, the duration and style of travel, prevention measures used and individual characteristics [1–3]. This risk is difficult to quantify, and new risk areas may emerge [4], while risk at traditional traveller destinations can decline [5, 6]. Even though malaria imported to non-endemic countries is a notifiable disease, the population at risk (*i.e.* the exact number of travellers to a specific destination) is often impossible to ascertain.

Therefore, it has been proposed to use data on malaria endemicity in the indigenous population and then extrapolate the risk to travellers [7], although many travellers will have a lower risk of malaria than local populations in endemic areas because they have shorter exposition periods of exposure and higher standards of accommodation. The exceptions here are travellers visiting friends and relatives (VFR) who have a malaria risk comparable to or higher than the local communities.

With regard to tolerability, several studies quantify the rate of adverse events in short-term travellers using malaria chemoprophylaxis but the methodologies used are rarely comparable so that incidence rates can differ enormously between studies. The recent Cochrane Review of the tolerability of antimalarial drugs and the comments on this review highlight the need for more targeted and controlled studies to address this topic [8]. Data on the long-term use and tolerability of drugs for chemoprophylaxis are lacking. There is also a lack of data on the safety and efficacy of strategies for vulnerable populations such as pregnant/lactating women and small children.

2 Low Risk of Malaria Infection

One of the most difficult questions in malaria prophylaxis today is how to advise travellers who visit low-risk areas, often areas with unstable transmission and a changing malaria epidemiology. Few studies report the absolute risk for specific areas, but these data are necessary to allow a rational decision on whether to recommend chemoprophylaxis or not, balanced against the risk of side effects. Travellers should not be exposed to a "substantial risk" (discussed in the next section) of adverse events from malaria chemoprophylaxis in areas where the risk of malaria infection is very low.

Can we quantify the risk of contracting malaria and weigh this against the risk of adverse events from chemoprophylaxis?

A study from Sweden found the risk of malaria to travellers in West Africa, South Africa, South America and Thailand to be 302, 46, 7.2 and 2 per 100,000 visitors, respectively [9].

6 Prophylaxis for Long-Term Visitors and Frequent Visitors

The risk of malaria cumulates over time, but the increased risk cannot just be attributed to longer exposure alone [10]. Long-term travellers, defined as persons staying permanently in an area for 6 months or longer and frequent travellers to endemic regions, behave differently from short-term visitors. Long-term travellers frequently discontinue chemoprophylaxis prematurely because they perceive the risk to be lower than expected. They also believe that they can effectively manage an infection and they are worried about side effects from long-term use of malaria chemoprophylaxis [19, 20]. Use of counterfeit drugs (*i.e.* fake drugs containing no or sub-therapeutic doses of active compounds) is particularly an issue for long-term travellers who often buy local supplies of dubious quality because they are cheaper and expatriates should be encouraged to bring adequate quantities of antimalarial medications with them [20]. The consequences of travellers using counterfeit drugs are far-reaching; untreated *P. falciparum* has a high mortality rate, sub-therapeutic dosages lead to inadequate prophylactic doses, a risk of increased adverse events due to excessive dosage or potentially toxic contaminants and finally a loss of faith in genuine medicines.

The health personnel advising long-term travellers should understand these aspects and accept that these are important issues, which must be addressed.

The key to managing malaria prevention in long-term and frequent travellers is to provide the travellers with knowledge and understanding of malaria so that they can take more responsibility for their own health compared to the short-term traveller.

Long-term travellers to high-risk areas should take malaria chemoprophylaxis even if this is necessary over several years. Mefloquine is the best documented drug used by long-term travellers and, if well tolerated, can be used for prolonged periods (*i.e.* there is no upper time limits for the use of mefloquine) [21]. It has a simple weekly dosing schedule that encourages adherence [22] and toxic accumulation in the body does not occur during long-term use [23]. Doxycycline (a tetracycline compound) is used for treatment of skin infections for months, but side effects, especially vaginal candidiasis in women is a problem in long-term users. The more expensive monohydrate form of doxycycline is considered to have superior tolerability to the older hyclate form of the drug. A third option is the use of atovaquone/proguanil [24]. The CDC has no upper limit on the duration of intake of atovaquone/proguanil [16], and recent United Kingdom guidelines consider atovaquone/proguanil to be safe for continuous use of up to 12 months [25].

Overall, long-term travellers must have knowledge about malaria as a disease including key symptoms such as fever, adequate use of SBET, the problems of counterfeit drugs and the need to identify reliable health care facility in case of emergency.

Mosquito bite prophylaxis is even more relevant for the long-term traveller and comprehensive information about screening, impregnated bednets, coils, repellents, insecticides and the biting habits of mosquitoes is required. Systemic use of

Table 1 Guidelines for preventing malaria in long-term travellers and travellers with frequent visits to malaria-endemic areas

Inform about the disease; symptoms, diagnosis and need for rapid treatment
Discuss access to qualified medical assistance at destination and the need to identify qualified medical staff before the traveller becomes ill
Inform about alternatives to continuous chemoprophylaxis
Provide detailed advice on methods to prevent mosquito bites
Inform about the use of standby emergency treatment, SBET, and emphasise the need for medical assessment despite the use of SBET
Inform about the problems with counterfeit drugs in many malaria-endemic countries, and that drugs for both chemoprophylaxis and SBET should be purchased before arriving at the destination
Rapid diagnostic tests can be recommended for selected travellers but require comprehensive pre-travel instruction

impregnated bednets can reduce the risk of malaria by 50% [26]. In selected cases, the long-term traveller may be trained to use the rapid malaria diagnostic kits, but this option needs to be restricted to persons staying in isolated places and thorough pre-travel instruction is essential [19, 27].

Health personnel advising travellers must recognise that this group is unlikely to take chemoprophylaxis continuously for years, and it will increase the credibility of health care workers if travel health advisors acknowledge this and try to prepare realistically this group for their long-term travel.

Suggestions for advice to long-term travellers and frequent visitors are summarised in Table 1. The drug of choice for SBET depends on the use of prophylaxis, if any, and the expected drug susceptibility pattern at the destination and the guidelines of the traveller's country of origin, and as a rule, the drug used for treatment should not be the same as that used for prophylaxis. Malaria breakthrough while compliant and on prophylaxis could be due to resistance of the *P. falciparum* parasite to the chemoprophylactic drug, which is why the drug used for prophylaxis should not be re-used for treatment. There is also the possibility of reaching potentially toxic drug levels if the prophylaxis and treatment drug are the same.

7 Prophylaxis Against *P. vivax* and *P. ovale*

The main goal of malaria chemoprophylaxis is to prevent *P. falciparum*, which is primarily responsible for malaria fatalities. However, *P. vivax*, *P. ovale P.malariae* and *P. knowlesi* can cause serious febrile illness in non-immune individuals. The clinical presentation of *P. vivax* and *P. ovale* cannot be distinguished from *P. falciparum* malaria [28]. The burden of infection is considerable in indigenous populations, and *P. vivax* is the most prevalent malaria type in Southeast Asia and

South America and is also found in East Africa [29]. *P. vivax* is susceptible to most antimalarial drugs although isolates with reduced sensitivity to chloroquine and primaquine are found in parts of Indonesia, Papua New Guinea, Iraq and Afghanistan [1, 30, 31]. There is a decreased susceptibility of *P. vivax* in Thailand to sulfadoxine/pyrimethamine [32].

A study from Europe of 518 imported cases of *P. vivax* found that 60% were admitted to hospital in an average of 4 days after the start of symptoms, and seven had severe complications: hepato-splenomegaly (3 patients), spleen-rupture (1 patient), pancytopenia (1 patient), macrohaematuria (1 patient) and psychosis (1 patient) [33].

P. vivax and *P. ovale* develop hypnozoites when the host is infected, which may relapse later and cause malaria symptoms long after the traveller has returned home. The hypnozoites are only susceptible to primaquine. One study found that the first *P. vivax* attack was seen approximately 3 months after leaving the malarious area regardless of whether the traveller had taken prophylaxis or not [34]. Prevention of relapsing *P. vivax* can only be achieved by treating presumptively post travel with a course of primaquine or by using primaquine as a chemoprophylaxis during travel. So far, neither of these options has been extensively used and primaquine is not registered in most countries as a primary prophylaxis. Terminating a stay in a *P. vivax* endemic area with a 2-week course of primaquine without knowing whether the traveller is infected is not attractive for practitioners or patients, as primaquine has some side effects (primarily causing methaemoglobinemia). Use of primaquine requires that the individuals are tested for G6PD deficiency. The recommended adult dose for "anti-relapse treatment" based on clinical trials and expert opinion is 30 mg given daily for 14 days, starting on return from a malarious region and taken together with a blood schizonticide. This is based on evidence from the 1950s showing that primaquine's activity against hypnozoites is enhanced when given with chloroquine. The adult dose for primary prophylaxis is 30 mg daily starting 1 day before travel and continuing for 7 days after return [35].

One study used primaquine as prophylaxis alone in an area with a high risk of *P. falciparum* infections and found breakthroughs in four users out of 106 [36]. Primaquine has only limited activity against blood stage *P. falciparum* and if parasites emerge from the liver into the blood, primaquine alone will be insufficient as chemoprophylaxis.

Using primaquine as the drug of choice for *P. vivax* in endemic areas raises the concern of selecting for primaquine resistance, which so far has been found only in Irian Jaya, Papua New Guinea and a single report from Iraq and Afghanistan [31].

As long as the absolute risk of *P. vivax* and *P. ovale* infection is not known, the best strategy is still to inform travellers of the risk of a late onset attack of either parasite after return and cessation of chemoprophylaxis.

8 Malaria Prevention for Pregnant and Breastfeeding Women

A significant proportion of travellers are women of childbearing potential who need evidence-based advice on the use of antimalarials in the peri-conception period, during pregnancy and breastfeeding. Malaria during pregnancy is hazardous for the mother, the foetus and the neonate and is an important cause of maternal and child morbidity and mortality [37]. *P. falciparum* is responsible for the main burden of malarial disease in pregnant women [38]. Other malaria species do not parasitize placental blood to the same extent and hence have less impact [39]. The clinical features of *P. falciparum* malaria in pregnancy depend to a large extent on the immune status of the woman, which in turn is determined by her prior exposure to malaria. Non-immune travellers have little or no immunity and are prone to episodes of severe malaria leading to stillbirths, spontaneous abortions or even maternal death. Mosquito bite protection is essential and research shows that *N,N*-diethyl-3-methylbenzamide (DEET) is effective, safe and has a low risk of accumulation in the foetus [38]. Pyrethroid insecticide-treated nets are safe and have been shown to substantially reduce the risk of placental malaria [33, 39]. Bednets are particularly indicated in rooms with no air-conditioning.

Malaria chemoprophylaxis in pregnancy is complex. Due to ethical and safety restrictions, few antimalarials have been evaluated for pregnant travellers, and there is also a dearth of information on drug disposition in the pregnant woman (Table 2). Chloroquine can be used but widespread resistance limits this option. Doxycycline and primaquine are contra-indicated. Due to insufficient data, atovaquone/proguanil is not recommended, although proguanil is considered safe in pregnancy and no teratogenicity has been observed in animal studies using atovaquone [40]. Mefloquine is an option for pregnant women who cannot defer travel and who need chemoprophylaxis for chloroquine-resistant malaria-endemic areas. Some authorities now allow the use of mefloquine in all trimesters, others advise against using the drug in the first trimester apart from exceptional circumstances. A recent trial of chloroquine prophylaxis for *P. vivax* malaria in pregnant women in Thailand found no effect on maternal anaemia or birth weight [41]. However, in areas with predominantly *P. vivax* malaria, infection during pregnancy contributes to maternal morbidity and mortality [42].

With regard to breastfeeding, chloroquine, hydroxychloroquine and mefloquine are considered compatible with breastfeeding and atovaquone/proguanil can be used if the breastfed infant weighs more than 5 kg. Proguanil is excreted into human milk in small quantities. Infants who are breastfed do not receive adequate concentrations of any antimalarials and require their own chemoprophylaxis [13].

Table 2 Antimalarials for chemoprophylaxis and stand-by emergency treatment in pregnancy

Antimalarial	Chemoprophylaxis in pregnancy	Emergency self-treatment in pregnancy	Comments
Atovaquone/proguanil	No data, should not be used	No data, should only be used if no other options are available	Not recommended due to lack of safety data. Inadvertent use in pregnancy probably safe but few data available
Chloroquine (hydroxychloroquine)	Can be used	Can be used	Regarded as safe but resistance is widespread
Proguanil	Can be used	Not used for treatment	Supplement with folic acid is recommended. Should be used only in combination with chloroquine
Doxycycline	Not recommended	Not recommended	May cause bone malformation and discoloured teeth
Mefloquine	Can be used after the first trimester. Some authorities (WHO, CDC) allow the use of mefloquine in the first trimester if the risk of malaria is high and travel cannot be deferred	Can be used after 16th gestational week or if no other options are available	Regarded as safe after 16th gestational week based on post-marketing surveillance. Inadvertent use in the peri-conception period or during pregnancy is not considered an indication for a termination
Artemisinins	Not used for prophylaxis	Few data. Can only be used if no other options are available	One small study found no adverse impact on the pregnant mother or the foetus
Quinine	Not used for prophylaxis	Can be used	Drug of choice in *P. falciparum* malaria. Combination with clindamycin is recommended
Primaquine	Contra-indicated	Treatment of *P. vivax* hypnozoites should be deferred until after the pregnancy	Use of primaquine in pregnancy would necessitate G6PD testing for mother and foetus

9 Malaria Prevention for Small Children

Imported malaria case numbers in children are increasing with the rise in travel among children and changing profiles of immigrants, particularly settled immigrants VFR in malaria-endemic countries. The use of DEET containing insect repellents is recommended for children older than 2 months and an alternative repellent, picaridin, can be recommended for children older than 2 years. Chloroquine is safe for children of all ages and weights but the use of this option is limited by widespread resistance to the drug. Mefloquine can be used for children >5 kg and atovaquone/proguanil prophylaxis (as paediatric tablets) can be used for children >5 kg according to new CDC guidelines [16]. The manufacturer, the World Health Organisation [18] and some European authorities sanction the use of this combination only for children weighing more than 11 kg. Coartem®, an artemisinin combination, in dispersible tablet form with cherry flavour, has been available since 2009 in Switzerland and some African countries for the treatment of infants (personal communication Novartis, Switzerland), but cannot be used for prophylaxis. Doxycycline is for children aged >8 years (in the UK, allowed for children aged over 12 years). Table 3 shows the currently recommended doses for malaria prophylaxis in children.

When possible, deferral of travel is recommended for pregnant and breastfeeding women and also for young children.

Table 3 Antimalarial chemoprophylaxis for children

Antimalarial	Chemoprophylaxis	Dosing	Comments
Atovaquone/proguanil	[a]>5 kg CDC >11 kg manufacturer some European countries	–*Daily* –Paediatric tablets**	–Palatable –Expensive
Chloroquine (hydroxychloroquine)	All ages and weights	5 mg base/kg *weekly*	Limited use due to resistance
Proguanil	All ages and weights	3 mg/kg per day	Only in combination with chloroquine
Doxycycline	Children > 8 years	1.5 mg salt/kg *daily*	Contra-indicated for children under 12 years
Mefloquine	>5 kg	5 mg/kg *weekly*	Bitter taste
Primaquine	Children > 4 years WHO CDC specifies no lower age limit	0.5 mg/kg base *daily*	–G6PD testing essential – Last choice

[a]New 2007
**Specified by the manufacturer

References

1. Baird JK, Purnomo BH, Bangs MJ, Subianto B, Patchen LC, Hoffman SL (1991) Resistance to chloroquine by *Plasmodium vivax* in Irian Jaya, Indonesia. Am J Trop Med Hyg 44:547–552
2. Chen LH, Keystone JS (2005) New strategies for the prevention of malaria in travelers. Infect Dis Clin North Am 19:185–210
3. Schlagenhauf P (ed) (2008) Traveler's malaria, 2nd edn. BC Decker, Hamilton, ON
4. Jelinek T, Behrens R, Bisoffi Z, Bjorkmann A, Gascon J, Hellgren U, Petersen E, Zoller T, Andersen RH, Blaxhult A for the TropNetEurop (2007) Recent cases of falciparum malaria imported to Europe from Goa, India, December 2006-January 2007. Euro Surveill 12: E070111.1
5. Schmid S, Chiodini P, Legros F, D'Amato S, Schöneberg I, Liu C, Janzon R, Schlagenhauf P (2009) The risk of malaria in travelers to India. J Travel Med 16:194–199
6. Behrens RH, Carroll B, Hellgren U, Visser LG, Siikamaki H, Vestergaard LS, Calleri G, Janisch T, Myrvang B, Gascon J, Hatz C (2010) The incidence of malaria in travellers to South-East Asia: is local malaria transmission a useful risk indicator? Malar J 9:266
7. Schlagenhauf P, Petersen E (2008) Malaria chemoprophylaxis: strategies for risk groups. Clin Microbiol Rev 21:466–472
8. Schlagenhauf P (2010) Cochrane Review highlights the need for more targeted research on the tolerability of malaria chemoprophylaxis in travellers. Evid Based Med 15:25–26
9. Askling HH, Nilsson J, Tegnell A, Janzon R, Ekdahl K (2005) Malaria risk in travelers. Emerg Infect Dis 11:436–441
10. Phillips-Howard PA, Radalowicz A, Mitchell J, Bradley DJ (1990) Risk of malaria in British residents returning from malarious areas. Br Med J 300:499–503
11. Kofoed K, Petersen E (2003) The efficacy of chemophrophylaxis against malaria with chloroquine plus proguanil, mefloquine and atovaquone plus proguanil in travellers from Denmark. J Travel Med 10:150–154
12. Behrens RH, Carroll B, Beran J, Bouchaud O, Hellgren U, Hatz C, Jelinek T, Legros F, Mühlberger N, Myrvang B, Siikamäki H, Visser L for TropnetEurop (2007) The low and declining risk of malaria in travellers to Latin America: is there still an indication for chemoprophylaxis? Malar J 6:114
13. Schlagenhauf P, Tschopp A, Johnson R, Nothdurft HD, Beck B, Schwartz E, Herold M, Krebs B, Veit O, Allwinn R, Steffen R (2003) Tolerability of malaria chemoprophylaxis in non-immune travellers to sub-Saharan Africa: multicentre, randomised, double blind, four arm study. Br Med J 327:1078–1085
14. Chen LH, Wilson ME, Schlagenhauf P (2007) Controversies and misconceptions in malaria chemoprophylaxis for travelers. JAMA 297:2251–2263
15. Arguin P, Schlagenhauf P (2008) Malaria deaths in travelers. In: Schlagenhauf P (ed) Travelers malaria, 2nd edn. BC Decker, London
16. CDC (2011) Travellers health, Yellow Book 2011. (http://www.cdc.gov)
17. Schlagenhauf P (2007) Pregnant? Breastfeeding? Traveling with infants? Malaria prevention challenges for mothers. Abstract PL03.03 P34. In: Proceedings of the 10th conference of the international society of travel medicine, Vancouver, Canada, 20–24 May 2007
18. WHO (2011) International travel and health. (http://www.who.int/ITH)
19. Chen LH, Wilson ME, Schlagenhauf P (2006) Prevention of malaria in long-term travelers. JAMA 296:2234–2244
20. Toovey S, Moerman F, van Gompel A (2007) Special infectious disease risks of expatriates and long-term travelers in tropical countries. Part I: Malaria. J Travel Med 14:42–49
21. Sonmez A, Harlak A, Kilic S, Polat Z, Hayat L, Keskin O, Dogru T, Yilmaz MI, Acikel CH, Kocar IH (2005) The efficacy and tolerability of doxycycline and mefloquine in malaria prophylaxis of the ISAF troops in Afghanistan. J Infect 51:253–258

22. Wells TS, Smith TC, Smith B, Wang LZ, Hansen CJ, Reed RJ, Goldfinger WE, Corbeil TE, Spooner CN, Ryan MA (2006) Mefloquine use and hospitalizations among US service members, 2002–2004. Am J Trop Med Hyg 74:744–749
23. Hellgren U, Angel VH, Bergqvist Y, Forero-Gomez JS, Rombo L (1991) Plasma concentrations of sulfadoxine-pyrimethamine, mefloquine and its main metabolite after regular malaria prophylaxis for two years. Trans R Soc Trop Med Hyg 85:356–357
24. Genderen PJ, van Koene HR, Spong K, Overbosch D (2007) The safety and tolerance of atovaquone/proguanil for the long-term prophylaxis of *Plasmodium falciparum* malaria in non-immune travelers and expatriates. J Travel Med 14:92–95
25. Swales CA, Chiodini PL, Bannister BA for the Health Protection Agency Advisory Committee on Malaria Prevention in UK Travellers (2007) New guidelines on malaria prevention: a summary. J Infect 54:107–110
26. Marbiah NT, Petersen E, David K, Lines J, Magbity E, Bradley DB (1998) Control of clinical malaria due to *Plasmodium falciparum* in children in an area of rural Sierra Leone with perennial transmission: a community-wide trial of lambdacyhalothrin-impregnated mosquito nets alone and/or with fortnightly dapsone/pyrimethamine/placebo chemoprophylaxis. Am J Trop Med Hyg 58:1–6
27. Schlagenhauf P, Adamcova M, Regep L, Schaerer MT, Rhein HG (2010) The position of mefloquine as a 21st century malaria chemoprophylaxis. Malar J 9:357
28. Bottieau E, Clerinx J, Van Den Enden E, Van Esbroeck M, Colebunders R, Van Gompel A, Van Den Ende J (2006) Imported non-*Plasmodium falciparum* malaria: a five-year prospective study in a European referral center. Am J Trop Med Hyg 75:133–138
29. Mendis K, Sina BJ, Marchesini P, Carter R (2001) The neglected burden of *Plasmodium vivax* malaria. Am J Trop Med Hyg 64(1–2 Suppl):97–106
30. Schwartz IK, Lackritz EM, Patchen LC (1991) Chloroquine-resistant *Plasmodium vivax* from Indonesia. N Engl J Med 324:927
31. Spudick JM, Garcia LS, Graham DM, Haake DA (2005) Diagnostic and therapeutic pitfalls associated with primaquine-tolerant *Plasmodium vivax*. J Clin Microbiol 43:978–981
32. Pukrittayakamee S, Imwong M, Looareesuwan S, White NJ (2004) Therapeutic responses to antimalarial and antibacterial drugs in vivax malaria. Acta Trop 89:351–356
33. Mühlberger N, Jelinek T, Gascon J, Probst M, Zoller T, Schunk M (2004) Epidemiology and clinical features of vivax malaria imported to Europe: sentinel surveillance data from TropNetEurop. Malar J 3:5
34. Elliott JH, O'Brien D, Leder K, Kitchener S, Schwartz E, Weld L, Brown GV, Kain KC, Torresi J for the GeoSentinel Surveillance Network (2004) Imported Plasmodium vivax malaria: demographic and clinical features in nonimmune travelers. J Travel Med 11:213–217
35. Hill DR, Baird JK, Parise ME, Lewis LS, Ryan ET, Magill AJ (2006) Primaquine: report from CDC expert meeting on malaria chemoprophylaxis I. Am J Trop Med Hyg 75:402–415
36. Schwartz E, Regev-Yochay G (1999) Primaquine as prophylaxis for malaria for nonimmune travelers: a comparison with mefloquine and doxycycline. Clin Infect Dis 29:1502–1506
37. Desai M, ter Kuile FO, Nosten F, McGready R, Asamoa K, Brabin B, Newman RD (2007) Epidemiology and burden of malaria in pregnancy. Lancet Infect Dis 7:93–104
38. McGready R, Ashley EA, Nosten F (2004) Malaria and the pregnant traveller. Travel Med Infect Dis 2:127–142
39. Lengeler C (2004) Insecticide-treated bed nets and curtains for preventing malaria. Cochrane Database Syst Rev 2:CD000363
40. Pasternak B, Hviid A (2011) Atovaquone-proguanil use in early pregnancy and the risk of birth defects. Arch Intern Med 171:259–260
41. Villegas L, McGready R, Htway M, Paw MK, Pimanpanarak M, Arunjerdja R, Viladpai-Nguen SJ, Greenwood B, White NJ, Nosten F (2007) Chloroquine prophylaxis against vivax malaria in pregnancy: a randomized, double-blind, placebo-controlled trial. Trop Med Int Health 12:209–218
42. Rodriguez-Morales AJ, Sanchez E, Vargas M, Piccolo C, Colina R, Arria M, Franco-Paredes C (2006) Pregnancy outcomes associated with *Plasmodium vivax* malaria in northeastern Venezuela. Am J Trop Med Hyg 74:755–757

Malaria Diagnostics: Lighting the Path

David Bell and Mark D. Perkins

Abstract Drugs cure malaria, while diagnostics reduce the use of drugs. Thus, they not only impose a risk, demanding strong health service structures, but also deliver clear advantages in drug targeting and evidence-based management of fever. Often in the past, the costs of delivering accurate diagnostic results were seen as a reason for remaining blind, contributing to a stagnation in malaria and febrile disease management. The advent of new diagnostic technologies and demands of declining transmission rates have brought diagnosis to center stage in malaria management, and promise to catalyze reform in the management of febrile illness and wider health service delivery.

On an October day in 1880, Laveran first identified a microscopic parasite in the blood of a febrile patient in Algeria. In doing so, he opened our eyes to the possibility of addressing malaria as a specific disease with a specific cause, as many of his contemporaries were doing with other human ailments. Malaria is almost uniquely distinguished in that many eyes were later closed again; across the swathes of the malaria-endemic world, the ability to distinguish malaria from other causes of tropical fever syndrome was deliberately ignored. In sub-Saharan Africa in particular, despite the lack of pathognomonic symptoms, building on Laveran's discovery required a level of effort deemed unmanageable or unnecessary. Essentially, all malaria-like fevers were to be managed as one; a legacy that is still influencing African health systems today.

D. Bell • M.D. Perkins (✉)
Foundation for Innovative New Diagnostics, Avenue de Budé 16, Geneva 1202, Switzerland
e-mail: mark.perkins@finddiagnostics.org

H.M. Staines and S. Krishna (eds.), *Treatment and Prevention of Malaria*,
DOI 10.1007/978-3-0346-0480-2_15, © Springer Basel AG 2012

1 Evolution of Diagnostics in Malaria Control

Malaria shares symptoms with several other common diseases that exact significant mortality and/or morbidity, such as the early stages of acute respiratory tract infection, meningitis, typhoid, and typhus [1–3]. Algorithms to distinguish malaria reliably on clinical grounds lack accuracy [4–8]. Managing each case of suspected malaria on an evidence base requires a significant degree of sophistication in health systems that until recently was considered unreachable, at least with the resources available. As a result, malaria diagnosis was largely guesswork, and many antimalarial drugs went to patients with other diseases. Not only was this practice wasteful of antimalarial drugs and dangerous for those misdiagnosed, but it also ensured continued ignorance about the epidemiology of malaria and the effectiveness of control efforts: malaria prevalence is highly heterogeneous but fever is quite uniformly common. The potential value of *excluding* malaria opened up by Laveran, thereby allowing a more focused treatment approach to other diseases, was largely lost.

Across much of Asia and South America, and the endemic areas of Europe and North America from which malaria is now largely eliminated, health systems did sometimes rise to the challenge of creating much of the diagnostic infrastructure necessary to allow targeted use of the slowly evolving stable of antimalarial drugs. Vertical structures to support microscopy were often associated with campaigns that were successful in eliminating malaria in Europe and North America, and in greatly reducing transmission in Asia and South America [9–12]. However, it is only in the past few years that guidelines for malaria management have fully embraced an evidence-based treatment policy such as is standard practice for most other infectious diseases, and in doing so opened the possibility of addressing a plethora of other diseases often exacting a similar or greater mortality in the same populations [13]. The recent increased use of parasite-based diagnosis is in part responsible for the steady reduction in reported incidence of malaria in areas where it was long considered the major public health problem; fever no longer equals malaria [14]. A review of published clinical studies, which tend to be carried out in high transmission areas, shows that malaria as a cause of fever has substantially declined across sub-Saharan Africa (Fig. 1) in the last 20 years: the median fraction of all patients with presumptive malaria found to be parasitemic was 22% in studies published since the year 2000 [15]. In many areas of sub-Saharan Africa previously thought to support high malaria transmission, incidence and prevalence are now known to be low [14]. In some areas, especially where effective control measures have been put in place, the decline in reported malaria that accompanies the introduction of diagnostics is dramatic. Data such as that reported from Livingstone district in Zambia (Fig. 2) demonstrate the impossibility of measuring the effectiveness of control measures in the absence of parasite-based diagnostics. As the proportion of febrile disease caused by malaria declines, the imperative for parasite-based diagnosis to target the use of antimalarial drugs increases further.

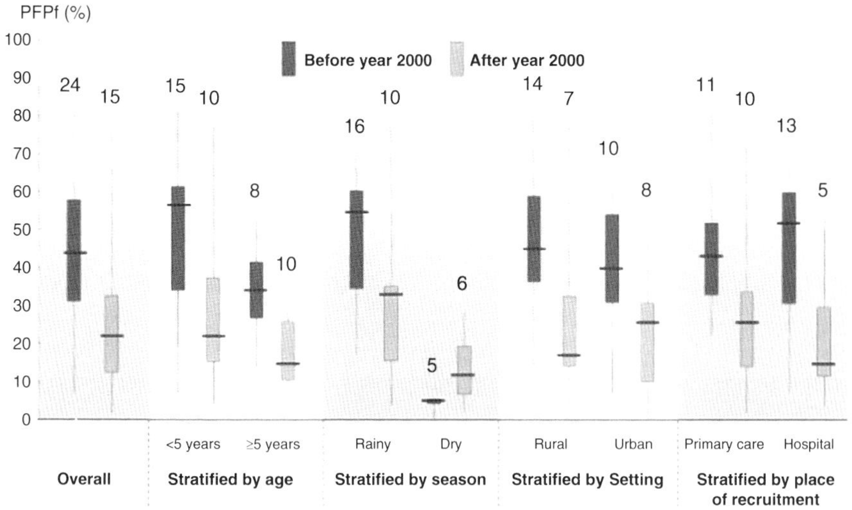

Fig. 1 Comparison between the proportions of fevers associated with *Plasmodium falciparum* parasitaemia (PFPf) in years ≤2000 and >2000, stratified by baseline characteristics (*numbers above plots* corresponds to the number of studies involved). Taken from [15]

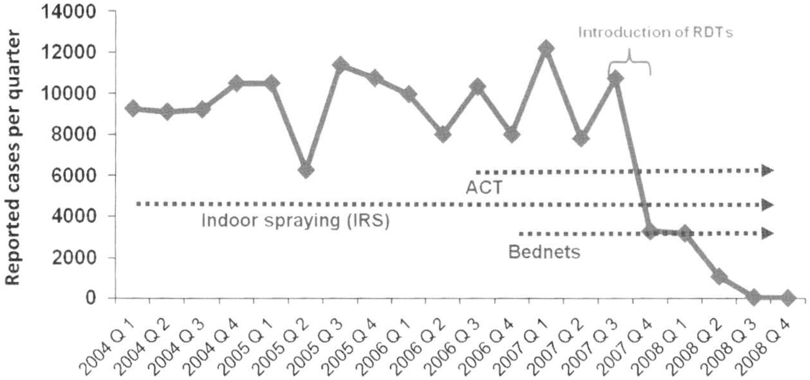

Fig. 2 Reported incidence of malaria in Livingstone district, Zambia, 2004–2008. Until late 2007, malaria was largely diagnosed clinically, masking the true incidence of disease and the effectiveness of control measures. Courtesy of the Zambia National Malaria Control Program

The diagnostic infrastructure established in the health systems of many Asian and South American countries likely had a significant impact in improving overall health system capacity and drug delivery. The structures set up to train, supply, and quality assure microscopists in countries such as India, Thailand, and Peru established supply lines and levels of management invaluable for delivery of a range of services, and a cadre of personnel trained sufficiently to use them [12]. While microscopy coverage often remained patchy and confined mostly to the

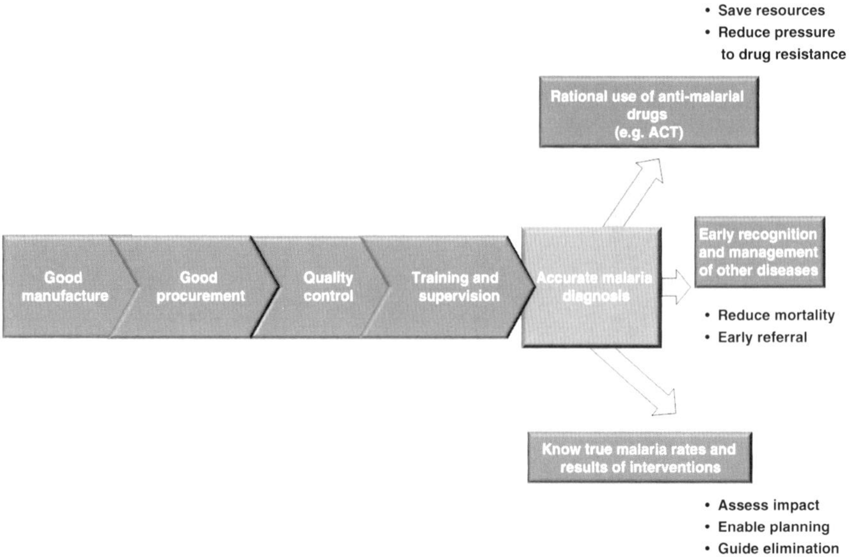

Fig. 3 Diagnostic effort and outcomes. Common to both microscopy and RDTs used at point of care, a well systematic quality control structure to ensure accuracy and provide confidence among users is expected to impart benefits beyond the management of malaria cases. Courtesy of the Foundation for Innovative New Diagnostics (FIND)

public health sector, the effort and infrastructure put in place to support diagnosis provided gains in malaria control well beyond case management (Fig. 3). It also strengthened advocacy for resources with some expectation that they would be directed to those who required them. The implementation of diagnostics required dedication of resources on the part of malaria control programs; antimalarial drugs, particularly chloroquine, and sulphadoxine–pyrimethamine, remained relatively cheap up to the last decade, and well below the costs of diagnosis.

Several factors have influenced the move, over the last decade, toward parasite-based diagnosis as an essential element in case management. First, the costs of antimalarial therapy rose dramatically with the introduction of artesunate-based combination therapies on a wide scale, particularly in Africa. As shown by several cost–benefit models and analyses, it was no longer economically viable to hand out antimalarial therapy to a large section of the population that had some other disease [16, 17]. Second, the advent of commercial lateral flow tests for malaria in the 1990s offered an alternative to microscopy that required significantly less infrastructure to support it. Evidence of the reproducible accuracy of some of the products in both laboratory and field settings has enabled confidence in their use [18–20]. Third, the rapid increase in funding available for malaria control programs over the past decade through mechanisms such as the Global Fund to Fight AIDS, Tuberculosis and Malaria has enabled Ministries of Health, especially in sub-Saharan Africa, to contemplate investments in malaria management not previously possible.

The rise of drug resistance, first to chloroquine and sulphadoxine–pyrimethamine, and more recently concern over the possible development of resistance to artemisinin-based combination therapies (ACTs), has further fuelled interest in diagnostic-based therapy. While an uninfected person will not add to pressure toward parasite resistance by inappropriately ingesting an antimalarial drug, the concern is that prolonged circulation of partner compounds, such as those accompanying artemisinin-based compounds in ACTs, could expose parasites from an infection arising after the one being treated to subtherapeutic doses and select for resistant strains.

In 2009, the World Health Organization (WHO), in response to increasing evidence of the safety and impact of diagnostics, updated its malaria control recommendations to include parasite-based diagnosis prior to the prescription of antimalarial drugs.

2 Utilization of Diagnostics

Although malaria parasites cause an acute, often recurring febrile illness, a significant number of individuals living in endemic areas may have asymptomatic parasitemia (Fig. 4). Such asymptomatic parasite carriage may occur as a result of clinical immunity to blood stage carriage, particularly of *Plasmodium falciparum* and *P. malariae*, or carriage of liver stages of *P. vivax* or *P. ovale*. For clinical purposes, laboratory methods for malaria diagnosis can therefore usefully be

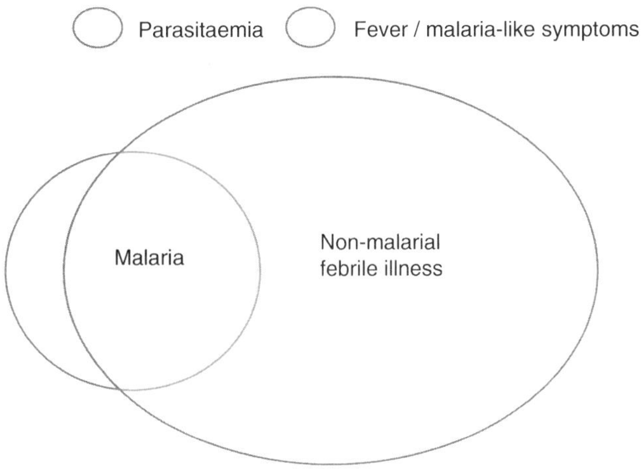

Fig. 4 The relationship between malaria parasitemia and malaria-like febrile illness in an endemic population. While most patients with malaria parasitemia are likely to have current or recent fever, a certain proportion can be asymptomatic and so do not present for diagnosis through fever case management. The clinical picture of febrile patients is indistinguishable for that of a number of other febrile diseases. Courtesy of the Foundation for Innovative New Diagnostics (FIND)

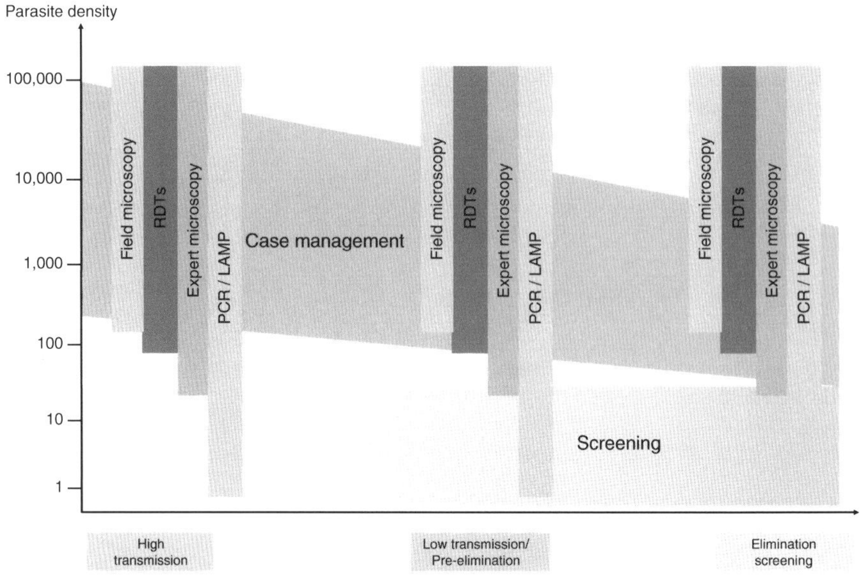

Fig. 5 The utility of different diagnostic methods will depend on the epidemiological setting of use, and on the indication – for case management or active screening. Courtesy of the Foundation for Innovative New Diagnostics (FIND)

divided into medical tests for case management – determining whether a febrile patient has malarial disease – and surveillance tools for determining the prevalence of parasites in a given population. Surveillance testing, especially in areas of low transmission and in elimination campaigns, needs to detect very low parasite densities (Fig. 5). Tests to detect other metabolic or pathological states relevant to malaria treatment are also required to varying extent to ensure safe case management; apart from routine biochemical monitoring to manage severe malaria, an example is the identification of glucose-6-phosphate dehydrogenase (G6PD) deficiency to allow safe use of 8-hydroxyquinolones in management of *P. vivax* and *P. ovale* (Box 1).

Box 1. *P. vivax*, Latent Parasites and Tafenoquine

An interesting case of the dependence of drug management on diagnostics, and the relative emphasis afforded the two areas of research and development, is seen in the currently rising interest in improved management and eventual elimination of *P. vivax*. In common with *P. ovale*, the latent liver-stage of *P. vivax* is not susceptible to drugs effective in clearing blood stages, which prevents acute treatment from clearing the parasite. Currently, the 8-aminoquinoline, primaquine, is the only available drug that can achieve radical cure, requiring a course conventionally of 14 days duration. Primaquine is further limited by the potential toxicity of the 8-aminoquinolines in cases of glucose-6-phosphate dehydrogenase (G6PD) deficiency. A newer drug in the same class, tafenoquine, has a longer half-life than primaquine, which could greatly simplify dosing. However, concern over the toxicity of a long-acting drug in G6PD-deficient patients is a major obstacle to tafenoquine introduction. In this context, it is perhaps surprising the large investment over more than 20 years in the development and evaluation of the drug has not been matched by a far more modest investment in the development of field-ready G6PD screening tests to accompany its use.

2.1 *Diagnostics for Case Management*

Light microscopy has remained essentially unchanged since the time of Laveran. Early improvements in staining, the advent of binocular microscopes, and light-emitting diode (LED) illumination have all made the technician's life easier and improved accuracy. However, despite notable exceptions, sometimes unsustained, in countries such as Thailand, India, Sri Lanka, and endemic areas of the Americas, in most countries microscopy has largely remained a tool limited to reference laboratories and hospitals. Because significant investment in human resources is required to maintain good malaria microscopy services, the extension of microscopy to lower levels of the health system has often resulted in a significant decline in performance standards [21–26].

From the 1970s to 1990s, a handful of new detection methods were developed to enhance or build on conventional light microscopy. Such methods, including fluorescent microscopy of acridine orange stained thick films or of centrifuged microhematocrit tubes [27–29], have thus far had limited impact. This is partially because they removed some advantages of light microscopy (including accurate species determination and utility of the equipment for a broad range of diseases), but more importantly because they failed to address the largest barrier to the success of microscopy in most malaria-endemic settings; the limited technical and human resources infrastructure available at the place where diagnostics are most needed – within walking distance of a sick patient.

Nucleic acid amplification [*e.g.*, polymerase chain reaction (PCR)] and enzyme linked immunoassays (ELISA) for plasmodial antigens added to the armamentarium of reference methods available at the top of national laboratory systems, but failed to offer a solution to the ease-of-use limitations of light microscopy for routine case management.

The development of lateral flow immunochromatographic test strips to detect plasmodial antigens in the early 1990s finally brought a challenge to the dominance of microscopy, and the potential to overcome many of its weaknesses in case management. Because of their simplicity, these immunochromatographic tests, commonly referred to as rapid diagnostic tests (RDTs) or "dipsticks," have been favored for field use over the other alternatives. Since the first RDT was evaluated

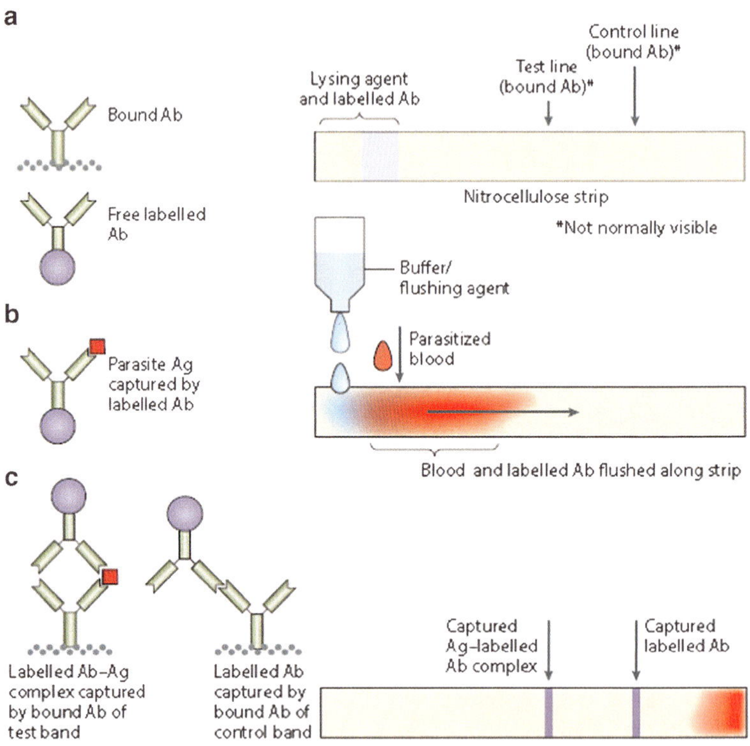

Fig. 6 Composition and detection mechanisms of lateral flow immunochromatographic tests for malaria. Courtesy of the WHO – Regional Office for the Western Pacific

in 1993, a steadily expanding number of malaria RDTs has appeared on the market, and others are under development [30]. All are based on the detection of parasite-derived protein in whole blood. Blood is lysed and drawn along a nitrocellulose test strip by a combination of capillary action and flushing by a buffer reagent. The protein is bound during testing by dye-labeled specific antibodies (colloidal gold is commonly used), and the labeled antigen–antibody complex is captured by a line of antibody bound to the strip (Fig. 6). Tests may contain multiple lines of antibody to detect species-differentiating *Plasmodium* proteins. A control line further along the strip, composed of anti-immunoglobulin antibody, also captures dye-labeled antibody and confirms that proper wicking has occurred.

Lateral flow tests are relatively simple to manufacture, given the availability of the main components including antibodies and nitrocellulose film. However, manufacturing in large quantities presents challenges in maintaining consistent quality. As a result, the number of available products mushroomed, but with a high variability in performance. More than 100 million RDTs are financed in the public sector in 2011 [31]. The diversity in quality, limited resources of the manufacturers involved, and lack of reference standards necessitated the development of an independent quality assurance program to guide procurement and provide the confidence necessary for reliance on these tools [32, 33].

WHO and the Foundation for Innovative New Diagnostics (FIND) and partners have collaborated to establish a quality assurance system that includes product testing, to determine the comparative performance of commercial tests, and lot testing, to check the quality of specific manufactured lots before they are distributed. Manufacturers operating under appropriate ISO standards are invited annually to submit products for testing against a highly characterized panel of blood samples with clinically relevant concentrations of *P. falciparum* or *P. vivax*. To date, four annual rounds of product testing have been initiated on 168 products submitted by more than 30 manufacturers. The reports of this product testing show a wide variability in the capacity of different tests to detect low parasite densities and in thermal stability – the latter an essential requirement for remote area transport and storage [18, 19]. They also show that there are many tests that perform very well. Since product testing has begun, the quality of RDT manufacturing has increased, as evident from significant performance improvements in tests that have been resubmitted for product testing after manufacturing changes.

2.2 Risks of Parasite-Based Diagnosis for Case Management

Efficacious antimalarial drugs cure malaria, while diagnostics, when replacing symptom-based diagnosis, lead to withdrawal of drugs from those who would otherwise have received them. This puts a huge imperative on quality of result before routine parasite-based diagnosis is contemplated. When accurate, diagnosis leads to gains in drugs saved, quality of data for disease tracking, and the potential to reduce mortality through the improved early management of other diseases as malaria is excluded [13]. When inaccurate, diagnostic use may increase malaria mortality as a result of missed diagnoses.

Conversely, parasite-based diagnosis may lead to inappropriate assumptions that malaria parasitemia is a cause of illness. Although symptomatic malaria parasitemia always deserves to be treated, parasitemia may occur as a secondary cause of fever or as an incidental infection in a fever caused by other etiology. As with all diagnostic testing, test results should not be interpreted in the absence of clinical assessment.

All diagnostic tests have limits; malaria RDTs detect a certain threshold of parasite density, or more precisely an antigen concentration equivalent to this. This detection threshold must be low enough that clinically significant disease is not missed. The ratio of antigen concentration to parasite density varies widely, as does the severity of symptoms at any given parasite density vary, depending on the level of immunity. A threshold of approximately 200 parasite/μL is considered a reasonable cut-off that must be detected reliably to ensure safe management [34] for the four common species of malaria infection humans. *P. knowlesi* may have a lower parasitemia threshold for causing symptoms.

Concern over the ability of tests to detect a sufficiently low threshold (adequate clinical sensitivity), together with the ingrained belief that "fever equals malaria," is probably largely responsible for the slow uptake of parasite-based diagnosis, whether by RDTs or microscopy. Evidence of poor adherence to diagnostic results is reported in antimalarial drug dispensing in several studies [35–37]. However, large reductions in drug dispensing are now being seen on a national scale, where RDT introduction has been systematically accompanied by programs to ensure both quality of the tests and training in their use [38, 39]. Moreover, studies on treatment withholding on negative test results have found that this is safe, even in young children, when the RDTs are confirmed to be working [40]. In order to establish this level of confidence, and safety, however, the structure of implementation and associated health systems development needs to be taken seriously and sufficiently resourced (Fig. 7).

The overall health benefits of parasite-based diagnosis and introduction of RDTs at a village or community level are difficult to quantify and are expected to vary in differing epidemiological situations, but may include improved adherence to antimalarial therapy promoted by surety of diagnosis, and better management of nonmalarial fever. Where early appropriate action is taken to identify and manage other etiologies of fever in patients with parasite-negative results, major gains in mortality reduction, especially for severely ill patients, may accrue [13, 16, 41], but the means and ability

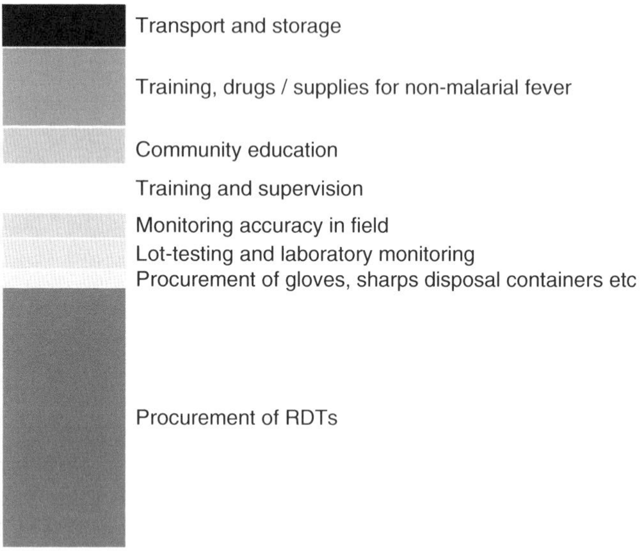

Fig. 7 The costs of implementation of malaria RDTs in a typical health system go well beyond the costs of RDT procurement. While the relative cost of the various areas of implementation will vary widely between programs, all must be adequately resourced. Courtesy of the WHO – Regional Office for the Western Pacific

to address these are often lacking. Together with exposing the need for the development of capacity for managing these nonmalarial febrile illnesses, RDT implementation, such as microscopy, requires considerable investment in health services to ensure adequate delivery and utilization of results. While this may slow implementation and impose new strains on resource-poor health systems, it also provides an opportunity to build structures to support community disease management that have spin-offs for other disease programs and other areas of health-care delivery.

2.3 Diagnostics for Screening and Surveillance

2.3.1 Nucleic Acid Detection

A nested PCR is the current gold standard for the detection of parasitemia, detecting less than one parasite per microliter [42, 43]. Quantitative PCR offers the potential to determine the concentration of circulating DNA, and therefore estimates of circulating parasite density and to a lesser extent parasite load. The applications of PCR-based methods are limited to well-equipped laboratories with specifically trained technicians, and are further limited by cost. Avoidance of contamination (leading to false-positive results) requires a high standard of laboratory practice. PCR capacity is limited in most malaria-endemic countries, and considerable resources would be required to establish and maintain this capacity. Restriction to well-equipped laboratories limits its utility for clinical care by preventing feedback timely enough for case management in most endemic areas, though the development of systems that automate and integrate sample processing, in place for some other diseases, may increase its accessibility [44].

Alternative molecular methods, which are inherently simpler than PCR, have been developed and applied to malaria. One such method is loop mediated isothermal amplification (LAMP), which operates at a single temperature and yields a turbid or fluorescent endpoint that can be detected by the naked eye. A sensitive malaria test using LAMP has been developed targeting mitochondrial DNA, and a report on a noncommercial early version of this assay published [45]. Such an approach has potential to reduce the training and infrastructure requirements of molecular diagnosis, making proximal implementation possible, where results could be rapidly available. This could be very useful for surveillance and active case finding for low-density parasitemia, and for monitoring parasite presence in drug efficacy monitoring and trials and vaccine trials. A LAMP assay that has sufficient throughput to be used for surveillance testing of large numbers of individuals has not yet been developed.

2.3.2 Serological Tests for Antibody Detection

Antibody detection, currently available in ELISA and RDT formats, can readily demonstrate infection with malaria. These tests are inappropriate for case management because they cannot reliably differentiate between past and current infection, and because antibodies may not be detectable in blood stage infections of very recent onset. They do, however, have a potential role in tracking the epidemiology of malaria. Parasite prevalence data provide a snapshot in any given season of the epidemiology of disease, whereas antibody responses represent transmission intensity over several years, reducing seasonal or annual bias. Age-adjusted rates of immune responses may be used to estimate the force of infection [46, 47]. As humoral immune responses are very sensitive measures of infection, they may have potential, little used till now, to guide stratification of malaria risk, where transmission is very low.

Detection of antisporozoite antibodies has been suggested as a surrogate for detecting individuals with a high likelihood of carriage of *P. vivax* hypnozoites (evidence of infection), and could therefore be used to guide the use of 8-aminoqunimolines for clearance of liver-stage *P. vivax* and *P. ovale*.

3 Conclusions

Parasite-based diagnosis has a large and growing role in malaria control and the case management of febrile illness, thanks to the development of simple and reliable assays and a mechanism to monitor their quality, and thanks to improved health funding and a clearer understanding of the benefits of parasite-based diagnosis. The expanding role of rapid testing in malaria control programs has in many areas transformed local understanding of the true prevalence of malaria and has, where properly supported by health-worker training, saved millions of courses of unwarranted malaria therapy. These benefits come with a cost, not just only for the commodities used, but also for the complexity that knowledge brings. Confronting a patient with an unknown cause of fever is much more complicated for health workers than simply dispensing, needed or not, antimalarial drugs. Having spent decades treating all fever in the tropics as malaria, health systems will have to adapt to maximize the advantages that knowing before treating can bring.

References

1. Cunha BA (2004) Osler on typhoid fever: differentiating typhoid from typhus and malaria. Infect Dis Clin North Am 18:111–125
2. Wright PW, Avery WG, Ardill WD, McLarty JW (1993) Initial clinical assessment of the comatose patient: cerebral malaria vs. meningitis. Pediatr Infect Dis J 12:37–41
3. Madhi SA, Klugman KP (2006) Acute respiratory infections. In: Jamison DT, Feachem RG, Makgoba MW et al (eds) Disease and mortality in Sub-Saharan Africa. World Bank, Washington, DC
4. Bassett MT, Taylor P, Bvirakare J, Chiteka F, Govere EJ (1991) Clinical diagnosis of malaria: can we improve? Trop Med Hyg 94:65–69

5. Redd SC, Kazembe PN, Luby SP, Nwanyanwu O, Hightower AW, Ziba C, Wirima JJ, Chitsulo L, Franco C, Olivar M (1996) Clinical algorithm for treatment of *Plasmodium falciparum* malaria in children. Lancet 347:223–227
6. Luxemburger C, Nosten F, Kyle DE, Kiricharoen L, Chongsuphafajaiidhi T, White NJ (1998) Clinical features cannot predict a diagnosis of malaria or differentiate the infecting species in children living in an area of low transmission. Trans R Soc Trop Med Hyg 92:45
7. Mwangi TW, Mohammed M, Dayo H, Snow RW, Marsh K (2005) Clinical algorithms for malaria diagnosis lack utility among people of different age groups. Trop Med Int Health 10:530–536
8. Quigley MA, Armstrong Schellenberg JR, Snow RW (1996) Algorithms for verbal autopsies: a validation study in Kenyan children. Bull World Health Organ 74:147–154
9. Hackett L (1937) Malaria in Europe: an ecological study. Oxford University Press, London
10. Ackerknecht EH (1945) Malaria in the Upper Mississippi Valley 1760–1900. Bull Hist Med (suppl) viii+142
11. Deane LM (1988) Malaria studies and control in Brazil. Am J Trop Med Hyg 38:223–230
12. Najera JA, Gonzalez-Silva M, Alonso PL (2011) Some lessons for the future from the Global Malaria Eradication Programme (1955–1969). PLoS Med 8:e1000412
13. Black RE, Cousens S, Johnson HL, Lawn JE, Rudan I, Bassani DG, Jha P, Campbell H, Walker CF, Cibulskis R, Eisele T, Liu L, Mathers C; Child Health Epidemiology Reference Group of WHO, UNICEF (2010) Global, regional, and national causes of child mortality in 2008: a systematic analysis. Lancet 375:1969–1987
14. WHO (2010) World Malaria Report. World Health Organization, Geneva
15. D'Acremont V, Lengeler C, Genton B (2010) Reduction in the proportion of fevers associated with *Plasmodium falciparum* parasitaemia in Africa: a systematic review. Malar J 9:240
16. Shillcutt S, Morel C, Goodman C, Coleman P, Bell D, Whitty CJ, Mills A (2008) Cost-effectiveness of malaria diagnostic methods in sub-Saharan Africa in an era of combination therapy. Bull World Health Organ 86:101–110
17. Lubell Y, Reyburn H, Mbakilwa H, Mwangi R, Chonya S, Whitty CJ, Mills A (2008) The impact of response to the results of diagnostic tests for malaria: cost-benefit analysis. Br Med J 336:202–205
18. WHO (2009) Malaria rapid diagnostic test performance, Round 1. World Health Organization, Geneva. ISBN 978 92 4 159807 1
19. WHO (2010) Malaria rapid diagnostic test performance, Round 2. World Health Organization, Geneva. ISBN 978 92 4 159946 7
20. Lot testing results. http://www.finddiagnostics.org/about/what_we_do/successes/malaria_rd-t_lot_testing_results/index.html
21. Kachur SP, Nicolas E, Jean-François V, Benitez A, Bloland PB, Saint Jean Y, Mount DL, Ruebush TK 2nd, Nguyen-Dinh P (1998) Prevalence of malaria parasitemia and accuracy of microscopic diagnosis in Haiti, October 1995. Rev Panam Salud Publ 3:35–39
22. Durrheim DN, Becker PJ, Billinghurst K (1997) Diagnostic disagreement–the lessons learnt from malaria diagnosis in Mpumalanga. S Afr Med J 87:1016
23. Kain KC, Harrington MA, Tennyson S, Keystone JS (1998) Imported malaria: prospective analysis of problems in diagnosis and management. Clin Infect Dis 27:142–149
24. Coleman RE, Maneechai N, Rachaphaew N, Kumpitak C, Miller RS, Soyseng V, Thimasarn K, Sattabongkot J (2002) Comparison of field and expert laboratory microscopy for active surveillance for asymptomatic *Plasmodium falciparum* and *Plasmodium vivax* in western Thailand. Am J Trop Med Hyg 67:141–144
25. O'Meara WP, McKenzie FE, Magill AJ, Forney JR, Permpanich B, Lucas C, Gasser RA Jr, Wongsrichanalai C (2005) Sources of variability in determining malaria parasite density by microscopy. Am J Trop Med Hyg 73:593–598
26. Kilian AH, Metzger WG, Mutschelknauss EJ, Kabagambe G, Langi P, Korte R, von Sonnenburg F (2000) Reliability of malaria microscopy in epidemiological studies: results of quality control. Trop Med Int Health 5:3–8

27. Anthony RL, Bangs MJ, Anthony JM, Purnomo (1992) On-site diagnosis of *Plasmodium falciparum*, *P. vivax*, and *P. malariae* by using the Quantitative Buffy Coat system. J Parasitol 78:994–998
28. Parija SC, Dhodapkar R, Elangovan S, Chaya DR (2009) A comparative study of blood smear, QBC and antigen detection for diagnosis of malaria. Indian J Pathol Microbiol 52:200–202
29. Adeoye GO, Nga IC (2007) Comparison of Quantitative Buffy Coat technique (QBC) with Giemsa-stained Thick Film (GTF) for diagnosis of malaria. Parasitol Int 56:308–312
30. Shiff CJ, Premji Z, Minjas JN (1993) The rapid manual ParaSight-F test. A new diagnostic tool for *Plasmodium falciparum* infection. Trans R Soc Trop Med Hyg 87:646–648
31. RBM, MMV: Tracking progress in scaling-up diagnosis and treatment for malaria. Geneva, Roll back Malaria Partnership, Medicines for Malaria Venture. 2009. http://www.rbm.who.int/partnership/wg/wg_procurementsupply/docs/ACT_RDTscaleUpAssess-ment_MMVRBMaugust2010.pdf
32. WHO-TDR-FIND (2010) Methods manual for laboratory quality control testing of malaria rapid diagnostic tests, Version Six. 2010. World Health Organization, Geneva
33. WHO-FIND-TDR-USCDC (2010) Methods manual for product testing of malaria rapid diagnostic tests (version 3). World Health Organization, Geneva
34. WHO (2009) Parasitological confirmation of malaria diagnosis. Report of a WHO technical consultation, Geneva, 6–8 Oct 2009. ISBN 978 92 4 159941 2
35. Bisoffi Z, Sirima BS, Angheben A, Lodesani C, Gobbi F, Tinto H, Van den Ende J (2009) Rapid malaria diagnostic tests vs. clinical management of malaria in rural Burkina Faso: safety and effect on clinical decisions. A randomized trial. Trop Med Int Health 14:491–498
36. Zurovac D, Njogu J, Akhwale W, Hamer DH, Larson BA, Snow RW (2008) Effects of revised diagnostic recommendations on malaria treatment practices across age groups in Kenya. Trop Med Int Health 13:784–787
37. Chandler CI, Chonya S, Boniface G, Juma K, Reyburn H, Whitty CJ (2008) The importance of context in malaria diagnosis and treatment decisions – a quantitative analysis of observed clinical encounters in Tanzania. Trop Med Int Health 13:1131–1142
38. D'Acremont V, Kahama-Maro J, Swai N, Mtasiwa D, Genton B, Lengeler C (2011) Reduction of anti-malarial consumption after rapid diagnostic tests implementation in Dar es Salaam: a before-after and cluster randomized controlled study. Malar J 10:107
39. Thiam S, Thior M, Faye B, Ndiop M, Diouf ML, Diouf MB, Diallo I, Fall FB, Ndiaye JL, Albertini A et al (2011) Major reduction in anti-malarial drug consumption in Senegal after nation-wide introduction of malaria rapid diagnostic tests. PLoS One 6:e18419
40. D'Acremont V, Malila A, Swai N, Tillya R, Kahama-Maro J, Lengeler C, Genton B (2010) Withholding antimalarials in febrile children who have a negative result for a rapid diagnostic test. Clin Infect Dis 51:506–511
41. Msellem MI, Martensson A, Rotllant G, Bhattarai A, Stromberg J, Kahigwa E, Garcia M, Petzold M, Olumese P, Ali A et al (2009) Influence of rapid malaria diagnostic tests on treatment and health outcome in fever patients, Zanzibar: a crossover validation study. PLoS Med 6:e1000070
42. Snounou GS, Viriyakosol S, Zhu XP, Jarra W, Pinheiro L, do Rosario VE, Thaithong S, Brown KN (1993) High sensitivity of detection of human malaria parasites by the use of nested polymerase chain reaction. Mol Biochem Parasitol 61:315–320
43. Mixson-Hayden T, Lucchi NW, Udhayakumar V (2010) Evaluation of three PCR-based diagnostic assays for detecting mixed *Plasmodium* infection. BMC Res Notes 3:88
44. Boehme CC, Nabeta P, Hillemann D, Nicol MP, Shenai S, Krapp F, Allen J, Tahirli R, Blakemore R, Rustomjee R et al (2010) Rapid molecular detection of tuberculosis and rifampin resistance. N Engl J Med 363:1005–1015
45. Polley SD, Mori Y, Watson J, Perkins MD, González IJ, Notomi T, Chiodini PL, Sutherland CJ (2010) Mitochondrial DNA targets increase sensitivity of malaria detection using loop mediated isothermal amplification. J Clin Microbiol 48:2866–2871

46. Snow RW, Molyneux CS, Warn PA, Omumbo J, Nevill CG, Gupta S, Marsh K (1996) Infant parasite rates and immunoglobulin M seroprevalence as a measure of exposure to *Plasmodium falciparum* during a randomized controlled trial of insecticide-treated bed nets on the Kenyan coast. Am J Trop Med Hyg 55:144–149
47. Drakeley CJ, Corran PH, Coleman PG, Tongren JE, McDonald SL, Carneiro I, Malima R, Lusingu J, Manjurano A, Nkya WM et al (2005) Estimating medium- and long-term trends in malaria transmission by using serological markers of malaria exposure. Proc Natl Acad Sci USA 102:5108–5113

Index

H.M. Staines and S. Krishna (eds.), *Treatment and Prevention of Malaria*,
DOI 10.1007/978-3-0346-0480-2, © Springer Basel AG 2012

Printed by Printforce, the Netherlands

Michael Blake
Zwischen Gerechtigkeit und Gnade

Michael Blake

Zwischen Gerechtigkeit und Gnade

Eine Ethik der Migrationspolitik

Aus dem Englischen von Thorben Knobloch

wbg Academic

Die Originalausgabe ist auf Englisch bei Oxford University Press unter dem Titel
Justice, Migration & Mercy erschienen.

Die Deutsche Nationalbibliothek verzeichnet diese Publikation
in der Deutschen Nationalbibliographie;
detaillierte bibliographische Daten sind im Internet über
www.dnb.de abrufbar.

wbg Academic ist ein Imprint der wbg.

Die Herausgabe des Werkes wurde durch die Vereinsmitglieder der wbg ermöglicht.

Lektorat: Sophie Dahmen, Karlsruhe
Satz: Arnold & Domnick GbR, Leipzig
Einbandgestaltung: Harald Braun, Helmstedt
Gedruckt auf säurefreiem und alterungsbeständigem Papier
Printed in Europe

Besuchen Sie uns im Internet: www.wbg-wissenverbindet.de

ISBN 978-3-534-27254-9

Elektronisch sind folgende Ausgaben erhältlich:
eBook (PDF): 978-3-534-74630-9
eBook (epub): 978-3-534-74631-6

Inhaltsverzeichnis

Vorwort

Es war der fünfte September 2017, als ich einen Vortragssaal im Süden Seattles betrat, meine rechte Hand hob und ein Bürger der Vereinigten Staaten von Amerika wurde. Am selben Tag verkündete Präsident Donald Trump das Ende des Programms *Deferred Action for Childhood Arrvials* (DACA) – ein Programm, das undokumentierte Immigrantinnen und Immigranten, die als Kinder in die Vereinigten Staaten gebracht worden waren, vor der Abschiebung schützen soll – zugunsten einer nicht näher spezifizierten, zukünftigen Regelung. Während ich darauf wartete, meinen Eid zu leisten, verkündete der damalige Justizminister Jeff Sessions, dass die Regierung sich dazu entschieden hatte, selbst den geringen Schutz aufzuheben, den DACA bis dahin geboten hatte. Diese Entscheidung erschien mir damals wie heute als zugleich ungerecht und grausam. Mein Weg zur amerikanischen Staatsbürgerschaft war lang und mit den üblichen bürokratischen Hürden versehen, aber er verlief doch so reibungslos, wie es sich für einen solchen Prozess nur wünschen lässt, und am fünften September endete er schließlich damit, dass ich Staatsbürger der Vereinigten Staaten wurde. Für viele andere Menschen jedoch stellte der fünfte September eine andere Art von Ende dar: Ihnen wurde eine Möglichkeit genommen, sich ein für sie wertvolles Leben aufzubauen, und in der Folge mussten sie sich einer neuen, unsicheren Realität stellen. Als ich nach der Zeremonie nach Hause kam, begann ich dieses Buch zu schreiben.

Diese Tatsachen führe ich aus zwei Gründen an. Der erste ist die Anerkennung der simplen Wahrheit, dass Migration und Migrationspolitik sich stetig verändernde Gegenstände politischer Kontroversen sind. So unterscheidet sich die heutige politische Landschaft von derjenigen, die ich zu Beginn meiner Reflektionen über dieses Thema vorfand, und zum Zeitpunkt der Veröffentlichung dieses Buches wird sie sich weiter verändert haben. Ich habe versucht, diesen Tatsachen nach Möglichkeit dadurch Rechnung zu tragen, dass ich einen moralischen Rahmen für die Bewertung neuer Gesetze und Vorhaben in dieser sich stetig verändernden politischen Landschaft suche; und ich hoffe, dass diese Überlegungen uns auch dann leiten oder zumindest an unsere besten ethischen Werte erinnern können, wenn sich die spezifischen Details der Migrationspolitik ändern. Der zweite Grund besteht darin, anzuerkennen, dass ich eine bestimmte Position innerhalb des Kosmos der Migration einnehme: Ich bin selbst ein Migrant, allerdings mit einer weitaus unproblematischeren Migrationserfahrung als viele andere Migrantinnen dieser Welt. Ich habe versucht, ehrlich zu sein – sowohl im Hinblick darauf, was ich als die richtigen ethischen Werte ansehe, als auch auf meine eigene Migrationserfahrung. In Staaten wie den USA wurde die Migrationspolitik oft zum Schutze von Personen wie mir und zum Nachteil für Menschen gestaltet, die eine vulnerablere Position innehaben – Menschen ohne meinen (relativen) Wohlstand, meine ethnische Identität oder meinen Bildungshintergrund. Anlass zum Verfassen dieses Buches war unter anderem der Gedanke, dass Migration nicht notwendig als ein Menschenrecht angesehen werden muss, um zu zeigen, dass wir die Migrationspolitik eines Staates moralisch kritisieren können und auch sollten. Kurz, es bedarf keiner Verteidigung offener Grenzen, um scharfe Kritik an der Art gegenwärtiger Grenzregime zu üben. Es ist allerdings sinnvoll, zu betonen, dass ich allein aus meiner eigenen Perspektive schreiben kann und jene Grenzregime für mich bedeutend weniger gewaltsam sind als für viele andere.

Dieses Buch umfasst mehrere Jahre des Nachdenkens über Migration und ich schätze mich glücklich, eine Gruppe von Kolleginnen und Menschen um mich gehabt zu haben, die mir auf diesem Weg geholfen haben. Selbstverständlich sind nur einige wenige von ihnen mit allem einverstanden, was ich sage; alle haben jedoch dazu beigetragen, meine Überlegungen voranzubringen, sei es durch Gespräche, Kritik oder Fürsorge. Meine Studentinnen Patrick Taylor Smith, Amy Reed-Sandoval, Mitch Kaufmann, Julio Covarrubias und Stephen Blake Hereth waren so freundlich, meine Gedanken während ihrer Entstehung zu begleiten. Die University of Washington ist ein außergewöhnlicher Ort für Arbeit und Lehre und besonders Stephen Gardiner, Jamie Mayerfeld, Michael Rosenthal, Bill Talbott und Andrea Woody bin ich dankbar für ihre Unterstützung in den letzten Jahren. Eine große Gruppe von Forscherinnen und Freundinnen haben mir dabei geholfen, meine Gedanken zu den hier besprochenen Problemen zu vertiefen; besonders dankbar bin ich dabei Gillian Brock, Joseph Carens, Luara Ferracioli, Javier Hidalgo, Adam Hosein, Stephen Macedo, Pietro Maffetone, José Jorge Mendoza, David Owen, Ryan Pevnick, Mathias Risse, Alex Sager, Grant Silva, Sarah Song, Anna Stilz, Christine Straehle, Laura Valentini,

Christopher Heath Wellman, Shelly Wilcox, Caleb Yong und Lea Ypi. Verschiedene Versionen der Ideen aus den letzten beiden Buchkapiteln wurden bei Kolloquien der University of Durham, der University of British Columbia und der Princeton University präsentiert; ich danke dem Publikum an all diesen Universitäten. Die Struktur dieses Buches wurde zum ersten Mal im Rahmen einer Summer School über Migration des Instituts für Philosophie I der Ruhr-Universität Bochum dargelegt; ich möchte Volker Heins, Thorben Knobloch, Corinna Mieth, Andreas Niederberger, Christian Neuhäuser, Itzik Pazuelo und Janelle Pötzsch für die Gespräche und die philosophische Gemeinschaft danken. Schließlich bin ich auch Peter Ohlin für seine Unterstützung dieses Projekts sehr dankbar.

Teile des vierten Kapitels dieses Buchs sind bereits in „Immigration, Jurisdiction, and Exclusion", 41 (2), *Philosophy and Public Affairs* (2013) sowie „Immigration and Political Equality", 45 (4), *San Diego Law Review* (2008) erschienen. Ich danke diesen Zeitschriften für die Erlaubnis, das Material in diesem Kontext verwenden zu dürfen. Dieses Buch wurde zum Teil durch einen Zuschuss des National Endowment for the Humanities im Rahmen der Initiative The Common Good unterstützt; ich möchte daher auch dem NEH für die Unterstützung meiner Arbeit danken. Das Manuskript wurde schließlich während eines Aufenthalts am Helen Riaboff Whitely Center der Friday Harbor Laboratories der University of Washington fertiggestellt und ich danke dem Personal des Whiteley Center dafür, einen idealen Ort für meine Studien als auch für das Verfassen des Manuskripts geschaffen zu haben. Zuletzt stehe ich, wie immer, zutiefst in der Schuld meiner Frau Melissa Knox und meiner Kinder, Eloise und Gus, die weiterhin den Grund für jeglichen Optimismus darstellen, der mir im Hinblick auf den Zustand dieser Welt und ihre Zukunft geblieben ist.

Zwischen Gerechtigkeit und Gnade.
Eine Ethik der Migrationspolitik

1
Über Moral und Migration

Als ich begann, dieses Buch zu schreiben, wollte ich nicht weniger als eine endgültige Moraltheorie für das Phänomen der Migration entwickeln. Es ist mir nicht gelungen. Tatsächlich erscheint es mir mittlerweile so, als ob der Gegenstand der Migration schlicht zu *umfassend* ist, als dass er auf angemessene Weise in einem einzigen Buch behandelt werden könnte. Wenn Menschen sich zwischen Ländern bewegen, erschaffen sie neue Dinge: ein neues Leben für sich selbst, neue Beziehungen, neue Normen sowohl im Wirtschaftsleben als auch in der Zivilgesellschaft, neue Lebensformen. Migration spiegelt das gesamte Spektrum menschlicher Erfahrungen und ist daher schlicht zu komplex, um ihre verschiedenen Elemente gleichzeitig innerhalb einer einzelnen Theorie zu erfassen. Nehmen Sie zum Beispiel die vielen Themen, zu denen sich eine vollständige Moraltheorie der Migration verhalten müsste:

(1) – **Migration und Geschichte**: Die Welt besteht aus Ländern, deren Wohlstand und Macht immer noch durch das gemeinsame Erbe von Kolonialismus und Ausbeutung bestimmt werden. Wir sind nicht mit Immigrantinnen aus hypothetischen Ländern konfrontiert, sondern aus ganz bestimmten Gesellschaften, die aus eben jener gemeinsamen und brutalen Geschichte hervorgegangen sind. In diesem Zusammenhang ist auch die Tatsache zu beachten, dass viele Menschen aus ehemaligen Kolonialstaaten in die wohlhabenderen Gesellschaften ehemaliger Kolonialmächte zu migrieren versuchen. Eine vollständige Moraltheorie der Migration würde sich damit auseinandersetzen, ob und inwiefern das koloniale Erbe besondere Verpflichtungen begründet, derartige Migrationsbewegungen zu erlauben.[1] Wer sich der Auseinandersetzung mit diesen Dingen zu entziehen versucht, riskiert es, eine inkohärente Perspektive auf das Phänomen der Migration einzunehmen. Wie Kishore Mahbubani schreibt, sind die Bürgerinnen weniger entwickelter Länder

> „wie hungrige und kranke Passagiere auf einem leckgeschlagenen, überfüllten Boot, das gerade in gefährliche Gewässer abdriftet, in denen viele der Reisenden umkommen werden. Der Kapitän des Bootes ist ein rauer Typ –

> manchmal fair, manchmal nicht. An den Ufern steht eine Gruppe reicher, gut genährter und wohlmeinender Zuschauer. Sobald diese Zuschauer beobachten, wie einer der Passagiere geschlagen, eingesperrt oder gar um seine Meinungsfreiheit gebracht wird, entern sie das Boot, um einzugreifen und die Passagiere vor dem Kapitän zu schützen. Aber die Passagiere bleiben hungrig und krank. Sobald sie versuchen, ans Ufer und in die Arme der Wohltätigen zu schwimmen, werden sie mit Bestimmtheit auf das Boot zurückgebracht, wo ihre ursprünglichen Sorgen unvermindert fortbestehen.“[2]

Die Bürgerinnen der reichsten Länder der Welt – oder zumindest politische Theoretikerinnen aus diesen Ländern – tendieren zu der Überzeugung, dass sie zu einem gewissen Grad dazu verpflichtet sind, dabei zu helfen, das globale Erbe des Kolonialismus zu überwinden und nehmen dabei für sich das Recht in Anspruch, diejenigen zu kritisieren, die sich an Tyrannei und Ausbeutung im Ausland beteiligen.[3] Selten fragen sie sich aber, ob den Armen dieser Welt nicht auch durch die Einwanderung in Länder geholfen werden könnte, in denen sie weniger arm wären. Zusammenfassend gesagt würde eine vollständige Moraltheorie der Migration uns helfen zu verstehen, ob und inwiefern Migration eine Antwort auf den Kolonialismus darstellen kann.[4]

(2) – **Migration und globale wirtschaftliche Gerechtigkeit**. Die Welt weist derzeit hinsichtlich Macht und Wohlstand bedeutende Unterschiede zwischen den Staaten auf. Über die vergangenen Jahrzehnte hinweg haben politische Philosophen versucht, die Konzepte von Recht und Gerechtigkeit auf diese globale Kluft zwischen Wohlstand und Armut anzuwenden.[5] Selbst wenn diese Ungleichheiten nicht durch den Kolonialismus und seine Folgen verursacht worden sein sollten, können wir uns doch immer noch fragen, ob Migrationsrecht und -politik nicht eine Rolle bei der Überwindung globaler wirtschaftlicher Ungleichheit spielen sollten. Zumindest könnte überlegt werden, welche Rolle der Migrationspolitik bei der *Aufrechterhaltung* dieser Ungleichheiten zukommen könnte. Der Immigration and Nationality Act der Vereinigten Staaten von 1965 hatte beispielsweise zur Folge, dass sich die legale Einwanderung aus Mexiko in die Vereinigten Staaten halbierte – und das zu einem Zeitpunkt, als globale wirtschaftliche Entwicklungen gleichzeitig die Zahl der Arbeitsplätze in Mexiko zurückgehen ließen. In der Folge stieg die Zahl der illegal in den USA arbeitenden mexikanischen Bürgerinnen.[6] Gleichzeitig nahmen auch die Gewinne US-amerikanischer Unternehmen

zu, da diese nun frei über eine Arbeiterschaft verfügen konnten, deren gesellschaftliche Position es nicht zuließ, Arbeitsrechte einzufordern.[7] Nur ein vergleichsweise geringer Anteil dieser Gewinne floss zurück nach Mexiko. Auf der „richtigen" Seite des Rio Grande geboren zu sein scheint demnach so etwas wie das moderne Äquivalent einer glücklichen Geburt innerhalb feudaler Gesellschaften darzustellen, also die zufällige Verteilung unverdienter Vorteile an einige glückliche Personen qua Geburt.[8] Eine vollständige Moraltheorie der Migration würde uns daher helfen, zu verstehen, wie Staaten bei der Ausgestaltung ihrer Migrationspolitik durch Forderungen globaler Verteilungsgerechtigkeit eingeschränkt werden könnten.

(3) – **Migration und innerstaatliche wirtschaftliche Gerechtigkeit**. Wenn Menschen migrieren, treten sie oft dem Arbeitsmarkt ihrer neuen Gesellschaft bei; die damit verbundenen Effekte sind komplex und widerstehen jeglicher Vereinfachung.[9] Es gibt allerdings einige häufiger auftretende Effekte, die einer moralischen Betrachtung wert sein könnten. Der erste betrifft die Tendenz einheimischer Personen, in Zeiten wirtschaftlicher Schwierigkeiten Immigrantinnen als die *Ursache* der ökonomischen Misere anzusehen. Wie die US Commission on Civil Rights es prägnant formulierte: jede neue Generation von Immigrantinnen wird letztendlich von der Generation abgelehnt, die vor ihr kam.[10] Der zweite Effekt betrifft hingegen die simple Tatsache, dass die Anwesenheit von Einwanderern auf dem Arbeitsmarkt sowohl das Arbeitsangebot für die einheimische Bevölkerung als auch die Löhne für verschiedene Tätigkeiten beeinflusst.[11] Von vielen dieser Entwicklungen profitieren alle Teile der Gesellschaft, von einigen allerdings nicht, und dieser Umstand betrifft vor allem Menschen mit niedrigen Bildungsabschlüssen, die im Wettbewerb mit Immigranten um gering entlohnte Arbeit und Gelegenheitsjobs stehen.[12] Sofern wir uns um die wirtschaftliche Gerechtigkeit zwischen reichen und armen Bürgerinnen eines demokratischen Staates sorgen, könnten wir also gute Gründe haben, diese Entwicklungen kritisch zu hinterfragen – genauso wie für die Frage, ob eine vollständige Moraltheorie der Migration es überhaupt vorschreiben würde, unsere Migrationspolitik in Abhängigkeit dieser Problematik auszugestalten.

(4) – **Migration und ethnische Gerechtigkeit.** Die Migrationsbewegungen in die Vereinigten Staaten nach dem Immigration and Nationality Act von 1965 spiegeln nicht explizit Ideen ethnischer Über- oder Unterlegenheit wider. Allerdings besitzt die Migrationspolitik durchaus das Potential, ethni-

sche Hierarchien abzubilden und zu verstärken. So trug die frühere amerikanische Migrationspolitik deutlich rassistische Züge. Der Immigration Act von 1924 wurde mit dem „wissenschaftlichen" Rassismus Madison Grants gerechtfertigt, der auch für Gesetze gegen die sogenannte „Rassenmischung" eintrat. Noch vor diesem Gesetz hatte der Chinese Exclusion Act explizit „jede Person der chinesischen oder mongolischen Rasse" vom Betreten der Vereinigten Staaten zum Zwecke der Einwanderung ausgeschlossen.[13] Neuere Formen des Ausschlusses offenbaren ihren Rassismus weniger eindeutig; dennoch können sie rassistischen, verurteilungswürdigen Neigungen entspringen oder diese sogar noch verschärfen. So wurde beispielsweise Sheriff Joe Arpaio dafür schuldig gesprochen, eine gerichtliche Anordnung missachtet zu haben, die ihn an der gezielten Schikanierung von Menschen mexikanischer Abstammung hindern sollte: Das Justizministerium hatte herausgefunden, dass Lateinamerikanerinnen und -amerikaner von Arpaios Streifen vier- bis neunmal häufiger als Weiße kontrolliert wurden.[14] In den Augen vieler hatte Arpaio bestimmte Gesetze gegen irregulären Aufenthalt dazu genutzt, Menschen mexikanischer Abstammung, gleich ob Bürgerinnen der USA oder nicht, zu schikanieren und auszugrenzen. Kurzum: Durch ihre Migrationspolitik können Gesellschaften ausdrücken, wer am Rande der Gesellschaft steht und wer nicht; mittels der Maßnahmen zur Kontrolle staatlicher Grenzen kann ausgedrückt werden, wer innerhalb dieser Grenzen von Bedeutung ist.[15] Sogar die Idee der *illegalen* Einwanderung an sich kann dazu genutzt werden, Immigrantinnen den Status als Menschen abzusprechen. So ist es sicherlich kein Zufall, dass die in Europa und den USA so erfolgreichen populistischen Bewegungen dazu neigen, Immigrantinnen als Überträger von Krankheiten zu dämonisieren.[16] In einer Vielzahl sozialer Kontexte können wir das Phänomen beobachten, dass Migration mit ethnischer Herkunft verbunden wird und in der Folge Immigrantinnen sowohl aufgrund ihrer ethnischen Herkunft als auch ihres Status *als* Immigrantinnen weniger Respekt zuerkannt wird. So beschreibt Swetlana Alexijewitsch die Worte eines tadschikischen Aktivisten wie folgt:

> „Zwei junge Tadschiken wurden mit einem Krankenwagen von einer Baustelle ins Krankenhaus gebracht. [...] Sie lagen die ganze Nacht in der kalten Aufnahme, und niemand kümmerte sich um sie. Die Ärzte hielten mit ihrer Meinung nicht hinterm Berg: „Was wollt ihr Schwarzärsche denn alle

> hier?" [...] Mit einem Polizeigeneral habe ich mich lange unterhalten. Er war kein Idiot, kein Kommisskopf, sondern ein gebildet wirkender Mann. ‚Wissen Sie', sagte ich zu ihm, ‚bei Ihnen arbeitet ein echter Gestapo-Mann. Ein richtiger Foltermeister, alle haben vor ihm Angst. Die Obdachlosen und Gastarbeiter, die ihm in die Hände fallen, macht er zu Invaliden.' [...] Doch er sah mich nur an und lächelte. ‚Nennen Sie mir seinen Namen. Ein toller Mann! Wir werden ihn befördern, ihn auszeichnen ... Wir machen die Bedingungen für euch hier absichtlich unerträglich, damit ihr möglichst schnell wieder verschwindet. In Moskau gibt es zwei Millionen Gastarbeiter, so viele Zugereiste kann die Stadt nicht verkraften. Ihr seid zu viele.'"[17]

Der Offizier ist ehrlicher als viele andere, aber ich vermute, dass der Kern seiner Auffassung – *es gibt zu viele von euch hier* – vielen Menschen nicht unvertraut ist, auch wenn sicherlich nur wenige bereit wären, auf dieser Grundlage Folter zu rechtfertigen. Eine vollständige Moraltheorie der Migration könnte uns allerdings mit den Mitteln ausstatten, die schlimmsten Versionen dieser Auffassung zu verstehen und zurückzuweisen; oder, sollte das fehlschlagen, uns Werkzeuge an die Hand geben, um unseren Widerspruch kundzutun.

(5) – **Migration und moralische Phänomenologie**. Eine vollständige Moraltheorie der Migration würde nicht einfach nur theoretisch erörtern, wie wir politisch auf Immigrantinnen reagieren sollten; sie würde uns auch dazu befähigen, zu verstehen, was es heißt, ein Migrant *zu sein* und welche Bedeutung dieser Erfahrung der Migration aus Perspektive der Ethik zukommt.[18] Eine Migrantin zu sein bedeutet, auf eine bestimmte Art und Weise Mensch zu sein: ein Mensch, der sich dem Aufbau eines neuen Lebens verschrieben hat, an einem neuen Ort, mit anderen, neuen Menschen. Jede neue Möglichkeit bringt es jedoch mit sich, dass eine andere Möglichkeit aufgegeben werden muss; und der Preis für ein neues Leben ist, dass das alte Leben mehr oder weniger aus dem Blickfeld verschwindet. Dieser Umstand erklärt, warum Abschiebungen tatsächlich so gewalttätig sind, wie sie uns erscheinen. In seiner Ablehnung der Alien and Sedition Acts von 1798 schrieb James Madison wortgewandt über die Bedeutung des Verlustes der eigenen Heimat:

> „Wenn die Ausweisung eines Fremden aus einem Land, das ihm ursprünglich unter Verheißung größten Glücks eine Zuflucht angeboten hat – ein

> Land, in dem er womöglich die zärtlichsten Verbindungen aufgebaut hat; ein Land, in dem er womöglich sein gesamtes Eigentum investiert und Eigentum der dauernden, beweglichen oder vorübergehenden Art erworben hat; ein Land, in dem er unter den Gesetzen einen größeren Anteil der Segnungen persönlicher Sicherheit und Freiheit genossen hat als er sich irgendwo sonst erhoffen konnte und wo er fast alle Voraussetzungen der Staatsbürgerschaft erfüllt hat … wenn eine solche Ausweisung keine Bestrafung, nicht eine der schlimmsten aller Bestrafungen, sein soll, so wird es schwierig, sich ein Verhängnis vorzustellen, das diesen Namen wirklich verdient."[19]

Daher schmerzen uns Abschiebungen, denn sie bedeuten den Verlust von etwas; einer Heimat, die vielleicht am besten als ein Teil des Selbst zu verstehen ist, das diese Heimat bewohnt. Durch Migration sind wir mit einer Situation konfrontiert, in der wir uns auf neue Weise mit den Mitgliedern unserer sozialen Welt auseinandersetzen müssen; und das mag wiederum erfordern, eine vollkommen neue Person zu werden – sowohl für andere als auch für uns. Während seines Exils in Belgien beschrieb Jean Améry diesen Umstand auf vortreffliche Weise. Was ist *Heimat*?, fragt Améry, und antwortet:

> „Heimat ist, reduziert auf den positiv-psychologischen Grundgehalt des Begriffs, Sicherheit. Denke ich zurück an die ersten Tage des Exils in Antwerpen, dann bleibt mir die Erinnerung eines Torkelns über schwankenden Boden. Schrecken war es allein schon, daß man die Gesichter der Menschen nicht entziffern konnte. Ich saß beim Bier mit einem großen, grobknochigen Mann mit viereckigem Schädel, der konnte ein solider flämischer Bürger sein, vielleicht sogar ein Patrizier, ebensogut aber auch ein verdächtiger Hafenbursche, der drauf und dran war, mir seine Faust ins Gesicht zu schlagen und sich meiner Frau zu bemächtigen. Gesichter, Gesten, Kleider, Häuser, Worte (auch wenn ich sie halbwegs verstand) waren Sinneswirklichkeit, aber keine deutbaren Zeichen. In dieser Welt war für mich keine Ordnung."[20]

Die Emigration verlangt den Aufbau eines neuen Selbst und auch wenn viele Menschen freiwillig auswandern, sollte doch stets bedacht werden, dass dieser Prozess schmerzhaft ist und notwendigerweise tiefgreifende Verluste mit sich bringt. Im Falle der unfreiwilligen Emigration sollten wir darüber hinaus auch den Grund der Auswanderung nicht aus den Augen verlieren und nicht vergessen, dass dieser Verlust durch einen anderen Akteur veranlasst wurde, der meint, ein solcher Verlust sei erträglicher als weiterhin in Gemeinschaft mit dem zur Emigration Gezwungenen zu leben. Wie ich zeigen werde, ist eine Abschiebung nicht immer moralisch falsch; aber sie bedeutet immer die teilweise Vernichtung dessen, was die Abgeschobenen aus ihrem Leben gemacht haben. Es ist daher kein Zufall, dass *Abschiebung* metaphorisch oft mit Auslöschung in Verbindung gebracht wird. Der Völkermord an den Armeniern begann mit der Verkündung des Tehcir-Gesetzes, das die Abschiebung von Armeniern erlaubte, die in „Spionage gegen den Staat" verwickelt waren.[21] Ganz ähnlich beschrieb die Wannsee-Konferenz den systematischen Mord an den Juden durch die euphemistische Formulierung *nach dem Osten transportiert*.[22] Eine umfassende Betrachtung der Moralität der Migration würde, so ist zu hoffen, uns dazu befähigen, zu verstehen, was bei der Migration auf dem Spiel steht: was in der Emigration verloren geht, was verloren geht, wenn sie unterbleibt, und wie wir die moralische Bedeutung der gewaltsamen Abschiebung aus der eigenen Heimat verstehen können.

Eine vollständige Betrachtung der durch Migration aufgeworfenen moralischen Probleme würde uns mit den Werkzeugen ausstatten, die für ein Verständnis all dieser Aspekte und ihres Verhältnisses zueinander vonnöten sind. Ein solches Unterfangen würde eine umfassende Analyse der komplexen Wege erfordern, auf denen das gemeinsame Vermächtnis von Gewalt und Gräuel unsere Welt sowohl in der Vergangenheit geformt hat als auch heute noch formt. Ich betrachte es als unwahrscheinlich, dass eine solche Theorie jemals entwickelt wird – ich bin mir allerdings sicher, dass, sollte eine solche Theorie demnächst veröffentlicht werden, sie nicht von mir stammen wird. In diesem Buch versuche ich etwas Bescheideneres. Ich möchte nur zwei Ziele verfolgen und dabei gebührend anerkennen, dass die damit verbundenen Überlegungen uns nur mit einem Teil der moralischen Erläuterungen versorgen, die wir benötigen.

Der erste Teil meines Vorhabens besteht darin, die in meinen Augen beste Rechtfertigung für ein staatliches Recht auf den Ausschluss von Migrantinnen darzulegen. Meiner Meinung nach muss diese Rechtfertigung vorgebracht worden sein, bevor eines der von mir beschriebenen spezifischeren moralischen Probleme adäquat analysiert werden kann. Wenn ein Staat manche Menschen daran hindern möchte, sein Territorium zu betreten, wird dieser Staat eine Erklärung dafür vorbringen müssen, warum er dazu berechtigt ist. Ich möchte in diesem Buch eine mögliche Version einer solchen Erklärung vorlegen – eine, von der ich glaube, dass sie gute Chancen hat, mit jenen Idealen der Gerechtigkeit im Einklang zu stehen, die die liberale politische Philosophie mit Leben füllen. Ihren Ausgang nehmen meine Überlegungen von Annahmen, die ich, wenn auch nicht als unkontrovers, so doch als potentiell allgemein attraktiv ansehe und aus denen ich im weiteren Verlauf einen spezifischen Ansatz für die Rechtfertigung eines staatlichen Rechts auf Ausschluss abzuleiten versuche. Das Verständnis dieses Ansatzes kann uns meiner Meinung nach bei der Navigation durch die schwierigen Gewässer der öffentlichen Debatte über Migration in liberalen Gemeinwesen helfen. Mein Vorschlag wird sowohl ein Recht auf Ausschluss rechtfertigen, als auch – gleich bedeutsam – die Grenzen dieses Rechts aufzeigen; es gibt demnach Menschen, deren Ausschluss ungerechtfertigt ist und Gründe, die nicht zur Rechtfertigung eines Ausschlusses angeführt werden können. Das Verständnis dessen, was die von mir vorgeschlagenen moralischen Ausführungen nicht erlauben, ist genauso wichtig wie das Verständnis dessen, was sie erlauben. Doch selbst wenn ich hinsichtlich dieser Aufgabe erfolgreich sein sollte, werde ich nur einen Teil des komplexen Gegenstands der Migration erfasst haben. Was ich hier sage, kann prinzipiell von bestimmten anderen Faktoren übertrumpft zu werden – so mag die individuelle Geschichte eines Landes es uns mitunter erlauben, zu zeigen, dass in einem bestimmten Fall verboten ist, was für gewöhnlich erlaubt sein mag. Auf diese Weise hat Michael Walzer gezeigt, dass die Vereinigten Staaten kein Recht haben, diejenigen zurückzuweisen, die durch die amerikanische Invasion in Vietnam zu Flüchtlingen wurden: Die militärische Geschichte der USA hatte sie „de facto bereits amerikanisiert, ehe sie die Küsten Amerikas erreichten."[23] Walzer zufolge haben politische Gemeinschaften das Recht, eine Vielzahl von Personen auszuschließen – aber aufgrund ihrer Geschichte können sie dieses Recht verlieren, zumindest mit Blick auf bestimmte Personen. Mein Ansatz ist offen für derlei Überlegungen, obwohl ich hinsichtlich ihrer theoretischen Grundlagen einige Bedenken habe.[24] Im Folgenden werde ich mich jedoch hauptsächlich auf das Recht auf Ausschluss und seine Beschränkungen fokussieren und mir die Bedenken über die Umstände, unter denen dieses Recht ungültig sein könnte, für eine andere Gelegenheit aufheben. Diese Entscheidung soll nicht bedeuten, dass solche, an partikularen Fällen orientierte Argumente von geringerer Bedeutung sind; vielmehr denke ich, dass sie oftmals genauso wichtig sind wie das umfassende moralische Recht auf Ausschluss, gegen das sie ins Feld geführt werden. Meine vorliegende Aufgabe besteht jedoch darin, dieses Recht zu verstehen, da dieses Verständnis eine wichtige Vorstufe für die daran anschließende, umfassendere moralische Analyse darstellt.

Der zweite Teil dieses Buches wird über die Idee der Gerechtigkeit hinausgehen und einen Raum für die Anwendung alternativer moralischer Konzepte im Bereich der Migrationspolitik eröffnen. Unsere Diskussionen über Migration werden häufig unter Bezug auf konkurrierende Gerechtigkeitsideale geführt. Uneinigkeit über die moralischen Fragen der Migration beruht in diesen Fällen auf der Uneinigkeit darüber, welche Personen welche Rechtsansprüche gegenüber bestimmten sozialen Institutionen geltend machen können. Diese Debatten – sowohl in der Politik als auch der Philosophie – sind wichtig und werden wahrscheinlich auch in der näheren Zukunft fortbestehen. Aber sie sind unvollständig. Im alltäglichen Leben können wir schreckliche Menschen sein, ohne dabei jemals aktiv irgendwelche Rechte zu verletzen. Wir können grausam, hartherzig, bösartig, kleinlich und so vieles mehr sein – ohne dabei notwendig etwas zu tun, was berechtigterweise als *ungerecht* bezeichnet werden kann. Ich werde in diesem Buch auf das Konzept der Gnade zurückgreifen, um die spezifische moralische Tugend zu erfassen, die besagt, dass wir mehr tun müssen, als bloß die Verletzung von Gerechtigkeitspflichten zu vermeiden. Eine Person, die nicht im Sinne der Gnade handelt, fügt keiner anderen Person irgendein Unrecht zu – zumindest nicht aus dem Blickwinkel der Gerechtigkeit – aber es kann ihr dennoch vorgeworfen werden, ein schlechtes moralisches Vorbild zu sein. Wie Jeffrie Murphy schreibt:

> „Die Tugend der Gnade zeigt sich, wenn eine Person, aus Mitgefühl für die schwierige Lage einer bei ihr in der Schuld stehenden Person, auf das mit dieser Schuld verbundene Recht verzichtet und sie somit von dieser Verpflichtung und der mit ihr verbundenen Last befreit. Menschen, die stets und ohne Rücksicht auf die Folgen ihrer Handlungen für andere auf ihrem Recht bestehen, sind schlicht unerträglich. Solchen Personen kann zwar nicht vorgeworfen werden, dass sie ungerecht handeln, aber Vorwürfe können ihnen in jedem Fall gemacht werden.“[25]

In den späteren Kapiteln möchte ich diese Ideen nicht einfach nur im Hinblick auf Personen, sondern auch hinsichtlich der Politik untersuchen. Ich werde zeigen, dass ein Staat selbst dann moralisch kritisiert werden kann, wenn er nicht gegen Gerechtigkeitsnormen verstößt, nämlich wenn er es nicht schafft, durch sein Handeln die Tugend der Gnade angemessen zu demonstrieren. Staaten sind, wie auch Individuen, nicht einfach nur dazu verpflichtet, Rechte zu achten, sondern auch anteilnehmend für andere Personen zu handeln – und zwar unabhängig davon, ob diese Personen aus der Idee der Gerechtigkeit ein Recht auf jene Anteilnahme ableiten können. Ich glaube, dass diese Art von Tugend aus mindestens zwei Gründen eine bedeutende Ergänzung zur Gerechtigkeitsdebatte im Bereich der Migration darstellt. Der erste Grund besteht darin, dass es Fälle zu geben scheint, in denen bestimmte Aspekte der Migrationspolitik nicht mit dem Adjektiv *ungerecht* am besten beschrieben werden, sondern mit *gnadenlos*: Eine solche Politik wird nicht aufgrund von Gerechtigkeitserwägungen infrage gestellt, sondern lässt ihre Urheber unzureichend achtsam gegenüber jenen moralischen Werten erscheinen, die sie gerade aufgrund ihres Auftrags zur Ausgestaltung der öffentlichen Ordnung besonders ernst nehmen sollten. Ich glaube, dass es für Personen, die an politischer Ethik interessiert sind, Gründe gibt, bestimmte Formen politischer Kritik, die sich klar von Gerechtigkeitserwägungen unterscheiden, zu verteidigen; und ich glaube auch, dass die Idee der Gnade eine wichtige Tugend darstellt, die in die öffentliche Debatte um Migrationspolitik einbezogen und in ihr respektiert werden sollte. Wie ich zeigen werde, kann die Kategorie der Gnade in einer Vielzahl moralischer Theorien als eigenständiges Modul eingeführt werden. So finden sich ähnliche Ideen beispielsweise in der christlichen Ethik, der kantischen Hilfspflicht und der feministischen Care-Ethik. Das Konzept ist stark, sowohl rhetorisch als auch philosophisch, und für diejenigen unter uns, die eine ethische Reform der Migrationspolitik anstreben, kann es daher auch im öffentlichen Diskurs von Nutzen sein. Der zweite, umstrittenere Grund für die Einführung dieser Idee besteht darin, dass anhand des Konzepts der Gnade auch solche politischen Maßnahmen verteidigt werden können, die auf Forderungen der Gerechtigkeit beruhen. Es gibt viele Wege, das Bösartige an bestimmten Formen der Migrationspolitik zu beschreiben und es gibt Konstellationen, in denen die Sprache der Gnade sich als stärker erweist denn die bisweilen trockene Sprache der Gerechtigkeit.

Manche Menschen werden vielen der hier angestellten Überlegungen ablehnend gegenüberstehen. Als ich die Idee der Gnade das erste Mal vortrug, kam die erste Frage von einer Zuhörerin, die das Konzept selbst als ehrverletzend ansah: Sie würde eher für Gerechtigkeit kämpfen, als um Gnade zu betteln. Diese Reaktion ist nicht vollkommen falsch. So ist beispielsweise das Konzept der Gnade in jenen Fällen der falsche Ausgangspunkt, in denen eine bestimmte Politik deshalb abgeschafft werden sollte, weil sie ungerecht ist. Diejenigen, die gegen Jim Crow[26] kämpften, hätten ihre Forderungen wohl kaum in Form eines Gnadengesuchs vorgebracht. (Und wenn diejenigen, die den Civil Rights Act von 1964 erlassen haben, diesen Akt unter Berufung auf die Idee der Gnade rechtfertigen, halten wir das entsprechend für leicht obszön.)[27] Hierauf kann ich bloß zwei Antworten geben. Die erste lautet, dass der Diskurs der politischen Ethik verarmt, wenn wir unsere Auseinandersetzungen allein in der Sprache der Gerechtigkeit führen. Sofern der liberale Staat ein gewisses Recht hat, ungebetene Immigrantinnen auszuschließen, wofür ich argumentiere, könnten wir Bedarf an einem ethischen Diskurs haben, der uns dabei hilft, zu verstehen, wie wir am besten mit der aus diesem Recht resultierenden Freiheit umgehen sollten. Darauf zu bestehen, dass der Kampf für Gerechtigkeit der einzig gangbare Weg ist, würde bedeuten, die wundervolle Komplexität unseres ethischen Vokabulars zu ignorieren. Meine zweite Erwiderung lautet, dass es manchmal nicht darum geht, was wir gerne tun würden, sondern darum, was letztlich effektiv ist. Mitunter sollten wir die Frage danach, was Gerechtigkeit verlangt, von der Frage unterscheiden, wie wir andere von einer gerechten Politik überzeugen können. Die Idee der Gnade, so mein Argument, kann uns sowohl dabei helfen die Ethik der Migrationspolitik zu verstehen, als auch dabei, eine gerechte Politik in einem zunehmend völkisch und feindselig gesinnten politischen Gemeinwesen zu befördern.

Einige werden sicherlich weitaus mehr als das Konzept der Gnade ablehnen. Manche werden die Frage stellen, ob es überhaupt Staaten oder Grenzen geben sollte.[28] Auf diese Art von Einwänden möchte ich auf zwei Weisen reagieren. Die erste Reaktion ist begrifflicher Natur: das Konzept der Einwanderung ergibt in meinen Augen bloß vor dem Hintergrund des modernen, westfälischen Staatensystems einen Sinn. Als ich von Boston nach Seattle zog, bin ich nicht migriert – ganz im Gegensatz zu meinem Umzug von Kanada in die Vereinigten Staaten. So wie ich den Begriff in diesem

Buch verstehe und benutze, verlangt Migration das Übertreten von Grenzen souveräner Staaten, weshalb es die Existenz souveräner Staaten voraussetzt. Nicht alle Verwendungen dieses Konzepts beinhalten diese Implikation. So beginnt Michael Fishers Geschichte der Migration mit der Beschreibung von Gudrid dem Wanderer, der sich 1002 unserer Zeitrechnung auf dem Gebiet des heutigen Kanadas „als ein Immigrant aus Island" niederließ.[29] Meiner Meinung nach ergibt es allerdings nur dann Sinn, ein Immigrant aus Island zu sein, wenn es ein Island *gibt*, weshalb die Existenz als Immigrant die Existenz von Staaten *wie* Island voraussetzt. Daher werde ich für den Zweck des vorliegenden Buches annehmen, dass das System der Staaten, oder etwas ihm sehr ähnliches, als Hintergrund unserer normativen Diskussion bis auf Weiteres erhalten bleibt. Der zweite Grund für meine Ablehnung solch radikaler Überlegungen ist hingegen eher philosophischer Natur: Ich verstehe mich als einen institutionellen Konservativen – was sich, wie ich gleich anmerken sollte, recht deutlich von einem reinen Konservativen unterscheidet.[30] Der institutionelle Konservative beginnt mit den Institutionen, die er oder sie vorfindet, und fragt danach, wie diese beschaffen sein müssten, um gerechtfertigt werden zu können. Eine solche Position unterscheidet sich stark vom Bestehen darauf, dass die gegenwärtigen Institutionen bereits gute oder angemessene Arbeit leisten. Sie unterscheidet sich auch davon, zu fragen, welche Institutionen wir haben könnten, würden wir sie noch einmal von Neuem errichten. Der Grund für meine Ablehnung der letzten Position ist einfach: Sofern wir das Vorgefundene den Forderungen der Gerechtigkeit entsprechend verändern können, sollten wir auch entsprechend handeln – wobei dieses *sollte* seine Kraft sowohl aus den konzeptionellen Schwierigkeiten eines vollkommenen Neuaufbaus dieser Institutionen, als auch aus den praktischen Problemen revolutionärer Umwälzungen zieht.[31] Sollte diese Aufgabe jedoch hoffnungslos sein – wenn unsere vorhandenen Institutionen also nicht an die Forderungen der Gerechtigkeit angepasst werden könnten –, dann haben wir gute Gründe dafür, zu radikaleren Lösungen zu greifen. In diesem Buch werde ich jedoch schauen, ob diese radikaleren Lösungen, zumindest prinzipiell, nicht doch vermieden werden können.

An diesem Punkt möchte ich jedoch einige Schwierigkeiten der mir selbst auferlegten Aufgabe einräumen. Das erste Problem besteht in meiner Identität: Ich bin zwar ein Migrant, aber, wie ich bereits sagte, ein ziemlich privilegierter Migrant. Ich wurde in eine relativ wohlhabende Gesellschaft geboren und bin in eine andere recht wohlhabende Gesellschaft eingewandert. Sollte Améry Recht haben, ist jeder Emigrant zu einem gewissen Grad vulnerabel, allerdings war und bin ich aufgrund meines relativen Wohlstands und meiner sozialen Identität weit weniger vulnerabel als andere. Ich erwähne diese Umstände, da die Ideen, von denen ich hoffe, dass sie viele Leserinnen und Leser motivieren, von einer bestimmten Person kommen und daher wohl sehr wahrscheinlich, auf für mich unvorhersehbare Weise, die Besonderheiten meiner Situation widerspiegeln. Darüber hinaus muss ich anerkennen, dass ich innerhalb eines von amerikanischen und europäischen Stimmen dominierten Diskurses schreibe. Mein Text wendet sich an Akademiker und Akademikerinnen innerhalb einer von Akademikern und Akademikerinnen gebildeten diskursiven Agenda und ich muss einräumen, dass die Menschen, auf deren Überlegungen ich hier antworte, ebenfalls als eher privilegiert zu bezeichnen sind. All dies erwähne ich, da ich überzeugt davon bin, dass der von uns gestaltete Dialog sehr stark von Inklusion, gleich welcher Art, profitieren würde. Zumindest wären die Argumente dieses Buches besser, wären sie jener Art von kritischer Auseinandersetzung ausgesetzt, die in unserem philosophischen Kosmos bloß schwerlich entsteht. Was ich hier sage, soll daher höchstens als ein Beitrag zu aktuellen Debatten über die Moral der Migration verstanden werden und ist ausdrücklich nicht darauf ausgerichtet, diese Debatten zu beenden.

Zum Schluss möchte ich noch auf einige Grenzen der vorliegenden Untersuchung eingehen. Der Fokus liegt auf Einwanderung, unter der ich die Überquerung einer internationalen Grenze zu einem anderen Land verstehe, die mit der Absicht eines zeitlich unbegrenzten Aufenthalts in eben jenem Land einhergeht. Durch diesen Fokus werde ich einige ziemlich wichtige politische Themen nicht oder nur am Rande behandeln. So werde ich beispielsweise nicht auf die Möglichkeit einer partiellen Mitgliedschaft in einer politischen Gemeinschaft eingehen, sei sie begrenzt im Hinblick auf die Rechte der Immigrantinnen oder die zeitliche Dauer ihres Aufenthalts.[32] Auch die schwierige Frage nach den Bedingungen des Übergangs von einer solch partiellen zu einer vollen Mitgliedschaft als Bürgerin werde ich nicht behandeln, also beispielsweise die Frage, ob Einbürgerungstests illiberal sind oder durch die legitimen Interessen des liberalen Staates gerechtfertigt werden können.[33] Und in diesem Zusammenhang: Welche Arten von Integrationsbemühungen können berechtigterweise von einer Person erwartet werden, die in eine neue Gesellschaft eingewandert ist? Das sind offensichtlich wichtige Fragen, die jedoch in diesem Buch nur im Vorübergehen behandelt werden. Schließlich werde ich auch nicht darauf eingehen, wie sich Migration auf die Verteilungsgerechtigkeit im Zielland oder die zurückgelassenen Menschen im Herkunftsland auswirkt.[34] Diese Themen sind ebenfalls wichtig – tatsächlich sind sie der Hauptbestandteil vieler Diskussionen über die Konsequenzen einzelner Reformen innerhalb der Migrationspolitik. Im Folgenden werde ich mich jedoch damit beschäftigen, was den Ausschluss ungebetener Immigrantinnen durch eine politische Gemeinschaft rechtfertigen könnte; daher werde ich mich nur insoweit mit der Ökonomie der Migration beschäftigen, als wirtschaftliche Faktoren sich auf die Fähigkeit einer Gesellschaft auswirken können, den Anforderungen und Garantien der liberalen politischen Philosophie gerecht zu werden.

Wir können daher nun zur Diskussion dessen übergehen, was ich unter diesen Anforderungen und Garantien verstehe. An dieser Stelle möchte ich einige Prämissen erläutern, auf denen meine späteren Argumente aufbauen werden. Ich gehe davon aus, dass diese Prämissen verteidigt werden können und habe auch an anderer Stelle versucht, einige von ihnen zu rechtfertigen. Im Kontext dieses Buches möchte ich sie jedoch einfach als Ausgangspunkte anbieten. Wer mit diesen Punkten nicht einverstanden ist, wird sicher auch den nachfolgenden Überlegungen nicht zustimmen. Erklärungen müssen jedoch an einem bestimmten Punkt enden und die meinigen enden hier, mit fünf Prämissen, anhand derer ich das Recht auf Ausschluss sowohl begründen als auch einschränken werde.

1 – **Autonomie**. Ich gehe davon aus, dass menschliche Wesen ein Recht auf Verhältnisse haben, unter denen sie als autonome Akteure leben können. Damit meine ich nichts metaphysisch Stärkeres, als dass sie dazu fähig sein sollen, sich ein für sie wertvolles Leben aufzubauen, in dem ihre eigene Handlungsfähigkeit (*agency*) eine bedeutende Rolle bei der Bestimmung ihrer Aktivitäten und der von ihnen verfolgten Pläne spielt. Ich nehme an, dass es sich dabei insofern um einen kosmopolitischen Wert handelt, als dass *alle* Menschen einen Anspruch auf Verhältnisse haben, die eine solche Autonomie ermöglichen. Alle Menschen haben daher ein Recht auf dasjenige Bündel von Institutionen, das eine solche Autonomie garantiert und gewährleistet. Institutionen wie beispielsweise Staaten können ihre moralische Aufmerksamkeit daher nicht bloß auf die Autonomie ihrer eigenen Bürgerinnen

beschränken. Sie sind vielmehr dazu verpflichtet, gemeinsam eine Welt zu errichten, in der alle Individuen mit den Rechten und Ressourcen ausgestattet sind, die sie für ein autonomes Leben benötigen. Dabei werde ich eine recht schlanke Idee von Autonomie verwenden, die bloß verlangt, dass, wie Joseph Raz es formulierte, die betroffenen Individuen zumindest als Teilautoren ihres eigenen Lebens verstanden werden können.[35] Doch selbst aus einem solchen Verständnis von Autonomie können bedeutende Schlüsse gezogen werden. Ich verstehe liberale politische Philosophie, zumindest in der Tradition von John Rawls, als einen Weg zur Bestimmung der Art von Politik, die eine solche Form von Autonomie am ehesten respektiert. In seinem Werk ging Rawls davon aus, dass Individuen mit einer bestimmten Vorstellung davon ausgestattet sind, was ihr Leben wertvoll macht – in Rawls' Worten also eine Idee vom Guten besitzen – und dass politische Maßnahmen gegenüber allen Individuen unter Bezug darauf gerechtfertigt werden müssen, wie sie den Einzelnen die Verwirklichung dieser Idee innerhalb eines gesellschaftlichen Kooperationszusammenhangs ermöglichen.[36] Mir ist bewusst, dass die Idee der Autonomie deutlich komplexer ist, aber ich hoffe, dass diese Komplexität zumindest für den Moment beiseitegelassen werden kann.

2 – **Pläne**. Darüber hinaus werde ich annehmen, dass die Ausübung von Autonomie die Fähigkeit voraussetzt, Pläne machen zu können.[37] Diese Idee lässt sich ebenfalls bei Rawls finden, für den das Gut, das wir verfolgen, uns nicht nur ermöglicht, in der Gegenwart zu handeln, sondern auch, unserem Handeln über die Zeit hinweg eine Struktur zu geben. Eine Person ist für Rawls daher „ein menschliches Leben, das nach einem Plan gelebt wird".[38] Die Idee eines Plans ist deshalb wichtig, weil sie es uns erlaubt, Autonomie mit unserer Erfahrung von Zeit zu verbinden. Hätten wir keinen Sinn für Erinnerungen oder Erwartungen, könnten wir zwar mitunter noch immer über eine Vorstellung vom Guten verfügen, sie wäre aber nicht mehr von der Art, wie sie dem menschlichen Leben, so wie wir es kennen, eigen ist. Wir erinnern uns an Wertvolles in der Vergangenheit und blicken voraus auf Wertvolles in der Zukunft; wir leben sozusagen in der Zeit und wissen um diese Tatsache, weshalb sich das Gute für uns Menschen von, sagen wir, dem Guten für einen Goldfisch unterscheidet. Daraus folgt jedoch, dass unsere Autonomie nicht nur einfach durch die Verweigerung von Freiheit zum gegenwärtigen Zeitpunkt eingeschränkt werden kann, sondern auch durch die Verweigerung der Ressourcen und Rechte, die für ein Handeln

über die Zeit hinweg vonnöten sind. Wir benötigen somit Verhältnisse, die uns potentiell dazu befähigen, sowohl unsere Pläne auszugestalten als auch zu verfolgen und dabei zugleich den Schranken Rechnung zu tragen, die aus der Tatsache einer geteilten sozialen Welt erwachsen.

3 – **Staaten**. Aus den vorangegangenen Überlegungen folgt ein Bedürfnis nach politischer Gesellschaft. Wir brauchen gewisse gefestigte Erwartungen, ohne die wir keine Sicherheit erfahren könnten, sei es im Hinblick auf unsere Person oder unser Eigentum. Ich werde daher annehmen, dass es einen Bedarf an Staaten gibt, worunter ich Institutionen verstehe, die zur Durchsetzung bestimmter kollektiver Normen mittels Zwang berechtigt sind. Mit anderen Worten, ich nehme an, dass eine politische Gesellschaft für Menschen wie uns notwendig ist, um die implizit im Konzept der Autonomie enthaltenen Planungshandlungen ausführen zu können. Ich behaupte nicht, dass diese politischen Gesellschaften die Formen annehmen müssen, die wir ihnen im Verlauf der Geschichte gegeben haben. Noch weniger behaupte ich, dass die existierenden Grenzverläufe in irgendeinem Sinne zwingend sind oder ihre jetzige Form natürlicherweise haben müssen. Allerdings meine ich, dass irgendeine Form politischer Gesellschaft moralisch notwendig ist, und dass dieser Umstand die Existenz der Staaten dieser Welt rechtfertigt, sofern diese Staaten ihre politische Macht auf gerechtfertigte Weise ausüben.

4 – **Rechtfertigung**. Da wir geschichtlich betrachtet in einer Welt angelangt sind, in der Staaten territorial verankert sind, können Staaten in dieser Welt unterschiedliche Menschen auf unterschiedliche Weise behandeln. So bin ich beispielsweise gerade nicht in Frankreich und habe auch keinen Anspruch auf die französische Staatsbürgerschaft, weswegen mich der juristische Apparat Frankreichs nicht direkt zu bestimmten Handlungen oder Unterlassungen zwingen kann. Hingegen bin ich ziemlich vulnerabel im Hinblick auf die Entscheidungen der politischen und rechtlichen Institutionen der Vereinigten Staaten. Diese können mein Eigentum an sich nehmen, mich ins Gefängnis stecken oder mich – im äußersten Falle – hinrichten. Aus diesen Tatsachen folgt jedoch, dass Staaten unterschiedlichen Personen gegenüber auf verschiedene Weisen dazu verpflichtet sind, das eigene Handeln zu rechtfertigen. In einem anderen Zusammenhang habe ich argumentiert, dass viele Normen der Verteilungsgerechtigkeit am besten als eine Art der Rechtfertigung politischen Zwangs gegenüber den diesem Zwang unterworfenen Personen zu verstehen sind; John Rawls' innerstaat-

liche Gerechtigkeitsprinzipien sind demnach im globalen Kontext nicht anwendbar, weil es dort keinen Staat gibt, dessen Zwangsentscheidungen durch diese Prinzipien legitimiert werden würden.[39] Im Kontext dieses Buches stelle ich keine solche Behauptung auf; vielmehr möchte ich allein die strukturelle Tatsache feststellen, dass Staaten, zum einen, andere Verpflichtungen gegenüber denjenigen haben können, die unter ihrer Zwangsgewalt stehen, als gegenüber denjenigen, für die das nicht der Fall ist und, zum anderen, dass diese Differenz aus der Idee des universellen Respekts vor der individuellen Autonomie abgeleitet werden kann und diesem Respekt daher nicht widerspricht.

5 – **Menschen- und Bürgerrechte**. Schließlich möchte ich eine strukturelle Differenz zwischen verschiedenen Arten moralischer Rechte behaupten. Manche Rechte sind eine direkte Folge des Prinzips der Autonomie. Das Recht, nicht gefoltert zu werden, sehe ich als eine Instanz dieser Form von Rechten an: Wird ein Mensch gefoltert, ist er nicht länger ein autonomer Akteur, vielmehr ist sein Körper selbst das Werkzeug einer anderen Person geworden.[40] Das internationale Menschenrecht auf Freiheit von Folter spiegelt daher offenkundig das kosmopolitische Recht auf autonome Handlungsfähigkeit wider. Andere Rechte hingegen existieren als Rechtfertigung für politische Herrschaft mittels Zwang gegenüber denjenigen autonomen Individuen, die unter diesem Zwang leben. In meinen Augen ist das Recht zu wählen von solcher Art. Für die Entscheidung, mir das Wahlrecht in Frankreich zu verweigern, ist es irrelevant, dass ich bloß aufgrund des Zufalls meiner Geburt keine französische Staatsbürgerschaft besitze. Da ich nicht unter dem Zwang des französischen Gesetzes stehe, habe ich auch keinen Anspruch auf einen Anteil an der Macht, dieses Gesetz zu verändern. Wir haben daher Gründe dafür, zwischen denjenigen Rechten zu unterscheiden, die ich schlicht aufgrund meiner Handlungsfähigkeit als Mensch habe und solchen, die mir als Ergebnis meiner Verpflichtung gegenüber dem Zwangssystem eines territorialen Staates zukommen. Diese Rechte sind nicht als vollkommen getrennt voneinander zu betrachten – insbesondere könnten wir ein Menschenrecht auf eine Gesellschaft haben, in der unsere Bürgerrechte anerkannt werden. Damit ist allerdings nicht gemeint, dass diese zwei Arten von Rechten sich nicht in mancherlei Hinsicht unterscheiden und ich werde mit der Annahme fortfahren, dass sie auf verschiedene Weisen funktionieren.

Die Darstellung dieser Ideen war offensichtlich bloß kurz und skizzenhaft, aber sie sollte für den Zweck dieses Buches genügen. Was bedeuten diese Überlegungen also mit Blick auf ein staatliches Recht auf Ausschluss? Ich glaube, dass wir von ihnen ausgehend zeigen können, dass einem liberalen Staat ein eingeschränktes Recht auf den Ausschluss ungebetener Immigrantinnen zukommt, die sich jenseits seiner Grenzen befinden. In diesem Buch werde ich versuchen, den Gedanken zu verteidigen, dass der liberale Staat nicht verpflichtet ist, denjenigen Personen Zutritt zum Zwecke der Einwanderung zu gewähren, deren Fähigkeit zum Aufbau eines wertvollen Lebens bereits im Herkunftsland geschützt wird. Innerhalb des von mir verteidigten Ansatzes besitzen Menschen eine Vielzahl moralischer Rechte. Sie haben Menschenrechte, ohne die sie kein selbstbestimmtes und insofern lebenswertes Leben im Sinne von Freiheit und Auswahlmöglichkeiten führen können. Zudem haben sie politische Rechte und Bürgerrechte, sodass sie innerhalb ihrer politischen Gesellschaft ihre eigenen Vorstellungen von politischer Herrschaft effektiv einbringen und für sie eintreten können. Diese politischen und Bürgerrechte mögen – und ich glaube, sie sollten – Rechte auf ein bestimmtes Prinzip von Verteilungsgerechtigkeit beinhalten, ohne die der politische Diskurs verzerrt oder politische Partizipation gänzlich unmöglich werden würde. Allerdings haben Menschen kein Recht darauf, Mitglied derjenigen politischen Gesellschaft zu werden, die ihnen am meisten zusagt, jedenfalls nicht im Rahmen meiner Argumentation. Individuen mögen ein Recht auf Einwanderung erwerben, sofern diese für die Gewährleistung ihrer Menschenrechte notwendig ist, doch diese Menschenrechte selbst beinhalten nicht das Recht, in ein Land der eigenen Wahl zu migrieren.

Dieser Ansatz mag nun eher konservativ erscheinen, denn die Verteidigung eines Rechts auf Ausschluss scheint zugleich auch all die verschiedenen Formen der Zurückweisung von Migrantinnen zu verteidigen, die derzeit von Staaten angewendet werden. Das wäre allerdings ein Fehlschluss. Der Zweck meiner Argumentation besteht nicht einfach in der Rechtfertigung jedweder Form von Ausschluss, sondern darin, zu untersuchen, welchen Einschränkungen ein Recht auf Ausschluss selbst dann unterliegt, wenn ich davon ausgehe, es bestmöglich verteidigt zu haben. Wenn ich überhaupt etwas bezwecken will, dann, dass dieses Buch die globale Wirklichkeit der Migrationspolitik infrage stellt. Gegenwärtig haben wir bereits offene Grenzen – für bestimmte Personen. Wohlhabende Einwohner entwickelter

Länder – besonders diejenigen mit finanzieller oder unternehmerischer Macht oder wertvollen, seltenen Formen von Humankapital – können sich mehr oder weniger frei diejenige Gesellschaft aussuchen, in der sie leben wollen. Die am stärksten marginalisierten Personen, diejenigen also, die in repressiven Regimen oder in schlimmer Armut leben, sind den strengsten und rauesten Formen von Grenzkontrollen ausgesetzt. Sollte das, was ich in diesem Buch schreibe, zutreffen, haben wir gute Gründe, diese Wirklichkeit in ihr Gegenteil zu verkehren. Die Menschen mit dem stärksten moralischen Recht auf Migration sind demnach genau diejenigen, deren sonstige Rechte am schlechtesten verteidigt werden. Auf der anderen Seite sind diejenigen Personen, die meinen Überlegungen zufolge am ehesten ausgeschlossen werden könnten, gegenwärtig genau diejenigen, die kaum ausgeschlossen werden. Es müssen folglich nicht direkt offene Grenzen verteidigt werden, um eine kritische Perspektive auf die derzeitig wirksamen Formen von Grenzkontrollen zu eröffnen. Ich hoffe daher, dass meine Gedanken Anklang finden – sowohl unter meinen Kolleginnen, von denen einige für ein Recht auf Ausschluss argumentieren, als auch unter meinen Mitbürgerinnen, von denen fast alle ein solches Recht auf Ausschluss verteidigen. Kurz gesagt: Denjenigen, die nicht für ein Menschenrecht auf Migration eintreten, möchte ich Gründe dafür präsentieren, dass ein Recht auf Ausschluss stärker eingeschränkt sein könnte, als es auf den ersten Blick erscheint – und dass ein solches Recht höchstens den Beginn unserer moralischen Analyse darstellen kann, nicht ihren Endpunkt.

Selbstverständlich handelt es sich bei all dem bloß um eine andere Art des Ausdrucks meiner Erwartung, dass *niemand* viel von dem halten wird, was ich im Folgenden sagen werde. Diejenigen Akademiker, die ein Recht auf Ausschluss verteidigen, werden die von mir vorgebrachten Grenzen dieses Rechts ablehnen. Diejenigen, die sich für ein Recht auf Migration aussprechen, werden es ablehnen, dass ich überhaupt *irgendeine* Art von Ausschluss verteidige. Der gegenwärtige politische Diskurs rund um das Thema Migration ist zudem mittlerweile so vergiftet, dass er an frühere – und schrecklichere – Debatten über Migration und Zugehörigkeit erinnert. Die momentane Atmosphäre tendiert dazu, radikale Positionen zu belohnen und moderaten Perspektiven zu misstrauen. (Es ist nun allerdings auch nicht so, als ob eine moderate Haltung jemals, selbst zu besten Zeiten, viele Freunde gehabt hätte. Wie Jim Hightower bemerkt, gibt es in der Mitte der Straße bloß gelbe Streifen und tote Gürteltiere.)[41] Und schließlich hat die Philosophie ein wenig den Ruf, eine Disziplin feindseliger Grausamkeit zu sein – und offen gesagt ist dieser Ruf wohlverdient. Ein Buch über die Moral der Migration kann daher energische, um nicht zu sagen wütende, Reaktionen erwarten – falls der Autor überhaupt das Glück hat, nicht ignoriert zu werden. Um diese Kritik, auf eine zugegebenermaßen recht unwirksame Art, abzuwenden, möchte ich an dieser Stelle noch bemerken, dass mich all die Menschen, mit denen ich in diesem Buch nicht übereinstimme, bedeutende Dinge über Migration und Philosophie gelehrt haben. Ich hätte meinen eigenen Ansatz letztlich nicht ohne die Auseinandersetzung mit ihren Argumenten entwickeln können. Allerdings ändert das selbstverständlich nichts an der Tatsache, dass ich mit diesen Denkerinnen uneins bin – und es wird im Folgenden meine Aufgabe sein, diese Uneinigkeit zu erläutern.

2
Gerechtigkeit und die Ausgeschlossenen, Teil 1: Offene Grenzen

Unsere öffentliche Debatte über Migration ist zunächst einmal eine Debatte über Ethik. Der Gedanke, dass bestimmte Formen der Migration ungerecht, also unfair, sind, zieht sich als roter Faden durch den jüngeren politischen Diskurs. Um ein prominentes Beispiel zu nennen: Der Wahlkampf Donald Trumps im Jahr 2016 begann mit seiner Behauptung, dass unter den mexikanischen Immigranten Vergewaltiger, Drogenschmuggler und ganz allgemein „bad hombres" seien.[1] Sein zentrales Wahlkampfversprechen – die Errichtung einer von Mexiko bezahlten Mauer entlang der Grenze zwischen den USA und ihrem südlichen Nachbarstaat – wurde mitunter mit der Behauptung gerechtfertigt, dass die mexikanischen Einwanderer auf unfaire Weise die Arbeitsplätze amerikanischer Arbeitnehmer klauen würden. Trumps Antrittsrede verdeutlichte diesen moralischen Unterton:

> „Von diesem Moment an heißt es America First. Jede Entscheidung über Handel, Steuern, Einwanderung oder Außenpolitik wird zum Vorteil amerikanischer Arbeiter und Familien getroffen. Wir müssen unsere Grenzen vor den Verwüstungen durch andere Länder schützen, die unsere Produkte herstellen, unsere Unternehmen klauen und unsere Arbeitsplätze zerstören. […] Wir werden uns unsere Arbeitsplätze zurückholen. Wir werden uns unsere Grenzen zurückholen. Wir werden uns unseren Wohlstand zurückholen."[2]

Trumps Analyse ist in ihrem Kern moralisch: Die derzeitige globale Ordnung ist unfair. Es ist unfair, dass Immigrantinnen die Arbeitsplätze und den Wohlstand der derzeitigen Einwohner der Vereinigten Staaten an sich nehmen dürfen. Ich denke, dass Trumps Idee von Eigentum am besten als eine bestimmte Form moralischen Anspruchs interpretiert werden kann: Diese Arbeitsplätze und dieser Wohlstand sind *berechtigterweise* das Eigentum der amerikanischen Bürger. Trump gewann die Präsidentschaftswahlen teilweise auch dadurch, dass er Hillary Clinton als „globalist“ darstellte, die aufgrund mangelnder Liebe zu Amerika dazu bereit war, diese gegenüber den Amerikanerinnen unfairen Verhältnisse zu billigen. Ob wir uns nun angesprochen fühlen von solch einem Populismus oder nicht – und es sollte deutlich werden, dass dies in meinem Falle nicht zutrifft –, so stellt er doch eine bestimmte Vorstellung davon dar, wie Gerechtigkeit in Bezug auf Migration hergestellt werden könnte: Die Vereinigten Staaten müssen ungebetene Immigrantinnen ausschließen, denn alles andere wäre unfair.

Wie diese Position beruht auch die Perspektive der Gegenseite auf einer bestimmten Vorstellung von Gerechtigkeit. Große Teile des politischen Widerstands gegen die Politik der Trump-Administration fußen auf dem Gedanken, dass bestimmte Maßnahmen zur Durchsetzung der Migrationspolitik unfair sind. Der rhetorische Slogan *Kein Mensch ist illegal* verweist beispielsweise darauf, dass bereits die Sprache, mit der wir Personen ohne Aufenthaltspapiere bezeichnen, ausgrenzend und entmenschlichend und somit unfair sein kann.[3] Auch auf einer praktischeren Ebene hat die Linke in den USA ihren Widerstand gegen bestimmte Methoden und Zwecke des Ausschlusses unter Verweis auf moralische Gründe gerechtfertigt. Sowohl Bürgerinnen als auch Bundesstaaten haben sich in rechtlichen Auseinandersetzungen gegen den sogenannten *travel ban*, das Einreiseverbot der Trump-Administration engagiert, dessen Zweck es war, die Migration aus bestimmen Ländern mit muslimischer Bevölkerungsmehrheit zu unterbinden. Ein gegen dieses Verbot gerichteter Brief argumentierte, dass es sich hierbei um Diskriminierung und somit um unfaire Maßnahmen handele.[4] Die wiederauflebende Sanctuary-Bewegung[5] vertrat die Meinung, dass Städte und Individuen den mit Abschiebungen beauftragten Beamten der Immigration and Customs Enforcement (ICE) die Unterstützung verweigern sollten, andernfalls würden sie sich, so ein Gründer der Bewegung, „der Ungerechtigkeit mitschuldig“ machen.[6] Darüber hinaus haben undokumentierte Amerikanerinnen begonnen, zu argumentieren, dass der Status als Undokumentierte mit einer Form sozialer Ausgrenzung einhergehen kann, die in ihren negativen Konsequenzen der historischen Marginalisierung von Amerikanerinnen afrikanischer Abstammung gleichkommt. Jim Crow hat sich in Juan Crow verwandelt – und den Aktivistinnen zufolge sollten undokumentierte Amerikanerinnen, die bereit sind, für soziale Gleichheit einzustehen, gegen diese

Entwicklung Widerstand leisten.[7] Auch diese Argumente gründen auf einer Vorstellung von Gerechtigkeit. Sie bestehen darauf, dass manche Formen der Exklusion unfair und daher ungerecht sind. Die Vereinigten Staaten müssten demnach diejenigen Exklusionspraktiken abschaffen, die eine solche Ungerechtigkeit sowohl produzieren als auch dulden.

Ich möchte anmerken, dass nichts von all dem allein auf den politischen Diskurs in den Vereinigten Staaten zutrifft. So sind große Teile der europäischen Politik von ähnlichen Auseinandersetzungen zwischen unterschiedlichen Gerechtigkeitsvorstellungen im Bereich der Migration bestimmt. In Deutschland hat sich das Problem der Offenheit gegenüber syrischen Flüchtlingen zu einem zentralen Konfliktpunkt zwischen Angela Merkels Christdemokraten und populistischen Bewegungen wie der Alternative für Deutschland entwickelt. In Italien gelangten zwei populistische Parteien – die Fünf-Sterne-Bewegung und die Lega – größtenteils durch die Ablehnung von Einwanderung an die Macht. Eine der ersten Handlungen dieser Koalition bestand darin, ein Boot zurückzuweisen, das 629 Asylsuchende vor dem Ertrinken gerettet hatte.[8] Wiederum nichts von all dem trifft allein auf Europa zu. Populistische Bewegungen sind in solch unterschiedlichen Gesellschaften wie Brasilien, Zimbabwe oder Myanmar entstanden und auch dort beruht ihr Machtgewinn zum Teil auf der Verurteilung gegenwärtiger Migrationsbewegungen als ungerecht.[9]

In öffentlichen Auseinandersetzungen über Migration findet sich häufig ein bestimmtes Muster moralischer Argumentation: Was die jeweils andere Seite vorschlägt, ist ungerecht, und zwar deshalb, weil eine bestimmte Gruppe von Menschen dadurch unfair behandelt würde. Trotz dieser strukturellen Ähnlichkeiten gelangen die verschiedenen Argumentationslinien offensichtlich zu ganz unterschiedlichen Schlussfolgerungen. Sie unterstützen jeweils radikal unterschiedliche Bündel von Migrationsrechten und entwerfen radikal unterschiedliche Visionen für die jeweilige politische Gemeinschaft. Aber diese so verschiedenartigen Schlussfolgerungen beruhen auf ähnlichen konzeptionellen Grundlagen. Es ist daher wenig überraschend, dass diese Argumente, wie David Miller es ausgedrückt hat, oft mehr Aufregung denn Aufklärung mit sich bringen.[10] Der Unterschied zwischen der Seite, die ein Recht auf Ausschluss rechtfertigt und derjenigen, die eine solche Position als illiberal brandmarkt, verweist nicht auf gänzlich verschiedene moralische Standpunkte, sondern auf subtilere Unterschiede in der Anwendung gleicher moralischer Werte.

Die öffentliche Debatte zum Thema Migration neigt hingegen eher selten zur Subtilität. Einwanderung hat sich in den vergangenen Jahren zu einem der hitzigsten – wenn nicht gar explosivsten – Themen unserer gemeinsamen öffentlichen Diskussion entwickelt. Wir sind, um es vereinfacht auszudrücken, sehr gut darin, aneinander vorbei zu argumentieren. Wir können zeigen, wie unsere eigene Vorstellung von Fairness und Gerechtigkeit die von uns getroffenen Entscheidungen darüber rechtfertigt, wer welche Rechte auf Einlass in unsere politische Gemeinschaft hat. Wir sind zugleich ebenso gut darin, diejenigen zu dämonisieren und zu verurteilen, die nicht unserer Meinung sind. Auf der anderen Seite sind wir allerdings ziemlich schlecht darin, eine offene und sorgfältige Diskussion darüber zu führen, was berechtigterweise gegen unsere Vorstellung von Fairness vorgebracht werden könnte oder darüber, was wir erwidern könnten, um unsere Vorstellung gegenüber gegensätzlichen Ideen von Fairness zu verteidigen.

Was kann die politische Philosophie angesichts dieser Verhältnisse anbieten? Es ist unwahrscheinlich, dass sie so etwas wie einen Konsens über das Verhältnis von Gerechtigkeit und Migration hervorbringt. Uneinigkeit ist eine unvermeidliche Folge von Freiheit und wir sollten Einstimmigkeit zu allerletzt bei solch komplexen und hitzigen Themen wie der Migration erwarten. Aber die Philosophie könnte uns zumindest Werkzeuge anbieten, um unsere Uneinigkeit besser zu verstehen. Sie kann uns mitunter etwas Klarheit über die Konzepte und Ideen verschaffen, die unserem gemeinsamen Diskurs über Gerechtigkeit im Hinblick auf Migration zugrunde liegen und uns somit einen Weg nach vorne weisen – und wenn schon nicht zur Einstimmigkeit, so mag uns dieser zumindest zu so etwas wie gegenseitigem Respekt führen. Präzision im Hinblick darauf, wie wir Ideen und Begriffen nutzen, könnte unsere Debatten über Gerechtigkeit weniger feindselig gestalten, und sei es allein durch den Zwang, genau darzulegen, wie wir – und diejenigen, die uns widersprechen – das Wesen der von uns vorgebrachten Behauptungen verstehen.

Wie also können wir den Begriff der Gerechtigkeit verstehen? In diesem Buch werde ich auf John Rawls' Überlegungen zurückgreifen, für den Gerechtigkeit die „erste Tugend" sozialer Institutionen, und somit auch politischer Staaten, darstellte.[11] Nach Rawls handelt es sich bei der Gerechtigkeit um einen formalen Begriff (*concept of justice*), der im Einklang mit einer Vielzahl möglicher Gerechtigkeitsvorstellungen (*conceptions of justice*) interpretiert werden kann, wobei jede dieser Vorstellungen sozusagen im Detail ausführt, wie der formale Begriff der Gerechtigkeit zu verstehen und anzuwenden ist. Für den Anfang genügt jedoch der formale Begriff, um zu verstehen, wie die Idee der Gerechtigkeit im Falle der Migration spezifischer gefasst werden könnte:

> „Man kann sich also natürlicherweise neben den verschiedenen Gerechtigkeitsvorstellungen einen Gerechtigkeitsbegriff denken, der aus der ihnen gemeinsamen Rolle besteht. Menschen mit verschiedenen Gerechtigkeitsvorstellungen können sich also immer noch darin einig sein, daß Institutionen gerecht sind, wenn bei der Zuweisung von Grundrechten und -pflichten keine willkürlichen Unterschiede zwischen Menschen gemacht werden, und wenn die Regeln einen sinnvollen Ausgleich zwischen konkurrierenden Ansprüchen zum Wohle des gesellschaftlichen Lebens herstellen."[12]

Dieser Begriff wird in Rawls' eigenem Werk auf die Grundstruktur einer Gesellschaft angewendet. Diese regelt ihm zufolge „wie die wichtigsten sozialen Institutionen Grundrechte und -pflichten und die Früchte der gesellschaftlichen Zusammenarbeit verteilen." Nach diesem Verständnis ist Gerechtigkeit der bedeutendste moralische Aspekt einer politischen Gesellschaft und zwar derart, dass keine andere Tugend Vorrang vor dem Wert der Gerechtigkeit hat; folglich kann ein ungerechtes System von Institutionen nicht durch die Erfüllung anderer politischer Tugenden gerechtfertigt werden.

All das ist Kennern der modernen politischen Philosophie wohlvertraut. Es gibt allerdings drei Aspekte, die ich an dieser Stelle hervorheben möchte. Der erste ist der, dass die Idee der Gerechtigkeit *Stringenz* hinsichtlich ihrer Anwendung verlangt. Durch die Bezeichnung der Gerechtigkeit als erste Tugend sozialer Institutionen betont Rawls die Idee, dass aufseiten derjenigen, die unter Ungerechtigkeit leiden, starke Ansprüche bestehen. Die Idee der Gerechtigkeit begründet also bestimmte und besonders starke Ansprüche auf die Korrektur ungerechter Verhältnisse. Der zweite Aspekt, den ich hervorheben möchte, besteht in der Beziehung der Gerechtigkeit zur Idee der *Gleichheit*. Ungerechtigkeit bedeutet demnach Ungleichheit hinsichtlich eines bestimmten Lebensbereichs; tatsächlich kann es als eine Tatsache im Begriff der Gerechtigkeit betrachtet werden, dass sie bestimmte Aspekte von Personen benennt, in Hinsicht auf die diese Personen das Recht haben, als moralisch gleichwertig behandelt zu werden.[13] Rawls' Verwendung des Konzepts der Willkür gibt diese Vorstellung von Gleichheit wieder: Wenn verschiedene Individuen den sozialen Institutionen in einer ähnlichen Situation gegenüberstehen, ist es ungerecht, wenn diese sozialen Institutionen sie aufgrund eines willkürlich bestimmten Fakts unterschiedlich behandeln. Zuletzt möchte ich den *institutionellen* Aspekt der Gerechtigkeit hervorheben. Die Sprache der Gerechtigkeit ist umfassend: Wir können Individuen wie auch persönliche Verpflichtungen oder persönliche Beziehungen als gerecht oder ungerecht bezeichnen. Die besondere Form der Gerechtigkeit, von der Rawls spricht, bezieht sich hingegen auf Menschen, die durch soziale – allgemeiner: politische – Institutionen verbunden sind.

Wie aber kann uns dieses weite Verständnis von Gerechtigkeit dabei helfen, zu verstehen, welche Formen sozialer Institutionen zu Recht als ungerecht verurteilt werden können? Wie bereits erwähnt, argumentiert Rawls, dass wir eine bestimmte Vorstellung der Gerechtigkeit entwickeln müssen, um zu verstehen, wie wir die abstrakten Bestimmungen des formalen Begriffs der Gerechtigkeit interpretieren und anwenden sollen. Für die Entwicklung einer solchen Vorstellung werden uns jedoch ein paar Hinweise gegeben. Es gibt einige provisorische Orientierungspunkte, die uns dabei helfen, die Sprache der Gerechtigkeit auf die Welt der Politik anzuwenden. Rawls folgt dabei der Logik Abraham Lincolns, der anmerkte, dass der Begriff der Ungerechtigkeit jeglichen Sinn verlieren würde, könnte mit ihr nicht die Sklaverei als ungerecht verurteilt werden.[14] Mit Blick auf unsere jüngere Geschichte könnten wir ähnliche Dinge über die Muster der Marginalisierung und Ausgrenzung in den Vereinigten Staaten zu Zeiten von Jim Crow sagen: In den USA zur Zeit der Rassentrennung wurden rechtlich im Prinzip gleichgestellten Personen völlig ungleiche Anteile an politischer Macht, materiellen Ressourcen und öffentlichem Respekt zuteil. Wie im Falle der Sklaverei könnten wir sagen: Sollten diese Verhältnisse nicht als ungerecht gelten, wie könnten wir irgendetwas als ungerecht bezeichnen? Es gibt keinen legitimen Grund zu glauben, dass eine solche Art von Ungleichheit mit Gerechtigkeit in Einklang zu bringen sei und jede Vorstellung der Gerechtigkeit, die Jim Crow als gerecht bezeichnet, könnte allein aufgrund dieser Tatsache zu Recht als ungerecht zurückgewiesen werden.

Mit diesen Gedanken im Hinterkopf stellt sich die Frage, wie der abstrakte Begriff der Gerechtigkeit uns im Hinblick auf bestimmte Formen der Migration und der Migrationspolitik weiterhelfen könnte. Gibt es möglicherweise auch hier Schlussfolgerungen, die so eindeutig richtig sind, dass sie uns als Maßstab für die Bewertung von Gerechtigkeitsvorstellungen im Bereich der Migration dienen könnten?

Es fällt schwer, anzunehmen, dass es irgendetwas im Bereich der Migrationspolitik gibt, das nicht durch *irgendwen* abgelehnt wird. Ich glaube jedoch, dass es einige Schlussfolgerungen gibt, die bloß schwerlich von irgendwem zurückgewiesen werden könnten, der behauptet, den Werten moralischer Gleichheit verpflichtet zu sein, die dem in diesem Buch verwendeten Begriff der Gerechtigkeit zugrunde liegen. Demzufolge gibt es tatsächlich einige Möglichkeiten, bestimmte Formen der Migrationspolitik als im gleichen Maße ungerecht – also tatsächlich eindeutig ungerecht – zu bezeichnen wie die Verhältnisse zu Zeiten von Jim Crow. Sollte dies zutreffen, müsste jede Vorstellung von Migrationsgerechtigkeit entweder diese Schlussfolgerungen anerkennen oder mit einer recht wagemutigen Geschichte darüber aufwarten, warum sie dazu nicht verpflichtet ist; und ich kann mir nicht einmal ausmalen, wie eine solche Geschichte aussehen sollte.

Der erste Orientierungspunkt bezieht sich auf die Idee der Gleichheit von Bürgerinnen gegenüber dem Staat. Migrationspolitik betrifft demnach nicht bloß diejenigen außerhalb der Grenzen; wie diese Politik wahrgenommen und durchgesetzt wird kann durchaus auch die Gleichheit derjenigen einschränken, die selbst keine Migrantinnen sind. Denken Sie beispielsweise daran, wie die Rassifizierung des Migrationsrechts das öffentliche Leben für Bürgerinnen lateinamerikanischer und nicht-lateinamerikanischer Herkunft beeinflussen kann. So schafften beispielsweise die von Sheriff Joe Arpaio angewendeten Praktiken zur Durchsetzung des Migrationsrechts ein Klima der Angst und Einschüchterung unter den Bürgerinnen lateinamerikanischer Herkunft von Phoenix, und zwar ganz unabhängig von ihrem aufenthaltsrechtlichen Status: Arpaio schikanierte mit seinen Kontrollen regelmäßig Autofahrerinnen und Autofahrer lateinamerikanischer Herkunft und richtete sogar eine Task Force zu dem Zwecke ein, undokumentierte Einwanderer aufzuspüren und zu bestrafen, was schließlich dazu führte, dass die betroffenen Bürgerinnen in den einfachsten Verrichtungen ihres Alltags eingeschränkt wurden.[15] Die soziale Funktion der wiederholten Erfahrung solcher Maßnahmen besteht letztlich in der Marginalisierung und Disziplinierung aller Einwohnerinnen, die aus Lateinamerika stammen. Amy Reed-

Sandoval bezeichnet die Maßnahmen von Arpaios Beamten berechtigterweise als ein Schauspiel der Ungleichheit, durch das die Betroffenen an ihren untergeordneten und marginalisierten Status erinnert werden sollen.[16] Eine Theorie der Gerechtigkeit der Migration wird daher anerkennen müssen, dass bestimmte Ausschluss- und Durchsetzungsformen zu Ungleichheiten *zwischen Bürgerinnen* führen können und diejenigen Ungleichheiten zwischen Bürgerinnen, die innerstaatlich als Ausweis für Ungerechtigkeit gelten, werden auf ähnliche Weise auch im Hinblick auf die Bewertung der Migrationspolitik als Maßstab dienen können.

Der zweite Orientierungspunkt bezieht sich auf den Gedanken, dass es Bereiche geben kann, in denen Nicht-Bürgerinnen einen Anspruch darauf haben, vor dem Gesetz ebenso behandelt zu werden wie Bürgerinnen. Wie ausgedehnt diese Bereiche sind, ist selbstredend Gegenstand kontroverser Diskussionen. So dient die Beschwörung der „Illegalität" beispielsweise oft dem Zweck, die Inanspruchnahme von Rechten durch Migrantinnen prinzipiell zu diskreditieren. (Der rechte Slogan *What part of illegal dont't you understand?* versucht gar, undokumentierten Migrantinnen unter Verweis auf ihren fehlenden legalen Aufenthaltstitel jeglichen moralischen Status abzusprechen.) Aber diese Beschwörung der Illegalität scheint nicht wirklich stark genug, um den Gedanken zu widerlegen, dass es Bereiche gibt, in denen selbst diejenigen außerhalb des Staatsgebiets – oder diejenigen, die ohne Recht innerhalb des Staates anwesend sind – das Recht auf die gleiche Behandlung wie Bürgerinnen haben. Ein einfaches Beispiel ist sicherlich der Schutz durch die Polizei: Unrechtmäßig in den USA lebende Personen sind zwar stets von Abschiebung bedroht, verlieren hierdurch jedoch nicht ihr Recht auf den Schutz vor Mord oder Raub.[17] Ein anderes Beispiel ist der Fall von Kindern, deren unrechtmäßige Präsenz korrekterweise als Ergebnis der Handlungen ihrer Eltern verstanden wird und nicht als Resultat einer Handlung der Kinder. Der Supreme Court bemerkte in *Plyler v. Doe*, dass sich daraus auch für diejenigen ein Recht auf Grundschulbildung ableitet, die sich ohne Aufenthaltsrecht in den USA aufhalten:

> „Zumindest sollten diejenigen, die unser Staatsgebiet insgeheim und unter Verletzung unseres Gesetzes betreten, darauf gefasst sein, die Konsequenzen ihres Handelns zu tragen, darunter, wenn auch nicht allein, die Abschiebung. Allerdings trifft dies auf die Kinder dieser illegal eingereisten Personen nicht

> auf gleiche Weise zu. Ihre Eltern können das eigene Verhalten den geltenden sozialen Regeln anpassen und es ist anzunehmen, dass sie das Hoheitsgebiet des jeweiligen Bundesstaates auch wieder verlassen können; aber die Kinder, die im vorliegenden Fall die Klägerinnen sind, können weder das Verhalten ihrer Eltern noch ihre eigene Lage beeinflussen. Selbst wenn der Bundesstaat es als angebracht ansehen sollte, das Verhalten der Eltern durch Handlungen gegen deren Kinder zu beeinflussen, steht eine Gesetzgebung, welche die Folgen elterlicher Verfehlungen gegen deren Kinder wendet, im Widerspruch zu fundamentalen Vorstellungen der Gerechtigkeit."[18]

Mit anderen Worten: Es gibt einige eindeutige Fälle, in denen Immigrantinnen durch eine Bestrafung ungerecht behandelt werden, selbst dann, wenn im Einzelfall kein Recht auf einen Aufenthalt in dem Land besteht, das diese Strafe aussprechen möchte. Dem Supreme Court zufolge gibt es also Maßnahmen, die wir nicht nutzen dürfen, selbst wenn sie ein effektives Mittel zur Prävention unerlaubter Einwanderung wären. Die jüngsten Anstrengungen der Trump-Administration, die Kinder von Immigrantinnen direkt an der Grenze von ihren Eltern zu trennen, scheinen diesen moralischen Punkt zu missachten: Kinder werden unfair behandelt, wenn sie zu Unrecht für die Handlungen ihrer Eltern verantwortlich gemacht werden und ihnen die Fürsorge durch ihre Eltern verweigert wird – und es ist ungerecht, sowohl diesen Kinder, als auch ihren Eltern gegenüber, sie diesem Leid auszusetzen.

Der letzte Orientierungspunkt, den ich hervorheben möchte, ist womöglich der wichtigste: Es gibt Personen, in deren Fall die meisten von uns glauben, dass ein Recht auf Einwanderung besteht. Der gegenwärtige Rechtsbegriff des Flüchtlings entwickelte sich aus dem Grauen des Zweiten Weltkriegs und der daran anschließenden Scham vieler Staaten über die Zurückweisung derjenigen, die den mörderischen Intentionen des Nazi-Regimes zu entfliehen versucht hatten.[19] Wir können und sollten uns fragen, wer genau zum Schutz unter dem Abkommen berechtigt ist, das die Grundlage des internationalen Flüchtlingsrechts bildet; ich werde diese Frage in Kapitel sieben behandeln. Aber ich denke, die Abscheu gegenüber der Entscheidung der Vereinigten Staaten, die Flüchtlinge der *St. Louis* zurückzuweisen, sollte uns in diesem Fall eine Lehre sein.[20] *Bestimmte* Gruppen von Menschen haben gewiss das Recht, vor bestimmten Formen von Grausamkeit und Übeln geschützt zu werden und eine Theorie der Gerechtigkeit der

Migration, die diese Tatsache nicht anerkennen möchte, wäre bereits alleine aufgrund dieses Mangels abzulehnen. Der Flüchtling hat ein *Recht* auf Zuflucht und wir können sagen, dass die Verweigerung dieses Recht einer unfairen Behandlung gleichkommt – die Sprache der Gerechtigkeit erlaubt es uns, einen solchen Anspruch festzustellen.

Diese Orientierungspunkte setzen unserer Diskussion über die Gerechtigkeit der Migration also einige Schranken. Allerdings bieten sie uns nicht viel. Wir benötigen noch mehr Orientierungshilfen, um die abstrakte Idee der Gerechtigkeit auf das Gebiet der Migration anwenden zu können. Denn was ich bisher dargelegt habe, bietet uns zwar einige Überlegungen hinsichtlich dessen, was wir nicht sagen können – es bietet uns allerdings sehr wenig hinsichtlich der Bestimmung dessen, was wir sagen *könnten*.

Ein Großteil dieses Buches wird versuchen, eine bestimmte Vorstellung der Gerechtigkeit für den Bereich der Migration zu entwickeln. Mit ihr soll zugleich ein Ansatz ausgearbeitet werden, um die Freiheiten von Staaten im Hinblick darauf zu bestimmen, welche Migrantinnen sie einlassen oder aber ausschließen dürfen. Diese Vorstellung zu entwickeln bildet den Gegenstand der folgenden drei Kapitel. Beginnen möchte ich die Aufgabe in diesem Kapitel jedoch damit, dass ich zunächst diejenigen Theorien der Gerechtigkeit der Migration untersuche – und schließlich widerlege –, die gegen *jegliches* Recht auf Ausschluss ungebetener Migrantinnen argumentieren. Für diese Theoretikerinnen, die oftmals als Verteidiger *offener Grenzen* bezeichnet werden, gleicht der Versuch, ein gerechtes System des Ausschlusses zu entwickeln, dem Vorhaben, ein gerechtes System der Rassentrennung auszuarbeiten: Ein solches Vorhaben ist zwar nicht hoffnungslos, aber so etwas wie ein Widerspruch in sich.

Die theoretischen Verteidiger offener Grenzen vertreten die extreme Version einer Seite der bereits weiter oben dargestellten Dialektik. Viele der Menschen, die davon abgestoßen waren, wie die Trump-Administration Immigrantinnen behandelt, haben versucht, diese als rassistisch und völkisch wahrgenommene Migrationspolitik zurückzuweisen. Nicht alle von ihnen glauben, dass Grenzen offen sein sollten. Viele sind einfach der Meinung, dass die Trump-Administration die Grenzen auf falsche Art und gegenüber den falschen Menschen schließt. Der Glaube, dass es Beschränkungen dessen gibt, wie der Staat an seinen Grenzen handeln darf, ist nicht identisch mit dem Gedanken, dass diese Grenzen offen sein sollten – obwohl die Trump-Administration oft versucht hat, jeglichen Widerstand gegenüber ihren migrationspolitischen Maßnahmen als Verteidigung offener Grenzen darzustellen. Die Befürwortung offener Grenzen ist also nicht die einzige Möglichkeit, dem populistischen Ansatz der Migrationspolitik etwas entgegenzusetzen. Tatsächlich findet sich gegenwärtig kaum ein politischer Akteur, der bereit wäre, die Möglichkeit offener Grenzen zu verteidigen. Nichtsdestotrotz ist der Gedanke, dass Staaten *niemanden* zu Recht ausschließen können, eine genauere Betrachtung wert. Sollte diese Idee erfolgreich verteidigt werden können, hätten wir gute Gründe, die scheinbar unvermeidlichen Einwanderungskontrollen infrage zu stellen. Wie Joseph Carens bemerkt, sollten wir uns nicht von der Tatsache abschrecken lassen, dass uns offene Grenzen in der gegenwärtigen Welt als unmöglich erscheinen, denn eine Welt ohne Sklaverei erschien einstmals ebenso fantastisch.

Wo aber sollten wir mit der Verteidigung des Gedankens, dass Grenzen offen sein sollten, beginnen? In meinen Augen sollten wir uns durch den Begriff der Gerechtigkeit selbst inspirieren lassen. Laut Rawls verträgt sich Gerechtigkeit nicht mit willkürlichen Unterscheidungen, insbesondere dann nicht, wenn dadurch die grundlegenden Rechte und der materielle Besitz einzelner Personen betroffen sind. Wir könnten also meinen, dass die Praxis des Ausschlusses in beiden Hinsichten problematisch ist. Wenn die Vereinigten Staaten einer Migrantin aus einem relativ armen Land die Aufnahme verweigern, verurteilen sie diese Person zu einem Leben ohne denjenigen Wohlstand, der oft (wenn auch nicht immer) Teil des US-amerikanischen Lebens ist. Wenn sie die Aufnahme einer Person verweigern, die unter einem repressiven Regime lebt, verurteilen sie diese Person zu einem Leben ohne politische Freiheit. Wenn sie es einer Person, die zu migrieren versucht, versagen, ein von ihr gewähltes Projekt oder eine von ihr gewählte Beziehung in den Vereinigten Staaten zu verfolgen, beschränken sie die Möglichkeiten dieser Person, ein wertvolles Leben für sich und ihre Nächsten aufzubauen. Zudem tun die Vereinigten Staaten all das aufgrund einer Grenze, die aus einem moralischen Blickwinkel selbst als willkürlich erscheint; niemand hat es verdient, auf der „richtigen" Seite einer bestimmten, auf dem Boden gezogenen Linie geboren zu werden. Darüber hinaus repräsentiert diese Linie nicht nur einfach materielle Ungleichheit, sondern auch eine Geschichte von Gräueln und kolonialer Ausbeutung: Viele der gegenwärtigen Zurückweisungen an der Grenze bestehen darin, dass ehemalige Kolonialmächte Menschen aus verarmten ehemaligen Kolonien die

Einwanderung verweigern.[21] Wir könnten, einfach anhand dieser Gedanken, eine Vorstellung der Migrationsgerechtigkeit entwerfen, in der alle Formen von Ausschluss mutmaßlich ungerecht sind. In seiner Zurückweisung aller Versuche, ein Recht auf Ausschluss zu rechtfertigen, bietet uns Chandran Kukathas eine anschauliche kleine Skizze der Möglichkeiten eines Arguments für offene Grenzen:

> „Wie also können Einschränkungen der Einwanderung, oder, allgemeiner, der Bewegungsfreiheit, gerechtfertigt werden? Es ist nicht leicht, solche Beschränkungen aus der Perspektive von Individuen oder Völkern allgemeingültig zu verteidigen. [...] Ein Argument für die Beschränkung individueller Bewegungsfreiheit wird immer auch ein Argument sein, weshalb eine bestimmte (manchmal begünstigte) Gruppe dabei geschützt werden sollte, die Renten zu genießen, die sie sich durch das große Glück gesichert hat, in einem bestimmten Teil der Welt statt einem anderen zu leben."[22]

Dieses breit angelegte Argumentationsmuster führt also an, dass der Gedanke eines (potentiell) gerechtfertigten Systems staatlichen Ausschlusses eine Illusion sei, denn das Ziel aller liberalen Theoretikerinnen sollte es sein, jegliche Formen ungerechtfertigter Hierarchien niederzureißen – und die Struktur des Aufenthaltsrechts, die manche Menschen von manchen Orten ausschließt, könnte schlicht eine solche Form ungerechtfertigter Hierarchien darstellen. Ich denke, dass die Idee offener Grenzen in diesem Sinne einfach als eine Ausweitung des umfassenderen Systems von Bürgerrechten verstanden werden kann. Wie bereits Roger Nett vor einer Generation argumentierte, könnte es sich bei dem letzten Bürgerrecht – das vermutlich am schwierigsten durchzusetzen sei – um das Recht handeln, sich frei über die Erdoberfläche bewegen zu dürfen.[23]

Ich möchte im weiteren Verlauf dieses Kapitels nun vier spezifischere Versionen des Arguments für offene Grenzen diskutieren. Dabei handelt es sich weniger um einzelne Argumente, als vielmehr um Gruppen von Argumenten, die Möglichkeiten darstellen, das skizzierte Argument gegen staatlichen Ausschluss genauer und rigoroser zu formulieren. Wie ich bereits bemerkt habe, bin ich nicht der Meinung, dass diese Argumente überzeugend sind. Mein Widerspruch beginnt damit, wie die verschiedenen Argumente die besondere Beziehung begreifen, die zwischen einem mit

Zwangsbefugnissen ausgestatteten Staat und denjenigen Personen besteht, die ihr Leben unter den Gesetzes jenes Staates gestalten. In meinen Augen tendieren die Argumente für offene Grenzen einfach ausgedrückt dazu, die besonderen Verpflichtungen zu ignorieren, die Staaten gegenüber denjenigen haben, die in direkter Reichweite der staatlichen Zwangsbefugnis ihr Leben fristen. Sie tendieren darüber hinaus zu der Überzeugung, dass jegliches Bürgerrecht zugleich auch ein Menschenrecht sein sollte. Im Folgenden werde ich versuchen, sowohl meinen Dissens mit diesen Theoretikerinnen herauszuarbeiten, als auch den Verdienst, den sie an der Formulierung meines eigenen Ansatzes haben. In meinen Augen besteht ein zu großer Teil der Philosophie darin, die Fehler anderer aufzuspüren und eine theoretische Abhandlung ist immer dort am schwächsten, wo sie auf die Lücken anderer Ansätze verweist.[24] Was ich hier vorbringe, soll daher nicht als Demonstration einer einwandfreien Widerlegung der verschiedenen Positionen für offene Grenzen verstanden werden – für solch ein Vorhaben sind diese Argumente sowohl zu komplex als auch zu gut. Vielmehr handelt es sich höchstens um die Darstellung von Überlegungen, aufgrund derer ich die verschiedenen Argumente nicht akzeptieren kann und das trotz der Schuld, in der ich bei ihren Urhebern stehe.

1. Gerechtigkeit und Ausschluss

Wie also können wir den Gedanken, dass ein Staat kein Recht darauf hat, ungebetene Einwanderer auszuschließen, präziser fassen? Ich sollte nebenbei anmerken, dass die von mir diskutierten Autoren oft bereit sind zu akzeptieren, dass ein Staat das Recht hat *manche* Personen auszuschließen – so beispielsweise diejenigen, die vorhaben, dem Staat oder seinen Einwohnerinnen zu schaden. Allerdings argumentieren sie dann, dass das Recht auf das Überqueren internationaler Grenzen sehr ähnlich dem Recht sei, sich *innerhalb* von Staaten zu bewegen oder sehr ähnlich dem Recht, seine Religion frei auszuüben: Abstrakt betrachtet ist es möglich, diese Rechte einzuschränken, allerdings nur in wirklich furchtbaren (und, so wird gehofft, untypischen) Situationen. Im allgemeinen Fall jedoch, in der eine Person Staatsgrenzen einfach aus denselben Gründen überqueren will, die sie auch dazu motivieren könnten, sich innerhalb dieser Grenzen zu bewegen, behandelt der Staat eine Person diesen Autoren zufolge ungerecht, wenn er sie von eben dieser

Überquerung abhält. Was kann zur Unterstützung des Gedankens ins Feld geführt werden, dass der Begriff der Gerechtigkeit eine solche Form des Ausschlusses verurteilen sollte?

Die vier von mir diskutierten Versionen eines möglichen Beweises bezeichne ich als die Argumente der *Willkürlichkeit,* der *Verteilungsgerechtigkeit,* der *Kohärenz mit bestehenden Bewegungsrechten* und der *Zwangsgewalt.* Die Grenzen zwischen den verschiedenen Argumenten sind allerdings eher durchlässig; so könnten diejenigen, die eine Art dieser Argumente vorbringen, auch zugleich all die anderen anführen. Allerdings werde ich die unterschiedlichen Positionen zum Zwecke der analytischen Klarheit als eigenständige Argumente diskutieren und versuchen, jedes möglichst vorteilhaft darzustellen. All diese Ansätze haben eine ähnliche Struktur: Sie nutzen die groben Umrisse des bereits dargestellten Gerechtigkeitsbegriffs um zu zeigen, dass der ausschließende Staat die durch ihn Ausgeschlossenen dadurch in ihren Rechten verletzt, dass er sie unfair behandelt. Zudem verwenden sie häufig eine ähnliche Strategie: Ihnen zufolge unterstützen die Normen und Prinzipien, die wir bereits als Teil liberaler Gerechtigkeit anerkennen, den Gedanken, dass ein moralisches Recht auf freien Grenzübertritt zu Recht als Teil des Liberalismus betrachtet wird. Somit haben all diese Überlegungen ein gewichtiges Argument: Menschen als gleichwertig und mit Respekt zu behandeln bedeutet, sie nicht gewaltsam davon abzuhalten, an den Ort ihrer Wahl zu gelangen.[25]

1.1 Willkürlichkeit

Eine Möglichkeit, für offene Grenzen zu argumentieren, beruht schlicht auf dem, was bereits weiter oben erwähnt wurde: Grenzen sind demnach unter moralischen Gesichtspunkten willkürlich, da sie im Allgemeinen eher die gemeinsame Vergangenheit kolonialer Gewalt denn irgendetwas von moralischer Bedeutung repräsentieren. Sie sind selbstverständlich auch in der Hinsicht willkürlich, dass niemand behaupten könnte, er habe es in irgendeinem bedeutsamen Sinne *verdient,* auf der begünstigten Seite einer Grenze geboren worden zu sein. Das weiter oben angeführte Zitat von Chandran-Kukathas stellt einen Ausdruck dieses Gedankens dar: Wenn es den derzeitigen Einwohnerinnen eines Ortes erlaubt ist, ungebetene Immigrantinnen auszuschließen, fungiert der Zufall des Geburtsortes als Grundprinzip der

Verteilung von Rechten. Joseph Carens' frühes Werk zu diesem Thema zieht eine Parallele zwischen der Staatsbürgerschaft in einem reichen, attraktiven Staat und den Privilegien, die im Feudalismus mit der Geburt verbunden waren: „ein vererbter Status, der die Lebenschancen massiv verbessert."[26] Das Argument der Willkürlichkeit hängt jedoch nicht von materiellen Ungleichheiten ab oder davon, dass ein Staat wohlhabender als andere Staaten ist. Diesem Argument zufolge genügt bereits die Tatsache willkürlicher Unterschiede in der Verteilung von Rechten – Ihnen ist es aufgrund ihres Geburtsortes oder ihrer Eltern erlaubt, in Deutschland zu leben, mir nicht – um die Überzeugung zu rechtfertigen, dass Ausschluss ungerecht sei.

Carens ist nicht allein mit dem Gedanken, dass bereits die Willkürlichkeit von Staatsgrenzen den Ausschluss von Migrantinnen als ungerecht ausweist. Philip Cole argumentiert auf ganz ähnliche Weise, dass Liberale nicht meinen können, das Recht auf Ausschluss stünde im Einklang mit der liberalen Vorstellung von Gerechtigkeit. Cole geht sogar insofern weiter als Carens, als dass er meint, die Willkürlichkeit von Staatsgrenzen könnte uns dazu zwingen, die Vorstellung liberaler Gerechtigkeit an sich zu überdenken:

> „Wie wir sehen werden, geben viele Autoren zu, dass der Liberalismus exklusive Mitgliedschaften prinzipiell und Einschränkungen der Einwanderung praktisch ablehnen sollte, und zwar aufgrund seiner Verpflichtung gegenüber der moralischen Gleichwertigkeit aller Menschen als solche und der daraus resultierenden Abscheu gegenüber willkürlichen Unterscheidungen zwischen ihnen. Aber angesichts dieses Eingeständnisses fahren sie damit fort, eine Rechtfertigung für den Ausschluss von Migrantinnen zu suchen, oftmals auf der Basis des Gedankens, dass es möglich sein muss, eine von liberal-demokratischen Staaten so weitgehend akzeptierte Praxis zu rechtfertigen. Aber wie wir sehen werden, stellt ein solches Unterfangen die liberale Theorie vor besondere Probleme. Der Zweck dieses Buches ist es, diese Probleme zu untersuchen."[27]

Coles Analyse zufolge begründet die Willkürlichkeit von Staatsgrenzen den Schluss, dass die Praxis staatlichen Ausschlusses ungerechtfertigt ist. Die Tatsache, dass wir nicht bereit sind, diesen Schluss zu akzeptieren, ist für Cole Beweis dafür, dass wir vielleicht eher einer neuen Form der Debatte bedürfen – nicht nur über Grenzen, sondern über politische Gerechtigkeit im weiteren Sinne.

1.2 Verteilungsgerechtigkeit

Das soeben dargestellte Argument gewinnt offensichtlich an Stärke, wenn anerkannt wird, dass Staatsgrenzen nicht nur Bürgerinnen und Migrantinnen trennen, sondern oft auch wohlhabende von armen Menschen. Der Gedanke, dass die Zurückweisung von Migrantinnen nicht nur in sich falsch ist, sondern auch, weil sie die distributiven Rechte der Armen dieser Welt untergräbt, bildet die Grundlage für ein weiteres, starkes und wichtiges Argument gegen ein Recht auf Ausschluss. Seine Begründung kann allerdings zwei verschiedene Formen annehmen: Sie kann zum einen auf Überlegungen der *wirtschaftlichen Gerechtigkeit*, zum anderen auf Überlegungen der *Chancengleichheit* beruhen.

Das Argument der *wirtschaftlichen Gerechtigkeit* führt die schlichte Tatsache globaler wirtschaftlicher Ungleichheit an und zeigt, wie verschiedene Arten des Ausschlusses potentieller Migranten durch die reicheren Länder Nordamerikas und Europas dazu tendieren, diese ungleiche Verteilung des globalen Wohlstands zu verstärken. Dieses Argument kann durch Prognosen hinsichtlich möglicher Wohlstandsgewinne durch offene Grenzen weiter verstärkt werden. Wie Michael Clemens gezeigt hat, könnte die Abschaffung staatlicher Grenzen wirtschaftliche Gewinne vom anderthalbfachen Umfang des gegenwärtigen globalen Bruttoinlandsprodukts schaffen.[28] Darüber hinaus haben Geldüberweisungen von Migrantinnen, die aus einkommensschwachen in einkommensstarke Länder migriert sind, bedeutende Auswirkungen auf die Entwicklung jener einkommensschwachen Länder; laut Schätzungen ist der Umfang dieser Überweisungen derzeit größer als die weltweiten Aufwendungen für Entwicklungshilfe.[29] Der Punkt ist jedoch nicht, dass der Ausschluss von Migrantinnen die Welt ärmer macht als sie sein könnte. Zentral ist vielmehr, dass das Recht auf Ausschluss es den Wohlhabenden effektiv erlaubt, ihren Wohlstand trotz legitimer Gerechtigkeitsansprüche des armen Teils der Weltbevölkerung zu bewahren.[30] Darauf bezieht sich Kukathas' Argument der Renten und Carens' Argument des Feudalismus – wir meinen, dass die Wohlhabenden dieser Welt eine bessere Rechtfertigung für ihren Reichtum anführen können sollten als bloßes Glück. Mehr als die meisten anderen Philosophen fügt Philip Cole alldem den Gedanken hinzu, dass unsere gemeinsame Erfahrung globaler Ungleichheit auch eine gemeinsame Geschichte kolonialer Gewalt und kolonialen

Grauens widerspiegelt. Es ist demnach nicht bloß so, dass wir die Armut derjenigen verstetigen, die das Pech hatten, in einem Entwicklungsland geboren worden zu sein. Vielmehr setzen wir bis zum heutigen Tag eben jene koloniale Ausbeutung fort, die ursprünglich bestimmt hat, welches Land sich entwickeln konnte und welches nicht.

Der zweite Aspekt der Verteilungsgerechtigkeit, den ich diskutieren möchte, ist das Prinzip der *Chancengleichheit*. Wir sind überzeugt davon, dass Chancengleichheit innerhalb von Staaten ein zentraler Wert ist – wie John Rawls in seinen eigenen Ausführungen zur Gerechtigkeit feststellt, wäre eine Gesellschaft ungerecht, in der es keine zuverlässige Gewährleistung der Chancengleichheit gibt. Allerdings schaffen Staatsgrenzen eine Welt radikal ungleicher Chancen, indem sie Menschen daran hindern, an den Ort ihrer Wahl zu gelangen. Diese Version des Arguments könnte unter der Überschrift „Gleiche Freiheit" diskutiert werden, wie etwa bei Kieran Oberman, oder schlicht als ein weiterer Fall ungleicher Chancen. Laut Carens ist der springende Punkt in diesem Fall jedoch, dass wir jeden Tag Zwangsgewalt einsetzen, um Menschen davon abzuhalten, ein besseres Leben für sich und ihre Familien aufzubauen. Oberman betont diesen Punkt besonders nachdrücklich: Ihm zufolge schulden wir Menschen nicht bloß einen adäquaten Umfang an Möglichkeiten zur Gestaltung des eigenen Lebens, sondern den größtmöglichen. Da es laut Oberman für Personen von essentiellem Interesse ist, am Ort ihrer Wahl ein Leben gemäß ihrer Wertvorstellungen aufzubauen, wäre es folglich allein globale Bewegungsfreiheit, die mit der von uns vorgeblich hochgeschätzten liberalen Vorstellung von Gerechtigkeit kompatibel ist.

1.3 Kohärenz mit bestehenden Bewegungsrechten

Das Argument gegen den staatlichen Ausschluss von Migrantinnen erfährt darüber hinaus auch dadurch Unterstützung, dass liberale Autorinnen bereits zu der Überzeugung tendieren, bestimmte Bewegungsrechte seien impliziter Bestandteil liberaler Gerechtigkeit. Es gibt mindestens zwei Arten von Bewegungsrechten, die in diesem Zusammenhang diskutiert werden, und aus beiden wird jeweils ein Recht auf das Überqueren von Staatsgrenzen abgeleitet. Wir können diese Überlegungen als Argument der *innerstaatlichen Bewegungsfreiheit* sowie als *Emigrations*-Argument bezeichnen.

Das erste Argument beginnt bei der simplen Tatsache, dass das innerstaatliche Recht auf Bewegungsfreiheit allgemein als ein zentrales Menschenrecht anerkannt wird; so ist es unter anderem Bestandteil bedeutender Dokumente des internationalen Rechts, wie beispielsweise der Allgemeinen Erklärung der Menschenrechte. Das Argument der innerstaatlichen Bewegungsfreiheit besagt allerdings bloß, dass die Gründe für die Bedeutung der *innerstaatlichen* Art von Bewegungsfreiheit ebenso im Falle der Bewegungsfreiheit im allgemeineren Sinne gelten. Carens nennt diese Beweisform in Anlehnung an David Miller die „Kranarm"-Strategie (*cantilever strategy*): Wir beginnen bei etwas, das bereits erfolgreich verteidigt wurde, ob im internationalen Recht oder gemäß unseres besten Verständnisses der Gerechtigkeit, und zeigen, dass ähnliche Argumente auch für eine über den ursprünglichen Anwendungsbereich hinausgehende Auslegung dieses Rechts vorgebracht werden können:

> „Wenn es für Menschen so bedeutsam ist, sich frei innerhalb ihres Staates zu bewegen, ist es für sie dann nicht ebenso wichtig, ein Recht darauf zu haben, Staatsgrenzen frei zu überqueren? Jeder Grund dafür, warum eine Person sich innerhalb eines Staates bewegen will könnte genauso gut auch ein Grund für die Bewegung zwischen Staaten sein: Eine Person möchte eine bestimmte Arbeitsstelle; verliebt sich in eine Person aus einem anderen Land; gehört einer Religion an, die bloß wenige Anhängerinnen in ihrem Heimatland, jedoch viele in einem anderen Land hat; oder möchte einfach bestimmte kulturelle Möglichkeiten nutzen, die ihr bloß in einem anderen Staat zur Verfügung stehen. Diese radikale Trennung, die Bewegungsfreiheit innerhalb von Staaten als ein Menschenrecht ansieht und zugleich Staaten frei über die Kontrolle ihrer Grenzen verfügen lässt, macht moralisch keinen Sinn. Wir sollten das Menschenrecht auf Bewegungsfreiheit ausweiten. Wir sollten die Freiheit zu migrieren, zu reisen und zu leben wo immer eine Person will, als ein Menschenrecht anerkennen."[31]

Ähnliche Gedanken lassen sich in den Überlegungen von Oberman, Christopher Freiman und Javier Hidalgo finden.[32] Jede uns verfügbare Rechtfertigung für die Gewährleistung eines Rechts auf innerstaatliche Bewegungsfreiheit, so das Argument, bietet uns zugleich auch Gründe dafür, sich zwischen Staaten frei bewegen zu dürfen. Wenn es also ungerecht ist, die interne Bewegungsfreiheit einzuschränken, so ist es ebenso ungerecht, ungebetene Migrantinnen abzuweisen.

Ein anderes internationales Recht, das für die Verteidigung offener Grenzen herangezogen werden könnte, ist hingegen das Recht, das eigene Land *verlassen* zu dürfen. Die meisten liberalen Autorinnen – wenn auch nicht ausnahmslos – sehen dieses Recht als moralisch bedeutsam an.[33] Diejenigen Länder, die es ihren Bürgerinnen schwer machen auszuwandern, geben uns einige empirische Gründe für die Annahme, dass Gerechtigkeit nicht mit Emigrationsbeschränkungen einhergehen kann. Die Frage ist allerdings, ob ein solches Recht auf Auswanderung ohne ein ähnliches Recht darauf, am Ort der Wahl einreisen zu dürfen, überhaupt sinnvoll gedacht werden kann. Der Gedanke ist also, dass das Recht auf Emigration, um überhaupt mehr als eine bloß formale Garantie zu sein, auch derjenigen Mittel bedarf, anhand derer es erst sinnvoll genutzt werden kann; und falls es keine andere, zur Aufnahme verpflichtete Gesellschaft gibt, verliert dieses Recht folglich seine Bedeutung. Lea Ypi beschreibt diese Überlegung treffend:

> „Die Asymmetrie zwischen Aus- und Einwanderung verweist auf einen ernsten moralischen Mangel innerhalb der Theorie und stimmt nicht mit dem allgemeinen Prinzip der Gerechtigkeit im Bereich der Migration überein. Entweder ist Bewegungsfreiheit von Bedeutung, oder sie ist es nicht."[34]

Ypi argumentiert, dass sich diese Spannung nur dadurch auflösen ließe, dass wir unsere Analyse der Migration grundlegend überdenken. Ich werde diese Gedanken hier allerdings nicht weiter vertiefen und mich stattdessen auf die Überlegung konzentrieren, dass es moralisch problematisch ist, ein Recht auf Auswanderung ohne ein komplementäres Recht auf Einwanderung zu verteidigen. Cole beschreibt ähnliche Ideen, indem er sich auf ein Ann Dummett zugeschriebenes Argument bezieht:

> „Die liberale Asymmetrie-Position behauptet, dass es ein Recht darauf gibt, einen Staat zu verlassen, aber kein Recht darauf, einen Staat zu betreten – und diese Position ist gleichbedeutend mit der Behauptung, dass das Recht, einen Staat zu verlassen, tatsächlich nicht das Recht beinhaltet, internationale Grenzen zu überqueren, denn das würde einem Recht gleichkommen, einen anderen Staat zu betreten. Wie kohärent aber ist ein Recht darauf, einen Staat verlassen zu dürfen, wenn es kein Recht darauf beinhaltet, Grenzen zu überqueren?"[35]

Dieser Überlegung nach ist eine Vorstellung von Gerechtigkeit inkohärent, der zufolge Menschen zwar ein moralisches Recht darauf haben, einen Staat zu *verlassen*, ihnen zugleich aber kein Recht darauf gewährleistet wird, einen anderen Staat *betreten* zu können; und ein angebliches Recht auf Ausschluss wiederum steht in Spannung, wenn nicht gar in offenem Widerspruch, zu einem solchen Recht auf Eintritt.

1.4 Zwangsgewalt

Das letzte von mir betrachtete Argumentationsmuster ist das Argument der *Zwangsgewalt.* Viele liberale Theoretikerinnen vertreten die Position, dass Personen bestimmte Rechte gegenüber Institutionen haben, die dazu befugt sind, Zwang über sie auszuüben. Insbesondere neigen wir zu der Annahme, dass die durch eine Institution Gezwungenen besonders starke Rechte auf eine Rechtfertigung der Politik und der Praktiken jener Institution haben; und dass bei dieser Rechtfertigung Personen, die sich in einer vergleichbaren Situation befinden, auch als gleichwertig betrachtet werden sollten. Diese Idee habe ich bereits an anderer Stelle verteidigt.[36] Im Kontext der Migration gewinnt sie ihre Bedeutung aufgrund der simplen Tatsache, dass Staatsgrenzen, wie Carens schreibt, üblicherweise von Menschen mit Schusswaffen bewacht werden.[37] Die Ausschlusspraxis, von der wir hier sprechen, ist keine bloße Abstraktion, sondern, einfach gesagt, sowohl potentiell, als auch viel zu oft tatsächlich, gewaltsam. Der Umstand, dass Staatsgrenzen regelmäßig mit dem Einsatz von Zwangsgewalt einhergehen, sollte uns veranlassen, darüber nachzudenken, durch welche Art von Garantien gegenüber den betroffenen Personen diese Form von Zwang potentiell legitimiert werden kann.[38] Arash Abizadeh hat argumentiert, dass die soeben beschriebenen Tatsachen ausreichen würden, um neue internationale Institutionen mit Entscheidungskompetenzen hinsichtlich der Migrationspolitik einzelner Länder auszustatten; ihm zufolge gibt es kein *einseitiges* Recht, ungewollte Immigrantinnen auszuschließen.[39] Carens äußert ähnliche Gedanken, wenn er die bedeutenden Effekte der Ausschlusspraktiken für das Leben der gewaltsam ausgeschlossenen Personen beschreibt:

> „Es ist schwer vorstellbar, dass es eine Form staatlichen Zwangs jenseits der Haft gibt, die tiefgreifendere Auswirkungen auf das Leben einer Person hat,

> als die Verweigerung der Einreise. Mit ihr gehen gewaltige Folgen hinsichtlich der Lebensentscheidungen einher, die eine Person im Anschluss noch treffen kann […] für viele Menschen hat selbst die fortdauernde Präsenz staatlicher Macht in ihrem Alltag keinen solch durchdringenden Einfluss auf ihr Leben wie die ursprüngliche Bestimmung darüber, wo sie hingehören und wo sie leben dürfen – oder nicht dürfen."[40]

Für Carens sind in diesen Fällen die Auswirkungen staatlicher Zwangsgewalt auf die von ihr betroffenen Personen so schwerwiegend, dass wir meinen sollten, eine angemessene Rechtfertigung könnte kaum gegeben werden. Um es kurz zu machen: Ein liberaler Staat muss seinen eigenen Überzeugungen gerecht werden. Im Falle eines der Zwangsgewalt an der Grenze unterworfenen, potentiellen Immigranten scheint eine Rechtfertigung jedoch nicht vorhanden zu sein. Demnach ist seine Zurückweisung als ungerecht zu bezeichnen und das angebliche Recht eines Staates auf den Ausschluss ungebetener Migrantinnen muss als Illusion begriffen werden.

2. Die politische Gerechtigkeit staatlichen Ausschlusses

Die soeben nachvollzogenen Argumente sind in meinen Augen sehr stark und stellen, wie bereits gesagt, bedeutende Herausforderungen für die gängigen Vorstellungen politischer Moral innerhalb unserer Gesellschaft dar, in denen das Recht auf Ausschluss für gewöhnlich außer Frage steht. Allerdings glaube ich auch, dass diesen Herausforderungen erfolgreich begegnet werden kann. Entsprechend bin ich nicht der Meinung, dass eines dieser Argumente – so stark sie auch sein mögen – erfolgreich Gründe dafür formuliert, staatlichen Ausschluss als inhärent ungerecht auszuweisen. Auch hier möchte ich nochmals erwähnen, dass ich nicht behaupte, jegliche Form von Ausschluss könnte gerechtfertigt werden – wie ich in den folgenden Kapiteln zeigen werde, ist das ausdrücklich nicht der Fall. Wir benötigen eine Theorie der Gerechtigkeit im Bereich der Migration, die es uns ermöglicht, richtige Entscheidungen darüber zu treffen, welche Formen des Ausschlusses ungerecht sind. In diesem Kapitel möchte ich jedoch zunächst lediglich darlegen, weshalb ich weiterhin denke, dass diejenigen, die *jegliche* Form staatlichen Ausschlusses als ungerecht bezeichnen, falsch liegen.

Meiner Kritik dieser Argumente liegt die Überzeugung zugrunde, dass keine dieser Überlegungen auf angemessene Weise die Tatsache der *Gebietshoheit (jurisdiction)* in Betracht zieht – hierunter verstehe ich die Tatsache, dass der Staat die Ausübung seiner Zwangsgewalt legitimerweise allein über diejenigen Individuen beanspruchen kann, die sich in einem bestimmten Teil der Welt aufhalten. In meinen Augen haben alle Personen, die sich innerhalb eines solchen Hoheitsgebiets aufhalten, eine besondere Beziehung sowohl zueinander, als auch zu dem Staat, der über dieses Gebiet herrscht. Diese Beziehung ist eine *politische* Beziehung und beinhaltet die Durchsetzung zwingender rechtlicher Regeln innerhalb eines bestimmten Territoriums; für diejenigen, die sich in diesem Gebiet aufhalten, erschaffen jene Regeln eine soziale Welt. Diese Zwangsbefugnis begründet jedoch die Notwendigkeit einer Rechtfertigung gegenüber genau jenen Individuen, die mit dem Staat in dieser, mit Zwang verbundenen, politischen Beziehung stehen. Damit ist allerdings nichts anderes gemeint, als dass diese Individuen moralische Rechte auf bestimmte Garantien dafür haben, *wie* diese Zwangsgewalt ausgeübt wird. Folglich gibt es Rechte, die am besten als Bürgerrechte, nicht als Menschenrechte, bezeichnet werden, und zwar aus dem Grund, dass diese Rechte nur im Kontext einer Gesellschaft von Bürgerinnen, die von einem mit Zwangsgewalt ausgestatteten Staat regiert werden, ihre korrekte Anwendung finden können. Um Missverständnissen vorzubeugen möchte ich betonen, dass solche Rechte hinsichtlich ihres Geltungsbereichs universal sind; sie gelten also für alle Menschen, oder sollten das zumindest. An anderer Stelle habe ich gezeigt, dass alle Menschen das Recht auf einen Staat haben, der ihnen eine adäquate Form demokratischer Mitsprache ermöglicht.[41] Wir können daher ein Menschenrecht *auf* eine Gesellschaft haben, in der wir mit Bürgerrechten ausgestattet sind. Wie ich in Kapitel 4 zeigen werde, ist allerdings nicht klar, weshalb wir, sobald wir über solche Bürgerrechte verfügen, ein Recht darauf haben sollten, in einer *bestimmten* Gesellschaft zu leben. An dieser Stelle möchte ich jedoch zunächst nur bemerken, dass derlei Bürgerrechte nur konkreten Menschen in konkreten politischen Gemeinschaften zukommen und nicht abstrakten Menschen jenseits solcher Gemeinschaften. Viele der Argumente, die gegen ein Recht auf Ausschluss vorgebracht werden, übersehen diesen politischen Kontext, aus dem heraus die von ihnen angeführten Rechte jedoch erst ihre moralische Bedeutung gewinnen.

2.1. *Willkürlichkeit*

Das Argument der Willkürlichkeit führt an, dass Rawls' Gerechtigkeitsbegriff die Verteilung grundlegender Rechte auf der Basis willkürlicher Unterscheidungen ablehnt; dieser Gedanke wird dann genutzt, um den Ausschluss von Migrantinnen als illegitim anzufechten. Das Problem mit diesem Argument besteht jedoch darin, dass die Idee der *Willkürlichkeit* uns zwar vorschreibt, Ungleichheiten nicht auf der Basis moralisch irrelevanter Fakten zu rechtfertigen – allerdings behauptet sie nicht, dass jede Form von auf Zufall beruhender Ungleichheit auch moralisch fragwürdig sein muss. Wir können das an einem Beispiel aus dem vorangegangenen Kapitel erkennen: Es ist nicht ungerecht, mir das Wahlrecht in Frankreich zu verweigern, und dass obwohl die Grenzen zwischen Frankreich und anderen Ländern letztlich, je nach Perspektive, Resultate des Zufalls oder moralischen Unrechts sind – und dasselbe gilt hinsichtlich des Umstands, dass ich zufällig in Kanada statt in Frankreich geboren wurde. Aus der der Tatsache, dass in diesem System Glück und Zufall eine große Rolle spielen, folgt daher nicht notwendig, dass die ungleiche Verteilung des Wahlrechts in einem *moralisch zu verurteilenden* Sinne zufällig ist. Die französische Bürgerin hat eine bestimmte Beziehung zu ihrer Regierung, die in meinem Falle nicht besteht, da ich weder in Frankreich geboren wurde, noch dort lebe. Diese Beziehung lässt die besonderen Rechte der französischen Bürgerinnen gegenüber ihrer Regierung als moralisch angemessen erscheinen, selbst wenn es bloßer Zufall ist, dass sie in Frankreich geboren wurde und ich in Kanada zur Welt kam. Die Ungleichheit zwischen uns ist folglich, selbst wenn sie ursprünglich auf reinem Zufall beruhte, aufgrund der Tatsache gerechtfertigt, dass die *politische Beziehung* zwischen Ihnen und Ihrem Staat sich von der meinigen zu meinem Staat unterscheidet.

Warum aber ist das von Bedeutung? Ich denke, diese Beobachtung ist deshalb wichtig, weil einige der Argumente gegen ein Recht auf Ausschluss davon auszugehen scheinen, dass *jeglicher* Anteil von Zufall eine Ungleichheit bereits als ungerecht qualifiziert. So argumentiert beispielsweise Carens in seinen früheren Arbeiten, dass wir John Rawls' Vorstellung der Gerechtigkeit als Fairness global anwenden sollten und folglich die von Rawls verteidigten Bewegungsreche entsprechend für die Welt als Ganze gelten sollten. Eine solche Vorgehensweise übersieht jedoch, dass Rawls seinen Ansatz auf die spezifische Frage begrenzen wollte, wie eine politische Gesellschaft ihre mit Zwang verbundenen Gesetze und Praktiken vor allem gegenüber den in ihrem Hoheitsgebiet anwesenden Personen rechtfertigen kann. Als Rawls seine Aufmerksamkeit dem internationalen Recht zuwandte, sprach er sich mit Nachdruck gegen solche Auslegungen seiner theoretischen Überlegungen aus, die in ihnen ein Argument für die Abschaffung jeglicher, auf Zufall beruhender Unterschiede erkennen wollten – womit er zugleich auch all jenen widersprach, die meinten, aus seiner Theorie folge die Abschaffung des staatlichen Rechts auf Ausschluss. Laut Cole ist die Idee willkürlicher Grenzen eine theoretische Peinlichkeit für den Liberalismus. Allerdings haben alle diejenigen, die – wie Rawls selbst – den Liberalismus als dezidiert *politische* Theorie charakterisieren, keinen Grund, peinlich berührt zu sein. Wir können uns weiterhin fragen, was der Liberalismus im Bereich der Migration von uns verlangt. Aber willkürlich gezogene Staatsgrenzen sind bloß für diejenigen eine Peinlichkeit, die dachten, der Liberalismus fordere die Abschaffung jeglichen Zufalls. Die meisten Liberalen haben allerdings nichts dergleichen im Sinne.

Mit Blick auf einen ähnlichen Kontext ist es überdies erwähnenswert, dass wir womöglich auch ganz allgemein sensibler für die Unterscheidung von Bürger- und Menschenrechten sein sollten. Cristina Rodríguez merkt in ihrer Geschichte der Bürgerrechtsbewegung in den Vereinigten Staaten an, dass die Aufhebung dieser Unterscheidung es schwieriger gestalten könnte, das Wesen der Ansprüche von Einwanderern akkurat zu beschreiben. Ihr zufolge sollte der Versuchung widerstanden werden, die Rechte von Migrantinnen schlicht als ihrem Wesen nach identisch mit den Rechten afrikanischstämmiger Amerikanerinnen zu betrachten:

> „Das Migrationsrecht entwickelte sich in Auseinandersetzung mit den Bewegungen für Bürgerrechte und bürgerliche Freiheiten der 1960er und 1970er Jahre und es bestehen tatsächlich bedeutende Ähnlichkeiten zwischen der Situation vieler Migrantinnen heutzutage und der Situation marginalisierter Gruppen, deren Kämpfe den Ausgangspunkt der Bürgerrechtsbewegung darstellten. Viele arme, nicht-weiße Migrantinnen erledigen essenzielle, aber harte Arbeit, wobei sie zum einen den Abschiebegesetzen ausgeliefert sind und zum anderen kaum die Möglichkeit besitzen, ihre Interessen in den politischen Prozess einzubringen. Aber so wichtig diese Gemeinsamkeiten auch

> sein mögen, liegen der Aufnahme von Migrantinnen und der Bürgerrechtsbewegung doch zwei recht unterschiedliche Vorstellungen von Gerechtigkeit zugrunde. *Während die Protagonistinnen der Bürgerrechtsbewegung die Anerkennung der vollen Staatsbürgerschaft forderten, wie sie ihnen bei ihrer Geburt durch den vierzehnten Verfassungszusatz garantiert wurde, ersuchen die Migrantinnen um Einlass in ein neues Gemeinwesen, welches keine vorhergehende Verpflichtung zu einer solchen Aufnahme eingegangen ist.*“[42]

Wie im politischen Aktivismus, so ist es auch in der Philosophie. Die Ansprüche marginalisierter Bürgerinnen, die von einem Staat regiert werden und Gleichheit vor eben jenem Staat einfordern, können nicht so einfach mit den Ansprüchen von Personen gleichgesetzt werden, die außerhalb dieses Staates stehen und nicht auf diese Weise regiert werden, sondern vielmehr darum ersuchen, auf eben diese Weise regiert *zu werden*. Eine angemessene Theorie der Gerechtigkeit im Bereich der Migration müsste diesen Unterschied anerkennen. Sie würde zudem ein Verständnis von Zufall und Willkürlichkeit explizieren, das diesem Unterschied Rechnung trägt, und die ungleiche Verteilung von Rechten nicht als moralisch verdächtig betrachten, wenn diese moralisch bedeutsame Unterschiede zwischen Personen widerspiegeln.

2.2 *Verteilungsgerechtigkeit*

Wir können mit diesen Gedanken nun im Hinblick auf das Argument der Verteilungsgerechtigkeit fortfahren. Zu Beginn möchte ich jedoch bemerken, dass ich selbst nicht über die Kompetenz verfüge, irgendwelche Aussagen bezüglich der empirischen Effekte treffen zu können, die der Ausschluss von Migrantinnen mit sich bringt; die Ökonomie der Migration ist, um es vorsichtig auszudrücken, ein kontrovers diskutiertes Feld und zudem sollte sich niemand an einen Philosophen wenden, wenn es um Fragen empirischer Zusammenhänge geht.[43] Aus einer philosophischen Perspektive heraus kann ich jedoch so viel sagen: Selbst wenn wir unter Bezug auf die Idee der Verteilungsgerechtigkeit bestimmte Migrationsrechte rechtfertigen könnten, müssten wir immer noch dasjenige Bündel von Rechten identifizieren, das, moralisch betrachtet, die besten Resultate erzielt – und es ist unklar, ob aus einer Politik der offenen Grenzen eben solche Ergebnisse folgen würden. So hat zum einen Peter Higgins angeführt, dass offene Grenzen dazu neigen könnten, diejenigen zu benachteiligen, die aufgrund von körperlichen Einschränkungen, Alter oder sozialer Marginalisierung weniger mobil sind als andere.[44] Eine auf der Idee der Verteilungsgerechtigkeit beruhende Theorie der Migration muss daher zwar auf eine radikale Veränderung des geltenden Migrationsrechts bestehen; allerdings ist nicht klar, ob diese Veränderung die Abschaffung eines Rechts auf Ausschluss umfassen muss.

An dieser Stelle können wir zwei grundsätzlichere Herausforderungen für das Argument der Verteilungsgerechtigkeit anführen. Die erste fragt recht simpel, warum wir uns überhaupt um internationale Ungleichheit kümmern sollten. Das bedeutet selbstredend nicht, dass ich der Meinung wäre, wir sollten uns überhaupt nicht um sie kümmern; ich habe hier keinen Grund angeführt, aus dem zu folgern wäre, dass internationale Ungleichheit für unsere Erwägungen keine Rolle spielen sollte – und ziemlich sicher scheint globale *Armut* relevant zu sein und zwar ganz unabhängig davon, ob der Kluft zwischen Arm und Reich eine besondere Bedeutung zukommt oder nicht. Die von mir angeführte Herausforderung besagt bloß, dass es einer Begründung bedarf, *warum* Ungleichheit von Bedeutung ist – und dass es in meinen Augen nicht selbstverständlich ist, dass die besten derzeit verfügbaren Antworten auf diese Frage schlicht vom innerstaatlichen auf den internationalen Raum übertragen werden können. Rawls stellte mit Nachdruck fest, dass sein strenges Prinzip der Verteilungsgerechtigkeit – das Differenzprinzip – nur *innerhalb* des Staates anwendbar sei; es kann daher nicht legitimerweise zwischen Staaten oder über sie hinweg angewendet werden. Damit möchte ich, wie gesagt, nicht den Eindruck erwecken, die Idee der Verteilungsgerechtigkeit zwischen Staaten wäre bedeutungslos. Vielmehr will ich bloß zeigen, dass es einer Rechtfertigung dafür bedarf, *warum* sie eine Rolle spielen sollte. Wenn uns gesagt wird, dass das Menschenrecht auf Bewegungsfreiheit anhand eines Rechts auf einen Anteil am globalen wirtschaftlichen Wohlstand begründet werden kann, dann ist die Frage berechtigt, was diesem Argument seine moralische Kraft verleiht.

Die letzte grundlegende Herausforderung für die Argumente der Verteilungsgerechtigkeit besteht jedoch in einer etwas anderen Frage: Falls die Idee der Verteilungsgerechtigkeit von Bedeutung ist, wie würde sie sich dann zu anderen Normen politischer Gerechtigkeit verhalten, darunter der Idee der Selbstbestimmung? Verteilungsgerechtigkeit, um es klar zu sagen, ist nicht die einzig gültige politische Norm, sei es auf innerstaatlicher oder internationaler Ebene. Andere Rechte – wie beispielsweise das Recht, seine eigenen Angelegenheiten selbstbestimmt regeln zu dürfen – sind ebenfalls von Bedeutung. Selbst wenn gezeigt werden könnte, dass offene Grenzen die globale Verteilung von Gütern gerechter gestalten würden, könnten wir daraus nicht folgern, dass offene Grenzen *verpflichtend* seien. Stellen Sie sich zur Veranschaulichung das folgende Szenario innerhalb eines Staates vor: Ein bestimmtes Land wird vernünftig und gerecht regiert, betreibt jedoch eine recht schlichte (wenngleich populäre) Finanzpolitik, durch die eine große Menge an Goldbarren in der Zentralbank des Landes gelagert wird. Stellen Sie sich nun vor, dass Robin Hood erscheint, das Gold aus der Zentralbank befreit und es den Armen des Landes gibt – im Ergebnis ist die Verteilung des Wohlstands nun derjenigen näher gekommen, die unserer besten Vorstellung liberaler politischer Gerechtigkeit entspricht. Ist das Land nun moralisch verpflichtet, den Armen das Gold zu überlassen? Ich vermute, dass die Meinungen hier auseinander gehen werden, aber für mich lautet die Antwort: Nein. Verteilungsgerechtigkeit ist eine wichtige Norm, aber so verhält es sich auch mit der Idee der Selbstbestimmung, und Robin Hood ist nicht dazu berechtigt zu entscheiden, wie die Finanzpolitik eines Landes gestaltet werden sollte. Staaten – oder zumindest demokratische Staaten – haben ein Recht auf Dummheit, so lange sie sich im Rahmen der Menschenrechte bewegen.

Daraus folgt jedoch, dass eine bestimmte Politik nicht bloß deshalb als *zwingend* gilt, weil sie zu einer gerechteren Verteilung von Gütern führen würde. Der Gedanke, dass offene Grenzen die Welt einer gerechten Verteilung näherbringen würden, stellt daher kein vollständiges Argument für offene Grenzen, sondern vielmehr erst dessen Ausgangspunkt dar.

Über all dies könnte sicherlich noch sehr viel mehr gesagt werden. Ich denke, die Befürworterinnen offener Grenzen werden in der Lage sein, weitere Argumente dafür zu entwickeln, warum der Verteilungsgerechtigkeit größeres Gewicht zukommen sollte als Werten wie der Selbstbestimmung oder dem mutmaßlichen Recht darauf, die Verantwortung für den Schutz der Menschenrechte eines Dritten zurückzuweisen, das ich in meinem eigenen Ansatz anführe. An dieser Stelle möchte ich jedoch nur darauf bestehen, dass solche Argumente vorgebracht werden *müssen,* um den auf der Idee der Verteilungsgerechtigkeit beruhenden Überlegungen Geltung zu verschaffen. Nun möchte ich allerdings mit der Untersuchung des Arguments der Chancengleichheit fortfahren, da es in meinen Augen die erfolgversprechendste Version des Arguments der Verteilungsgerechtigkeit darstellt. Hierzu werde ich Kieran Obermans Überlegungen genauer untersuchen, dessen Argumente sehr gut entwickelt und daher einer ausführlicheren Betrachtung wert sind.

Oberman formuliert sein Argument der Chancengleichheit als Antwort auf die Behauptung David Millers, wir hätten kein Recht auf die größtmögliche, sondern bloß auf eine angemessene Menge an Lebensmöglichkeiten. Hiergegen verteidigt Oberman die Idee, dass alle Menschen einen Anspruch auf die größtmögliche Menge an verfügbaren Möglichkeiten haben, was er als „vollen Umfang verfügbarer Optionen zur persönlichen Lebensgestaltung" bezeichnet.[45] Er argumentiert gegen Miller, dass eine Person den Anspruch auf diesen vollen Umfang aufgrund der *persönlichen* und *politischen* Interessen besitzt, die den Kern des eigenen Lebens bilden. Darunter fällt zum Beispiel, dass ich zur Umsetzung eines bestimmten Lebensplans in ein anderes Land ziehen oder zu dem Zweck migrieren möchte, mehr darüber zu erfahren, wie andere Länder Politik betreiben. Laut Oberman käme die Verweigerung des Rechts auf eine derart motivierte Migration daher der Beseitigung verfügbarer Optionen innerhalb einer politischen Gemeinschaft gleich:

> „Sollte das Menschenrecht auf Bewegungsfreiheit die Menschen bloß zu einem ‚adäquaten' Umfang von Optionen der Lebensgestaltung berechtigen, würde den Bewohnerinnen der Staaten, in denen eine größere Menge solcher Optionen zur Verfügung steht, kein Menschenrecht auf freie Bewegung über das gesamte Gebiet ihres eigenen Staates hinweg zukommen. Sollte beispielsweise Belgien einen ‚adäquaten' Umfang solcher Optionen bieten, die Vereinigten Staaten im Vergleich dazu aber eine Vielzahl mehr, stünde Personen innerhalb der Vereinigten Staaten deutlich mehr als eine ‚adäquate' Menge an Optionen zur persönlichen Lebensgestaltung zur Verfügung. Bestünde allerdings bloß ein Recht auf eine ‚adäquate' Menge solcher Optionen, könnten die Vereinigten Staaten ihr Staatsgebiet in hunderte Parzellen von der Größe Belgiens aufteilen und jede einzelne Grenze zwischen den Parzellen mit Wachpersonal und Stacheldraht ausstatten, ohne dadurch das Menschenrecht auf Bewegungsfreiheit zu verletzen."[46]

Eine solche Position wäre laut Oberman genauso einfältig wie der Gedanke, dass ein Verbot der jüdischen Religion so lange keine Verletzung der Religionsfreiheit darstelle, wie eine „angemessene" Auswahl anderer Religionen zur Verfügung stehe. Oberman legt nahe, dass in beiden Fällen die Chancengleichheit nur dann wirklich angemessen erfüllt ist, wenn jede Person ein Recht auf den *vollen* Umfang möglicher Optionen zur persönlichen Lebensgestaltung besitzt, was wiederum bedeutet, dass *jeglicher* Ausschluss mutmaßlich ungerecht ist.

Ein solcher Gedankengang scheint jedoch die moralische Struktur von Bürgerrechten misszuverstehen. Zunächst erscheint es schlicht falsch, zu denken, dass jede Person ein Recht auf den *maximalen* Umfang von Möglichkeiten hat. Stellen Sie sich zum Beispiel vor, dass ein bestimmtes Land vor der Wahl steht sich zu industrialisieren oder weiterhin eine landwirtschaftliche Lebensweise zu pflegen. Auch wenn ich nicht sagen kann, wie in einem solchen Falle zu entscheiden wäre, sehe ich es als unwahrscheinlich an, dass dabei allein die *Quantität* der Möglichkeiten eine Rolle spielen sollte. Die Tatsache, dass mit der Industrialisierung mehr Möglichkeiten für die persönliche Lebensführung einhergehen, stattet uns daher nicht mit hinreichenden Gründen dafür aus, ein mögliches Ausbleiben der Industrialisierung als ungerecht zu bezeichnen.

Was stattdessen zählt, ist nicht der Umfang an Möglichkeiten, sondern *warum* sie eben diesen Umfang besitzen. Nehmen sie beispielsweise Belgien. Sollten sich die Vereinigten Staaten tatsächlich in kleinere Einheiten aufteilen, jede von ihnen bewacht von Menschen mit Schusswaffen, wäre ich ebenfalls der Meinung, dass es sich hierbei um einen eher ungerechten Vorgang handelt. Aber der Grund für dieses Urteil hat *rein gar nichts* mit der Frage zu tun, ob die Möglichkeiten der persönlichen Lebensgestaltung in Belgien angemessen sind. Stattdessen ist ausschlaggebend, dass es einem Staat nicht gestattet ist, seine Bevölkerung so zu behandeln, wie die Regierung der Vereinigten Staaten die auf amerikanischem Boden anwesenden Personen behandelt, und diese Personen dann zugleich davon abzuhalten, sich frei auf eben jenem Boden zu bewegen. Wenn wir die Rechtfertigung der innerstaatlichen Bewegungsfreiheit stärker als die Rechtfertigung eines Bürgerrechts verstehen, was Oberman nicht zulässt, dann besteht die Ungerechtigkeit darin, dass die mir zur Verfügung stehenden Möglichkeiten persönlicher Lebensgestaltung auf den Umfang belgischer Optionen reduziert wurden *und zugleich* seitens des Staates darauf bestanden wird, mich von der Hauptstadt aus zu regieren. Adam Hosein hat ein ähnliches Argument vorgebracht: Im Hinblick auf innerstaatliche Bewegungsfreiheit ist nicht die Bewegungsfreiheit als solche von Bedeutung, sondern wie die Art der *Einschränkungen* dieser internen Bewegungsfreiheit vom Staat dazu genutzt werden könnte, Bürgerinnen ungerechtfertigterweise ungleich zu behandeln.[47] Um Obermans Beispiel abzuwandeln: Sollten die Vereinigten Staaten in eine Vielzahl souveräner Territorien von der Größe Belgiens aufgeteilt werden, jedes von ihnen geschützt durch Stacheldraht und bewaffnete Grenzposten, würde ich das bedauerlich finden. Aber meine Reaktion auf diesen Fall wäre eindeutig eine andere, wenn die Vereinigten Staaten *sowohl* das Recht für sich beanspruchen würden, mich mittels Zwangsgewalt zu regieren, *als auch* meine Bewegungsfreiheit auf amerikanischem Boden einschränkten. Der erste Fall scheint mir bedauerlich, aber nicht in sich ungerecht. Im Gegensatz dazu erscheint mir der zweite Fall tatsächlich als ungerecht. Wenn die Regierung der Vereinigten Staaten das Recht beansprucht, mich zu regieren, so kann sie dies nicht gerechterweise beanspruchen, während sie mich an der freien Bewegung innerhalb der USA hindert.

Ich denke, dass ähnliche Erwägungen auf die Überlegungen zur Religionsfreiheit zutreffen. Oberman und ich stimmen darin überein, dass ein Verbot der jüdischen Religion moralisch falsch wäre. (Tatsächlich denke ich, dass es sich hierbei um einen der bereits im vorherigen Kapitel erwähnten grundlegenden Orientierungspunkte handelt.) Die Rechtfertigung dieser Behauptung aber hat weit weniger mit der Zahl mir verfügbarer Religionen zu tun, als mit der Historie staatlicher Versuche, manche Bürgerinnen als moralisch minderwertig im Vergleich zu ihren Mitbürgerinnen zu behandeln. Ich denke, es besteht ein Recht darauf, dass die eigenen religiösen Überzeugungen als ebenso moralisch bedeutsam anerkannt werden wie diejenigen der Mitbürgerinnen, und dass daraus ein Recht auf Religionsfreiheit erwächst. Es ist folglich in der Tat ungerecht, wenn ein Staat eine bestimmte Religion verbietet – allerdings beruht diese Ungerechtigkeit darauf, wie der Staat uns in diesem Fall im Vergleich zu unseren Mitbürgerinnen behandelt. Auch hier besteht die Frage nicht darin, wie groß der Umfang an Optionen hinsichtlich des eigenen religiösen Glaubens ist; sie besteht darin, wie und warum ein mit Zwangsgewalt ausgestatteter Staat eine Option aus der Auswahl entfernt.

Anders ausgedrückt: Religiöse Intoleranz kann durchaus verurteilt werden ohne dabei annehmen zu müssen, dass ein Recht auf den größtmöglichen Umfang von Optionen hinsichtlich der Wahl der eigenen Religionsgemeinschaft besteht. Das genannte Beispiel ist jedoch auch in anderer Hinsicht lehrreich. Aus der Religionsfreiheit wird oft ein Recht auf Freiheit vor staatlichen Eingriffen in die Ausübung der jeweiligen Religion abgeleitet. Allerdings folgt aus der Religionsfreiheit im Allgemeinen kein Recht auf die für diese Ausübung notwendigen Mittel, oder auf die Mitwirkung von Personen, deren Beitrag für diese Ausübung notwendig erscheinen mag. Wenn meine Religion beispielsweise von mir verlangt, eine kostspielige Pilgerreise auf mich zu nehmen, ist nicht direkt ersichtlich, warum die Idee der Religionsfreiheit den Staat dazu verpflichten sollte meine Reise zu finanzieren.[48] Wenn mein Gottesdienst eine *Minjan* von zehn Leuten verlangt, aber bloß neun Personen Teil meiner religiösen Gemeinschaft sind, kann ich nicht dem Staat gegenüber darauf bestehen, mich mit der zehnten Person zu versorgen, sei es durch Einwanderung oder (vielleicht) Anreizen zur Konversion. Die Freiheit der Religionsausübung umfasst kein Recht auf die Nutzung fremder Körper oder Ressourcen.

Diese Überlegungen sind im Hinblick auf die Analyse der Religionsfreiheit wohl nicht übermäßig umstritten. Meiner Meinung nach sollten wir uns ihrer allerdings auch in der Diskussion über Migration bewusst sein. Oberman klingt an vielen Stellen so, als wären Migrationseinschränkungen ebenso zu betrachten wie Fälle, in denen die Ausübung einer Religion durch den Einsatz von Zwangsgewalt verhindert wird. In meinen Augen handelt es sich jedoch beim Ausschluss von Migrantinnen um eine komplexere Angelegenheit, weshalb er in manchen Fällen eher mit der Verweigerung verglichen werden sollte, einen bestimmten Vorteil zu gewähren. Wenn eine Person in ein anderes Hoheitsgebiet einwandert, erschafft sie – wie ich in Kapitel vier zeigen werde – neue Formen der Verpflichtung aufseiten derjenigen, die bereits in diesem Hoheitsgebiet leben. Möchte eine Person ein Hoheitsgebiet betreten, ist damit weit mehr verbunden als schlicht die Überquerung einer Linie auf dem Boden; vielmehr versucht sie durch ihren Grenzübertritt, neue Verpflichtungen aufseiten der bereits dort lebenden Bürgerinnen zu etablieren. Sie vom Grenzübertritt abzuhalten könnte daher in manchen Fällen am besten analog dazu verstanden werden, einer Person die für die Ausübung ihrer Religion notwendigen Mittel zu verweigern, statt analog dazu, ihr diese Ausübung mittels Zwang zu verwehren. Während wir zuweilen verpflichtet sind, solche Mittel bereitzustellen, muss doch jeweils gezeigt werden, warum dies der Fall sein sollte. Ein solcher Nachweis ist durchaus möglich. So denken viele von uns, dass, um einen aktuellen Fall aufzugreifen, die Hersteller von Hochzeitstorten ihre Dienste sowohl für hetero- als auch homosexuelle Hochzeiten zur Verfügung stellen sollten.[49] Aber auch hier muss das Argument Bezug auf eine Idee wie die der Gleichheit vor dem Staat nehmen. Im Falle der Hochzeitstorten kann daher vorgebracht werden, dass eine solche Verweigerung aufseiten der Bäcker zur Stigmatisierung homosexueller Beziehungen beitragen würde. Allerdings findet sich in dieser Überlegung nichts, was ein Menschenrecht auf eine Hochzeitstorte begründen würde. Sollten alle Bäcker auf einen Schlag aus den Vereinigten Staaten verschwinden, würden keinerlei Rechte verletzt. Wir können daher folgern, dass die Religionsfreiheit auch ohne ein Recht auf den größtmöglichen Umfang von Optionen verteidigt werden kann – oder, wie in diesem Fall, ohne ein Recht auf eine Hochzeitstorte.

Ein letzter Punkt: Oberman nennt als Grundlage seiner Zurückweisung eines Rechts auf Ausschluss bestimmte Interessen wie „[das] Treffen persönlicher Entscheidungen und politisches Engagement". Selbst wenn ich dazu bereit wäre, die erstgenannte Art von Interessen zu akzeptieren, muss ich doch zugeben, dass mich die zweite Art etwas irritiert. Um es einfach auszudrücken: Politik ist kein spaßiger Zeitvertreib und existiert auch nicht allein für das eigene, private Selbstverständnis, sondern beinhaltet notwendig den Einsatz von Zwang gegenüber anderen. Um Cheshire Calhoun zu paraphrasieren: Meine politischen Überzeugungen sind von anderer Art als meine persönlichen Überzeugungen; sie sind Überzeugungen im Hinblick darauf, was *wir* sein sollten, und nicht im Hinblick darauf, was *ich* sein sollte.[50] Diese unterschiedlichen Arten von Überzeugungen im Kontext der Migration als gleichwertig zu betrachten scheint in meinen Augen zu einer für das Problem ungeeigneten Lösung zu führen. Migration bedeutet, wie Kukathas schreibt, in einer neuen Gesellschaft „anzukommen, zu bleiben und teilzuhaben". Warum sollten wir ein Recht auf *all das* haben, wenn der Zweck solchen Handelns in diesem Fall bloß darin besteht, herauszufinden, wie Politik betrieben wird? Migration bedeutet, an einen neuen Ort zu ziehen und ihn zu seinem eigenen zu machen. Es erscheint seltsam, zu glauben, dass wir dieses Recht aufgrund der Bedeutung politischen Handelns haben sollten. Wenn die Grundlage des von Oberman angeführten Interesses grundsätzlich informativer Natur sein sollte, dann halte ich es für ungeeignet, um ein Recht auf Migration zu verteidigen. Wir denken im Allgemeinen nicht, dass ein Recht darauf, in eine Beziehung mit anderen zu treten, mit dem Erlangen von *Informationen* begründet werden kann. Ich mag ein großes Interesse daran haben, wie Universitäten geleitet werden und aus diesem Grunde auch ein starkes Interesse daran, zu erfahren, wie die University of Miami den Fachbereich Philosophie führt.[51] Allerdings scheint nichts davon ausreichend, um mir ein Recht auf eine Stelle an der University of Miami zuzusprechen. Aus der Tatsache, dass meine Anwesenheit in dieser Gemeinschaft mich mit für mich wertvollen Informationen versorgen würde, folgt nicht, dass die derzeitigen Mitglieder dieser Gemeinschaft eine Verpflichtung hätten, mich in ihrem Kreis willkommen zu heißen.

2.3 Kohärenz mit bestehenden Bewegungsrechten

Wir können die Diskussion dieses Argumentationsmusters mit Carens' Analyse innerstaatlicher Migration beginnen. Ihm zufolge müssen die Gründe für den starken Schutz innerstaatlicher Migration, von welcher Art auch immer sie sein mögen, auch für die Frage internationaler Mobilität gelten; die der innerstaatlichen Bewegungsfreiheit zugrunde liegenden Interessen müssten demnach ebenso stark auch im internationalen Raum berücksichtigt werden. Carens führt keine bestimmte Vorstellung von der Beschaffenheit dieser Interessen an – tatsächlich macht er diesbezüglich *bewusst* keine Ausführungen –, sondern bemerkt bloß, dass jede beliebige Begründung, die für Bewegungsfreiheit angeführt werden könnte, sowohl innerstaatlich als auch international gelten müsste.[52] Der zentrale Punkt besteht Carens zufolge darin, dass jedes Interesse, welches eine der beiden Formen der Bewegungsfreiheit rechtfertigt, ebenso auch für die andere Form gilt.

Die Antwort auf dieses Argument besteht jedoch in dem Hinweis darauf, dass das *Interesse* daran, mobil zu sein, bloß ein Teil der umfassenderen Rechtfertigung innerstaatlicher Bewegungsfreiheit ist. Der andere Teil besteht in der Frage, ob eine Person oder Institution, die uns dieses Recht verweigert, auch das moralische Recht dazu hat. Wir betrachten im Allgemeinen nicht nur die Interessen einer Seite, wenn wir moralische Rechte zu bestimmen versuchen. Ihr Interesse daran, ein öffentliches Gelände zu betreten, mag dem Interesse ziemlich ähnlich sein, aufgrund dessen Sie ein privates Gelände betreten möchten. Daraus folgt jedoch nicht, dass es keinen moralisch bedeutsamen Unterschied zwischen diesen Fällen gibt. Ebenso mag Ihr Interesse an Ihrem Abendessen von gleicher Art sein wie *mein* Interesse an Ihrem Abendessen. Allerdings denken wir nicht, dass wir beide aufgrund dieses Umstands einen gleichwertigen Anspruch auf dieses Abendessen hätten.

Wie aber können uns diese Ideen bei einer Antwort auf das „Kranarm"-Argument helfen? Um zu sehen, warum sich internationale von innerstaatlicher Bewegungsfreiheit unterscheidet, müssen wir uns bloß anschauen, was der Staat in beiden Fällen zu tun gedenkt, wenn er diese Freiheit einschränken will. Wie bereits erwähnt, scheinen sich die zwei Fälle in dieser Hinsicht zu unterscheiden. Im ersten Falle nutzt der Staat seine Zwangsgewalt, um Menschen am Eintritt in dasjenige Hoheitsgebiet zu hindern, in dem er der Souverän ist. Im zweiten Fall nutzt der Staat ebenfalls Zwang, allerdings um über einzelne Personen zu regieren und ihnen *zugleich* vorzuschreiben, dass sie kein Recht darauf besitzen, sich innerhalb seines Hoheitsgebiets frei zu bewegen. Aus Perspektive der Moral sind diese Szenarien nicht identisch. Selbst wenn wir in beiden Fällen ähnliche Gründe für unseren Wunsch, uns frei zu bewegen, anführen könnten – wir möchten vielleicht eine neue Arbeit antreten oder eine Liebesbeziehung führen – würden wir daraus nicht schließen, dass dem Staat in beiden Fällen die Einschränkung unserer Bewegungsfreiheit auf gleiche Weise untersagt ist. Wir gehen davon aus, dass es ein Bürgerrecht gegenüber dem Staat gibt, von solcherlei Einschränkungen frei zu sein, und dass es denjenigen zukommt, die der Staat zu regieren beansprucht. Kurzum: das Recht auf innerstaatliche Bewegungsfreiheit ist ein Bürgerrecht. Dieses Bürgerrecht mag wiederum auf einem Menschenrecht beruhen – vielleicht einem Menschenrecht darauf, von einem Staat regiert zu werden, der es unterlässt, die interne Bewegungsfreiheit mittels Zwang einzuschränken. Aber das ändert nichts daran, dass dieses Recht korrekterweise nur von denen in Anspruch genommen werden kann, die sich bereits in Reichweite der staatlichen Zwangsbefugnis befinden.

Wir können allerdings darüber hinaus auch die spezifischeren Argumente in Carens' „Kranarm"-Argument untersuchen. Carens nennt fünf Arten, auf die innerstaatliche und internationale Bewegungsfreiheit unterschieden werden können, um dann zu zeigen, dass keine dieser Unterscheidungen funktioniert. Die von ihm genannte Differenz, die meiner soeben ausgeführten Erwiderung am nächsten kommt, ist die Idee, dass Bewegungsfreiheit ein auf Mitgliedschaft beruhendes Recht ist bzw. ein Recht darauf, in einem Staat zu leben, dessen Rechtsprechung frei von Diskriminierung ist. Carens weist diese Unterscheidung zurück und es lohnt sich, genauer zu betrachten, wie auf sein Argument reagiert werden kann.

Zunächst führt Carens aus, dass die Geltung der einem Recht zugrunde liegenden Interessen als unabhängig von der Staatsbürgerschaft einer Person begriffen werden sollte. „Aus der Perspektive des Individuums", schreibt er, „ist innerstaatliche Bewegungsfreiheit aus einer Vielzahl von Gründen wichtig, die keinerlei Bezug zur politischen Mitgliedschaft haben."[53] Wir sollten daher annehmen, dass Nicht-Bürgerinnen wie auch Bürgerinnen ein Recht darauf haben, sich frei innerhalb eines Landes zu bewegen. Das Problem hiermit ist jedoch erneut, dass der für den Wunsch nach Bewegungsfreiheit ausschlaggebende Grund uns nicht alles darüber mitteilt, warum diese Freiheit von Bedeutung sein sollte. So könnte ich beispielsweise die Religionsfreiheit aus dem Grund befürworten, dass ich Kirchenarchitektur und Choräle liebe und diese daher gerne erhalten wissen würde. (Was, nebenbei gesagt, tatsächlich zutrifft.) Allerdings besteht in der Begründung der Religionsfreiheit *keinerlei* Bezug zu diesen individuellen Interessen. Stattdessen spielen dabei spezifischere Gründe bezüglich der Historie staatlicher Verurteilung und Marginalisierung von Mitgliedern missliebiger Glaubensgemeinschaften eine Rolle. Die Bedeutung der Religionsfreiheit beruht auf dieser Geschichte staatlicher Angriffe auf religiöse Praktiken. Wir verteidigen die Religionsfreiheit nicht einfach deshalb, weil das Interesse an Religion stark ist, sondern aufgrund einer Vergangenheit, in der die Menschen- und Bürgerrechte der Anhänger bestimmter Religionsgemeinschaften allzu oft missachtet wurden. Ebenso sollten wir die interne Bewegungsfreiheit aus Gründen verteidigen, die sich speziell auf die Legitimation politischen Zwangs beziehen – und nicht einfach auf die Stärke der Interessen, die von einem solchen Recht verteidigt würden.

Einen solchen Zusammenhang zwischen politischer Gleichheit und innerstaatlicher Bewegungsfreiheit weist Carens mit wenigen Sätzen zurück: Innerstaatliche Bewegungsfreiheit sei „ein zu umfassendes Recht", als dass ihr vorrangiger Zweck in der Vermeidung von Diskriminierungen bestehen sollte.[54] Ich denke nicht, dass dieses Argument funktioniert. Andere, tendenziell stärkere – und umfassendere – Rechte sind häufig in vergleichsweise eng gefassten Zwecken begründet. Nehmen Sie zum Beispiel die Verteidigung der Rede- und Meinungsfreiheit John Stuart Mills, der diese Freiheit mit der Suche nach Wahrheit in persönlichen und politischen Diskursen rechtfertigt. Diese Verteidigung der freien Rede schützt aber nicht nur komplexe und erkenntnisreiche Diskussionen über die politische Ethik, sondern auch Groschenromane, Katzenblogs und die Filme von Michael Bay. Mit anderen Worten: Selbst eine eng umrissene Rechtfertigung kann ein umfassendes Recht begründen, vor allem dann, wenn uns die natürliche Tendenz des Staates, die Grenzen seiner legitimen Herrschaft zu überschreiten, Sorge bereitet.

Ich möchte anmerken, dass Carens' Position eine unter den Befürworterinnen offener Grenzen übliche Vorstellung von der Bewegungsfreiheit darstellt; sie verstehen Bewegungsfreiheit als ein vorpolitisches Recht. Dieser Vorstellung nach ist es so, als ob wir so lange dazu berechtigt gewesen wären, uns frei über die Oberfläche der Erde zu bewegen, bis Staaten entstanden und uns Mauern und Hindernisse in den Weg stellten. Dabei handelt es sich selbstredend bloß um eine bestimmte Art, die moralische Grundlage der Bewegungsfreiheit zu verstehen, und meiner Meinung nach handelt es sich dabei nicht um die einzig mögliche Betrachtungsweise – ganz abgesehen davon, dass ich sie auch einfach nicht besonders ansprechend finde. Die Vorstellung von einem Staat, der unsere vorpolitischen Rechte auf Bewegungsfreiheit einschränkt, scheint mir keine angemessene Konzeption der normativen Situation vor der Entstehung des Staates darzustellen. In einer vorpolitischen Situation wäre es unsere erste Aufgabe, eine Welt zu schaffen, in der Politik überhaupt erst möglich wird. Das Recht, sich innerhalb eines Staates frei zu bewegen, besteht meiner Meinung nach zu Recht gegenüber jedem Staat und zwar aus Gründen politischer Rechtfertigung. Aber in meinen Augen gibt es keinen Grund zu meinen, die Bewegungsfreiheit selbst sei auf die Art moralisch relevant, wie Carens und andere es behaupten.

Ähnliche Überlegungen können in meinen Augen auch gegen das Argument vorgebracht werden, dass aus der Kohärenz mit einem Recht auf Emigration ein präsumtives Recht gegen Ausschluss folgt. Wie ich bereits an anderer Stelle gezeigt habe, hat ein Staat kein Recht darauf, seine Einwohnerinnen mittels Zwangsgewalt in seinem Hoheitsgebiet zu halten, wenn andernfalls diese Einwohnerinnen frei wären, den Zugriffsbereich des staatlichen Zwangs zu verlassen und einer anderen politischen Gemeinschaft beizutreten.[55] Der Staat sollte Personen nicht durch die Androhung von Waffengewalt dazu zwingen, weiterhin mit ihm in einer politischen Beziehung zu stehen. Daraus folgt jedoch nicht, dass es ihm niemals gestattet ist, durch die Androhung solcher Gewalt Menschen am *Beitritt* zu einer solchen Beziehung zu hindern. Es verhält sich hierbei genauso wie wenn ich meine Haustür abschließe, um Sie in meinem Haus zu halten, oder wenn ich mit dem Abschließen bezwecke, Sie aus meinem Haus herauszuhalten. Letzteres ist moralisch in vielen Fällen zulässig, ersteres wird hingegen allgemein als moralisch falsch angesehen.

David Miller nutzt diese Ideen, um die Kohärenz eines Systems an Rechten zu verteidigen, dem zufolge einer Person zwar ein Recht auf Auswanderung aus seinem Heimatland zukommt, sie jedoch zugleich kein Recht darauf besitzt, ein Land ihrer Wahl zu betreten.[56] Christopher Heath Wellman nutzt die Analogie des Rechts auf Heirat – also das Recht darauf, eine Person zu heiraten, die freiwillig zustimmt, *mich* zu heiraten. Das Recht zu heiraten gewährleistet folglich kein Recht auf die Ehepartnerin oder den Ehepartner eigener Wahl, da keine Person einen Anspruch auf die Körper oder die Arbeit anderer zur Umsetzung seines eigenen Lebensplans besitzt.[57]

In seiner Verteidigung der Inkompatibilität eines Rechts auf Auswanderung mit einem Recht auf Ausschluss weist Cole diese Analogien mit dem Argument zurück, dass eine Person auch ohne Eheschließung ein gutes Leben führen könne, ganz im Gegensatz zu einem Leben ohne Staat:

> „Niemand muss jemals in eine Ehe, einen Golfclub oder irgendeine der anderen Vereinigungen eintreten, auf die im Kontext der Migrationsdebatte häufig Bezug genommen wird. Aus diesem Grund ist es in diesen Fällen plausibel, anzunehmen, dass ein Recht auf Austritt nicht mit einem Recht auf Zutritt einhergeht, da hier der Austritt nicht notwendig von einem Zutritt andernorts abhängt. Eine Person kann vom Recht auf Austritt aus einer dieser Vereinigungen Gebrauch machen ohne einer anderen Vereinigung beizutreten, und dabei ist besonders auf den ‚Raum' außerhalb der einzelnen Vereinigungen zu verweisen, der frei betretbar ist und in dem jede Person ihre eigenen Lebensvorstellungen, so sie denn mag, gut verfolgen kann. Das ist jedoch auf drastische und bedeutsame Weise nicht der Fall, wenn es um Nationalstaaten geht. [...] Es gibt zwar einen ‚Raum' der Staatenlosigkeit, aber diesen will kein

> Mensch betreten – er ist zutiefst problematisch, gefährlich und macht es unmöglich, eigene Lebensvorstellungen überhaupt zu entwickeln. Während es also plausibel ist, ein Recht auf Austritt ohne ein damit einhergehendes Recht auf Zutritt im Falle von Vereinigungen wie der Ehe oder bei Golfclubs anzunehmen, da es hier für den Austritt nicht zwingend eines Zutritts zu einer anderen Vereinigungen bedarf, liegt der Fall bei Nationalstaaten gänzlich anders, da hier das Recht auf Auswanderung den Zutritt zu einem anderen Gemeinwesen voraussetzt. Daher ist es in diesem Fall plausibel, dass aus dem Recht auf Austritt auch ein Recht auf Zutritt folgt."[58]

Für Cole setzt die *Existenz* eines Rechts auf Austritt daher eine tatsächlich verfügbare Möglichkeit dieses Austritts voraus. Das erscheint jedoch seltsam: Ich kann durchaus die Freiheit haben, etwas zu tun, ohne dabei auch über die konkreten Mittel zu verfügen, die für die Nutzung dieser Freiheit vonnöten sind. Rawls' Unterscheidung zwischen der Freiheit und dem Wert der Freiheit dient exakt dem Zweck, diese Differenz hervorzuheben; wir wollen zwischen der Freiheit, etwas zu tun, und meiner Verfügung über die dafür notwendigen Mittel unterscheiden können. Es erscheint daher schlicht falsch, anzunehmen, dass das Recht auf Austritt nur dann existiert, wenn eine Person dieses Recht auch tatsächlich nutzen kann – so wie es falsch ist, zu meinen, dass mein Recht auf Religionsfreiheit nur dann existiert, wenn ich die von meinem Glauben geforderte teure Pilgerreise unternehmen kann. Diesen Umstand erkennt selbst Carens an: Das Recht auf freie Migration, so bemerkt er, ist es wert, um seiner selbst willen verteidigt zu werden – selbst für diejenigen, die (noch) nicht über die Mittel verfügen, dieses Recht auch tatsächlich zu nutzen.[59]

Wir können daher die moralische Relevanz eines Rechts auf Austritt auch in einer Welt verteidigen, in der der Nutzung dieses Rechts mitunter bedeutende Hindernisse entgegenstehen. Dafür muss sich der Liberalismus nicht schämen; vielmehr ist es ein weiterer Beweis dafür, dass liberale Theoretikerinnen zwischen Rechten und den für ihre Nutzung nötigen Mitteln unterscheiden sollten. Ich möchte an dieser Stelle noch einmal wiederholen, dass das Verständnis von Rechten häufig nicht einfach unter Bezugnahme auf individuelle Interessen verstanden wird, sondern mit Blick auf die Akteure, die gedenken, im Rahmen oder entgegen dieser Interessen zu handeln. Das Recht auf Emigration ist das Recht darauf, nicht aufgrund von *Zwang* in einer dauerhaften Beziehung mit dem eigenen Staat verbleiben zu müssen. Dem Staat sollte es, einfach ausgedrückt, nicht erlaubt sein, diese Beziehung mittels Gewalt aufrechtzuerhalten. Aber daraus folgt nicht, dass wir einen Anspruch auf die Mittel haben, die für die Migration zwischen Staaten benötigt werden.

2.4 Zwangsgewalt

Meiner Antwort auf dieses Argument möchte ich die Bemerkung voranstellen, dass das Konzept des Zwangs im Allgemeinen nicht als eigenständiger Grund für offene Grenzen angeführt wird. Anzuerkennen, dass wir an den Grenzen Zwangsgewalt ausüben, zielt vielmehr darauf ab, die Annahme zu erschüttern, Ausschluss wäre moralisch unproblematisch. Wir bedrohen Individuen an der Grenze und müssen ihnen daher vielleicht auch eine Rechtfertigung für unser Handeln anbieten. Diese Tatsachen nutzt Abizadeh, um auf der Unzulässigkeit der einseitigen Zurückweisung von Personen zu bestehen. Carens wiederum verweist in seinem Argument auf diesen Umstand, um die von mir getroffene Unterscheidung – zwischen denjenigen, die sich in Reichweite der Zwangsbefugnis des Staates befinden und denjenigen, die in diese Reichweite gelangen wollen – als moralisch falsch zu verwerfen. Schließlich wird über die Menschen an den Grenzen ebenso Zwangsgewalt ausgeübt und jener Zwang hat ebenfalls „tiefgreifende Folgen", die in ihrer Schwere den Folgen gleichkommen, die innerstaatlicher Zwang für die Bürgerinnen des ausschließenden Staates bedeutet.[60]

Allerdings können wir auch in diesem Fall fragen, ob es allein die Stärke des infrage stehenden Interesses ist, die ein politisches Recht begründen kann. Ein Staat schuldet verschiedenen Menschen unterschiedliche Dinge, und zwar abhängig davon, wie und wo er Zwangsgewalt über sie ausübt. Daraus folgt jedoch nicht, dass alle der von dieser Zwangsgewalt betroffenen Personen über dieselben Rechte angesichts dieses staatlichen Zwangs verfügen. Ganz klar ins Auge fällt dieser Umstand im innerstaatlichen Kontext. Wenn meine Frau durch den Staat bestraft wird, ergeben sich daraus unzweifelhaft auch bedeutende Folgen für mich selbst. Ich könnte in manchen Fällen durch die Haft meiner Frau in gleichem Maße geschädigt werden wie sie selbst. Aber daraus folgt nicht, dass die Strafe für meine Frau nicht insbesondere ihr gegenüber gerechtfertigt werden sollte. Während in diesem

Fall zwar wir beide in unseren Interessen betroffen sind, ist sie allein diejenige, die durch die Zwangsgewalt des Staates bestraft wird und dieser Umstand stattet sie mit Rechten aus, die mir nicht zukommen. Es ist daher, in anderen Worten, nicht die gefühlte *Bedeutung* dessen, was mir angetan wird, die den zentralen Teil unserer Diskussion darstellen sollte, sondern ob das, was mir angetan wird, gerechtfertigt ist.

Diese Ideen können wir nun auf den Einsatz von Zwangsgewalt an Staatsgrenzen anwenden. Stellen Sie sich vor, ich möchte nach England ziehen und wurde davon mit gewaltsamen Mitteln abgehalten – diese Zurückweisung macht es mir nun unmöglich, eine Reihe von Dingen zu erreichen, die ich gerne in meinem Leben hätte. Wir können uns darüber hinaus vorstellen, dass meine Zurückweisung durch das Vereinigte Königreich einer „richtungsweisenden Entscheidung“ mit „tiefgreifenden Folgen“ für mein restliches Leben entsprach. Das bedeutet jedoch nicht, dass mir in der britischen Politik dieselben Rechte zustehen wie denjenigen, die bereits im Vereinigten Königreich leben. Da das Leben letzterer bereits anhand des britischen Rechts durch die staatliche Zwangsgewalt strukturiert wird, verfügen sie auch über spezifische Rechte gegenüber den britischen politischen Institutionen, darunter, wie ich bereits gezeigt habe, das Recht auf Bewegungsfreiheit innerhalb des Vereinigten Königreichs. Meine Interessen statten mich nicht mit dem gleichen Bündel an Rechten aus. Eine andere Art über dieses Problem nachzudenken besteht in der Vorstellung, dass noch vor meiner Zurückweisung eine Katastrophe den britischen Staat zusammenbrechen lässt und das Vereinigte Königreich in einen Zustand permanenter Anarchie abrutscht. Das wäre offensichtlich kein ganz geringes Problem für meine Zukunftspläne. Aber als Außenstehender würde mir durch die Abwesenheit politischer Institutionen im Vereinigten Königreich kein *Unrecht* geschehen. Ich könnte hier in den Vereinigten Staaten verbleiben; ich wäre zwar enttäuscht, aber es kann kein Menschenrecht darauf geben, frei von Enttäuschungen zu leben. Mein Interesse an den politischen Institutionen Großbritanniens ist einfach mein – wie auch immer stark ausgeprägtes – Interesse daran, durch diese Institutionen regiert *zu werden*. Den gegenwärtigen Einwohnerinnen des Vereinigten Königreichs – denjenigen also, die bisher unter dem britischen Gesetz standen – würde hingegen durch einen solchen Zusammenbruch ein schwerwiegendes Unrecht angetan. Selbst wenn ich von tiefstem Herzen ein Bürger des Vereinigten Königreichs werden möchte, werde ich durch die Abwesenheit des Vereinigten Königreichs als politischer Gesellschaft nicht ungerecht behandelt. Dem Volk Großbritanniens allerdings würde ein drastisches Unrecht widerfahren. Diese Menschen haben ein Recht auf eine gewisse Form politischer Gemeinschaft, und die neu eingetretene Gesetzlosigkeit würde dieses Recht verletzen. Genau wie ich mögen sie Zukunftspläne im Vereinigten Königreich geformt haben, aber der Unterschied zwischen uns besteht in der Beziehung zu dem, was das Vereinigte Königreich *ist*. Wir können daher bei der Bestimmung der Rechte, die wir gegenüber bestimmten Institutionen haben, nicht einfach nur darauf schauen, wie der Ausschluss aus einer politischen Gemeinschaft unsere Zukunftspläne beeinflusst. In welchem Verhältnis wir zu diesen Institutionen stehen – was sie gegenüber uns zu tun gedenken– ist ein wichtiger Teil des Gesamtbilds und zwar ein Teil, der von den Verteidigerinnen offener Grenzen im Allgemeinen ignoriert wird.

Ich möchte zum Abschluss dieses Kapitels jedoch die Bedeutung des Konzepts der *Zwangsgewalt* für die Theoriebildung im Bereich der Migration anerkennen. Selbst wenn Grenzen nicht immer mit gewaltsamen Mitteln verteidigt werden, sind potentielle Migrantinnen doch *häufig* mit der Möglichkeit von Gewaltanwendung konfrontiert, wenn sie eine Linie zwischen zwei Hoheitsgebieten überschreiten wollen. Und diese Gewaltanwendung müssen wir auf irgendeine Weise rechtfertigen. In diesem Kapitel habe ich versucht zu zeigen, weshalb wir nicht denken müssen, dass eine solche Rechtfertigung unmöglich ist – warum also aus der Idee liberaler Gerechtigkeit nicht notwendigerweise die Öffnung aller Grenzen folgt. Allerdings benötigen wir eine Theorie, die uns zeigt, wie und in welchen Fällen und aus welchen Gründen wir diese Grenzen schließen dürfen. Es ist diese Aufgabe, der wir uns nun zuwenden können.

3
Gerechtigkeit und die Ausgeschlossenen, Teil 2: Geschlossene Grenzen

Sofern meine vorangegangenen Ausführungen korrekt sind, haben sie bisher nur so viel gezeigt: Nichts innerhalb des Begriffs der Gerechtigkeit zeichnet den staatlichen Ausschluss von Migrantinnen als an sich ungerecht aus. Wir können also nach einer theoretischen Begründung suchen, anhand derer es möglich ist, festzustellen, wann und warum es Staaten erlaubt ist, ungebetene Immigrantinnen zurückzuweisen. Das vorherige Kapitel hat die Frage offengelassen, wie genau eine solche Begründung aussehen könnte. Die Ausgeschlossenen haben, so nehmen wir an, ein Recht darauf, zu erfahren, warum ihr Ausschluss nicht ungerecht ist. Welchen Grund könnten wir ihnen nennen? Was kann einer Person gesagt werden, um zu erklären, dass sie kein Recht darauf hat, eine willkürlich gezogene Linie zwischen zwei Staaten zu überqueren?

Eine mögliche Antwort auf diese Frage besteht sicherlich in der Behauptung, dass wir gar keiner solchen Rechtfertigung bedürfen. Wie bereits in den vorherigen Kapiteln erwähnt, führen Rechtspopulisten oft das Argument ins Feld, dass bei der Bestimmung einer Ethik der Migration allein die Rechte und Interessen der Inländer zählen sollten. Präsident Trumps Rede zur Amtseinführung brachte ziemlich deutlich zum Ausdruck, dass unter seiner Führung die Interessen der derzeitigen Einwohnerinnen der Vereinigten Staaten die einzige Basis für politische Entscheidungen darstellen würden. Aus dieser Perspektive braucht es keine Rechtfertigung gegenüber ungebetenen Migrantinnen: Sie stehen *außerhalb*, und zwar sowohl juristisch als auch moralisch.

Diese Antwort ist jedoch nicht annähernd befriedigend. Es gibt manche Handlungen, die nicht zu rechtfertigen sind; und zwar unabhängig davon, welche Staatsbürgerschaft die betroffenen Personen haben. Wie bereits weiter oben erwähnt, hat die Trump-Administration die Trennung von Familien veranlasst, die an der südlichen Grenze der Vereinigten Staaten Zuflucht suchten – und in diesem Zusammenhang minderjährige Kinder in Sicherheitseinrichtungen untergebracht, die Gefängnissen ähneln. Die Eltern wurden dann dazu angehalten, ihre Asylanträge zurückzuziehen, wobei die Zusammenführung mit ihren Kindern als zusätzlicher Ansporn für ihr Einverständnis genutzt wurde. Die Ungerechtigkeit dieser Praxis, sowohl gegenüber den Eltern, als auch den Kindern, ist enorm; viele Beobachterinnen waren vor allem vom Leid schockiert, das den Kindern zugefügt wurde und zwar vor allem angesichts der Tatsache, dass diese Kinder in keiner Weise als moralisch verantwortlich für die Situation bezeichnet werden können, in der sie sich befinden.[1] Ein Moderator auf Fox News antwortete auf diese ablehnenden Reaktionen wie folgt:

> „Ob sie es mögen oder nicht, das sind nicht unsere Kinder. […] Es ist nicht so, als ob er das den Menschen aus Idaho oder Texas antun würde. Das sind Menschen aus einem anderen Land."[2]

Die Repliken darauf kamen schnell und wiesen die Aussage zu Recht strikt zurück. Es gibt Dinge, die nicht getan werden dürfen, gegenüber niemandem, und um zu bestimmen, was darunter zu verstehen ist, braucht es eine Theorie darüber, worauf Menschen sowohl an der Grenze, als auch beim Überschreiten dieser Grenze, einen Anspruch haben. Wir brauchen nicht viele theoretische Überlegungen, denke ich, um die Politik der Trennung von Familien an staatlichen Grenzen zu verurteilen. Aber die Tatsache, dass wir diese Politik verurteilen wollen, beinhaltet zugleich eine Zurückweisung der populistischen Perspektive: Der Umstand, dass eine Person aus einem anderen Land kommt, negiert nicht ihr Recht auf eine Rechtfertigung dessen, was ihr angetan wird.

Aus diesem Grund müssen wir untersuchen, was ein Staat in Übereinstimmung mit der liberalen Idee der Gerechtigkeit zur Rechtfertigung von Praktiken vorbringen kann, mit denen er einige der Menschen zurückweisen möchte, die seine territorialen Grenzen überschreiten wollen. Wie wir sehen werden, hat diese Rechtfertigung zwei Aspekte. Der erste würde den Ausgeschlossenen erklären, warum ihr Ausschluss selbst nicht moralisch falsch ist – warum also der sie ausschließende Staat in diesem Fall nicht ungerecht handelt. Der zweite Aspekt betrifft jedoch den Staat selbst, indem ihm aufgezeigt wird, in welchen Fällen eine solche Rechtfertigung nicht verfügbar ist. Dieser zweite Teil hilft uns bei der Orientierung in und Kritik an der staatlichen Migrationspolitik, indem er darlegt, wie bestimmte migrationspolitische Vorhaben des Staates besten Rechtfertigung widersprechen, die wir für eine Kontrolle der Migration vorbringen können. In den Kapiteln fünf und sechs werde ich zwei mögliche Formen der Einschränkung staatlicher Migrationspolitik diskutieren. Die erste Form wird sich mit der Gruppe von Menschen befassen, die aufgrund der Umstände in ihren Heimatländern nicht gerechterweise ausgeschlossen werden können. Die zweite Form beschäftigt sich damit, wie auch denjenigen ohne eigenständiges Recht auf Einwanderung ein Anspruch auf eine faire Behandlung zukommt, sofern ein Land sich dazu entscheidet, einigen von ihnen (aber nicht allen) ein Recht auf Einwanderung zuzusprechen.

In diesem Kapitel möchte ich jedoch zunächst bloß untersuchen, mit welchen Argumenten andere Autoren versucht haben zu zeigen, warum der staatliche Ausschluss von Immigrantinnen gerechtfertigt sein könnte. Auf Grundlage eines solchen Arguments könnten wir uns dann der Aufgabe zuwenden, zu bestimmen, wer ausgeschlossen werden darf und wer nicht – und welche Arten der Migrationspolitik mit dem liberalen Gerechtigkeitsbegriff in Einklang stehen würden. Allerdings hätten wir in diesem Fall tatsächlich erst damit begonnen zu verstehen, wie der Begriff Gerechtigkeit auf das Phänomen der Migration anzuwenden ist. Es werden auch weiterhin Fragen dahingehend bestehen, welche Güter eine bestimmte Migrationspolitik gerechterweise als Zweck verfolgen darf; selbst wenn also ein Staat eine große Anzahl an Menschen ausschließen darf, können wir uns fragen, ob damit auch einhergeht, dass alle sozialen Güter, wie z. B. der Erhalt einer bestimmten Kultur, zu Recht durch migrationspolitische Maßnahmen angestrebt werden dürfen. Es werden darüber hinaus noch eine Vielzahl weiterer Probleme bestehen bleiben, die sich nicht auf die Idee der Gerechtigkeit selbst beziehen, sondern auf den Zusammenhang zwischen dieser Idee und der spezifischen Geschichte bestimmter Länder. So verweist die Debatte um die Trennung von Familien an der südlichen Grenze der USA auf eine umfassendere Diskussion über die amerikanische Außenpolitik in Mexiko und Zentralamerika.[3] Zudem werden, selbst wenn ein Staat Immigrantinnen legitimerweise ausschließen darf, Fragen darüber bestehen bleiben, welche Mittel bei der Zurückweisung genutzt werden dürfen. Wir haben bereits die Trennung von Kindern und ihren Eltern besprochen; ebenso könnten wir betrachten, wie die bereits errichteten physischen Mauern an der Grenze zu Mexiko die Migrantinnen dazu gezwungen haben, für ihre Reise auf die gefährlichsten Teile der Wüste auszuweichen – was zur Folge hat, dass viele von ihnen auf dem Weg in die Vereinigten Staaten ums Leben kamen.[4] Wir könnten fragen, inwiefern solche Mittel zulässig sind, auch wenn wir ihren Zweck verteidigt haben. Alle diese Fragen sind jedoch erst dann von Bedeutung, wenn wir gezeigt haben, weshalb der Ausschluss ungebetener Migrantinnen überhaupt gerechtfertigt ist. Dieser Aufgabe möchte ich mich nun zuwenden.

Es gibt eine Vielzahl möglicher Argumente dafür, warum es einem Staat erlaubt sein sollte, ungebetene Migrantinnen auszuschließen. In diesem Kapitel werde ich jedoch bloß vier, meiner Meinung nach relativ überzeugende, Familien solcher Argumente untersuchen. Die ersten zwei Familien behandeln Güter, die durch staatlichen Ausschluss verteidigt werden könnten, sowie deren gerechte Verteilung. Die letzten zwei Familien hingegen behandeln Rechte, insbesondere die Rechte der Menschen, die bereits in einem bestimmten Staat leben.

Diese Argumentfamilien basieren, im weitesten Sinne, auf den Ideen des *Lands*; der *Solidarität*; des *Eigentums*; und der *Vereinigungsfreiheit*. Sie alle bieten eine mögliche Begründung, die gegenüber bestimmten Personen außerhalb des Staatsgebiets angeführt werden könnte um zu zeigen, dass ihr Ausschluss nicht ungerecht ist. (Und jede dieser Familien bietet auch eine mögliche Theorie darüber an, welche Personen *nicht* ausgeschlossen werden dürfen – in welchem Fall die von ihnen vorgeschlagene Rechtfertigung also nicht funktioniert.) Zwischen diesen Argumentfamilien gibt es in meinen Augen keine scharfe Abgrenzung; sie überlappen sich oft und wie auch im vorherigen Kapitel nutzen die Autorinnen verschiedenartige Ideen zur gleichen Zeit. Aber meiner Meinung nach stellt die von mir vorgeschlagene Klassifizierung einen guten Ausgangspunkt für unsere Untersuchung dar, weshalb ich davon ausgehe, dass die Unterscheidungen ausreichend klar sind.

Im Folgenden werde ich erklären, wie diese Argumente staatlichen Ausschluss rechtfertigen und warum sie dabei letztendlich nicht erfolgreich sind. Mein Argument lautet, dass all diese Erklärungsansätze die Verweigerung des Zutritts zu einer politischen Gemeinschaft jeweils unter Bezug auf etwas rechtfertigen, das ich als unwesentlich für eine solche

Art von Gemeinschaft erachte. Sie tragen der Tatsache nicht angemessen Rechnung, dass mit der Erlaubnis zum Betreten eines bestimmten Hoheitsgebietes auf verschiedene Weisen der Eintritt in das für dieses Hoheitsgebiet spezifische System politischer und rechtlicher Verpflichtungen einhergeht. Wenn ich beispielsweise die Grenze zu Mexiko überquere, erschaffe ich durch diesen Akt bestimmte Verpflichtungen zum Schutz meiner Grundrechte, die sowohl den mexikanischen Institutionen zufallen, als auch den derzeitigen Bewohnerinnen Mexikos, die diese Institutionen unterstützen und verwalten. Wie ich im folgenden Kapitel zeigen werde, ist es diese Tatsache, die Mexiko ein Recht darauf gibt, mich auszuschließen. Die bisherigen Ansätze haben in meinen Augen dem *politischen* Wesen des Ausschlusses nicht die angemessene Aufmerksamkeit zukommen lassen. Ich möchte jedoch betonen, dass meine Zurückweisungen sehr gut damit vereinbar sind, dass ich tief in der Schuld derjenigen stehe, die das Recht auf Ausschluss in ihren Werken verteidigt haben. Meine Ideen möchte ich daher erneut nicht als eine endgültige Widerlegung dieser alternativen Sichtweisen verstanden wissen, sondern als eine einfache Erklärung dafür, weshalb mich diese Alternativen nicht überzeugen.

Schließlich möchte ich noch bemerken, dass die hier diskutierten Ansätze die Debatte nicht erschöpfend abbilden. Andere Autorinnen haben das staatliche Recht auf Ausschluss anhand von Überlegungen gerechtfertigt, die nicht auf die hier untersuchten Arten von Argumenten reduziert werden können. So hat zum Beispiel Philip Cafaro Migrationsbeschränkungen unter Verweis auf einen drohenden ökologischen Kollaps gerechtfertigt: Angesichts des durchschnittlichen ökologischen Fußabdrucks eines US-Amerikaners sollten laut Cafaro nicht mehr US-Amerikaner als nötig die Erde bevölkern.[5] Stephen Macedo hat dafür plädiert, bestimmte Muster von Migrationsrechten auf Basis der Idee innerstaatlicher Verteilungsgerechtigkeit zu kritisieren.[6] Beide Ansätze sind wichtig und verdienen mehr Platz, als ich ihnen in diesem Kontext einräume. Meine einzige Antwort auf diese Ansätze besteht in der Bemerkung, dass ohne die Existenz eines Rechts darauf, etwas Bestimmtes zu tun, es auch kein Recht darauf geben kann, es für gute politische Zwecke zu tun. Argumente wie die von Cafaro und Macedo funktionieren daher nur, wenn wir bereits geklärt haben, warum und wie Staaten das Recht haben, Personen auszuschließen, die nicht bereits Mitglied der politischen Gemeinschaft sind. Kommt dem Staat nicht das Recht zu, auch nur irgendeine Person auszuschließen, dann kann er dieses Recht schließlich auch nicht dadurch erlangen, dass der Ausschluss die wirtschaftliche Situation gerechter oder nachhaltiger gestalten würde. Wir benötigen daher eine Erklärung dafür, warum ein Staat Nicht-Mitglieder ausschließen darf; und wir können die Suche nach dieser Erklärung damit beginnen, dass wir uns zunächst mit der Idee vom Land selbst als physischer Ressource befassen.

1. Land

Die ausgeschlossene Person wird daran gehindert, eine bestimmte Linie auf der Erdoberfläche zu überqueren. Diese Linie kann Verschiedenes markieren: eine politische Gemeinschaft, ein juristisches Hoheitsgebiet und sehr oft eine sprachliche oder kulturelle Gemeinschaft. Aber was sie unmittelbar markiert ist ein *Territorium* – die physische Tatsache also des Landes selbst. Wird eine potentielle Migrantin ausgeschlossen, so ist es ihr nicht gestattet, ein bestimmtes Stück Land zu nutzen. Dabei ist Land nicht das Resultat individuellen Handelns. Abgesehen von Fällen wie der Errichtung von *Seasteads* oder den von China erbauten künstlichen Inseln im Südchinesischen Meer existierte das Land, das wir nutzen, bereits vor der Inanspruchnahme durch irgendeine Person. Das Land wurde, um die Sprache früherer Philosophen zu nutzen, uns allen durch Gott gegeben und stellt daher ein Gut dar, auf das gewissermaßen alle einen gleichwertigen Anspruch haben. Viele Philosophen – darunter Immanuel Kant und Hugo Grotius – nahmen an, dass diese Tatsache bedeutende Implikationen für die Bestimmung politischer Rechte habe.[7] Jüngere Philosophinnen haben ähnliche Ideen entwickelt. So sah es beispielsweise Henry Sidgwick als offensichtlich an, dass ein staatliches Recht darauf, jedwede Person ausschließen zu dürfen, einhergeht mit „einem Vorbehalt im Falle von Staaten, die das Eigentum über große Flächen unbewohnten Landes beanspruchen".[8] Ganz ähnlich argumentierte Michael Walzer, dass das Recht auf Ausschluss nicht bestehen würde, falls das Land, aus dem bedürftige Nicht-Mitglieder ausgeschlossen werden, nur spärlich bewohnt sei:

> „Das Anrecht der weißen Australier auf die riesigen, leeren Räume des Subkontinents gründete auf nichts anderem als auf dem erstmals von ihnen darauf erhobenen und gegen die eingeborene Bevölkerung durchgesetzten

> Anspruch. Es scheint kein Anrecht zu sein, das sich gegenüber bedürftigen Menschen, die Einlaß begehren, mühelos verteidigen ließe. [...] Wenn wir also davon ausgehen, daß es tatsächlich überflüssiges Land gibt, dann würden die unabweisbaren Erfordernisse eine politische Gemeinschaft wie ‚das weiße Australien' dazu zwingen, eine Grundsatzentscheidung zu treffen. Ihre Mitglieder könnten, um ihre Homogenität zu bewahren, Land abtreten, oder sie könnten auf die Homogenität verzichten (und der Schaffung einer multirassischen Gesellschaft zustimmen), um das Land in seiner vollen Ausdehnung zu erhalten. Andere ihnen offenstehende Alternativen gäbe es nicht. Das weiße Australien könnte nur als Kleinaustralien überleben."[9]

Walzers Argument beruht in diesem Fall auf einer zuvor ausgeführten Theorie der Gerechtigkeit territorialer Inbesitznahme. Viele haben sich schon gefragt, ob „White Australia" überhaupt ein Existenzrecht zukommen sollte, besonders auf dem Land, das zuvor von den Aborigines bewohnt wurde.[10] Der Punkt, den ich an dieser Stelle untersuchen möchte, ist allerdings der umfassendere Gedanke, dass es das Vorhandensein „überschüssigen" Landes ist, das den Ausschluss von Immigrantinnen moralisch untersagt. Diesem Ansatz nach würde der Fakt, dass Land unterbevölkert ist, es mutmaßlich unmöglich machen, Migrantinnen abzuweisen; und umgekehrt könnte die Überbevölkerung eines bestimmten Territoriums ein stärkeres Recht auf Ausschluss begründen.

Ideen wie diese wurden in den letzten Jahren am besten im Werk von Matthias Risse entwickelt, der in seinem *On Global Justice* die Über- und Unterbeanspruchung von Land als einen für Gerechtigkeitserwägungen relevanten Grund untersucht. Risses Arbeit ist vielschichtig und verdient eine sorgfältige Betrachtung. An dieser Stelle werde ich mich allein auf die verschiedenen Arten konzentrieren, auf die Risse die Idee verteidigt, dass das *Land* selbst als eigenständige Grundlage für die Beurteilung der Gerechtigkeit von Migrationsregimen betrachtet werden sollte. Er beginnt damit, dass er einem Land einen bestimmten Wert für menschliche Zwecke zuordnet – im Falle des Staates S nennen wir diesen Wert V_S – und vergleicht diesen Wert mit der Population P, genannt P_S. Für jeden Staat können wir dann V_S und P_S vergleichen und ihr Verhältnis als *pro tanto*-Grund für die moralische Bewertung eines Migrationsregimes heranziehen:

> „Das Territorium von S ist relativ unausgelastet (oder schlicht unausgelastet), wenn VS/PS größer ist als der Durchschnitt dieser Werte über alle Staaten hinweg (sodass die durchschnittliche Person über ein Ressourcenbündel von höherem Wert verfügt als die Durchschnittsperson in einem durchschnittlichen Land). Es ist relativ überbeansprucht (oder schlicht überbeansprucht), wenn der Wert unterhalb dieses Durchschnitts liegt. Wenn V_S/P_S über dem Durchschnitt liegt, haben die Mitbesitzer andernorts einen *pro tanto*-Grund für Einwanderung in dem Sinne, dass unausgelastete Länder, so lange die Ansprüche dieser Mitbesitzer unerfüllt bleiben, nicht vernünftigerweise erwarten können, dass andere sich ihrer Einwanderungspolitik fügen. Es ist daher in diesem Falle eine Frage vernünftigen Handelns, dass ein Staat Einwanderung erlaubt. [...] Wenn ein Land sein Territorium auslastet, kann von Personen vernünftigerweise verlangt werden, seine Einwanderungspolitik zu akzeptieren (sofern keine davon unabhängigen Probleme im Hinblick auf diese Politik bestehen).“[11]

Risse nutzt diese Ideen, um die gegenwärtigen Debatten über Migration in den Vereinigten Staaten zu analysieren. Die USA hat eine Bevölkerungsdichte von 31 Personen pro Quadratkilometer, in Japan leben auf der gleichen Fläche 320 Personen. Laut Risse veranschaulicht dieser Vergleich, dass in den Vereinigten Staaten „Ressourcen in kritischem Maßstab nicht ausgelastet werden“, woraus für ihn folgt, dass die USA von undokumentierten Migrantinnen nicht erwarten können, ihrem Staatsgebiet fern zu bleiben. Die Bevölkerungsdichte – die Risse selbst als bloß „sehr grobe Orientierung“ hinsichtlich des Wertes von Land bezeichnet, auf dem sein Argument letztlich beruht – reicht demnach aus, um amerikanische Ansprüche darauf zurückzuweisen, mexikanische Bürgerinnen entweder bereits an der Grenze abweisen oder, nach ihrem unrechtmäßigen Betreten des Staatsgebiets, abschieben zu können.[12]

An diesem Punkt möchte ich gegen die moralische Relevanz von Land argumentieren. Das ist in gewissem Maße seltsam, da ich durch meine Argumentation gegen Risse zugleich eine frühere Version meiner Selbst kritisiere:[13] Manche der Ideen Risses bezüglich Migration wurden zuerst in einem von uns gemeinsam verfassten Artikel entwickelt. Seitdem bin ich jedoch zu der Überzeugung gelangt, dass wir es vermeiden sollten zu denken, das Land selbst wäre moralisch von zentraler Bedeutung für die Rechtfertigung von Ausschluss. Land ist, um es einfach auszudrücken, allein dann moralisch relevant, wenn es eine Voraussetzung dafür darstellt, eine autonome Existenz *auf diesem Land* zu realisieren. Es sind diese Existenzen, auf die unsere Aufmerksamkeit gerichtet sein sollte. Werden Individuen mit den Umständen und Ressourcen ausgestattet, anhand derer es ihnen möglich ist, ein für sie wertvolles Leben aufzubauen? Werden ihnen die Mittel an die Hand gegeben, die es für ein Leben als autonomer Akteur braucht? Sofern die Antwort auf diese Fragen ja lautet, scheint es nebensächlich, wie viel oder wenig Land dafür vonnöten ist. Wir brauchen eine gewisse Menge an Land, um ein wertvolles Leben aufbauen zu können; aber der Marktpreis dieses Landes – und die Anzahl der auf ihm lebenden Menschen – sagt uns diesbezüglich ziemlich wenig.

Eine mögliche Veranschaulichung dieses Arguments stellt der Aufstieg von Städten in den letzten Jahrzehnten dar. Zum ersten Mal in der Menschheitsgeschichte leben mehr Menschen in Städten als auf dem Land.[14] All das wurde möglich durch Elisha Otis' Erfindung einer Sicherheitsvorrichtung für Aufzüge.[15] Vor Otis' Erfindung war die Höhe von Gebäuden in der Praxis auf drei Stockwerke begrenzt. Nach Otis aber wurden mehrstöckige Gebäude – Wolkenkratzer, hoch aufragende Wohnkomplexe und so weiter – eine reale Möglichkeit; mit dem Resultat, dass sich der Zusammenhang veränderte, der zwischen der physischen *Landschaft* eines Orts und den *Flächen* besteht, auf denen dort gelebt und den eigenen Angelegenheiten nachgegangen werden kann. In Manhattan gibt es beispielsweise 46,5 Millionen Quadratmeter Bürofläche, von der sich nur wenig auf dem ursprünglichen *Land* befindet, dass uns gemeinsam von Gott gegeben wurde. Viel eher stehen wir auf Etagen, die selbst wiederum auf Etagen stehen, die dann irgendwann schließlich auf jenem ursprünglichen Land aufliegen. Zwei Dinge sind im Hinblick auf diesen Umstand wichtig. Zuerst, dass Manhattan an einem durchschnittlichen Wochentag fast vier Millionen Einwohner hat, was einer Bevölkerungsdichte von ca. 65.000 Personen pro Quadratkilometer entspricht.[16] Bei einer solchen Bevölkerungsdichte könnte die gesamte Weltbevölkerung auf einem Gebiet der Größe von Texas leben. Der zweite Fakt betrifft die Tatsache, dass Manhattan *funktioniert*. Sicherlich nicht perfekt und auch nicht für jede einzelne Person. Allerdings funktioniert Manhattan gut genug um zu zeigen, dass die Bevölkerungsdichte – oder die biophysischen Eigenschaften des Lands und sein Wert – für die Frage der

Migration ziemlich irrelevant zu sein scheint. Wäre Manhattan ein Staat und die Mongolei (Bevölkerungsdichte: ca. 2 Personen pro Quadratkilometer) auch weiterhin ein Staat, lägen den Prinzipien der Migrationsgerechtigkeit, welcher Art sie auch immer sein sollten, sehr wahrscheinlich nicht die ungleiche Bevölkerungsdichte dieser beiden Staaten zugrunde. Sofern Manhattan für die dort lebenden Menschen funktioniert, verfügen diese Menschen über das, worauf sie einen Anspruch haben. Der Gedanke, dass die Mongolei dazu verpflichtet wäre, Einwohnerinnen Manhattans aufzunehmen, scheint der moralischen Idee der *Überanspruchung* mehr moralische Macht zu verleihen, als ihr zusteht.

Selbstverständlich basiert Risses Argument eigentlich darauf, dass, da alle Menschen gemeinsam Eigentümerinnen der Erdoberfläche sind, es so etwas wie Verteilungsgerechtigkeit geben sollte, wenn es um die Verteilung des Landes geht, oder zumindest um den Wert dieses Landes. An diesem Punkt widersprechen wir uns in meinen Augen am stärksten. Risse denkt, dass Prinzipien der Verteilungsgerechtigkeit für Güter wie die Erdoberfläche gelten. Für mich hingegen ist es am wichtigsten, dass Menschen Teil einer politischen Gemeinschaft sind, in der ihre Grundrechte geschützt werden. Die Verteilung von Gütern wie *Land* scheint aus dieser Perspektive eher irrelevant. Dieser Unterschied wird besonders in Risses Kritik meines eigenen Ansatzes deutlich:

> „Ein Virus dezimiert die Bevölkerung von Syldavien, wobei er aus ungeklärter Ursache bloß Menschen in Syldavien befällt. Der Virus schlägt so schnell zu, dass die Menschen sterben bevor sie fliehen können. Es gibt bloß 50 Überlebende. Allerdings errichtete Syldavien bereits vor der Infektion ein elektronisches Grenzüberwachungssystem, inklusive Maschinen, die Menschen gewaltsam vom Staatsgebiet fernhalten. Verfügbare Robotertechnik erlaubt es den Überlebenden in Syldavia, ihren hohen Lebensstandard zu erhalten. Sie bewahren ihre stolzen institutionellen Traditionen durch wöchentliche Online-Beratungen, leben allerdings in kleinen Gruppen über entlegene Gebiete des Landes verteilt. Anfragen von Bordurianern, sich dort niederzulassen und am Wirtschaftsleben beteiligen zu dürfen, werden zurückgewiesen. Bordurianer, die diese Entscheidung nicht akzeptieren wollen, werden von Grenzrobotern überwältigt."[17]

Risse charakterisiert das Verhalten der Syldavianer als „unvernünftig“, und darin stimme ich ihm zu. Die Frage ist jedoch, wie dieser Begriff zu verstehen ist. Die von mir bevorzugte Art dieses Wort zu lesen ist die, dass sich die Syldavianer zwar als egoistisch und unangenehm erweisen, durch die Zurückweisung allerdings kein Bordurianer auf signifikante Weise *ungerecht* behandelt wird. Wenn Risse bloß meint, dass wir von den Bordurianern nicht erwarten sollten, dass sie sich an das sie ausgrenzende Gesetz halten, sind wir erneut einer Meinung; wie ich weiter unten zeigen werde, kann es Fälle geben, in denen wir den Bruch geltender Migrationsgesetze nicht mit einem mangelhaften moralischen Charakter gleichsetzen sollten, selbst wenn das infrage stehende Gesetz nicht ungerecht ist. Aber wenn wir bei Risse ein schärferes Argument sehen, nämlich, dass der Ausschluss der Bordurianer eine *Ungerechtigkeit* ihnen gegenüber darstellt, dann ist mein Widerspruch ebenfalls schärfer. Stellen Sie sich zum Beispiel vor, dass alle Staaten der Welt – Bordurien und Syldavien sind vielleicht die einzigen – florieren und weit über einer wie auch immer definierten Schwelle politischer Angemessenheit liegen, die Staaten legitimiert. Kurz gesagt: Bordurien und Syldavien geht es gut. Warum sollten wir aber nun angesichts einer Verringerung der Bevölkerung Syldaviens denken, dass aufgrund dieses Umstands irgendein Bordurianer nun irgendeinen Grund für einen Anspruch auf Einwanderung erhält? Damit dieses Argument funktioniert, müssten wir meiner Meinung nach davon ausgehen, dass es die Verteilung von Land selbst ist, und nicht die durch dieses Land ermöglichten Existenzen, die moralisch von Bedeutung ist. Das aber erscheint mir eher als Fetischismus. Land ist ein Gut und nichts anderes darüber hinaus. Risse verteidigt diese distributiven Konsequenzen selbstverständlich mit Verweis auf die Idee, dass alles Land gemeinsames Eigentum der Weltbevölkerung ist. Allerdings bezweifle ich, dass die Diskussion damit zu einem Ende findet. Ich selbst habe den gleichen und wechselseitigen Anspruch auf Land immer als Ausdruck eines gleichen und wechselseitigen Anspruchs auf die Art von autonomen Leben verstanden, die durch Land möglich wird. Ist dieser Anspruch erfüllt, erscheint mir Gleichheit im Hinblick auf andere Dinge – sei es die Bevölkerungsdichte, der Wert des Landes oder etwas anderes – als schlicht irrelevant.

Risses Analyse kann daher dahingehend kritisiert werden, dass sie dem Land selbst zu viel moralische Bedeutung beimisst und dabei die auf diesem Land errichteten politischen Gemeinschaften nicht angemessen einbezieht. Aus diesem Grund ist diese Version seines Ansatzes für mich nicht zufriedenstellend. Wir stimmen beide darin überein, dass die Syldavianer schreckliche Menschen sind. Sie sind egoistisch und gemein und ihr Mangel an politischer Tugend ist bemerkenswert. Aber wir müssen solche Arten von Urteilen von dem stärkeren Urteil unterscheiden, dass die Syldavianer den ausgeschlossenen Bordurianern ein Unrecht antun. Ich glaube daher, dass eine Theorie der Gerechtigkeit in Migrationsfragen, will sie zu angemessenen Schlüssen kommen, auf etwas anderes als auf diese Ideen zurückgreifen muss.

2. Solidarität

Viele der stärksten Verteidigungen des staatlichen Ausschlusses von Migrantinnen beginnen mit Tatsachen hinsichtlich dessen, was von den bereits an einem Ort lebenden Menschen geteilt wird. Die von mir im Folgenden untersuchten Ansätze gehen dabei nicht einfach nur von geteilten Eigenschaften aus, sondern betrachten, was Einwohnerinnen im Sinne ihrer Selbstbeschreibung oder Identität gemeinsam haben. Wir gehen oft davon aus, dass die Menschen an einem Ort nicht bloß eine Sprache oder Tradition teilen, sondern dass sie diese Gemeinsamkeiten in ihrem Alltag und der Gestaltung ihrer gegenseitigen Beziehungen *nutzen*. Wenn sie um eine Selbstbeschreibung gebeten werden – also erklären sollen, wer sie sind –, werden sie auf diese Gemeinsamkeiten verweisen und charakteristische moralische Einstellungen gegenüber denjenigen offenbaren, die jene Gemeinsamkeiten teilen. Die Bedeutung dieser Ideen für das Problem des Ausschlusses sollten klar ersichtlich sein: Sofern diesen Gemeinsamkeiten selbst eine moralische Bedeutung zukommt – oder sie die Voraussetzung für andere, für diese Menschen wesentliche, Güter darstellen – scheint der Ausschluss derjenigen, die diese Gemeinsamkeiten nicht aufweisen, moralisch vertretbar.

Es gibt selbstverständlich viele unterschiedliche Ansätze, in denen diese Perspektive entwickelt wurde. Manche Vertreter eines akademischen Nationalismus führen besonders starke Varianten ins Feld. Für diese Autoren beruht die Rechtfertigung von Ausschluss auf der moralischen Bedeutung, die der besonderen Beziehung zwischen den gegenwärtigen Mitgliedern einer Gruppe zukommt; dieser Beziehung selbst entspringt demnach ein eigener Wert. David Millers frühere Arbeiten stellen eine außergewöhnlich anschauliche Variante dieser an der Nation orientieren Perspektive dar; ihm zufolge besitzt die nationale Gemeinschaft einen enormen moralischen Wert, da sie auf verschiedene Weise Teil des moralischen Rahmens ist, innerhalb dessen sich einzelne Personen ihr eigenes Leben aufbauen.[18] In seinem späteren Werk argumentiert er, dass die Ethik der Migration unter Bezug auf die besonderen Werte bestimmter Gemeinschaften entwickelt werden sollte, deren Mitglieder sich um das Wohlergehen und den Fortbestand eben jener Gemeinschaften sorgen:

> „Dabei handelt es sich häufig um kollektive Werte, die mit der grundsätzlichen Gestalt und dem Charakter der Gesellschaft zu tun haben, in die die Einwanderer einreisen wollen – wie zum Beispiel mit der Größe der Gesamtbevölkerung, der Sprache oder den Sprachen, die ihre Bewohner sprechen, oder mit ihrer tradierten nationalen Kultur. Diese Dinge sind für die angestammten Bürger oft von größter Bedeutung. […] Die Menschen möchten das Gefühl haben, dass sie die zukünftige Gestalt ihrer Gesellschaft mitbestimmen können. Sie haben ein Interesse an politischer Selbstbestimmung, wozu auch die Fähigkeit zu der Entscheidung darüber gehört, wie viele Einwanderer kommen dürfen, welche von ihnen ausgewählt werden sollten, wenn diese Anzahl überschritten wird, und was von denen, die aufgenommen werden, zu Recht erwartet werden kann.“[19]

Für Miller reicht es aus, dass der Charakter einer nationalen Gemeinschaft für diejenigen von *Bedeutung* ist, die bereits an einem bestimmten Ort leben. Ausgehend von dieser Prämisse sollen wir auf das moralische Recht der Gemeinschaft schließen, den Charakter dieses Ortes durch Einwanderungskontrollen bestimmen zu dürfen.

Ich nutze an dieser Stelle die unspezifische Kategorie der *Solidarität*, um auf die verschiedenen Ansätze Bezug zu nehmen, die die Gemeinsamkeiten zwischen bestimmten Menschen als Quellen für Werte und Motivation eben jener Menschen betrachten. Millers Nationalismus kann daher als ein bestimmtes Verständnis von Solidarität verstanden werden: Die Verbindungen zwischen den derzeitigen Mitgliedern einer Gesellschaft sind bedeutsam und die teilweise durch diese Verbindungen geprägten Menschen haben ein Recht darauf, bei der Bestimmung des zukünftigen Charakters und Selbst-verständnisses dieser Gesellschaft eine Rolle zu spielen. Miller ist nicht der einzige Autor, der auf eine solche Vorstellung von nationaler Identität zurückgreift; auf jeweils unterschiedliche Weise begründen auch Michael Walzer und Will Kymlicka das Recht auf Ausschluss anhand der Bedürfnisse der bereits anwesenden Personen nach nationaler Deliberation und Solidarität.[20] Allerdings wird Solidarität auch von Autorinnen als Argument angeführt, deren politische Philosophie nicht als in diesem Sinne nationalistisch bezeichnet werden kann. So ist die Idee der nationalen Selbstbestimmung beispielsweise in Sarah Songs Argumentation für ein Recht auf Ausschluss von zentraler Bedeutung. Songs Vorstellung beruht stärker auf demokratischer Selbstbestimmung denn auf Nationalismus, aber auch hier spielt das Konzept der Solidarität eine Schlüsselrolle für die Rechtfertigung eines Rechts auf Ausschluss:

> „Wie ich zeigen werde, handelt es sich bei der politischen Gleichheit um eine konstitutive, bei der Solidarität hingegen um eine instrumentelle Bedingung der Demokratie. […] Wenn wir die normative Bedingung der politischen Gleichheit auf das demokratietheoretische Problem der Begrenzung anwenden, können wir demokratische Gründe dafür finden, warum der Demos durch den Territorialstaat begrenzt werden sollte."[21]

Für Song ist die Bedeutung der Solidarität von instrumentellem Charakter; sie ermöglicht demokratische Legitimität. Ihr zufolge sollte diese zentrale Bedeutung der Solidarität in der Diskussion um ein Recht auf Ausschluss anerkannt und bedacht werden. Demnach haben wir also Gründe dafür, solche identitätsstiftenden Arten von Gemeinsamkeiten als eine Vorbedingung demokratischer Selbstregierung zu bewahren.

Es kann sicherlich noch sehr viel mehr über Solidarität gesagt werden. Allerdings möchte ich mit der Bemerkung beginnen, dass es einige klassische Probleme bei der Anwendung solcher Konzepte gibt, zumindest in dem sehr starken Sinne, in dem sie von Nationalisten wie Miller verstanden werden. Kritiker haben darauf hingewiesen, in welcher Hinsicht diese Art von Nationalismus das Potential besitzt, diejenigen Einwohnerinnen einer bestimmten Gesellschaft zu entfremden, deren persönliche Eigenschaften von denjenigen Eigenschaften abweichen, die von der Elite dieser Gesellschaft hochgehalten werden. Darüber hinaus wurde darauf verwiesen, dass ein Recht auf die eigene Gemeinschaft in Spannung zu dem Recht außenstehender Personen steht, ein wertvolles Leben aufzubauen – weshalb zu untersuchen wäre, wie schwer diese Rechte im Vergleich zueinander wiegen. Hier sollte ich anmerken, dass Miller dieses Problem durchaus anerkennt und einigen Personen als Flüchtlingen Rechte auf Zuflucht innerhalb dieser Gemeinschaften zuspricht. Die umfassendere Frage war jedoch lange, ob nationale Gemeinschaften ein Recht auf die präsumtiven Befugnisse haben, die ihnen von akademischen Nationalisten zugeschrieben werden.

In diesem Kapitel möchte ich jedoch ein etwas anderes Argument in Bezug auf Solidarität und Migration vorbringen – ein Argument, das die Anwendung des Konzepts der Solidarität womöglich sowohl in seiner starken als auch schwachen Version untergräbt. Theoretikerinnen wie Miller und Song führen an, dass Solidarität *nützlich* für eine Gesellschaft sei; und sie sei aufgrund dieser Nützlichkeit ein Gut an sich oder ermögliche zumindest einen einfacheren Zugang zu den gerechten politischen Beziehungen einer Demokratie. Aber ich denke nicht, dass die Nützlichkeit von etwas – und sei es auch für einen noch so bedeutsamen moralischen Zweck – ausreicht, um zu zeigen, dass etwas moralisch erlaubt ist. Es scheint hingegen vielmehr so, als ob ein eigenständiges Argument nötig ist um zu zeigen, dass das jeweils infrage stehende Gut legitimerweise anhand des Ausschlusses von Migrantinnen verfolgt werden darf.

Zur Veranschaulichung können wir uns eine Gesellschaft vorstellen, die über keine besondere Form von Solidarität oder Gruppenidentität verfügt. Das ist tatsächlich nicht weit hergeholt, wenn wir bedenken, dass viele Staaten, vor allem in der Zeit nach dem Kolonialismus, unabhängig von jeglicher empfundenen Solidarität gebildet wurden. Diese Tatsache ist auf vielerlei Arten von Bedeutung. Laut Paul Collier könnte dieser Umstand erklären, warum manche afrikanischen Staaten sich langsamer entwickelten als ökonomische Theorien es erwarten lassen konnten.[22] Stellen Sie sich weiterhin vor, dass aus irgendwelchen Gründen eine solche, von Solidarität relativ freie Gesellschaft, letztlich tatsächlich funktioniert. Sie ist effizient, fair und die demokratischen Institutionen der politischen Entscheidungsfindung funktionieren nicht schlechter als andernorts. Diese Gesellschaft ist eine bunt gemischte Gruppe von Individuen, die wenig gemeinsam haben und sich gegenseitig auch nicht sonderlich mögen – aber sie arbeiten dennoch für den Erhalt dieser Gesellschaft, da sie denken, dass diese Arbeit getan werden sollte. Mich interessiert nicht, ob dieses Beispiel eine realistische Möglichkeit darstellt, sondern vielmehr, welche Rechte einer solchen Gesellschaft im Falle ihrer Existenz zukommen sollten. Ich meine, dass, sofern irgendeine Gesellschaft ein Recht auf Ausschluss hat, dieses Recht ebenso auch allen anderen rechtsstaatlichen Gesellschaften zukommen sollte, und zwar unabhängig vom Umfang an gefühlter Solidarität, der in ihnen herrscht. Wenn die von uns imaginierte, vielfältige Gesellschaft zu diesen Garantien fähig ist, dann scheint es so, als ob sie einen gleichwertigen Anspruch auf das Recht auf Ausschluss hat. Es wäre doch zutiefst abwegig, sollte sie dieses Recht gerade aufgrund der Tatsache ihrer Diversität verlieren. Warum sollte eine vielfältige Gesellschaft, deren Mitglieder nicht viel gemein haben, mit weniger politischen Rechten ausgestattet sein als konventionellere Gemeinwesen? Mir fällt keine Antwort ein, bei der die moralische Gewichtung nicht an der falschen Stelle ansetzt. Die Erscheinungsformen der Solidarität sind meiner Meinung nach tatsächlich politisch nützlich, allerdings sollte bei der Rechtfertigung eines Rechts auf Ausschluss das Konzept der Politik selbst den Ausschlag geben – und nicht Solidarität. Anschaulicher ausgedrückt: Flaggen und Hymnen sind nützlich um Politik zu betreiben, aber eine Gesellschaft, die ihre Flaggen und Hymnen verloren hätte, würde dadurch nicht auch ihr Recht auf Ausschluss verlieren.

Hier sollte ich anmerken, dass insbesondere Song eine komplexe Vorstellung von Solidarität hat, der zufolge sich der Gedanke der Solidarität und die Tatsache starker Diversität nicht widersprechen müssen.[23] In der Tat kann Solidarität für Song auch auf einem gemeinsamen Respekt vor der Vielfalt innerhalb eines Staates beruhen. Aber auch in diesem Fall ist das Recht auf Ausschluss noch abhängig davon, dass wir etwas in unseren Einstellungen gemeinsam haben – in diesem Fall ist es die gemeinsame Einstellung gegenüber dem, was nicht geteilt wird. Wenn wir ein Recht auf Ausschluss rechtfertigen wollen, müssen wir jedoch weiter gehen und herausfinden, warum genau eine politische Gemeinschaft ein Recht auf Maßnahmen hat, die Migrantinnen ausschließen, und ich glaube, dieses Recht würde sich nicht auf Solidarität, sondern auf Politik gründen. Wie ich im nächsten Kapitel ausführen werde, kann Solidarität überaus nützlich sein – aber diese Tatsache, wie schwierig oder schädlich sie auch sein mag, sollte nicht für die Rechtfertigung eines Rechts auf Ausschluss herangezogen werden.

Ich möchte mit der Bemerkung enden, dass ich Autorinnen wie Song und Miller viel verdanke. Die von ihnen aufgestellten Theorien sind bemerkenswert – und wie sich zeigen wird, schulde ich beiden viel. Vor allem Songs vom Konzept des Hoheitsgebiets ausgehende Perspektive steht in enger Verbindung zu meinem eigenen Ansatz. Wenn ich mit diesen Autorinnen nicht übereinstimme, so liegt das allein daran, dass ich nicht der Meinung bin, dass Solidarität – oder irgendeine andere Form geteilter psychologischer Tatsachen – eine Rechtfertigung für ein Recht auf Ausschluss begründen kann.

3. Eigentum

Beide bisher besprochenen Argumentfamilien haben bei der Bedeutung bestimmter Güter für deren Besitzer begonnen. Wie ich versucht habe zu zeigen, reicht keine der Ideen vollständig aus, um ein Recht auf Ausschluss zu begründen. Rechtfertigungen, die mit dem Wert des Lands arbeiten, gewichten das Land selbst zu stark gegenüber dem Wert der autonomen Leben, die auf diesem Land existieren. Rechtfertigungen, die vom Wert der Solidarität ausgehen, messen den gemeinsamen Einstellungen im Vergleich zu den politischen Beziehungen, die sich mit ihrer Hilfe herausbilden, zu viel Wert bei. Wir können daher damit fortfahren, die Ansätze ins Auge zu fassen, die das Recht auf Ausschluss nicht anhand bestimmter Güter, sondern mittels eines bestimmten Systems von Rechten individueller Personen begründen wollen.

Argumente auf Basis von Eigentumsrechten sind besonders starke Varianten dieser Herangehensweise, denn schließlich enthält die Idee der Eigentumsrechte den Gedanken, dass manche Menschen Rechte über bestimmte Dinge haben, die anderen nicht zukommen. Wenn ich das Eigentum an meinem Haus habe, bedeutet das im Allgemeinen, dass ich entscheiden kann, was mit diesem Haus geschehen soll: ob es verkauft, repariert oder zerstört wird und so weiter. Von größerer Bedeutung aber ist, dass ein solches Eigentumsrecht in den meisten Fällen auch bedeutet, dass ich entscheiden kann, Sie aus meinem Haus herauszuhalten. Eigentumsrechte werden darüber hinaus oft in Verbindung mit Arbeit verstanden. Insbesondere John Lockes Theorie des Eigentumsrechts spricht denjenigen Recht auf Land zu, die an dessen Verbesserung mitgewirkt haben. Es liegt daher nahe, diese soeben ausgeführten Ideen zu einer bestimmten Perspektive auf politische

Institutionen zu verbinden und zu argumentieren, dass die bereits an einem Ort anwesenden Menschen Eigentumsrechte an den Institutionen haben, die diesen Ort verwalten, und zwar schlicht deshalb, weil sie am Aufbau und Erhalt dieser Institutionen mitgewirkt haben. Wir können diejenigen außerhalb der Staatsgrenzen gewissermaßen deshalb ausschließen, weil wir – und nicht sie – Eigentumsansprüche an den Institutionen erworben haben, die unser Leben bestimmen.

Ryan Pevnick hat mehr als irgendwer sonst dafür getan, diese Ideen auf die politische Moral der Migration anzuwenden. Sein Ansatz fußt auf dem Gedanken, dass es eine intergenerationale Gemeinschaft gibt, die daran gearbeitet hat, die Institutionen eines bestimmten Ortes aufzubauen; diese Gemeinschaft hat ein spezifisches Bündel an Eigentumsrechten aufgrund der Tatsache erworben, dass diese Gemeinschaft Arbeit in den Aufbau der Institutionen investiert hat. Die fragliche Gemeinschaft muss, darin ist Pevnick klar, generationsübergreifend verstanden werden, was bedeutet, dass die von uns genutzten Institutionen von früheren Mitgliedern der Gemeinschaft aufgebaut wurden und von uns an die nachfolgenden Generationen weitergegeben werden. Die Institutionen werden uns demnach dann zu eigen, wenn wir unseren Teil zu ihrem Aufbau und Erhalt beitragen:

> „Wie eine Familienfarm ist auch der Aufbau staatlicher Institutionen ein geschichtliches Projekt, das sich über Generationen hinweg erstreckt und in das die Individuen hineingeboren werden. So, wie der Wert einer Farm größtenteils durch beständige Verbesserungen geschaffen wird, ist auch der Wert der Mitgliedschaft in einem Staat zum großen Teil das Ergebnis der Arbeit und Investitionen der Gemeinschaft. Die Bürgerschaft erhält durch Besteuerung Ressourcen und investiert diese in wertvolle öffentliche Güter: grundlegende Infrastruktur, Verteidigung, Aufbau und Erhalt eines effektiven Marktes, ein Bildungssystem, und so weiter. […] Dies alles sind Güter, die einzig aufgrund der Arbeit und Investitionen der Mitglieder der Gemeinschaft existieren."[24]

Diese Überlegungen erlauben es uns also, zwischen Mitgliedern und außenstehenden Personen zu unterscheiden und bieten uns zudem eine Rechtfertigung für den Ausschluss letzterer. Die außenstehenden Personen haben aufgrund der Tatsache, dass sie nicht am Erhalt der infrage stehenden Institutionen mitgearbeitet haben, kein Recht darauf, diese Institutionen zu kontrollieren und somit auch kein Recht darauf, sich darüber zu beschweren, wenn sie von den Eigentümerinnen dieser Institutionen ausgeschlossen werden. Da der Einlass außenstehender Personen diesen eine proportionale Kontrolle über die Institutionen zusprechen würde, wäre ihr potentieller Ausschluss demnach nicht ungerecht.

Pevnicks Ansatz ist komplex und liefert eine starke Verteidigung des Rechts auf Ausschluss. Allerdings glaube ich nicht, dass sie durchweg funktioniert. In meinen Augen hat sein Ansatz zwei Probleme. Das erste besteht darin, dass die Theorie des auf Arbeit beruhenden Eigentums bloß eine schwache Beschreibung dessen darstellt, was wir unter einer politischen Gemeinschaft verstehen sollten. Das zweite Problem betrifft den Umstand, dass die Theorie durch ihre Verbindung gegenwärtiger Rechte mit vergangener Arbeit Schlussfolgerungen zu rechtfertigen scheint, die auf eklatante Weise das Prinzip moralischer Gleichheit verletzen. Ich werde die Bedenken in dieser Reihenfolge besprechen.

Das von Pevnick entwickelte Argument besteht aus zwei Schritten: Wenn wir bestimmte Institutionen aufgebaut haben, können wir als deren Eigentümer betrachtet werden; und wenn wir erst einmal Eigentum an ihnen erworben haben, können wir ungebetene Immigrantinnen von der Nutzung dieser Institutionen ausschließen. Beide Schritte scheinen ein paar Probleme zu beinhalten. Der zweite Schritt argumentiert, dass der Kern eines Eigentumsrechts das Recht auf Ausschluss sei. Das ist jedoch nicht notwendigerweise der Fall. Wie Generationen von Jura-Studentinnen lernen mussten, beinhaltet das Konzept des Eigentums ein Bündel an Rechten, wovon keines der konzeptionelle Kern des Konzepts selbst ist. Wichtiger ist jedoch, dass wir Anlass dazu haben, den ersten Schritt detaillierter zu untersuchen. Ihm zufolge kann behauptet werden, wir besäßen Eigentum an einem Gegenstand, wenn wir gemeinsam an seiner Produktion gearbeitet haben. Das mag für viele Dinge der Fall sein, allerdings ist unklar, ob es für *alle* Dinge gilt. Susan Moller Okin hat auf bemerkenswerte Weise herausgearbeitet, wie absurd es ist, die von Frauen geleistete Reproduktionsarbeit durch die Brille der Locke'schen Eigentumstheorie zu betrachten. Ihrem Argument zufolge wird Frauen kein Eigentum an ihren Kindern zugesprochen, obwohl sie Arbeit in die Existenz dieser Kinder investiert haben.[25] Was für Kinder gilt, kann allerdings ebenso auf politische Institutionen angewendet werden. Auch in deren Fall ist die Arbeit, die in den Aufbau der Institutionen investiert wurde, nicht der einzige moralisch bedeutsame Faktor. So könnten wir auch die Beziehungen zwischen den Menschen innerhalb des Staates und ihre Beziehung zu diesem Staat für unsere Betrachtungen heranziehen. Wir könnten diese Beziehungen selbst – und nicht die historische Erzählung darüber, wie diese Beziehungen zustande gekommen sind – als moralisch grundlegend ansehen.

Zur Veranschaulichung möchte ich den Fall von Pullman, Illinois, besprechen. Diese Stadt wurde von George Pullman gegründet, der sich selbst zugleich als Eigentümer der Stadt verstand. Walzer beschreibt die Situation wie folgt:

> „Pullman (der Besitzer) baute nicht nur Fabriken und Wohnheime, wie es etwa fünfzig Jahre zuvor in Lowell, Massachusetts, geschehen war. Er errichtete Villen, Reihenhäuser und Mietblöcke für sieben- bis achttausend Menschen, dazu (unter kunstvollen Arkaden gelegene) Läden und Büroräume sowie Schulen, Sportanalgen, Spielplätze, einen Markt, ein Hotel und eine Bibliothek, ein Theater und sogar eine Kirche: kurz, eine Modellstadt, eine auf dem Reißbrett geplante Stadtgemeine. Und jedes Stück darin gehörte ihm."[26]

Pullman sah die von ihm errichtete Stadt also als sein Eigentum an und handelte entsprechend. Seine eigenen Ansichten waren in jeder Hinsicht entscheidend, sei es in Fragen der Religion, der Bildung, des Geschäftsbetriebs oder sogar des Kleidungsstils. Wenn die Einwohnerinnen sich über seine Einschränkungen beschwerten, verwies er einfach auf die Tatsache, dass er diesen Ort erschaffen hatte und er noch immer vollständig sein Eigentum sei; sollten seine Ansichten einem Arbeiter missfallen, sei dieser stets frei, die Stadt zu verlassen.

Inwiefern ist dieser Fall nun aber für Pevnicks Analyse relevant? An ihm zeigen sich zwei unterschiedliche Möglichkeiten, ein Bündel sozialer Institutionen zu betrachten. Die erste Perspektive beginnt beim Aufbau dieser Institutionen und argumentiert, dass sie eine Form von Eigentum sind; die Geschichte ihrer Entstehung – also wessen Arbeit sie ins Leben gerufen hat – erlaubt es uns, zu bestimmen, wer ausgeschlossen werden darf und welche Arten von Ausschluss zulässig sind. Die andere Perspektive beginnt dort, wo ich gerne beginnen würde, nämlich bei den Beziehungen zwischen den Menschen und ihren Institutionen, und fragt, was den Ausschluss von diesen Institutionen rechtfertigen könnte – allerdings beruht die Begründung in diesem Fall nicht auf in der Vergangenheit erworbenem Eigentum, sondern auf den gegenwärtig bestehenden Beziehungen. Welche Rechte Pullman hat, ist demnach abhängig von den eigenen Ansichten bezüglich dieser Unterscheidung. In der Realität wurde selbstverständlich verfügt, dass Pullman seine Modellstadt verkaufen musste. Dem Supreme Court of Illinois zufolge steht das Eigentum an einem solchen Bündel von Institutionen „im Gegensatz zu guter Politik, sowie zu der Theorie und dem Geist unserer Institutionen."[27] Mit anderen Worten: Für das Gericht war demnach nicht die Arbeit ausschlaggebend, die der Unternehmer Pullman investiert hatte, sondern die gegenwärtigen Beziehungen, die in der Stadt Pullman existierten.

Etwas ähnliches kann meiner Meinung nach auch über den Staat selbst gesagt werden, wenn er ungebetene Migrantinnen auszuschließen gedenkt. Der Staat ist nicht einfach nur ein Ding, sondern die Bezeichnung für ein komplexes Muster menschlicher Beziehungen. Wenn wir versuchen, Ausschluss auf Basis dieser Beziehungen zu rechtfertigen, haben wir allen Grund, diese Beziehungen selbst genauer zu betrachten. Es mag dabei relevant sein, wie diese Beziehungen ursprünglich entstanden sind, aber diese Relevanz ist zwangsläufig komplizierter Natur. Unser Handeln geschieht über die Zeit hinweg und wir nehmen unser Wirken durch die von uns gemachten Pläne wahr. Aus diesem Grund sollten wir die Vergangenheit betrachten, wenn wir die Gegenwart verstehen wollen. Allerdings kann ein solches Verständnis auch ohne Rückgriff auf die Idee von einer intergenerationalen Gemeinschaft oder die von dieser Gemeinschaft in der Vergangenheit investierte Arbeit erreicht werden. Tatsächlich erscheint es als konzeptioneller Fehler, eine Gemeinschaft für die Rechtfertigung eines Rechts auf Ausschluss anzuführen. Sollte ein Recht auf Ausschluss zu verteidigen sein, scheint es von einem neuen Staat ebenso in Anspruch genommen werden zu können wie von einem bereits länger bestehenden Staat. Angenommen, ein Staat würde plötzlich um Mitternacht entstehen, dann würde ihm vom Moment seiner Existenz an ein Recht auf Ausschluss zukommen. Die Bewohnerinnen innerhalb der Grenzen dieses Staats müssten demnach nicht warten, bis sie Arbeit in die Institutionen dieses Staates investiert haben, bevor sie ein solches Recht in Anspruch nehmen könnten.

Diese Überlegungen führen uns zu einem weiteren Problem von Pevnicks Ansatz: Der Verweis auf die Geschichte – und auf die Arbeit – trägt das Potential moralischer Abwegigkeit in sich. Wenn, erstens, die intergenerationale Gemeinschaft die Eigentumsrechte innehat, besteht beständig die Sorge, dass denjenigen mit einer engeren Beziehung zu dieser historisch gewachsenen Gemeinschaft auch mehr Rechte im Hinblick auf die weitere Ausgestaltung dieser Institutionen zukommen. Nehmen Sie zum Beispiel die soziale Macht, die in den Vereinigten Staaten damit einhergeht, die eigene Abstammung auf einen Passagier der *Mayflower* zurückführen zu können. Laut dem *Chicago Tribune* stellen Personen mit einer solchen Abstammung „die Spitze des amerikanischen Establishments in der Finanzwirtschaft, dem Rechtswesen, der Literatur, der Unterhaltung und nahezu jedem anderen Feld" dar, unter ihnen sieben Präsidenten.[28] Wenn das Recht darauf, Teil einer bestimmten Gesellschaft zu sein, den Platz widerspiegelt, den eine Person innerhalb eines generationenübergreifenden Aufbauprozesses einnimmt, scheint das kaum außergewöhnlich; offensichtlich haben in einem solchen Fall diejenigen, deren Familien länger vor Ort waren, ein Recht auf größere Macht an eben diesem Ort. Allerdings gehe ich davon aus, dass die meisten von uns die Macht, welche den Nachfahren der *Mayflower* zukommt, wahrscheinlich eher als illegitimes Privileg betrachten. Sollte ein Migrant über Nacht eingebürgert werden, scheint er nach Pevnicks Ansatz weniger Mitspracherechte hinsichtlich, beispielsweise, der Migrationspolitik zu besitzen, als eine innerhalb der Grenzen der Gemeinschaft geborene und aufgewachsene Bürgerin, und diese wiederum würde über weniger Rechte verfügen als die Nachfahren der *Mayflower*-Pioniere. In meinen Augen führt das zu tiefgreifenden Problemen. Ein zentraler Bestandteil der Einbürgerung ist schließlich, dass die Eingebürgerten sich als allen anderen Bürgerinnen gleich an Rechten gegenüber dem Staat verstehen dürfen. Darüber hinaus ignoriert ein Ansatz, der das Recht auf Meinungsäußerung an die investierte Arbeit koppelt, dass Menschen auf vielfältige Weisen nicht in der Lage sein können zu arbeiten. Kinder, ältere Menschen oder Menschen mit Behinderungen haben allesamt verschiedene Einschränkungen im Hinblick darauf, auf welche Weise und wie einfach sie für die staatlichen Institutionen arbeiten können. Die Idee gleicher Staatsbürgerschaft scheint jedoch zu fordern, dass alle Bürgerinnen hinsichtlich des Rechts auf politische Meinungsäußerung gleich behandelt werden.

Tatsächlich erkennt Pevnick selbst an, dass sein Ansatz die Möglichkeit bietet, ungewollte Kinder aus der Gesellschaft auszuschließen, woraufhin er die politische Gleichheit als sekundäre Grundlage für die Beurteilung der Ansprüche von Bürgerinnen einführt.[29] In meinen Augen verweist die Notwendigkeit solcher ad-hoc Prinzipien mitunter darauf, dass ein Problem im Ansatz selbst vorliegt. Indem er seine Theorie mit geleisteter Arbeit und der Geschichte einer Gemeinschaft beginnt, spricht Pevnick der Geschichte die falsche Art von Bedeutung für die gegenwärtige Diskussion zu und kommt zu Ergebnissen, die er mit der Idee gleicher Staatsbürgerschaft korrigieren muss. Es wäre meiner Meinung nach besser, mit dem Staat in seiner gegenwärtigen Gestalt zu beginnen – verstanden als eine Beziehung zwischen bestimmten Personen, ganz unabhängig von der Arbeit, die sie in der Vergangenheit aufgewendet haben –, und zu schauen, welche Rechte auf Ausschluss aus dieser spezifischen Perspektive abgeleitet werden können.

4. Vereinigungsfreiheit

Dieses Kapitel endet mit der Untersuchung einer besonders starken Verteidigung des Rechts auf Ausschluss. Diese Verteidigung, entwickelt von Christopher Heath Wellman, begründet das Recht auf Ausschluss mit dem Recht auf Vereinigung, dass den Mitgliedern eines Staates zukommt. Wellmans Ansatz ist bewundernswert schlicht. Er beginnt mit drei ziemlich unstrittigen Prämissen, von denen die erste lautet, dass Menschen ein Recht darauf haben, sich mit bestimmten anderen Personen zu verbinden. Tatsächlich ist dieses Recht der Ursprung vieler Dinge, die uns besonders wichtig sind: Wir gestalten unser Dasein mit bestimmten anderen Menschen und gestalten auf diesem Wege ein für uns wertvolles Leben. Wellmans zweite Prämisse lautet, dass dieses Recht nicht bloß ein Recht auf Vereinigung darstellt, sondern ebenso ein Recht darauf, unerwünschte Verbindungen zurückzuweisen. Wenn eine bestimmte Vereinigung allen offen stehen muss – selbst denjenigen, die den Zielen dieser Vereinigung feindlich gesinnt sind –, kann sie ihre Aufgabe in den Leben derjenigen, die von dieser Vereinigung profitieren, nicht mehr erfüllen. Die letzte von Wellman angeführte Prämisse lautet, dass dieses Recht nicht allein individuellen natürlichen Personen, sondern auch Vereinigungen mehrerer Personen und anderen Körperschaften zukommt – darunter auch dem Staat.[30]

Dieses Argument ist in der Darstellung simpel, hinsichtlich seiner Schlussfolgerungen jedoch wenig kompromissbereit. Für Wellman ist das Recht darauf, eine unerwünschte Vereinigung mit ungebetenen Immigrantinnen vermeiden zu dürfen, tatsächlich ein sehr starkes Recht; der Staat kann mehr oder weniger all diejenigen ausschließen, mit denen seine Einwohnerinnen keine Verbindung eingehen wollen. Wellman erkennt zwar die Ansprüche von Flüchtlingen an, argumentiert aber, dass der Staat seine Pflichten gegenüber bedürftigen Migrantinnen auch durch andere Mittel als das der Einwanderung nachkommen kann – so mitunter auch durch Entwicklungshilfe. Die simple Tatsache, dass Menschen möglicherweise auf ihren Straßen oder in ihren Schulen keine Einwanderer sehen wollen, genügt demnach, um ein Recht auf Ausschluss zu begründen. Wellman zufolge müssen wir diese Entscheidung nicht rechtfertigen, genauso wie ich auch keiner Person eine Rechtfertigung schulde, mit der ich nicht befreundet sein oder die ich nicht heiraten möchte. Migration bedeutet Vereinigung, und Vereinigung gehört ins Reich der Freiheit.

Wellmans Ansatz sieht sich intensiver Kritik ausgesetzt.[31] An dieser Stelle möchte ich nur ein paar der Gründe aufzeigen, weshalb ich seine Methodologie nicht in Gänze akzeptieren kann. Zuerst bin ich mir nicht sicher, ob das Recht auf Vereinigungsfreiheit so stark ist, wie Wellman behauptet – so haben beispielsweise Gerichte in den Vereinigten Staaten oft befunden, dass andere Rechte von größerem Gewicht sind als das Recht darauf, frei von ungewollten Verbindungen zu sein. Die Jaycees wurden zum Beispiel dazu gezwungen, ihre Vereinigung in den USA auch für Frauen zu öffnen – eine Tatsache, die für den Supreme Court darauf beruhte, dass gleichen Bürgerrechten ein größeres Gewicht zukommt als der Vereinigungsfreiheit.[32] Einige Kommentatoren haben zudem angeführt, dass die von Wellman angewandte Methodik verschiedene Arten von Beziehungen unterschiedlich abbilden müsste, was zu komplizierten Ergebnissen führt. So hat beispielsweise Matthew Lister dahingehend argumentiert, dass familiäre Verbindungen vor Gericht im Allgemeinen als bedeutsamer angesehen werden als andere Arten von Verbindungen; falls also meine Bedürfnisse, meine Frau zu sehen, mit Ihrem Bedürfnis in Konflikt gerät, meiner Frau nicht auf dem Marktplatz begegnen zu müssen, würden Ihre Präferenzen wohl gegenüber meinen eigenen zurückstehen müssen.[33]

Zum Abschluss dieser Diskussion möchte ich darauf hinweisen, dass Wellmans Ansatz meiner Meinung nach nicht korrekt wiedergibt, was beim Eintritt einer neuen Person in meine Gesellschaft tatsächlich passiert. Wellman behauptet, dass meine Entscheidung, eine Verbindung mit dieser Person zu vermeiden, respektiert werden sollte. Aber ich bin mir nicht sicher, ob Verbindung in diesem Fall der richtige Begriff für den infrage stehenden Vorgang ist. In jedem halbwegs großen Staat ist es unwahrscheinlich, dass ich einen bestimmten Einwanderer tatsächlich treffe. Womöglich lebe ich so behütet oder so fern von Ballungsräumen, dass ich noch nicht einmal von deren Existenz weiß. Wellmans Ansatz scheint jedoch auf einem stärkeren Konzept der *Vereinigung* zu beruhen. Für ihn besteht der Unterschied zwischen meiner Beziehung zu meiner Frau und meiner Beziehung zu einer bestimmten Migrantin schlicht in der Intensität dieser Beziehung; mit beiden pflege ich Umgang, allerdings tue ich dies im Falle meiner Frau häufiger und aufgrund von Zwecken, die sich in meinem Umgang mit der Migrantin nicht finden. Ich hingegen denke, dass die beiden Beziehungen von verschiedener *Art* sind. Ich *verbinde* mich mit meiner Frau: Ich sehe sie regelmäßig, rede mit ihr und so weiter und ich werde gereizt, wenn dem nicht so ist. Mit der Immigrantin hingegen teile ich etwas Bedeutsames – nämlich die Verpflichtung gegenüber einem bestimmten Staat. Diese Gemeinsamkeit ist moralisch relevant; tatsächlich ist sie bedeutsam genug, um einen plausiblen Ansatz für ein Recht auf Ausschluss zu entwickeln, wie ich im nächsten Kapitel zeigen werde. Aber zu meinen, dass diese beiden Beziehungsformen bloß ihrer Intensität nach voneinander verschieden sind, scheint mir einen fundamentalen Unterschied zu übersehen. Meiner Meinung nach sollten wir die Verbindung, die durch eine gemeinsame Unterwerfung unter das Gesetz entsteht, für sich genommen analysieren, statt sie bloß als ein Beispiel aus der Kategorie persönlicher Verbindungen zu betrachten. Es ist diese geteilte rechtliche Verbindung, die ich im nächsten Kapitel analysieren werde, und nicht die empfundene emotionale Verbindung, die mich zweifellos an meine Frau bindet (und die im Falle der Migrantin nicht existiert.)

In meinen Augen liegt Wellman richtig, wenn er sich die Beziehungen zwischen den bereits an einem Ort lebenden Personen anschaut, und diese Beziehungen anschließend für die Formulierung einer Rechtfertigung nutzt, die gegenüber ungebetenen Migrantinnen vorgebracht werden kann. Allerdings liegt er meiner Meinung nach im Hinblick auf das, was jene ungebetenen Migrantinnen zu tun gedenken, nicht ganz richtig. Wir sollten daher eine spezifisch *politische* Idee des Ausschlusses betrachten und uns dabei insbesondere auf das konzentrieren, was von den bereits an einem Ort lebenden Personen geteilt wird. Diese Idee werde ich im folgenden Kapitel behandeln.

Meine Kritik an Wellmans Ansatz war kurz, was teilweise auch daran liegt, dass mir vieles richtig erscheint; wie sich zeigen wird, kann mein eigener Ansatz mitunter als eine verbesserte Version seiner Überlegungen interpretiert werden. In meinem Argument ist die Verbindung, die zur Rechtfertigung staatlichen Ausschluss herangezogen wird, selbstverständlich keine persönliche, sondern eine politische. Die Struktur unserer Argumente ist jedoch bemerkenswert ähnlich. Was ich über Wellman zu sagen habe, gleicht allerdings dem, was ich auch über die bereits zuvor diskutierten Autorinnen geäußert habe. Viele dieser Leute haben starke Theorien der Gerechtigkeit im Bereich der Migration entwickelt, die jedoch auf den falschen Grundlagen fußten: auf Fakten in den Köpfen anstatt auf den Beziehungen von Recht und Politik; auf Zuneigung statt auf der politischen Gemeinschaft, für die wir diese Zuneigung empfinden; auf dem Land statt auf der Politik, die mit und auf diesem Land betrieben wird. Wir haben daher Grund dazu, einen neuen Ausgangspunkt für die Rechtfertigung des Rechts auf Ausschluss zu suchen, und dieser Aufgabe widmet sich das nächste Kapitel.

4
Gerechtigkeit, Gebietshoheit und Migration

In den vorangegangenen Kapiteln habe ich mich zum einen gegen diejenigen Autorinnen gewandt, denen zufolge jeglicher Ausschluss von Immigrantinnen als an sich ungerecht zu betrachten ist. Wie ich gezeigt habe, achten all diese Ansätze zu wenig auf die Unterschiede zwischen Bürger- und Menschenrechten. In meinen Augen liegt ihr Fehler darin, den Unterschied, den eine politische Gemeinschaft ausmacht, nicht wahrzunehmen. Allerdings habe ich auch gegen diejenigen argumentiert, die ein Ausschlussrecht mit Verweis auf Land, Solidarität, Eigentumsrechte oder das Recht auf Vereinigungsfreiheit verteidigen. Hier hatte ich beanstandet, dass die ausschließende Instanz bloß ungenügend als politische Gemeinschaft begriffen wird. Es ist der Staat, verstanden als politische Einheit, der ungebetene Migrantinnen ausschließt; und es ist folglich die politische Natur des Staates, die ein Recht auf Ausschluss begründen sollte.

Es sollte daher nicht überraschen, dass ich nun für eine bestimmte Vorstellung des Rechts auf Ausschluss eintreten möchte, in der dieses Recht aus dem politischen Wesen des Staates abgeleitet wird. Die von mir vertretene Vorstellung staatlichen Ausschlusses hängt entsprechend nicht von der Existenz psychischer Einstellungen oder einem Konzept privater Vereinigungen ab, sondern von der spezifischen Form einer politischen Gemeinschaft, in der sich diejenigen Personen wiederfinden, die sich in einem Staat versammelt haben. Dabei handelt es sich, wie ich bemerken will, um eine *schwache* Konzeption des Staates. Einer solchen Konzeption zufolge stellt der nationale Charakter eines Staats keine eigenständige Quelle an Wert dar. Der von mir verteidigte Ansatz ist prinzipiell offen dafür, auch solchen politischen Gemeinschaften ein Recht auf Ausschluss zuzusprechen, die keinen eigenen nationalen Charakter haben; es reicht demnach aus, dass ein Staat auf gewisse Weise politisch organisiert ist; ich verlange keine substantielleren Eigenschaften. Darüber hinaus werde ich es vermeiden, mich in wesentlichen Punkten irgendwie auf generationsübergreifende Gemeinschaften, Land oder nationale Selbstbestimmung zu beziehen. Stattdessen werde ich schlicht versuchen, das Recht auf Ausschluss aus dem politischen Wesen des Staates abzuleiten.

Ausschlaggebend ist bei diesem Argument, dass der Staat eine territoriale und rechtliche Gemeinschaft darstellt, wobei das Territorium ein Hoheitsgebiet ausweist, in dem die Gesetze des Staates effektiv wirken. Daraus folgt, dass die Person, die ein Hoheitsgebiet betritt, die Einwohnerinnen dieses Territoriums dazu verpflichtet, den rechtlichen Schutz der Grundrechte auch auf sie auszuweiten. Dieser Vorgang begrenzt jedoch die Freiheit der gegenwärtigen Einwohnerinnen des Hoheitsgebiets und aus diesem Grunde sind die Einwohnerinnen berechtigt, ungebetene Migrantinnen am Betreten des Territoriums ihres Staates zu hindern. Dieses Recht auf Ausschluss übertrumpft jedoch nicht in allen Fällen die Rechte potentieller Migrantinnen, denn diese haben ein Recht auf Umstände, unter denen ihre Rechte geschützt werden. Daher wird es zahlreiche Fälle geben, in denen einem liberalen Staat der Ausschluss ungebetener einwanderungswilliger Personen nicht gestattet ist, da der Liberalismus die Einwohnerinnen dieses Staates in solchen Fällen dazu verpflichtet eine rechtliche Beziehung mit den Immigrantinnen einzugehen. Ich schlage daher vor, dass ein Recht auf Ausschluss zwar existiert, es allerdings nicht all diejenigen Ausschlusspraktiken rechtfertigen kann, die von den wohlhabenden Gesellschaften derzeit angewendet werden.

Ich werde dieses Argument in drei Schritten entwickeln. Im ersten Schritt werde ich eine bestimmte Methodik beschreiben, anhand derer ich versuchen werde, das Recht auf Ausschluss aus der im internationalen Recht niedergelegten Struktur des Menschenrechtsschutzes abzuleiten. Im zweiten Schritt werde ich zu zeigen versuchen, wie diese Methode zum einen ein präsumtives Recht auf Ausschluss einwanderungswilliger Personen rechtfertigen, sowie, zum anderen, auch ein Recht auf den Einsatz bestimmter Formen von Gewalt zur Durchsetzung jenes Rechts begründen kann. Der dritte Schritt wird schließlich darin bestehen, zwei mitunter problematische Implikationen dieses Ansatzes zu untersuchen. Aus diesen Implikationen folgt nicht zwingend, dass der hier vorgestellte Ansatz falsch ist, allerdings zeigen sie, dass das in ihm verteidigte Recht auf Ausschluss nicht die derzeit von manchen Staaten angewendeten Ausschlusspraktiken rechtfertigen kann. Den abschließenden Teil bildet eine Diskussion dreier Arten von Einwänden gegen die Struktur meines Arguments. Diese Einwände – basierend auf den Ideen der *Ausweisung*, *Reproduktion* und *Freiheit* – sind durchaus stark, aber ich hoffe, dass mein im Folgenden dargestelltes Argument die Auseinandersetzung mit ihnen übersteht.

1. Politische Gemeinschaft und Menschenrechte

Wir sollten aufgrund des Gesagten mit einer Analyse dessen beginnen, was meiner Meinung nach von allen Personen innerhalb einer Gesellschaft geteilt wird. Das ist in meinen Augen die Verpflichtung gegenüber der spezifischen Hoheitsgewalt des Staates. Was auch immer sonst ein Staat sein mag – ein Ort für eine Kultur, für eine bestimmte Identität oder für ein besonderes historisches Projekt –, so ist er in seinem Kern doch ein auf der Idee der *Gebietshoheit* basierendes Projekt, und zwar derart, dass er durch die Machtbefugnis definiert wird, die er über ein bestimmtes Gebiet innehat. Sofern ein Staat ein Recht auf Ausschluss besitzen und ihm dieses aufgrund seines Daseins als Staat zukommen soll, muss der hoheitsrechtliche Aspekt seines Wesens – also die Rechte, die ihm allein aufgrund seiner Existenz als Staat notwendig zukommen – den Ausschlag geben. Falls ein Staat jedoch über ein solches Recht verfügt, kann er es selbstredend für die Verteidigung anderer Güter nutzen – vielleicht für die Förderung einer Kultur oder für Handlungen aus Respekt vor den Wünschen seiner Mitglieder im Hinblick darauf, mit wem sie sich verbinden möchten. Die erste Frage lautet jedoch, ob ein solches Recht überhaupt existiert.

Beginnen möchte ich mit der Bestimmung dessen, was Staaten sind und zwar im Hinblick auf die grundlegendsten juristischen Begriffe, um dann zu analysieren, inwiefern diese juristischen Begriffe für die Begründung eines Rechts auf Ausschluss herangezogen werden können. Leitend ist dabei die Idee, einen Staat in seinen grundlegendsten Aspekten zu beschreiben – also hinsichtlich derjenigen Eigenschaften, ohne die wir den infrage stehende Objekt überhaupt nicht als Staat beschreiben würden –, um dann zu sehen, ob wir das Recht auf Ausschluss aus dieser Beschreibung entwickeln können. Diese Methode ähnelt derjenigen, die Robert Nozick für seine Rechtfertigung des Staats genutzt hat. Laut Nozick sollten wir die günstigste Version einer vor-politischen Situation konstruieren und dann untersuchen, was in einem solchen Fall zur Entstehung eines legitimen politischen Staats führen würde. Nozicks Rechtfertigungsansatz folgt also der Idee, den Staat als eine

Eigenschaft auszuweisen, die aus der bestmöglichen vor-politischen Situation entsteht: Wenn wir Gründe haben, diesen Naturzustand zu verlassen, hätten wir demnach auch Gründe dafür, uns mit zumindest einigen Formen politischer Organisation im Hier und Jetzt zu versöhnen.[1] Mein eigener Ansatz ist weniger ambitioniert. Ich gehe davon aus, dass Staaten existieren und dass sie gewisse Charakteristika aufweisen, ohne die wir sie nicht als Staaten bezeichnen würden – selbstverständlich *ohne* dabei ein Recht auf Ausschluss bereits vorauszusetzen. Daran anschließend möchte ich fragen, welche Aspekte dieser charakteristischen Eigenschaften als Gründe für ein kollektives Recht auf den Ausschluss potentieller Migrantinnen herangezogen werden könnten. Sollte sich herausstellen, dass aus diesem Material ein Recht auf Ausschluss hervorgeht, könnten wir gute Gründe für den Schluss haben, dass ein deontisches Recht auf Ausschluss verteidigt werden kann.

Dieses Vorgehen hat einige Vorzüge; insbesondere stellt uns das internationale Recht Material zur Verfügung, das wir für die Definition des Begriffs „Staat" heranziehen können. Zumeist wird in diesem Zusammenhang die Montevideo-Konvention von 1934 zitiert, die den Staat anhand von vier Eigenschaften definiert: eine ständige Bevölkerung, ein definiertes Staatsgebiet, eine Regierung sowie die Fähigkeit, in Beziehungen mit anderen Staaten treten zu können.[2] Die letzte Eigenschaft ist für uns in diesem Zusammenhang nicht weiter relevant – ganz im Gegensatz zu den ersten drei. Die dieser Definition zugrunde liegende Idee ist also, dass ein Staat nicht ohne die gleichzeitige Erfüllung dreier Eigenschaften existieren kann: einer Regierung, die dazu fähig ist Zwangsgewalt auszuüben; eines bestimmten Teils der Erdoberfläche, auf dem diese Zwangsgewalt ausgeübt werden kann; und schließlich einer bestimmten Gruppe von Menschen, über die jene Zwangsgewalt ausgeübt wird. Die letzte Eigenschaft zielt offensichtlich darauf ab, die unbewohnten oder unbewohnbaren Teile der Erdoberfläche von diesen Überlegungen auszuschließen; die Antarktis ist kein Staat und wird (von ziemlich extremen klimatischen Veränderungen einmal abgesehen) niemals ein Staat sein. Für unsere Zwecke ist vor allem von Bedeutung, dass sich ein Staat durch die effektiv über einen bestimmten Teil der Erdoberfläche ausgeübte Rechtshoheit konstituiert. Dieses Prinzip scheint, selbst für die Verfasser der Montevideo-Konvention grundlegend zu sein: Ihr zufolge müssen alle Einwohnerinnen eines Hoheitsgebiets, Fremde wie Bürgerinnen, „unter dem gleichen Schutz des Rechts" stehen.[3] Das Fundament eines Staates bildet demnach ein Bündel von Institutionen, das dazu fähig ist, all diejenigen Individuen effektiv zu regieren, die sich innerhalb eines bestimmten Hoheitsgebiets aufhalten.[4]

Wir können von dieser Vorstellung ausgehend mit unserer Analyse des Rechts auf Ausschluss beginnen. Gemäß diesem Ansatz existiert ein Staat überall dort, wo eine Regierung effektiv dazu imstande ist, politische und rechtliche Kontrolle über ein bestimmtes Hoheitsgebiet auszuüben. Das Recht auf Ausschluss ist nicht bereits Teil dieser Vorstellung, da es sich dabei genau um das handelt, was wir aus den genannten Eigenschaften abzuleiten versuchen. Mein Ansatz geht also allein von der Existenz territorialer Staaten aus und eine solche Annahme erscheint mir legitim. Tatsächlich meine ich, dass die Frage der Einwanderung an sich bloß unter einer solchen Prämisse Sinn ergibt, denn gäbe es nur eine globale Regierung, die über alles bewohnbare Land herrschte, wäre das Konzept der Einwanderung nicht sinnvoll anwendbar.

Wir gehen also davon aus, dass Staaten die Eigenschaften aufweisen, die sie zu Staaten machen. Darüber hinaus können wir annehmen, dass Menschen bestimmte Rechte schlicht aufgrund der Tatsache zukommen, dass sie Menschen sind. An dieser Stelle möchte ich vorsichtig sein mit dem, was ich sage; ich denke nämlich nicht, dass wir bisher irgendeine bestimmte, endgültige Antwort darauf haben, welche Menschenrechte tatsächlich durch rechtliche Institutionen verteidigt werden sollten. Theoretikerinnen und Praktikerinnen werden auch weiterhin uneins darüber sein, welche Rechte wir als universal ansehen sollten.[5] Es mag jedoch in Ordnung sein, über diese Schwierigkeiten hinwegzugehen, da ich eher die Struktur als den Inhalt der Menschenrechte untersuchen möchte. Obwohl Menschenrechte als Rechte definiert sind, die Menschen aufgrund ihrer Existenz als Menschen zukommen, bedeutet dieser Umstand nicht, dass alle Menschenrechte von allen Menschen auf gleiche Weise gegenüber allen menschlichen Institutionen vorgebracht werden können. In einer Welt, die in unterschiedliche Rechtsgebiete aufgeteilt ist, erlegen die Menschenrechte unterschiedlichen politischen Gemeinschaften auch unterschiedliche Verpflichtungen auf. Insbesondere kann hier auf die übliche Dreiteilung menschenrechtlicher Verpflichtungen verwiesen werden, nach der die Menschenrechte durch Staaten zu *achten*, zu *schützen* und zu *gewährleisten* sind.[6] Diese drei Arten von Verpflichtung erfordern es, dass Staaten gegenüber unterschiedlichen Personen auf unterschiedliche Weise handeln. Zunächst sind Staaten global gesehen dazu verpflichtet, die Menschenrechte zu *achten*; ein legitimer Staat darf die Menschenrechte anderer demnach nicht verletzen, unabhängig davon, ob sich die betroffenen Personen in seinem Hoheitsgebiet befinden oder nicht. Darüber hinaus sind Staaten jedoch auch dazu verpflichtet, die Menschenrechte derjenigen zu *schützen*, die sich in ihrem Hoheitsgebiet aufhalten. Damit ist gemeint, dass die Individuen innerhalb des Hoheitsgebiets eines Staates ein Recht darauf haben, das dieser Staat ihre Menschenrechte verteidigt und schützt. Diese Schutzpflicht verlangt selbstverständlich auch die Einrichtung politischer Institutionen, die dauerhaft in der Lage sind, jene Rechte zu schützen und ihre Verletzungen zu ahnden. Dies wird als die Verpflichtung verstanden, die Menschenrechte der innerhalb des Hoheitsgebiets eines Staates anwesenden Personen zu *gewährleisten*.

An dieser Stelle möchte ich betonen, dass die erste Pflicht zwar universell, die anderen zwei jedoch ausdrücklich bloß lokal Geltung beanspruchen können. Der Staat steht unter einer universellen Verpflichtung, die Verletzung von Menschenrechten zu vermeiden, ob diese Verletzungen nun innerhalb seines Hoheitsgebiets stattfinden würden oder nicht. Aber der Staat hat keine damit korrespondierende universelle Pflicht, die Rechte von Menschen aufgrund ihres Daseins als Menschen zu schützen oder zu gewährleisten. Der Staat ist hingegen bloß dazu verpflichtet, die Rechte mancher Menschen zu schützen und zu gewährleisten, nämlich derjenigen, die sich innerhalb seines Territoriums befinden. Diese Einschränkung scheint nicht schon an sich gegen die liberale Forderung der moralischen Gleichheit von Personen zu verstoßen – vielmehr wird durch sie die Idee der Gleichheit in einer Welt territorial organisierter Staaten erst handhabbar. Aus diesem Grund ist ein Anschlag in Frankreich auf einen französischen Staatsbürger unzweifelhaft eine Menschenrechtsverletzung und sollte ebenso unzweifelhaft von allen Staaten bedauert werden. Aber die Vereinigten Staaten sind in einem solchen Fall nicht dazu verpflichtet, ihre institutionellen Kapazitäten für die Verteidigung des Rechts dieses französischen Bürgers, frei von Anschlägen zu leben, aufzubringen. (Tatsächlich würde es die französische Regierung wohl eher als problematisch betrachten, sollten die USA Institutionen einrichten, die sich französischen Straftätern widmen.) Die Vereinigten Staaten dürfen ihre institutionellen Kapazitäten allein dem Schutz und der Gewährleistung der Rechte derjenigen widmen, die sich auf amerikanischem

Boden befinden. Der Grund dafür liegt nicht darin, dass französische Leben in den Augen der Vereinigten Staaten von geringerem Wert sein könnten als amerikanische Leben, denn schließlich ist es für die Zeit und den Aufwand, den die Institutionen der USA im Falle eines Angriffs auf eine Person innerhalb ihres Hoheitsgebiets investieren, unerheblich, ob es sich dabei um einen französischen Touristen oder einen US-Amerikaner handelt – gesetzt den Fall, dieser Aufwand ist gerechtfertigt. Die USA dürfen ihre Institutionen derart nutzen, da ihre Hoheitsrechte Beschränkungen unterliegen, die sie dazu autorisieren und verpflichten, die Menschenrechte bloß innerhalb eines bestimmten Gebiets der Erdoberfläche zu schützen und zu gewährleisten.[7] Diejenigen, die am amerikanischen System teilhaben, sind darüber hinaus dazu autorisiert und auch verpflichtet, das institutionelle System bei der Erfüllung dieser Aufgabe zu unterstützen. Wenn wir, wie ich, glauben, dass wir als Individuen eine allgemeine Pflicht dazu haben, gerechte Institutionen aufrechtzuerhalten, dann haben diejenigen, die innerhalb des Staatsgebiets der USA leben und arbeiten, eine sowohl rechtliche als auch moralische Verpflichtung, die Menschenrechte ihrer (auf das Hoheitsgebiet bezogenen) Nachbarinnen zu schützen und zu gewährleisten.[8]

Mit diesen Überlegungen im Hinterkopf können wir nun fortfahren zu untersuchen, was genau passiert, wenn eine Person die Grenzen eines Staates überschreitet. Nehmen Sie für den Moment einmal an, dass eine Person einfach von Frankreich aus in die Vereinigten Staaten schwimmt. Sie kommt an Land an und befindet sich nun innerhalb des Hoheitsgebietes der USA, wie weiter oben bereits beschrieben. Sollten die Vereinigten Staaten sie ausweisen wollen, müssten sie dafür, meiner Meinung nach, Gründe anführen, die sich auf das soeben entworfene Bild der Hoheitsgewalt beziehen. Dabei könnten wir beispielsweise mit dem Hinweis darauf beginnen, dass diese französische Staatsbürgerin in Frankreich bereits unter Institutionen lebte, die moralisch verpflichtet waren, ihre Menschenrechte zu schützen und zu gewährleisten. Darüber hinaus war sie auch Teil einer Bevölkerung, die legal und moralisch dazu verpflichtet war, diese Rechte mittels dieser Institutionen zu schützen. (Ich sollte hier bereits anmerken, dass die genannten Eigenschaften Frankreichs mit großer Wahrscheinlichkeit nur auf wenige andere Länder der Welt zutreffen; dieser Umstand wird vor allem dann von Bedeutung sein, wenn wir die Beschränkungen des Rechts auf Ausschluss betrachten, die aus dem auf der Gebietshoheit beruhenden Ansatz folgen. Durch das Verlassen des

Hoheitsgebiets Frankreichs hat diese Person zu einem gewissen Grad das Recht darauf aufgegeben, dass ihre Menschenrechte durch die französischen Institutionen geschützt und gewährleistet werden. Sie gibt dieses Recht selbstverständlich nicht in Gänze auf; sie hat jedes Recht dazu, nach Frankreich zurückzukehren und mitunter auch manche Rechte darauf, im Falle einer Verhaftung in den Vereinigten Staaten die Leistungen des französischen Konsulats in Anspruch zu nehmen. Allerdings kann sie, solange sie sich innerhalb des Hoheitsgebiets der Vereinigten Staaten aufhält, nicht mehr die französischen Institutionen für den direkten Schutz und die Gewährleistung ihrer grundlegenden Menschenrechte in Anspruch nehmen. Sie hat nun jedoch Anspruch darauf, dass ihre Menschenrechte durch die Vereinigten Staaten geschützt und gewährleistet werden – allein die Tatsache ihrer Anwesenheit innerhalb des Hoheitsgebiets reicht aus, um diese Institutionen zur Verteidigung ihrer Recht zu verpflichten. Darüber hinaus sind auch die Personen in den Vereinigten Staaten moralisch und rechtlich dazu verpflichtet, die Rechte dieser Person zu verteidigen – eine Verpflichtung, die sie vor ihrer Anwesenheit auf dem Territorium der USA nicht hatten. Die Bevölkerung ist nun also dazu verpflichtet, so zu handeln, dass die Rechte dieser Person effektiv geschützt und gewährleistet werden. So sind sie unter anderem dazu verpflichtet, ihren Beitrag zur Finanzierung der Polizei zu leisten, die die körperliche Sicherheit der Person im Falle eines Angriffs schützen soll, oder dazu, als Geschworene vor dem Gericht zu dienen, das den Angreifer verurteilen würde – sie sind also dazu verpflichtet solche Institutionen zu schaffen und zu erhalten, die dem Schutz der grundlegenden Menschenrechte dieser Person angemessen sind. Es sollte dabei erwähnt werden, dass diese Verpflichtung schlicht dem Fakt der Anwesenheit dieser Person entspringt; sie bedarf demnach keines speziellen rechtlichen Status innerhalb dieses Hoheitsgebiets. Das internationale Recht verlangt, dass die Reichweite staatlicher Institutionen all diejenigen umfasst, die sich innerhalb des jeweiligen staatlichen Hoheitsgebiets aufhalten.[9] Diese Überlegungen hinsichtlich des Geltungsbereichs staatlicher Rechtshoheit finden sich auch im innerstaatlichen Recht wieder: So meinen beispielsweise die Vereinigten Staaten, dass das Recht auf Grundschulbildung all jenen zukommt, die sich innerhalb ihres Hoheitsgebiets aufhalten, und zwar unabhängig davon, ob ihr Aufenthalt legal ist oder nicht. Dieser Punkt wird auf instruktive Weise auch im Fall *Plyler v. Doe* angeführt, den wir bereits in Kapitel zwei betrachtet haben:

> „Der Schutz des vierzehnten Verfassungszusatzes erstreckt sich auf jeden, Bürger oder Ausländer, der den Gesetzen eines Bundesstaates unterworfen *ist*, und reicht bis in jede Ecke des Gebiets dieses Bundesstaats. Dass der Eintritt einer Person in einen Bundesstaat oder die Vereinigten Staaten ursprünglich rechtswidrig war, und sie deshalb ausgewiesen werden kann, negiert nicht die schlichte Tatsache ihrer Präsenz innerhalb des Territoriums eines Bundesstaats. Im Falle einer solchen Anwesenheit ist sie in vollem Umfang den Verpflichtungen unterworfen, die ihr durch die zivil- und strafrechtlichen Gesetze dieses Bundesstaats auferlegt werden. Bis zu dem Zeitpunkt, an dem sie das Hoheitsgebiet verlässt – sei es freiwillig oder unfreiwillig und in Übereinstimmung mit der Verfassung sowie den Gesetzen der Vereinigten Staaten – steht ihr der gleiche Schutz durch die Gesetze des Bundesstaats zu.“[10]

Aus alledem ergibt sich in meinen Augen ein starkes Bündel an Verpflichtungen aufseiten der gegenwärtigen Einwohnerinnen eines bestimmten Hoheitsgebiets hinsichtlich des Schutzes sowie der Gewährleistung der Menschenrechte all jener Personen, und ausschließlich jener Personen, die sich innerhalb dieses Hoheitsgebiets befinden. Die Bewohnerinnen eines Hoheitsgebiets sind nicht dazu verpflichtet, die Rechte von Personen im Ausland zu schützen und zu gewährleisten – genauer gesagt sind sie hierzu nicht verpflichtet, wenn die Rechte dieser Personen durch die in ihrem Land herrschenden Institutionen angemessen geschützt werden. Demnach sind amerikanische Bürger gegenüber Bürgern von Somalia mitunter dazu verpflichtet, beim Aufbau politischer Institutionen in Somalia zu helfen oder es den somalischen Bürgern zu erlauben, das Hoheitsgebiet der Vereinigten Staaten zu betreten, sodass sie durch die US-amerikanischen Institutionen geschützt werden. Allerdings kommt den Einwohnerinnen der USA keine Verpflichtung zu, diesen Schutz auch auf in Frankreich lebende französische Bürgerinnen auszuweiten und wenn ein französischer Bürger die Vereinigten Staaten betritt, legt er den derzeitigen Bewohnerinnen der USA daher ein neues Bündel von Verpflichtungen auf. Sofern es ein Recht auf den Ausschluss ungebetener Immigrantinnen gibt, muss dieses Recht meinem Ansatz zufolge aus diesen Tatsachen erwachsen.

2. Das Recht auf Ausschluss

Wie also könnte ein Staat, der auf Grundlage dieser Begriffe verstanden wird, begründen, dass er ein Recht auf die Zurückweisung einer einwanderungswilligen Person hat? Dieses Argument wird in meinen Augen seinen Ausgang von der Tatsache nehmen müssen, dass die ein Hoheitsgebiet betretende Person denjenigen eine Verpflichtung auferlegt, die sich bereits innerhalb dieses Hoheitsgebiets aufhalten: primär eine Verpflichtung, Institutionen zu schaffen und zu unterstützen, die dazu fähig sind, die Rechte von Neuankömmlingen zu schützen, sowie eine Pflicht, sich innerhalb dieser Institutionen so zu verhalten, dass die Verteidigung dieser Rechte auch tatsächlich gewährleistet wird. Sofern wir ungebetene potentielle Migrantinnen zu Recht ausschließen dürfen, wird der Grund dafür sein, dass wir ein gewisses Recht darauf haben, diese neue Verpflichtung zurückzuweisen. Ich sollte bereits im Vorfeld darauf hinweisen, dass dieses Argument sich nicht auf die *Kosten* dieser Verpflichtung bezieht, sondern auf die bloße Tatsache, dass diese Verpflichtung existiert – mich beschäftigt die Frage, ob wir ein Recht darauf haben, frei einer Verpflichtung zu sein, die uns vorschreibt, auf eine bestimmte Weise gegenüber bestimmten Personen zu handeln und nicht, ob ihre Anwesenheit uns monetäre Kosten auferlegt. (Wir könnten annehmen, dass der ungebetene französische Auswanderer einen finanziellen Segen für die Vereinigten Staaten darstellt, da er sehr wahrscheinlich Steuern zahlen und nicht auf Leistungen des Sozialsystems angewiesen sein wird.) Gibt es ein Recht darauf, eine Person schlicht deshalb auszuschließen, weil ihre Anwesenheit uns zu bestimmten Handlungen verpflichten würde?

In meinen Augen ist das durchaus plausibel, sofern wir die Idee der Freiheit ernst nehmen. Eine Person, die mir eine Verpflichtung auferlegt, schränkt meine Freiheit ein, wenn auch auf sehr begrenzte Weise. Im Falle einer gesetzlichen Verpflichtung ist meine Freiheit in dem Sinne eingeschränkt, dass ich eine bestimmte Handlung nun nicht mehr ohne negative Konsequenzen ausführen kann, denn schließlich beruht das Wesen zwangsbewährter Gesetze auf ihrer Fähigkeit, bestimmte Handlungsmöglichkeiten auszuschließen (wie beispielsweise die Option, Ihr Fahrrad zu stehlen *und* dafür nicht ins Gefängnis zu kommen). Im Falle einer moralischen Verpflichtung ist mein moralisches Recht eingeschränkt, etwas Bestimmtes zu tun: Wenn ich beispielsweise die moralische Verpflichtung auferlegt bekomme, für Ihren Welpen zu sorgen, habe ich moralisch kein Recht mehr darauf, den Abend mit etwas anderem als einem Spaziergang mit Ihrem Hund zu verbringen. Philosophen, die sich mit Verpflichtungen beschäftigen, neigen dazu die Moralität des *Verpflichtet-Seins* zu betrachten, also wie wir die Struktur der Pflicht, auf eine bestimmte Weise zu handeln, verstehen sollten, und wie die verschiedenen Arten von Verpflichtungen zusammenwirken. Meines Wissens haben nur wenige die Moralität der Art von Handlungen untersucht, *die andere verpflichten.*[11] Das ist durchaus nachvollziehbar, denn die meiste Zeit kümmern wir uns vorrangig darum, wie wir die verschiedenen Verpflichtungen verstehen und befolgen sollen, die den normativen Hintergrund unseres Alltagslebens bilden. Aber wir sollten uns auch um die Moral von Handlungen kümmern, die andere zu etwas verpflichten. Meiner Meinung nach ist es möglich, das folgende Prinzip zu verteidigen, auch wenn ich in unserem Kontext keine adäquate Rechtfertigung anführen kann: Wir haben ein präsumtives Recht darauf, frei von Verpflichtungen zu sein, die andere uns ohne unsere Zustimmung auferlegen. Dabei handelt es sich lediglich um ein präsumtives Recht und es ist durchaus möglich, dass wir in vielen Fällen eine bereits bestehende Verpflichtung haben, die uns aufgrund ihrer Stärke dazu nötigt, eine neue Verpflichtung einzugehen. Aber der Nachweis einer solchen bestehenden Verpflichtung muss erbracht werden, andernfalls handelt es sich bei der neuen Verpflichtung schlicht um eine ungerechtfertigte Freiheitseinschränkung.

Zur Veranschaulichung können wir ein beliebtes Beispiel von Judith Jarvis Thomson umfunktionieren. Stellen Sie sich vor, dass Thomsons libertäre Einwände fehlschlagen, und wir tatsächlich einer – legalen und moralischen – Pflicht unterworfen sind, unseren Blutkreislauf mit einem bedürftigen Violinisten zu teilen, falls eine solche Unterstützung für sein Überleben unabdingbar ist.[12] Der Violinist kann uns zudem schlicht durch eine Berührung mit seinen Fingerspitzen erfolgreich dazu verpflichten, ihm diese Unterstützung zu gewähren. (Angenommen, dass eine solche Berührung ansonsten erlaubt ist; ein bloßes Berühren mit den Fingerspitzen gilt in dieser vorgestellten Welt also nicht als Körperverletzung.) Stellen Sie sich schließlich vor, dass der Violinist nun mit einer Person verbunden ist und ihm von dieser ausreichend geholfen wird; allerdings möchte er nun stattdessen mit Ihrem Blutkreislauf verbunden werden. Hat der Violinist ein Recht darauf, Sie zu berühren und dadurch zugleich zu verpflichten, ihn mit dem zu versorgen, worauf er einen moralischen Anspruch hat? Mir ist nicht ersichtlich, warum dies der Fall sein sollte – denn was auch immer es ist, auf das er einen Anspruch hat, er erhält es in diesem Fall bereits von der Person, mit der er verbunden ist. Sie sind nicht dazu verpflichtet, diejenige Person zu werden, die die Ansprüche des Violinisten erfüllen muss; tatsächlich wären Sie sogar dazu berechtigt, bis zu einem gewissen Grad Zwang auszuüben, um sich vor den Fingerspitzen des Violinisten zu schützen. Dabei handelt es sich selbstverständlich nicht um einen Freifahrtschein – es gibt viele Dinge, die Sie nicht tun dürfen um Ihr Recht darauf zu verteidigen, nicht zur Unterstützung des Violinisten verpflichtet zu werden. Aber mir scheint, als hätten Sie ein gewisses Recht auf die Verteidigung ihrer Freiheit, ein Leben zu führen, in dem Sie nicht zur Unterstützung eines bestimmten Violinisten verpflichtet werden, zumindest dann, wenn die Rechte des Violinisten bereits adäquat von einer anderen Person geschützt werden. Der Violinist hat demnach ein Recht auf Schutz, aber nicht auf den Schutz durch eine Person seiner Wahl.

Allerdings unterscheidet sich der Fall einer Immigrantin offensichtlich in vielerlei Hinsicht vom Fall des Violinisten. Der Violinist ist eine Last; die Immigrantin womöglich nicht – die Beziehung, die zu ihr entsteht, mag mit der Forderung nach einer bestimmten Handlung durch eine bestimmte Person verbunden sein, mitunter ist dies aber niemals der Fall. Einwanderungen sind oft durch gute Gründe motiviert, seien es Gründe familiärer Art, ökonomischer Suffizienz, oder gar des bloßen Überlebens; im Gegensatz dazu scheint der Violinist aus einer bloßen Laune heraus von Person zu Person zu springen. Der Violinist bietet seinen Unterstützerinnen zudem keinerlei Vorteil; die Immigrantin hingegen bietet dem Aufnahmeland im Allgemeinen einige solcher Vorteile – zumindest ihre Arbeitskraft und die daraus resultierenden Steuereinnahmen. Der Violinist kann folglich eher verurteilt werden als die Immigrantin. Sollten wir den Vergleich daher nicht einfach verwerfen?

In meinen Augen sollten wir diese Unterschiede eher so verstehen, dass sie für unser moralisches Urteil über den *Charakter* der Person, die Pflichten auferlegt, von Bedeutung sind, statt die Idee aufzugeben, das Auferlegen von Verpflichtungen selbst sei moralisch problematisch. Die Gründe der Immigrantin für ihre Einwanderung sind für uns in vielen Fällen vollkommen nachvollziehbar und respektabel. (Tatsächlich ist es möglich, wie ich in Kapitel acht zeigen werde, dass wir kein moralisches Recht darauf haben, denjenigen einen Mangel an Tugend vorzuwerfen, die eine staatliche Grenze ohne Recht auf Einreise überqueren; wir mögen ein Recht darauf haben, sie aufzuhalten, allerdings folgt daraus nicht, dass wir auch dazu berechtigt sind, sie als Monster zu betrachten.) Aus den Unterschieden zwischen der Immigrantin und dem Violinisten folgt also, dass wir vorsichtig damit sein sollten, unser negatives moralisches Urteil über den Violinisten direkt auf die Immigrantin zu übertragen. Vieles von dem, was uns am Violinisten stört, könnte durch Umstände bedingt sein, die im Falle der Immigrantin nicht zutreffen.

Wir sollten daher Vorsicht walten lassen, wenn wir die Immigrantin mit der gleichen Härte verurteilen wie den Violinisten. Die Quelle unseres moralischen Unbehagens scheint jedoch identisch zu sein. Selbst wenn das Verhalten des Violinisten für uns beschwerlicher und launenhafter sein mag als das der Immigrantin, so sind doch beide Verhaltensweisen aus demselben Grunde falsch: Sie legen denjenigen eine Verpflichtung auf, die keine bestehende Verpflichtung haben, derart verpflichtet zu werden. Die Last durch die Immigrantin mag weniger gewichtig und stärker reziproker Natur sein als die durch den Violinisten verursachten Lasten, aber in beiden Fällen besteht kein Recht, diese Art von Beziehung einem moralischen Akteur ohne dessen Zustimmung aufzuerlegen. Es steht mir auch nicht zu, eine auferlegte Verpflichtung unter Verweis auf die Geringfügigkeit der mit ihr verbundenen Kosten zu rechtfertigen – so wie es mir auch nicht zusteht, einen Einbruch in Ihr Haus mit dem Verweis auf die Tatsache zu rechtfertigen, dass es dort nach meinem Einzug wahrscheinlich ordentlicher wäre. Derjenige, der einbricht und stiehlt oder einbricht und wertvolle Erbstücke zerstört, ist härter zu verurteilen als eine Person, die einbricht und aufräumt; allerdings wird in keinem Fall etwas moralisch Richtiges getan, da die Einwilligung der Hausbesitzerin fehlt.[13] Das gleiche gilt in meinen Augen auch im Hinblick auf den Violinisten und die Immigrantin: Wir mögen über ersteren mitunter härter urteilen, aber wir sind im Recht, wenn wir in beiden Fällen urteilen, dass etwas moralisch Falsches getan wurde.

An diesem kritischen Punkt möchte ich einen Einwand von Michael Kates und Ryan Pevnick besprechen. Ihnen zufolge erlaubt es der von mir präsentierte Ansatz, diese neue, durch die Immigrantin auferlegte Verpflichtung auf zwei Art zu verstehen. Sollte die Verpflichtung gegenüber der einzelnen Person bestehen, die die Grenzen überquert, würde es sich um eine neue Verpflichungs*instanz* handeln, woraufhin Kates und Pevnick meinen, dass sich das von mir beschriebene moralische Problem bloß dann einstellt, wenn diese Instanz tatsächlich Kosten verursacht. Sollte es sich bei dieser neuen Verpflichtung hingegen um eine bestehende Verpflichtung handeln – hilf jedem innerhalb deines Hoheitsgebiets –, dann stellt die Verpflichtung gegenüber dem Neuankömmling gar keine der Art nach neue Verpflichtung dar, in welchem Falle die für den Ausschluss herangezogene Begründung fehlschlagen würde:

> „Wenn aber die Einwohnerinnen eines Staates durch Einwanderung keine wirklich neuen Arten von Verpflichtungen auf sich nehmen, erscheint es eher so, als ob Einwanderung die Position der Einwohnerinnen dadurch verändert, dass neue Instanzen bereits bestehender Arten von Verpflichtungen eingeführt werden. Das Problem ist jedoch, dass die Einführung neuer Instanzen bereits bestehender Verpflichtungen die Situation der gegenwärtigen Einwohnerinnen eines Staates allein dadurch zu verändern scheint, dass sie die Kosten der Erfüllung jener Pflichten verändert, die sich die Bürgerinnen bereits gegenseitig schulden. Wenn also die infrage stehenden neuen Verpflichtungen bloß Instanzen bereits existierender Arten von Verpflichtungen sind, ist unklar, was Blake mit der Behauptung meint, sein Argument beruhe auf neuen Verpflichtungen anstatt auf Kosten. Denn nochmals: Diese ‚neuen' Verpflichtungen verändern allein dadurch die an die derzeitigen Bewohnerinnen eines Staates gerichteten Ansprüche, indem sie die Kosten für die Unterstützung derjenigen Institutionen verändern, zu deren Unterstützung die Einwohnerinnen ohnehin bereits verpflichtet sind."[14]

Kates und Pevnick sehen also nur zwei Möglichkeiten: Entweder sind die Verpflichtungen nicht wirklich neu, woraufhin sich durch die Ankunft eines Neuankömmlings nichts von besonderer moralischer Bedeutung verändert – oder diese Verpflichtungen sind allein deshalb interessant, weil sie Kosten verursachen, woraufhin Argumente hinsichtlich der durch den Neuankömmling tatsächlich verursachten Kosten ausreichen würden, um jeden potentiellen Einwand gegen die Einreise der Immigrantin zurückzuweisen.

Ich denke, dass es sich dabei um eine falsche Beschreibung der moralischen Landschaft handelt. Viele moralische Pflichten können auf verschiedenen Abstraktionsstufen beschrieben werden: Ich kümmere mich um die Nachbarschaft, in der mein Haus steht, was eine Möglichkeit darstellt, meiner Pflicht zur Nachbarschaftshilfe nachzukommen, was wiederum ein Weg ist, meine Pflicht, die Welt ein wenig besser zu hinterlassen, als ich sie vorgefunden habe, zu erfüllen, was schließlich mitunter als eine Möglichkeit der Erfüllung meiner Pflicht, dem Wort Gottes Folge zu leisten, verstanden werden mag. Selbst wenn es zutreffen sollte, dass das beste Verständnis meiner Pflichten gegenüber dem Neuankömmling unter der bestehenden Verpflichtung „kümmere dich um die Menschen um dich herum" subsumiert werden könnte, wäre es falsch zu denken, dass sie nicht auch *zugleich* individuelle – und neue – Pflichten gegenüber dieser bestimmten Person umfasst. Diese zwei Beschreibungen meiner Verpflichtung gegenüber der Immigrantin sind also durchaus miteinander kompatibel. Stellen Sie sich zur Veranschaulichung vor, Sie würden durch Adoption ein Kind in Ihre Familie aufnehmen. Es wäre gewiss überaus seltsam zu behaupten, dass Sie keine neuen Verpflichtungen gegenüber diesem Kind hätten – dass Sie also bloß eine bestehende Verpflichtung hätten, sich allgemein um diejenigen Kinder zu kümmern, die Ihnen gegenüber schutzbedürftig sind. Diese bestehende Verpflichtung lässt sich jedoch am besten unter Verweis auf die *besonderen* Handlungen beschreiben, die Sie diesem *besonderen* Kind schulden. Darüber hinaus kann die Verpflichtung in diesem Fall auch nicht allein anhand der spezifischen Kosten verstanden werden, die durch die Erziehung dieses Kindes anfallen. Stellen Sie sich beispielsweise vor, dass in Ihrer Familie genug Geld für die Erziehung dieses Kindes vorhanden ist – fünf Kinder können vielleicht genauso günstig leben wie vier, ohne dass sich dafür der Lebensstandard verändern müsste. Auch in diesem Fall wäre es äußerst seltsam, zu behaupten, dass der Einzug des Kindes in Ihren Haushalt kein neues Bündel an Pflichten in Ihr Leben bringen würde. Diese Pflichten sind in der Tat schlicht aufgrund der Tatsache neu, dass Sie nun Dinge für dieses Kind tun müssen, die zuvor nicht in Ihren Aufgabenbereich fielen. Es ist nebensächlich, ob dies nun mit Kosten verbunden ist. Der Punkt ist stattdessen, dass *Sie* nun die Person sind, die diese Aufgaben zu erfüllen hat.

Aus diesen Gründen bleibe ich bei der Meinung, dass die Schaffung neuer Bündel von Verpflichtungen in diesem Fall moralisch bedeutsam ist und zwar unabhängig von den durch dieses Bündel verursachten Kosten – und dass selbst dann, wenn diese neuen Verpflichtungen als aus bestimmten, abstrakteren Pflichten abgeleitete Verbindlichkeiten verstanden werden können. Meiner Meinung nach ist von Bedeutung, dass eine Beziehung geschaffen wurde, in der ich dazu verpflichtet bin, bestimmte Dinge zu tun und mich darum bemühen sollte, sie gut zu tun.

Können wir nun also anhand dieses Gedankengangs das Recht auf Ausschluss rechtfertigen? Zumindest für manche Fälle ist das nun möglich. Die potentielle Immigrantin, die in ein bestimmtes Hoheitsgebiet einwandern will, bürdet mit ihrer Handlung den gegenwärtigen Einwohnerinnen dieses Hoheitsgebiets ein Bündel von Verpflichtungen auf. Diese Verpflichtungen schränken wiederum die Freiheit der Einwohnerinnen dadurch ein, dass ihnen die besondere Verpflichtung auferlegt wird, zum Schutz der Rechte dieser Immigrantin auf bestimmte Weise zu handeln. In Reaktion darauf können legitime Staaten es Immigrantinnen untersagen, ihr Staatsgebiet zu betreten, und zwar aus dem Grund, dass die Einwohnerinnen dieser Staaten ein Recht darauf haben, die ihnen durch die potentiellen Einwanderer auferlegten Verpflichtungen zurückzuweisen. Dieses generelle Recht erlegt wiederum den potentiellen Einwanderern die Pflicht auf, einen bestimmten Grund dafür anzugeben, weshalb die gegenwärtigen Einwohnerinnen eine Verpflichtung haben sollten, sich gegenüber ihnen verpflichten zu lassen. Sofern ein solcher Grund fehlt, scheint es mir, als ob Staaten das Recht dazu hätten, verhältnismäßige Formen von Gewalt anzuwenden, um die potentiellen Immigrantinnen vom Betreten ihres Hoheitsgebiets abzuhalten – da es die simple Tatsache der Anwesenheit innerhalb dieses Hoheitsgebiets ist, aus der die Verpflichtung zum Schutz der Grundrechte der Immigrantinnen erwächst. Der jeweilige Staat kann entscheiden, für welche Zwecke er dieses Recht in Anspruch nimmt: Er mag damit die kulturelle Einheit, die Solidarität oder ein anderes bestimmtes Gut verteidigen. Das Recht auf Ausschluss selbst ist jedoch aus dem generellen Recht abgeleitet, unerwünschte Verpflichtungen zurückweisen zu dürfen, wenn keine gesonderte Verpflichtung besteht, diese einzugehen. Das Recht auf Ausschluss geht folglich auf individuelle Rechte zurück. Aus diesem Grund könnten Staaten, sofern ihnen dieses Recht auf Ausschluss faktisch noch nicht zukommen sollte, auf Basis der Existenz dieser individuellen Rechte für die Plausibilität eines allgemeinen Rechts auf Ausschluss argumentieren.

Zur Veranschaulichung kann ich erneut meine eigene Geschichte anführen: Ich zog im Alter von 22 Jahren für meine Doktorarbeit in Philosophie von Kanada in die Vereinigten Staaten.[15] Als ich amerikanischen Boden betrat, verpflichteten sich die USA mir gegenüber auf bestimmte Weise. Sie hätten sich aufgrund dieser Tatsache dafür entscheiden können, mich auszuschließen. Mit Sicherheit hatten sie keine Pflicht, mir gegenüber eine Verpflichtung zum Schutz meiner Rechte einzugehen, da diese Rechte bereits in Kanada ausreichend geschützt wurden. Ich bin sehr froh, dass die Vereinigten Staaten sich für eine Einwanderungspolitik entschieden haben, die mir die Einreise erlaubte und schließlich auch die Möglichkeit gab, amerikanischer Staatsbürger zu werden. Ich glaube jedoch nicht, dass ich ein bestimmtes *Recht* darauf hatte, die Vereinigten Staaten zu betreten. Die Entscheidung, mir ein Studium an einer amerikanischen Universität zu erlauben, war eine Ermessensentscheidung und mir wäre im Falle eines Ausschlusses kein Unrecht widerfahren.[16]

All das ist natürlich bloß der erste Schritt eines Arguments. Selbst wenn legitime Staaten ein Recht darauf haben sollten, einwanderungswillige Personen auszuschließen, muss noch einiges getan werden, um die Konturen dieses Rechts zu bestimmen. Denn selbst wenn wir über ein Recht auf Ausschluss verfügen, ist es stets möglich zu fragen, ob dieses Recht eine bestimmte Einwanderungspolitik rechtfertigt. So könnten wir fragen, ob das Recht auf Ausschluss bestimmte Formen seiner Durchsetzung legitimiert; beispielsweise könnte die Militarisierung der südlichen Grenze der USA moralisch problematisch sein, auch wenn das Recht besteht, Migrantinnen an der Grenze zurückzuweisen.[17] Wir könnten darüber hinaus auch fragen, ob ein liberaler Staat dazu berechtigt ist, all jene Güter zu verfolgen, die weiter oben zur Verteidigung dieses Rechts angeführt wurden.[18] Schließlich könnten wir auch fragen, ob es fair ist, wenn ein Staat dieses Recht einseitig ausübt.[19] Für den Moment möchte ich diese Probleme jedoch beiseitelassen und mich auf zwei beunruhigende Konsequenzen des von mir entwickelten Ansatzes konzentrieren. Die aus ihnen folgenden Einwände will ich nicht nur aufgrund ihrer Plausibilität diskutieren, sondern auch deshalb, weil sie uns in meinen Augen etwas über die Grenzen des von mir verteidigten Rechts auf Ausschluss sagen können.

3. Implikationen: Föderalismus und Zuflucht

Den ersten Einwand können wir relativ schnell abhandeln. Er führt an, dass Staaten nicht die einzigen Einheiten sind, die über ein Hoheitsgebiet definiert werden: Föderale Untereinheiten, inklusive Provinzen, Gemeinden und dergleichen, haben allesamt ein begrenztes territoriales Gebiet, in dem sie fähig sind, bindende Gesetze zu erlassen. Problematisch wird es diesem Einwand zufolge dann, wenn diese Untereinheiten im gleichen Sinne wie souveräne Staaten dazu berechtigt sein sollen, ungebetene Immigrantinnen auszuschließen. Es bedarf keiner besonderen Erwähnung, dass eine solche Schlussfolgerung die Validität meines Arguments infrage stellen würde: Eine Theorie des Rechts auf Ausschluss, die ein solches Recht gleichermaßen der Stadt Seattle, dem Bundesstaat Washington und den Vereinigten Staaten von Amerika zusprechen würde, wäre nicht wirklich reich an Erkenntnissen.

Die passende Antwort auf diesen Einwand besteht, meiner Meinung nach, in dem Hinweis darauf, dass die Struktur eines föderalen Staates notwendigerweise mit einem bestimmten konstitutionellen Vorhaben einhergeht, das wiederum festlegt, wie die politischen Machtverhältnisse zwischen den föderalen Untereinheiten und der Regierung dieser Föderation ausgestaltet sein sollten. Unter diesen Umständen sind die föderalen Einheiten womöglich nicht in der Lage, ungebetene Migrantinnen auszuschließen – nicht, weil es ihnen prinzipiell untersagt wäre, sondern weil der besondere Zweck der rechtlichen Verbindung dieser Untereinheiten es ihnen nicht gestattet. In meinen Augen ist das die korrekte Beschreibung des gegenwärtig in den USA herrschenden Verständnisses des Rechts auf Bewegungsfreiheit; dieses Recht wurde auch vom Supreme Court an die Bedeutung des föderalen Projekts der Bildung einer politischen Einheit gekoppelt:

> „Das bedeutet jedoch nicht, dass der gesetzgeberischen Aktivität der Bundesstaaten keine Grenzen gesetzt sind. Tatsächlich gibt es solche Grenzen. Die bedeutsamste dieser Grenzen besagt, dass es einem Bundesstaat untersagt ist, sich den allen Bundesstaaten gemeinsamen Schwierigkeiten dadurch zu entziehen, dass er die Bewegung von Menschen und Eigentum über seine Grenzen hinweg unterbindet. Es ist häufig der Fall, dass sich ein Bundesstaat unter dem Druck der Ereignisse dadurch eine vorübergehende Atempause verschaffen möchte, dass er seine Tore zur Außenwelt schließt. Aber in den

Worten von Richter Cardozo: ‚Die Verfassung wurde vor dem Hintergrund einer weniger engstirnigen politischen Philosophie ausgearbeitet. Sie wurde mit dem Gedanken verfasst, dass die Bevölkerungen der verschiedenen Bundesstaaten entweder zusammen schwimmen oder zusammen untergehen müssen, und dass auf lange Sicht Wohlstand und Heil nur in der Einheit, nicht der Teilung liegen.'"[20]

Dieser Analyse zufolge muss das Recht auf Ausschluss, das den einzelnen föderalen Untereinheiten jeweils zukommt, hinter dem Vorhaben einer gemäß der Verfassung regierten politischen Gemeinschaft zurückstehen. Die angemessene Antwort auf den ursprünglichen Einwand lautet also, dass die föderalen Untereinheiten kein Recht auf Ausschluss haben, sofern dieses Recht der Errichtung einer gemeinsamen politischen Gemeinschaft entgegensteht. An der Vorstellung, dass der Staat Washington, oder gar die Stadt Seattle, in einem anderen institutionellen Kontext ein Recht auf den Ausschluss von Migrantinnen haben könnte, ist nichts inhärent falsch. (Der Roman *Snow Crash* beschreibt exakt ein solches Phänomen.)[21] Allerdings haben sie unter den gegenwärtig herrschenden Institutionen kein Recht darauf, aus anderen Landesteilen einreisende, potentielle Einwohnerinnen auszuschließen, selbst wenn diese den gegenwärtig anwesenden Einwohnerinnen signifikante Kosten aufbürden sollten.[22] Der Grund hierfür liegt in dem auf Bundesebene zum Zweck der gemeinsamen Selbstregierung eingerichteten politischen Projekt, dem zufolge nur die Bundesregierung das Recht auf Ausschluss innehat. Die bundesstaatlichen Einheiten haben hingegen kein solches Recht: Gemäß der von mir verwendeten Begriffen sind sie dazu verpflichtet, sich von denjenigen verpflichten zu lassen, die ihr Gebiet betreten wollen.

Diese Antwort eröffnet allerdings eine weitere Möglichkeit, den von mir verteidigten, auf der Gebietshoheit beruhenden Ansatz zu kritisieren – oder genauer gesagt: die Fähigkeit dieses Ansatzes, die gegenwärtig von Staaten genutzten Ausschlusspraktiken zu rechtfertigen. Die soeben von mir formulierte Antwort bestand in der Behauptung, dass manche Formen politischer Projekte von den unter ihnen versammelten hoheitlichen Einheiten verlangen, die ihnen zukommenden Ausschlussrechte aufzugeben. Der Kritiker kann nun einwenden, dass solche Projekte auch auf internationaler Ebene existieren. Sollte das jedoch zutreffen, könnte die Inanspruchnahme eines

Rechts auf Ausschluss potentieller Migrantinnen durch Staaten als ähnlich illusorisch wie seine Inanspruchnahme durch die Stadt Seattle betrachtet werden.

Es gibt zwei Versionen dieses Arguments, eine stärkere und eine schwächere. Die stärkere argumentiert global und führt an, dass mittlerweile eine den Ausschluss von Migrantinnen untersagende Form politischer Gemeinschaft auf globaler Ebene existiert. Die schwächere Version argumentiert hingegen auf der Ebene einzelner Staaten und wendet ein, dass bestimmte Formen politischer und rechtlicher Verflechtung zwischen einzelnen Staaten den Ausschluss innerhalb dieses Aggregats von Staaten untersagen würde. Mir erscheint die stärkere Version als bloß schwer zu verteidigen: Die gegenwärtig auf globaler Ebene existierenden Institutionen sind derart schwach und für ihre Legitimation sowie Durchsetzung derart auf souveräne Staaten angewiesen, dass die Annahme, ihre Vernetzung sei ausreichend weit fortgeschritten um den Ausschluss durch Staaten zu untersagen, wenig plausibel erscheint. Damit habe ich selbstverständlich kein vollständiges Bild davon gezeichnet, wie globale Institutionen beschaffen sein müssten um einen Grad an Integration zu erreichen, der ein Verbot staatlichen Ausschlusses rechtfertigen könnte. Ich habe schlicht angenommen, dass die Vereinigten Staaten hinreichend stabil sind und die Vereinten Nationen nicht. Es ist daher gut möglich, dass ein Argument entwickelt wird, dass meine Annahmen widerlegt. Allerdings bleibe ich diesbezüglich skeptisch.

Die schwächere Version ist hingegen plausibler. Auch hier ist es schwierig, ein finales Urteil zu fällen, solange eine substantiellere Theorie darüber fehlt, welche Formen politischer Ordnung ein staatliches Recht auf Ausschluss einschränken würden. Allerdings erscheint es mir plausibel, dass manche Formen politischer Integration eine politische Gemeinschaft von ausreichender Macht und Reichweite schaffen, sodass die Untereinheiten dieser Gemeinschaft ihr Recht auf den Ausschluss ungebetener Migrantinnen verlieren. Das am weitesten fortgeschrittene Projekt einer solchen transnationalen Gemeinschaft kann wohl in Europa gefunden werden, das eine Reihe politischer Institutionen besitzt, die so etwas wie eine Gemeinschaft zum Zwecke der Politik bilden: ein europäisches Parlament, den Europäischen Gerichtshof und so weiter.[23] Aus dieser Form der Integration könnte folgen, dass die Staaten der Europäischen Union nicht mehr länger das Recht besitzen, ungebetene Migrantinnen zurückzuweisen, die aus anderen Staaten der EU einwandern. Sollte das zutreffen, hätten die Staaten der EU ein System politischer Institutionen geschaffen, das unter anderem der Einrichtung einer gemeinsamen Staatsbürgerschaft bedarf; das Recht auf Bewegungsfreiheit könnte daher, wie in den USA, aus den Anforderungen dieser gemeinsamen politischen Identität hervorgehen. Es ist daher für Staaten durchaus möglich, ihr Recht auf Ausschluss zu verlieren, sofern sie freiwillig die Einrichtung stabiler politischer Institutionen betreiben, innerhalb derer sie als Bestandteil fungieren. Meiner Meinung nach hat dieser Prozess in Europa zu einem solchen Grade stattgefunden, dass die Bewegungsfreiheit innerhalb der Europäischen Union zu einem moralischen Imperativ geworden ist – was darüber hinaus auch dann der Fall wäre, wenn die EU dies nicht anerkennen würde. Ich bin nicht sicher, ob die Europäische Union das nächste Jahrhundert überstehen wird; dabei handelt es sich jedoch nicht um philosophische Einsichten, sondern um Pessimismus (und Ehrlichkeit) hinsichtlich der Herausforderungen, vor denen diese Organisation steht. Nun bin ich jedoch weniger an der Verteidigung eines Rechts auf Bewegungsfreiheit innerhalb der EU, als an der Anerkennung der Stärke des vorgebrachten Einwands interessiert: Es trifft meiner Meinung nach zu, dass wir uns in einer Welt mit robusten transnationalen Institutionen nicht vollständig mit einem staatlichen Recht auf Ausschluss begnügen können. Zumindest darf die Frage gestellt werden, ob solche Institutionen ausreichen, um diesem Recht bestimmte Grenzen zu setzen. In Erwiderung auf diese Frage steht dem kosmopolitischen Kritiker des Rechts auf Ausschluss eine bestimmte argumentative Struktur zur Verfügung: Er kann anführen, dass, selbst wenn dieses Recht abstrakt verteidigt werden könnte, es unter den besonderen Umständen der uns gegenwärtig bekannten Welt nicht in Anspruch genommen werden kann. Während ich vermute, dass der kosmopolitische Kritiker und ich nicht darin übereinkommen werden, was genau diese Einschränkung des Rechts auf Ausschluss begründet, bin ich zumindest zufrieden, dass wir nun eine gemeinsame Beschreibung des Gegenstands unserer Auseinandersetzung gefunden haben: Wir wollen klären, ob die von uns geteilten Institutionen solcher Art sind, dass sie uns dazu verpflichten können, uns gegenüber Migrantinnen verpflichten zu lassen, die um Zutritt zu unserem Hoheitsgebiet ersuchen.

Der zweite Einwand ist simpler, aber womöglich auch bedeutsamer. Er verweist schlicht auf den Inhalt meines Arguments: Wir können ungebetene

Immigrantinnen ausschließen, da sie bereits einen angemessenen Schutz ihrer Rechte in ihren Heimatländern genießen und uns nun zum Schutz dieser Rechte verpflichten wollen. Das Argument gilt also nur, wenn in dem Land, aus dem eine Person auswandern will, die Rechte dieser Person auch *tatsächlich* adäquat geschützt werden. Das galt sicherlich bei meiner Emigration aus Kanada und trifft zweifellos auf viele andere potentielle Migrantinnen zu. Es ist jedoch unwahrscheinlich, dass dies auf den Großteil der Migrantinnen zutrifft, von denen viele Umständen zu entkommen versuchen, in denen (recht kaltblütig ausgedrückt) ihre Rechte bloß ungenügend geschützt werden. In solchen Fällen würde meine Rechtfertigung selbstverständlich scheitern. Wir dürfen die ungebetenen Immigrantinnen nun nicht mehr ausschließen, da die von uns angeführten Gründe – dass wir nämlich keine Verpflichtung haben, uns dazu verpflichten zu lassen, die Rechte dieser Personen zu schützen – nicht mehr zutreffen. Die Migrantinnen würden beim Betreten unseres Territoriums schlicht in den Genuss des Rechtsschutzes kommen, auf den sie einen Anspruch haben. Sofern wir nun Gewalt anwenden wollen, um diese Personen am Eintritt in unser Staatsgebiet zu hindern, nutzen wir die Gewalt schlicht dazu, sie zurück in eine moralisch untragbare Situation zu zwingen. Ein solcher Einsatz von Gewalt ist unmöglich zu rechtfertigen. Entsprechend könnten wir annehmen, dass der von mir entwickelte Ansatz – abgesehen von seiner allgemeinen Verteidigung eines Rechts auf Ausschluss – tatsächlich weitaus radikaler ist, als er zunächst erscheint. Er würde zu einer radikalen Überarbeitung des Flüchtlings- und Asylrechts verpflichten und zwar in dem Sinne, dass wir keine Gewalt für den Ausschluss von Migrantinnen einsetzen dürfen, deren Herkunftsländer die Menschenrechte nur unzureichend beachten.

Auch dieser Einwand richtet sich gegen eine Lesart meines Ansatzes, die ihn als eine unbedingte Verteidigung gegenwärtiger Ausschlusspraktiken versteht; allerdings möchte ich diese Überlegungen nicht vollständig zurückweisen. Vielmehr möchte ich zeigen, inwiefern dieser Einwand möglicherweise ein paar Einschränkungen unterliegt, wobei ich der Meinung bin, dass er dennoch grundsätzlich korrekt ist. Tatsächlich glaube ich, dass wir, sofern wir uns auf die Rechtfertigung staatlicher Gewalt konzentrieren, wahrscheinlich zu einer Vorstellung von Flüchtlings- und Asylrecht gelangen werden, der zufolge wir einer großen Zahl potentieller Migrantinnen die Aufnahme nicht verweigern dürfen. Eine solche Perspektive auf Einwanderung wäre meiner Meinung nach stärker als der verwandte Ansatz von Wellman, der es legitimen Staaten erlaubt, sich das Recht auf Ausschluss dadurch zu erkaufen, dass in angemessenem Umfang Ressourcen für die Entwicklungshilfe bereitgestellt wird.[24] Meinem Ansatz zufolge ist das nicht einmal prinzipiell möglich. Wir können die Anwendung von Gewalt gegen eine Person nicht durch Verweis auf einen eventuell daraus resultierenden Nutzen für eine andere Person rechtfertigen; die Begründung muss vielmehr so sein, dass die betroffene Person sie akzeptieren kann, ohne sich dabei über Gebühr mit den Interessen anderer zu identifizieren. Das ist der Kern von Rawls' Lehre, dass Personen als getrennt voneinander zu betrachten sind. Diesem Gedanken zufolge ist es falsch, von einer Person, die durch eine Institution, deren Zwang sie unterliegt, schlecht behandelt wird, zu verlangen, dieser Behandlung aufgrund des vergleichsweise größeren Nutzens für andere Personen zuzustimmen.[25] Der Vorschlag, dass wir die Anwendung von Gewalt gegenüber einer schutzlosen Person durch den Verweis auf andernorts geleistete Hilfe rechtfertigen können, scheint ein Schlag ins Gesicht für diese Idee zu sein. Wir können Gewalt an den Grenzen nur dann rechtfertigen, wenn wir sie gegenüber Menschen anwenden, deren Rechte bereits auf angemessene Weise in ihren Heimatländern geschützt werden. Sie gegen andere Personen zu verwenden erscheint schlicht und einfach als die Verteidigung eines illegitimen *status quo* und ganz unabhängig davon, als wie gerecht sich unsere übrige Außenpolitik darstellen sollte, ist ein solches Vorgehen moralisch nicht vertretbar.

Die Einschränkung der Rechte von Immigrantinnen aus unterdrückten Ländern muss demnach auf eine Weise geschehen, die mit ihrem Recht in Einklang steht, als moralisch gleichwertige Personen behandelt zu werden. Gibt es einen Weg, dies zu tun? Ich denke, es gibt prinzipiell zwei solcher Wege, obwohl keiner von beiden denjenigen, die den internationalen *status quo* erhalten wollen, viel Trost spenden wird. Der erste Weg beginnt bei der einfachen Tatsache, dass die Verpflichtung, sich verpflichten zu lassen, nicht nur einfach einem individuellen Staat, sondern allen legitimen Staaten gemeinsam zukommt. Um auf den modifizierten Fall des Violinisten zurückzukommen: Ein sterbender Violinist mag ein Recht darauf haben, dass eine Person ihm die für ein würdiges Leben notwendige Unterstützung zukommen lässt, aber er hat kein Recht darauf, sich diese Person auszusuchen. Nichts in der Vorstellung von grundlegenden Rechten, die diesem Ansatz zugrunde liegen, legt nahe, dass die *allgemeine* Verpflichtung zum Schutz der Rechte von Menschen bloß durch die Wünsche der Schutzberechtigten in eine *besondere* Verpflichtung – also eine Verpflichtung, die einer bestimmten Person auferlegt wird – umgewandelt werden kann. Daraus folgt in meinen Augen, dass unser Verständnis von Ausschluss komplexer sein muss. Wir sind nicht dazu berechtigt, zum Zweck des Ausschlusses individueller Personen Gewalt anzuwenden, sofern diese unser Hoheitsgebiet aus Ländern erreichen, die die Rechte dieser Personen bloß unzureichend schützen. Dabei wird jedoch nicht behauptet, dass es unser Hoheitsgebiet sein muss, in dem diese Personen letztendlich ihr Leben werden führen können. Wenn die Bürde der Flüchtlingsbewegungen von allen wohlhabenden Staaten der Welt gemeinsam getragen werden sollte, dann steht es einem bestimmten Staat vollkommen offen, vorzubringen, dass ein anderer Staat sich weigert, den ihm zufallenden Anteil zu tragen. Auf diese Weise wird eine nachgeordnete moralische Analyse der fairen Verteilung der Lasten der Migrationspolitik möglich. Darüber hinaus eröffnet sich auch die Möglichkeit legitimer internationaler Abkommen über die Zurückweisung von Personen, die auf dem Weg zu dem von ihnen gewählten Zufluchtsort bereits rechtsstaatliche Länder durchquert haben. Das Dublin-Abkommen schreibt beispielsweise vor, dass das Asylgesuch einer Person von demjenigen Staat zu bearbeiten ist, an dessen Grenze die Person erstmals das Gebiet der Europäischen Union betreten hat.[26] Dieses Abkommen mag unfair gegenüber den peripheren Grenzstaaten der EU sein, die einen proportional größeren Anteil an den Kosten der Flüchtlingsbewegungen zu tragen haben; aber es kann nicht behauptet werden, dass es unfair gegenüber den Flüchtlingen selbst sei.

Die zweite Art von Einschränkung, der der oben genannte Einwand unterliegt, ist komplexer – und auch potentiell gefährlicher. Diese Einschränkung geht auf die Idee zurück, dass wir unter bestimmten Umständen von Menschen erwarten können, dass sie gewisse Kosten für die Erhaltung gerechter Institutionen in Kauf nehmen. Das ist in dieser Allgemeinheit eine offensichtlich recht unstrittige Idee. Wir akzeptieren beispielsweise, dass Personen dazu verpflichtet sind, zur Unterstützung eines gerechten Staates Steuern zu zahlen. Im gleichen Sinne können wir aber auch anführen, dass Personen dazu verpflichtet sind, solche Kosten auch bei der Einrichtung gerechter Institutionen zu tragen. So könnten wir beispielsweise verlangen, dass die Bewohnerinnen eines undemokratischen Staats einige der Kosten tragen, die beim Übergang dieses Staates zu einer legitimen Demokratie anfallen würden. Zudem können diese Kosten auch in anderen Dingen denn finanzieller Belastung bestehen. So fordern wir in bestimmten Fällen, dass Personen im Namen der Gerechtigkeit persönliche Risiken in Kauf nehmen. (Die meisten Theorien eines gerechten Krieges beinhalten schließlich die Idee, dass Gerechtigkeit ein Ideal von hinreichender Bedeutung darstellt, um Personen dem Risiko eines gewaltsamen Tods auszusetzen.) Wenn also eine Person aus einem Land auswandern will, das ihre Rechte nicht respektiert, sollten wir mitunter das Recht haben, darauf zu bestehen, dass sie in ihrem Heimatland bleibt, um bei der Verbesserung der dortigen Situation zu helfen.[27] Gibt uns das hinreichende Gründe, um darauf bestehen zu können, dass Personen, die aus einem ihre Rechte missachtenden Land fliehen, zurückkehren und bei dessen Verbesserung helfen sollten?

Ich möchte diese Ideen nicht gänzlich zurückweisen. Allerdings glaube ich nicht, dass sie ausreichen, um ein Recht auf Ausschluss potentieller Migrantinnen aus Ländern mit unterdrückerischen Regierungen zu rechtfertigen. Der Grund ist in meinen Augen der recht einfache Gedanke, dass es unfair wäre, wenn eine aus einem solchen Land fliehende Person mehr Pflichten gegenüber diesem Staat hätte als eine Person, die nicht Teil dieses Staates ist.[28] Die Beziehung zwischen diesem Staat und dem Emigranten ist kaum von der anhaltenden, wertvollen Art, aus der für gewöhnlich die Pflichten der an einer politischen Beziehung beteiligten Parteien entspringen. Sollten wir das nicht bereits einfach aufgrund der Tatsache anerkennen, dass der Staat ungerechtfertigt Zwang ausübt, könnten wir darauf verweisen, dass der Emigrant aktiv versucht, diese Beziehung zu verlassen. Darauf zu bestehen, dass eine solche Beziehung Pflichten begründet, kommt in diesem Falle der Vorstellung gleich, dass ein willkürlicher Fakt – der bloße Fakt der Geburt – ausreicht, um Verpflichtungen zu begründen. Eine solche Vorstellung ist jedoch moralisch unzulässig. So könnte uns die Migrantin fragen, warum von ihr verlangt werden kann, eine größere Laste bei der gemeinsamen Aufgabe der Demokratisierung ihres Heimatlandes zu tragen. Sie ist zwar dort geboren, aber das scheint kaum ausreichend, um ihr eine größere Verpflichtung als denjenigen zuzuschreiben, die das Glück hatten, in einem wohlhabenderen Staat geboren worden zu sein. (Sie mag mehr Wissen haben und aufgrund dessen auch besser bei dieser Aufgabe helfen können, allerdings erscheint eine Vorstellung von Pflichten ungerecht, bei der die Stärke von Verpflichtungen von bestimmten persönlichen Fähigkeiten abhängt.) Meiner Meinung nach besteht die einzige Möglichkeit für einen Erfolg diese Arguments darin, alle Parteien – sowohl die derzeitigen Bewohnerinnen der wohlhabenden Staaten als auch die potentiellen Migrantinnen – als moralisch gleichwertig anzusehen, wonach jedem die gleiche Pflicht zukommt, die demokratischen Fähigkeiten des infrage stehenden Staates zu fördern. In diesem Falle könnte von potentiellen Migrantinnen legitimerweise verlangt werden, einen bestimmten Teil der Kosten des Wiederaufbaus ihres Herkunftsstaates zu tragen. Aber es würde gleichfalls auch von den Einwohnerinnen der rechtsstaatlichen Länder verlangt werden können, etwas zu der gemeinsamen Aufgabe beizutragen, eine dem Rechtsschutz verpflichtete Welt einzurichten. Wie ich bereits sagte, ist die Idee einer spezifischen Pflicht gegenüber dem eigenen Heimatland gefährlich, da sie uns dazu verleitet, diejenigen, die von einem bestimmten Ort kommen, als der Verbesserung dieses Ortes besonders verpflichtet anzusehen. Eine gemeinsame Verantwortung muss allerdings auch wirklich aufgeteilt werden, und zwar auf eine gerechte Weise, zwischen allen Bürgerinnen der Welt. Es scheint daher schwer vorstellbar, dass wir diese Ideen tatsächlich heranziehen könnten, um potentielle Migrantinnen auszuschließen – zumindest bis wir an den Punkt gelangen, an dem wir selbst Kosten tragen müssten, die den Kosten der Migrantinnen gleichkommen. Aber ich denke, diesen Punkt überschreitet derzeit keiner der wohlhabenderen Staaten.

Als Ergebnis all dieser Überlegungen hat sich das Recht eines Staates, Personen aus unterentwickelten Ländern und solchen mit unterdrückerischen Regierungen auszuschließen, als relativ schwach herausgestellt. In meinen Augen ist das kein Mangel meines Ansatzes; es erscheint mir als richtig, dass ein Staat, der sich anschickt, Gewalt zu nutzen, um Menschen von seinem Hoheitsgebiet fern zu halten, auch Verantwortung für die Rechte derer trägt, gegen die er diese Gewalt richtet. Allerdings habe ich bei dieser Diskussion viele bedeutsame Fragen beiseitegelassen. Wir könnten beispielsweise fragen, welche Menschenrechte es genau sind, die durch internationales Recht sowie durch Einwanderungsgesetze geschützt werden sollten. Ebenso können wir fragen, welche anderen Überlegungen jenseits des Schutzes von Menschenrechten das staatliche Recht auf Ausschluss beeinflussen sollten.[29] Diese Fragen werde ich im nächsten Kapitel genauer besprechen. Allerdings kann uns das, was ich hier angeführt habe, bereits Anlass zu der Annahme geben, dass das Recht auf Ausschluss existiert – und dass dieses Recht bei weitem nicht die Formen von Einwanderungspolitik rechtfertigen kann, die derzeit von den wohlhabenden Gesellschaften dieser Welt betrieben werden. Wir haben als Kollektiv ein Recht darauf, manche ungebetenen Einwanderer auszuschließen. Wir sollten es uns aufgrund dieses Umstands jedoch nicht bequem machen: Vieles von dem, was wir derzeit tun, ist zutiefst ungerecht und sollte von uns auch entsprechend betrachtet werden. Die vorgebrachte Verteidigung eines Rechts auf Ausschluss sollte daher als ein Aufruf zu Reformen und nicht als quietistisches Plädoyer verstanden werden. Es ist uns in der Tat erlaubt, manche Migrantinnen auszuschließen, allerdings sollten wir daraus nicht folgern, dass wir sie mittels der Maßnahmen ausschließen dürfen, die wir derzeit einsetzen.

4. Einwände: Ausweisung, Reproduktion, und Freiheit

Ich möchte dieses Kapitel mit der Untersuchung einiger besonders wichtiger Einwände schließen, die Kritiker gegen meinen Ansatz vorgebracht haben. Drei dieser Einwände möchte ich im Folgenden besprechen: Wir können sie als Argumente der *Ausweisung*, *Reproduktion* und *Freiheit* bezeichnen.

Das Argument der Ausweisung stammt von Kates und Pevnick. Ihnen zufolge scheint mein Ansatz, da er das Wesen des Staates nicht unter Verweis auf eine generationenübergreifende Gemeinschaft bestimmt, prinzipiell zu erlauben, jedwede unerwünschte Person auszuweisen. Schließlich hat eine Migrantin, die aus einer Gesellschaft kommt, in der ihre Menschenrechte geschützt sind, keinen Anspruch auf ein Leben in einer Gesellschaft ihrer Wahl; sie hat bereits all das, worauf sie einen Anspruch hat. Warum sollten wir dasselbe nicht auch über Bürgerinnen sagen können, die wir einfach so ausweisen wollen?

> „Stellen Sie sich vor, dass Sie in einem Land geboren werden, das ihren Akzent oder Musikgeschmack unausstehlich findet. Während das Gemeinwesen dieses Landes Ihre Rechte aus einem Gefühl rechtlicher Verantwortung heraus verlässlich und effektiv schützt, würde es Sie dennoch gerne loswerden. Stellen Sie sich nun vor, dass irgendein anderes Land herantritt und behauptet, es wäre froh, Ihre Rechte zu gewährleisten und hätte darüber hinaus auch die Kapazitäten, die es hierfür braucht. Plötzlich erscheint es so, als ob Sie in

derselben Situation wie der Einwanderer sind, da Sie nun über eine absolut adäquate Ersatz-Option verfügen (zumindest in dem Sinne, dass ein anderes Land fähig und willens ist, Ihre Recht zu schützen). Aufgrund all dessen entscheidet sich schließlich Ihr Heimatland dazu, Sie auszuweisen – im sicheren Wissen, dass dieses neue Land Ihre Rechte verlässlich schützen und dabei auch keine unerwünschten neuen Verpflichtungen eingehen wird (da es Sie mit offenen Armen empfängt). Für gewöhnlich denken wir jedoch nicht, dass eine solche Ausweisung akzeptabel wäre. Aber gibt es irgendetwas in Blakes Ansatz, dass es ihm erlauben würde, eine solche Maßnahme abzulehnen?"[30]

Ich denke, die Antwort lautet ja. In meinen Augen verlangt die Idee der Menschenrechte, dass wir Menschen als sich in der Zeit bewegende Akteure wahrnehmen: Wir machen Pläne und sehen den Erfolg dieser Pläne als ein Anzeichen des Erfolgs unseres Lebens. Um mich der Sprache von Kieran Oberman zu bedienen: Möglichkeiten und Verbundenheit sind verschiedene Dinge.[31] Das Recht darauf, nicht ausgewiesen zu werden, geht demnach darauf zurück, dass es für Menschen von großer Bedeutung ist, auf den Beziehungen und Plänen aufbauen zu können, die sie an einem bestimmten Ort entwickelt haben. Ich werde in Kapitel sieben zeigen, unter welchen Umständen Abschiebungen moralisch vertretbar sein können, und zwar um bestimmte Formen des Ungehorsams gegenüber dem Gesetz zu ahnden. Allerdings muss auch in diesen Fällen gezeigt werden, dass diese Abschiebung mit dem Respekt vor den Menschenrechten des Abgeschobenen kompatibel ist und dabei handelt es sich um eine hohe Hürde. Immerhin werden die Bindungen des Abgeschobenen fast vollständig zerstört, und diese Art von Schaden erfordert eine außergewöhnlich starke Form der Rechtfertigung. Was Kates und Pevnick mir jedoch zuschreiben wollen, ist ein pauschales staatliches Recht darauf, die Bindungen *einer jeden Person* zunichte zu machen – und ich denke, dass meine Theorie darauf schlicht und einfach antworten kann, dass *das* nicht der Fall ist. Dabei möchte ich anmerken, dass dieser Schluss ohne den Vorschlag von Kates und Pevnick möglich ist, das Recht auf Ausschluss aus der besonderen Geschichte eines Gemeinwesens abzuleiten. Es gibt in meinem Ansatz nichts, was uns dazu nötigt, die leidige Frage zu verfolgen, was nun unter der Geschichte einer bestimmten Gemeinschaft genau zu verstehen ist oder wie wir das Wesen einer generationenübergreifenden Gemeinschaft zu verstehen haben. Alles, was wir behaupten müssen, ist das: Für Geschöpfe wie uns unterscheidet sich die Zukunft von der Vergangenheit.

Kates, Pevnick und ich stimmen daher vollkommen darin überein, dass die unerwünschte Bürgerin nicht ausgewiesen werden sollte. Die von ihnen vorgebrachte Rechtfertigung bezieht sich jedoch auf etwas, das jenseits der Handlungsfähigkeit dieser Bürgerin zu finden ist. Für Kates und Pevnick müssen wir uns eine selbstbestimmte Gemeinschaft vorstellen, die unabhängig von den individuellen Mitgliedern dieser Gemeinschaft über Generationen hinweg existiert und die Verantwortung hinsichtlich der staatlichen Institutionen trägt. Meiner Meinung nach müssen wir nicht auf die Behauptung der Existenz einer solchen Gemeinschaft zurückgreifen, um erklären zu können, warum die Ausweisung in dem von Kates und Pevnick beschriebenen Falle falsch ist. Wir können dies anhand der Geschichte der *Person* erklären und müssen dafür nicht auf die Geschichte der Gemeinschaft verweisen. Ein solches Vorgehen scheint mir zwei Vorteile zu haben. Zum einen ist es metaphysisch um einiges schlichter: Wir können uns darauf einigen, dass die jeweilige Person existiert, selbst wenn wir hinsichtlich der Existenz der Nation uneins sein mögen. Zum anderen vermeidet dieser Ansatz die im vorherigen Kapitel diskutierten Probleme der Marginalisierung. Das Recht auf Schutz vor Ausweisung unter Berufung auf eine historische und über Generationen gewachsene Gemeinschaft zu begründen hieße, diejenigen zu gefährden, die sich nicht so leicht mit dieser Gemeinschaft identifizieren können. Meinem Ansatz zufolge ist das mutmaßliche Recht einer Bürgerin darauf, nicht ausgewiesen zu werden, immer von gleicher Bedeutung, ob sie nun ganz klar ein Teil dieser historischen Gemeinschaft oder ihr erst heute Morgen als Bürgerin beigetreten ist.

Neben diesem Einwand wurde auch das Problem der *Reproduktion* gegen meinen Ansatz vorgebracht. Viele Kommentatoren hatten das Gefühl, dass das Recht, Immigrantinnen auszuweisen, auch ein Recht darauf implizieren müsste, die Reproduktion meiner Mitbürgerinnen zu verhindern. In diesem Falle wäre der hier behandelte Ansatz mit Sicherheit einer *reductio ad absurdum* überführt. Jan Brezger und Andreas Cassee sehen in diesem Problem einen verhängnisvollen Einwand gegen jeden Ansatz, der einem Staat das Recht auf Ausschluss zuspricht:

> „Wenn der Zugang zu den mit der Mitgliedschaft in einem Staat verbundenen Vorteilen allein durch Verweis auf die Vereinigungsfreiheit, die Freiheit von neuen Verpflichtungen oder bestimmte Eigentumsrechte reguliert werden soll, folgt daraus, dass die derzeitige Bürgerschaft den Zutritt nicht nur den potentiellen Einwanderinnen, sondern auch dem Nachwuchs der einheimischen Bürgerinnen verwehren könnte.“[32]

In diesem Zusammenhang ist es sicherlich wichtig darauf hinzuweisen, dass das von mir verteidigte Prinzip ein Recht darauf enthält, frei davon zu sein, dass mir Verpflichtungen auferlegt werden – *außer* wenn ich eine Verpflichtung habe, mich verpflichten zu lassen. Es kann viele Fälle geben (und es gibt sicherlich viele Fälle), in denen ich nicht frei darüber entscheiden kann, verpflichtet zu werden. Das Flüchtlingsrecht ist beispielsweise solch ein Fall: Diejenigen, deren Menschenrechte nicht verteidigt werden, können uns zum Schutz ihrer Rechte verpflichten und darauf bestehen, dass wir in politische Beziehungen zu ihnen treten. Die Reproduktion erscheint mir ebenso als ein Fall, in dem andere berechtigterweise beanspruchen, uns verpflichten zu können.

Allerdings können wir über diesen Fall in meinen Augen noch mehr sagen. Es wird allgemeinhin als wahr angesehen, dass Ihre Reproduktion mir – aufgrund meiner Verpflichtungen Ihnen gegenüber – nicht schadet. Aber wir müssen bestimmen, warum dies so sein sollte. Es erscheint zumindest mir *nicht* unplausibel, dass die Anwesenheit neuer Menschen in meiner sozialen Welt mich auf moralisch signifikante Weise betreffen kann. Ist es tatsächlich falsch zu denken, dass Neuankömmlinge – seien es Migrantinnen oder Neugeborene – mir gegenüber Ansprüche erheben könnten? Menschen sind, offen gesagt, moralisch anspruchsvoll und ihre moralischen Rechte begründen Ansprüche uns gegenüber. In meinen Augen kann die Abneigung gegenüber dem Gedanken, dass wir unsere Mitbürgerinnen an der Fortpflanzung hindern dürfen, nicht auf der Idee beruhen, dass ihre Reproduktion sich nicht auf uns auswirkt. Denn *selbstverständlich* wirkt sie sich auf uns aus. Die einzige Frage ist, ob dieser Umstand die Grundlage für ein Recht darauf darstellen kann, Reproduktion zu verhindern.

Das kann er jedoch mit Sicherheit nicht; der Grund dafür liegt allerdings nicht darin, dass Babys keine Ansprüche an uns haben, sondern darin, dass die Mittel, die uns zur Verfügung stehen, um Menschen von der Reproduktion *abzuhalten*, offensichtlich und auf entsetzliche Weise ungerecht sind. Würden wir Personen zu medizinischen Eingriffen zwingen, um ihre Fortpflanzung präventiv zu unterbinden, wäre das zutiefst ungerecht. Aber es wäre deshalb ungerecht, weil es sich um einen unfreiwilligen körperlichen Eingriff handelt und nicht deshalb, weil das betroffene Kind mich und meine Freiheit nicht weiter berührt. In vielen anderen Lebensbereichen geben wir bereitwillig zu, dass die Anwesenheit von Menschen uns bestimmte Dinge erschwert; auf gleiche Weise können wir auch in diesem Kontext argumentieren. Die wahre Grundlage für die Freiheit zur Reproduktion ist nicht die Tatsache einer neu ins Leben tretenden Person, sondern der körperliche Eingriff, den die Prävention dieses Eintritts verlangen würde.[33]

In meinen Augen folgt aus alledem so etwas wie eine dreistufige Verteidigung des Verbots, die Reproduktion von Mitbürgerinnen zu verhindern. Die erste Stufe besteht in der Anerkennung der Behauptung, dass aufgrund der besonderen Beziehung zwischen der Frau und ihrem eigenen Körper der hierfür notwendige körperliche Eingriff mutmaßlich falsch ist. Die Frau hat das Recht darauf, ihren Körper frei von politischen Eingriffen zu nutzen – sowohl aus Gründen der Geschlechtergerechtigkeit als auch aufgrund grundlegender Menschenrechte. Die zweite Stufe besteht in der Anerkennung der Tatsache, dass das neugeborene Kind ungemein vulnerabel ist. Neugeborene sind, milde ausgedrückt, ungeheuer schlecht darin, für sich selbst zu sorgen. Sie sind mit ihrer Geburt sozusagen Flüchtlinge, die ein Recht auf eine Gemeinschaft haben, in der ihre Rechte geschützt werden. Zwar mag das Kind auf verschiedene Arten geschützt werden können, aber es scheint doch so, als ob das Recht der Eltern, diese Beschützer zu *sein*, als besonders wichtig dafür angesehen werden kann, die langfristige Entwicklung des Kindes sicherzustellen. Die dritte Stufe besteht hingegen darin, von dem Menschenrecht auf den eigenen Körper und die eigenen Eltern abzusehen und auf die soziale Bedeutung von Elternschaft zu blicken. Das Recht queerer Paare sowie von Adoptiveltern darauf, als gleichwertig angesehen zu werden, bedeutet, dass wir eine Gesellschaft einrichten sollten, die alle Formen der Kindererziehung mit den Rechten ausstattet, die nötig sind, um das Kind in seiner Entwicklung zu schützen und zu führen.[34] In meinen Augen würden wir aufgrund von mir angestellten Überlegungen daher in eine Situation gelangen, die der heutigen ziemlich stark ähnelt: In einer Welt, in der die Rechte von Eltern verteidigt werden und in der mein eigenes Recht, nicht verpflichtet zu werden, keine Aussicht darauf hat, das Recht auf Reproduktion einzuschränken. Allerdings folgen daraus keinerlei Probleme hinsichtlich des staatlichen Rechts auf Ausschluss. Die genauere Begründung dafür, warum wir die Schwangerschaften anderer nicht abbrechen dürfen, sagt uns nichts bezüglich der Frage, ob wir unsere Staatsgrenzen schließen dürfen oder nicht.

Das letzte Problem, mit dem wir uns auseinandersetzen müssen, ist das Problem der *Freiheit*. Andy Lamey hat behauptet, dass, sollte das Recht auf der Idee der Freiheit beruhen, wir dieses Recht verlieren könnten, sofern entweder unsere Freiheit nicht berührt wird oder die Kosten des Verzichts auf die Inanspruchnahme dieser Freiheit minimal sind:

> „Es gibt einen Schweregrad, ab dem automatisch die Pflicht entsteht, den Immigranten aufzunehmen; allerdings gibt es auf der anderen Seite keinen gleichwertigen Schwellenwert hinsichtlich der Geringfügigkeit der durch einen Grenzübertritt entstehenden Freiheitseinschränkung, der ebenso automatisch die Einwanderung erlauben würde. Die Aufnahmegesellschaft eines liberalen Staates kann jederzeit einen Immigranten zurückweisen, selbst wenn die Kosten für seine Aufnahme minimal wären oder sie sogar von dessen Ankunft profitieren würde. […] Es fällt schwer, diese Asymmetrie als gerechtfertigt anzusehen. Im einen Fall sind die Kosten vom Adressaten einer potentiellen Verpflichtung zu tragen, im anderen Fall von demjenigen, der die Verpflichtung trägt. Es erscheint als konsistenter mit dem Prinzip moralischer Gleichheit, das die Rechtsprechung begründet, zu behaupten, dass die Kosten für beide Parteien in die Begründung einer Verpflichtung einfließen müssen. Aus einer solchen Perspektive wäre Raum für den Gedanken, dass die Kosten einer Pflicht mitunter so geringfügig sind, dass wir keine Gründe haben, sie zurückzuweisen.“[35]

In einem gewissen Sinne glaube ich, dass Lamey Recht hat: Wo die Kosten einer Aufnahme gering sind – und die Vorteile für die Immigrantin signifikant –, haben wir gute moralische Gründe, diese Migrantin aufzunehmen. Aber mir scheint, dass diese Überlegungen Vorstellungen von Gnade widerspiegeln, die ich später diskutieren werde. Die Idee der Gerechtigkeit hingegen blickt darauf, ob der ausgeschlossenen Person Unrecht getan wurde; und wie zuvor scheint es schlicht falsch zu behaupten, dass wir deshalb *verpflichtet* wären in eine bestimmte Beziehung einzutreten, weil diese Beziehung uns keinerlei Kosten auferlegt. (Ich muss mit Ihnen heute Abend nichts ins Kino gehen, selbst wenn Sie mir beweisen würden, dass ich einen solchen Besuch tatsächlich mehr genießen würde, als zu Hause zu bleiben.) Warum sollten wir meinen, dass einer Person durch meine Weigerung, mit ihr in eine bestimmte Beziehung zu treten, ein Unrecht geschieht? Eine mögliche Art, Lamey zu lesen, besteht sicherlich darin, anzunehmen, dass der Freiheitsverlust dadurch *kompensiert* wird, dass mir durch die Anwesenheit der Migrantin keinerlei Kosten entstehen. Diese Vorstellung von Freiheit scheint jedoch schlicht falsch. Im Allgemeinen gehen wir nicht davon aus, dass ich mich Ihrer moralischen Erwartungen aufgrund der Tatsache entledigen kann, Ihnen zu einem späteren Zeitpunkt Geld anbieten zu können. Wenn ich die Freiheit habe, nicht in eine Beziehung mit Ihnen zu treten, dann ist das eine Freiheit, die ich ganz unabhängig davon habe, ob ich mit dem Verzicht auf diese Freiheit gut beraten wäre. Kurzum: Rechte gelten unabhängig von den durch sie verursachten Kosten – wenn ich ein Recht habe etwas nicht zu tun, löst die Tatsache, dass ich es ohne Kosten tun könnte, dieses Recht nicht auf.

Lameys letzte Überlegung betrifft den Gedanken, dass eine Person dadurch vor Ausschluss geschützt werden könnte, dass sie freiwillig auf ihre eigenen Rechte verzichtet. Schließlich sind es die freiheitseinschränkenden Auswirkungen der Rechte von Einwanderern, die mir das Recht geben, diese Personen auszuschließen – wenn eine Immigrantin nun aber erklärt, mein Handeln zu keinem Zeitpunkt in Anspruch zu nehmen, scheint es unzulässig, sie auszuschließen:

> „Aus welchen Gründen genau aber wäre es für sie falsch, auf ihre Rechte zu verzichten? Es gibt Fälle, in denen wir es Individuen erlauben, ein Gebiet zu besiedeln, in dem effektiv kein Zugang zu einer Regierung besteht, die dazu fähig wäre, ihre Rechte zu schützen. [...] Auf diese Weise befinden sich Menschen, die eine Weltreise per Boot machen, jenseits der effektiven Reichweite von Staaten und somit zugleich in einer Position, in der praktisch gesehen kein Gesetz zuhanden ist, um sie zu schützen. Es scheint nicht an sich, sondern höchstens bedingt unmoralisch, wenn sich eine Person freiwillig jenseits des effektiven Schutzes ihrer Menschenrechte stellt. Wenn das der Fall ist, warum sollten wir es informierten Individuen nicht gestatten, sich innerhalb eines existierenden Staates jenseits des Rechts zu stellen?“[36]

Dieser Einwand ist raffiniert; aber ich denke, er schlägt fehl, wenn auch nur, weil ich mit Kant den Gedanken teile, dass im Naturzustand die erste Pflicht einer jeden Person darin besteht, alles Notwendige zu tun, um diesen Naturzustand zu *verlassen*.[37] Ob es nun einer Person erlaubt sein sollte, den grundlegendsten Schutz ihrer Menschenrechte freiwillig aufzugeben, kann ich hier nicht abschließend beurteilen; aber ich bin überzeugt, dass es für jeden Staat zutiefst unmoralisch wäre, einer Migrantin das Recht auf ein Leben innerhalb seiner Staatsgrenzen unter der Bedingung anzubieten, dass sie jeglichen Schutz ihrer Rechte aufgibt. In vormodernen Rechtssystemen bestand die härteste Form der Bestrafung nicht selten in der Ächtung – einem Status, in dem es unmöglich war, zum Schutz des eigenen Lebens oder der eigenen Rechte auf die politische Gemeinschaft zurückzugreifen. Das Konzept des *homo sacer* – des Menschen, der geopfert werden kann – entstammt dem schrecklichen Wesen dieser Bestrafung. Ich würde es eher bezweifeln, dass wir irgendeiner Person ein Leben als *homo sacer* anbieten können und dabei noch im Einklang mit dem moralischen Imperativ der Menschenrechte stehen. Was die Seereisen und andere Formen des Lebens außerhalb des Schutzes durch den Staat anbelangt: Ich denke, es sollte hierbei betont werden, dass die Entdecker der Weltmeere trotz allem recht strikte – in der Tat recht brutale – Regierungssysteme an Bord ihrer Schiffe unterhielten.[38] Nehmen Sie zum Beispiel den Kodex von Kapitän Bartholomew Roberts, dem Piratenkapitän der *Royal Fortune*. Kapitän Roberts' Piratenkodex kam nichts weniger als einer Verfassung gleich, die der Regulierung der Beziehungen unter den reisenden Männern diente. Nehmen Sie die ersten beiden Klauseln dieses Kodex als Beispiel für den Inhalt dieses (umfangreichen) Dokuments:

> „I. Jedermann hat Stimmrecht in wichtigen Angelegenheiten; hat denselben Anspruch auf wann immer erbeutete Nahrungsmittel oder starken Schnaps, es sei denn ein Mangel (nichts ungewöhnliches unter ihnen) macht es erforderlich, um des Wohles aller willen, eine Beschränkung zu verhängen.
>
> II. Jedermann soll aufgrund einer Liste abwechselnd an Bord von Prisen (mitgebrachte Schiffe, Anm. d. Ü.) gesetzt werden, denn man gestand ihnen (über ihren normalen Anteil hinaus) bei dieser Gelegenheit einen Posten Leinenzeug zu: Wenn aber jemand die Gesellschaft auch nur um einen Dollar in Edelmetall, Juwelen oder Geld betrog, so wurde er mit Aussetzung bestraft.

> Wenn einer lediglich einen anderen beraubte, so gaben sie sich damit zufrieden, die Ohren oder die Nase des Schuldigen aufzuschlitzen und ihn an Land zu setzen, zwar nicht an einem unbewohnten Ort, doch immerhin irgendwo, wo er sicher sein konnte, großen Härten ausgesetzt zu sein."[39]

Aufgrund dieser Aussagen meine ich, dass es wohl vergleichsweise wenige Plätze auf dieser Welt gibt, an denen wir es gutheißen würden, dass Menschen außerhalb jeglicher zwingender Regierungsstruktur leben. Eine Person, die tatsächlich außerhalb der Reichweite zwingenden Rechts steht, ist nicht fähig, ihre Recht zu verteidigen, einmal abgesehen von der Verteidigung durch ihre eigenen Hände; und nur über diese Form der Verteidigung zu verfügen ist in der Tat eine schlechte Sache. Auch wenn ich es an dieser Stelle nicht beweisen kann, denke ich, dass es für einen Staat unzulässig ist, einer Person eine solch schlechte Form des Lebens anzubieten. (Mit Sicherheit würde niemand ein solches Angebot akzeptieren, außer einer Person, deren Rechte bereits ernsthaft bedroht sind; und diese Personen sind wiederum zu weit mehr berechtigt, als ihnen dieser Handel anbietet.) Ich bin auf jeden Fall nicht überzeugt davon, dass uns die von Lamey vorgebrachten Überlegungen einen Grund dafür geben, unsere Haltung hinsichtlich der Rechte von Migrantinnen zu überdenken, von denen im Übrigen sehr wenige – wie wir uns vorstellen können – darauf aus sind, ihre Rechte aufzugeben und sich in Ächtung zu begeben.

Den Schluss dieses Kapitels soll ein Hinweis auf die Bescheidenheit dessen bilden, was ich zu verteidigen versucht habe. Die von mir vertretene Vorstellung von Ausschluss beruht nicht auf Nationalität, Territorium, einer generationenübergreifenden Gemeinschaft oder etwas, das metaphysisch anspruchsvoller wäre als das bloße Faktum einer politischen Gemeinschaft. Ich glaube, dass diese Idee von Ausschluss am erfolgreichsten bei der Erklärung ist, wie und warum ein liberaler Staat Migrantinnen gerechterweise ausschließen darf. Allerdings habe ich bisher nicht gezeigt, wie diese Ideen für das Verständnis der spezifischeren Problemstellungen herangezogen werden können, die oft den Hintergrund unserer Debatten über Migrationspolitik bilden. Dieser Aufgabe wende ich mich nun zu.

5
Zwang und Zuflucht

Das vorangegangene Kapitel beschrieb einen bestimmten Ansatz zur Rechtfertigung von Einwanderungsbeschränkungen. Gemäß dieses Ansatzes müssen wir dabei nicht auf Ideen wie eine nationale Identität, soziale Solidarität oder irgendein anderes besonderes Gut zurückgreifen, das anhand solcher Beschränkungen verteidigt werden könnte. Stattdessen schauen wir, ob die Menschenrechte der potentiellen Migrantin bereits in ihrem Herkunftsland angemessen geschützt werden. Sofern dies der Fall ist, haben andere Staaten ein präsumtives Recht, diese Migrantin auszuschließen. Demnach haben Menschen also ein Recht auf eine politische Gemeinschaft, die ihre Rechte schützt, jedoch kein Recht darauf, Mitglied in einer von ihnen gewählten Gemeinschaft zu sein. Diejenigen, die bereits innerhalb eines bestimmten Landes leben, haben eine Recht darauf, die Entstehung einer besonderen Beziehungen zu den potentiellen Neuankömmlingen zu verweigern – und eine solche Beziehung würde notwendigerweise entstehen, sollten die potentiellen Migrantinnen die Staatsgrenze tatsächlich überqueren und dadurch in das Hoheitsgebiet des Staates eintreten. Im Falle eines adäquaten Rechtsschutzes in den Herkunftsländern haben die übrigen Länder der Welt also ein präsumtives Recht darauf, Migrantinnen auszuschließen.

Im Falle vieler Menschen ist jedoch die erste Prämisse dieses Prinzips nicht erfüllt. Die Welt ist voll tiefgreifender Ungerechtigkeiten. Überall auf dem Globus sind Menschen mit verschiedenen Formen des Unrechts konfrontiert: Sie werden verfolgt, sind verarmt oder marginalisiert, leiden Hunger, werden zu Unrecht eingesperrt oder schlicht missachtet. Sie sind menschlichen Boshaftigkeiten genauso unterworfen wie Naturkatastrophen und einer Unzahl anderer, unerwarteter Verheerungen. Die Welt ist also voller Menschen, auf die die von mir vertretene Rechtfertigung des Rechts auf Ausschluss nicht zutrifft. Was kann uns mein Ansatz angesichts dieser Tatsachen sagen?

Der naheliegendste Schluss besteht darin, dass Menschen, die in gravierender Ungerechtigkeit leben, ein Recht darauf haben, Grenzen zu überqueren, oder dass, genauer gesagt, die Staaten dieser Welt moralisch nicht berechtigt sind, diese Personen mittels Zwangsgewalt auszuschließen. Dieser offenkundige Schluss scheint einfach – der von mir verteidigte Ansatz impliziert demnach, wenn schon nicht offene Grenzen, dann zumindest offene Grenzen für viele der verarmten und marginalisierten Bewohnerinnen dieses Planeten. In diesem Schluss steckt einiges an Wahrheit, denn mit Sicherheit erlaubt es uns der von mir auf Basis der Gebietshoheit entwickelte Ansatz nicht, irgendetwas in der Form der gegenwärtig angewendeten Ausschlusspraktiken zu rechtfertigen. Allerdings glaube ich auch, dass die Implikationen meiner Herangehensweise etwas komplexer sind, als sie zunächst erscheinen. Wie genau mein Ansatz das staatliche Handeln an den Grenzen – und auch über diese Grenzen hinaus – einschränkt, wird der Gegenstand dieses Kapitels sein.

Der Gedanke, dass es Menschen gibt, die von den gewöhnlichen Regeln des Ausschlusses nicht betroffen sind, hat sowohl in der Theorie als auch in der Praxis eine lange Tradition. So umfasste die mittelalterliche Praxis des Kirchenasyls (sanctuary) Elemente kirchlicher sowie politischer Macht und bot manchen Personen einen Weg, um mit kirchlichen Mitteln gegen die Legitimität der ihnen drohenden, weltlichen Strafen Einspruch zu erheben.[1] War die Gewährung von Asyl ursprünglich ein religiöser Akt, so war er doch schlicht genug, als dass auch Staaten den Verfolgten einer anderen Nation einen solchen Schutz bieten konnten – und während der auf die Reformation folgenden Kriege begannen sowohl Katholiken als auch Protestanten, fremde Staaten als mögliche Zufluchtsorte zu betrachten. Selbstverständlich diskutierten auch Philosophen darüber, wie die moralische Grundlage dieser staatlichen Gewährung von Asyl zu verstehen sei. Bereits im siebzehnten Jahrhundert entwickelten Autoren wie Hugo Grotius, Samuel von Pufendorf und Christian Wolff komplexe Analysen hinsichtlich der Frage, wie und wann Staaten sich der Ausweisung derjenigen widersetzen sollten, die behaupteten, in ihren Herkunftsländern unter Ungerechtigkeit zu leiden. Grotius' Überlegungen sind für den Gedankengang dieses Buches besonders relevant. Sein Argument bestand darin, dass Asyl denjenigen zugesprochen werden sollte, die unter „unverdienter Feindseligkeit leiden" (*immerito odio laborant*) – womit gemeint war, dass ihnen eine ungerechte Bestrafung in ihren Heimatländern drohte.[2] Grotius führte aus, dass ein Staat berechtigterweise darauf verzichten könne, seine Zwangsgewalt zur Ausweisung eines Fremden einzusetzen, falls diesem in seinem Herkunftsland ein Unrecht widerfahren würde. Diese Vorstellung von Asyl hatte bis in die frühe

Moderne hinein Bestand. So verweigerte beispielsweise Thomas Jefferson die Ausweisung von Flüchtlingen nach Spanien auf Grundlage der Überlegung, dass wer für eine Reform ungerechter politischer Verhältnisse kämpfte, oftmals des Hochverrats beschuldigt wurde. Laut Jefferson verdienten jedoch diejenigen, die sich, nachdem sie im Namen der Gerechtigkeit gekämpft hatten, aus Spanien in die Vereinigten Staaten flüchteten, Schutz statt Ausweisung. „Wir sollten uns davor hüten", so Jefferson, „einen in seinem Kampf gescheiterten Patrioten, der bei uns Zuflucht sucht, dem Henker auszuliefern."[3] Diese Vorstellung gründet auf der Idee, dass die Patriotin bloß dann Schutz verdient, wenn ihr eine ungerechte Bestrafung im Heimatland droht. Wie bereits zuvor stellt die Vorstellung, dass ein gerechter Staat seine Zwangsmacht nicht dazu nutzen sollte, eine Person in ungerechte Umstände zu zwingen, einen zentraler Bestandteil der moralischen Begründung dar.

Die Idee des Asyls war von jeher komplex – sowohl in moralischer als auch in rechtlicher Hinsicht –, und es wurde im Laufe des zwanzigsten Jahrhunderts zunehmend komplizierter. Der Erste und Zweite Weltkrieg hinterließen überall auf der Welt Narben und vertrieben Menschen, denen diese weltweiten Konflikte jeden Schutz nahmen. Die Geschichte dieser Entwicklung kann ich hier nicht einmal im Ansatz adäquat wiedergeben. Stattdessen werde ich mich schlicht den Dokumenten zuwenden, die als internationale Antwort auf die Schrecken dieses Jahrhunderts dienen sollten: dem Abkommen über die Rechtsstellung der Flüchtlinge aus dem Jahr 1951 (Genfer Flüchtlingskonvention), zusammen mit seinem Protokoll von 1967. Diese Dokumente enthalten die moderne rechtliche Definition des Begriffs Flüchtling und eine bestimmte Vorstellung dessen, was den um Zuflucht ersuchenden Personen nicht angetan werden darf. Die Genfer Flüchtlingskonvention war zum einen dazu bestimmt, staatliches Recht zu leiten; zum anderen aber auch eine Aussage über international geltende moralische Standards und als solche bemerkenswert schlicht.[4] Sie verpflichtet ihre Unterzeichner zur Anerkennung des Grundsatzes, dass diejenigen nicht zurückgewiesen werden dürfen, denen eine bestimmte Form von Unheil in ihren Heimatländern droht. Ein Flüchtling wird als eine Person definiert, die sich

> „aus der begründeten Furcht vor Verfolgung wegen ihrer Rasse, Religion, Nationalität, Zugehörigkeit zu einer bestimmten sozialen Gruppe oder wegen ihrer politischen Überzeugung sich außerhalb des Landes befindet, dessen

> Staatsangehörigkeit sie besitzt, und den Schutz dieses Landes nicht in Anspruch nehmen kann oder wegen dieser Befürchtungen nicht in Anspruch nehmen will; oder die sich als staatenlose infolge solcher Ereignisse außerhalb des Landes befindet, in welchem sie ihren gewöhnlichen Aufenthalt hatte, und nicht dorthin zurückkehren kann oder wegen der erwähnten Befürchtungen nicht dorthin zurückkehren will."[5]

Personen, die unter diese Definition fallen, haben eine Vielzahl an Rechten – das bedeutendste unter ihnen ist dabei das Recht auf *Non-Refoulement* (Nichtzurückweisungsprinzip), dem zufolge sie davor geschützt sind, mittels staatlicher Zwangsgewalt in diejenigen Umstände zurückgebracht zu werden, aus denen sie aufgrund von Verfolgung geflohen sind.[6] Die Unterzeichnerstaaten des Abkommens und seines Protokolls stimmen darüber hinaus zu, den Schutzsuchenden weitere Rechte zu gewähren, darunter den Zugang zu einer Anhörung, in der die Gültigkeit ihres Asylanspruchs fair bewertet werden kann.

Es ist nur schwer möglich, eine akkurate Zusammenfassung des durch die Genfer Flüchtlingskonvention begründeten Rechtsregime zu geben. Das Verständnis dieses Regimes bedarf sowohl eines Verständnisses internationalen Rechts, als auch einer Sensibilität dafür, wie dieses Recht von Individuen und Staaten ausgelegt und angewendet wird, sowie eines Bewusstseins dafür, wie internationales Recht durch einzelstaatliche Politik geformt wird und zugleich auf die Politik der Einzelstaaten zurückwirkt. In diesem Kapitel möchte ich von vielen dieser Aspekte absehen und mich stattdessen auf einen bestimmten Teil des Regimes konzentrieren. Von besonderer Bedeutung ist dabei, dass der *Verfolgung* eine gewichtige Rolle innerhalb der durch die Genfer Flüchtlingskonvention begründeten internationalen Moral zukommt.Um das Recht, Grenzen zu überqueren, in Anspruch nehmen und darauf bestehen zu können, dass Staaten eine moralische Pflicht haben, mir dies auch zu erlauben, muss ich zeigen können, dass ich verfolgt werde. Es ist demnach beispielsweise nicht genug, dass ich unter Hunger, willkürlicher Gewalt oder dem politischen Zusammenbruch meines Heimatlandes leide. Dieser besondere Fokus auf die Idee der Verfolgung wird seit Bestehen der Genfer Flüchtlingskonvention regelmäßig kritisiert, sowohl innerhalb der Philosophie als auch in der juristischen Praxis. Dokumente wie die *Cartagena Declaration on Refugees* von 1984 haben das Konzept des Flüchtlings um Personen erweitert, die von „allgemeiner Gewalt, ausländischer Aggression, internen Konflikten, massiven Menschenrechtsverletzungen oder anderen Umständen, welche die öffentliche Ordnung auf schwerwiegende Weise gestört haben“, bedroht sind.[7] Auch innerhalb der Philosophie wurde die zentrale Bedeutung des Konzepts der Verfolgung kritisiert.[8] Die Idee des Flüchtlings ist mittlerweile derart mit der Genfer Flüchtlingskonvention verbunden, dass manche, wie Alexander Betts, vorgeschlagen haben, das Konzept selbst abzuschaffen. Stattdessen, so Betts, sollten wir uns auf „Überlebensmigrantinnen“ konzentrieren, deren Leben von der Rettung durch eine Gesellschaft außerhalb ihres Heimatlandes abhängen.[9] Andere – darunter Matthew Price und Matthew Lister – argumentierten, dass das Abkommen einige bedeutende moralische Weisheiten enthielte: Nur der Begriff der Verfolgung, so ihr Argument, ist in der Lage, das spezifische moralische Unrecht adäquat zu erfassen, das allein durch die Gewährung von Aufenthaltsrechten andernorts auf angemessene Weise gelindert werden kann.[10]

Ich werde an dieser Stelle nicht direkt auf die Frage eingehen, ob das Abkommen moralisch haltbare Ideale widerspiegelt oder ob, wie die Kritiker behaupten, es bloß einer besonderen historischen Konstellation entspringt. Stattdessen will ich mich dieser Frage dadurch annähern, dass ich untersuche, welche Arten von Einschränkungen des Rechts auf Ausschluss aus dem von mir verteidigten Ansatz der Gebietshoheit folgen. Hierfür werde ich zunächst zwei verschiedene Arten skizzieren, auf die dieser Ansatz den Staat bei seinen Entscheidungen leiten kann. Anschließend werde ich spezifischer analysieren, wie es möglich ist, dass dieser Ansatz zu solchen Einschränkungen führt. Schließlich werde ich mich an einer ersten Beschreibung der Verpflichtungen versuchen, die aus diesem Ansatz folgen, woraus sich wiederum ein kritischer Ausgangspunkt für die Kritik oder Verteidigung bestimmter rechtlicher Institutionen gewinnen lässt.

Aus diesem Grund werde ich im vorliegenden Kapitel drei Dinge tun. Beginnen werde ich damit, verschiedene Lesarten eines Rechts auf Grenzübertritt zu diskutieren. Anhand einer genaueren Betrachtung der Einschränkungen eines Rechts auf Ausschluss können wir zwischen einem *positiven* und einem *negativen* Recht auf Bewegungsfreiheit unterscheiden, abhängig davon, was dieses Recht von dem Staat verlangt, gegenüber dem es in Anspruch genommen wird. Daran anschließend werde ich mit einer Untersuchung der Bedingungen fortfahren, unter denen die Überlegungen des auf der Gebietshoheit basierenden Ansatzes gegen einen Staat vorgebracht werden können, der Zwangsgewalt zum Schutz seiner Grenzen einzusetzen gedenkt. Das Konzept der Verfolgung, so mein Argument, ist dabei nicht hinreichend, um die Gruppe derjenigen zu definieren, gegen die keine Zwangsgewalt an der Grenze eingesetzt werden darf. Der abschließende Teil dieses Kapitels wird jedoch über den Einsatz von Zwangsgewalt an staatlichen Grenzen hinausgehen und fragen, gemäß welcher Prinzipien ein Staat Ressourcen für flüchtende Personen bereitstellen sollte, die bereits um Unterstützung ersuchen, bevor sie an der Grenze dieses Staates angelangt sind. Meinen Überlegungen zufolge gibt es Personen, die nicht nur einfach ein Recht darauf haben, vor Zurückweisung geschützt zu sein, sondern auch einen Anspruch darauf, während ihrer Flucht unterstützt zu werden – worunter in meinen Augen auch fällt, verhältnismäßige Gewalt für ihre Verteidigung einzusetzen. Nach meinem Argument könnte das Konzept der *Verfolgung* eine Rolle bei der Bestimmung spielen, wer Anspruch auf eine solche Unterstützung erheben kann. Dadurch mag der Fokus der Genfer Flüchtlingskonvention auf das Konzept der Verfolgung legitimiert werden – allerdings nur dann, wenn das Abkommen so verstanden wird, dass es in einem umfangreicheren Bündel von Prinzipien wirkt, dem zufolge mit Zwang verbundene Sanktionen gegenüber Verfolgerstaaten legitim sind. Wir können die Genfer Flüchtlingskonvention also durchaus verteidigen, allerdings nur, wenn wir dazu bereit sind, einigen strikteren Geboten gerecht zu werden, die uns durch später entstandenes rechtliches Material auferlegt worden sind.

1. Positive und negative Migrationsrechte

Wir können damit beginnen, zwei verschiedene Lesarten eines Rechts auf Grenzübertritt zu untersuchen.[11] Einem möglichen Verständnis zufolge sind wir im Falle der Existenz eines solchen Rechts moralisch dazu berechtigt, bei einem versuchten Grenzübertritt keiner Gewalt ausgesetzt zu sein. Diese Variante können wir als *negatives* Recht auf Migration bezeichnen. Es hält andere davon ab, mich durch Zwang an einer Bewegung im physikalischen Raum zu hindern, durch die ich Linien überschreiten würde, die das Hoheitsgebiet eines Staates begrenzen. Eine Vielzahl bedeutsamer moralischer Rechte lässt sich auf diese Weise am besten beschreiben. Das Recht auf Pressefreiheit gewährleistet, dass ein Staat eine Zeitung nicht schließen und ebenso wenig ihre Herausgeber festnehmen darf. Es gewährleistet hingegen einem Autor nicht das Recht, dass seine Texte durch eine Zeitung publiziert werden – und wenn sich keine Zeitung findet, die seine Texte veröffentlichen will, wird seine Freiheit durch diesen Umstand nicht eingeschränkt. Ähnlich verhält es sich mit dem Recht auf anwaltliche Vertretung in einem Zivilprozess, welches mir garantiert, dass der Staat meine Vertretung vor Gericht durch einen Rechtsanwalt nicht verhindern darf. Dieses Recht wird jedoch nicht missachtet, wenn ich mir keinen Anwalt leisten kann und auch nicht, wenn mein Fall so aussichtslos ist, dass kein Anwalt mich vertreten will.

Meist meinen wir aber etwas Stärkeres, wenn wir über Migrationsrechte reden. Ein *positives* Recht auf Migration würde demnach nicht nur einfach die Abwesenheit gewaltsamer Einschränkungen meiner Pläne bedeuten, sondern bestimmte Handlungen zur Beförderung dieser Pläne fordern. Ein solches Verständnis würde bedeuten, dass der Staat etwas zum Erfolg meines Vorhabens beiträgt – was auch den Einsatz von Gewalt *gegen* diejenigen umfassen könnte, die meine Migrationspläne verhindern wollen. Auf diese Weise funktioniert beispielsweise das US-amerikanische Recht auf anwaltliche Vertretung in einem Strafprozess: Ich bin nicht einfach nur dazu berechtigt, frei von staatlichen Eingriffen einen Anwalt zu beauftragen; vielmehr habe ich auch einen Anspruch darauf, dass mir ein Anwalt bereitgestellt wird, sollte ich mir keinen leisten können.[12]

All das ist selbstverständlich nicht neu. Der Unterschied zwischen positiven und negativen Rechten ist uns schon länger vertraut, wenn er auch nicht unumstritten ist. Allerdings sollten wir hinsichtlich dieser Unterscheidung sensibel sein, da wir zu der Überzeugung tendieren, dass es einen moralischen Unterschied zwischen einem positiven und einem negativen Recht gibt. Sie unterscheiden sich darin, was sie von uns verlangen und als wie schwerwiegend wir ihre Verletzung einschätzen sollten. Auf diesen moralischen Intuitionen beruht die Arbeit von Thomas Pogge. Er argumentiert, dass wir globale Armut nicht als ein Versagen begreifen sollten, unsere positiven Hilfspflichten gegenüber den Armen zu erfüllen – was selbstredend bloß die logische Folge des positiven Rechts der Armen auf unsere Hilfe darstellen würde. Wir unterliegen ihm zufolge vielmehr der negativen Pflicht, globale Armut nicht zu *verursachen*, und es ist diese negative Pflicht, die wir kollektiv verletzen. Die Grundlage von Pogges Arbeit ist der weit verbreitete moralische Gedanke, dass die Verletzung anderer einem strengeren moralischen Verbot unterliegt als das Versagen, eine solche Verletzung zu verhindern; das Verletzen eines negativen Rechts stellt somit ein größeres moralisches Unrecht dar als die Verletzung eines positiven.[13]

Meiner Meinung nach kann eine ähnliche moralische Intuition unsere Reaktion auf unterschiedliche Interpretationen des Rechts auf Grenzübertritt erklären. Das Recht, Grenzen zu überqueren, könnte positiv in dem Sinne verstanden werden, dass es einen Anspruch gegenüber anderen Akteuren – inklusive Staaten – umfasst, bei der Überquerung internationaler Grenzen – sofern der Zweck in der Suche nach Zuflucht besteht – unterstützt zu werden. Es könnte allerdings auch in dem Sinne als negativ verstanden werden, dass der Staat bloß dazu verpflichtet ist, nicht gewaltsam in den Akt der Grenzüberquerung einzugreifen. Intuitiv scheint es schlimmer, einer Person im Falle der Grenzüberquerung mit Gewalt zu drohen, als ihr die Mittel vorzuenthalten, die sie für die Überquerung dieser Grenze benötigen würde. Kieran Oberman folgt dieser Intuition. Zwar folgt aus seinem Ansatz ein Recht auf Grenzüberquerung, allerdings folgt daraus für hin nicht, dass jede Person mit einem Flugticket für die von ihr favorisierte Gesellschaft ausgestattet werden muss. Wir können daher ein Recht auf Einwanderung verteidigen, ohne gleichzeitig behaupten zu müssen, dass die Migrantin mit den für die Ausübung dieses Rechts notwendigen Mitteln ausgestattet werden sollte.

Somit können wir also zwischen der Freiheit vor gewaltsamen Eingriffen in unsere Pläne und dem moralischen Recht auf positive Unterstützung bei der Umsetzung dieser Pläne unterscheiden. Ersteres mag eher mager und inkonsequent erscheinen – obwohl an dieser Stelle anzumerken ist, dass unter anderem Joseph Carens den Gedanken verteidigt hat, dass das Recht auf Einwanderung auch dann noch von Wert sein kann, wenn eine Person nicht über die zur Nutzung dieses Rechts notwendigen Mittel verfügt.[14] Das negative Recht auf Einwanderung würde demnach seine Stärke auch dann nicht verlieren, wenn einzelne Personen sich nicht die Transportmittel leisten könnten, die für das Erreichen dieser Grenze notwendig wären. Eine negative Pflicht mag also weniger belastend sein als eine positive Pflicht, denn wir sind ihr zufolge nicht dazu verpflichtet, Migranten bei ihrem Vorhaben zu unterstützen, sondern bloß dazu, dieses Vorhaben nicht zu verhindern. *Was* die negative Pflicht allerdings fordert, scheint sie mit besonderer Strenge zu fordern. Der Staat, der diese negativen Pflichten verletzt, schadet bestimmten Personen, und diese Personen haben in der Folge besonders starke Ansprüche auf Wiedergutmachung durch diesen Staat.

Wir können das soeben Gesagte in meinen Augen sehr gut an unseren Reaktionen auf bestimmte Fälle der Migrationspolitik festmachen. Nicht alle Formen von Migrationspolitik, die der Verringerung von Einwanderung dienen, erreichen ihren Zweck durch die gewaltsame Verhinderung von Grenzübertritten. Insbesondere Sanktionen gegen Beförderungsunternehmen funktionieren so, dass Luftfahrt- und Bootsunternehmen finanzielle Strafen drohen, sollten sie Migrantinnen befördern, die keine legalen Einreisedokumente für das Zielland besitzen.[15] Mit diesen Instrumenten drohen den Beförderungsunternehmen finanzielle Strafen, um von der Beförderung von Personen abzusehen, die keine Einreiseerlaubnis für den jeweils infrage stehenden Staat bekommen haben. Diese Vorgehensweise resultierte, angesichts der großen Menge verzweifelter Menschen, die ihren von Gewalt und Gräuel gezeichneten Herkunftsländern zu entkommen versuchen, selbstverständlich in einer Zunahme von Schleppern. Folglich gibt es heute einen florierenden Sekundärmarkt für Migrationsdienstleistungen, der für die Migrantinnen mit höheren Kosten und während ihrer Reise auch mit deutlich größeren Risiken für Unfälle und Gewalt einhergeht. Um beispielsweise von der Halbinsel Bodrum auf die griechische Insel Kos zu gelangen, müssen entweder 19 oder 1000 Dollar bezahlt werden – je nachdem, ob die Person ein Visum besitzt oder nicht. Ganz ähnlich ist es im Falle einer Bootsüberfahrt von Libyen auf die italienische Insel Lampedusa, die entweder 200 oder 2000 Dollar kostet, abhängig davon, ob die „richtigen" Dokumente vorliegen.[16]

Diese Maßnahmen sollen offensichtlich dazu dienen, die Zahl der Personen, die an den Grenzen der wohlhabenden Staaten ankommen, zu verringern. Erreicht eine Person die Staatsgrenze, hat sie ein Recht auf Prüfung des Asylantrags durch den jeweiligen Staat; aus diesem Grund sind die europäischen Staaten sehr daran interessiert, die Zahl ankommender Menschen zu verringern – vor allem in Anbetracht der Tatsache, dass viele dieser Personen einen Anspruch auf Schutz gemäß der Genfer Flüchtlingskonvention geltend machen könnten.[17] Sanktionen für Beförderungsunternehmen sind daher vielen Menschen ein Dorn im Auge, denn sie scheinen schutzbedürftigen Personen – darunter vielen Flüchtlingen im Sinne der Genfer Flüchtlingskonvention – die Suche nach effektivem Rechtsschutz zu erschweren. Aber verletzen diese Sanktionen tatsächlich das Recht dieser Individuen auf Grenzübertritt? Die Antwort auf diese Frage hängt in meinen Augen von unserem Verständnis davon ab, was sie den betroffenen Migrantinnen bei ihrem Versuch, die Grenze zu übertreten, tatsächlich antun.

Einer möglichen Perspektive zufolge sind diese Sanktionen gewaltsam, und zwar gegenüber den Migrantinnen selbst, denn schließlich sind sie letztlich das Ziel dieser Zwangsmaßnahme. Selbst wenn der Zwang zunächst gegenüber den Beförderungsunternehmen ausgeübt wird, besteht sein eigentlicher Zweck doch darin, das Reisen nach Europa zu erschweren oder vollkommen unmöglich zu machen. Wir können diese Sanktionen demnach in ihrer Struktur mit Formen gewaltsamer Zurückweisung an der Grenze vergleichen. Teresa Bloom und Verena Risse bieten eine solche Analyse; ihnen zufolge sollten die mit Zwangsgewalt bewehrten Sanktionen moralisch genauso bewertet werden wie die gewaltsame Zurückweisung einer Migrantin an der Grenze. Der Zwang sei demnach in diesem Falle „versteckt", weshalb die Sanktionen eher als eine Form von Verschleierung denn als moralisch distinkte Tatsache zu verstehen seien:

> „Im Jahr 1990 weigerten sich vier große Luftfahrtunternehmen (Lufthansa, Swissair, Iberia und Alitalia) die gegen sie durch das Vereinigte Königreich erhobenen Strafen zu zahlen und begründeten dies damit, dass sie durch diese Praxis aufgefordert würden ‚als Einwanderungsbeamte zu handeln'. Das britische Innenministerium antwortete darauf, dass die Airlines im Falle unterlassener Zahlungen ihre Landerechte verlieren könnten. *Dabei handelt es sich, wie bereits weiter oben gezeigt, eindeutig um den Versuch, eine gewisse Form von Zwang auszuüben, unabhängig von der umfassenderen Natur der Beziehung zwischen dem jeweiligen Staat und den Beförderungsunternehmen, und dieser Zwang bezieht sich auf die Staatsgrenze.* Allerdings richtet sich diese Form staatlichen Zwangs nicht direkt gegen die potentiellen Migrantinnen, sondern wird durch eine andere Konstellation zwingender und gezwungener Akteure verschleiert."[18]

Sollten diese Überlegungen zutreffen, gebe es keinen moralisch bedeutsamen Unterschied zwischen der gewaltsamen Zurückweisung von Migrantinnen an der Grenze und der zwangsweisen Zurückweisung einer Migrantin durch die Mitarbeiterinnen eines Luftfahrtunternehmens aufgrund angedrohter

Sanktionen. Beide Formen der Zurückweisung wären demnach Akte von Zwangsgewalt, die sich letztendlich gegen die Migrantin selbst richten; und sofern diese über ein negatives Recht auf Grenzüberquerung verfügt, verletzen diese Sanktionen jenes Recht ebenso evident wie eine Waffe, die an der Grenze auf die Migrantin gerichtet wird.

Dabei handelt es sich jedoch nicht um die einzige Möglichkeit, die Sanktionen gegen Luftfahrtunternehmen zu verstehen. Aus einer anderen Perspektive könnte diese Maßnahmen so beschrieben werden, dass darin gar kein Zwang gegenüber Migrantinnen involviert ist: Selbstverständlich sind diese Sanktionen auch aus dieser Perspektive mit Zwang gegenüber den Luftfahrtunternehmen verbunden, allerdings zwingen sie die Unternehmen diesem Verständnis zufolge bloß dazu, die für eine erfolgreiche Reise notwendigen Mittel zu verweigern – und verletzen somit das Recht auf Migration allein dann, wenn es als ein positives Recht verstanden wird. Die Verweigerung notwendiger Mittel ist schließlich nicht an sich ein gewaltsamer Akt. Wenn die Fähre zwischen Bodrum und Kos nicht mehr fahren würde, da sich der Betrieb nicht mehr rentiert, wäre es seltsam, diesen Umstand als gewaltsam gegenüber den Einwohnerinnen der Halbinsel Bodrum anzusehen. Das Beförderungsunternehmen hat keine Verpflichtung, die Fähre anzubieten, zumindest nicht unter gewöhnlichen Umständen; sollte es aufhören, diese Dienstleistung anzubieten oder sie gar nicht erst angeboten haben, wäre keiner Person etwas widerfahren, was sich als Gewalt bezeichnen ließe. Allerdings ist unklar, ob sich an diesem Urteil etwas ändert, wenn es der Staat ist, der den Transport zumindest einer großen Anzahl potentieller Passagiere unprofitabel *macht*, indem er das Unternehmen finanziell für den Transport jener Personen bestraft. Die Sanktionen schränken das Beförderungsunternehmen selbstredend gewaltsam ein; die Bußgelder werden nicht bloß angedroht. Aber diese Form des Zwangs scheint nicht transitiv. Aus seiner Ausübung gegenüber dem Unternehmen scheint kein Zwang gegenüber den Reisenden selbst zu folgen.

Ich möchte an diesem Punkt klarstellen, dass es mir nicht um eine Verteidigung der Sanktionen selbst geht. Wie im Folgenden ersichtlich werden sollte, denke ich, dass diese Sanktionen in vielen Fällen ungerecht sind. Mein Punkt ist vielmehr, dass die Ungerechtigkeit dieser Maßnahmen es mitunter nötig machen könnte, das Recht auf Mobilität als ein positives Recht zu verstehen, welches die positive Pflicht für Dritte umfasst – seien es Beförderungsunternehmen oder Staaten –, die Mittel für das Erreichen eines Zufluchtsstaats bereitzustellen. Wir empfinden zu Recht großes Mitgefühl für die syrischen Migrantinnen, die von Bodrum nach Kos übersetzen wollen. Aus diesem Mitgefühl folgt jedoch nicht, dass wir die Sanktionen gegen Beförderungsunternehmen als Sanktionen gegen die Migrantinnen selbst betrachten müssen. Stattdessen könnten sie als bloße Verweigerung derjenigen Mittel verstanden werden, die für eine Nutzung des Rechts auf Einwanderung erforderlich wären. Unsere Abneigung gegenüber diesen Sanktionen lässt sich in meinen Augen anhand der Idee eines positiven Rechts auf Migration erklären, dem zufolge eine Migrantin nicht nur ein Recht darauf hat, frei von Zwangsgewalt die staatlichen Grenzen zu überqueren, sondern auch auf Hilfe dabei, an dieser Grenze anzukommen.

Hier möchte ich anmerken, dass auch andere Autoren die Frage diskutiert haben, ob alle Maßnahmen zur Prävention von Migration zugleich als Einsatz von Zwangsgewalt zu betrachten sind. So haben Arash Abizadeh und David Miller in einer wichtigen Debatte versucht, die Frage zu beantworten, ob alle Einschränkungen der Bewegungsfreiheit zu Recht als gewaltsam betrachtet werden sollten.[19] Millers Argument besteht, vereinfacht gesagt, darin, dass es sich bei vielen Maßnahmen zur Einschränkung von Einwanderung nicht um Zwang, sondern Prävention handelt. Demnach nehmen wir den Migrantinnen durch unsere Maßnahmen nicht alle Optionen, sondern bloß eine Option – und nicht alle Arten, ihnen diese Option zu nehmen, sind in sich gewaltsam. So würde beispielsweise eine unüberwindbare Mauer Grenzübertritte effektiv verhindern und zugleich nicht als gewaltsam beschrieben werden können. Was ich hier sage, stimmt in Teilen mit Millers Argument überein; wir haben also Grund dafür, nicht illegitimerweise anzunehmen, dass alle Arten der Bewegungseinschränkungen moralisch mit Gewaltandrohungen gleichzusetzen sind.[20] Demnach sind nicht alle Bewegungseinschränkungen notwendig als gewaltsam zu verstehen; die Entfernung einer Möglichkeit, die Grenze eines Staates zu erreichen, sollte folglich aus moralischer Perspektive nicht auf den Einsatz von Zwangsgewalt an der Grenze selbst reduziert werden.

Ich werde auf diesen Fall am Ende noch einmal zurückkommen. Wie uns diese Diskussion zeigt, sollten wir sehr vorsichtig mit der Frage umgehen, welche Maßnahmen wir als Formen zwingender Eingriffe in die Bewegungsfreiheit ansehen und welche einer Verweigerung gleichkommen, die für dieser Bewegungsfreiheit notwendigen Mittel bereitzustellen. Zwang ist moralisch bedeutsam und diejenigen, die ihn einsetzen, müssen ihn gegenüber den von diesem Zwang betroffenen Personen rechtfertigen. Personen, die ein negatives Recht auf Einwanderung geltend machen könnten, dürfen einem solchen Zwang nicht ausgesetzt werden; er könnte ihnen gegenüber nicht gerechtfertigt werden und folglich haben sie einen Anspruch darauf, frei von staatlichen Zwangsmaßnahmen die Grenze zu überqueren. Sollten Personen jedoch über ein positives Recht auf Einwanderung verfügen, könnten sie einen Anspruch auf mehr denn die bloße Unterlassung von Zwangsmaßnahmen formulieren, nämlich einen Anspruch auf unsere Unterstützung. Indem wir im Folgenden untersuchen, wie der auf der Gebietshoheit basierende Ansatz diese zwei Arten von Rechten behandelt, können wir zugleich auch genauer bestimmen, wie dieser Ansatz die Moral der Flucht versteht.

2. Zwang an der Grenze: Das negative Recht auf Zuflucht

Beginnen wir mit der Vorstellung, dass eine Gruppe von Menschen an der Grenze eines Staates steht. Diese Menschen sind aus einem Land geflohen, in dem ihre Recht bloß unzureichend geschützt werden. Wie auch immer wir die Idee eines angemessenen Schutzes der Menschenrechte fassen: Das Land, aus dem diese Menschen fliehen, bietet ihn nicht. Wir könnten diese Menschen in unser Land lassen ohne dabei unsere gerechten Institutionen zu gefährden, die unsere Menschenrechte achten, schützen und gewährleisten. Haben wir irgendein Recht, diese Menschen auszuschließen? Die Antwort scheint nein zu lauten; mit Sicherheit fallen diese Menschen nicht unter das im letzten Kapitel beschriebene Prinzip zur Rechtfertigung staatlichen Ausschlusses. Würden wir Waffen nutzen, um diese Menschen während ihres Grenzübertritts zu bedrohen, käme das dem Einsatz von Gewalt zur Aufrechterhaltung moralisch unzulässiger Umstände gleich. Wir würden ihnen nicht bloß die Mittel für ihre Einwanderung verwehren, sondern durch unseren angedrohten Zwang auch aktiv dafür sorgen, dass sie weiterhin in einer moralisch unrechtmäßigen Form politischer Beziehung verbleiben. Mit anderen Worten: Wir würden mittels Gewalt eine Welt realisieren, in der die grundlegenden Rechte mancher Menschen verletzt werden; und das scheint eindeutig einen moralisch illegitimen Einsatz von Gewalt darzustellen. Im vorherigen Kapitel habe ich zwei mögliche Einschränkungen dieser Schlussfolgerung erwähnt: Wir könnten verlangen, dass einzelne Personen einen gerechten Anteil der Lasten übernehmen, die bei der Reform eines ungerechten Staates entstehen, und wir könnten anführen, dass Personen nicht immer einen Anspruch darauf haben, durch den von ihnen favorisierten Staat gerettet zu werden. Diese Einschränkungen scheinen jedoch nicht auszureichen, um den grundlegenden Schluss zu widerlegen, dass ein Staat Personen nicht rechtmäßig ausschließen kann, die seine Grenzen auf der Flucht vor einem Staat erreichen, der ihre Menschenrechte bloß unzureichend schützt.

Die rechtliche Idee des Non-Refoulement spiegelt Ideen wie diese wider. Staaten handeln dann falsch, wenn sie ihre Zwangsbefugnis nutzen um flüchtende Menschen in ihre Heimatländer abzuschieben und diese Länder ihnen keine angemessene politische Gemeinschaft bieten. Wie wir die Idee „moralischer Angemessenheit" an dieser Stelle verstehen sollten, ist eine komplizierte Angelegenheit, sowohl aus philosophischer als auch juristischer Perspektive. Rechtlich gesehen war der Geltungsbereich der Idee des Non-Refoulement von den Verfassern der Genfer Flüchtlingskonvention ursprünglich auf die Gruppe derjenigen begrenzt worden, die vor Verfolgung fliehen; den Unterzeichnerstaaten des Abkommens war es demnach untersagt, solche Flüchtlinge zurück in die Länder zu schicken, aus denen sie geflohen waren. Der Personenkreis, dem ein Recht auf Non-Refoulement zukommt, ist jedoch größer geworden und umfasst nun auch eine Vielzahl von Personen, deren Vulnerabilität nicht nur auf Verfolgung, sondern auf anderen Arten des Unrechts beruht – darunter willkürliche Gewalt, Zusammenbruch der Gesellschaft und andere Formen inadäquaten Rechtsschutzes.[21]

Eine diesen Prinzipien verpflichtete Moral des Flüchtlingsschutzes verlangt von uns sicherlich nicht, dass wir all die Personen, die derart von Ungerechtigkeit bedroht sind, mit Mitteln ausstatten, um die Grenzen unseres Staates zu erreichen. Aber welche Gründe können wir dafür anführen? Schließlich erscheint es zunächst so, als ob jeder Person, die mit einer gleichen Form von Unrecht konfrontiert ist, auch ein gleicher Anspruch auf Rettung vor diesem Unrecht zukommt – dass also alle Personen, sei es nun an unserer Grenze, innerhalb des sie unterdrückenden Landes oder andernorts, mit den notwendigen Mitteln ausgestattet werden sollten, um Zuflucht in unserem Land zu suchen. Warum also sollten wir nicht glauben, dass das Recht auf Zuflucht jederzeit und notwendigerweise ein positives Recht auf

Hilfe darstellt? Warum sollten beispielsweise die Vereinigten Staaten nicht ihre beträchtliche militärische Macht einsetzen, um Menschen weltweit aus moralisch verwerflichen Situationen auszufliegen und mit dem Recht auszustatten, in den USA zu leben?

Es gibt in meinen Augen viele mögliche Antworten auf diese Frage. Eine besteht darin, dass die Flüchtlinge an der Grenze unseres Staates gewissermaßen in eine Eins-zu-Eins-Beziehung mit uns treten; sie ersuchen um *unsere* Hilfe und nicht um die eines anderen Staates, der ebenso gut fähig wäre, sie zu retten; und sie werden diese Hilfe erhalten, sofern *wir* auf Zwangsmaßnahmen gegen sie verzichten. In anderen Fällen richten Flüchtlinge ein Gesuch an die Staaten der Welt als Kollektiv; wenn sie jedoch an unserer Grenze stehen, richtet sich ihre Bitte direkt an *uns*. Aus dieser ersten Antwort folgt eine weitere: Die Flüchtlinge wollen eine bestimmte *Form* der Hilfe – nämlich das Recht auf Eintritt in unsere politische Gemeinschaft – statt Hilfe im allgemeineren Sinne. Den in ihren Ländern Geschundenen und Marginalisierten kann im Prinzip auf vielerlei Weisen geholfen werden; so könnten auch Entwicklungshilfe, humanitäre Interventionen oder andere Formen staatlicher Eingriffe ihr Leben verbessern. Allerdings ersucht die Person an der Grenze um eine bestimmte Form von Hilfe – um die Hilfe, die mit dem Überqueren der Grenze einhergeht, nämlich ein Leben innerhalb des jeweiligen politischen Hoheitsgebiets zu beginnen. Die wichtigste Antwort auf die vorangehende Frage ist jedoch die grundlegendste: Es gibt, zumindest scheinbar, ein moralisches Ungleichgewicht zwischen der Drohung, auf eine Person zu schießen, und der Verweigerung positiver Unterstützung. Ersteres scheint für die meisten von uns schwieriger zu rechtfertigen. Sollte das zutreffen, könnte es richtig sein, den Bedürftigen an unserer Grenze andere Rechte zuzusprechen als denjenigen, die von dieser Grenze weiter entfernt sind. Es wäre demnach moralisch falsch, eine solche Person durch die Androhung körperlicher Gewalt vom Überqueren unserer Landesgrenze abzuhalten und das selbst dann, wenn andere, auf gleiche Weise bedürftige Personen kein Recht darauf haben, dass wir sie mit den Mitteln für die Reise in unser Land ausstatten.

An dieser Stelle müssten selbstverständlich noch weitere bedeutende Probleme behandelt werden. Selbst wenn einer Person das Recht zukommt, die Grenzen unseres Staates zu überqueren, müsste noch untersucht werden, welche Ansprüche sie durch den Eintritt in unser Land erwirbt.[22] Die

Unterscheidung zwischen permanenter Einwanderung und eher temporären Schutzformen scheint bedeutsam und wurde zugleich bisher nur unzureichend theoretisch aufgearbeitet. Wir könnten die Grenze auch als einen *Ort* untersuchen – insbesondere dann, wenn Menschen auf See vom Militär eines Staates daran gehindert werden, die Küste dieses Staates zu erreichen. Diese Form der Migrationskontrollen wurde bereits durch die Regierungen der USA als auch Australiens eingesetzt, jeweils begleitet von öffentlicher Kritik.[23] Wir könnten schließlich auch fragen, ob der Schutz des Flüchtlings auf dem Territorium desjenigen Staates gewährleistet werden muss, der um Hilfe ersucht wird. So hat wiederum Australien damit begonnen, Asylantragsteller nach Vanuatu zu transportieren: Die australische Regierung hat behauptet, ihre einzige Verpflichtung bestehe darin, Zuflucht zu gewähren; es müsse sich dabei aber nicht notwendigerweise um Zuflucht innerhalb der Grenzen Australiens handeln. Wir könnten an dieser Stelle fragen, wie es bereits viele getan haben, ob ein solches Vorgehen, entweder prinzipiell oder in diesem besonderen Fall, moralisch gerechtfertigt werden kann. An dieser Stelle möchte ich diese Probleme jedoch beiseitelassen und stattdessen eine recht grundlegende Frage stellen: Wie können wir die Gruppe von Personen definieren, die ein Recht darauf hat, nicht mittels Zwangsgewalt vom Übertreten einer Grenze abgehalten zu werden? Wer genau sind diese Menschen, denen der Eintritt nicht verwehrt werden darf, wenn sie an den Grenzen eines staatlichen Hoheitsgebiets stehen?

In meinen Augen wird eine solche Definition nicht ohne Bezug auf die Genfer Flüchtlingskonvention von 1952 auskommen können. In ihr wurde festgehalten, dass eine bestimmte Gruppe von Menschen bestimmte Ansprüche hat – und es steckt, wie ich zeigen werde, eine gewisse moralische Weisheit in dem Gedanken, dass die Mitglieder dieser Gruppe etwas gemeinsam haben. Allerdings erscheint die Menge derjenigen Personen, die nicht vom Grenzübertritt abgehalten werden dürfen, größer als die durch die Konvention definierte Gruppe zu sein. Es gibt viele Arten, ungerecht behandelt zu werden. Eine Person kann selbstverständlich aufgrund ihrer Zugehörigkeit zu einer bestimmten Gruppe verfolgt werden, aber genauso gut kann sie aus den eigenwilligsten Gründen brutal behandelt werden, willkürlicher Gewalt unterworfen sein, Opfer staatlicher Korruption sein oder schlicht hungern. Es scheint plausibel, dass all diese Phänomene eine politische Beziehung beschreiben, die zum einen durch Zwang charakterisiert ist und zum anderen am zutreffendsten als ungerecht bezeichnet werden kann. Mit Sicherheit kann gesagt werden, dass in solchen Beziehungen Menschen derjenigen moralischen Rechte beraubt werden, die ihnen gerechterweise zustehen. Wie ich gezeigt habe, können wir den Einsatz von Zwangsgewalt zur Prävention von Grenzübertritten in dem Fall rechtfertigen, dass dieser Übertritt die bereits im Land anwesenden Menschen in eine bestimmte Form von Beziehung zwingt. Aber wir können anhand dieses Arguments nicht den Einsatz von Gewalt zu dem Zwecke rechtfertigen, dass eine Person in einer moralisch unzulässigen politischen Beziehung *verbleibt*. Selbst wenn ich keine starke Verpflichtung hätte, Ihre Flucht aus einer solchen Beziehung aktiv zu unterstützen, so könnte ich doch nicht behaupten, richtig gehandelt zu haben, wenn ich Gewalt einsetze, damit Sie in solchen Umständen verbleiben. Wenn wir Zwangsgewalt zur Zurückweisung von Migrantinnen einsetzen, in deren Herkunftsländern solche Ungerechtigkeiten herrschen, machen wir jedoch genau das. Diese moralischen Tatsachen werden bei den gegenwärtigen Bemühungen, Grenzen zu sichern, häufig übersehen. So hat die Trump-Administration beispielsweise unlängst angekündigt, dass sie Personen, die vor Bandengewalt in Zentralamerika fliehen, keine legitimen Ansprüche auf Asyl mehr zuerkennt.[24] Diese Entscheidung geht jedoch notwendigerweise mit dem Einsatz US-amerikanischer Zwangsgewalt zu dem Zweck einher, Menschen in Umstände zurückzuschicken, in denen ihre grundlegendsten Rechte – darunter das Recht auf Leben – gefährdet werden. Dabei handelt es sich um genau jene Akte, von denen ich behaupte, dass sie ein Staat nicht gerechterweise ausführen darf. Indem Gewalt genutzt wird, um illegitime Zustände aufrechtzuerhalten, versagt der Staat nicht nur Rettung, sondern erhält aktiv eine Beziehung von Gewalt und Ungerechtigkeit aufrecht.

Da ich die Trump-Administration erwähnt habe, wäre es vielleicht an der Zeit, kurz über die Grenzmauer zu sprechen, die einen der Eckpfeiler in Donald Trumps Präsidentschaftswahlkampf darstellte. Wir könnten meinen, dass die Mauer die soeben ausgeführten Überlegungen verkompliziert, da eine Mauer schließlich keine Gewalt ausübt. In Millers Worten handelt es sich bei ihr eher um ein Mittel der Prävention denn des Zwangs. Sollte ich Recht haben und es den Vereinigten Staaten nicht gestattet sein, Zwangsgewalt gegen eine Gruppe von Schutzsuchenden anzuwenden, die weiter gefasst ist als in der Genfer Flüchtlingskonvention von 1951, dürfen die Vereinigten Staaten diese Gruppe dann folglich auch nicht durch Aufrichtung eines Hindernisses ausschließen?

Ich denke, dass diese Frage zum Teil dadurch beantwortet werden kann, dass es sich bei der Mauer nicht einfach nur um eine Mauer handelt, sondern um ein bestimmtes Mittel, dass bei der zwangsweisen Durchsetzung des Grenzregimes eingesetzt wird. Wie David Bier es pointiert ausgedrückt hat: Eine Mauer oder ein Zaun kann keine illegale Einwanderung verhindern, „da eine Mauer oder ein Zaun niemanden festnehmen kann."[25] Trumps Mauer mag es für die Grenzbeamtinnen einfacher machen, diejenigen zu bedrohen, die die Grenze überschreiten, aber sie ist kein Ersatz für sie, was sich auch an der Tatsache zeigt, dass der Plan der Trump-Administration für die Errichtung der Mauer auch 8,5 Milliarden US-Dollar für neue Grenzsoldatinnen und -beamtinnen umfasst.[26] (Ebenso sollte erwähnt werden, dass dieselbe Regierung in Reaktion auf eine sich den Vereinigten Staaten nähernde Karawane von Migrantinnen den Einsatz „tödlicher Gewalt" an der Grenze autorisiert hat.)[27] Aus philosophischer Perspektive könnten wir fragen, was es bedeuten würde, wenn wir die Grenze eines bestimmten Staates ohne den Einsatz von Zwangsmitteln effektiv schließen könnten. Wenn also die Vereinigten Staaten, wie wir festgestellt haben, derzeit eine vor Gewalt in Zentralamerika fliehende Person nicht ausschließen darf, dürften sie es dann tun, wenn sie von einer unüberwindbaren Mauer umgeben wären? Immerhin würden die Mittel, durch die jene Migrantin ausgeschlossen würde, keinen Zwang mehr beinhalten. In meinen Augen würde unsere Antwort in diesem Fall davon abhängen, wie wir die Bedeutung physischer Präsenz gewichten. Die Person an unserer hypothetischen, perfekten Mauer wendet sich schließlich immer noch an *uns* und diese Tatsache könnte ausreichen, um uns die Pflicht aufzuerlegen, uns dieser bestimmten Person gegenüber für unser Handeln zu rechtfertigen. In unserer jetzigen Welt brauchen wir uns um diesen hypothetischen Fall nicht zu kümmern: In dieser Welt gehen Grenzen mit Waffen einher, unabhängig davon, ob sie auch mit Mauern befestigt sind.

Es könnte nun jedoch eingewandt werden, dass die von mir beschriebene Gruppe anspruchsberechtigter Personen viel zu groß sei. Die Welt ist offensichtlich voll von Ungerechtigkeiten und es gibt keine Gesellschaft auf dieser Welt, die rundum gerecht wäre, sowie eine Vielzahl von Staaten, die noch nicht einmal ansatzweise als gerecht bezeichnet werden kann. Haben all diese Menschen ein Recht darauf, beispielsweise die Vereinigten Staaten oder Griechenland zu betreten, sofern sie die Mittel haben, diese Reise anzutreten? Würde das nicht einen Großteil der Weltbevölkerung mit starken Rechten auf Schutz in den wenigen Staaten ausstatten, in denen überhaupt Zuflucht gesucht werden kann?

Dieser Eindruck ist meiner Meinung nach falsch. Die Gruppe ist zwar groß, aber nicht so groß, wie sie zunächst erscheinen mag. Dafür gibt es zwei Gründe. Zunächst kann uns, wie im vorangegangenen Kapitel bereits diskutiert, bis zu einem gewissen Grad eine Verantwortung für die Ungerechtigkeit in unserem eigenen Land zugesprochen werden. Die Pflicht, ungerechte Gesellschaften in einen Zustand der Gerechtigkeit zu überführen, lastet auf allen Menschen, und auch wenn diese Pflicht von den Wohlhabenden und Privilegierten mehr Anstrengungen verlangt als sie derzeit bereit sind aufzubringen, wird sie nichtsdestotrotz weiterhin auch für diejenigen gelten, die sich innerhalb einer ungerechten Gesellschaft befinden. Es ist für einen Staat zumindest möglich zu behaupten, dass manche Menschen auf gerechte Verhältnisse in ihrem eigenen Land hinwirken müssen, sofern sie die dafür nötigen Ressourcen und die nötige Macht besitzen. Dieses Prinzip kann natürlich missbräuchlich verwendet werden; aber das würde bloß einen solch egoistischen und heuchlerischen Gebrauch dieses Prinzips anfechten und nicht das Prinzip selbst.

Der zweite Grund ist aus philosophischer Perspektive jedoch von größerem Interesse. Es gibt in meinen Augen eine Beziehung zwischen der Idee der Toleranz und der Idee der Zuflucht. Im Hinblick auf die Außenpolitik liberaler Staaten haben Philosophen argumentiert, dass wir unsere eigenen moralischen Überzeugungen mit einer gewissen Form prinzipiengeleiteten Respekts für die falschen Überzeugungen anderer in Einklang bringen müssen. John Rawls gibt ein Beispiel für ein solches Argument in seiner Besprechung achtbarer hierarchischer Gesellschaften, deren Gesetze trotz fehlender demokratischer Legitimation eine am Gemeinwohl orientierte Vorstellung der Gerechtigkeit darstellen und der Bevölkerung genuin moralische Pflichten auferlegen.[28] An anderer Stelle habe ich argumentiert, dass wir hinsichtlich unserer Überzeugung, die Wahrheit über die Forderungen der Gerechtigkeit gefunden zu haben, zurückhaltend sein sollten.[29] Diese Zurückhaltung ist nicht gleichzusetzen mit Relativismus; aus ihr folgt beispielsweise nicht, dass demokratische und totalitäre Regierungsformen gleichwertig sind, solange sie nur irgendwie zur Geschichte einer bestimmten Gemeinschaft „passen". Wir können – und ich glaube, wir sollten – darauf bestehen, dass die Demokratie der Tyrannei moralisch vorzuziehen ist. Was von uns stattdessen gefordert wird, ist die Anerkennung der Tatsache, dass wir hinsichtlich unserer Überzeugungen davon, was Demokratie *verlangt*, sehr wahrscheinlich voreingenommen sind und falsch liegen. Wir neigen dazu, diese Ideen bloß im Kontext unserer Außenpolitik zu untersuchen; aber wir können sie mitunter auch für unser Verständnis davon heranziehen, wer ein Recht auf Zuflucht hat.

Zur Veranschaulichung können die Uneinigkeiten zwischen den Vereinigten Staaten und Kanada betrachtet werden. Für Kanada ist die öffentliche Gesundheitsversorgung ein moralisch notwendiger Teil legitimer Herrschaft; die USA sehen das anders. Die Vereinigten Staaten glauben, dass die Todesstrafe moralisch legitim ist; Kanada sieht das anders. Angesichts dieser Uneinigkeiten akzeptieren Staaten im Allgemeinen so etwas wie eine Idee der Toleranz. Die Vereinigten Staaten gehen nicht davon aus, dass sie Kanada zur Abschaffung seiner öffentlichen Gesundheitsversorgung drängen dürfen, und Kanada mahnt die USA kaum dazu, die Todesstrafe abzuschaffen.[30] Rawls' Analyse beinhaltet den Gedanken, dass sowohl Kanada als auch die Vereinigten Staaten – und im Übrigen auch manche nicht-demokratischen Staaten – vernünftige Formen der Organisation einer Gesellschaft gefunden haben und folglich innerhalb der internationalen Staatengemeinschaft mit einem prinzipiengeleiteten Respekt behandelt werden sollten. Mein eigener Ansatz reflektiert diese Idee durch die Überlegung, dass wir die Existenz einer richtigen Antwort hinsichtlich der Gesundheitsversorgung (oder der Todesstrafe) von unserer Überzeugung, diese richtige Antwort zu kennen, unterscheiden sollten. Aber Rawls und ich – und viele andere – würden darin übereinstimmen, dass es etwas wie prinzipiengeleitete Toleranz für die falschen Antworten geben sollte, zu denen andere Länder gelangt sind. Wir können glauben, dass wir Recht haben, aber wir sollten nicht darauf bestehen, wenn wir unsere Außenpolitik rechtfertigen.

Darüber hinaus behaupte ich jedoch, dass ähnliche Ideen auch unsere Migrationspolitik beeinflussen können. Stellen Sie sich einen kanadischen Bürger vor, der an der Grenze zu den Vereinigten Staaten um Zuflucht bittet, wobei er als Grund angibt, dass die kanadische Gesundheitspolitik eine ungerechte Einschränkung seiner Vertragsfreiheit darstellt. Wir können annehmen, dass die US-amerikanische Regierung eine sehr ähnliche Überzeugung vertritt, und mit Sicherheit repräsentiert die Gesetzeslage der USA den Gedanken, dass die Marktfreiheit das Recht auf medizinische Versorgung übertrumpft. Aber die Vereinigten Staaten würden Kanada Unrecht tun, sollten sie in die kanadische Gesellschaft mit dem Ziel eingreifen, das kanadische Gesetz den US-amerikanischen Freiheitsvorstellungen anzupassen. Die Vereinigten Staaten sind dazu verpflichtet, das kanadische Gesetz als Ausdruck einer Vorstellung von Gesundheitspolitik anzusehen, die zwar fehlgeleitet sein mag, aber dennoch Respekt verdient. Aus diesem Respekt folgt jedoch, dass das kanadische Volk nicht von seinen gesetzlichen Fehlern „errettet" werden sollte, sei es durch Intervention oder die Gewährung von Asyl. Das der kanadischen Gesundheitsversorgung zugrunde liegende Prinzip ist, schlimmstenfalls, tolerabel und das heißt sowohl, dass Kanada ein Recht auf Toleranz hat als auch, dass die aus Kanada fliehende Person nicht aus einem moralisch unhaltbaren Zustand flieht.

Wir können diesen Schluss auch mittels Rawls' Analyse internationaler Gerechtigkeit verteidigen. Rawls beruft sich auf den von Philip Soper stammenden Gedanken, dass ein Rechtssystem nicht vollständig gerecht sein muss, um die unter ihm lebenden Personen zum Befolgen seiner Vorschriften verpflichten zu können.[31] Ein derart verpflichtendes Rechtssystem tut denjenigen innerhalb dieses Systems kein Unrecht an, weshalb die unter ihm lebenden Personen auch kein Recht darauf haben, vor diesem System gerettet zu werden. Es wäre gewiss seltsam, sollte diesen Personen eine Pflicht zum Befolgen der Gebote des Rechtssystems zukommen und externe Personen zugleich berechtigt sein, eben jenes System zu untergraben. Das gleiche könnte dann auch für die Migrationspolitik gelten. Wir können demnach annehmen, dass das Rechtssystem einer anderen Gesellschaft auf einer falschen Vorstellung von Gerechtigkeit beruht und dennoch meinen, dass sie eine Vorstellung der Gerechtigkeit darstellt, die prinzipiell Respekt verdient. Mein eigener Ansatz kommt zu einem ähnlichen Schluss: Sofern ein Staat gemäß eines Prinzips öffentlicher Moral regiert wird, das – ohne systematische Täuschung oder die Aufgabe jeglicher Selbstachtung – als ein Prinzip der Gerechtigkeit verstanden werden kann, hat er ein präsumtives Recht auf prinzipielle Toleranz. Sollte dem so sein, wäre die Gruppe derjenigen, die ein Recht auf die Überquerung von Staatsgrenzen haben, doch nicht so groß wie zunächst gedacht. Wenn Kanada also ein Recht darauf hat, seine Einwohnerinnen weiterhin gemäß seiner angeblich verfehlten Gerechtigkeitsvorstellung zu regieren, wird diesen Einwohnerinnen kein Unrecht getan.

Ich denke, dass etwas Ähnliches über viele der Länder gesagt werden kann, die Rawls als achtbare hierarchische Gesellschaften bezeichnen würde, obwohl ich dieses Argument hier nicht vollständig entwickeln kann. Mein Argument ist daher schlicht struktureller Natur, nämlich dass die Gruppe der Personen mit einem Recht auf Grenzübertritt meinem Ansatz zufolge zwar groß ist, aber nicht so groß, wie es anfänglich den Anschein erweckt haben mag. Diese Perspektive müsste um ein Konzept der Toleranz ergänzt werden, das ich in diesem Kontext nicht entwickeln kann. Darüber hinaus möchte ich darauf hinweisen, dass ich in meinen Ausführungen zwar die Möglichkeit systemischer, nicht jedoch individueller Vulnerabilität diskutiert habe. Es ist schließlich möglich, dass individuelle Personen einen berechtigen Anspruch auf Schutz haben, selbst wenn ihrem Herkunftsland ein Recht auf Toleranz zukommt, und zwar schlicht aufgrund der Tatsache, wie diese Länder *sie*, als einzelne Personen, behandeln. Für den Moment möchte ich jedoch auf den primären Gedanken zurückkommen, dass die Gruppe derjenigen, die nicht mittels Zwangsgewalt an der Grenze zurückgewiesen werden dürfen, größer ist, als in der Genfer Flüchtlingskonvention von 1951 angenommen wurde. Es gibt viele Menschen auf dieser Welt, die vielfältigen Formen von Unrecht zu entkommen versuchen, und die Grenze zwischen tolerabel und intolerabel, wie auch immer wir sie ziehen wollen, spiegelt mit Sicherheit nicht die Ideen wieder, die der Genfer Flüchtlingskonvention zugrunde liegen.

3. Zwangsgewalt jenseits von Grenzen

An dieser Stelle möchte ich auf diese Konvention zurückkommen und untersuchen, ob sie auf moralischen Idealen beruht, die es wert sind, verteidigt zu werden. Die Genfer Flüchtlingskonvention beginnt mit dem Gedanken, dass Verfolgung eine eigenständige Kategorie moralischen Unrechts darstellt. Hierauf gibt es mindestens zwei mögliche Reaktionen. Die erste Reaktion besagt, dass die Konvention falsch liegt oder, etwas freundlicher, dass sie ein Kind ihrer Zeit ist. Sie reagiert auf bestimmte Formen des Unrechts, die die Welt in den Jahren vor ihrer Entstehung zerrüttet haben – aber es kann nicht davon ausgegangen werden, dass sie eine zeitlose Form moralischer

Wahrheit darstellt. Der zweiten Reaktion zufolge stellt Verfolgung tatsächlich einen besonderen Tatbestand dar und verlangt daher besondere Schutzformen. Moralisch gesprochen unterscheidet sich Verfolgung demnach von den vielen anderen Formen moralischen Unrechts, die wir bei der Suche nach Schutz anführen könnten, und verlangt daher von den Staaten der Welt besondere Aufmerksamkeit.

Ich möchte eine bestimmte Version dieser zweiten Reaktion verteidigen. In meinen Augen handelt es sich bei Verfolgung um ein besonderes Unrecht – auf das die Genfer Flüchtlingskonvention allerdings nur dann eine angemessene moralische Antwort gibt, wenn ihre Garantien deutlich stärker verstanden werden als gemeinhin üblich. Um zu zeigen, was damit gemeint ist, müssen wir sowohl über die Frage, was Staaten an ihren Grenzen moralisch gestattet werden kann, als auch über die damit verbundenen Begriffe von Toleranz und Gerechtigkeit hinausgehen. Wir müssen stattdessen verstehen, wann es Staaten moralisch erlaubt ist, zur Unterstützung bei der Überwindung staatlicher Unterdrückung Zwangsmaßnahmen gegen andere Staaten anzuwenden. Die Konvention richtet ihren Fokus meiner Meinung nach zu Recht auf das Problem der Verfolgung – aber sie sollte als ein Bündel rechtlicher und moralischer Garantien für die Verfolgten verstanden werden, das in seiner Gesamtheit die Ausübung von Zwang gegen die Verfolger von Schutzbedürftigen autorisiert.

Meiner Lesart zufolge spiegelt die Konvention diejenigen Werte wider, die erst später unter dem Titel der *Responsibility to Protect* (Schutzverantwortung; R2P) kodifiziert wurden.[32] Die der R2P zugrunde liegenden Dokumente wurden zwischen den Jahren 2001 und 2005 verfasst und argumentierten, dass jeder Staat primär dafür verantwortlich ist, seine eigenen Bürgerinnen vor den vier Formen schwerer Menschenrechtsverletzungen zu schützen, als da sind Genozid, Kriegsverbrechen, ethnische Säuberung und Verbrechen gegen die Menschlichkeit. Für unsere Zwecke ist der Umstand von Bedeutung, dass R2P allen Staaten eine „residuale Verantwortung" auferlegt, die Rechte betroffener Personen zu verteidigen, „wenn ein bestimmter Staat offensichtlich nicht willens oder nicht fähig ist, seiner Schutzverantwortung selbst nachzukommen, oder jene Verbrechen oder Gräuel selbst verübt." Nach meiner Lesart der Genfer Flüchtlingskonvention kann sie – fünfzig Jahre nach ihrem Inkrafttreten – dadurch moralisch kohärent interpretiert werden, dass sie als eine Form von R2P verstanden wird, die sich auf eine bestimmte Personengruppe bezieht. Allerdings kann selbst diese Form von Kohärenz nur dann verteidigt werden, wenn die Konvention als Teil eines breiteren Arsenals von Reaktionen auf staatliches Fehlverhalten verstanden wird.

Wir können für dieses Argument mit der Frage beginnen, was an der Verfolgung durch einen Staat besonders ist. Wird ein Staat zum Verfolger, verkehrt sich die normative Welt: Die eigentlich zum Schutz der eigenen Rechte verpflichtete Institution bedroht nun eben diese Rechte. Matthew Lister verweist auf diesen besonderen Aspekt der Verfolgung und argumentiert, dass die Genfer Flüchtlingskonvention auf dem Gedanken beruht, dass den Verfolgten nicht dort geholfen werden kann, wo sie sich gerade befinden; wir können diese Hilfe nicht vermittelt über die lokale Regierung leisten, wenn diese Regierung selbst der Ursprung des Problems ist.[33] Zwar erscheint mir das korrekt, aber ich möchte über diese vernünftigen Bedenken hinaus betonen, dass die Einzigartigkeit der Verfolgung genau darin zu bestehen scheint, wie sie die soziale Welt der Verfolgten verändert. Dieses Unrecht ist in dreierlei Hinsichten besonders. Zuerst hinsichtlich der simplen Tatsache, dass der Staat nicht einfach nur seine Aufgaben inadäquat oder gar nicht erfüllt, sondern seine Zwangsgewalt aktiv einsetzt, um den infrage stehenden Personen Unrecht zu tun. Der zweite Aspekt ist eher empirischer Natur und beruht auf dem Gedanken, dass Verfolgung oftmals in Gräueltaten endet. Wenn das Ziel der Genfer Flüchtlingskonvention darin besteht, eine bestimmte Kategorie von Unrecht hervorzuheben, könnte die Bemerkung relevant sein, dass ein Unrecht, das mit Verfolgung beginnt, häufig im Genozid endet. Das ist sicherlich eine Frage empirischer Untersuchungen; aber es erscheint zumindest plausibel, dass Verfolgungen deshalb ein besonderes moralisches Gewicht zukommt, weil sie uns etwas über den weiteren Verlauf der Ereignisse sagen können. Der letzte, den Tatbestand der Verfolgung auszeichnende Aspekt besteht darin, wie er uns mit anderen Menschen und Gemeinschaften verbindet. Wie bereits gezeigt, ruht die Last, die Welt zu einem gerechten Ort zu machen, auf allen Schultern, und unter bestimmten Umständen könnte es durchaus fair sein, von Menschen zu verlangen, dass sie mit dieser Aufgabe in ihren eigenen Gesellschaften beginnen. Wenn die politischen Organe dieser Gesellschaften allerdings selbst zum Ursprung von Verfolgung werden, scheint es – milde ausgedrückt – unfair, zu verlangen, dass die Verfolgten weiterhin mit ihren Verfolgern zusammen an der

Einrichtung einer gerechten Gesellschaft arbeiten sollen. Eine Person kann sich durch Verfolgung sozusagen *entfremden*. Jean Améry beschreibt dieses Phänomen sehr anschaulich, wenn er sich daran erinnert, wie er auf seiner Flucht vor der Verfolgung durch die Nazis die Stimme einer Person aus seiner Heimatstadt Wien hörte:

> „Ich hatte lange diesen Tonfall nicht mehr vernommen, und darum regte sich in mir der aberwitzige Wunsch, ihm in seiner eigenen Mundart zu antworten. Ich befand mich in einem paradoxen, beinahe perversen Gefühlszustand von schlotternder Angst und gleichzeitig aufwallender familiärer Herzlichkeit, denn der Kerl, der mir in diesem Augenblick zwar nicht gerade ans Leben wollte, dessen freudig erfüllte Aufgabe es aber war, meinesgleichen in möglichst großer Menge einem Todeslager zuzuführen, erschien mir plötzlich als ein potentieller Kamerad. […] In diesem Augenblick begriff ich *ganz* und für immer, daß die Heimat Feindesland war und der gute Kamerad von der Feindheimat hergesandt, mich aus der Welt zu schaffen."[34]

Diese paradoxe Mischung an Reaktionen – mein Heimatland ist mir feind geworden – kann möglicherweise erklären, was genau das Unrecht der Verfolgung einzigartig macht. Die Verfolgung macht aus einer Person einen Fremden, während sie doch, rechtlich betrachtet, immer noch zu Hause ist. Der Staat, dem eigentlich eine Schutzpflicht mir gegenüber zukommt, hat sich entschieden, die eigene Person zum Feind zu erklären und entsprechend zu behandeln.

Anhand dieser Ideen können wir in meinen Augen erklären, was die Kategorie des *Verfolgten* moralisch auszeichnet. Wie aber können wir diese Ideen nutzen? In meinen Augen nicht dazu, diejenige Gruppe von Personen zu bestimmen, die frei von Zwangsgewalt die Grenze eines Staates passieren darf. Stattdessen können wir sagen, dass es sich bei der Gruppe der Verfolgten gerade um diejenigen Personen handelt, die ein *Anrecht* auf den Einsatz von Zwangsgewalt haben – allerdings in diesem Falle nicht gegen sich selbst, sondern gegen die Institutionen und Akteure, die ihre Bewegungsfreiheit einschränken wollen. Das Abkommen bestimmt demnach einen moralisch signifikanten Aspekt einer bestimmten Personengruppe; was diese Personen jedoch untereinander gemein haben, sollte nicht als ein Recht auf Bewegungsfreiheit verstanden werden, sondern als ein Recht auf *Schutz* –

wobei Schutz hier sowohl ein eigenständiges Recht auf Migration meint, als auch einen gewissen Anspruch auf ein gewaltsames Eingreifen durch die anderen Staaten der Welt. Das mag mitunter anachronistisch klingen, allerdings geht die Genfer Flüchtlingskonvention der Idee von R2P immerhin auch ein halbes Jahrhundert voraus. Vielleicht ist es aber auch kein Zufall, dass die Genfer Flüchtlingskonvention aus Kriegen hervorging und insbesondere aus einem Krieg, der teilweise zur Verteidigung der Idee geführt wurde, dass Menschen gewisse Rechte auch gegenüber ihrem eigenen Staat zukommen. Die Konvention ist in ihrer Geltung nicht auf Zeiten von Krieg und Gewalt beschränkt, aber es könnte dennoch wahr sein, dass die Frage der Gewalt – in jedem Fall aber die Frage der Zwangsgewalt – bei der Interpretation des Abkommens nicht ignoriert werden sollte.

Als Beleg können wir die moralische Logik der Konvention selbst heranziehen. Das Abkommen spricht Verfolgten Migrationsrechte zu, wenn diese sich außerhalb des Hoheitsgebiets ihres Herkunftslands befinden. Allerdings erscheint das Abkommen als moralisch inkohärent, wenn es nicht mit der substanziellen Garantie einhergeht, dass andere Länder sich für die Rechte dieser verfolgten Personen einsetzen, ob sie nun das Hoheitsgebiet ihrer Verfolger verlassen haben oder nicht. Sofern es eine Schutzverantwortung gibt, muss diese Verantwortung darin bestehen *die Verfolgten zu schützen.* Allerdings kann die Gewährung von Migrationsrechten nur einen Teil der verfolgten Personen schützen. Laut R2P müssen Staaten gewisse Kosten auf sich nehmen, um diejenigen zu unterstützen, denen es nicht möglich ist, das für die Gräueltaten verantwortliche Land zu verlassen. Die Konvention muss daher mit einer gewissen Pflicht einhergehen, Zwangsmaßnahmen gegen die Verfolgerstaaten einzusetzen. Ohne diese Pflicht gäbe es einen Unterschied in den Rechten, der bloß unzureichend durch den Verweis auf einen Unterschied zwischen den Personen begründet werden könnte.

Wie aber würde eine kohärente Vorstellung von der Pflicht zum Schutz verfolgter Personen aussehen? In meinen Augen müsste sie eine Form positiver Unterstützung für diejenigen beinhalten, die nicht fähig sind den Staat zu verlassen, der sie verfolgt – unabhängig davon, ob die Mittel zur Ausreise fehlen oder der Staat diese Ausreise effektiv verhindert. Wir können uns daher vorstellen, dass die Konvention allein dann moralisch kohärent wäre, wenn sie um die Möglichkeit gewaltsamer Interventionen in den Verfolgerstaaten ergänzt würde. Darüber hinaus können wir uns vorstellen, dass es einen legitimen Anspruch auf positive Unterstützung aufseiten derjenigen gibt, die einen Ort aufgrund fehlender Mittel nicht verlassen können. Sich beiden Maßnahmen zu verweigern scheint jedoch der Einstellung gleichzukommen, positive Handlungen zur Unterstützung derjenigen, die am stärksten der Rettung vor Gräueltaten und Verfolgung bedürfen, vermeiden zu wollen – einer Einstellung also, die wohl eher Eigennutz denn ein Streben nach Gerechtigkeit auszudrücken scheint.

Wir können also den Fokus des internationalen Gesetzes auf Verfolgung rechtfertigen, solange wir Verfolgung auch als Grundlage für ein Bündel internationaler Maßnahmen ansehen, das nicht nur ein Verbot von Zurückweisungen an staatlichen Grenzen, sondern auch positive Maßnahmen zur Unterstützung der Verfolgten umfasst. Die moderne rechtliche Konzeption des Flüchtlings kann folglich nur dann als Ausdruck valider moralischer Überlegungen angesehen werden, wenn diese Konzeption von den Staaten der Welt mehr fordert als sie gegenwärtig zu tun bereit sind. Die Genfer Flüchtlingskonvention könnte daher schlicht als ein Aspekt der Ansprüche verstanden werden, die R2P an die Staaten dieser Welt richtet – sie sagt uns, was den entkommenen Verfolgten nicht angetan werden darf. Wir sollten die Konvention also nicht derart verstehen, dass sie die Grenze dessen festlegt, was wir den Verfolgten schulden, die sich noch innerhalb des Verfolgerstaats befinden. Ihnen wird sehr viel mehr geschuldet, als es die Konvention oder die derzeitige Praxis vorsehen – in Form von aktiver Unterstützung und, im äußersten Fall, militärischem Zwang.

Ich möchte dieses Kapitel mit einer Bemerkung darüber abschließen, wie bescheiden sich selbst diese umfassendere Lesart ausnimmt. In den vorangehenden Absätzen habe ich die der Genfer Flüchtlingskonvention zugrunde liegende Moral teilweise verteidigt; aber ich sollte nicht so verstanden werden, dass ich zu stark auf der Einzigartigkeit von Verfolgung bestehe. Es mag durchaus der Fall sein, dass andere Formen von Unrecht gleichfalls zu solch koordinierten internationalen Anstrengungen verpflichten. So scheint beispielsweise der Klimawandel als potentiell genauso verheerend für eine Gesellschaft wie Menschenrechtsverletzungen. Eine Gesellschaft, der wortwörtlich das Wasser bis zum Hals steht, hat schließlich kaum Aussichten auf eine blühende Zukunft.[35] Ich möchte auch anmerken, dass die Welt Situationen gegenüberstehen mag, in denen nichts – oder zumindest nichts, was effektiv oder moralisch zulässig wäre – getan werden kann, um diejenigen verfolgten

Personen zu schützen, die sich noch innerhalb des Verfolgerstaats befinden. Moralische Tragödien sind ein unvermeidlicher Teil des Lebens und es könnte Zeiten geben, in denen das Beste, was wir tun können, darin besteht, den wenigen Menschen zu helfen, die es geschafft haben zu fliehen. Es mag, mit anderen Worten, also Zeiten geben, in denen es uns nicht möglich ist, den Verfolgten zu helfen und zugleich gegen die Verfolger vorzugehen. Aber ich würde dennoch darauf bestehen, dass wir es – manchmal – *versuchen* sollten. Selbst wenn denen, die sich außerhalb ihrer Heimatländer befinden, einfacher geholfen werden kann, folgt daraus nicht, dass sie unsere Hilfe aus moralischer Perspektive mehr *verdient* hätten.

Allerdings ist meine Betrachtung des Problems der Schutzgewährung nicht ganz zufriedenstellend. Mein Ansatz besteht aus zwei Teilen. Ein Prinzip trifft auf diejenigen zu, die an den Grenzen eines rettenden Staates anlangen, ein anderes regelt die Ansprüche derjenigen, denen dies nicht möglich war. Das Konzept der Verfolgung mag nun für letztere Gruppe nützlich sein, aber ich denke nicht, dass es der ersten Gruppe viel nützt. Es gibt allerdings eine tieferliegende Logik in diesem Ansatz. Das erste Prinzip regelt, was der Staat hinsichtlich der ihm zur Verfügung stehenden Zwangsmittel nicht tun darf, um Menschen an der Überquerung seiner Grenzen zu hindern. Wir können diesem Prinzip zufolge keine Zwangsgewalt einsetzen, um Menschen zurück in eine moralisch unzulässige Beziehung zu zwingen. Das zweite Prinzip formuliert, was der Staat mittels positiver Unterstützung – und, im Grenzfall, Gewalt – zu tun verpflichtet ist, um denen zu helfen, die in ihren Heimatländern verfolgt werden. Es gibt also manche Menschen, die nicht bloß schlicht ein Recht darauf haben, frei von Zwangsgewalt zu sein, sondern darauf, dass Zwangsgewalt zu ihren Gunsten ausgeübt wird; und die Konvention wird am besten als ein Teil der umfassenderen Aufgabe gelesen, die Gruppe dieser Menschen zu bestimmen.

Diese Analyse lässt naturgemäß viele Probleme außen vor. So habe ich nicht besprochen, wie wir die Lasten aufteilen sollten, die aus einer solchen Vorstellung von Gerechtigkeit resultieren würden. Auch habe ich nicht ansatzweise diskutiert, wann genau ein Staat zu tolerieren ist oder welche Art von Rechten diejenigen beanspruchen können, die aus einem Verfolgerstaat gerettet wurden. Mein Hauptanliegen bestand vielmehr darin zu zeigen, dass selbst die Frage, wer über ein Recht auf Bewegungsfreiheit verfügt, komplexer ist als es zunächst erscheint; sie beansprucht einen Fokus darauf, was dieses Recht tatsächlich voraussetzt und hat mitunter mehr als eine Antwort. Zum Ende dieses Kapitels möchte ich auf das bereits diskutierte Problem der Sanktionen für Beförderungsunternehmen zurückkommen. Eine erste, durch viele von uns vertretene Antwort lautet, dass die Sanktionen Zwang bedeuten und daher das Prinzip des Non-Refoulement verletzen. Allerdings glaube ich, dass das nicht ganz richtig ist. Die Sanktionen sind kein Zwang gegenüber den Migrantinnen, sondern gegenüber den Beförderungsunternehmen, das ihnen die für die Migration notwendigen Mittel potentiell bereitstellt. Das tatsächliche Problem mit diesem Sanktionsregime besteht stattdessen darin, dass es denjenigen Menschen die Mittel nimmt, die nicht bloß ein Recht darauf haben, solche Mittel für die Flucht vor Gräueltaten in Syrien zu erwerben, sondern sogar ein Recht darauf, dass die anderen Länder der Welt alles dafür tun, um sie vor diesen Gräueltaten zu *retten*. Wenn die Staaten der Welt angemessen auf die Schrecken in Syrien reagieren würden, müssten sie nicht nur von den Sanktionen gegen Beförderungsunternehmen zurücktreten, sondern *damit beginnen die Beförderung selbst durchzuführen* und, was sicherlich umstrittener ist, versuchen mithilfe von Zwangsmaßnahmen in jeden Versuch des syrischen Staates einzugreifen, der den Zugang zu einer solchen Beförderung zu verhindern versucht. Die meisten Staaten haben sich selbstverständlich nicht für ein solches Vorgehen entschieden und werden es wohl auch nicht tun. Dieses Kapitel soll jedoch mit einer positiven Bemerkung enden. Die kanadische Regierung setzte im Jahr 2016 Militärflugzeuge ein, um syrischen Flüchtlingen dabei zu helfen, den Nahen Osten zu verlassen und ein neues Leben in Kanada zu beginnen. Die sogenannte *Operation Provision* war möglicherweise nicht genug; sie stellte nur 25.000 Syrerinnen eine neue Heimat zur Verfügung und konzentrierte sich auf Personen, die bereits aus Syrien geflohen waren.[36] Die Regierung Bashar Al-Assads fährt darüber hinaus, unterstützt durch die russische Regierung, mit ihren Gräueltaten fort. Die *Operation Provision* spiegelte jedoch zumindest die Tatsache wider, dass die von Verfolgung Bedrohten nicht nur einen Anspruch auf die Überquerung von Grenzen, sondern auch auf unsere Unterstützung bei diesem Grenzübertritt haben. Wir haben allen Anlass dazu, diese kleinen moralischen Handlungen in einer Welt wertzuschätzen, die solche Maßnahmen allzu oft unmöglich macht.

6
Auswahl und Zurückweisung: Über Migration, Ausschluss und das Veto des Heuchlers

Im vorherigen Kapitel habe ich eine bestimmte Vorstellung davon verteidigt, welche Maßnahmen Staaten nicht zustehen. Sie dürfen Migrantinnen nicht ausschließen, wenn diese dadurch weiterhin unter Bedingungen leben müssten, in denen ihre Grundrechte nicht respektiert werden. Staaten dürfen diesen Ausschluss auch nicht mit Zwang gegenüber solchen Menschen durchsetzen, die sich in ihren Herkunftsländern einem mangelnden Rechtsschutz ausgesetzt sehen. Auch wenn diese Gruppe anhand eines Konzepts prinzipieller Toleranz bestimmt werden muss, handelt es sich hierbei doch um eine bedeutende Einschränkung. Zudem ist dies alles, wie ich bereits betont habe, bloß ein Ausschnitt der umfassenderen Einschränkungen, die staatlichem Handeln im Ausland durch die Grundrechte ausländischer Personen auferlegt werden. Eine umfassendere Vorstellung von Schutz, in der Staaten nicht nur Zwang vermeiden, sondern diese Zwangsmacht im Ausland zur Verteidigung der Rechte von Verfolgten einsetzen, könnte ebenfalls ein notwendiger Teil unseres moralischen Arsenals sein.

Allerdings liegt nicht jedem Fall von Migration eine Missachtung der Menschenrechte zugrunde. Menschen migrieren aus allen möglichen Gründen: aus Liebe, für Geld oder ihre berufliche Entwicklung oder einfach aufgrund des Wunsches, das Leben an einem anderen Ort zu beenden als dort, wo es begonnen wurde. Für den Moment nehme ich an, dass keine dieser Personen ein bestehendes moralisches Recht darauf hat, an dem Ort eingelassen zu werden, den sie sich für ihre Einwanderung ausgesucht hat. Manchen in dieser Gruppe könnten zwar besondere Ansprüche auf Einwanderung zukommen; darauf werde ich jedoch erst im nächsten Kapitel eingehen. Hier und jetzt können wir jedoch annehmen, dass kein solcher moralischer Anspruch besteht. Nehmen wir also an, bestimmte Personen möchten einfach in ein bestimmtes Land einwandern, wobei es sich um mehr Menschen handelt, als dieses Land aufnehmen möchte. Allerdings ist es für dieses Land durchaus möglich, einige Einzelpersonen aus dieser Gruppe aufzunehmen. Wie können – im Einklang mit den Prinzipien des moralischen Egalitarismus, die die liberale politische Theorie so attraktiv machen – Personen aus dieser Gruppe ausgewählt werden? Welche Arten von Unterscheidungskriterien sind erlaubt, wenn keiner einzelnen Kandidatin ein Recht auf Einlass zukommt?

Neben dieser Frage werde ich in diesem Kapitel auch untersuchen, was daraus folgt, wenn die gefundenen Antworten das Handeln liberaler politischer Gemeinschaften einschränken sollten. Die erste Frage behandelt also, welche Arten von Prinzipien für die Unterscheidung der Ansprüche von Migrantinnen akzeptabel sind. Seit langer Zeit bereits ziehen Staaten manche Migrantinnen anderen einwanderungswilligen Personen vor. Einige der dafür herangezogenen Gründe scheinen zutiefst unethisch zu sein, wie beispielsweise der offensichtlich rassistische Chinese Exclusion Act aus dem Jahr 1882. Andere scheinen hingegen moralisch erlaubt, so wie die Extrapunkte, die Kanada oder Australien für ein niedriges Lebensalter oder einen relativ hohen Bildungsstand vergeben.[1] Um die Legitimität eines solchen Vorgehens zu bestimmen, benötigen wir jedoch einen theoretischen Rahmen, anhand dessen die Gerechtigkeitsansprüche derjenigen beschrieben werden können, die nicht ausgewählt wurden und folglich kein Recht auf Einwanderung erhalten haben. Wie ich zeigen werde, gibt es signifikante Einschränkungen hinsichtlich der Kriterien, die ein Staat zur Unterscheidung zwischen Migrantinnen heranziehen kann, selbst wenn keiner einzelnen Migrantin ein Recht auf Einlass zukommt.

Die daran anschließende Frage beginnt hingegen mit der Einsicht, dass diese Schlussfolgerung, unabhängig von ihrem Wahrheitsgehalt, wahrscheinlich recht unpopulär sein wird. Eine von politischen Philosophen bloß selten gestellte Frage lautet, was passiert, wenn die Lücke zwischen dem, was gerecht ist, und dem, was politisch möglich ist, zu groß wird. In meinen Augen ist es durchaus möglich, dass viele der gegenwärtigen Einwohnerinnen demokratischer Gemeinschaften selbst gegen zaghafte Reformen der Migrationspolitik, die bestehende Ungerechtigkeiten bloß mindern, Widerstand leiste würden – darunter Formen von Widerstand, die den Fortbestand des Gemeinwesens selbst gefährden. In solchen Situationen gelangen wir meiner Meinung nach an die Grenzen dessen, was die politische Philosophie leisten kann und zwar aus dem Grund, dass wir in solchen Situationen bestehende Ungerechtigkeiten nicht beseitigen können ohne zugleich diejenigen Institutionen zu gefährden, die uns den gegenwärtigen Grad an Gerechtigkeit sichern. Die Welt könnte demnach nicht nur ungerecht, sondern auch tragisch sein und zwar in dem Sinne, dass wir keine gerechten Wege hin zu einer vollkommen gerechten Welt identifizieren können und daher bestimmen müssen, welche Arten von Ungerechtigkeit am ehesten toleriert werden können.

Bevor ich allerdings auf diese Fragen eingehen kann, muss ich zunächst den Gedanken verteidigen, dass es bedeutende Einschränkungen hinsichtlich der Auswahlkriterien für Migrantinnen gibt. Im ersten Teil dieses Kapitels werde ich diskutieren, wo wir bei dieser Untersuchung beginnen könnten. Der zweite Teil wird dann vier verschiedene, für Staaten potentiell attraktive, Auswahlprinzipien diskutieren und untersuchen, wie diese vom Standpunkt der Gerechtigkeit aus zu bewerten sind. Der abschließende Teil des Kapitels wird noch einmal auf die Bedenken hinsichtlich der realen Möglichkeit einer gerechten Migrationspolitik eingehen und die Überlegung verteidigen, dass wir in der Migrationspolitik mit einer tragischen Entscheidung konfrontiert sein könnten: Einer Entscheidung zwischen Gerechtigkeit und dem Erhalt der Institutionen, durch die Gerechtigkeit erst möglich wird.

1. Einwanderung und Gleichheit

Ich habe vorausgesetzt, dass wir eine Gruppe potentieller Migrantinnen betrachten, in der keine Person über ein bestehendes Recht auf Einlass verfügt. Keine von ihnen ist demnach in einer Situation, wie sie im vorherigen Kapitel beschrieben wurde: Ihre Rechte werden in ihren Herkunftsländern angemessen geschützt und sie behaupten auch nicht, ein Recht auf Schutz durch den Staat zu haben, dem sie beitreten wollen. Sie würden durch eine Zurückweisung also nicht auf die Art ungerecht behandelt, wie eine Person, deren Menschenrechte bedroht sind. Es könnte nun behauptet werden, dass wir damit am Ende unserer ethischen Untersuchung angelangt sind: Wenn diese Personen durch einen Ausschluss nicht falsch behandelt werden, warum kann der ausschließende Staat dann nicht einfach gemäß seinen eigenen Vorstellungen handeln?

Der Grund liegt meiner Meinung nach darin, dass bestimmte *Muster* der Gewährung von Vorteilen falsch sein können, selbst wenn keine der beteiligten Personen ein bestehendes Recht auf diese Vorteile hat. Würden die Vereinigten Staaten ein Programm auflegen, dem zufolge ausschließlich alle weißen, männlichen Bürger ein Auto geschenkt bekommen, erschiene uns dieses Programm als unfair und daher als zutiefst ungerecht. Dieses Beispiel unterscheidet sich offensichtlich in dem Sinne vom Fall der Migration, dass es bei bestimmten politischen Beziehungen beginnt: Diejenigen Bürgerinnen, die kein Auto erhalten, fühlen sich zu Recht als moralisch minderwertig behandelt, da sie sich in einer besonderen politischen Beziehung – nämlich der von Mitbürgerinnen – befinden, in der sie ein Recht darauf haben, als Gleiche behandelt zu werden. Ähnliche Gedanken könnten aber auch jenseits der besonderen Beziehungen von Mitbürgerinnen gelten. Demnach wäre es möglich, dass ein bestimmtes Entscheidungsmuster selbst dann unfair und somit moralisch unzulässig ist, wenn die involvierten Personen nicht die Art von politischer Beziehung teilen, die den Hintergrund eines Großteils unserer politischen Theorie bilden.

In diesem Kapitel werde ich daher davon ausgehen, dass die um Einlass bittenden Migrantinnen in spezifischen politischen Beziehungen stehen – sowohl untereinander als auch gegenüber dem Staat, dem sie beitreten wollen. Beginnen werde ich mit der Feststellung, dass das Ideal moralischer Gleichheit innerhalb dieser Beziehungen genauso Bestand haben muss wie in den Beziehungen zwischen Mitbürgerinnen. Um liberal zu sein, muss ein liberaler Staat die moralische Bedeutung aller Personen anerkennen – also nicht nur seiner Bürgerinnen, sondern aller Personen im Allgemeinen. Andernfalls würde die Geltung der moralischen Garantien des Liberalismus auf illegitime Weise in ihrer Reichweite eingeschränkt werden, was dann in der Tat bedeuten würde, die Idee feudaler Geburtsprivilegien implizit zu einem Teil der liberalen Gleichheit zu machen. Allerdings geht mit der Idee moralischer Gleichheit kein spezifisches Bündel an politischen Rechten einher. Moralische Gleichheit verlangt vielmehr, dass in unterschiedlichen Situationen auch unterschiedliche Bündel an Rechten und Pflichten gelten.

Der moralische Egalitarismus kann also in der Tat mit verschiedenen Mustern der Gewährung von Rechten einhergehen; leben Menschen in unterschiedlichen institutionellen Beziehungen, ist es durchaus angemessen, daraus auf unterschiedliche Rechte und Pflichten zu schließen, die jene Idee der Gleichheit zum Ausdruck bringen, die diesen institutionellen Beziehungen eigen ist. Eine stichhaltige moralische Analyse des Phänomens der Migration wird diese Tatsachen einbeziehen müssen. Potentielle Immigrantinnen stehen somit in einer bestimmten, sehr spezifischen institutionellen Beziehung zu dem Staat, in den sie einwandern wollen, die sich jedoch von derjenigen der gegenwärtigen Bürgerinnen zu diesem Staat unterscheidet. So sind einwanderungswillige Personen beispielsweise noch nicht in das staatliche System des Zivil- und Strafrechts eingebunden, sind nicht dazu verpflichtet, Steuern an diesen Staat zu entrichten und auch nicht dazu berechtigt, die staatlichen Stellen zur Schlichtung privater Streitigkeiten anzurufen.

Entsprechend wäre es ein Fehler darauf zu bestehen, dass die Migrationsrechte potentieller Immigrantinnen identisch mit den Bewegungsrechten der derzeitigen Mitglieder der politischen Gemeinschaft sein sollten.

Allerdings ist die Beziehung potentieller Migrantinnen zu dem infrage stehenden Staat, auch wenn sie nicht von der Art wie diejenige der derzeitigen Bürgerinnen ist, doch auch nicht identisch mit der Beziehung von Ausländern zu diesem Staat im allgemeineren Sinne. Selbst wenn die Welt derart verbunden wäre, dass jede Person Beziehungen zu irgendeinem fremden Staat hätte, hätten potentielle Migrantinnen eine ganz spezifische Beziehung zu dem Staat, in den sie einwandern wollen. Durch einen freiwilligen Akt haben sie sich, zumindest für eine einzelne Entscheidung, in die Reichweite der Zwangsbefugnis dieses fremden Staates begeben. Die einwanderungswilligen Personen sind freiwillig an eine staatliche Grenze gekommen, ob nun im wortwörtlichen oder im rechtlichen Sinne durch einen Bewerbungsakt, und haben zugestimmt, dass das rechtliche System des Staates über ihre Bewerbung auf Mitgliedschaft entscheiden darf.

Diese Tatsache kann uns als Ausgangspunkt für das Verständnis der Rechte potentieller Migrantinnen dienen. Der bloße Fakt, dass sie um eine Begünstigung ersuchen, auf die sie noch keinen Anspruch haben, und sich dafür freiwillig in eine politische und mit Zwangsgewalt bewehrte Beziehung begeben, bedeutet nicht, dass der infrage stehende Staat ein Recht darauf hat, frei über den Einsatz dieser Zwangsgewalt zu entscheiden. Vielmehr kommen wir unter der Annahme, dass es sich bei der Beziehung potentieller Migrantinnen um eine Form politischer Beziehungen *sui generis* handelt, zu dem Schluss, dass ein gerechter Staat eine Verpflichtung hat, solche Personen als einander gleich zu behandeln, und zwar aufgrund der allgemeineren Pflicht solcher Staaten, bei der Ausübung ihrer Zwangsgewalt die betroffenen

Personen als moralisch gleich zu betrachten. Der Staat ist allerdings nicht dazu verpflichtet, diese potentiellen Migrantinnen seinen Bürgerinnen politisch gleichzustellen; so ist es beispielsweise legitim, potentiellen Migrantinnen das Wahlrecht zu verweigern oder ihnen zu untersagen, zivilrechtliche Institutionen anzurufen, bevor ihre Einwanderung gebilligt wurde. Aufgrund der unterschiedlichen institutionellen Umstände, in denen sich die Bürgerinnen und die potentiellen Migrantinnen befinden, verletzt nichts an diesen Maßnahmen das Prinzip der moralischen Gleichheit. Allerdings verletzt der Staat mit Sicherheit seine Pflichten, wenn er potentielle Migrantinnen untereinander moralisch ungleich behandelt. Die von ihm angeführten Gründe für eine Unterscheidung zwischen den verschiedenen potentiellen Einwanderern müssen daher von solcher Art sein, dass sie selbst von denjenigen nicht zurückgewiesen werden können, deren Antrag auf Einwanderung abgelehnt wird. Diesem Ansatz zufolge hat der Staat also ein generelles Recht darauf, potentielle Migrantinnen auszuschließen; der Egalitarismus verlangt nicht nach offenen Grenzen. Wenn der Staat aber nun unter den potentiellen Migrantinnen einige Personen auswählen und diesen die Einwanderung gewähren will, muss er auf Gründe zurückgreifen, die die moralische Gleichheit aller potentiellen Migrantinnen widerspiegeln – Gründe also, die letztendlich von den Ausgeschlossenen selbst akzeptiert werden könnten. Das ist alles, was die Idee politischer Gleichheit unter potentiellen Einwanderern verlangen sollte.

Was aber würde all das in der Praxis bedeuten? Es ist hilfreich, an dieser Stelle die Idee genauer zu betrachten, dass es Gründe gibt, die nicht vernünftigerweise zurückgewiesen werden können.[2] Wollen sie diese Prüfung bestehen, müssen die vorgebrachten Gründe die Interessen der Betroffenen als voneinander getrennte und unantastbare moralische Personen ernst nehmen. Was das im jeweils konkreten Fall bedeuten würde, ist stets schwer zu bestimmen und wird immer vom spezifischen institutionellen Kontext abhängen, in dem das Prinzip der Gleichheit angewendet werden soll. Soll beispielsweise das staatliche Handeln gegenüber den Bürgerinnen des Staates gerechtfertigt werden, würde ein eher umfangreiches Bündel an Rechten und Pflichten vonnöten sein, darunter die Gewährleistung einer gewissen materiellen Gleichheit. Was im Kontext potentieller Immigrantinnen verlangt wird, ist im Gegensatz dazu einfacher zu bestimmen, da in diesem Fall nicht ein System politischer, mit Zwangsgewalt ausgestatteter Institutionen gerechtfertigt werden muss, sondern bloß eine einzige mit Zwang verbundene Entscheidung hinsichtlich der Zulassung einer Person. Was wir in diesem Fall also suchen, sind nicht die politischen Garantien gleicher Staatsbürgerschaft, sondern eine weniger anspruchsvolle Form gleicher Behandlung, die dem entsprechenden Kontext angemessen ist. Die Rechtfertigung muss daher so sein, dass sie die moralische Gleichheit der betroffenen Personen ernst nimmt und von ihren Adressatinnen akzeptiert werden kann, ohne dass diese dabei ihrer eigenen moralischen Ungleichheit zustimmen müssten.

Hierzu könnte noch deutlich mehr gesagt werden; aber ich möchte vorschlagen, mit dieser breit angelegten Vorstellung davon, wie die von einem Staat vorgeschlagenen Prinzipien der Auswahl potentieller Migrantinnen bewertet werden sollen, fortzufahren. Ein solches Auswahlprinzip muss die Migrantinnen als gleich hinsichtlich ihrer moralischen Würde behandeln – und in meinen Augen wird ein Prinzip diesen Test nicht bestehen, wenn eine Person seine Gültigkeit nicht akzeptieren kann, ohne dabei zugleich die eigene moralische Minderwertigkeit in Kauf nehmen zu müssen. Wenn wir also mit der vorgebrachten Rechtfertigung nur dann einverstanden sein können, wenn wir unserer eigenen moralischen Minderwertigkeit zustimmen, haben wir es mit einem Prinzip zu tun, dass den grundlegendsten Test der Fairness nicht besteht. Ich behaupte nicht, dass es sich hierbei um die einzige Möglichkeit handelt, wie ein solches Auswahlprinzip die Forderungen der Gerechtigkeit missachten kann, aber ich glaube, dass viele ungerechte Prinzipien unter Rückgriff auf die soeben ausgeführten Überlegungen als ungerecht ausgewiesen werden können.

Nehmen Sie zum Beispiel die von der Trump-Administration angeführten Gründe, um die Verstärkung der Sicherheitsvorkehrungen an der südlichen Grenze der USA zu rechtfertigen. Als Präsidentschaftskandidat begann Donald Trump mit der Behauptung, dass Mexiko Kriminelle „schicken“ würde, darunter Drogendealer und Vergewaltiger. Als Präsident verfolgte er dieses Thema mit der Behauptung weiter, dass mexikanische Bandenmitglieder (und womöglich alle Mexikanerinnen) „Tiere“ seien.[3] Die Trump-Administration behauptete, dass eine solche Sprache vor dem Hintergrund der Verbrechen von Gangs wie MS-13 legitim sei.[4] Allerdings hat diese Rechtfertigung verstärkter Sicherheitsvorkehrungen an der Grenze nur sehr wenig mit den Fakten, die über Kriminalität vorliegen, zu tun. Die verfügbaren sozialwissenschaftlichen Erkenntnisse weisen darauf hin, dass Migrantinnen, inklusive solcher Personen ohne Aufenthaltspapiere, weniger Verbrechen begehen als Nicht-Migrantinnen. Die tatsächliche Rechtfertigung für die Verstärkung der Grenzsicherheit findet sich in der emotionalen Anziehungskraft der vorgebrachten Gründe – in einer Welt, in der Migrantinnen als mutmaßlich gefährlich betrachtet werden, und mexikanische Migrantinnen als mutmaßlich gefährlicher gelten als norwegische. Die Trump-Administration hat es sich gleichfalls zur Aufgabe gemacht, Veranstaltungen mit sogenannten „Angel Families" auszurichten – Familien, in denen ein Angehöriger durch einen undokumentierten Migranten ermordet wurde. Der Bezug auf die Angel Families beruht dabei nicht auf einem unvoreingenommenen Glauben, dass die Präsenz undokumentierter Migranten zu Mord führt. Vielmehr stellte Präsident Trump in einer Diskussion über die Angel Families schlicht fest, dass undokumentierte Personen mit höherer Wahrscheinlichkeit Verbrechen begehen:

> „Ich höre das immer wieder. ‚Nein, diese Menschen sind weniger kriminell als die Menschen, die in diesem Land leben.' Ihr habt das gehört, Leute. Oder? Ihr habt das gehört. Ich höre es so oft. Und ich sage, ‚Ist das möglich?' Die Antwort ist, es stimmt nicht. Wenn man das hört scheint es, als ob sie bessere Menschen sind als die, die wir hier haben – als unsere Bürger. Es ist nicht wahr."[5]

Stattdessen scheint diese Rechtfertigung des Ausschlusses mit dem Gedanken zu beginnen, dass bestimmte Personen an sich schlechter sind als andere und eher bereit dazu, einen Mord zu begehen – unabhängig von der empirischen Beweislage. Einer Person, die auf der Grundlage eines solchen emotionalen Appells ausgeschlossen wird, ist es in meinen Augen unmöglich, die Überlegungen hinter dieser Aussage zu akzeptieren, es sei denn, sie betrachtet sich selbst als moralisch minderwertig.

Selbstverständlich besteht die Geschichte des Einwanderungsrechts größtenteils aus genau solchen Verurteilungen. So versuchten beispielsweise die Vereinigten Staaten in der Vergangenheit, Menschen chinesischer Abstammung von der Einwanderung auszuschließen und während des Wahlkampfs im Jahre 1888 war der dauerhafte Ausschluss chinesischer Personen Teil des Programms beider Parteien. Die Republikanische Partei wurde dabei besonders deutlich:

> „Wir erklären unsere Ablehnung der Einwanderung ausländischer und chinesischer Arbeitskräfte, denen unsere Zivilisation und Verfassung fremd sind; zudem verlangen wir die strenge Durchsetzung bereits existierenden Rechts gegen eine solche Einwanderung und befürworten eine unverzügliche Gesetzgebung, die solche Arbeitskräfte aus unserem Land fernhält.“[6]

Der Gedanke, dass chinesischen Arbeiterinnen die Demokratie sowie die Zivilisation der Vereinigten Staaten „fremd“ seien, war weit verbreitet und beruhte, wie bereits erwähnt, nicht auf Tatsachen, sondern allein auf der Behauptung, dass chinesische Migrantinnen ausgeschlossen werden sollten, da sie eben chinesisch seien. Es ist bemerkenswert, wie unempfänglich diese Art von Ausschlussprinzip für Fakten ist. Was wir über die Entscheidungen einer Person vorhersagen können ist, wenn überhaupt, eine Frage der empirischen Wissenschaften; aber die in diesem Fall den Ausschluss rechtfertigenden Prinzipien geben bloß vor, auf empirischen Tatbeständen zu beruhen. Stattdessen sind sie ein Ausdruck des Glaubens, dass manche Menschen sich schlicht weniger dazu eignen, ein Teil der Zukunft eines bestimmten Landes zu sein. Diejenigen, die aufgrund dieses Glaubens ausgeschlossen würden, können allerdings zu Recht behaupten, dass ein solches Prinzip ihr Recht auf eine angemessene moralische Behandlung verletzt. Sie könnten dieses Prinzip nicht unterstützen, ohne dabei ihre eigene moralische Unzulänglichkeit zu billigen; und diejenigen, die sich auf dieses Prinzip berufen, behandeln die derart Ausgeschlossen daher unfair.

Wir könnten noch mehr über diese Art von Prinzipien sagen, aber diese Ausführungen mögen für den Anfang unserer Diskussion genügen. Lassen Sie uns nun vier mögliche Prinzipien untersuchen, die ein Staat zur Auswahl potentieller Migrantinnen nutzen könnte.

2. Auswahlprinzipien

Es gibt sicherlich eine Vielzahl solcher Prinzipien und ich werde an dieser Stelle nicht versuchen, alle von ihnen zu besprechen. So erlauben viele Staaten die Einwanderung, wenn ein bestimmter finanzieller Betrag aufgebracht wird. Eine Person kann zum Beispiel auf Dominica gegen eine „Spende“ von 100.000 Dollar an den National Transformation Fund das Recht auf die Staatsbürgerschaft erwerben oder, in den USA, durch ein Investment von einer Millionen US-Dollar in ein US-amerikanisches Unternehmen einen permanenten Aufenthaltstitel in den Vereinigten Staaten erlangen.[7] Viele Staaten nutzen zudem Auswahlprinzipien, die machtpolitische Erwägungen widerspiegeln; so haben die USA über die vergangenen vier Dekaden hinweg Kubanerinnen Einwanderungsrechte gewährt, um dem kubanischen Regime „den Stempel des Versagens aufzudrücken".[8] Die Zahl der Auswahlprinzipien ist so groß wie der menschliche Einfallsreichtum; aber ich glaube, dass die moralische Legitimität eines jeden Prinzips durch den Vergleich mit einem der vier folgenden Prinzipien bestimmt werden kann.

2.1. *Zufall und Eigenwilligkeit*

Lassen Sie uns mit der Untersuchung der beiden Prinzipien beginnen, die auf Glück zu basieren scheinen. Teile des Migrationsrechts beinhalten Zufallselemente; so haben unter anderem die Vereinigten Staaten eine Diversity Lottery, in der Bewerberinnen aus Ländern, die bloß wenige Migrantinnen in die USA entsenden, die Chance auf ein Einwanderungsrecht bekommen. Dieses Vorgehen ist nicht identisch mit der Idee, alle Einwanderungsrechte anhand einer Lotterie zu verteilen; die Diversity Lottery beruht auf einer bestimmten Vorstellung davon, welche Art von Diversität von Bedeutung ist, weshalb sie sowohl Ausdruck einer bestimmten politischen Philosophie, als auch des Prinzips einer zufälligen Verteilung von Einwanderungsrechten ist. Allerdings könnten wir uns eine Welt vorstellen, in der der Zufall eine deutlich größere Rolle spielt. So hat beispielsweise Alexander Guerrero Argumente für eine stärkere Nutzung von Zufallselementen hinsichtlich administrativer Fragen der Politik entwickelt.[9] Könnten wir ähnliche Ideen auch für die Steuerung der Migration heranziehen?

Aus dem Einsatz von Zufallselementen könnte gute oder schlechte Politik entstehen – in dem gewöhnlichen Sinne, dass gute Politik in der Bestimmung des besten Weges für das Erreichen eines legitimen öffentlichen Ziels besteht. Allerdings könnte ein solches Vorgehen wohl kaum als *unfair* bezeichnet werden. Die Lotterie scheint, sofern sie die üblichen Prüfungen auf Fairness der Wahrscheinlichkeitsverteilung besteht, keine bestimmte Teilnehmerin zu benachteiligen. Die Tatsache, dass manchen von uns dieser Einsatz von Zufälligkeit missfallen wird, mag daran liegen, dass manche der an dieser Lotterie teilnehmenden Personen ein *Recht* darauf haben, zu gewinnen – eine solche Lotterie könnte beispielsweise mitunter Menschen ausschließen, die einen Anspruch auf Asyl geltend machen können. In manchen Fällen lehnen wir die zufällige Verteilung eines Guts aus dem Grund ab, dass bestimmte Personen einen Anspruch auf dieses Gut haben. (Ich sollte erwähnen, dass das gleiche auch hinsichtlich schlechter Dinge gelten mag; so missfiel vielen Leuten David Lewis' versuchsweise Verteidigung zufälliger Bestrafungen, da sie meinten, ein Krimineller hätte ein *Recht* auf eine Bestrafung.)[10] Wenn wir jedoch davon ausgehen, dass diejenigen Personen, die über solche Rechte verfügen, bereits angemessen behandelt wurden und somit nur diejenigen verbleiben, die solche Rechte nicht in Anspruch nehmen können, scheint kein Einwand mehr vorzuliegen. Eine zufällige Verteilung mag also zumindest in diesem spezifischen Kontext fair sein.

Interessanter wird es jedoch, wenn wir bedenken, dass Zufälligkeit bisweilen nicht auf der bewussten Einführung von Zufallselementen beruht, sondern auf den eigenwilligen Vorlieben von Personen in verantwortlichen Positionen. Wäre ich ein rothaariger Mann und Ezekiah Hopkins würde die Liga der rothaarigen Männer gründen, um Menschen wie mir zu helfen, so erschiene es von meinem Standpunkt aus so, als ob ein zufälliges Ereignis mir eine günstige Gelegenheit zugespielt hätte. Ähnliche Dinge scheinen auch im Falle des Migrationsrechts zuzutreffen. Zwischen 1992 und 1994 wurde, im Rahmen eines durch den Abgeordneten Bruce Morrison eingerichteten Programms, 48.000 irischen Bürgerinnen das Recht auf Einwanderung in die Vereinigten Staaten zugesprochen.[11] Morrison fügte die seinem Programm zugrunde liegende Formulierung in den Immigration and Nationality Act von 1990 ein und scheint dies vor allem aufgrund der Zuneigung für sein ehemaliges Heimatland getan zu haben. (Es sollte dabei hervorgehoben werden, dass dem gesamten Vereinigten Königreich im Rahmen dieses Programms bloß 6000 Visabefreiungen zugesprochen wurden.) Ist es möglich, diese Vorgehensweise aufgrund ihrer Ähnlichkeit mit einer rein zufälligen Prozedur zu verteidigen?

In meinen Augen ist es durchaus möglich, dass willkürliche Unterscheidungen nicht inhärent ungerecht sind – sofern sie Menschen nicht auf unzulässige Weise hierarchisieren. Allerdings müssen mindestens drei Dinge bedacht werden. Zunächst ist es so, dass viele willkürliche Entscheidungen in der Tat auf moralisch unzulässige Überzeugungen zurückgeführt werden können. Ezekiah Hopkins' Vorliebe für rothaarige Männer mag vielleicht unproblematisch sein. Damit möchte ich jedoch nicht sagen, dass dem so ist; wir könnten einen Fall konstruieren, in dem Hopkins' scheinbare Vorliebe für rothaarige Männer von denjenigen als eine Form sozialer Ausgrenzung erlebt wird, die keine roten Haare haben. Mehr Sorgen bereitet mir jedoch Morrisons Visa-Programm und zwar deshalb, weil es den spezifischen Vorlieben und Zuneigungen eines Mitglieds des US-Kongresses entsprungen ist, also einer Person mit außergewöhnlich viel Macht. Es erstaunt kaum, dass Morrison sich dafür entschied, Irland statt beispielsweise Gambia zu helfen; immerhin gab es bisher viel mehr Politiker irischer Abstammung in den Vereinigten Staaten als solche mit gambischen Wurzeln. Hierin scheint jedoch genau das Problem zu liegen. Eine scheinbar harmlose Entscheidung zugunsten eigenwilliger Interessen kann viel zu sehr wie eine Gesetzgebung erscheinen, die pauschal die partikularen Interessen mächtiger Personen fördert und verteidigt. Die potentielle Legitimität zufallsbasierter Entscheidungsregeln kann also kaum zur Verteidigung einer Regel herangezogen werden, die es mächtigen Personen erlaubt, bei der Verteilung von Gütern ihre eigenwilligen Interessen zu verfolgen. Was mit solchen Präferenzen beginnt, kann durchaus in ungerechtfertigter Ungleichheit enden.

Doch Eigenwilligkeit und Zufälligkeit unterscheiden sich noch in einer zweiten Hinsicht: Das Prinzip der Eigenwilligkeit legt viel Macht in wenige Hände. Selbst wenn alle beschriebenen Probleme ausgeräumt wären – wenn also keine Fehlallokation von Macht bestünde und wir uns nicht darum sorgen müssten, dass hinter den eigenwilligen Präferenzen möglicherweise rassistische Interessen stehen –, könnten wir immer noch der Meinung sein, dass es schlicht nicht gut ist, so viel Macht in so wenige Hände zu legen. Um diesen Gedanken auszuführen, müssen wir über den Politischen Liberalismus hinausgehen und die Idee der Dominanz heranziehen, wie sie sich bei Theoretikerinnen des Republikanismus findet.[12] Allerdings habe ich dafür weder die Zeit noch die nötige Kompetenz. In meinen Augen hat jedoch der Gedanke, dass eine solche Konzentration von Macht in den Händen einer einzigen Person moralisch unzulässig ist, einen gewissen Wahrheitsgehalt. Ich bin mir nicht sicher, ob dem tatsächlich so ist; es ist durchaus möglich, dass ein Entscheidungsträger seine Wahl anhand von Prinzipien trifft, die keine Person marginalisieren oder dämonisieren – und in einem solchen Falle liegen wir mitunter mit der Behauptung richtig, dass die von dieser Person getroffene Entscheidung keine Ungerechtigkeit darstellt, da sie nicht unfair ist. Allerdings bin ich offen dafür, noch weiter darüber nachzudenken, in welcher Hinsicht auf Eigenwilligkeiten beruhende Auswahlprozesse inhärent falsch sein könnten – und zwar nicht aufgrund der aus ihnen resultierenden Güterverteilung, sondern aufgrund der Verteilung von Macht zwischen den Entscheidungsträgerinnen und denjenigen, die von deren Entscheidungen betroffen sind.

Der letzte Punkt, den es zu bedenken gilt, spiegelt hingegen etwas wider, was ich bereits weiter oben angemerkt habe – dass nämlich einige unserer Reaktionen gegen eigenwillige Prinzipien eigentlich eine Reaktion gegen die Anwendung solcher Prinzipien in Kontexten darstellt, in denen manche Menschen einen Anspruch auf das infrage stehende Gut haben. Wenn ich einen auf Gerechtigkeit beruhenden Anspruch auf etwas habe – sei es das Recht auf Einwanderung, das Recht zu wählen oder etwas anderes –, dann wird mir Unrecht getan, wenn mir dieses Gut aufgrund der Funktionsweise von Institutionen vorenthalten wird. Ein solches Unrecht besteht in meinen Augen ganz unabhängig davon, ob die dem Unrecht zugrunde liegende Entscheidung daher rührt, dass ich schlecht behandelt werden soll oder aus einem eigenwilligen Entscheidungsprozess resultiert, in dem ein bestimmter Akteur das Gut an manche Menschen verteilt (und nicht an mich). In beiden Fällen besteht mein primärer Einwand darin, dass mir nicht zugeteilt wurde, was mir zusteht. Es erscheint mir jedoch schlimmer, die eigenen Rechte aufgrund eigenwilliger Entscheidungen verweigert zu bekommen; eine solche Entscheidung scheint, in den Worten Avishai Margalits, sowohl ungerecht als auch entwürdigend zu sein.[13] Aber diese Reaktion bezieht sich darauf, dass der Zufall oder eigenwillige Präferenzen im falschen Kontext eingesetzt wurden und beeinträchtigt somit nicht den möglichen Nutzen dieser Prinzipien selbst.

2.2 *Identität*

Wenn ich von Identität spreche, dann meine ich die Aspekte des Selbst, die der Verortung dieses Selbst in der sozialen Welt dienen: das eigene soziale Geschlecht, die eigene ethnische Identifikation, die eigene Religion und so weiter. Die Kategorie der Identität umfasst alle Aspekte des Selbst, die diesem Zweck dienen – allerdings möchte ich an dieser Stelle zwei Punkte betrachten, die in den jüngsten politischen Debatten an Relevanz gewonnen haben: die ethnische sowie die religiöse Identität.

Wir können mit dem Aspekt der Ethnie beginnen. Nicht, weil dieser Aspekt einfacher wäre – das ist er nicht –, sondern weil Diskriminierung und Unterdrückung aufgrund der ethnischen Abstammung gemeinhin (und zu Recht) als die kanonischen Formen illiberalen Handelns angesehen werden. Sollte im Bereich innerstaatlicher Politik ein Prinzip als rassistisch ausgewiesen werden, reicht uns das im Allgemeinen aus, um zu sagen, dass es den Test politischer Moralität nicht bestanden hat. Laut Joseph Carens können wir das gleiche auch im Kontext der Migration sagen: Ein rassistisches Ausschlussprinzip ist unrechtmäßig und zwar genauso, wie es ein rassistisches Gesetz im Inland wäre.[14]

Hierin stimme ich mit Carens überein, aber wir können noch mehr vorbringen, um diesen Schluss zu rechtfertigen. Zunächst ist anzumerken, dass ein rassistisches Ausschlussprinzip aus einem ganz bestimmten Grund unrechtmäßig sein kann, der genauerer Betrachtung wert ist. Selbst wenn wir von der Migrantin selbst absehen und einfach auf die Rechte der Bürgerinnen eines bestimmten Staates schauen würden, hätten wir Gründe für die Behauptung, dass der Staat bei Maßnahmen zum Ausschluss von Migrantinnen nicht vollkommen frei handeln darf. Was ein Staat an seinen Grenzen unternimmt, wird auch von den Menschen innerhalb seiner Grenzen wahrgenommen, weshalb eine bestimmte Form der Migrationspolitik sich nicht nur deshalb als unrechtmäßig herausstellen könnte, weil sie Migrantinnen unrechtmäßig behandelt, sondern auch diejenigen, die mit den derart Ausgeschlossenen in einer sozialen Beziehung stehen.

Präsident Trumps Dämonisierung mexikanischer Immigrantinnen betrifft nicht nur einfach diese Personengruppe, sondern eine große Zahl an Menschen, die als mexikanisch oder migrantisch *wahrgenommen* werden. Ganz abgesehen von den Rechten der Migrantinnen selbst, wird das Recht von Personen lateinamerikanischer Abstammung darauf, bei der Entstehung und Durchsetzung von Gesetzen als moralisch gleichwertig behandelt zu werden – darunter das Recht, von der Polizei als moralisch Gleiche behandelt zu werden – durch die von der Trump-Administration verkündeten Formen der Grenzkontrollen gefährdet. Amy Reed-Sandoval entwickelte das Konzept der „sozial Undokumentierten“ um dieses Problem zu analysieren; demnach kann eine Person zwar alle Dokumente besitzen, um von den eigenen Rechten Gebrauch machen zu können, aber dennoch durch die Migrationspolitik der Behörden substantiell ungleich behandelt werden.[15] Präsident Trumps

Begnadigung von Sheriff Joe Arpaio brachte beispielsweise die Tatsache zum Ausdruck, dass die Trump-Administration durchaus den Willen dazu hat, die Rechte von Bürgerinnen lateinamerikanischer Abstammung zu untergraben, wenn sie dadurch Härte gegenüber undokumentierten Personen demonstrieren kann.[16]

Die Struktur meines Arguments beginnt mit der Gleichheit von Bürgerinnen. Eine bestimmte Ausschlusspraxis ist demnach unrechtmäßig, sofern sie einer Gruppe von Bürgerinnen den Eindruck vermittelt, sie würden politisch nicht in gleicher Weise beachtet und respektiert. Mit anderen Worten: Wenn der Eindruck entstünde, dass wir uns über eine Gesellschaft freuen würden, in der sie nicht präsent wären. Ein Staat, der eine solche Botschaft vermittelt, kann sicherlich nicht als legitim betrachtet werden.

Wir können dieses Argument noch erweitern und untersuchen, was darüber hinaus vorgebracht werden kann, wenn wir die Rechte der Migrantinnen selbst in Betracht ziehen. So kann beispielsweise die Wahrnehmung der rassistischen Prinzipien durch die benachteiligten Mitglieder einer ethnischen Gruppe den Grund dafür darstellen, solche Prinzipien als unrechtmäßig zu bezeichnen, und zwar ganz unabhängig von den Auswirkungen dieser Prinzipien auf die Bürgerinnen, die jener ethnischen Gruppe angehören. Für manche Philosophinnen stellt die Ethnie eine Kategorie dar, die gar nicht unabhängig von Macht und Marginalisierung begriffen werden kann, weshalb eine Ausschlusspraxis, die sich auf Prinzipien der ethnischen Abstammung beruft, notwendigerweise unzulässige Botschaften vermittelt. Nehmen Sie zum Beispiel noch einmal den Chinese Exclusion Act und seine Beschreibung der chinesischen Staatsangehörigen als Menschen, „denen unsere Zivilisation und Verfassung fremd sind". Es ist schwer vorstellbar, wie eine hierdurch ausgeschlossene Person diesen Satz interpretieren sollte *ohne* ihn als eine Aussage über die eigene moralische Minderwertigkeit zu verstehen. Keine chinesische Person wäre verpflichtet, diese Aussage als Rechtfertigung zu akzeptieren. Tatsächlich scheint es einer derart ausgeschlossenen Person, die den dieser Begründung zugrunde liegenden Gedanken unterstützt, auf unangenehme Weise an Selbstachtung zu mangeln.

Das Problem besteht nun jedoch darin, dass nicht alle Auswahlprinzipien ihren Rassismus so offen zur Schau stellen wie der Chinese Exclusion Act. An dieser Stelle gewinnt das juristische Konzept der Scheinbegründung (*pretext*) an Bedeutung. Selbst wenn ein bestimmtes Gesetz anhand eines neutralen Prinzips gerechtfertigt werden könnte, es also von den betroffenen Personen unterstützt werden könnte, ohne dabei die eigene moralische Minderwertigkeit in Kauf nehmen zu müssen, sind wir demnach nicht immer dazu verpflichtet, es in diesem Sinne zu verstehen – nämlich dann nicht, wenn offensichtlich sein sollte, dass der eigentliche Zweck des Gesetzes in der Stigmatisierung oder Verurteilung einer bestimmten Gruppe besteht. Auch hier bietet sich die Geschichte der Marginalisierung chinesischer Personen zur Veranschaulichung an. Der Supreme Court entschied im Fall *Yick Wo v. Hopkins* im Jahre 1886, dass ein Gesetz das Recht auf gleichen Schutz durch das Gesetz verletzen könne, selbst wenn es auf den ersten Blick neutral erschiene, und zwar aufgrund der Rolle, die Formulierung als auch Anwendung dieses Gesetzes bei der Marginalisierung einer bestimmten Gruppe gespielt haben. Das infrage stehende Gesetz war eine Verordnung der Stadt San Francisco, die es kommerziellen Wäschediensten untersagte, ihr Geschäft ohne eine spezielle Zulassung in Holzbauten zu führen. Zu diesem Zeitpunkt arbeiteten viele chinesische Wäschedienste in solchen Holzbauten und das Gesetz wurde durch die städtische Polizei zur Drangsalierung chinesischer Arbeiterinnen genutzt. Es wurde behauptet, das Gesetz diene Sicherheitszwecken, aber die Art seiner Anwendung offenbarte, dass es sich dabei bloß um eine Scheinbegründung handelte. In Wahrheit war die Ablehnung einer bestimmten Ethnie der Beweggrund für das Gesetz sowie seine Anwendung:

> „[Während die Kontrolleure die Ausstellung von Genehmigungen an] 200 andere [verweigerten], die ebenfalls darum baten, alle von ihnen chinesische Personen, wurde 80 anderen, unter ihnen kein einziger Chinese, eine Genehmigung für die gleiche Geschäftstätigkeit unter ähnlichen Bedingungen gewährt. Die Tatsache der Diskriminierung wird zugegeben. Für dieses Vorgehen wird kein Grund genannt und daher kann der Schluss nicht vermieden werden, dass kein Grund außer einer ablehnenden Einstellung gegenüber der Ethnie und Nationalität der Antragstellerinnen besteht, was aus der Perspektive des Rechts nicht gerechtfertigt werden kann. Die Diskriminierung ist daher illegal und die sie ausführende öffentliche Verwaltung verweigert das Prinzip gleichen Rechtsschutzes, was eine Verletzung des 14. Verfassungszusatzes darstellt."[17]

Das von *Yick Wo v. Hopkins* ausgehende Versprechen wurde in den darauffolgenden Jahren zwar nicht angemessen erfüllt; allerdings zeigt es auf moralischer Ebene, dass selbst ein auf den ersten Blick neutrales Prinzip mitunter nicht ausreicht, um einen bestimmten Rechtsakt zu rechtfertigen. Wir sind uns oft der wahren Zwecke eines Gesetzes bewusst und sehen deutlich, auf welche Art es diejenigen marginalisiert und schädigt, gegen die es sich wendet. Wie Oliver Wendell Holmes es ausgedrückt hat: Selbst ein Hund versteht, ob er aus Versehen angerempelt oder getreten wird.[18]

All das zeigt uns, dass es für ein bestimmtes Ausschlussprinzip zumindest möglich ist, den Test politischer Moralität selbst dann nicht zu bestehen, wenn eine scheinbar neutrale Begründung für dieses Prinzip angeführt werden könnte. Das Zusammenspiel von Ethnie und Migration ist komplex und wirkmächtig – und ich kann an dieser Stelle bloß anerkennen, wieviel Arbeit für die Untersuchung dieses Zusammenhangs nötig wäre. Es könnte noch viel darüber gesagt werden, auf welche verschiedenen Arten ein bestimmtes Ausschlussprinzip die von mir vorgeschlagenen Tests nicht besteht, das Prinzip also zu Recht zurückgewiesen werden kann, da es seine Adressatinnen dazu nötigen würde, der eigenen moralische Minderwertigkeit zuzustimmen. Für den Moment möchte ich jedoch mit der allgemeinsten Schlussfolgerung enden: Nämlich, dass der Bezug auf eine ethnisch begründete Abneigung im Bereich der Migrationspolitik moralisch unzulässig ist, selbst wenn diese Abneigung durch das Feigenblatt einer Scheinbegründung getarnt wird.

Diese Überlegungen bringen uns zum Problem der Religion. Ich möchte mich in meiner Analyse insbesondere darauf konzentrieren, welche Rolle die Religion – sowie die Nutzung von Scheinbegründungen – bei dem von der Trump-Administration verteidigten sogenannten Travel-Ban gespielt hat, dessen Ziel darin bestand, die Einreise von Menschen aus bestimmten Ländern mit muslimischer Bevölkerungsmehrheit zu verhindern. Dieses Verbot hat – und hatte – eine Vielzahl von Problemen, von denen eines der bedeutendsten den Fakt betrifft, dass er schlecht ausgearbeitet wurde und keine Unterscheidung zwischen Personen traf, die dauerhafte Einwohner der Vereinigten Staaten waren und solchen, die bisher noch nicht einmal die Vereinigten Staaten besucht hatten.[19] Dieser juristische Fehler ist moralisch bedeutsam; ein Gesetz, das unverständlich ist, kann letztlich auch nicht befolgt werden. Aber die wichtigste moralische Frage hinsichtlich des Travel-Ban ist einfach: Gibt es irgendeinen Weg, eine solche Art von Politik zu verteidigen?

Zur Beantwortung dieser Frage können wir für den Moment die überarbeitete Version des Travel-Ban beiseitelassen, die ihn als eine Antwort auf globale Sicherheitsprobleme darzustellen versuchte. Während seiner Kampagne forderte Trump einen „vollständigen und totalen Stop" muslimischer Einwanderung in die Vereinigten Staaten. Die Gründe, aus denen diese Version des Travel-Ban moralisch falsch ist, ähneln den bereits weiter oben ausgeführten Überlegungen. Erstens würde der Bann von muslimischen Bürgerinnen der Vereinigten Staaten so verstanden werden, dass er sie als untauglich für die amerikanische Staatsbürgerschaft darstellt. Indem er Muslime unter Generalverdacht stellt, artikuliert der Travel-Ban eine Vorstellung von amerikanischer Staatsbürgerschaft, die von muslimischen Amerikanerinnen berechtigterweise zurückgewiesen werden könnte. Zweitens würden sich muslimische Migrantinnen durch eine solche Politik sehr wahrscheinlich als moralisch minderwertig bezeichnet fühlen. Der Travel-Ban behauptet, dass muslimische Migrantinnen mutmaßlich eher zur Kriminalität neigen als andere Migrantinnen; zumindest forderte Donald Trumps Wahlkampfteam, den Travel-Ban für Muslime solange aufrechtzuerhalten, „bis die Repräsentanten unseres Landes herausfinden können, was zur Hölle los ist." Diese Darstellung des Banns scheint mir mit großer Sicherheit eine Umkehr der Unschuldsvermutung im Falle von Muslimen zu bedeuten, die nun solange verdächtig sind, bis ihre Unschuld bewiesen ist – eine Vermutung, die durch den Vorschlag Donald Trumps, ein Melderegister für Muslime einzurichten, bloß weiter verstärkt wurde. All das scheint die Botschaft transportieren zu wollen, dass Menschen eines bestimmen Glaubens als besonders ungeeignet für die Vereinigten Staaten betrachtet werden.

Das führt uns nun aber zu der Version des Travel-Ban, die letztendlich von der Trump-Administration erlassen wurde. Diese Version des Banns beginnt nicht mit Gründen, die sich auf eine bestimmte Religion beziehen, wie es noch die Variante aus Trumps Wahlkampf tat. Stattdessen wird hier eine globale Sicherheitslage beschrieben, in der bestimmte Länder als Feinde der Vereinigten Staaten dargestellt werden. Es gab mindestens zwei Arten, diese Version des Travel-Ban zu lesen. Die erste basiert auf den Begrifflichkeiten, die der Travel-Ban selbst verwendet und in denen keinerlei Bezug auf Religion gleich welcher Art gemacht wird. Im Fall *Trump v. Hawaii* nutzte der Supreme Court diese Lesart:

„Die Verkündung dient ausdrücklich legitimen Zwecken: Den Zutritt von Personen zu verhindern, die nicht ausreichend überprüft werden können, sowie andere Länder dazu zu veranlassen, ihre Sicherheitskontrollen zu verbessern. Der Text enthält keinerlei Aussagen über eine bestimmte Religion. Der Kläger und die abweichende Meinung betonen dennoch, dass die Bevölkerung von fünf der sieben Länder, die derzeit in der Verkündung genannt werden, mehrheitlich muslimisch sind. Allerdings erlaubt diese Tatsache allein noch nicht den Schluss auf eine feindselige Gesinnung gegenüber einer bestimmten Religion, da das Gesetz nur acht Prozent der weltweiten muslimischen Bevölkerung betrifft und auf Länder beschränkt ist, die bereits zuvor entweder vom Kongress oder vorhergehenden Regierungen als nationales Sicherheitsrisiko eingestuft wurden.“[20]

Richterin Sotomayor hingegen argumentierte in ihrer abweichenden Meinung für eine Lesart des Banns, die den von mir verteidigten Begrifflichkeiten näherkommt:

„Angesichts der aktenkundigen Beweise würde ein vernünftiger Beobachter darauf schließen, dass dieses Gesetz anti-muslimisch motiviert war. Das alleine reicht aus, um zu zeigen, dass die Kläger mit ihrer Klage gemäß des ersten Verfassungszusatzes wahrscheinlich erfolgreich sein werden. Die Mehrheit ist jedoch anderer Meinung, wobei sie die Tatsachen ignoriert, Präzedenzfälle falsch auslegt und gegenüber dem Schmerz und Leid die Augen verschließt, den dieses Gesetz über unzählige Familien und Individuen bringt, darunter viele amerikanische Staatsbürgerinnen. Da diese verstörenden Folgen unserer Verfassung sowie unseren Präzedenzfällen entgegenstehen, widerspreche ich.“[21]

An dieser Stelle lasse ich selbstverständlich bestimmte Fragen hinsichtlich der Interpretation der Verfassung außen vor; Richter Roberts argumentierte in seiner Entscheidung beispielsweise insbesondere damit, dass das Gericht aufpassen müsse, das Urteil der Exekutive nicht durch das Urteil des Gerichts zu ersetzen – und das Gericht der Exekutive daher bei der Bestimmung des dem Travel-Ban zugrunde liegenden Prinzips den Vortritt lassen sollte. Jenseits solcher Rücksichtnahme weigert sich der Supreme Court jedoch, den Travel-Ban anhand jener Begrifflichkeiten zu lesen, die am ehesten geeignet sind, um zu verstehen, wie er von muslimischen Amerikanerinnen erlebt wird. Die Scheinbegründung der nationalen Sicherheit sollte uns nicht ausreichen, um den ziemlich offensichtlichen Schluss zu ignorieren, dass die wahre Motivation des Travel-Ban Feindseligkeit und nicht Sicherheit ist; und wenn das, was ich hier gesagt habe, richtig ist, kann dies nicht als moralisch legitim betrachtet werden.

All das folgt natürlich den bereits diskutierten Überlegungen hinsichtlich ethnischer Unterscheidungskriterien. Allerdings gibt es in diesem Fall einige Besonderheiten, die einer näheren Betrachtung wert sind. Zunächst ist es nicht ausgeschlossen, dass wir unter bestimmten Voraussetzungen tatsächlich in einer Welt enden, in der Sicherheitsrisiken stark mit religiösen Überzeugungen korrelieren. (Ein Beispiel wäre die Sekte Ōmu Shinrikyōdie, die für die Sarin-Attacke in Tokyo im Jahre 1995 verantwortlich war und deren Ziel unter anderem darin bestand, das Ende der Zeit herbeizuführen.)[22] Im Unterschied zur Ethnie sind kanonische Glaubenssätze und Verhaltensregeln wesentliche Merkmale von Religionen, weshalb die Entstehung einer Glaubensgemeinschaft, deren Anhängerinnen mit hoher Wahrscheinlichkeit bestimmte Verbrechen begehen, nicht unmöglich erscheint. Zwar glaube ich nicht, dass wir in einer solchen Situation leben und bin darüber hinaus auch der Meinung, dass diejenigen, die das Gegenteil behaupten, mit hoher Wahrscheinlichkeit schlechte Absichten verfolgen – wahrscheinlich ist ihre Behauptung eines nationalen Notstands eher auf politisch motivierte Verleumdungen als auf Sicherheitsbedenken zurückzuführen. Nichtsdestotrotz ist es sinnvoll zu fragen, wie unsere Antwort in einer solchen Situation aussehen würde. Aus der Tatsache, dass die religiöse Identität uns in diesem Fall nützliche Informationen hinsichtlich der kriminellen Neigungen von Personen gibt, folgt jedoch nicht, dass wir diese Informationen nutzen sollten. Randall Kennedys Analyse des Racial Profiling scheint mir an dieser Stelle hilfreich: Selbst wenn wir die ethnische Zugehörigkeit einer Person für die Ermittlungsarbeit nutzen könnten, sollten wir es aus dem Grund nicht tun, dass die langfristigen Nachteile hinsichtlich der Idee gleicher Staatsbürgerschaft stärker wiegen könnten als der kurzfristige Nutzen für die Polizeiarbeit. Eine Gesellschaft, in der afrikanisch-stämmige Amerikanerinnen regelmäßig von der Polizei kontrolliert werden, ist laut Kennedy eine Gesellschaft, in der sich das soziale Gewebe zwischen den Bürgerinnen und der Polizei aufzulösen beginnt – was ihm zufolge weitaus schlimmer wäre, als auf eine maximal erfolgreiche Strafverfolgung zu verzichten.[23] Etwas ähnliches können wir auch im Hinblick auf Migration sagen. Selbst wenn wir Informationen über die ethnische Zugehörigkeit einer Person nutzen könnten, um gewalttätige Personen auszuschließen, würden wir damit die Fähigkeit marginalisierter Bürgerinnen beschädigen, sich selbst als Teil der Gesellschaft wahrnehmen zu können – und letztendlich mag das moralisch bedeutsamer sein als ein Maximum an Sicherheit zu erreichen. Ein Teil des Arguments aus dem vorhergehenden Kapitel lautete, dass es neben dem Eigennutz noch andere, moralische Gründe für staatliches Handeln gibt. So müssen Staaten Kosten in Kauf nehmen, um Rechte zu schützen. Das gleiche gilt auch in diesem Fall: Wir können nicht annehmen, dass die effizienteste Art, für Sicherheit zu sorgen, notwendigerweise auch moralisch zulässig ist.

Zuletzt wird die Diskussion dadurch noch ein Stück komplexer, dass wir, angesichts der Interpretation religiöser Präferenzen im Falle der Migrationspolitik, anerkennen müssen, dass nicht alle Formen solcher religiösen Präferenzen moralisch identisch sind. In einer Zeit, in der es einen bedeutenden weltweiten Vorbehalt gegenüber einer bestimmten Religion gibt, könnte beispielsweise eine *Bevorzugung* von Mitgliedern dieser Religion eher Gerechtigkeit denn Voreingenommenheit bedeuten. So übten in den 1970er Jahren die Vereinigten Staaten Druck auf die Sowjetunion aus, damit diese die Auswanderung von Juden nach Israel erlaubte, und rechtfertigten diese Entscheidung mit der vorausgegangenen *Feindschaft* der Sowjetunion gegenüber ihren jüdischen Bürgerinnen.[24] Die Entscheidung, Anhängerinnen einer bestimmten Religion zu bevorzugen, kann folglich auch etwas anderes als Feindseligkeit ausdrücken, dann nämlich, wenn die Welt, in der diese Entscheidung getroffen wird, die Mitglieder dieser Religion schlecht behandelt. Allerdings gibt es auch hier Raum für Bedenken, denn es ist eine Sache, von dieser Methode Gebrauch zu machen, eine andere jedoch, von ihr guten Gebrauch zu machen. So forderte beispielsweise Ted Cruz, dass sich die Vereinigten Staaten, angesichts des Ausmaßes der Verfolgung von Christen an verschiedenen Orten der Welt, auf die Zulassung christlicher Migrantinnen konzentrieren sollten.[25] Selbst wenn die Struktur dieses Arguments richtig sein sollte, scheint es doch inhaltlich falsch: Christen werden in der Tat verfolgt, aber so widerfährt es auch Muslimen, Buddhisten und einer Vielzahl anderer Menschen – und wir könnten bloß dann eine Bevorzugung von Christen rechtfertigen, wenn es uns möglich wäre zu zeigen, dass sie auf eine ganz andere Art verfolgt würden als die Anhängerinnen anderer Religionen und wir zugleich wie kein anderes Land dazu in der Lage wären, ihnen zu helfen. Zwar handelt es sich dabei um empirische Fragen, aber in meinen Augen ist es eher unwahrscheinlich, dass diese Argumente erfolgreich verteidigt werden können.

2.3 *Fähigkeiten*

Viele Länder tendieren dazu Immigrantinnen auszuwählen, die über Fähigkeiten verfügen, die unter den bereits anwesenden Personen bloß spärlich vorhanden sind. Oft stehen denjenigen, die über seltene und wertvolle Fähigkeiten verfügen, Türen offen, die ihren weniger befähigten Landsleuten verschlossen bleiben. Das Gewicht, das dem Kriterium der Fähigkeiten bei der Auswahl von Migrantinnen zukommt, unterscheidet sich je nach Land. So legt beispielsweise Kanada mehr Wert auf Fähigkeiten, die in der Bevölkerung bloß knapp vorhanden sind, als die Vereinigten Staaten, die eher Familienzusammenführungen priorisieren. Zudem unterscheidet sich in den verschiedenen Gesellschaften auch, was als wichtige Fähigkeit angesehen wird. Allerdings stellen alle Länder auf die ein oder andere Weise gesetzliche Instrumente bereit, anhand derer bestimmten Menschen mit seltenen und wertvollen Fähigkeiten die Einwanderung ermöglicht wird. Trotz der Bedeutung der Familienzusammenführung, erleichtern auch die USA Personen mit solchen Fähigkeiten die Einwanderung, unter anderem mit einer speziellen Visa-Kategorie für sogenannte „Ausländer mit außergewöhnlichen Fähigkeiten."[26]

All das ist vor allem aus politischer Perspektive interessant – was aber kann über diese Tatsachen vom Standpunkt der Gerechtigkeit aus gesagt werden? Ist etwas ungerecht, so muss es letztendlich ungerecht gegenüber einer bestimmten Person sein. Wenn wir also die Behauptung verteidigen wollen, dass die Bevorzugung qualifizierter Personen ungerecht ist, müssen wir verstehen, wer zu einer solchen Behauptung berechtigt wäre – wer könnte in diesem Fall behaupten, unfair behandelt zu werden?

In meinen Augen gibt es hier drei Möglichkeiten: Die Unfairness könnte von den potentiellen Migrantinnen empfunden werden, die nicht ausgewählt wurden; von denjenigen, die bereits im Zielland anwesend sind; oder von denjenigen, die im Auswanderungsland zurückgelassen wurden. Die letzte

Möglichkeit wird oft unter dem Begriff des „Brain Drain“ (der Abwanderung von hochqualifizierten Arbeitskräften) diskutiert und ich habe bereits an anderer Stelle argumentiert, dass dies in der Tat eine Form von Ungerechtigkeit darstellen kann. In diesem Zusammenhang habe ich auch gezeigt, dass eine sich entwickelnde Gesellschaft ein Recht darauf hat, das eigene Humankapital dadurch zu erhalten, dass es die Auswanderung erschwert.[27] An dieser Stelle möchte ich kurz bemerken, dass wir die Entscheidung, qualifizierten Personen die Einwanderung zu erlauben, durchaus mit Verweis auf die Auswirkungen anfechten könnten, die eine solche Entscheidung auf die armen Bürgerinnen des Auswanderungslandes hat – dabei sollten wir allerdings aus zwei Gründen sehr vorsichtig sein. Der erste Grund besteht darin, dass die empirischen Effekte im Bereich der Migration außergewöhnlich komplex sind und die Entscheidung, die Grenzen für qualifizierte Personen zu schließen, den qualifizierten wie auch den unqualifizierten Personen im potentiellen Auswanderungsland schaden könnte. Der zweite Grund ist der, dass wir ziemlich vorsichtig mit dem Gedanken sein sollten, dass die Last globaler Transformation hin zu mehr Gerechtigkeit von den qualifizierten Bürgerinnen der Entwicklungsländer zu tragen ist.

Wir können uns nun stattdessen dem Gedanken zuwenden, dass die Bevorzugung qualifizierter Personen unfair gegenüber den Bürgerinnen des Einwanderungslandes ist und eine solche Bevorzugung aus diesem Grund zurückweisen. Das ist der Gedanke hinter der bereits eingangs erwähnten Behauptung Donald Trumps, die Vereinigten Staaten hätten ihre eigenen Bürgerinnen bloß unzureichend geschützt. Die Struktur dieser Rechtfertigung muss jedoch deutlicher herausgearbeitet werden. Stellen Sie sich zum Beispiel vor, die Behauptung sei wahr, dass die Einwanderung einer bestimmten Gruppe qualifizierter Arbeiterinnen – sagen wir qualifizierter Informatikerinnen, die im Rahmen des H1-B Visaprogramms in die USA kommen – auf die Löhne der lokalen Informatikerinnen drückt. Was genau ist in einem solchen Falle unfair? In meinen Augen wäre es ein schlechtes Argument, zu behaupten, dass eine solche Migrationspolitik schlicht deshalb unfair sei, weil ihr Nettoeffekt in einem Rückgang lokaler Löhne bestünde. Zwar mögen diese Lohneinbußen der Beginn eines Arguments sein, allerdings sicher nicht dessen Schluss: Wir haben in diesem Sinne kein Menschenrecht auf einen gleichbleibenden Lohn. Ich denke, dass ein besseres Argument mit dem Gedanken beginnen könnte, dass eine solche Migrationspolitik die

Verteilungsgerechtigkeit – insbesondere hinsichtlich der Kluft zwischen den Wohlhabenden und den weniger Begüterten – beeinträchtigen kann. Wenn Einwanderung zu niedrigeren Löhnen führt, dabei auch noch die Solidarität unter den Arbeiterinnen verringert und zugleich den Reichtum und die Macht von Unternehmen vergrößert, wäre eine Theorie der Gerechtigkeit, die all diese Entwicklungen verurteilt, wohl auch dazu berechtigt, die Politik zu verurteilen, die jenen qualifizierten Migrantinnen ursprünglich die Einwanderung gestattet hat. (Um die Sache noch komplizierter zu machen: Es könnte auch der Fall sein, dass Unternehmen von undokumentierten Arbeitskräften profitieren; manche Unternehmen scheinen gerade deshalb darauf aus zu sein, solche Personen anzustellen, weil diese durch ihren rechtlichen Status besonders vulnerabel sind.)[28] Auf diese Weise könnte das Prinzip der Verteilungsgerechtigkeit eine demokratische Entscheidung gegen die Zulassung bestimmter Migrantinnen begründen.

Ich möchte nicht behaupten, dass dies falsch sei, allerdings würde ich gerne zwei mäßigende Einwände gegen diese Art von Argumenten anführen. Erstens sollte ein Argument, dass mit dem Prinzip der Verteilungsgerechtigkeit beginnt, in Betracht ziehen, dass die Kraft dieses Prinzips auf der globalen und der innerstaatlichen Ebene unterschiedlich ist. Ich glaube nicht, dass dem Prinzip der Verteilungsgerechtigkeit im Verhältnis zwischen Staaten die gleiche Bedeutung zukommt wie innerhalb eines Staates; daraus folgt jedoch nicht, dass ihm im internationalen Raum keinerlei Bedeutung zukommt. Eine Theorie, die das Recht auf Ausschluss mit Forderungen der Verteilungsgerechtigkeit begründet, müsste daher bestimmen, wie Reichtum und Armut jeweils auf globaler sowie lokaler Ebene zu verstehen sind und wie die diesen Ebenen entsprechenden Bedenken gegeneinander abgewogen werden sollten – es könnte also schwieriger sein, als wir glauben, eine einzelne Lösung für das Problem zu finden, wie die Verteilungsgerechtigkeit mittels unserer Migrationspolitik befördert werden könnte.

Daran anschließend verweist der zweite mäßigende Einwand auf die empirische Komplexität der ökonomischen Analyse. Wie bereits erwähnt, sind die tatsächlichen Auswirkungen jedweder Form des Ausschlusses von Migrantinnen nicht problemlos zu bestimmen. Gleiches gilt für jedes Aufnahmeprogramm, wie beispielsweise die Bevorzugung qualifizierter Arbeitskräfte. Um den Zusammenhang zwischen diesen Programmen und den Phänomenen der Verteilungsgerechtigkeit sicher bestimmen zu können, müssten wir sehr viel darüber wissen, wie Einwanderung auf Löhne wirkt – und zwar meiner Meinung nach mehr, als wir derzeit wissen. Allerdings können wir, wie bereits erwähnt, diese Überlegungen, nicht für den Ausschluss wahrhaft verzweifelter Personen anführen, es sei denn, wir sind willens, die eher unplausible Behauptung aufzustellen, dass dem Prinzip der Verteilungsgerechtigkeit der Vorrang vor dem nackten Überleben gegeben werden sollte.

Dies führt uns zur letzten Hinsicht, in der die Bevorzugung qualifizierter Migrantinnen ungerecht sein könnte: Sie könnte unfair gegenüber denjenigen unter den potentiellen Migrantinnen sein, die nicht über solche Qualifikationen verfügen. An dieser Stelle ist das Argument in meinen Augen mit großen Schwierigkeiten konfrontiert. An Gesetzen wie dem Travel-Ban ist die Tatsache moralisch zu verurteilen, dass sie ihre Feindseligkeit in ein Gewand scheinbar neutraler Interessen kleiden, wie beispielsweise Sicherheitsinteressen. Qualifizierten Personen den Vorrang zu geben stellt hingegen tatsächlich ein solch neutrales Interesse dar. So scheint beispielsweise eine Präferenz für qualifizierte Ärzte nicht notwendig auf Überlegungen zu beruhen, die in irgendeiner Weise mit Feindseligkeit oder Annahmen hinsichtlich einer moralischen Minderwertigkeit anderer Migrantinnen einhergehen. Es ist selbstverständlich möglich, dass unsere *Wahrnehmung* davon, wer über solche Qualifikationen verfügt, von Vorurteilen beeinflusst wird. Die Erlaubnis, diejenigen zu bevorzugen, die eine bestimmte Arbeit gut erledigen, umfasst nicht die Erlaubnis, Vorurteile in die Prognosen über zukünftige Leistungen einfließen zu lassen. (Eine Familiengeschichte kann hier als Beispiel dienen: Meine Mutter war in den 1960er Jahren Informatikerin und half während ihrer Zeit an den Jet Propulsion Labs dabei, die Mariner-Mission zum Mars zu programmieren. IBM lud sie aufgrund ihrer Arbeit zu einem Vorstellungsgespräch ein, teilte ihr anschließend jedoch mit, dass sie nicht eingestellt würde, da die Firma keine Prognosen hinsichtlich der zukünftigen Leistung einer weiblichen Informatikerin aufstellen könne.) Zudem kann eine Gesellschaft auch Vorurteile bezüglich der Frage haben, welche Tätigkeiten wirklich notwendig sind. So ist es beispielsweise durchaus möglich, dass wir uns bei der Auflistung der zur Einreise berechtigenden Qualifikationen des Sexismus schuldig machen. Wie Lori Watson gezeigt hat, besteht zudem die Möglichkeit, dass diese Qualifikationen auf globaler Ebene anhand von Vorurteilen verteilt sind.[29] Handeln Gesellschaften vorhersehbar sexistisch, kann prognostiziert werden, dass in die Ausbildung von Jungen mehr Zeit und Geld investiert wird als in die Ausbildung von Mädchen und der scheinbar neutrale Vorzug gut ausgebildeter Personen somit letztendlich, global gesehen, Sexismus verstärkt – darüber hinaus sogar noch von ihm profitiert. All das stimmt und die Überwindung des globalen Sexismus mag internationale Kooperation zwischen den verschiedenen Staaten erfordern – allerdings folgt aus all dem nicht, dass der Widerstand gegen diese Probleme bei der *Migrationspolitik* beginnen muss. Wenn ich in einer rassistischen Gesellschaft, in der allein weiße Männer Chirurgen sein dürfen, eine Blinddarmoperation benötige, ist es in meinen Augen zulässig, dass ich einen weißen Chirurgen mit der Operation beauftrage. Die Pflicht, gegen Rassismus vorzugehen, umfasst in diesem Fall nicht die Verpflichtung so zu tun, als ob die Fähigkeiten eines Chirurgen nicht selten oder nicht wertvoll seien. Genauso verhält es sich mit den Fähigkeiten, anhand derer Gesellschaften an ihren Grenzen zwischen Migrantinnen wählen.

Die Bevorzugung qualifizierter Migrantinnen kann also mitunter dazu tendieren, globale und gelegentlich auch innerstaatliche Ungerechtigkeiten zu verschärfen – aber es kann nicht behauptet werden, dass dieses Auswahlprinzip an sich auf die Art ungerecht ist, wie es Präferenzen für eine bestimmte ethnische oder religiöse Gruppe sind. Eine Migrantin, die zugunsten einer besser qualifizierten Migrantin zurückgewiesen wurde, wird nicht in dem weiter oben ausgeführten Sinne unfair behandelt. Vielmehr wird ihr ein Grund für ihren relativen Nachteil genannt, dem sie aller Voraussicht nach zustimmen kann – ohne sich dabei, ob offen oder versteckt, selbst als moralisch minderwertig ansehen zu müssen.

3. Das Veto des Heuchlers

An dieser Stelle lohnt es sich Bilanz zu ziehen. Die von mir verteidigten Prinzipien erlauben dem Staat einigen Spielraum bei der Entscheidung, welche Migrantinnen er aufnehmen möchte. Diese Prinzipien sind dem staatlichen Handeln gegenüber zumindest freundlicher eingestellt als solche Prinzipien, die offene Grenzen einfordern. Allerdings erscheinen sie im Vergleich zu der Realität, in der ich dieses Buch schreibe, als eher restriktiv. Die Gruppe an Menschen, die ein Recht auf die Überquerung von Grenzen hat – darunter auch die Grenze zwischen den Vereinigten Staaten und Mexiko – ist in meinem Ansatz bedeutend größer als unter den gegenwärtig geltenden Gesetzen. Tatsächlich könnten, sofern die Argumente des vorangegangenen Kapitels zutreffen, die Vereinigten Staaten – wie auch alle anderen Länder – eine Pflicht haben, ihre Zwangsgewalt gegen ungerechte und bösartige Regierungsformen im Ausland einzusetzen, und ein Teil dieser Anstrengungen könnte in der Pflicht bestehen, Möglichkeiten dafür zu schaffen, dass die in diesen Staaten unterdrückten Personen auch tatsächlich Zuflucht in den Vereinigten Staaten finden können. Zudem darf nach meinem Ansatz kein Land Migrantinnen anhand ethnischer oder religiöser Präferenzen auswählen, sei es offen oder unter dem Deckmantel einer Scheinbegründung. Zuletzt darf kein Land bei der Entscheidung, welche qualifizierten Personen einwandern dürfen, allein die inländischen Auswirkungen dieser Entscheidung betrachten. Wie ich gezeigt habe, folgt aus der Befürwortung des Prinzips der Verteilungsgerechtigkeit, dass sowohl globale als auch lokale Effekte der Migration hinsichtlich der Verteilungsgerechtigkeit einbezogen werden müssen. Was ich hier schreibe, ist mit Blick auf die Freiheiten des Staates also eher restriktiv. Selbst wenn es für viele liberale Philosophinnen übermäßig konservativ erscheinen mag, würde es als Programm einer politischen Partei von vielen Bürgerinnen wohl eher als übermäßig liberal angesehen werden.

Dieser Umstand birgt jedoch ein bestimmtes Problem. Damit meine ich kein philosophisches Problem – ob meine Überlegungen richtig oder falsch sind, hängt davon ab, was über die von mir vorgebrachten Argumente gesagt werden kann. Das Problem besteht vielmehr darin, dass meine Argumente zugleich richtig und in den uns bekannten Gesellschaften unmöglich umzusetzen sein könnten.

Das vergangene Jahrzehnt hat einige unangenehme Wahrheiten über liberale Demokratien offenbart. Während John Rawls – im Rückblick recht unbekümmert – von der Annahme ausging, dass denjenigen, die in einer liberalen Demokratie erzogen wurden, vertraut werden könnte, die Normen jener liberalen Demokratien zu bewahren, waren wir in den vergangenen zehn Jahren Zeugen einer erstaunlichen Wiedergeburt von Rechtspopulismus und Autoritarismus.[30] Schwache, aber dennoch existierende Demokratien wie in Ungarn oder der Türkei begannen sich so zu verändern, dass sie heute fast autoritären Ein-Parteien-Regimen gleichen. Während es übertrieben wäre, in Donald Trump einen Diktator zu sehen, hat er doch eine

Vorliebe für autoritäre Methoden gezeigt – und für autoritäre Herrscher, wie seine Freundschaften zu Rodrigo Duterte und Wladimir Putin zeigen. Meiner Meinung nach beginnen wir langsam zu erkennen, wie falsch Francis Fukuyama lag, als er 1989 das Ende der Geschichte prognostizierte. Nach dem Zerfall der Sowjetunion hätten wir das Wiederaufleben eines hässlichen und völkischen Phänomens sehen sollen und nicht eine globale Einigung auf die Regierungsform der liberalen Demokratie. Das vergangene Jahrzehnt war Zeuge einer wachsenden Bereitschaft, People of Color und ausländischen Personen die Schuld für stagnierende Löhne zu geben; eine Perspektive, die vor allem von weißen Bürgerinnen der Arbeiterklasse wohlhabender Nationen eingenommen wird. Frederick Harris und Robert Lieberman heben hervor, dass der Aufstieg autoritärer Politik in den Vereinigten Staaten genau zu dem Zeitpunkt stattfand, als die weiße Arbeiterklasse mit einem Absinken ihres Lebensstandards konfrontiert wurde; mehr als die Hälfte dieser Menschen stimmt darüber hinaus der Behauptung zu, dass Rassismus gegen Weiße genauso weit oder gar weiter verbreitet sei als Rassismus gegen Nicht-Weiße.[31]

Bei alledem handelt es sich offensichtlich nicht um philosophische Erwägungen. Allerdings können wir zeigen, dass diese Entwicklungen für die Philosophie durchaus von Bedeutung sind, wenn wir uns die Zusammenhänge genauer ansehen, die zwischen wachsender Diversität, die mit Migration einhergeht, und dem Aufstieg jener populistischen Bewegungen bestehen, die zum Widerstand gegen Migration aufrufen – sowie der Rückkehr derjenigen an die Macht, die meinten, sie verloren zu haben. Die Vereinigten Staaten sind dafür ein gutes Beispiel. So ist es kein Zufall, dass Donald Trump seinen Wahlkampf damit begann, mexikanische Migrantinnen anzuprangern. Dabei handelt es sich jedoch keineswegs bloß um ein US-amerikanisches Phänomen. Nehmen Sie zum Beispiel die skandinavischen Länder, die von liberalen Nordamerikanern wie mir traditionellerweise als Hochburgen der sozialen Gerechtigkeit betrachtet werden. In den vergangenen zehn Jahren hat die Migration aus dem Mittleren Osten und Nordafrika in diese Länder zugenommen. Das Ergebnis war eine deutliche Zunahme gesetzlicher Regelungen und politischer Parteien, die sich gegen Migrantinnen und deren Lebensformen wandten. In Dänemark wurde beispielsweise offiziell damit begonnen, einige migrantisch geprägte Gegenden als „Ghettos“ zu klassifizieren, verbunden mit der Anordnung, dass die dort aufwachsenden Kinder mehrere Stunden pro Woche getrennt von ihren Eltern verbringen sollten.[32] In Schweden konnte sich die nationalistische Sverigedemokraterna – eine von Neonazis gegründete Partei, die sich für eine Verringerung der Einwanderung nach Schweden einsetzt – am Machtkampf beteiligen.[33] Wir könnten in der Tat eine ziemlich lange Liste von Parteien und Politikerinnen aufstellen, bei denen die Angst vor dem nicht integrierbaren Ausländer die Grundlage ihres politischen Programms bildet.

Wo aber liegt hier das Problem? Meiner Meinung nach könnte aus diesen Zusammenhängen eine tragische Situation für die liberale Theorie erwachsen. Ich möchte diese Idee anhand von drei Thesen diskutieren – zwar beruhen diese Thesen auf empirischen Prämissen, allerdings gehe ich davon aus, dass ich damit die tatsächlichen Verhältnisse akkurat abbilde. Die erste These besagt, dass wir in den vergangenen zehn Jahren vermehrt Belege dafür gesehen haben, dass viele Menschen schlicht keine Politik mit Ausländern machen *wollen*. Sicherlich wird die Linie zwischen Einheimischen und Ausländern stets unterschiedlich gezogen; dass sie aber weiterhin existiert, und dass trotz unseres, zumindest den Worten nach, geteilten Bekenntnisses zum moralischen Egalitarismus, scheint kaum bestreitbar. Die zweite These besagt, dass ein Anstieg sozialer Diversität zu einem Rückgang des sozialen Vertrauens führen kann, wenn er sich vor dem Hintergrund eines erheblichen, gefühlten Widerstands gegen diese Diversität vollzieht. Das Pew Research Center hat über ein halbes Jahrhundert hinweg Daten über das Vertrauen in den Vereinigten Staaten gesammelt. Im Jahr 1958 gaben über 70% der Teilnehmerinnen an, dass sie der Regierung der Vereinigten Staate vertrauen, wohingegen im Jahr 2015 diese Zahl auf 20% gesunken war. Im Jahr 1972 meinte fast die Hälfte aller US-Amerikanerinnen, dass den meisten Menschen vertraut werden könne, wohingegen diese Zahl heute knapp unter 30% liegt und kontinuierlich abfällt.[34] Es wäre nun zu einfach, all das auf eine zunehmende gesellschaftliche Vielfalt zurückzuführen, aber die Zunahme von Diversität ist unzweifelhaft ein Faktor – ebenso wie das rassistische Ressentiment weißer Personen, die sich zunehmend als Opfer von Programmen zur Förderung ethnischer Minderheiten, inklusive Migrantinnen, sehen. Zumindest nutzen rechte Politikerinnen diesen Faktor vermehrt für ihre politischen Zwecke. Dies führt uns zur dritten und letzten meiner Thesen: Ein Verlust des sozialen Vertrauens ist deshalb zutiefst beunruhigend, weil das politische System für seine Erhaltung auf die freiwillige Kooperation einer großen Zahl von Menschen angewiesen ist. Der Rückzug selbst weniger Menschen aus diesem Kooperationszusammenhang kann das politische System zerstören oder es zumindest darin beeinträchtigen, seine Aufgaben auf angemessene Weise zu erfüllen. Gene Sharp, der gefeierte Theoretiker des zivilen Ungehorsams, beschreibt diese Idee treffend: Ungerechte Gesellschaften können ihm zufolge durch eine vergleichsweise kleine Anzahl von Menschen in die Knie gezwungen werden, wenn diese sich den Methoden und Normen verweigern, die der infrage stehende Staat eingerichtet hat. Das Problem ist jedoch, dass diese Beobachtung nicht nur dann gilt, wenn sich Aktivistinnen verweigern. Der Staat kann durch Nationalisten ebenso leicht in die Knie gezwungen werden wie durch Menschenrechtsanwälte.

All das legt meiner Meinung nach nahe, dass eine Moraltheorie der Migration sich komplizierter gestalten könnte, als wir zunächst angenommen haben. Es ist durchaus möglich, dass eine Umsetzung meiner Vorschläge zur Zunahme populistischer und gegen Migrantinnen gerichteter Bewegungen beitragen könnte. Zumindest bin ich davon überzeugt, dass das bei den stärkeren Forderungen von Kieran Oberman und Joseph Carens der Fall wäre. An einer Stelle schreibt Carens in abwertendem Ton über die Folgen steigender Einwanderungszahlen für die Innenpolitik:

> „Wir wissen, wie Einwanderung funktioniert. Trotz der vereinzelt zu vernehmenden rhetorischen Phrase, dass das Boot voll sei, kann kein demokratischer Staat in Europa oder Nordamerika behaupten, dass er nicht viel, viel mehr Migrantinnen aufnehmen könnte, als er derzeit tut, ohne dabei zusammenzubrechen oder gar ernsthaft Schaden zu nehmen."[35]

Das ist jedoch nicht der Punkt. Es geht nicht darum, ob das Boot voll ist, sondern ob diejenigen, die momentan an Bord dieses Bootes sind, auch weiterhin *rudern* werden.

Diese Bedenken gehen mit manchen Einwänden einher, die bereits David Miller vorgebracht hat. Während er allerdings in den einzelnen lokalen Gemeinschaften die legitimen Grenzen der Ethik sieht, betrachte ich die beschriebenen Verhaltensweisen schlicht als ein dem Menschen eigenes Krankheitsbild. Terry Pratchett schrieb einmal, dass der Mensch einen natürlichen Konstruktionsfehler hätte – eine Tendenz dazu, sich in den Knien zu beugen.[36] Dem würde ich einen weiteren Defekt hinzufügen: Die

Versuchung zu glauben, dass das Wort *Fremder* seiner Funktion nach dem Wort *Dieb* entspricht.

Ich bin nicht die einzige Person mit derlei Bedenken. Hillary Clinton drückte ähnliche Sorgen aus, als sie angesichts der Ankündigung Angela Merkels, sich politisch zurückzuziehen, folgendes sagte:

> „Ich denke, dass Europa die Migration in den Griff bekommen muss, denn die Migration hat dieses Feuer entfacht [...] Ich bewundere den sehr generösen und mitfühlenden Ansatz, der besonders von führenden Politikerinnen wie Angela Merkel verfolgt wurde, aber ich denke, es kann fairerweise gesagt werden, dass Europa seinen Teil geleistet hat und eine ganz klare Botschaft vermitteln muss – ‚es wird uns nicht möglich sein, weiterhin Zuflucht und Hilfe zu gewähren' –, denn wenn wir keinen guten Umgang mit dem Problem der Migration finden, wird dieses Problem das politische Gemeinwesen weiter aufwühlen."[37]

Damit will Clinton nicht diejenigen verteidigen, die jenes Gemeinwesen aufwühlen; es ist vielmehr gut denkbar, dass sie das Wiederaufleben des Rechtspopulismus in Deutschland ganz klar ablehnt. Stattdessen stellt sie bloß fest, dass Deutschland vor einer Art tragischer Entscheidung stehen könnte. Es kann den am schlechtesten Gestellten dieser Welt Gerechtigkeit widerfahren lassen – allerdings auf Kosten eben jener Institutionen, die eine solche Hilfe erst möglich gemacht haben. Oder es kann die liberale Demokratie bewahren – allerdings nur zu dem Preis, Menschen auszuschließen, die sehr gut begründete Schutzansprüche haben. Laut Clinton ist es jedoch keine Option, *gar nichts* aufgeben zu müssen. Deutschland kann den Liberalismus bewahren oder ihm folgen – allerdings nicht beides zugleich.

Sollte all das stimmen, haben die Deutschen – wie auch der Rest von uns – bloß eine tragische Wahl. Mit tragisch meine ich nicht bloß schwierig oder lästig; stattdessen meine ich die Tatsache, dass es keinen moralisch richtigen Weg zu geben scheint – also keinen Weg ohne moralische Kosten solchen Ausmaßes, dass wir zögern sollten, bevor wir sie akzeptieren. Auf der einen Seite könnten wir einfach feststellen, dass die Idee liberaler Gerechtigkeit eine ethnische oder religiöse Diskriminierung von Immigrantinnen verbietet. Wir könnten also schlicht sagen: Möge Gerechtigkeit walten, auch wenn der Himmel darüber einstürzt. (Ein philosophischer Freund drückte das in einer Antwort auf meine Ideen sehr gut aus: Unsere Aufgabe ist es, zu sagen, was ethisch wahr ist – nicht, uns um Idioten zu kümmern.) Das ist kohärent, könnte uns aber möglicherweise auch teuer zu stehen kommen. Im Ernstfall könnte es die infrage stehende Gesellschaft schwer belasten, wenn eine große Zahl an Menschen nicht mehr daran glaubt, dass dem Erhalt der Gesellschaft ein großer moralischer Wert zukommt. Wie ein altes Sprichwort feststellt, ist die Verfassung kein Selbstmordkommando; wir dürfen sie nicht in einem Sinne auslegen, der zum Zerfall der Einheit der Vereinigten Staaten führt. Das gleiche könnte jedoch auch über das Projekt liberaler Gerechtigkeit gesagt werden.

Wir könnten also eine andere Taktik wählen. Wir könnten bezüglich der Migration schlicht nicht das tun, was das Prinzip der Gerechtigkeit verlangt, da ein solches Handeln die Fähigkeit unserer Institutionen beeinträchtigen könnte, sich selbst zu erhalten und auch weiterhin, so gut es ihnen möglich ist, Gerechtigkeit herzustellen. Das würde allerdings einem Vetorecht für Heuchler gleichkommen, das den Rassisten und Nationalisten das Recht zuspräche, die Regierung dadurch von einem gerechten Handeln abzuhalten, dass sie – implizit oder anderweitig – damit drohen, die Gesellschaft unregierbar zu machen. Das ist offensichtlich inakzeptabel. Wir möchten nicht, dass sich die liberale Gerechtigkeit den Forderungen der Illiberalen beugt. Allerdings erscheint ein Liberalismus, der diese Menschen nicht berücksichtigt, gleichsam trocken und theoretisch – wie eine Spielzeug-Theorie, die in manchen Fällen nützlich sein mag, aber unfähig ist, uns in der unübersichtlichen Wirklichkeit unserer Welt Orientierung zu bieten. Wir können zwar meinen, dass Rassisten und Nationalisten falsch liegen, aber wir können nicht behaupten, dass sie nicht *existieren* – andernfalls sind unsere Theorien zur Irrelevanz verdammt.

Dieser Umstand stellt in meinen Augen eine tragische Situation dar. Keine der verfügbaren Optionen ist attraktiv– und keine lässt die entscheidende Person ohne den Makel einer ernsthaften Verfehlung. Wir könnten selbstverständlich versuchen, das tragische Element abzumildern. Wir könnten unsere Infrastruktur ausbauen, inklusive des Bildungswesens, so dass wir nicht mehr länger versucht sind, diejenigen außerhalb unserer Gemeinschaft als Wettbewerber zu sehen. Oder wir könnten versuchen Bildungsinstitutionen zu errichten, die gegen Nationalismus und Rassismus arbeiten. All das mag die Tragik mildern, aber ich denke nicht, dass wir sie vollends beseitigen können. Beide Lösungen scheinen die Aversion der Menschen gegen Ausländer im Reich der Vernunft zu verorten. Sie versuchen, die materiellen Umstände so zu verändern, dass die Menschen keine Gründe mehr haben, Migrantinnen gegenüber feindlich gesinnt zu sein; oder sie versuchen zu zeigen, dass eine solche Feindseligkeit irrational ist. Das Problem liegt allerdings meiner Meinung nach darin, dass ein Großteil dieser feindlichen Gesinnung nicht der Vernunft entspringt, sondern einer weit weniger kognitiven Quelle: Wir sind feindselig gegenüber dem, was unbekannt, schrill oder unangebracht ist und wir sind genau dann am feindseligsten, wenn wir uns unserer eigenen Verwundbarkeit am stärksten bewusst sind. Sollte es eine Möglichkeit geben, dieses Dilemma zu überwinden, wird sie nicht dem Reich der Vernunft entstammen, sondern der simplen Tatsache, dass das, was uns zunächst noch unbekannt ist, schließlich vertraut werden kann. Nehmen Sie zum Beispiel den recht bemerkenswerten Wandel der Einstellungen gegenüber homosexuellen Frauen und Männern in der zweiten Hälfte des 20. Jahrhunderts. Weit verbreiteter Hass und Missbilligung wichen etwas, das sozialer Toleranz nahekommt. Das mag natürlich nicht genug sein, aber es ist dennoch eine außergewöhnliche Entwicklung in Anbetracht ihres Ausgangspunkts. Die Erklärung für diesen Sinneswandel scheint jedoch nicht bei Argumenten, sondern bei der Tatsache der Sichtbarkeit zu beginnen: Als mehr und mehr Menschen sich öffentlich zu ihrer Homosexualität bekannten, gewöhnten sich mehr und mehr Menschen an Homosexualität und fühlten sich damit auch wohl.[38]

Falls es also eine Lösung für das soeben skizzierte tragische Dilemma geben sollte, müsste sie in meinen Augen auf solche Ideen zurückgreifen. Vertrautheit und Zeit könnten uns letztendlich akzeptieren lassen, was uns anfangs fremd und seltsam erschien. Die Empörung über den Chinese Exclusion Act liegt nicht nur darin begründet, dass er unmoralisch ist, sondern auch darin, dass er auf einer nachweisbar falschen Annahme beruht: Migrantinnen aus China waren der Verfassung gegenüber *nicht* feindlich gesinnt und die Urheber jenes Gesetzes hätten das selbst damals schon wissen müssen. Wir könnten also unsere Migrationspolitik gemäß den Forderungen der Gerechtigkeit gestalten und hoffen, dass wir später ganz ähnlich auf die Ungerechtigkeiten der Gegenwart blicken werden, darunter den Travel-Ban. Was wir allerdings nicht vermeiden können, ist der Gedanke, dass dieses Vorgehen ein gewisses Risiko mit sich bringt: Wir hoffen, um es offen zu sagen, dass Zeit und Vertrautheit stärker sind als Rassismus und Bigotterie. In meinen Augen ist diese Wette es wert, eingegangen zu werden; aber ich behaupte nicht, dass es letztendlich nicht doch bloß eine Wette ist, die durchaus auch verloren werden kann. Wir können damit schließen, dass das Projekt der liberalen Gerechtigkeit in einer Welt wie der unseren ein risikoreiches Unterfangen darstellt. Um eine gerechte Welt zu schaffen, müssen wir nicht nur tugendhaft sein, sondern auch auf unser Glück hoffen.

7
Menschen, Orte und Pläne: Über Liebe, Migration und Aufenthaltspapiere

Im letzten Kapitel habe ich untersucht, was Staaten gegenüber denjenigen, die ihre Grenze überschreiten wollen, nicht tun dürfen. Meinem Argument zufolge verlangt der Liberalismus nicht nach offenen Grenzen, weshalb Staaten manche der ungebetenen potentiellen Immigrantinnen ausschließen dürfen. Es gibt allerdings auch Personen mit einem Recht darauf, die Grenzen zu übertreten, und zwar schlicht aufgrund dessen, was ihnen innerhalb der Grenzen ihrer Herkunftsländer widerfährt. Sie verfügen nicht über das, worauf sie aufgrund ihres Daseins als menschliche Wesen ein Recht haben, und aus diesem Grund haben sie einen Anspruch gegenüber den übrigen Staaten der Welt auf etwas, das der Idee des Asyls sehr nahekommt. Diesen anderen Staaten ist es demnach zumindest nicht erlaubt, Zwang einzusetzen, um diese Gruppe von Migrantinnen am Grenzübertritt zu hindern. Die Ansprüche dieser Migrantinnen sind jedoch allgemeiner Natur, was bedeutet, dass sie sich an die Gruppe der Staaten dieser Welt als Ganzes richten und fordern, dass ihr Schicksal für alle Staaten ein Anlass zu sorgendem Handeln sein sollte. In diesem Zusammenhang werden Personen vor allem anhand jener Vulnerabilitäten und Gefahren betrachtet, die allen Menschen gemeinsam sind oder ihnen drohen können.

Manche Migrantinnen können allerdings besondere Ansprüche gegenüber bestimmten Staaten formulieren. Manchmal entstehen solche Ansprüche aus dem, was ein Staat den Migrantinnen selbst oder aber Personen, die diesen Migrantinnen sehr ähnlich sind, angetan hat. Weiter oben habe ich bereits Michael Walzers Argument erwähnt, dass durch den Vietnamkrieg vertriebene Vietnamesen ein Recht auf Einwanderung in die Vereinigten Staaten hätten. Laut Walzer wurde diese spezifische Gruppe von Migrantinnen aufgrund der Folgen US-amerikanischer Gewalt in ihrem Heimatland zu US-Amerikanerinnen.[1] In anderen Fällen wird ein solch spezifischer

Anspruch hingegen nicht aufgrund der Handlungen eines Staates postuliert, sondern aufgrund des Handelns der *Migrantinnen* oder aufgrund dessen, was sie gerne tun würden. In diesem Kapitel möchte ich mich vor allem mit dieser zweiten Form partikularer Ansprüche befassen. Manche Migrantinnen behaupten, ein Recht auf eine besondere Beziehung zu einer Gesellschaft zu haben, der sie (zumindest rechtlich gesehen) nicht angehören. Vielleicht versuchen sie zu bewahren, was sie sich in der Vergangenheit aufgebaut haben, oder wollen zukünftig ein neues Leben aufbauen. Es gibt selbstverständlich viele mögliche Formen solcher Forderungen. In diesem Kapitel werde ich nur drei dieser Formen untersuchen: Forderungen, die auf einer besonderen *Affinität* zu einem bestimmten Ort, den Rechten und Interessen *undokumentierter Immigrantinnen* oder den Geboten der *Familienzusammenführung* beruhen.

Die These dieses Kapitels ist diesen Ansprüchen gegenüber vor allem negativ: Vergleichsweise wenige dieser spezifischen Forderungen können als Grundlage für Gerechtigkeitsansprüche betrachtet werden. Staaten können daher in vielen Fällen den Neuankömmlingen die Aufnahme verweigern, selbst wenn die vorgebrachten besonderen Bindungen an den jeweiligen Staat sehr stark sein sollten. Die derart ausgeschlossenen Migrantinnen werden nicht unfair und folglich auch nicht ungerecht behandelt. Allerdings folgt hieraus nicht, wie ich betonen möchte, dass die Verweigerung der Aufnahmen in vielen solcher Fälle moralisch löblich sei. Wie ich im abschließenden Kapitel dieses Buches zeigen werde, sollte ein moralisch angemessen handelnder Staat einige politische Instrumente entwickeln, um solchen spezifischen Forderungen mit einem Recht auf Einwanderung zu entsprechen. Ein gutes Gemeinwesen sollte demnach ein legales Recht auf Familienvereinigung, ein rechtliches Programm für die Einwanderung zum Zwecke der Nutzung lokaler Ressourcen und Möglichkeiten, sowie ein Amnestieprogramm zur Legalisierung bereits länger im Land lebender undokumentierter Personen einrichten. Diese gesetzlich verankerten Rechte werden jedoch nicht durch Normen der Gerechtigkeit begründet, sondern beruhen auf der eigenständigen Norm der Gnade. Während manche derer, die um Amnestie ersuchen, auch gemäß den Normen der Gerechtigkeit einen entsprechenden Anspruch haben, sollte sie auch einer Vielzahl derer zugesprochen werden, die nicht über einen solchen Anspruch verfügen. In meinen Augen zeigen wir uns hartherzig und ungnädig, wenn wir bloß versuchen, ungerechte Handlungen zu vermeiden. Vor der Entwicklung dieses positiven Arguments möchte ich jedoch auf den folgenden Seiten zunächst einige negative Überlegungen ausführen. Demnach stellen viele, uns instinktiv sympathische Forderungen nach einem Recht auf Einwanderung keine Ansprüche im Sinne der Gerechtigkeit dar. Sollte mir nun Hartherzigkeit vorgeworfen werden, möchte ich erneut betonen, dass das, was ich im Folgenden schreibe, nicht darauf abzielt, die moralische Bedeutung der diskutierten Forderungen zu untergraben. Vielmehr soll es die Behauptung widerlegen, dass solche Forderungen ausschließlich oder idealerweise anhand der Idee der Gerechtigkeit interpretiert werden sollten.

1. Affinität zu einem Ort

Wir können mit der konzeptionell einfachsten, aber auch unrealistischsten der von mir genannten partikularen Forderungen beginnen. Stellen Sie sich vor, dass eine Person in ein neues Land auswandern möchte, da sie in diesem Land Umstände vorfindet, die für ihre Vorstellung von einem gelungenen Lebens von größter Bedeutung sind. Ihre Pläne setzen mitunter die physische Beschaffenheit des Ziellandes voraus; ihre Pläne sind in diesem Sinne *verortet*, wie Anna Stilz schreibt, wobei diese verorteten Lebenspläne sich auf ein bestimmtes Hoheitsgebiet beziehen könnten, für das die Person keine rechtliche Aufenthaltserlaubnis besitzt.[2] Wenn ich beispielsweise leidenschaftlich gerne ein Ozeanologe werden möchte, allerdings nur über die Staatsbürgerschaft eines Landes verfüge, das keinerlei Zugang zum Meer bietet, könnte ich um das Recht ersuchen, in ein bestimmtes Land mit einem solchen Zugang einwandern zu dürfen. Andere Personen hingegen könnten die besonderen Institutionen und Beziehungen nutzen wollen, die ein bestimmtes Land bietet. Vielleicht würde es mir dort erlaubt sein, meinen religiösen Pflichten entsprechend meiner persönlichen Vorstellungen nachzukommen. Vielleicht *möchte* ich aber auch einfach nur an einen anderen Ort ziehen. Schriftsteller möchten zum Beispiel oft in Paris leben und wir könnten diese Tatsache selbst als Grundlage für den Anspruch betrachten, dass Paris für den Zuzug solcher Personen offen sein sollte. In einem weniger romantischen Sinne wollen viele Akademikerinnen aufgrund von Bildungs- oder Karrierechancen in andere Länder migrieren. So wollte ich für das Verfassen meiner Doktorarbeit aus dem spezifischen Grund nach Kalifornien ziehen, dass Debra Satz dort tätig war (und ist), weshalb sowohl meine Ausbildung als auch meine Karriere gelitten hätten, wäre ich an diesem Umzug gehindert worden.[3]

Wir werden uns später damit beschäftigen, welchen Unterschied es macht, falls ich bereits damit begonnen habe, mir ein Leben an einem bestimmten Ort aufzubauen. Im folgenden Abschnitt werde ich jedoch bloß untersuchen, welche Ansprüche ich aufgrund meines Wunsches formulieren kann, dass mein zukünftiges Leben eine bestimmte Form annimmt. Rufen wir uns erneut Kieran Obermans Unterscheidung zwischen Verbundenheit und Möglichkeit ins Gedächtnis, wobei Verbundenheit sich auf das bezieht, was bereits aufgebaut wurde und Möglichkeit auf das, was bisher bloß einen Wunsch darstellt. An dieser Stelle untersuchen wir eine (bloße) Möglichkeit. Könnten wir daraus, dass ich ein starkes Bedürfnis habe, nach Kalifornien zu ziehen – wenn ich also tatsächlich zeigen könnte, dass mein Leben ohne ein Recht darauf, nach Kalifornien zu ziehen, fundamental schlechter wäre als im Falle der Existenz eines solchen Rechts – folgern, dass ich ein moralisches Recht auf diesen Umzug habe?

Meine Antwort lautet, wenig überraschend, nein. Die Diskussion ähnelt dem, was bereits in Kapitel zwei über Obermans Ansatz gesagt wurde. Eine Person hat allein dann ein *Recht* auf eine solche Art von Einwanderung – es wäre dann also ungerecht, diese Person an der Einwanderung zu hindern –, wenn sie ein Recht auf das größtmögliche Bündel von Möglichkeiten und Chancen hätte; und wie ich bereits weiter oben dargelegt habe, denke ich nicht, dass ein solches Recht existiert. Vor allem habe ich nicht in jedem Fall ein Recht darauf, bei der Umsetzung meiner persönlichen Ziele von anderen unterstützt zu werden. Durch meinen Umzug nach Kalifornien habe ich nicht nur eine Verbindung zu Debra Satz aufgebaut (und im Übrigen auch zu den Studierenden und Mitarbeiterinnen am Institut für Philosophie). Darüber hinaus habe ich auch eine besondere Verbindung zu den Menschen aufgebaut, die den Gesetzen des Staates Kalifornien unterworfen sind – ganz zu schweigen von den Menschen, die den Gesetzen der Vereinigten Staaten unterworfen sind. Dabei hat die Intensität meines Wunsches, Philosoph zu werden, keinerlei Verpflichtungen aufseiten der Bürgerinnen Kaliforniens begründet, mich einzulassen. Wenn sie mir die Einwanderung verweigert hätten – wenn sie sich beispielsweise dazu entschieden hätten, einen Gesetzgeber zu wählen, der das F-1 Visa-Programm einstellt – hätte ich nicht behaupten können, dass sie auf ungerechte Weise meine Freiheit hinsichtlich meines philosophischen Denkens oder meines Bildungswegs eingeschränkt hätten.

Zudem möchte ich anmerken, dass es, ganz unabhängig von der Bedeutung dieser Freiheiten, im Allgemeinen nicht ungerecht ist, wenn von mir bisweilen verlangt wird, meine Wünsche an die gegebenen Umstände anzupassen. Wir können das anhand geschichtlicher Entwicklungen nachvollziehen. Die Vereinigten Staaten des 21. Jahrhunderts bieten beispielsweise Ausrufern, Hirten oder Lampenanzündern keine besonderen Möglichkeiten für die Ausübung ihres Berufes. Und es stellt keine Verletzung meiner Berufsfreiheit dar, wenn mir gesagt wird, dass ich meine Ambitionen an die gegenwärtig verfügbaren Möglichkeiten anpassen solle. Das gleiche scheint nun jedoch nicht nur in zeitlicher, sondern auch in räumlicher Hinsicht zu gelten. Sollte es eine politische Gemeinschaft geben, in der es möglich wäre, als Ausrufer zu arbeiten, würde aus dieser Tatsache nicht folgen, dass ich ein moralisches Recht darauf hätte, Teil dieser Gemeinschaft zu werden – wie gerne auch immer ich diesen Beruf ausüben würde. Die Intensität meiner subjektiven Wünsche in Bezug auf diese Berufe begründet demnach keine Pflicht aufseiten der Bürgerinnen dieser Gemeinschaft, mit mir in eine politische Beziehung zu treten.

All das ist uns selbstverständlich recht vertraut, schließlich wiederholt es bloß, was ich bereits zuvor gesagt habe. Es ist jedoch nützlich, sich diese Schlussfolgerungen ins Gedächtnis zu rufen, bevor wir uns komplexeren – aber realistischeren – Fällen partikularer Migrationsansprüche zuwenden. In allen Fällen jedoch müssen die vorgebrachten Forderungen, welcher Art auch immer sie sind, auf anderen Faktoren als der bloßen Intensität von Wünschen beruhen. Sofern es Gerechtigkeitserwägungen sein sollen, die ein Recht auf Einwanderung begründen, muss also mehr gesagt werden, als dass dieses Recht *begehrt* wird.

2. Undokumentierte Immigrantinnen

Unter dem Begriff *undokumentierter Immigrantinnen* verstehe ich Personen, die in einer politischen Gemeinschaft leben, ohne dabei über ein Recht auf Aufenthalt in dieser Gemeinschaft zu verfügen. Die Rechte undokumentierter Immigrantinnen sind, um es vorsichtig zu formulieren, Gegenstand einiger politischer Kontroversen, zumindest über die letzte

Dekade hinweg. Die rechtspopulistischen Bewegungen in Nordamerika, Europa und Australien haben immer wieder die Bedeutung der Sicherheit an den Grenzen hervorgehoben und dabei undokumentierte Einwanderer als gierige Ausländer beschrieben, die den verdienstvollen Einheimischen die Arbeit wegnehmen. Der Aufstieg des Populismus hat dabei seinen eigenen Gegenentwurf in Form einer sich entwickelnden politischen Bewegung für die Rechte undokumentierter Immigrantinnen hervorgebracht, die insbesondere in den USA für das Recht auf soziale Gleichheit einsteht. „Wir sind undokumentiert und unerschrocken", so ihr Slogan.

Dieser recht grobe Überblick verdeckt selbstverständlich die Vielfalt an Positionen, wie auch die Debatte über die Rechte undokumentierter Immigrantinnen selbst dazu tendiert, viele potentiell relevante Unterscheidungen zu übersehen. Undokumentierte Immigrantinnen existieren schließlich nicht als bloße Abstraktionen, sondern als konkrete Individuen innerhalb konkreter Gesellschaften, von denen eine jede ihre eigene Geschichte und ihren eigenen Umgang mit eben dieser Geschichte hat. Zudem bezeichnet die abstrakte Kategorie der undokumentierten Immigrantinnen schlicht die Tatsache, dass Personen ohne legale Erlaubnis innerhalb einer politischen Gemeinschaft anwesend sind, und unterscheidet daher selbst nicht zwischen Fällen, die wahrscheinlich recht unterschiedliche moralische Reaktionen hervorrufen. Nehmen Sie auf der einen Seite zum Beispiel Angela Luna, eine Bürgerin der Vereinigten Staaten, die im Jahr 2004 ihr japanisches Visum um zwei Wochen überzog und sich dadurch in den Augen der japanischen Regierung eines unerlaubten Aufenthalts und somit eines Verbrechens schuldig machte. Luna wurde über mehrere Stunden in einem japanischen Gefängnis festgehalten und anschließend wurde ihr untersagt, in den nächsten fünf Jahren nach Japan einzureisen. Sie selbst bezeichnete die Strafe als überzogen:

> „Wir wurden wie Kriminelle behandelt. […] Ich bin wütend, vor allem da Vieles ein Ausdruck politischen Willens und kein Fehler im bürokratischen System war. Die Bestrafung war dem Verbrechen sicherlich nicht angemessen."[4]

Im selben Jahr ertranken mindestens 23 undokumentierte chinesische Arbeiterinnen in Morecambe Bay beim Sammeln von Herzmuscheln. In Morecambe Bay herrschen tückische Gezeiten. Da die chinesischen Arbeiter kaum Englisch sprachen, verstanden sie die Warnungen der anderen Muschelsammlerinnen nicht, die sie auf die Gefahr hinwiesen, von der Küste abgeschnitten zu werden. Die chinesischen Arbeiter hatten Menschenhändler mit Verbindungen zur Snakehead Gang bezahlt, um von der Provinz Fujian an die Küste Englands gebracht zu werden – bloß um dann festzustellen, dass sie nach ihrer Ankunft quasi wie Sklaven behandelt wurden. Bei Fujian handelt es sich um eine der ärmsten Provinzen Chinas, weshalb die geschmuggelten Männer nach einem Einkommen für ihre in Armut verbliebenen Familien suchten. Mit Li Hua lebt momentan einer der wenigen Überlebenden dieses Ereignisses im Rahmen eines Zeugenschutzprogramms im Vereinigten Königreich. Er war bereit, vor Gericht gegen die Köpfe der Gang auszusagen, denen die Arbeiter ausgeliefert waren, weshalb sein Leben nach wie vor in Gefahr ist.[5]

Ein Problem besteht nun darin, dass unter die Kategorie undokumentierter Immigrantinnen sowohl Li Hua als auch Angela Luna fallen – zwei Personen, für die wir in meinen Augen einen vollkommen unterschiedlichen Grad an Sympathie hegen. Li Hua bezahlte Menschenhändler, um in das Vereinigte Königreich zu gelangen, da in Fujian drückende Armut und ein Mangel an Alternativen zur Auswanderung herrschten. Angela Luna hingegen entschied sich eher willkürlich dafür, ohne Erlaubnis in einem Land zu bleiben und dadurch, zumindest für zwei Wochen, als undokumentierte Immigrantin zu leben. Wenn wir eine angemessene Theorie der Gerechtigkeit für die Personengruppe der Undokumentierten entwickeln wollen, wird sie die Tatsache miteinbeziehen müssen, dass die Kategorie *undokumentierter* Immigrantinnen für uns nicht ausreicht, um belastbare moralische Aussagen treffen zu können.

Ein Unterschied zwischen diesen beiden Fällen besteht offensichtlich im Hinblick auf die simple Tatsache der *Zeit*. Angela Luna verbrachte bloß zwei Wochen undokumentiert in Japan, wohingegen Li Huas undokumentierter Aufenthalt deutlich länger andauerte. Joseph Carens greift in seinen Überlegungen zu einer Ethik der Amnestie auf diesen Gedanken zurück. Ihm zufolge ist das Recht darauf, in einem Land bleiben zu dürfen – selbst dann, wenn die Person ohne ein Recht auf Einwanderung in dieses Land gekommen ist – durch die Tatsache der von ihm so genannten *sozialen Mitgliedschaft* begründet. Demnach bauen sich die an einem Ort lebenden Personen ein eigenes Leben an eben diesem Ort auf, und mit der Zeit lassen sich dieser

Ort und dieses Leben immer weniger klar voneinander unterscheiden. Die Bindungen, die wir aufbauen, gewinnen für unsere Identität eine solch zentrale Bedeutung, dass sie bewahrt werden sollten – selbst dann, wenn wir ursprünglich kein Recht hatten, diese Bindungen aufzubauen:

> „Innerhalb der Spanne eines menschlichen Lebens sind 15 Jahre eine lange Zeit. In 15 Jahren wachsen Bindungen: zu Ehegattinnen und Lebenspartnern, Söhnen und Töchtern, Freundinnen und Nachbarn, Mitarbeiterinnen, Menschen, die wir lieben und solchen, die wir hassen. [...] Wir schlagen in 15 Jahren tiefe Wurzeln und diese Wurzeln sind auch dann von Bedeutung, wenn es uns ursprünglich nicht erlaubt war, uns an diesem Ort niederzulassen. Die moralische Bedeutung der eigenen sozialen Mitgliedschaft sollte die Bedeutung überwiegen, die der Durchsetzung unserer Einwanderungsgesetze zukommt."[6]

Um sein Argument zu stützen, beschreibt Carens den Fall von Miguel Sanchez, dem in seiner Heimatstadt Armut und Polizeigewalt drohten und der nun ohne einen legalen Aufenthaltstitel in den Vereinigten Staaten lebt.[7] Laut Carens hat Sanchez ein Recht auf Verbleib, da das von ihm gegenwärtig geführte Leben – aufgrund der beständig drohenden Abschiebung – moralisch inakzeptabel sei. Sanchez sei faktisch ein Mitglied der Vereinigten Staaten und was er sich in diesem Land aufgebaut hat, müsse respektiert werden. Carens schließt daraus, dass eine Abschiebung nach Ablauf einer bestimmten Aufenthaltszeit – er schlägt drei bis fünf Jahre vor – inhärent ungerecht sei.

Dieses Argument ist stark, aber durchaus nicht die einzig mögliche Verteidigung. So hat beispielsweise Adam Hosein argumentiert, dass neben der Aufenthaltsdauer auch die individuelle Handlungsfähigkeit ein zentraler Faktor für die Begründung der Rechte undokumentierter Immigrantinnen darstelle.[8] Ein bedeutender Teil seiner Überlegungen besteht in dem Argument, dass ein Staat seine mit Zwangsgewalt bewehrten Gesetze gegenüber all denjenigen rechtfertigen muss, die durch diese Zwangsgewalt potentiell betroffen sein könnten – ob sie nun im Besitz von Aufenthaltspapieren sind oder nicht. Aus einem solchen Respekt vor der Handlungsfähigkeit undokumentierter Einwanderer folgt für Hosein, dass der Staat ihnen die Möglichkeit bieten muss, Pläne machen und in dem Wissen leben zu können, dass ihnen nicht permanent die plötzliche und unerwartete Zerstörung all dessen droht, was sie sich aufgebaut haben. Hosein teilt mit Carens den Gedanken, dass ein Leben unter dem Damoklesschwert der Abschiebung vermutlich ungerecht ist, da keine Person mit der ständigen Drohung leben sollte, dass ihr all das genommen wird, was sie liebt. Wo Carens allerdings schlicht den Schmerz der Abschiebung betont, argumentiert Hosein damit, dass die beständige Drohung der Abschiebung gegenüber undokumentierten Personen ungerecht sei. Keine Person, so Hosein, könne gerechterweise mit sogenannten willkürlichen Einwanderungsrechten ausgestattet werden; diesem Gedanken zufolge handelt es sich um eine Verletzung der Rechte einer Person, wenn es ihr zunächst gestattet wurde, an einen Ort zu ziehen und sie sich anschließend der permanenten Drohung ausgesetzt sieht, abgeschoben zu werden. Hosein erkennt an, dass die Immigrantin auf diese Rechte auch verzichten kann – allerdings hat ein solcher Verzicht allein dann moralische Bedeutung, wenn er fair und frei erfolgt ist. So war beispielsweise Li Huas Entscheidung zur Auswanderung weder fair noch frei. Auch könnten wir, so Hosein weiter, eine Abschiebung nicht einfach mit dem Rauswurf einer Person gleichsetzen, die unbefugt unser Eigentum betreten hat. Der Staat beanspruche, Menschen durch Zwangsgewalt zu regieren, wodurch er sich prinzipiell von Grundbesitzern unterscheide; aus diesem Grund könne er den ihm zukommenden Anspruch auf die Ausübung dieser Zwangsgewalt solange nicht erheben, wie er sich das Recht vorbehält, bereits seit langer Zeit ansässige Einwohnerinnen abzuschieben.

Es gibt noch eine Reihe weiterer Argumente, die zur Verteidigung eines Rechts auf Amnestie für seit Langem ansässige, undokumentierte Einwanderinnen vorgebracht werden können. Hosein selbst diskutiert beispielsweise Argumente, die auf den volkswirtschaftlichen Beitrag dieser Personengruppe verweisen oder den moralischen Imperativ anführen, gegen Kastenstrukturen vorzugehen. Im Kontext dieses Buches werde ich mich jedoch nur auf die soeben dargestellten Argumente der *Autonomie* sowie der *sozialen Mitgliedschaft* konzentrieren, da diese in meinen Augen die besten Aussichten auf Erfolg haben. Ich denke jedoch nicht, dass eines dieser Argumente die Forderung rechtfertigen kann, dass undokumentierten Personen – nach einer angemessenen Zeit des Aufenthalts – ein Recht auf Verbleib zusteht.

Ich möchte jedoch mit der Bemerkung beginnen, dass ich, wie Hosein, Li Huas Auswanderung als Resultat wahrhafter Verzweiflung betrachte – im

Gegensatz zu Angela Lunas Fall, bei der es sich eher um eine freiwillige Entscheidung handelte. Li Huas Armut mag in der Tat so gravierend gewesen sein, dass ihm ein Recht auf Einwanderung zur Suche nach wirtschaftlichen Chancen zustand, um für die Versorgung seiner Familie aufkommen zu können. Obwohl ich das nicht wirklich beweisen kann, denke ich doch, dass ihm dieses Recht mit großer Wahrscheinlichkeit zukam, insbesondere angesichts der schrecklichen Bedingungen, die er für seine Flucht in Kauf nahm. Allerdings bin ich mir weniger sicher als Hosein, dass es prinzipiell nicht möglich ist, zu behaupten, wir hätten durch einen Grenzübertritt ohne entsprechendes Einwanderungsrecht freiwillig unsere Rechte darauf verwirkt, bestimmte Pläne machen zu können.

Stellen Sie sich beispielsweise vor, Angela Luna wäre für fünf Jahre statt für zwei Wochen in Japan geblieben. (Das ist kein abstruser Fall: Thailand verkündete erst vor Kurzem die Festnahme eines deutschen Bürgers, der die Frist seines Touristenvisums um 19 Jahre überschritten hatte.)[9] Was wäre in diesem Falle falsch daran, Angela Luna abzuschieben? Die Abschiebung wäre sicherlich unerfreulich, so viel ist klar. (Genauso unerfreulich war es wohl auch für den deutschen Mann, der in sein Heimatland zurückkehren musste.) Aber die Tatsache, dass meine Pläne zunichte gemacht werden, stellt keine Ungerechtigkeit dar, sofern diese Pläne auf Beiträge anderer Menschen angewiesen sind, zu denen diese Menschen wiederum nicht verpflichtet sind. Stellen Sie sich zum Beispiel vor, dass ich ein hübsches Wandgemälde an der Seite ihrer Scheune anbringe. Dieses Gemälde ist ein Meisterwerk und die Linie zwischen ihm und mir ist (in einem metaphorischen Sinne) nicht klar zu ziehen. Stellen Sie sich zudem vor, dass Sie mir das Malen auf Ihrer Scheune nicht erlaubt haben. Sie beginnen nun damit, mein Meisterwerk zu übermalen. Werde ich dadurch ungerecht behandelt? Ich kann nicht sagen, warum das der Fall sein sollte. Sie sind in meinen Augen ungnädig, aber nicht ungerecht. Ich wusste, dass sie jedes Recht dazu haben, dieses Werk zu übermalen, oder zumindest hätte ich es wissen sollen. Zudem handelt es sich dabei nicht um eine Bestrafung. Sie erteilen mir keine Lektion oder fügen mir irgendwelche Schmerzen zu. Der Schmerz kommt schlicht dadurch zustande, dass Sie etwas tun, wozu Sie jedes Recht haben: mich und mein Werk von dem auszuschließen, was Ihnen gehört, und nicht mir. Wenn Sie mich durch vage Versprechen, mein Werk zu erhalten, dazu veranlasst hätten, Ihr Gelände zu betreten, könnte ich mitunter daraus einen Anspruch auf den

Erhalt meines Gemäldes ableiten. Aber ich denke, dass selbst ein solcher Anspruch kaum gelten würde – es sei denn, Sie hätten deutlich mehr getan als mir ein paar vage Versprechen zu machen. Wenn Sie und ich uns der Tatsache bewusst waren, dass Sie mir kein Recht auf das dauerhafte Anbringen meines Werks an Ihrer Scheune gegeben haben, denke ich nicht, dass mir durch die Zerstörung meines Werkes irgendein Unrecht widerfahren würde.

Etwas Ähnliches gilt in meinen Augen auch für künstlerisch (oder monetär) sehr wertvolle Werke. Im Jahre 2009 wurde Banksys *One Nation Under CCTV* – das auf der Seite eines Regierungsgebäudes angebracht wurde – auf Veranlassung des Westminster City Council übermalt. Der Vorsitzende des Rats gab an, dass der Rat „jedes Recht" auf die Zerstörung des Kunstwerks habe, da Banksy keine Genehmigung für das Anbringen des Kunstwerks an der Seite des Gebäudes gehabt hätte.[10] Viele von uns mögen sich vielleicht wünschen, dass der Rat diese Genehmigung nachträglich erteilt hätte; aber ich denke nicht, dass der Vorsitzende bei der Bewertung der Rechte des Rates falsch lag.

Was im Falle von Eigentum gilt, scheint jedoch auch auf andere Formen menschlicher Beziehungen anwendbar. Wenn ich unsere Beziehung durch eine Verletzung Ihrer Rechte herstelle, kann sie zu Recht als eine von Beginn an beschädigte Beziehung betrachtet werden. Stellen Sie sich beispielsweise vor, dass ich durch gewiefte Täuschungen an meine Stelle als Professor gelangt bin – ich habe meinen Abschluss gefälscht, den Packen glühender Empfehlungsschreiben selbst geschrieben und so weiter. Nun zeigt sich nach fünf Jahren, dass ich ziemlich gut darin bin, die Aufgaben eines Professors zu erfüllen: Ich veröffentliche Artikel, unterrichte, nehme an akademischen Sitzungen teil und so weiter. Falls mein Betrug nach diesen fünf Jahren entdeckt werden sollte, ist es meiner Meinung nach ziemlich offensichtlich, dass ich entlassen werden sollte, und das auch verdient hätte. Mir stand diese Stelle niemals zu und die Tatsache, dass diese Entlassung meine Zukunftspläne beeinträchtigt, erscheint mir als größtenteils irrelevant.[11] Aus Hoseins Argument scheint hingegen zu folgen, dass das Bedürfnis dauerhafter Planungssicherheit so bedeutsam ist, dass selbst unrechtmäßige Beziehungen auf Dauer gestellt werden sollten – und das ist in meinen Augen schlicht falsch.

Man könnte nun natürlich erwidern, dass diese Beispiele am eigentlichen Punkt vorbeigehen. Das zentrale Element in Hoseins Argument sei demnach nicht, dass ein Anspruch darauf besteht, alle Vorhaben auf Dauer verfolgen zu dürfen, sondern dass eine Regierung, die sich als gerecht versteht, Umstände schaffen muss, unter denen es möglich ist, die eigenen Vorhaben dauerhaft zu verfolgen. Allerdings sagen uns die genannten Beispiele meiner Meinung nach bloß, dass diese Pläne, sofern sie ohne eine entsprechende Rechtsgrundlage zustande gekommen sind, durchaus zunichte gemacht werden können, ohne dass dabei ein Unrecht geschieht. Ein Staat, der Angela Luna abschiebt, tut dies ohne sie ungerecht zu behandeln, selbst wenn er sie erst nach einigen Jahren aus dem Land verweisen würde. Hosein könnte einwenden, dass ein derart handelnder Staat Angela Luna dennoch ungerecht behandelt – da er es ihr unmöglich macht, ihre Pläne auf Dauer zu verfolgen. Darauf gibt es nun mindestens zwei Antworten. Zum einen ist der Staat fähig – und in der Tat sogar verpflichtet –, Angela Luna gesetzlichen Schutz zu gewähren, während sie sich innerhalb seines Hoheitsgebietes aufhält. Sie ist vorübergehend ein Mitglied der politischen Gemeinschaft dieses Staates, auch wenn sie sich ohne Aufenthaltsrecht innerhalb seiner Grenzen befindet; um sein Handeln ihr gegenüber rechtfertigen zu können, muss der Staat ihre Menschenrechte schützen und gewährleisten, was allerdings nicht bedeutet, dass diese Rechte ein Wahlrecht oder ein Recht auf unbefristeten Aufenthalt umfassen. Zum anderen kann der Staat auch schlicht anführen, dass es in Lunas Macht *stand,* ihre Pläne auf Dauer zu verfolgen. Sie hätte in den Vereinigten Staaten bleiben und ihre Pläne dort verfolgen können. Was sie sich in Japan aufgebaut hat, wurde – gleich dem was ich mir an der Seite Ihrer Scheune aufgebaut habe – ohne eine entsprechende Rechtsgrundlage errichtet und darf daher zerstört werden. Wenn Angela Luna – oder mir – das nicht gefällt, wäre es bloß nötig gewesen nicht zu nehmen, was ihr – oder mir – nicht zusteht.

Wir können an diesem Punkt auf Carens' Analyse sozialer Mitgliedschaft zurückkommen. Carens beschreibt, in welchen Hinsichten eine beständig drohende Abschiebung ziemlich schmerzhaft sein kann: Die betroffene Person kann nicht einfach reisen, hat Angst bei Verkehrskontrollen und so weiter. Diese schmerzhaften Erfahrungen sind jedoch nicht notwendigerweise aufgrund dieses Schmerzes auch zugleich ungerecht. Carens nutzt zur Illustration das Beispiel von Miguel Sanchez, der, wie bereits geschildert, in seiner mexikanischen Heimat seinen Lebensunterhalt nicht bestreiten konnte und aus diesem Grund ohne Erlaubnis in die Vereinigten Staaten emigrierte.

Miguel Sanchez' Vulnerabilität innerhalb der Vereinigten Staaten ist laut Carens ungerecht, denn warum sollte er ein Leid erdulden müssen, dem seine einheimischen Nachbarn nicht ausgesetzt sind?

Es mag durchaus stimmen, dass Miguel Sanchez über die Ansprüche verfügt, die Carens ihm zuspricht; aber ich glaube, dass er uns hierfür mehr über die Umstände sagen muss, aus denen heraus Miguel Sanchez emigriert ist. Um zu sehen, warum das der Fall ist, lassen Sie uns den realen Fall von Morgan untersuchen, einem undokumentierten kanadischen Migranten, der sich in den Vereinigten Staaten aufhält. Er lebt nach eigenen Angaben seit mittlerweile fast zehn Jahren in Portland, bekommt sein Gehalt bar auf die Hand und bezahlt keine Einkommenssteuer. Er ist sich der Gefahr einer Abschiebung bewusst, zieht es aber vor, dieses Risiko aufgrund seiner Affinität zu Portland, Oregon, einzugehen. Er gibt zu, bei Verkehrskontrollen nervös zu sein: „Ich bin immer ein wenig aufgekratzt." Aber er hegt keine Pläne, die Vereinigten Staaten wieder in Richtung Kanada zu verlassen. Seine US-amerikanischen Freunde argumentieren zudem, dass er ein Recht darauf habe, in den USA zu bleiben, und klingen in ihrer Begründung ganz ähnlich wie Carens:

> „[Morgans] Freundin Amber Whittenberg sagt, dass Morgans Gründe [für einen Verbleib] sogar noch schlichter sind. „Es liegt daran, dass er hier zu Hause ist," sagt sie. „Er ist hier, weil Portland sein Zuhause ist." [...] Sein Freund Eric Roser sagt, dass Morgan sich seine Träume auch in Kanada erfüllen könnte. „Im Endeffekt geht es um die Tatsache, dass er hier sein will", sagt Roser. „Hat er deshalb weniger Rechte als beispielsweise ein Flüchtling? Ja. Aber bedeutet dies, dass er nicht hier sein sollte? Ich denke nicht. Ich denke vielmehr, dass es hier genug Platz für ihn gibt."[12]

Meine eigene Reaktion auf Morgan ist jedoch eine ganz andere. In meinen Augen hat Morgan schlicht gewettet: Er wird sich ein Leben in Portland aufbauen und wenn ihm das rechtlich nicht gestattet ist, wettet er darauf, nicht abgeschoben zu werden. Dabei handelt es sich um das, was Ronald Dworkin *kalkuliertes Glück* genannt hat: Eine Person geht ein Risiko ein und der Liberalismus erlaubt es ihr zum einen, die daraus resultierenden Gewinne zu behalten, verpflichtet sie jedoch zum anderen auch dazu, die möglichen Kosten zu tragen.[13] Laut Dworkin sollte der Liberalismus die Verluste von

Menschen nur dann kompensieren, wenn ihnen etwas aus *reinem Zufall* widerfährt, wenn ihnen also schlechte Dinge widerfahren oder sie Leiden ausgesetzt sind, die durch bestimmte Umstände oder die Entscheidungen Dritter entstanden sind. Der Liberalismus sollte jedoch keine Kompensation leisten, wenn Personen eine Wette eingehen und sich letztlich nicht die Situation einstellt, auf die sie gesetzt haben. Der Professor ohne Doktortitel geht ebenfalls ein Risiko ein und es scheint (zumindest mir) ziemlich eindeutig, dass die Kosten dieses Risikos zu Recht von ihm getragen werden müssen. So ist es auch im Fall von Morgan. Innerhalb des Liberalismus wird es zumindest schwerfallen, zu zeigen, dass Morgan aufgrund des Leids einer Abschiebung *ungerecht* behandelt würde. Selbst wenn er dabei psychische Schmerzen empfinden würde, die seine einheimischen Nachbarn nicht ertragen müssen, würde sich in diesem Unterschied keine Unfairness manifestieren: Er ging ein bestimmtes Risiko ein und seine Nachbarn nicht.

All das zielt selbstverständlich nicht darauf ab, zu zeigen, dass undokumentierte Immigrantinnen keine Gerechtigkeitsansprüche formulieren können. Diese Überlegungen nötigen uns stattdessen dazu, herauszufinden, wie diese Forderungen korrekt begründet werden können – und eine Grundlage für diese Ansprüche zu finden, die über die bloße Aufenthaltsdauer hinaus auf eine Analyse der Umstände abzielt, unter denen die Entscheidung zum Grenzübertritt ursprünglich getroffen wurde. Ich glaube, dass die Beispiele von Morgan und Angela Luna uns zwingen, genauer über das Wesen der vorgebrachten moralischen Ansprüche nachzudenken. Allerdings glaube ich *nicht*, dass die Situationen aller – oder gar der meisten – undokumentierten Personen mit den Umständen in den Fällen von Morgan oder Angela Luna vergleichbar ist. Sie ähneln vielmehr derjenigen von Miguel Sanchez oder Li Hua. Das aber ist genau der Punkt: Wir müssen in unserer Analyse etwas tiefer gehen, bevor wir angemessen bestimmen können, was die Gerechtigkeit in den verschiedenen Fällen von uns verlangt.

Lassen Sie uns also schauen, was in Morgans Fall tatsächlich zutrifft. Er ist ein autonomer Akteur, oder zumindest können wir ihn als die Art Geschöpf betrachten, der im allgemeinen Verantwortung für ihre Entscheidungen zugesprochen werden kann. Er kommt aus Umständen, die aus einer moralischen Perspektive vollkommen angemessen sind. Zwar gefällt ihm Kanada anscheinend nicht besonders, aber eine derart begründete Auswanderung kommt nicht einer Flucht aufgrund gravierender Armut oder der Verweigerung grundlegender Menschenrechte gleich. Zudem würden im Falle der Abschiebung zwar seine persönlichen Pläne vereitelt, jedoch nicht seine Fähigkeit, Pläne zu machen und zu verfolgen. Er würde sehr vieles verlieren, aber er wäre weiterhin dazu fähig, als ein Mensch unter Menschen zu leben; und das in einer Gesellschaft, die ihm angemessene Möglichkeiten dafür bietet, seine persönlichen Pläne zu entwickeln.

In meinen Augen fehlen alle drei Aspekte von Morgans Fall sehr häufig, wenn wir die Situation einzelner undokumentierter Immigrantinnen betrachten – was zur Folge hat, dass die soeben dargestellte Rechtfertigung von Abschiebungen für diese Personen schlicht nicht gilt. Zunächst ist Morgan ein vollwertiger moralischer Akteur. Das trifft nicht auf alle Menschen zu. So nehmen insbesondere Kinder eine eigenständige – und wenig theoretisch bearbeitete – Position innerhalb unseres moralischen Denkens ein. Sie sind zwar Geschöpfe mit Rechten, verfügen jedoch zugleich nicht im vollen Maße über diejenigen kognitiven Fähigkeiten, die es uns erlauben würden, ihnen ihre Entscheidungen vorzuhalten. Zwar unterscheiden sie sich untereinander hinsichtlich des Grades, zu dem sie über diese Fähigkeiten verfügen – ein 15-jähriger Jugendlicher ähnelt kaum einem 15 Tage alten Säugling. Allerdings wird zu Recht in keinem der beiden Fälle davon ausgegangen, dass sie über diejenige moralische Freiheit verfügen, die wir erwachsenen Personen im Allgemeinen zuschreiben. Meiner Argumentation zufolge wäre es nicht unfair, Morgan abzuschieben, da er die Entscheidung getroffen hat, sein Glück auf die Probe zu stellen. Sollte er Pech haben, ist der daraus resultierende Schmerz aus diesem Grund nicht unfair ihm gegenüber. Im Falle einer Person, die zum Zeitpunkt ihres Grenzübertritts zwar kein Recht auf Einwanderung hatte, allerdings noch ein Kind war, können wir diese Argumentation jedoch nicht anwenden. Kindern werden, sowohl rechtlich als auch moralisch, korrekterweise nicht die gleichen Freiheiten wie Erwachsenen zugesprochen: Wir lassen sie keinen Alkohol trinken, nicht für Lohn arbeiten, nicht dem Militär beitreten und so weiter. Darin spiegelt sich die einfache Tatsache wider, dass Kinder nicht über die Art von Handlungsfähigkeit verfügen, um solche Entscheidungen treffen oder um ihnen solche Entscheidungen vorhalten zu können. Und dennoch dachte der Abgeordnete Steve King in der Diskussion um das Deferred Action for Childhood Arrivals (DACA) Programm, dass es im Falle undokumentierter Immigrantinnen, die ursprünglich als Kinder, und somit nicht als vollwertige moralische Akteure, die Grenze überquert hatten, angemessen wäre, Kinder für das Begehen von Verbrechen verantwortlich zu machen:

> „Ich glaube, dass wir bloß die illegale Einwanderung verstärken, wenn wir eine Amnestie für diejenigen 12 bis 20 Millionen illegaler Einwanderer diskutieren, die heute in den Vereinigten Staaten leben [...] Amnestie begnadigt Menschen, die die Einwanderungsgesetze gebrochen haben und belohnt sie mit dem Ziel ihres Verbrechens: der Staatsbürgerschaft."[14]

Meiner Meinung nach ist das moralisch betrachtet ein wenig absurd: Wenn ein fünfjähriges Kind versuchen würde, der Armee beizutreten, würde King hoffentlich nicht willens sein, dieses Kind die Konsequenzen seiner „Wahl" tragen zu lassen. Das gleiche sollte allerdings auch im Hinblick auf die „Wahl" eines Kindes gelten, widerrechtlich einzuwandern. Morgan können die Konsequenzen seiner Entscheidung aufgebürdet werden, so wie beispielsweise auch, wenn er sich dazu entscheiden würde, der Armee beizutreten. Im Falle eines Kindes ist das nicht möglich.

Aus diesem Grund ist das DACA-Programm meiner Meinung nach besonders gut zu rechtfertigen. Ungeachtet der allgemeinen Analyse hinsichtlich der Rechte undokumentierter Immigrantinnen erscheint mir die Tatsache, dass die von DACA profitierenden Personen keine eigene Entscheidung zum Bruch der Gesetze getroffen haben, für die von ihnen vorgebrachten spezifischen Ansprüche als moralisch relevant. Zudem würde ich anmerken, dass Kinder ein Recht auf deutlich mehr als die bloße Anerkennung ihrer Unfähigkeit zu bestimmten „Entscheidungen" haben. So haben sie ein Recht auf die Anwesenheit bestimmter Personen, die das Handeln des Kindes führen und formen. In diesem Fall besteht eine Analogie zwischen dem Staat und den Eltern. Menschliche Wesen sind vulnerabel und haben ein Recht darauf, dass ihre grundlegendsten Menschenrechte geschützt werden. Wir haben, um es mit Hannah Arendt zu sagen, ein Recht darauf, Rechte zu haben – was bedeutet, ein Recht auf diejenigen sozialen und politischen Institutionen zu haben, die es braucht, um unsere Rechte mittels Zwang durchsetzen zu können.[15] Das Kind hat jedoch nicht nur ein Recht auf einen Staat, sondern auch auf Eltern und Betreuerinnen. Die Obama-Administration erließ sowohl DACA als auch das Programm Deferred Action for Parents of Americans (DAPA), dem zufolge es US-amerikanischen minderjährigen Kindern garantiert wurde, dass ihre undokumentierten Eltern nicht abgeschoben werden.[16] Die Grundlage hierfür bilden jedoch nicht die Rechte der Eltern, sondern die Rechte der Kinder. Das Kind, dessen Eltern abgeschoben werden, wird in seinem Recht auf Betreuung und Schutz auf gleiche Weise verletzt wie Erwachsene, die staatenlos werden. DAPA wurde kurz nach seiner Einführung auf gerichtlichem Wege gestoppt, aber es drückt einen starken moralischen Gedanken aus: Das Kind hat, rechtlich wie moralisch betrachtet, ein Recht auf bestimmte Personen, die es lieben und betreuen. Der Staat sollte diese spezifische Verletzlichkeit des Kindes dadurch anerkennen, dass er es vor dem Verlust seiner Eltern schützt. Auch in diesem Fall besteht ein Gerechtigkeitsanspruch: Das Kind wird, gelinde gesagt, ungerecht behandelt, wenn es der spezifischen Personen beraubt wird, die für die Entwicklung seiner Handlungsfähigkeit verantwortlich sind.

In meinen Augen zeigen diese Tatsachen, weshalb die Politik der Trump-Administration, die Kinder von Immigrantinnen an der Grenze von ihren Familien zu trennen, so verabscheuungswürdig ist. Diese Kinder werden in Gebäuden untergebracht, deren Architektur an Gefängnisse erinnert.[17] Die implizite Logik dieser Gebäude besteht darin, dass das Kind etwas falsch gemacht hat, denn mit Sicherheit kann aus der Perspektive des Kindes die Erfahrung des Eingesperrt- oder Eingekerkert-Seins bloß als Bestrafung verstanden werden. Dieses Vorgehen missachtet jedoch die Tatsache, dass das Kind nichts falsch gemacht hat, da es – als Kind – nicht über die kognitiven Voraussetzungen verfügt um *eine solche Art von Verbrechen* zu begehen. Was die Trump-Administration also missachtet, sind die Bedingungen der Handlungsfähigkeit dieser Kinder – und angesichts des zu erwartenden Schadens dieser Maßnahme für die kognitive Entwicklung dieser Kinder auf lange Sicht, erstreckt sich diese Missachtung auch auf ihre zukünftige Handlungsfähigkeit.

Hierbei handelt es sich also um eine Gruppe von Menschen, die sich klar von Morgan unterscheidet. Das Kind hat keine Entscheidung getroffen und daher kann ihm auch keine Entscheidung vorgehalten werden. Im Falle erwachsener undokumentierter Personen müssen wir uns hingegen die Bedingungen genauer anschauen, unter denen sie ihre Entscheidung zur Auswanderung getroffen haben. In dieser Hinsicht unterscheidet sich auch Miguel Sanchez von Morgan. Insbesondere bei Miguel könnte angeführt werden, dass er gar keine Wahl hatte. Sein Gesetzesverstoß resultierte aus ungerechten Umständen oder zumindest aus Umständen, die dem sehr nahekommen, was wir als ungerecht bezeichnen. Diese Tatsachen können herangezogen werden, um einen Anspruch Miguels und ähnlich situierter Menschen zu formulieren.

Eine recht offensichtliche Möglichkeit ein solches Argument zu formulieren, besteht darin, Carens' Beschreibung von Miguel wörtlich zu nehmen. Miguel, schreibt Carens, „ist nicht imstande, genug zu verdienen", um seinen Lebensunterhalt in seinem Heimatland Mexiko bestreiten zu können. Diese Aussage bezieht sich vermutlich auf lebensnotwendige Güter. Obwohl Carens hier keine näheren Angaben macht, scheint es unplausibel, anzunehmen, dass Miguel Schulden aufgrund besonders kostspieliger Vorlieben hat. Die bereits zuvor in diesem Buch entwickelte Logik besagt jedoch, dass staatliche Grenzen legitimerweise nicht gegenüber Personen geschlossen werden können, deren Grundrechte in ihrem Heimatland verletzt werden. In meiner Besprechung der Genfer Flüchtlingskonvention lag unser Fokus zwar darauf, was der Herkunftsstaat bei der Verletzung dieser Recht selbst aktiv tut – schließlich setzt das Konzept der Verfolgung die aktive Entscheidung eines Akteurs voraus, einer bestimmten Person Schaden zuzufügen. Aber ich habe auch argumentiert, dass das Recht auf Grenzübertritt selbst solchen Personen zukommt, deren Lebensumstände nicht mit Verweis auf das Konzept der Verfolgung beschrieben werden können. So sind beispielsweise Personen, denen der Hungertod droht, zur Rettung ihres Lebens zumindest dazu berechtigt, sich ungehindert über die Erdoberfläche bewegen zu können. Jeder Staat, der Gewalt anwendet, um diese Personen in die Situation zurückzudrängen aus der sie ursprünglich geflohen sind, würde Gewalt zum Zwecke der Aufrechterhaltung eines moralisch zu verurteilenden Zustands einsetzen; und ich habe gezeigt, dass eine solche Form des Zwangs moralisch unzulässig ist. Diese Tatsachen haben jedoch nicht nur an, sondern auch innerhalb der Grenzen bestimmte Folgen. Wenn der Grenzübertritt einer Person nicht gerechterweise hätte verhindert werden dürfen, dann kann die diesem Grenzübertritt zugrunde liegende Entscheidung auch nicht herangezogen werden, um das aus ihr resultierende Leben einem Risiko auszusetzen. Einfacher ausgedrückt: Wenn wir kein moralisches Recht hatten, die Migrantin mittels Zwang an ihrem Grenzübertritt zu hindern, dann haben wir jetzt auch kein Recht, sie mittels Zwangs aus unserer Gesellschaft abzuschieben.

Das alles scheint mir klar zu sein – oder zumindest scheint es eindeutig aus der Theorie zu folgen, die ich im Rahmen dieses Buches verteidige. Es wird jedoch meiner Meinung nach interessanter, wenn wir den Satz „nicht imstande, genug zu verdienen" in einem umfassenderen Sinne lesen. Wir könnten ihn auch so verstehen, dass er nicht nur die Unmöglichkeit erfasst, lebensnotwendige Dinge zu erwerben – was auch immer wir darunter verstehen –, sondern auch die Unmöglichkeit, das zu erwerben, was einer Person *zusteht*. Stellen Sie sich zum Beispiel vor, dass Miguel in jeder plausiblen Theorie globaler Verteilungsgerechtigkeit der Anspruch auf einen Lebensstandard zukommt, der höher ist als derjenige, den er innerhalb seines Heimatlandes erreichen kann. Stellen Sie sich des Weiteren vor, dass es ihm sehr einfach möglich ist, die Grenze zu den Vereinigten Staaten zu überqueren und seinen Lebensstandard dadurch zu verbessern. Miguel scheint also durch seine Einreise in die Vereinigten Staaten die Welt etwas näher an den Zustand zu bringen, in dem sie, gemäß unseren besten Vorstellungen von Gerechtigkeit, sein sollte. Er hat eine moralisch zu verurteilende Form von Armut hinter sich gelassen, ohne dabei irgendeine andere Person schlechter zu stellen (soweit unsere Annahme). Geben uns diese Umstände genügend Gründe, um anzunehmen, dass Miguel ein Recht darauf hat, die Grenze zu den Vereinigten Staaten zu überqueren und dass er nun, nach einiger Zeit in den USA, auch ein Recht darauf hat, innerhalb dieser Grenzen verbleiben zu dürfen?

Ich glaube, die Antwort lautet nein. Wie ich im Rest des Buches zeigen werde, würden die USA sich im Falle einer Abschiebung von Miguel als moralisch defizitär offenbaren. Sie würden – zumindest gemäß meinem Verständnis von Gnade– einen beanstandenswerten Mangel an Gnade zeigen. Aber es würde sich dabei nicht um eine ungerechte Handlung gegenüber Miguel handeln. Miguel versucht, seine Entscheidung zum Grenzübertritt durch den Verweis auf die Forderungen globaler Verteilungsgerechtigkeit zu rechtfertigen. Das Problem ist jedoch, dass die politische Gemeinschaft, die er betritt – in diesem Falle die Vereinigten Staaten – ein moralisches Recht darauf hat, für sich selbst zu entscheiden, wie sie diesen Forderungen gerecht werden möchte. Das bereits in Kapitel zwei erwähnte Beispiel von Robin Hood unterstützt dieses Argument. Miguel ist schlicht nicht derjenige Akteur, dem zu entscheiden zukommt, auf welchem Wege die Vereinigten Staaten ihre globalen Pflichten erfüllen sollten. Ich denke, dass Miguel in einer ähnlichen Position ist wie Robin Hood. Er hat nicht das Recht, die Welt durch das Brechen von Gesetzen gerechter zu gestalten und selbst wenn es stimmen sollte, dass die Welt nach diesem Gesetzesbruch gerechter wäre, wovon ich ausgehe, bedeutet das nicht, dass sein Handeln gerechtfertigt war. Miguel kann, mit anderen Worten, nicht behaupten, sein Gesetzesbruch sei rechtens gewesen und entsprechend können wir in diesem Fall nicht sagen, dass es falsch im Sinne von ungerecht wäre, ihm das, was er sich aufgebaut hat, wieder zu nehmen.

Ich werde im nächsten Kapitel mehr über diese Art von Gesetzesbruch sagen. An dieser Stelle möchte ich jedoch bemerken, dass nichts in meinen Überlegungen impliziert, man könne Miguel aufgrund seiner Handlungen einen mangelhaften Charakter zuschreiben. Das ist tatsächlich nicht der Fall. Aber es ist eine Tatsache, dass unsere moralische Reaktion auf Gesetzesverstöße uns nichts darüber sagt, ob Miguel das Gesetz, welches ihn vom Grenzübertritt abhalten sollte, auch zu befolgen hatte.

Schließen möchte ich mit einem letzten möglichen Unterschied zwischen Miguel und Morgan. Morgans persönliche Pläne würden durch eine Abschiebung nach Kanada zunichte gemacht. Es würde im schwerfallen, seine Freundschaften in den Vereinigten Staaten zu erhalten, die Sehenswürdigkeiten in Portland zu betrachten, an die er sich gewöhnt hat, und so weiter. Aber er wird sehr wahrscheinlich weiter handlungsfähig bleiben – wir können davon ausgehen, dass er über die Fähigkeiten verfügt und in der Lage ist, sich ein neues Leben aufzubauen. Das gleiche kann jedoch in Miguels Fall nicht angenommen werden – und wohl auch nicht, so meine Vermutung, in einer Vielzahl anderer Fälle. Wenn dem so ist, würde die Entscheidung für eine Abschiebung bedeuten, nicht nur das zunichte zu machen, was durch die Handlungsfähigkeit Miguels entstanden ist, sondern auch diese Handlungsfähigkeit selbst – und eine solche Entscheidung können wir in meinen Augen nicht rechtfertigen. Selbst wenn es uns in manchen Fällen erlaubt sein sollte, Menschen die Kosten ihrer Entscheidungen aufzubürden, schrecken wir doch im Allgemeinen davor zurück, Menschen sich selbst *vernichten* zu lassen. Eine solche Aussage ist sicherlich umstritten und Liberale sind sich uneinig darüber, bis zu welchem Grad wir eine freiwillige Entscheidung respektieren sollten, die zur Auflösung der eigenen Fähigkeit führt, überhaupt eine Entscheidung treffen zu können. Aber die meisten von uns würden einen solchen Fall zumindest als verschieden von den hier angestellten Überlegungen ansehen – weswegen wir in diesem Fall annehmen können, dass aus der Tatsache, dass eine Abschiebung mehr als nur Schmerz verursachen würde, ein Gerechtigkeitsanspruch erwächst.

Einige Beispiele mögen verdeutlichen, dass wir Menschen nicht gerechterweise in Umstände abschieben können, in denen sie buchstäblich ausgelöscht würden, ganz unabhängig davon, wie sich ihr Grenzübertritt gestaltet hat. So wurde beispielsweise Thomas Mann bei seiner Ankunft in Frankreich 1933 mitgeteilt, dass eine Rückkehr nach Deutschland sein Leben gefährden würde.[18] In manchen Fällen käme die Entscheidung für eine Abschiebung buchstäblich der Todesstrafe gleich. Wie ich gezeigt habe, können wir eine Abschiebung in einem solchen Falle nicht gerechterweise betreiben, ganz gleich, wie die jeweilige Person die Grenzen überquert hat. Aber ich denke, dass hier noch mehr gesagt werden kann. Es gibt manche Menschen, für die der Wiederaufbau eines eigenen Lebens schlicht so beschwerlich wäre, dass eine Abschiebung moralisch der Vernichtung dieser Person gleichkäme. So können beispielsweise drastische Unterschiede hinsichtlich der Sprache und Kultur den Prozess der Entwicklung von Plänen und Projekten nahezu unmöglich machen. Solche Bedenken scheinen von größerer Bedeutung für ältere Personen zu sein, da mit dem Alter unsere Fähigkeit für einen Neubeginn – ebenso wie unsere Fähigkeit, die Sprache oder Normen eines Ortes (erneut) zu lernen – auf natürliche Weise zu schwinden scheinen. Carens diskutiert in seiner Verteidigung solcher Ideen den Fall von Marguerite Grimmonds, die im Alter von zwei Jahren nach Schottland kam und mit achtzig Jahren beinahe in die Vereinigten Staaten abgeschoben worden wäre.[19] Wir könnten anführen, dass Grimmonds aufgrund der Tatsache nicht abgeschoben werden sollte, dass ihre ursprünglich ohne Erlaubnis erfolgte Einreise nach Schottland nicht als ihre Entscheidung angesehen werden kann. Aber wir könnten auch darauf hinweisen, dass die Abschiebung einer 80-jährigen Frau der Vernichtung ihrer Handlungsfähigkeit gleichkommt. Ähnliche Überlegungen könnten auch im aktuellen Fall von Francis Anwana angestellt werden, einem 48-jährigen tauben Mann mit kognitiven Einschränkungen, dem nun die Abschiebung nach Ghana droht, nachdem er als Kind in die Vereinigten Staaten gebracht worden war. (Die Abschiebung droht ihm alleine deshalb, weil die Organisatoren in seiner Gemeinschaftsunterkunft keinen Antrag auf Verlängerung seines Visums gestellt hatten.) Anwana war es möglich, ein wertvolles Leben in den Vereinigten Staaten aufzubauen; und es ist unwahrscheinlich, dass ihm ein solches Leben auch im Falle einer „Rückkehr" nach Ghana möglich wäre, wo er weder eine Familie, noch ein System sozialer Unterstützung vorfinden würde.[20]

Ich möchte dieses Argument nicht über Gebühr beanspruchen; schließlich bewegen sich Menschen regelmäßig zwischen völlig verschiedenen Ländern und in diesem Zusammenhang wissen wir auch von 80-jährigen, die freiwillig ihre Sachen packen und migrieren. Allerdings ist dabei von zentraler Bedeutung, dass es sich in manchen dieser Fälle um Anstrengungen handelt – und diese Anstrengungen können solche Ausmaße annehmen, dass eine Abschiebung der Vernichtung der abzuschiebenden Person gleichkäme.

3. Liebe und Familie

Wir können dieses Kapitel mit einer Untersuchung der Familienzusammenführung schließen. Die Vereinigung von Familien ist ein zentraler Teil des Narrativs, das dem US-amerikanischen Einwanderungsrecht zugrunde liegt: Die Vereinigten Staaten sind traditionellerweise außergewöhnlich großzügig bei der Gewährung von Einwanderungsrechten für Angehörige US-amerikanischer Staatsbürgerinnen oder dauerhafter Einwohnerinnen der USA. Allerdings haben wir in den letzten Jahren auch die Entstehung eines Gegennarrativs erlebt, demzufolge die Familienzusammenführung der ungerechtfertigten Einwanderung einer endlosen Zahl von Ausländerinnen Vorschub leisten würde; insbesondere die Trump-Administration hat damit begonnen, Ausdrücke wie *Kettenmigration* zu verwenden, um ein angebliches Übermaß der US-amerikanischen Bereitschaft zur Unterstützung von Familien auszudrücken. Beide Erzählungen beruhen jedoch jeweils auf einer bestimmten Vorstellung davon, warum die Zusammenführung von Familien eine gute Sache ist und es lohnt sich, genauer zu bestimmen, was die Familie tatsächlich *ist* und ob aus diesem Konzept Gerechtigkeitsansprüche abgeleitet werden könnten. Erneut versuchen wir also herauszufinden, ob die Ansprüche von Familienmitgliedern am besten als Gerechtigkeitsforderungen verstanden werden sollten, oder doch eher auf einem anderen moralischen Konzept wie beispielsweise der Gnade beruhen.

Die Idee der Familie selbst kann unterschiedlich interpretiert werden.[21] Wenn wir von Familie sprechen, beziehen wir uns oft auf zwei ähnliche, jedoch voneinander zu unterscheidende Ideen. Die erste ist die der *Abhängigkeit*: Ein Kind ist abhängig von seinen Eltern und die Zusammenführung einer Familie könnte unter Verweis auf den moralischen Imperativ begründet werden, dass sich das Kind zu einer eigenständigen Person entwickeln können soll. Die zweite Idee ist die der *Zuneigung*: Die Familie ist ein Ort der Liebe, oder etwas ihr ähnlichem, und wir sind durch die Idee der Gerechtigkeit dazu verpflichtet, Beziehungen zu respektieren, die eine solche Liebe ermöglichen. Wenden wir uns diesen Ideen nun nacheinander zu.

Die aus der Idee der Abhängigkeit resultierenden Ansprüche sind in meinen Augen recht offensichtlich. Erwachsene haben, wie bereits weiter oben beschrieben, ein Recht auf solche Umstände, in denen sie fähig sind, autonom zu handeln. Kinder haben darüber hinaus ein Recht auf Bedingungen, in denen ihre entstehende Handlungsfähigkeit gestärkt und entwickelt wird. Das bedeutet – wie Carens betont –, dass es Orte und Institutionen braucht, in denen Kinder und Jugendliche dazu befähigt werden, als politische Akteure innerhalb ihrer Gesellschaften zu wirken.[22] Die bedeutendste Folge dieser Überlegungen ist jedoch, dass es bestimmte Personen im Leben dieser Kinder geben muss, die sie anleiten und ihre sich entwickelnde Handlungsfähigkeit fördern. Darüber hinaus folgt aus diesen Überlegungen auch, dass die Beziehung zwischen den Eltern und ihren Kindern nicht aufgelöst werden kann, ohne die Kosten einer illegitimen Vernichtung der Handlungsfähigkeit der Kinder in Kauf zu nehmen. Ein Kind hat das Recht auf die Anwesenheit seiner Eltern. Wie ich bereits gezeigt habe, unterstützen diese Überlegungen die moralische Bedeutung gesetzlicher Maßnahmen wie DACA oder DAPA – obwohl ich anmerken würde, dass selbst diese Programme sehr wahrscheinlich ungenügend sind, denn wie Hosein betont, bedarf Handlungsfähigkeit der Möglichkeit, Pläne zu machen – und sowohl DACA als auch DAPA stellen bloß eine vorübergehende Aussetzung von Abschiebungen dar. Die Tatsache, dass ich Hoseins Analyse hinsichtlich der Rechtfertigung von Abschiebungen nicht teile, bedeutet also nicht, dass wir uns auch darüber uneins sein müssen, wie bedeutend die Fähigkeit ist, Pläne machen zu können.

Darüber hinaus würde ich anführen, dass die aus der Idee der Abhängigkeit folgenden Ansprüche potentiell auch über das Kind hinaus ausgeweitet werden können. Das Kind bedarf besonderer Fürsorge für seine Handlungsfähigkeit und das gleiche gilt auch für ältere Menschen sowie diejenigen, die unter kognitiven Einschränkungen leiden und prinzipiell für all die Personen, deren Handlungsfähigkeit nicht bloß einer Rechtsgemeinschaft bedarf, sondern auch der Anwesenheit bestimmter Personen, die eine bestimmte Funktion in ihrem Leben übernehmen. Ich würde argumentieren, dass es sich dabei um Gerechtigkeitsansprüche von gleicher Stärke wie im Falle minderjähriger Kinder handelt. Wir würden falsch handeln, sollten wir Personen abschieben, die für die effektive Ausübung ihrer individuellen Handlungsfähigkeit bestimmter Personen bedürfen, oder solche Personen, deren Handlungsfähigkeit wir uns nicht sicher sein können.

Es wird jedoch komplexer, wenn wir uns den Forderungen zuwenden, die der Idee der Liebe entspringen. Wenn es also eine Forderung der Gerechtigkeit sein sollte, dass den Partnerinnen einer Liebesbeziehung ein Recht auf Zusammenführung zukommt, müsste es an dieser Liebesbeziehung etwas Besonderes geben. Wir würden demnach falsch handeln, sollten wir es einer solchen Form der Beziehung erschweren, sich zu entwickeln und zu wachsen. Darüber hinaus würden wir im Falle der Beeinträchtigung einer solchen Liebesbeziehung auch größeres Unrecht tun als im Falle von Beziehungen zwischen Freundinnen, Ko-Autorinnen oder Arbeitskolleginnen. Immerhin haben diese letztgenannten Beziehungsformen auch keinen speziellen Status innerhalb des Migrationsrechts, wohingegen familiären Beziehungen – und insbesondere der Beziehung zwischen Ehegatten – eine außerordentliche Achtung zuteil wird. Die Frage ist jedoch, ob wir bei Liebesbeziehungen eine spezifische Eigenschaft identifizieren können, die einen solchen besonderen Status rechtfertigen würde. Luara Ferracioli hat argumentiert, dass wir keine moralisch angemessene Rechtfertigung dieses Status finden können und dass die Achtung gegenüber Liebesbeziehungen nicht auf philosophische Reflektionen, sondern vielmehr historische Zufälle zurückzuführen sei:

> „Sobald wir anerkennen, dass es einen weiten Bereich besonderer Beziehungen gibt, die einem liberalen Ethos zufolge gedeihen sollten, wird offensichtlich, dass die Migrationsgesetze liberaler Staaten nicht willkürlich eine dieser Beziehungen auf Kosten der anderen bevorzugen dürfen. Das Fazit lautet daher, dass das Migrationsregime so ausgestaltet werden sollte, dass entweder alle Arten besonderer Beziehungen, die dem menschlichen Leben Bedeutung verleihen, berücksichtigt werden, oder aber gar keine. Der momentane philosophische und praktische Konsens, dass familiäre und romantische Bezie-

hungen einen *einzigartigen* Status innerhalb des Migrationsrechts innehaben sollten, kann nach gründlicher Überlegung nicht gerechtfertigt werden."[23]

Diese Überlegungen stellen die erste Herausforderung für den Gedanken dar, dass die Zusammenführung von Familien einen Gerechtigkeitsanspruch darstellt. Sollten sich die Ansprüche der Liebe oder der biologischen Verwandtschaft aus moralischer Perspektive nicht von Ansprüchen anderer Beziehungsverhältnisse unterscheiden, könnte der liberale Staat Ehegattinnen auch nicht gerechterweise Freundinnen vorziehen. Wenn ich mit meinem besten Freund zusammenleben möchte und Sie mit ihrem Ehepartner oder ihrer Ehepartnerin, würde ich demnach von einem Staat unfair – und daher ungerecht – behandelt, wenn er Ihnen, aber nicht mir, das zuspräche, was wir uns beide wünschen.

Ich denke, dass es sich dabei um starke Bedenken hinsichtlich der Rechtfertigung von Familienzusammenführungen handelt und ich glaube, dass noch weitere Bedenken angeführt werden können. Ein weiterer Einwand besteht darin, dass wir, sobald wir uns von der Idee der Abhängigkeit lösen, auf ein komplexes Verhältnis zwischen dem Gut, das wir verfolgen – Liebe oder etwas ihr ähnliches – und den von uns genutzten Kategorien stoßen. Warum sollte es das Gesetz beispielsweise vorsehen, die Einwanderung der Schwiegermutter eines US-amerikanischen Staatsbürgers zu fördern, nicht aber die Einreise seiner Cousine? Die Antwort wird vermutlich lauten, dass die Beziehung einer Schwiegermutter zu ihrem Schwiegersohn oder ihrer Schwiegertochter wahrscheinlich auf etwas wie gegenseitiger Zuneigung beruht, während die Cousine im Stammbaum weiter „entfernt" liegt und somit wahrscheinlich auch die gegenseitige Zuneigung weniger stark ausgeprägt ist. Es gibt allerdings keinen Grund, anzunehmen, dass dies im individuellen Fall dann auch den Tatsachen entspricht, denn schließlich ist Zuneigung ein Phänomen, das sich recht schwer vorhersagen lässt, und darüber hinaus auch die Angewohnheit hat, jene Kategorien zu ignorieren, die wir zum Zwecke ihrer Einhegung geschaffen haben.

Die Trump-Administration musste sich mit diesen Schwierigkeiten bei der Bestimmung der Gruppe von Personen auseinandersetzen, die eine „*bona-fide*-Beziehung" zu einer US-Bürgerin oder einer dauerhaften Einwohnerin haben. Die Regierung erklärte, dass das Beziehungsverhältnis einer Schwiegermutter den *bona-fide*-Kriterien entsprechen würde, viele andere Beziehungsverhältnisse jedoch nicht. Der neunte Gerichtsbezirk hielt das für eine willkürliche Unterscheidung und untersagte der Regierung die Anwendung der von ihr festgelegten Regeln:

> „Einfach ausgedrückt bietet die Regierung keine überzeugende Erklärung dafür, warum das Beziehungsverhältnis der Schwiegermutter offensichtlich eine *bona-fide*-Beziehung im Sinne der vorherigen Rechtsprechung des Supreme Court darstellen soll, die eines Großelternteils, eines Enkels, einer Tante, eines Onkels, einer Nichte, eines Neffen oder eines Cousins hingegen nicht."[24]

Diese Überlegungen sind korrekt – allerdings übersieht der Richter hier, dass dieses Problem auch für alle anderen Beziehungsverhältnisse gilt. Wenn es keine klare Linie gibt, anhand derer die Schwiegermutter vom Cousin unterschieden werden kann, gibt es vermutlich auch keine klare Linie zwischen diesem Cousin und einer entfernter verwandten Cousine und schließlich auch nicht zwischen dieser Cousine und einer anderen Person, die derart entfernt mit uns verwandt ist, dass wir nicht einmal den Grad dieser Verwandtschaft genau bestimmen können und so weiter. Mit anderen Worten: Es fehlen ziemlich viele klare Linien. Wenn wir allerdings einen Gerechtigkeitsanspruch begründen wollen – wenn wir also die Einwanderung der Schwiegermutter als Forderung der Gerechtigkeit darstellen wollen, nicht aber die einer entfernten Cousine – brauchen wir eine solch klare Linie, um die Unterschiede zwischen diesen moralischen Rechten erklären zu können. Das Gesetz arbeitet notwendig mit arbiträren Unterscheidungen. So akzeptieren wir, dass wir in den USA erst mit 16 Jahren Auto fahren oder mit 21 Jahren Alkohol trinken dürfen, da legale Rechte solch klarer Linien bedürfen. Die Grundlagen moralischer Ansprüche müssen jedoch robuster sein und wenn von uns gefordert wird, die Cousine anders als die Schwiegermutter zu behandeln, dann ist diese Cousine berechtigt zu wissen, warum das der Fall sein sollte.

All dies soll die Annahme hinterfragen, dass Familienmitglieder einen einzigartigen und besonders starken Anspruch auf Zusammenführung haben. Hierfür könnten wir auch noch den Gedanken anführen, dass die Zurückweisung eines ausländischen Ehegatten nicht wirklich einer erzwungenen Vereitelung der Liebe oder einer liebevollen Beziehung durch den Staat gleichkommt. Vielmehr verweigert er dieser Liebe eine Heimstatt. Beispielsweise würden es die Vereinigten Staaten für mich weitaus komplizierter machen, wenn sie meiner ausländischen Ehegattin kein Einwanderungsrecht zusprechen würden. Aber ich kann nicht behaupten, dass sie mich dadurch davon abhalten würden, eine liebevolle Beziehung zu meiner Ehegattin zu pflegen, was beispielsweise der Fall wäre, wenn es als illegal erklärt werden würde, Personen eines bestimmten Geschlechts zu heiraten (oder eine Liebesbeziehung mit ihnen zu führen).[25] Die Versagung der für eine Beziehung notwendigen Mittel ist nicht gleichbedeutend mit einem zwangsweisen Eingriff in diese Beziehung. Nehmen wir zum Beispiel an, dass ich mich in eine Frau verlieben würde, die an der University of Miami arbeitet, ich aber nicht nach Miami ziehen kann, es sei denn, um an der University of Miami zu arbeiten (da meine Fähigkeiten sich auf das Philosophieren beschränken), diese aber kein Interesse daran hat, mich einzustellen. Greift die University of Miami in unsere Beziehung ein, wenn sie mir eine Arbeitsstelle verweigert? Ich denke, die Antwort lautet nein; sie versagt mir zwar die für diese Beziehung notwendigen Mittel, allerdings hatte sie auch keinerlei Verpflichtung, mir diese zur Verfügung zu stellen. Genauso stellt es sich auch im Falle von Ehegattinnen mit ausländischer Staatsbürgerschaft dar. Der Staat, der den ausländischen Ehegatten die Einwanderung verweigert, erschwert zwar die Fortsetzung dieser Ehe. Aber es kann nicht behauptet werden, dass er diese Ehe per Zwang verhindert – und diese Tatsache macht den Gedanken, dass gegenseitige Zuneigung Gerechtigkeitsansprüche begründen kann, in meinen Augen noch schwieriger nachzuvollziehen.

Eine letzte Überlegung möchte ich noch anstellen. Es wird oft von uns verlangt, unsere Wünsche und Begehren so anzupassen, dass sie mit den Rechten anderer und den uns zur Verfügung stehenden Mitteln kompatibel sind. Es fällt schwer, zu denken, dass dies auch auf die Liebe zutreffen sollte, was allerdings nicht bedeutet, dass dem nicht tatsächlich so sein könnte. In meinen Augen sind wir oft dazu aufgefordert, uns der verschiedenen Arten bewusst zu sein, auf die unsere persönlichen Beziehungen durch die Rechte anderer eingeschränkt sein könnten. Wir sind in solchen Fällen nicht bloß Opfer unserer Herzen – während Woody Allen bekanntlich behauptete, dass das Herz nun einmal will, was es will, würden viele von uns darauf gerne entgegnen, dass vom Herzen durchaus erwartet werden kann, sich zu kontrollieren. So erwarten wir von denjenigen, die Doktorandinnen betreuen, ihre Emotionen und Handlungen unter Kontrolle zu haben. Sich in einen Doktoranden zu verlieben entschuldigt nichts. Auch erwarten wir von Soldatinnen oder Diplomatinnen, dass sie nach langen Auslandsaufenthalten nach Hause zurückkehren, ungeachtet der Freunde, die sie dabei zurücklassen. Wir sind, um es kurz zu fassen, verpflichtet, die Verantwortung dafür zu übernehmen, wen und wie wir lieben. Wie kann meine Beziehung zu meiner Ehefrau demnach als eine Art Trumpf angesehen werden, um Rechte einzufordern, die anderen Formen menschlicher Beziehungen vorenthalten werden?

Ich möchte an dieser Stelle klar sagen, dass ich nicht gegen ein Gesetz argumentiere, dass Ehegattinnen oder Familienmitgliedern Einwanderungsrechte zuspricht. Ich würde nur ungern in einem Land ohne ein solches Gesetz leben wollen. Allerdings habe ich gezeigt, dass es unklar ist, wie ein solches Gesetz aus der Idee der Gerechtigkeit abgeleitet werden könnte. Die Liebe, die wir für unsere Ehepartnerinnen empfinden, mag für sie und uns selbst von besonderer Art sein. Allerdings ist mir schlicht nicht klar, weshalb sie auch vom Standpunkt der Gerechtigkeit aus als besonders gelten sollte.

Die beste Begründung einer solchen Forderung wurde in meinen Augen von Matthew Lister formuliert, der für sein Argument auf die Wünsche und Interessen der Bürgerinnen des potentiellen Einwanderungslands statt auf die der ausländischen Person zurückgreift.[26] Lister führt an, dass das Recht einer Bürgerin auf ihren Ehepartner sich der Art nach von dem Recht ihrer Mitbürgerinnen unterscheidet, sich nicht mit diesem ausländischen Ehegatten auseinandersetzen zu müssen. Listers Argument zufolge tendieren wir zu der Annahme, dass eine Ehebeziehung tiefer reicht als die Verbindungen zwischen Mitbürgerinnen. Demnach wäre mein Interesse, mit meiner Frau zusammenzuleben, gewichtiger als Ihr Interesse, meiner Frau nicht im Supermarkt begegnen zu müssen.

In meinen Augen könnte Lister hier nicht ganz falsch liegen, zumindest scheint mir sein Argument die stärkste potentielle Grundlage für die Forderung eines speziellen Status ehelicher Verbindungen darzustellen. Es lohnt sich jedoch, an dieser Stelle darauf hinzuweisen, dass selbst wenn Lister richtig läge, Ferraciolis Problem weiterhin in den Fällen bestehen würde, in denen andere Beziehungen den Antragstellerinnen ebenso wichtig sind wie eine eheliche Beziehung. Daher denke ich auch nicht, dass Lister vollkommen Recht hat – trotz der Stärke seines Arguments. Stellen Sie sich zum Beispiel vor, dass ich mich in eine Frau aus Italien verliebe, die noch nie einen Fuß in die Vereinigten Staaten gesetzt hat. Das mag vielleicht unrealistisch sein, aber es ist theoretisch möglich. Nehmen Sie nun an, dass die Vereinigten Staaten dieser Frau den Zutritt verweigern. Stellen Sie sich zuletzt noch vor, ich würde gemäß Listers Überlegungen argumentieren, dass ich durch diese Entscheidung in meinen Rechten verletzt werde, da meine Liebe für mich von größerem Wert ist als das, was meine Mitbürgerinnen durch den Ausschluss meiner Frau zu erreichen versuchen – was auch immer das sein mag. Es scheint mir als hätten meine Mitbürgerinnen in diesem Fall eine Antwort parat: Du hast das Recht auf eine Gesellschaft in der du gerecht behandelt wirst, weshalb wir dich durch unsere Gesetze nicht als moralisch minderwertig behandeln dürfen. Wir können daher deine Ehe nicht für illegal erklären. Aber wir behandeln dich nicht ungerecht, wenn wir der Ausländerin, die du liebst, kein Aufenthaltsrecht gewähren. Wir verurteilen nicht deine Liebe oder behaupten, dass sie weniger wert sei als die unsere. Vielmehr verweigern wir etwas, zu dessen Gewährung wir nicht verpflichtet sind, sei es gegenüber dir oder irgendwem anders. Wir stehen deiner Liebe nicht im Wege; aber wir haben uns auch nicht dazu entschieden, dir das zu gewähren, worum du gebeten hast – oder irgendwem anders, dessen Ehegattin nicht bereits Teil unserer politischen Gemeinschaft ist.

Das könnte eine erfolgreiche Strategie sein. Ich denke, dass sie funktioniert, obwohl ich offen für Argumente bin, die mich vom Gegenteil überzeugen. Die soeben geschilderte Antwort erscheint jedoch ohne Frage als hartherzig – und sie ist es in der Tat. Angesichts dieser Hartherzigkeit gibt es mindestens drei Antworten. Die erste besagt, dass die Antwort sowohl hartherzig als auch ungerecht sei: Sie sei ungerecht, da die Stärke des Bedürfnisses nach meiner Frau sich der Art nach von anderen Bedürfnissen unterscheide, oder da es für mich derart schädlich sei, von meiner Frau getrennt zu sein, dass es meine Handlungsfähigkeit einschränke. Der zweiten Antwort zufolge seien die meisten Ehen nicht von der Art wie die von mir illustrierte Ehe mit einer Frau aus Italien. Diese Ehe sei zwar alles in allem möglich – allerdings gingen die meisten Ehen damit einher, dass die Ehegattinnen für eine längere Zeit im gleichen Land gelebt haben, sodass sie sowohl Möglichkeit als auch Verbundenheit im Sinne Obermans in sich vereinen. Das mag der Wahrheit entsprechen; aber wir müssten auch in diesem Falle sagen können, warum diese Tatsachen von Bedeutung sein sollten – warum also diese Unterschiede von der Art sind, dass sie Einwanderungsrechte als Gerechtigkeitsansprüche begründen können. Der letzten Antwort zufolge liegt das moralische Problem der Reaktion meiner Mitbürgerinnen nicht in einer angeblichen Ungerechtigkeit, sondern in ihrer Herzlosigkeit. Dabei handelt es sich offensichtlich um meine eigene Position. Ich kann mich demnach darüber beschweren – und zwar vehement –, dass meiner Frau kein Recht auf Einwanderung gewährt wird. Aber nicht alle moralischen Beschwerden fußen auf dem Begriff der Gerechtigkeit und unser konzeptionelles Vokabular ist reichhaltig genug, um meine Beschwerde präzise zu formulieren. In den nächsten drei Kapiteln werde ich versuchen, genau das zu tun.

8
Reziprozität, die Undokumentierten und Jeb Bush

Das vorangegangene Kapitel endete mit dem Gedanken, dass nicht alle moralischen Beschwerden in der Sprache der Gerechtigkeit formuliert werden müssen. Wir können in meinen Augen die beschriebenen politischen Entscheidungen als bösartig bezeichnen – als Ausdruck von Grausamkeit und Hartherzigkeit – ohne damit notwendigerweise zu implizieren, dass eine Partei ungerecht behandelt wurde. In den folgenden drei Kapiteln möchte ich daher ein Konzept der politischen Gnade entwickeln, das uns meiner Meinung nach eine Möglichkeit aufzeigt, wie diese Art von Beschwerden vorgebracht werden könnten. Ich möchte diese Aufgabe jedoch aus einem anderen Modus heraus beginnen und zunächst über die Moralität undokumentierter Migration sprechen – und zwar nicht vom Standpunkt des Staates aus, sondern ausgehend vom Migranten und unserer Bewertung seiner Handlungen. Wie ich zeigen werde, sollten wir zwischen dem, was wir rechtmäßig zu verhindern versuchen dürfen, und unserer moralischen Haltung gegenüber denjenigen, die unsere Gesetze umgehen möchten, unterscheiden. Wir dürfen demnach versuchen, Menschen vom Überqueren unserer Grenzen abzuhalten – mitunter sogar recht bedürftige Personen –, ohne dabei jedoch zugleich berechtigt zu sein, diejenigen, die sie dennoch überqueren, als moralisch defizitär zu betrachten. Das scheint ein billiger Trost zu sein, aber ich denke, dass diese Überlegungen überraschende Implikationen haben, sowohl im Hinblick auf unsere Migrationspolitik, als auch dahingehend, welche Zwecke wir mit ihr verfolgen sollten.

Als eine Art Einführung in dieses Thema möchte ich an die Worte von Jeb Bush im Rennen um die Präsidentschaftskandidatur der Republikaner erinnern:

> „Aber so wie ich die Sache sehe – und ich werde das sagen, es wird aufgezeichnet und so sei es. So wie ich die Sache sehe, kommt eine Person, die in

> unser Land einreist, da sie nicht legal einreisen kann, aus dem Grund in unser Land, da ihre Familien – da ist der Vater, der seine Kinder liebt – sich darum sorgten, dass die Kinder kein Essen auf dem Tisch haben. Und sie wollten sichergehen, dass ihre Familien intakt bleiben und daher überquerten sie die Grenze, da sie keine anderen Möglichkeiten hatten, um durch Arbeit für ihre Familien zu sorgen. Ja, sie haben das Gesetz gebrochen, aber das war keine Straftat. Es war ein Akt der Liebe."[1]

Wir haben uns in diesem Buch bisher nicht sonderlich auf die moralischen Pflichten der Migrantinnen konzentriert und vor allem auch nicht darauf, ob eine Migrantin die Pflicht hat, das sie ausschließende Gesetz zu befolgen. Das ist auf der einen Seite nachvollziehbar, denn der Staat ist weitaus mächtiger als der Migrant und wir sorgen uns zumeist um diejenigen Akteure, die den größten Schaden anrichten können. Auf der anderen Seite stellt das jedoch ein Problem dar. Große Teile unseres öffentlichen politischen Diskurses beginnen, wenn es um die Moral der Migration geht, auf der Ebene des einzelnen Migranten. Rufen Sie sich in Erinnerung, welchen argumentativen Rahmen wir im ersten Kapitel als Ausgangspunkt nahmen: Die politische Rechte der Vereinigten Staaten nutzt die rituelle Beschwörung der Illegalität als unmoralische Daseinsform zu rhetorischen Zwecken: „Welchen Teil von *illegal* verstehen Sie nicht?"[2] Die Linke ihrerseits antwortet mit der Zurückweisung jeglichen Zusammenhangs von Legalität und Moralität: „Kein Mensch ist *illegal*".[3]

In diesem Kapitel möchte ich ein paar vorläufige Überlegungen dazu anbieten, wie der Bruch von Gesetzen, deren Zweck im Ausschluss von Migrantinnen besteht, moralisch verstanden werden sollte. Hierzu werde ich eine Vorstellung davon präsentieren müssen, was uns im Allgemeinen einen moralischen Grund dazu gibt, einer rechtlichen Vorschrift Folge zu leisten. Dabei handelt es sich um eine recht umfangreiche Aufgabe mit einer ehrwürdigen Geschichte, der ich in diesem Kontext nicht vollständig werde Rechnung tragen können. Stattdessen werde ich mich in meiner Analyse auf zwei Ansätze konzentrieren, die zeigen, wie und weshalb ein bestimmtes Rechtssystem uns zur Gefolgschaft verpflichten kann. Andrew Lister folgend, werde ich unterscheiden zwischen einem Verständnis einer solchen Verpflichtung in der Tradition von *Hume*, dem zufolge diese Pflicht auf dem moralischen Imperativ der Reziprozität bei der Verteilung der Früchte und Lasten sozialer Kooperation beruht; und einem Ansatz in der Tradition *Kants*, dem zufolge eine solche Verpflichtung stattdessen auf abstraktere Vorstellungen wie der Pflicht zur Unterstützung gerechter Institutionen zurückzuführen ist.[4] Meine erste These lautet dann, dass die Kant folgenden Überlegungen vielversprechender sind als die auf Hume zurückgehenden Argumente. Der Grund hierfür ist, dass uns die Argumente nach Hume zwar zu zeigen erlauben, dass eine Migrantin ein sie ausschließendes Gesetz nicht befolgen muss, allerdings um den Preis, dass wir Formen der Verletzung internationalen Rechts erlauben müssten, die wir zu Recht verurteilen. Meine zweite These lautet hingegen, dass selbst die Überlegungen nach Kant uns nicht einfach diejenigen verurteilen lassen, die ohne legale Erlaubnis einwandern. Wie ich zeigen werde, müssen wir zwischen einer Pflicht zur Unterstützung gerechter Institutionen und den Kosten unterscheiden, die einer Person durch die Erfüllung dieser Pflicht auferlegt werden. Es könnten demnach Umstände existieren, in denen wir zwar auf einer Pflicht zur Befolgung des Gesetzes bestehen dürfen, allerdings zugleich anerkennen sollten, dass die Kosten dieser Befolgung im Einzelfall derart hoch sein können, dass jede Person es als schwer oder gar unmöglich ansehen würde, sie auch tatsächlich tragen zu können. Unter solchen Umständen haben wir meinen Überlegungen zufolge kein Recht, diejenigen moralisch zu verurteilen, die der infrage stehenden Pflicht nicht nachkommen. Wir begehen demnach keinen moralischen Fehler, wenn wir darauf bestehen, dass eine Pflicht zur Befolgung des Gesetzes existiert, allerdings begehen wir einen bedeutenden moralischen Fehler, sollten wir denjenigen, die sie nicht befolgen, zugleich einen moralisch defizitären Charakter zuschreiben. Dies führt letztendlich dazu, dass ein legitimer Staat korrekt handelt, wenn er auf seinem Recht auf den Ausschluss bestimmter Migrantinnen besteht *und zugleich* denjenigen, die unerlaubt eingewandert sind, keinen besonderen moralischen Defekt zuschreibt.

Um zu diesem Schluss zu gelangen werde ich allerdings einige Annahmen treffen müssen. Dazu zählt die Prämisse, dass der ausschließende Staat von der Art ist, dass er verbindliche Vorschriften aufstellen kann – wie auch immer wir also die Schwelle politischer Zumutbarkeit bestimmen, unterhalb derer Staaten die Befolgung ihrer Gesetze nicht mehr erwarten dürfen: Der im Folgenden betrachtete Staat liegt in jedem Fall über dieser Schwelle. Das bedeutet denn auch, dass es eine Gruppe von Staaten gibt, die dazu fähig ist, verschiedenen Akteuren anhand von gesetzlichen Bestimmungen gültige moralische Pflichten aufzuerlegen, und dass der von uns betrachtete Staat zu dieser Gruppe zählt. Darüber hinaus werde ich selbstverständlich auch annehmen, dass die zuvor in diesem Buch ausgeführten Argumente Bestand haben und es somit kein allgemeines Recht auf die Überquerung staatlicher Grenzen gibt. Wir nehmen also an, dass die Argumente von Joseph Carens und Kieran Oberman fehlschlagen und der Versuch, potentielle Migrantinnen auszuschließen, nicht inhärent ungerecht ist. Zudem gehe ich davon aus, dass das infrage stehende Gesetz zum einen tatsächlich anordnet, dass manche der potentiellen Migrantinnen nicht einwandern dürfen, und zum anderen darauf besteht, dass diejenigen, die dennoch kommen, einen Gesetzesverstoß begehen.[5] Zusätzlich werde ich annehmen, dass wir in unserer Betrachtung solche Personen ausnehmen, die einen besonderen und nicht einen allgemeinen Anspruch gegenüber dem jeweiligen Staat geltend machen können; wir sehen also sowohl von den Fällen bereits lange anwesender undokumentierter Immigrantinnen ab, als auch von den Fällen derjenigen, die einen moralischen Anspruch gegenüber dem infrage stehenden Staat aufgrund seiner besonderen Geschichte oder Taten im Ausland geltend machen können.[6] Immerhin kann selbst ein Staat, der ein Recht auf Ausschluss besitzt, dieses Recht aufgrund bestimmter Handlungen verlieren, woraufhin bestimmte Personen wiederum ein Recht hätten, selbst gegen den Willen dieses Staates einwandern zu dürfen. Schließlich werde ich *nicht* annehmen, dass das den Ausschluss vorschreibende Gesetz selbst gerecht ist, sondern bloß davon ausgehen, dass der Staat die Art von Instanz ist, die zur Befolgung dieses Gesetzes verpflichten kann – und zwar unabhängig davon, ob dieses Gesetz selbst gerecht ist.

Kurzum: Ich treffe eine Menge faktischer wie auch ethischer Annahmen. Viele Menschen werden meinen, dass eine oder mehrere dieser Prämissen falsch sind, entweder in jedem Fall oder aber in einem bestimmten politischen Kontext. Das sollte allerdings kein Problem darstellen; mein Ziel in diesem Kapitel besteht nicht darin zu zeigen, dass potentielle Migrantinnen eine allgemeine Pflicht haben ein Gesetz zu befolgen, welches ihnen die Einwanderung untersagt, sondern wie diese Verpflichtung verstanden und behauptet und schließlich auch, wie sie selbst unter idealen Bedingungen eingeschränkt werden könnte. Ich möchte also schauen, welche Grenzen unseren Erwartungen gegenüber potentiellen Migrantinnen selbst unter idealen Bedingungen gesetzt werden könnten. Im Falle der Existenz solcher

Grenzen wäre es uns zumindest möglich, den Raum für weitergehende normative Untersuchungen auszuloten.

Lassen Sie uns also annehmen, dass all die von mir beschriebenen Prämissen einer näheren Betrachtung wert sind. Wir sind demnach mit einem Staat konfrontiert, der aufgrund seiner Beschaffenheit dazu berechtigt ist, Verpflichtungen durch Vorschriften zu schaffen. Ein solcher Staat schreibt demnach Personen, die sich außerhalb seines Hoheitsgebiets befinden, Folgendes vor: Ihr sollt diese Grenze nicht übertreten. Welchen Grund könnten die adressierten Personen haben, dieser Vorschrift Folge zu leisten?

1. Hume und Reziprozität

Es ist eine bemerkenswerte Eigenschaft des Einwanderungsgesetzes, dass diejenigen, denen gegenüber es Autorität beansprucht, an seiner Genese keinen Anteil haben. Sofern Reziprozität eine Bedingung moralisch legitimer Gesetzgebung ist und nach Fairness in der Verteilung der Früchte und Lasten der sozialen Kooperation verlangt, könnte diese Tatsache problematische Folgen haben. So könnten wir zu einer gewissen Radikalität in Bezug auf die Legitimität der Idee staatlichen Ausschlusses verleitet werden. Wie bereits weiter oben dargestellt, nutzt Arash Abizadeh die soeben dargestellte Problematik, um die Unzulässigkeit einer durch einzelne Staaten einseitig bestimmten Ausschlusspraxis zu demonstrieren.[7] Selbst diejenigen unter uns, die nicht so weit gehen, könnten zu der Annahme verleitet werden, dass die Missachtung des Prinzips der Reziprozität die Pflicht zum Befolgen der Gesetze untergräbt. Die derart Ausgeschlossenen könnten demnach einfach sagen: Wenn dieses Gesetz mir Lasten auferlegt, ohne mir dabei auch einen gewissen Vorteil zu bieten – ganz zu schweigen von einer substantiellen Mitsprache im Prozess der Gesetzgebung –, welcher Grund wird mir dann genannt, um die freiwillige Befolgung dieses Gesetzes anordnen zu können? Anders ausgedrückt: Warum sollte ich ein Gesetz befolgen, das zwar behauptet, mich zu seiner Befolgung verpflichten zu können, mir aber weder eine politische Stimme noch irgendwelche Vorteile gewährt?

Ich sollte anmerken, dass die Pflicht zur Befolgung des Gesetzes gemäß dieser Vorstellung letztlich auf der Fairness im Hinblick auf die Verteilung der Früchte und Lasten der im Rahmen dieses Gesetzes organisierten gesellschaftlichen Kooperation beruht. Zum näheren Verständnis dieser Überlegungen können wir die Idee der Reziprozität untersuchen, wie sie dem Kontraktualismus von John Rawls zugrunde liegt. Rawls' Kriterium der Reziprozität besagt, dass wir die von uns vorgeschlagenen politischen Regeln nur dann als vernünftig betrachten können, wenn wir „annehmen können, dass es für die anderen zumindest vernünftig ist, ihnen als freie und gleiche Bürgerinnen zuzustimmen."[8] Dieses Kriterium der Reziprozität leitet uns bei der Entwicklung sozialer und politischer Regeln, von denen wir erwarten, dass sie gerecht und über die Zeit hinweg stabil sind – darunter auch die Regeln hinsichtlich der Arten von Ungleichheit, die gerechterweise unter Bürgerinnen akzeptiert werden. Rawls' eigenes Verteilungsprinzip – das Differenzprinzip – wird von ihm unter Berufung auf diese Vorstellung von Reziprozität verteidigt; es rechtfertigt Ungleichheiten gegenüber den am schlechtesten Gestellten auf eine Weise, die die moralische Gleichheit aller Bürgerinnen akzeptiert und widerspiegelt.[9] Diese Ideen werden von Rawls allerdings nicht dazu genutzt um eine Pflicht zur Befolgung des Gesetzes zu begründen. Wie ich in Abschnitt zwei zeigen werde, verteidigt Rawls in seinem Spätwerk vielmehr eine natürliche Pflicht zu Unterstützung gerechter Institutionen. Allerdings können wir seine Ideen nutzen, um eine solche Pflicht gegenüber dem Gesetz zumindest in einer gerechten Gesellschaft, die dem Prinzip der Reziprozität folgt, direkt zu begründen: Du sollst das Gesetz befolgen, da es sowohl dein wechselseitiges Recht beachtet, bei der Entstehung des Gesetzes angehört zu werden, als auch die Früchte der Kooperation auf faire Weise verteilt.[10]

Unabhängig davon, ob diese Vorstellung von Verpflichtung attraktiv erscheint, kann sie keine Verpflichtung begründen, Gesetze im Bereich der Migration zu befolgen – oder zumindest nicht so leicht. Schließlich ist dieses Gesetz gegen diejenigen gerichtet, die der Definition nach keine Mitglieder der infrage stehenden Gesellschaft sind. Wovon sie ausgeschlossen sind, ist die Mitgliedschaft selbst. Daher wird eine Vorstellung von Verpflichtung, die mit der Idee einer reziproken politischen Gemeinschaft beginnt, kaum von analytischer Hilfe für die Beantwortung der Frage sein, ob und warum Einwanderungsgesetze zu befolgen sind, denn es geht ja gerade um den Eintritt *in* diese Gemeinschaft wechselseitiger Kooperation, den dieses Gesetz verweigert.

Hier könnte nun sicherlich schlicht der Punkt erreicht sein, an den wir gelangen wollten. Denn schließlich kann diese spezifische Vorstellung von Reziprozität sowohl für die Rechtfertigung sozialer Institutionen als auch für ihre Kritik herangezogen werden. Sie mag uns Gründe dafür geben, diese Institutionen zu respektieren und ihren Geboten zu folgen, wenn sie diese Art von Reziprozität verkörpern – und auch Gründe, sie zu verurteilen, sollte dies nicht der Fall sein. In eine ähnliche Richtung argumentiert auch Jeffrie Murphy, wenn er behauptet, dass die Theorie der Bestrafung krimineller Handlungen zwar formal korrekt sei, unter den gegebenen Umständen aber nicht zur Rechtfertigung der weitverbreiteten Inhaftierung ethnisch und ökonomisch marginalisierter Gruppen herangezogen werden kann:

> „Nehmen Sie ein Beispiel: Ein Mann wurde wegen eines bewaffneten Überfalls verurteilt. Während der Befragung erfahren wir, dass es sich um eine verarmte Person schwarzer Hautfarbe handelt, deren gesamtes Leben vom frustrierenden Ausschluss aus der herrschenden sozio-ökonomischen Struktur geprägt war – keine Arbeit; kein Nahverkehr, wenn es denn mal Arbeit gab; unterdurchschnittliche Ausbildung für die eigenen Kinder; schreckliche Wohnbedingungen und unzureichende Gesundheitsversorgung für die gesamte Familie; erniedrigende, verspätete und unangemessene Sozialleistungen; Belästigungen durch die Polizei ohne effektiven Schutz durch eben jene Polizei bei Verbrechen in der Nachbarschaft und schließlich ein fast vollkommener Ausschluss aus dem politischen Prozess. Im Wissen um all diese Umstände, wollen wir da immer noch – wie so viele es tun – davon sprechen, dass diese Person ihre Strafe erleidet, ‚um ihre Schuld gegenüber der Gesellschaft zu begleichen'? Sicherlich nicht. Welche Schuld?“[11]

Murphy möchte mit diesem Argument zeigen, dass der Staat mitunter nicht dazu berechtigt ist Personen zu bestrafen; allerdings könnte ein ähnliches Argument auch lauten, dass die infrage stehende Person keine Verpflichtung hat, den Gesetzen der Gesellschaft Folge zu leisten oder zumindest, dass sie keine Verpflichtung hat, dem Gesetz bloß deshalb Folge zu leisten, *weil es das Gesetz ist.* (Sie mag, schlicht aufgrund der Rechte anderer, frei von der Bedrohung durch Waffen zu leben, verpflichtet sein, einen bewaffneten Überfall zu vermeiden.) Derart könnten wir auch im Falle einer potentiellen Migrantin argumentieren. Sollten wir behaupten, dass das Gesetz, welches ihrem Ausschluss zugrunde liegt, für sie bisher von Vorteil gewesen sei und dass sie daher eine Schuld zum Befolgen jenes Gesetzes erworben hätte, könnte sie berechtigterweise fragen: Welche Schuld?[12]

Diese Argumente gehen aber vielleicht zu weit. Manche Kritikerinnen der Rawls'schen Vorstellung der Gerechtigkeit als Fairness meinen, dass seine Analyse der Gerechtigkeit diejenigen, die nicht an der Grundstruktur der Gesellschaft teilhaben, bei Gerechtigkeitserwägungen ausblendet, sofern diese Erwägungen nicht modifiziert oder eindeutiger formuliert werden. Da zum Beispiel schwer eingeschränkte Menschen mitunter nicht fähig sind, an der Marktwirtschaft oder an der politischen Entscheidungsfindung teilzuhaben, könne demnach eine Theorie der Gerechtigkeit, die sich auf die faire Verteilung der Früchte der sozialen Kooperation konzentriert, mitunter nichts über die Gerechtigkeit politischer Maßnahmen sagen, die jene Menschen betreffen. Da sie nicht an dem Prozess teilnehmen, der die zu verteilenden Güter hervorbringt, hätten sie gemäß dieser Gerechtigkeitsvorstellung auch keinen Anspruch darauf, angehört zu werden, wenn über die Verteilung dieser Güter durch die sozialen Institutionen diskutiert wird.[13] An diesem Punkt sollten wir meinen, über genügend Gründe zu verfügen, um Rawls' Theorie beiseitezulegen oder zumindest zu verlangen, dass sie eindeutiger formuliert oder überarbeitet wird.[14] Derart eingeschränkten Personen kommen korrekterweise gewisse Rechte hinsichtlich der Bestimmung einer gerechten Güterverteilung zu, unabhängig davon, wie sie bei der Genese dieser Güter mitgewirkt haben. Das scheint allerdings auf *Verpflichtungen* im gleichen Maße zuzutreffen wie auf *Rechte*. Wir sind versucht zu sagen, dass diejenigen, denen ein Gesetz keinerlei Vorteile verschafft, auch keine Verpflichtung zum Befolgen dieses Gesetzes haben. Allerdings kommt eine solche Aussage zu sehr der Idee nahe, dass die Moralität eines Gesetzes sich nur mit den Interessen und Verpflichtungen derjenigen beschäftigt, die fähig dazu sind, an den Aktivitäten mitzuwirken, die durch dieses Gesetz koordiniert werden – eine Schlussfolgerung, die wir zu Recht vermeiden wollen. Wenn wir weiterhin darauf bestehen wollen, dass manche Personen, die nicht an der politischen Gesellschaft teilhaben, nichtsdestotrotz ein Recht darauf haben, durch diese politische Gesellschaft gehört zu werden, dann müssten wir auch akzeptieren, dass diese Menschen gewisse Verpflichtungen haben.

Wir können diese Überlegungen weiter stützen, indem wir uns ansehen, inwiefern wir selbst dazu verpflichtet sein könnten, die Gesetze fremder Gesellschaften zu respektieren. Es scheint zum Beispiel Fälle zu geben, in denen ich dazu verpflichtet bin, das Gesetz eines Landes zu befolgen, in dem ich niemals war und dessen Dasein mir bisher keinerlei Vorteile beschert hat. Nehmen wir zum Beispiel die australischen Gesetze zur Regulierung von Videospielen. Das Spiel *Blitz: The League* ist momentan für den Import und Verkauf in Australien gesperrt. Es zeigt eine hemmungslose American Football League, in der die Nutzung von Steroiden nicht nur möglich, sondern auch förderlich für die Leistung des eigenen Spielcharakters ist.[15] (Es versteht sich von selbst, dass dieses Spiel in den Vereinigten Staaten problemlos erworben werden kann.) Nehmen wir nun an, dass ich der Meinung bin, Australiens Entscheidung, dieses Spiel zu zensieren, sei sinnlos und regressiv und ich mich deshalb dazu entscheide, das entsprechende Gesetz dadurch zu unterwandern, dass ich *Blitz: The League*-Kopien einschmuggle und anschließend kostenlos verteile. Mache ich damit etwas falsch? Vielleicht dann, wenn diese Spiele ihren Weg in die Hände von Kindern finden; aber die Frage bleibt selbst dann bestehen, wenn ich dies verhindern könnte. (Vielleicht schmuggle ich die Kopien nach Australien und gebe sie dann ein paar erwachsenen, leicht zu amüsierenden australischen Freunden.) Sollte ich das tun, könnte mir in meinen Augen immer noch der Vorwurf gemacht werden, die politische Gemeinschaft Australiens nicht angemessen zu respektieren. Diese politische Gemeinschaft ist schließlich mindestens so demokratisch wie meine eigene und mir könnte ein unzulässiger Mangel an Respekt gegenüber dem reziproken Prozess der politischen Entscheidungsfindung in Australien vorgeworfen werden, wenn ich die diesem Prozess entspringenden Gesetze schlicht als einfältig zurückweise und mich somit ihrer Wirkung auf mich zu entziehen versuche. Ich missachte das Gesetz, demgemäß es verboten ist, Kopien von *Blitz: The League* nach Australien zu bringen und es könnte angenommen werden, dass ich dadurch andere dazu ermutige das Gesetz zu missachten, das den Verkauf dieses Spiels innerhalb Australiens verbietet. Meiner Meinung nach habe ich eine Verpflichtung, die Gesetze Australiens zu befolgen und zwar nicht, weil sie weise oder bedeutend sind – im vorliegenden Fall trifft meiner Meinung nach beides nicht zu –, sondern aus einem gewissen Grad an Respekt gegenüber der politischen Gemeinschaft, aus der diese Gesetze hervorgegangen sind.

Daraus folgt allerdings, dass ich eine gewisse Verpflichtung haben könnte, nicht nur solche Gesetze zu befolgen, die mir Reziprozität im Prozess der Gesetzgebung oder dessen Ergebnissen garantieren, sondern auch die Gesetze (irgend)eines anderen Landes, und zwar aufgrund der Art und Weise, wie dieses Land eine solche Art von Reziprozität für seine eigenen Mitglieder realisiert. Ich mag verpflichtet dazu sein, Gesetze zu befolgen, die mich in der Gesetzgebung als Gleichen behandeln – gleichermaßen kann ich aber auch dazu verpflichtet sein, die Gesetze *anderer* Gemeinschaften zu respektieren; zumindest, wenn diese Gemeinschaften von einem bestimmten Charakter sind. Letzteres kann jedoch nicht direkt aus meinen eigenen Ansprüchen auf wechselseitige Rechtfertigung politischer Macht abgeleitet werden; wir müssen diese Verpflichtung damit begründen, den Prozess zu respektieren, durch den Menschen verantwortliche und gerechtfertigte Staaten einzurichten versuchen. Wir sollten daher das Argument zurückweisen, dass es angeblich keine Verpflichtung zur Befolgung von Einwanderungsgesetzen gibt, weil diese Gesetze die Migrantinnen bei der Verteilung der Vorteile sozialer Kooperation nicht als Gleiche anerkennen. Wie wir nun gesehen haben, setzt eine solche Pflicht nicht notwendig voraus, dass dieses Gesetz für uns tatsächlich von Vorteil ist, ganz gleich ob dieser Vorteil nun im Vergleich zu einem gegenwärtigen Zustand oder einem bestimmten moralischen Minimum verstanden wird.

Es gibt selbstverständlich eine Vielzahl möglicher Reaktionen auf diese Überlegungen. Eine mögliche Reaktion besteht darin, in den sauren Apfel zu beißen und zu erklären, dass diejenigen, die vom Prozess der Gesetzgebung ausgeschlossen sind, auch keine Verpflichtung haben, das aus ihm entspringende Gesetz zu befolgen; allerdings ist das, wie ich bereits gezeigt habe, mit einigen Kosten verbunden. Eine andere mögliche Reaktion besteht darin, zu behaupten, dass die zurückgewiesenen Migrantinnen tatsächlich an einer Art von kooperativer Aktivität mit dem sie zurückweisenden Staat teilhaben, was sie wiederum dazu berechtigt, Fragen hinsichtlich der Fairness dieser Kooperation zu stellen. Es gibt diesem Ansatz zufolge einen Kooperationszusammenhang in den Beziehungen zwischen Staaten, innerhalb dessen sich Staaten freiwillig wechselseitig Respekt für ihre jeweilige territoriale Souveränität und politische Aktivität zusprechen; wir könnten demnach die Pflicht zur Befolgung von Einwanderungsgesetzen tatsächlich damit begründen, dass sich die Staaten an diesem bestehenden Projekt beteiligen.[16] Selbst wenn dieser Sachverhalt zutreffen sollte, bestünde jedoch das Problem, dass ein solches Argument eher Staaten denn Individuen verpflichten würde. Staaten haben mutmaßlich ein Interesse daran, dass sich andere Staaten nicht ihres Territoriums bemächtigen oder ihre politische

Unabhängigkeit beeinträchtigen; damit ließe sich eine gewisse Pflicht zur Befolgung der Regeln des internationalen Rechts begründen, sofern sich dieses Recht durch das Prinzip der Reziprozität rechtfertigen ließe. Allerdings begründet in diesem Falle nichts eine Verpflichtung aufseiten der individuellen Migrantinnen, ein sie zurückweisendes Gesetz zu befolgen. Die hier zur Debatte stehende Wechselseitigkeit besteht zwischen Staaten, nicht aber zwischen Personen. Wir könnten nun behaupten, dass die einzelne Person eine Pflicht hat, diese Form der Reziprozität zu respektieren; allerdings würde diese Pflicht auch in diesem Falle nicht auf der Idee der Reziprozität selbst beruhen, sondern auf einer davon unabhängigen Pflicht dazu, reziproke Beziehungen zwischen anderen Personen zu respektieren und zu fördern. Wir sind also wieder bei dem Gedanken angelangt, dass wir nicht deshalb dazu verpflichtet sind, die Gesetze einer fremden Gesellschaft zu befolgen, weil diese politische Gemeinschaft uns mittels wechselseitiger Rechtfertigung als moralisch Gleiche behandelt, sondern weil wir dazu verpflichtet sind, die Fähigkeit einer *anderen* Gemeinschaft, auf reziproke Art Politik zu betreiben, zu schützen und zu bewahren.

Es scheint daher angemessen, dass wir uns den zweiten Grund für eine mögliche Pflicht zur Befolgung von Einwanderungsgesetzen ansehen: Die kantische Vorstellung, der zufolge wir dazu verpflichtet sind, gerechte Institutionen zu schützen und zu bewahren. Wie ich zeigen werde, kann dieser Ansatz potentiellen Migrantinnen eine gewisse Pflicht auferlegen, einem sie ausschließenden Gesetz Folge zu leisten. Sollte ich richtig liegen, dann hängt eine solche Pflicht jedoch von bestimmten Umständen ab – und selbst wenn eine solche Pflicht begründet werden könnte, würde sie mitunter nicht als Rechtfertigung für die Art von Einwanderungspolitik dienen können, die Staaten wie die USA derzeit verfolgen. Diesen Überlegungen werde ich mich nun zuwenden.

2. Kant und natürliche Pflichten

Wie bereits erwähnt, begründete Rawls die Pflicht zur Befolgung des Gesetzes nicht direkt mit dem Gedanken der Reziprozität; die Pflicht, den Gesetzen eines gerechten Staates Folge zu leisten, kommt uns ihm zufolge nicht aufgrund fairen Umgangs zu, sondern aufgrund einer natürlichen Pflicht, gerechte Institutionen zu unterstützen.[17] Die von mir zuvor diskutierte Idee der Reziprozität hat immer noch einen Platz in Rawls' Theorie eines gerechten Staates; allerdings wird sie selbst nicht herangezogen, um den verpflichtenden Charakter staatlicher Gebote zu erklären. Rawls ist der prominenteste Verteidiger dieses Ansatzes, allerdings bei weitem nicht der einzige; eine Reihe von Theoretikerinnen haben in letzter Zeit weitere Varianten dieser Pflicht entwickelt. Zwischen diesen Ansätzen gibt es selbstverständlich bedeutsame Unterschiede und jeder von ihnen arbeitet wichtige Unterscheidungen hinsichtlich der Handlungen heraus, die eine solche Pflicht gebietet.[18] An dieser Stelle möchte ich von diesen Unterschieden jedoch absehen und stattdessen fragen, was aus einer solchen Pflicht im Allgemeinen für potentielle Migrantinnen folgen würde. Sofern eine Migrantin durch den Bruch eines sie zurückweisenden Gesetzes ein Unrecht begeht, scheint es plausibel, anzunehmen, dass dieses Unrecht am besten unter Verweis auf eine natürliche Pflicht zum Befolgen der Gebote gerechter Institutionen verstanden werden kann. Eine Person, die das Gesetz bricht, vermindert die Autorität der Institutionen, die jenes Gesetz hervorgebracht haben; sofern wir also das Gesetz ungestraft brechen können, nehmen wir ihm insgesamt etwas von seinem Wesen als Gesetz. Ein Rechtsbruch kann verschiedene konkrete Effekte mit sich bringen; ein weit verbreiteter Ungehorsam gegenüber dem Gesetz schließlich kann dafür sorgen, dass das Vorhaben deliberativer Politik selbst unmöglich wird.[19] Aber wir müssen für unsere Argumentation keine solch konkreten Effekte nachweisen; es genügt zu sagen, dass diejenige Person, die das Gesetz bricht, die allgemeinere Pflicht verletzt hat, die gerechten Institutionen zu fördern und zu bewahren, die jenes Gesetz hervorgebracht haben. Wir können etwas ähnliches auch im Falle der Migration sagen. Weit verbreiteter Ungehorsam gegenüber einem Gesetz, das Migrantinnen ausschließt, könnte demnach negative Effekte auf unsere Fähigkeit haben, miteinander Politik zu betreiben.[20] Aber wir müssten noch nicht einmal zeigen, dass es tatsächlich so kommen würde, denn es wäre bereits ausreichend, wenn wir zeigen, dass die Person, die ein solches Gesetz bricht, die Autorität der diesem Gesetz zugrunde liegenden gerechten Institutionen mindert und somit eine bestehende Pflicht gegenüber diesen Institutionen verletzt.

Diese Vorstellung von einer Pflicht zur Befolgung der Gesetze ist zwar plausibler, allerdings ebenfalls nicht ohne Schwierigkeiten. Die erste und offensichtlichste Schwierigkeit besteht darin, dass eine solche Pflicht höchstens *prima facie* Geltung beanspruchen kann. So wäre eine solche Pflicht sicherlich mit Formen des Rechtsbruchs kompatibel, die sich an legitimen Prinzipien orientieren, wie beispielsweise zivilem Ungehorsam. Für den Moment möchte ich solche Formen des Rechtsbruchs jedoch außen vor lassen und mich stattdessen darauf konzentrieren, was diese *prima facie*-Pflicht unter bestimmten Umständen zu leisten vermag. Meine Behauptung wird sein, dass sie weniger stark ist, als sie zunächst erscheint. Sie wird uns vermutlich nicht die klare und einfache moralische Verurteilung irregulärer Migrantinnen erlauben, die vor allem im rechten politischen Spektrum verbreitet ist. Ich möchte diese Schlussfolgerung im Folgenden anhand dreier Fallbeispiele verteidigen, die ich ihrer moralischen Bedeutung nach in absteigender Reihenfolge besprechen werde.

a. *Abraham* wird in seinem Heimatland Opfer einer Verletzung internationaler Menschenrechte. Er hat die Möglichkeit, über die Grenze in ein benachbartes Land zu laufen, das seine Recht schützen würde, das jedoch nichtsdestotrotz ein Gesetz verabschiedet hat, das ihm dies verbietet.

Ich nehme es als ziemlich offensichtlich an, dass Abraham unter Berücksichtigung aller Umstände keine Pflicht dazu hat, die Grenze nicht zu überqueren. Es scheint in meinen Augen einen einfachen Grund zu geben, den Abraham zur Verteidigung seiner Entscheidung anführen kann. Die Verletzung, die dem gerechten Staat durch das Brechen des Gesetzes widerfahren würde, wiegt um einiges leichter als die Verletzung, die Abraham durch die Misshandlung seines Körpers erfahren musste. Mir fällt es schwer, Gründe dafür anzuführen, und ich vermute, dass jeder Grund, den ich anführen könnte, letztendlich weniger überzeugend wäre als die Schlussfolgerung selbst. Eine diesem Schluss widersprechende Person würde das abstrakte Wohlbefinden des Staates über die fortdauernde Existenz einer konkreten, über Rechte verfügenden Person stellen. Derlei Überlegungen waren es, die dem modernen Asylsystem überhaupt erst zu seiner Existenz verholfen haben, von dem oft angenommen wird, dass es aufgrund der Gräuel des Zweiten Weltkriegs entstand. Staaten ist es rechtlich verboten, Menschen wie Abraham zurückzuweisen. Wie auch immer ein staatliches Recht auf Ausschluss begründet wird, ist es moralisch doch weniger bedeutsam als das Recht von Personen, bestimmten Formen absoluten Grauens zu entkommen. Wenn wir all das nicht aus der Perspektive des Staates, sondern aus derjenigen Abrahams betrachten, können wir uns vorstellen, dass er zwei Pflichten unterliegt: Zum einen der Pflicht, die Rechte des (ansonsten) gerechten Staates zu respektieren, der ihn ausschließen möchte, und zum anderen die Pflicht, sein eigenes Recht darauf zu respektieren, als ein über Rechte verfügendes Individuum überleben zu können. Darauf zu bestehen, dass Abraham die erste Pflicht der zweiten vorziehen sollte, scheint fehlgeleitet, um das Mindeste zu sagen. Tatsächlich würde einem Abraham, der die erste Pflicht befolgen würde, zu Recht der Vorwurf mangelnder Selbstachtung gemacht werden.

Diese Überlegungen werden jedoch strittiger, je mehr wir uns von der Art von Menschenrechtsverletzungen entfernen, die dem modernen Recht auf Asyl zugrunde liegen. Sehen Sie sich zur Verdeutlichung nun einen Fall an, in dem die gefährdeten Rechte von umstrittenerer Art sind und bei denen derzeit nicht davon ausgegangen wird, dass sie ein Recht auf Asyl begründen können: Das Recht darauf, frei von undemokratischen und tyrannischen Formen der Regierung leben zu können.

> b. *Bobby* ist in seinem Heimatland reiner Tyrannei ausgesetzt. Er kann über die Grenze hinweg in ein Land laufen, dass seine Recht schützt, das jedoch nichtsdestotrotz ein Gesetz verabschiedet hat, das ihm dies verbietet.

Dieser Fall ist in meinen Augen aufgrund einer Vielzahl von Gründen komplizierter – nicht zuletzt deshalb, weil das Recht auf eine demokratische Regierung als Quelle internationalen Rechts umstrittener ist.[21] Gemäß internationalem Recht kann Bobby kein Recht auf Eintritt in ein Land zum Zwecke des Schutzes vor seiner eigenen Regierung geltend machen; bloße Tyrannei ist demnach nicht genug, um die prinzipielle Annahme außer Kraft zu setzen, dass Staaten den Zutritt zu ihrem Territorium kontrollieren dürfen. Es sollte allerdings angemerkt werden, dass das rechtliche Konzept des *subsidiären Schutzes* eine Möglichkeit bietet, Bobby und Personen, die sich in ähnlichen Situationen befinden, einen gewissen, begrenzten Schutz vor Zurückweisung zu bieten – obwohl diese Form des Schutzes weit hinter dem zurückbleibt, was durch die Verträge des internationalen Schutz- und Asylsystems gewährleistet wird.[22]

Unsere Frage ist jedoch eher von moralischer denn juristischer Natur; und gemäß den von mir in diesem Buch und an anderer Stelle vorgebrachten

Argumenten würde ich behaupten, dass Bobby ein Recht darauf hat, frei von der Art von Tyrannei zu leben, mit der er in seinem Heimatland konfrontiert ist, weshalb andere Staaten kein Recht darauf haben, Zwang einzusetzen, um ihn in dieser ungerechten Form politischer Beziehung zu halten. Aus Bobbys Perspektive müssen wir diese Frage jedoch überhaupt nicht stellen, sondern bloß fragen, ob und inwiefern er das ihn zurückweisende Gesetz als verbindlich ansehen sollte. Ob dieses Gesetz nun ungerecht ist, spielt in dieser Betrachtung keine zentrale Rolle, denn schließlich können wir dazu verpflichtet sein, ein ungerechtes Gesetz zu befolgen, wenn es einem ansonsten gerechten politischen System entspringt. Folgt also aus Bobbys allgemeiner Pflicht gegenüber dem möglichen Aufnahmestaat auch dann eine Pflicht zur Befolgung einzelner Vorschriften, wenn diese seine Zurückweisung gebieten?

Ich denke, dem ist nicht so. Stattdessen glaube ich, dass wir es hier mit zwei verschiedenen Vorstellungen davon zu tun haben, wie gerechte Institutionen unterstützt und gefördert werden sollten. Eine Vorstellung besagt, dass die einzelnen Vorschriften dieses Staates befolgt werden sollten. Die andere Vorstellung hingegen fordert, dass die Reichweite dieser Institutionen vergrößert werden sollte und zwar dadurch, dass sie Zugriff auf einen größeren Teil der menschlichen Welt erhalten. Wir unterscheiden nur selten zwischen diesen zwei Vorstellungen unserer Pflicht gegenüber staatlichen Institutionen. Im Allgemeinen nehmen wir die Gruppe von Personen, über die der Staat regiert, als gegeben hin und fragen nur, was wir als Mitglieder dieser Gruppe tun müssen, um diesen Staat zu erhalten. In Fragen der Migration können wir jedoch nicht von dieser Prämisse ausgehen. Wir könnten in diesem Fall den Zweck der gerechten Institutionen dadurch befördern, dass wir ihre Autorität auf einen größeren Teil der Welt ausweiten, statt auf ihre Vorschriften zu hören – im Kern würde das bedeuten, diesen Institutionen mehr Menschen zuzuführen, über die sie regieren können. Wir könnten also annehmen, dass die Pflicht, gerechte Institutionen zu fördern, dadurch erfüllt werden kann, dass wir einen größeren Teil der Menschheit unter diesen Institutionen versammeln. Sollte dem so sein, scheint Bobby jedoch mit einem Konflikt konfrontiert zu sein, der aus der natürlichen Pflicht entspringt, gerechte Institutionen zu bewahren und zu fördern: Er könnte entweder den Vorschriften der ihn zurückweisenden Institutionen Folge leisten oder – indem er eben jene Vorschriften ignoriert – die Reichweite dieser

Institutionen vergrößern, indem er ihnen eine weitere Person zuführt, für die diese Institutionen autoritativ gelten. Wenn wir also dazu verpflichtet sind, stärkere Institutionen aufzubauen, besteht ein Weg möglicherweise darin, die Anzahl an Personen zu vergrößern, über die diese Institutionen herrschen und es steht Bobby zumindest offen zu behaupten, dass seine Entscheidung zum Grenzübertritt genau das bedeuten würde.

Dieses Argument sieht zugegebenermaßen ein wenig wie ein Trick aus. Zumindest enthält es den Hauch eines inneren Widerspruchs: Ich soll diese gerechten Institutionen respektieren und bewahren, indem ich ihren Versuch ignoriere, mich zurückzuweisen! Trotzdem denke ich, dass diese Überlegungen Bobbys Entscheidung zur Einwanderung tatsächlich rechtfertigen können. Wenn wir möchten, können wir dieses Argument auch unter Verweis auf die moralische Nichtigkeit politischer Beziehungen innerhalb einer Tyrannei umformulieren. So könnten wir anführen, dass die Pflicht zur Gerechtigkeit als eine Pflicht gesehen werden kann, die Verbreitung und Macht totalitärer Regime zu verringern. Davon ausgehend könnten wir schließlich annehmen, dass eine Möglichkeit, diese Pflicht zu erfüllen, darin besteht, Menschen aus den Fängen solch tyrannischer Regime zu befreien; und es gibt keinen prinzipiellen Grund, warum die Person, die ich vor der Tyrannei bewahre, nicht *ich selbst* sein kann.

An dieser Stelle könnte selbstverständlich noch viel mehr gesagt werden; aber ich denke, dass diese Ideen zumindest den Ansatz einer Rechtfertigung für Bobbys Entscheidung darstellen können, das ihn zurückweisende Gesetz zu missachten. Demzufolge kann er vorbringen, dass die Pflicht, gerechte Institutionen zu bewahren und zu befördern, sich nicht bloß auf die Pflicht beschränkt, jede einzelne Vorschrift dieser Institutionen zu befolgen; wenn Menschen Grenzen überqueren ist es durchaus möglich, dass die Verletzung einer bestimmten Vorschrift moralisch weniger bedeutsam ist als die Schaffung einer neuen und gerechten politischen Beziehung. Sobald Bobby eine gerechte Gesellschaft betritt, erwirbt er, schlicht aufgrund seiner Anwesenheit innerhalb des Hoheitsgebiets dieser Gesellschaft, bestimmte Rechte und die Welt ist daher, sozusagen aufgrund seines Grenzübertritts, gerechter geworden.

Der dringlichste – und wohl auch umstrittenste – Fall betrifft jedoch nicht verwerfliche Tyranneien, sondern schlicht ökonomische Benachteiligung. Ich muss an dieser Stelle klar feststellen, dass wir nicht über Fälle von Armut reden, die so gravierend sind, dass sie die Fähigkeit einschränken, in Würde zu leben. Es kann Fälle von Armut geben, die so gravierend sind, dass sie wie die bereits soeben diskutierten Fälle zu behandeln sind – ethisch und mitunter auch rechtlich. Stattdessen werde ich mich hier bloß mit der Art wirtschaftlicher Ungleichheit befassen, die dazu führt, dass ein Individuum die eigenen Lebensperspektiven durch die Überquerung staatlicher Grenzen drastisch verbessern kann.

> c. *Carla* ist arm, allerdings droht ihr in ihrem Heimatland aufgrund dieser Armut kein verfrühter Tod und auch kein gravierender, vermeidbarer Schaden. Sie hat die Möglichkeit, die Grenze in ein Land zu überqueren, das ihre Rechte schützen würde, und würde durch diesen Grenzübertritt ihren Lebensstandard drastisch verbessern. Allerdings hat dieses Land ein Gesetz erlassen, das ihr den Grenzübertritt verbietet.

Carlas Fall ist dem Fall von Miguel Sanchez offensichtlich sehr ähnlich. Carla, und alle ähnlich situierten Personen, werden oft als „Wirtschaftsmigrantinnen" bezeichnet; dieser Begriff wird manchmal verurteilt, da er herunterzuspielen scheint, in welchem Maße materielle Armut ebenso zerstörerisch wirken kann wie staatliche Verfolgung.[23] An dieser Stelle möchte ich jedoch behaupten, dass Carlas Fall keiner jener Fälle zu sein scheint, in denen es eine angemessene Rechtfertigung dafür gibt, das zurückweisende Gesetz zu brechen – zumindest, solange keine anderen moralisch relevanten Umstände bekannt sind. Das ausschließende Gesetz benachteiligt Carla zwar; allerdings benachteiligen alle Gesetze *irgendwen* und im Allgemeinen sehen wir diesen Umstand nicht als eine Tatsache an, die den Bruch eines Gesetzes rechtfertigt. Das Gesetz mag ungerecht sein; mitunter würde die beste uns zur Verfügung stehende Theorie globaler Gerechtigkeit diese Form von Armut als illegitim ausweisen. Diese Tatsache scheint jedoch ebenfalls nicht ausreichend, um ein Recht darauf zu begründen, die Gesetze eines gerechten Staates zu brechen. So betrachten wir moralische Perfektion gemeinhin nicht als Voraussetzung, um Pflichten auferlegen zu können; selbst wenn also die Verteilung des Wohlstands in den Vereinigten Staaten ungerecht ist – und ich kenne sehr wenige Menschen, die sich nicht über zumindest *einige* Aspekte dieser Verteilung beschweren –, folgt daraus nicht, dass die durch die USA geschaffenen Eigentumsrechte nicht berechtigterweise einen gewissen Grad an Respekt verdienen. Sollten die Vereinigten Staaten also eine gewisse

Form von Respekt beanspruchen dürfen, besteht die angemessene Antwort auf ungerechte Verhältnisse nicht im Bruch von Gesetzen, sondern in politischem Engagement. Selbst wenn wir in einer vollkommen gerechten Gesellschaft eine deutlich andere Eigentumsverteilung vorfinden würden, folgt daraus nicht, dass Sie in *dieser* Gesellschaft einfach mein Auto an sich nehmen dürften. Ebenso gibt die Tatsache, dass die Welt derzeit ungerecht ist, und diese Ungerechtigkeit zudem Carla betrifft, Carla nicht das Recht – jedenfalls noch nicht –, das sie zurückweisende Gesetz zu brechen.

Diese Ergebnisse sind, gelinde gesagt, unbefriedigend. Es gibt allerdings viele Wege, die wir nun einschlagen können. Wir könnten den Gedanken zu widerlegen suchen, dass Carla dazu verpflichtet ist, das sie zurückweisende Gesetz zu befolgen. So könnten wir zum Beispiel anführen, dass Carla nicht aus jenem wohlhabenden Land ausgeschlossen werden darf, da der Reichtum dieses Landes auf globaler Ausbeutung und Kolonialismus beruht, die wiederum die Ursachen für Carlas gegenwärtige Armut sind.[24] Allerdings ist unklar, ob derlei Tatsachen Carlas Gesetzesbruch rechtfertigen könnten. Solche Überlegungen würden uns dabei helfen zu verstehen, warum Carlas Ausschluss möglicherweise ungerecht ist, allerdings könnten wir auf diesem Wege dem Staat nicht die Fähigkeit absprechen, verbindliche Vorschriften zu erlassen. Alternativ könnten wir uns auch der Idee des zivilen Ungehorsams gegenüber dem zurückweisenden Staat zuwenden. Allerdings wird ziviler Ungehorsam, zumindest gemäß den meisten theoretischen Ansätzen, als Erfüllung einer bestimmten Pflicht gegenüber diesem Staat verstanden – schließlich sind diejenigen, die am zivilen Ungehorsam teilnehmen, keine Anarchistinnen. Ich für meinen Teil denke, dass wir in eine andere Richtung gehen sollten. Dabei möchte ich nicht über eine Pflicht zur Befolgung des Gesetzes sprechen, sondern über die Kosten der *Erfüllung* dieser Pflicht. Diese Diskussion wird uns zu Überlegungen zurückführen, die in ähnlicher Form bereits im Zusammenhang mit der Idee der Reziprozität besprochen wurden. Ich schlage demnach vor, dass wir nicht erwarten können, dass eine Pflicht erfüllt wird, sofern die mit der Erfüllung dieser Pflicht verbundenen Lasten so erheblich sind, dass wir nicht annehmen können, wir selbst – oder irgendwer – seien in der Lage, ihren Anforderungen gerecht zu werden.

Beginnen möchte ich dieses Argument mit einer Unterscheidung zwischen dem, was eine Pflicht von uns verlangt, und den konkreten Folgen einer der Pflicht entsprechenden Handlung für unsere eigene Situation. Eine besonders starke Art jener Unterscheidung findet sich in der Geschichte *Fräulein Smillas Gespür für Schnee*. In dieser Geschichte wird Smilla – einer grönländischen Inuk – mit Haft gedroht. Da sich die Lebensform der Inuit vor dem Hintergrund unendlicher Weiten und entfernter Horizonte entwickelt hat, stellt Gefangenschaft einen besonderen Schrecken für Smilla dar, genauso wie für alle ihr ähnlichen Personen:

> „‚Eingesperrtsein', sagt er langsam; ‚in einem kleinen schalltoten Raum ohne Fenster, ist, so habe ich mir sagen lassen, besonders unangenehm, wenn man in Grönland aufgewachsen ist.' Er hat nichts Sadistisches an sich. Er weiß nur genau und vielleicht ein bißchen melancholisch über seine Druckmittel Bescheid. In Grönland gibt es keine Gefängnisse. Der größte Unterschied zwischen der Gesetzgebung in Dänemark und in Nuuk besteht darin, daß man in Grönland Gesetzesübertretungen, für die man in Dänemark mit Haft oder Gefängnis bestraft wird, weit häufiger mit Geldbußen ahndet. Die grönländische Hölle ist nicht die schwefelschwappende europäische Klippenlandschaft. Die grönländische Hölle ist der geschlossene Raum. Ich erinnere mich an meine Kindheit, als seien wir nie in Innenräumen gewesen. Es war undenkbar für meine Mutter, längere Zeit am selben Ort zu wohnen. Mir geht es mit meiner räumlichen Freiheit wie – nach meiner Beobachtung – Männern mit ihren Hoden."[25]

Wären wir Anhänger einer Theorie, die Strafe als Vergeltung begreift, wäre Smillas Gefangenschaft eine angemessene Reaktion auf ihr Verbrechen; vor Gericht haben wir keinen Grund, danach zu fragen, worin ihre *besondere* Abneigung gegenüber der Gefangenschaft besteht. Wenn Smilla das gleiche Verbrechen wie Dennis begeht – ein ehemaliger Mönch mit einer Vorliebe für kleine Räume –, können wir sie berechtigterweise für die gleiche Zeit einsperren; Dennis wird zwar weniger leiden als Smilla, allerdings besteht unser Ziel bei dieser Form der Bestrafung darin, Gerechtigkeit herzustellen, und zwar im Sinne des Verwirkens eines zu Unrecht erworbenen Vorteils, und nicht im Sinne eines perfekt austarierten psychischen Schmerzes. An Smillas Bestrafung scheint nichts ungerecht zu sein, selbst wenn die meisten von uns hoffen mögen, dass in dem sie betreffenden Strafrechtssystem nicht allein die Tugend der Gerechtigkeit den Ausschlag gibt. Aus dem Gesagten folgt jedoch, dass Gerechtigkeit sich auf das bezieht, was richtigerweise zu tun ist; die daraus resultierenden psychischen Belastungen einzelner Personen spielen dabei größtenteils keine Rolle. Angesichts Smillas Geschichte könnten wir jedoch zu glauben beginnen, dass die Idee der Gnade, wie sie im Strafrecht zu finden ist, einer näheren Betrachtung wert ist. Demnach hätten wir gute moralische Gründe, Smilla nicht derart zu bestrafen, selbst wenn ihr durch ihre Gefangenschaft kein Unrecht widerfahren würde.

Ich denke, dass etwas Ähnliches nicht nur hinsichtlich der Bestrafung eines Gesetzesbruchs, sondern auch über die Befolgung des Gesetzes gesagt werden kann. Es ist unzweifelhaft wahr, dass verschiedene Menschen unterschiedliche Belastungen empfinden, wenn sie das Gesetz befolgen. Eine Person, die kein Interesse daran hat, anderen ins Gesicht zu schlagen, wird es als recht einfach empfinden, einem Gesetz zu folgen, das Körperverletzungen untersagt. Einer Person hingegen, die *jeden* schlagen möchte, wird es große persönliche Schwierigkeiten bereiten, diesem Gesetz Folge zu leisten. Dieser Umstand selbst gibt uns selbstverständlich keinen besonderen Grund zu meinen, es sei jeder Person freigestellt das Gesetz zu befolgen; es gibt gute Gründe für seine Existenz und wir wissen, dass die Kosten seiner Befolgung ungleich verteilt sein werden. Besteht die Pflicht dem Gesetz Folge zu leisten, folgt aus der Tatsache einer ungleichen Verteilung der aus ihr resultierenden Belastungen also nicht, dass die Pflicht nicht mehr für alle gleichermaßen gilt.

Wir können diese Analyse nun auf den Fall von Carla anwenden. Stellen Sie sich vor, dass Carla ihr Einkommen durch den Grenzübertritt verdreifachen würde. Dabei handelt es sich keinesfalls um einen abstrusen Gedanken: Obwohl Menschen selten allein aus wirtschaftlichen Gründen migrieren, ist es doch unzweifelhaft ein Teilmotiv und diejenigen, die beispielsweise aus Mexiko in die Vereinigten Staaten auswandern, vervielfachen ihr Einkommen tatsächlich um den Faktor drei bis sechs.[26] Die Kosten, die *für Carla* aus der Befolgung des Gesetzes resultieren, sind also, gelinde gesagt, signifikant, und das insbesondere, sobald wir anerkennen, dass ihr Einkommen nicht bloß ihr zugutekommt, sondern auch denjenigen unter ihren Familienmitgliedern, die nicht am Markt teilnehmen – darunter sowohl Kinder als auch ältere Menschen. Eine mögliche Art, auf diese Tatsachen zu reagieren besteht darin, das Gesetz, das Carla zurückweist, als ungerecht zu bezeichnen. Unsere gegenwärtige Frage beschäftigt sich jedoch nicht mit der Gerechtigkeit dieses Gesetzes, sondern damit, *ob die Verpflichtung Carlas gegenüber dem Gesetz aufgehoben ist*. Wenn meine Überlegungen korrekt sind, muss die Antwort nein lauten. Carla wird zwar durch dieses Gesetz belastet, allerdings folgt aus diesen Lasten nicht, dass das Gesetz für sie nicht mehr gilt.

Das alles klingt ziemlich hartherzig und sollten wir dem nichts mehr hinzufügen, könnte unsere Antwort zu Recht als unmenschlich bezeichnet werden. Ich denke jedoch, dass wir noch ein Stück weiter gehen können. Selbst wenn Carla eine Verpflichtung hat, dem Gesetz Folge zu leisten, könnten wir fragen, welche moralischen Konsequenzen aus einer solchen Verpflichtung folgen. Es könnte schließlich durchaus möglich sein, dass diese Pflicht durch bestimmte Verpflichtungen, die anderen gegenüber bestehen, übertrumpft wird; so ist es zumindest möglich, dass Carlas Verpflichtung gegenüber einem Kind oder einem Elternteil berechtigterweise Vorrang vor der Pflicht hat, das sie ausschließende Gesetz zu befolgen. (In diesem Gedanken spiegelt sich die Aussage von Jeb Bush.) Allerdings bin ich mehr an der Frage interessiert, wie diese Überlegungen über Kosten unsere Bewertung von Carlas moralische *Charakter* beeinflussen könnten. Aus dieser Perspektive könnten Ideen wie Kosten und Reziprozität nun wieder zu ihrem Recht kommen; zwar nicht als Grundlage für eine Pflicht, das Gesetz zu befolgen, aber als Prinzipien zur Bewertung des moralischen Charakters derjenigen, die dieses Gesetz brechen.

Es gibt an dieser Stelle mindestens zwei Möglichkeiten. Eine Möglichkeit besteht darin, sich zu fragen, wie hoch die relativen Kosten der Befolgung einer solchen Pflicht für die Einwohnerinnen wohlhabender sowie ärmerer Gesellschaften jeweils ausfallen. Das bestehende Migrationsrecht erlaubt es den reichen wie auch den armen Ländern, externe Personen auszuschließen, so wie es das Gesetz sowohl reichen als auch armen Personen verbietet, unter Brücken zu schlafen. Die daraus resultierenden Belastungen sind für wohlhabende Personen offensichtlich deutlich geringer als für arme Personen. Im gleichen Sinne haben wohlhabende Personen im Großen und Ganzen weniger Gründe zu migrieren, und es gestaltet sich für sie deutlich einfacher, sollten sie sich doch einmal dafür entscheiden.[27] Es ist möglich, dass diese Art mangelnder Reziprozität dazu führt, dass wir ein größeres Verständnis dafür aufbringen sollten, wenn arme Menschen die infrage stehende Verpflichtung zur Befolgung des Gesetzes brechen – und dass wir daher mitunter auch nicht auf einen deformierten moralischen Charakter dieser Personen schließen sollten –, als wenn ein solcher Bruch durch eine wohlhabende Person erfolgen würde.

Eine andere mögliche Herangehensweise besteht hingegen darin, zu argumentieren, dass es manche Lasten gibt, angesichts derer es fast jeder normalen Person unmöglich wäre, eine bestimmte Pflicht zu erfüllen. Ein möglicher Ausgangspunkt für diese Überlegung ist der berühmte Fall *Regina v. Dudley and Stephens*. Dieser Fall, der den meisten Jura-Studentinnen bekannt sein dürfte, behandelt den Mord und Kannibalismus an einem Schiffsjungen durch ältere Seeleute an Bord eines schiffbrüchigen britischen Boots. Die Richter waren in diesem Fall mit der Frage konfrontiert, ob eine Notsituation einen Mord rechtfertigen kann; ob also die Tatsache, dass der Mord für die Rettung der Seeleute notwendig war, ausreicht, um diesen Mord als rechtlich erlaubt zu verteidigen. Die Antwort lautete, das dem nicht so sei. Allerdings bin ich mehr an einem seltsamen Eingeständnis der Richter interessiert, nämlich dass sie selbst wie die Seeleute gehandelt hätten, wären sie an Bord dieses Boots gewesen. Ihrer Ansicht nach folgt daraus jedoch nicht, dass die Tat rechtmäßig war:

> „Das Urteil darf nicht so verstanden werden, dass über die Ablehnung, eine Versuchung als Entschuldigung für ein Verbrechen anzuerkennen, vergessen worden wäre, wie schrecklich die Versuchung war, wie schlimm das Leiden und wie hart es ist, in solchen Fällen klar zu urteilen und rein zu handeln. *Wir sind oft versucht, Standards anzulegen, die wir selbst nicht erreichen würden und Regeln festzulegen, die wir selbst nicht erfüllen könnten.* Aber eine Person hat kein Recht, eine Versuchung als Entschuldigung gelten zu lassen, auch wenn sie selbst ihr mitunter nachgegeben hätte, noch darf sie es dem Mitgefühl für den Kriminellen erlauben, die rechtliche Definition des Verbrechens zu ändern oder abzuschwächen.“[28]

Es handelt sich hier um eine bemerkenswerte Aussage, die feststellt, dass es moralische Prüfungen geben kann, bei denen wir davon ausgehen können, dass jeder normale Mensch – selbst ein Richter – sie nicht bestehen würde. Allerdings wird die infrage stehende Pflicht dadurch nicht geschmälert. Sind die Lasten der Befolgung einer Pflicht zu groß, verlangt die Situation vielleicht nach Gnade – die Richter im Fall *Dudley* sahen es so – aber die Befolgung selbst ist dennoch geboten.[29]

Das könnte als zutiefst seltsam erscheinen. Sollte dem so sein, so entstammt dieses Gefühl wohl dem Gedanken, dass die Moral der Anleitung unserer Handlungen dienen sollte und wir es mitunter schwierig finden, auf der einen Seite zu behaupten, dass etwas getan werden *sollte,* und zugleich davon ausgehen sollen, dass es selbst von unserer Meinung nach guten Menschen nicht getan *werden wird.* Aber es scheint zumindest für mich eine richtige Beobachtung, dass die Welt uns mit moralischen Prüfungen konfrontieren kann, an denen wir vorhersehbar scheitern werden.[30] Ich vermute, dass die meisten von uns sich letztendlich dazu entscheiden würden, ein unschuldiges Kind zu schlagen, wenn die Alternative dazu in einem abgrundtiefen Schmerz für uns selbst bestehen würde. Daraus folgt nicht, dass wir monströs wären, sondern bloß, dass wir menschlich sind. Es bedeutet auch nicht, dass das Kind irgendwie sein Recht darauf verloren hätte, frei von körperlicher Gewalt zu leben. Der Schlag mag vielleicht zu entschuldigen sein – was nichts anderes bedeutet, als dass die gewöhnliche Art bestrafender Reaktionen in diesem Fall unangemessen wäre, allerdings bleibt diese Art von Handlung selbst falsch. Der Erzähler in Jonathan Littells *Die Wohlgesinnten* beschreibt eine solche Situation sehr gut:

> „Es sei noch einmal gesagt, der Klarheit wegen: Ich will hier nicht behaupten, ich sei an diesem oder jenem nicht schuldig. Ich bin schuldig, ihr seid es nicht, wie schön für euch. Trotzdem könntet ihr euch sagen, dass ihr das, was ich getan habe, genauso hättet tun können. [...] Wenn ihr in einem Land und in einer Zeit geboren seid, wo nicht nur niemand kommt, um eure Frau und eure Kinder zu töten, sondern auch niemand, um von euch zu verlangen, dass ihr die Frauen und Kinder anderer tötet, dann danket Gott und zieht hin in Frieden. Aber bedenkt immer das eine: Ihr habt vielleicht mehr Glück gehabt als ich, doch ihr seid nicht besser. Denn solltet ihr so vermessen sein, euch dafür zu halten, seid ihr bereits in Gefahr."[31]

Die Überlegungen des Erzählers mögen eigennützig erscheinen, allerdings müssen sie deshalb nicht zwangsläufig auch falsch sein. Die Welt mag mit Situationen aufwarten, in denen gewöhnliche Menschen daran scheitern werden, die Gebote der Ethik zu befolgen. Die Idee, dass solche Situationen nicht existieren, findet ihren Ursprung in Privilegien, nicht in der Moral.

Die Dinge verhalten sich in Carlas Fall vielleicht anders – sicherlich ist die Pflicht, ein zurückweisendes Gesetz zu befolgen, schwächer als die Pflicht, Mord oder Kannibalismus zu vermeiden, um es milde auszudrücken.

Allerdings ist die Struktur des Arguments identisch. Die mit der Befolgung dieser Pflicht verbundenen persönlichen Kosten für Carla können so groß werden, dass wir von uns selbst erwarten würden zu scheitern, wären wir in derselben Situation. Aber dieser Fehler ist nach wie vor ein moralischer Fehler; die Handlung wird nicht bloß dadurch richtig, da sie unvermeidlich ist. Allerdings macht diese Unvermeidlichkeit einen *gewissen* moralischen Unterschied. Es ist uns nicht gestattet, andere als moralisch defizitär zu verurteilen, wenn sie in einer Prüfung versagen, die wir selbst auch nicht bestehen würden. Wir können demnach andere nicht als besonders unmoralisch bezeichnen, wenn wir uns dem durch sie begangenen Verbrechen an ihrer Stelle ebenfalls schuldig machen würden.

Der Gedanke, dass es eine legitime Grenze unserer Erwartungen dahingehend geben sollte, in welchem Maße Menschen allein aus Pflichtmotiven handeln werden, hat eine lange Tradition innerhalb der politischen Philosophie. Rawls diskutiert sie als Einschränkungen der Verpflichtung; ihm zufolge ist eine politische Gesellschaft nur dann gerecht, wenn sie auf Prinzipien beruht, von denen wir uns vorstellen können, dass jeder sie als motivierend akzeptieren kann.[32] So können wir nicht erwarten, dass eine Gesellschaft fortbesteht, wenn sie auf Prinzipien beruht, die von manchen Menschen große Opfer zum Nutzen anderer erwarten. Aus diesem Gedanken entspringt Rawls' Ablehnung des Konsequentialismus, der ihm zufolge ein solche exzessive Opferbereitschaft verlangen würde. Diese Ideen könnten wir nun vielleicht heranziehen um der Schlussfolgerung zu entgehen, dass Carla eine Pflicht zur Befolgung des sie ausschließenden Gesetzes hat. Dieses Gesetz könnte demnach seinen verpflichtenden Charakter verlieren, wenn seine Befolgung psychologisch unplausibel wäre. Ich denke jedoch, dass ein solcher Schluss voreilig wäre. Rawls' Idee soll eine Grundlage bereitstellen, auf der die politische Zwangsbefugnis, das Herzstück moderner Verfassungsstaaten, gegenüber denjenigen gerechtfertigt werden kann, die eben jener Gewalt unterworfen sind. Mit anderen Worten: Er will herausfinden, wie Menschen über die Zeit hinweg miteinander Politik betreiben können. Seine Einschränkungen der Verpflichtung setzen einer konstitutionellen Demokratie daher einige Grenzen bei dem, was für diese Aufgabe von den Bürgerinnen verlangt werden kann. Aus diesen Überlegungen folgt jedoch nicht, dass der Einsatz von Zwang aufgrund der bloßen Tatsache falsch ist, dass der Bruch eines Gesetzes zu erwarten ist. Es steht uns zumindest offen, zu betonen, dass wir darauf schauen sollten, ob das Carla ausschließende Gesetz für sich genommen gerecht oder ungerecht ist; und dass die Verpflichtung, es einzuhalten, selbst dann bestünde, wenn der Bruch dieses Gesetzes sowohl nachvollziehbar als auch vorhersehbar ist.

3. Verpflichtung und Migration: Über Jeb Bush und Donald Trump

An diesem Punkt könnte die Frage aufkommen, was nun gewonnen wurde. Ich habe den Gedanken verteidigt, dass ein zurückweisendes Gesetz zu Recht von vielen Menschen missachtet werden darf – unter anderem von denjenigen, deren Heimatländer daran scheitern, ihnen die Rechte zu gewähren, die ihnen aufgrund ihrer Existenz als Menschen zukommen. Darüber hinaus habe ich argumentiert, dass dieses Gesetz, gleich ob gerecht oder ungerecht, auch weiterhin eine Vielzahl von Menschen zu seiner Einhaltung verpflichtet – wir aber zugleich nicht erwarten können, dass die Verpflichtung in diesem Falle Menschen mit einer gewöhnlichen moralischen Sensitivität auch tatsächlich motiviert. Selbst wenn es also eine gewisse Verpflichtung gibt, das zurückweisende Gesetz eines fremden Landes zu befolgen, können wir einen Grad an Verbesserung des eigenen Lebensstandards annehmen, jenseits dessen *jeder* die Entscheidung treffen würde, dem zurückweisenden Gesetz nicht Folge zu leisten. Es offenbart sich also, in anderen Worten, kein minderwertiger moralischer Charakter in der Entscheidung, eine relativ arme Gesellschaft zu verlassen und ohne rechtliche Grundlage in eine wohlhabendere Gesellschaft einzuwandern. Jeb Bush hat in den eingangs zitierten Überlegungen diesen Punkt gut getroffen. Die Migrantin hat zwar das Gesetz gebrochen, was moralisch nicht unerheblich ist; allerdings war dieser Bruch des Gesetzes nicht von der Art, dass daraus auf einen schlechten Charakter der Migrantin geschlossen werden könnte. Tatsächlich ist das Gegenteil der Fall: Migration findet in den überwiegenden Fällen statt, damit bestimmte andere Personen davon profitieren, nämlich diejenigen, die abhängig von der Migrantin sind – hieraus folgt denn auch der Gedanke, dass es sich bei Migration um einen Akt der Liebe handeln kann. Bush nimmt mit seiner Analyse einen seltsamen Platz innerhalb des öffentlichen Diskurses ein: Er argumentiert, dass eine Gesellschaft eine Migrantin zu Recht zurückweisen darf, und dass ihre Handlungen zugleich als nachvollziehbar, wenn nicht gar lobenswert, betrachtet werden sollten.

Diese Rede war verhängnisvoll für Bushs Wahlkampf: Donald Trumps Wahlkampfteam veröffentlichte ein Video mit dem Titel „Act of Love", in dem Polizeifotos undokumentierter Immigranten und eine Einblendung ihrer Gewalttaten zusammen mit Bushs Worten gezeigt wurden.[33] Trumps Team hatte einen recht schonungslosen Blick auf Migrantinnen, ob mit oder ohne Aufenthaltspapiere. Sein Wahlkampf versprach die berüchtigte Mauer entlang der Grenze zu Mexiko und eine neue „Deportation Force", deren Aufgabe es sein sollte, alle undokumentierten Immigrantinnen abzuschieben – Trump wiederholte bekanntlich immer wieder den Satz „Sie müssen gehen", wenn er über diese Menschen sprach.[34]

Das alles sind bekannte Tatsachen, aber welche Bedeutung haben sie für unsere Überlegungen? Es ist verlockend, anzunehmen, Jeb Bush habe inkohärent argumentiert und Trump äußere sich, wenn auch moralisch ungeheuerlich, so doch zumindest kohärent. Aber ich denke, das wäre falsch. Trumps Analyse vermengt zwei verschiedene Fragen: zunächst die Frage danach, was Staaten tun dürfen, um undokumentierte Einwanderer daran zu hindern, ihre Hoheitsgebiet zu betreten, und dann die Frage, wie wir den moralischen Charakter derjenigen beurteilen sollten, die zu Unrecht einwandern. Trumps Perspektive ist simpel: Wir können tun was auch immer nötig ist, da es sich bei diesen Menschen um (wie er sagt) „bad hombres" handelt. Bushs Analyse hingegen besagt, dass die Frage, zu was ein Staat zur Verhinderung von Einwanderung berechtigt ist, sich von der Frage unterscheidet, was wir über diejenigen denken sollten, die es dennoch schaffen einzuwandern. In meinen Augen ist Bushs Vorstellung zumindest kohärent, wenn es auch durchaus möglich ist, sie zu kritisieren. Ein Land kann manche Migrantinnen zurückweisen ohne dabei zu meinen, dass diejenigen, die dennoch irregulär einreisen, moralisch zu verurteilen seien. Die soeben angestellten Überlegungen können erklären, warum dem so ist. Sofern die Vereinigten Staaten eine Gesellschaft sind, die dazu fähig ist, verbindliche Verpflichtungen zwischen Menschen aufgrund von Gesetzen herzustellen, können die gegen undokumentierte Immigrantinnen gerichteten Gesetze – falls sie nicht durch bedeutendere moralische Erwägungen übertrumpft werden – die Bürgerinnen fremder Länder dazu verpflichten, nicht einzuwandern. Diejenigen, die nicht zurückgehalten wurden, begehen jedoch keinen moralischen Fehler, den wir unter denselben Umständen nicht auch gemacht hätten. Die Moral zwingt uns dazu, anzuerkennen, dass die Kosten, die diese irregulären Migrantinnen bei einem Verzicht auf die Einwanderung tragen würden, von einer Höhe sind, die wir im Namen einer solchen moralischen Pflicht ebenfalls nicht zu tragen bereit wären. Migration ohne gesetzliche Grundlage ist kein Akt moralischer Verderbtheit, sondern etwas, das jeder normale Mensch tun würde, befände er sich in der Situation der irregulär einwandernden Person.

Was folgt nun aus all dem? Die bedeutendste Schlussfolgerung besteht in meinen Augen darin, dass durchaus kohärent zugleich für ein Recht auf Ausschluss und gegen die moralische Verurteilung irregulärer Migrantinnen argumentiert werden kann. Ersteres mag legitim sein oder nicht; diese Frage sollte an anderer Stelle verhandelt werden. Sollte ein solches Recht jedoch existieren, können wir mit ihm einige der Aspekte unserer Migrationspolitik rechtfertigen, die eine Einwanderung ohne Papiere erschweren sollen. Der zweite Teil der Schlussfolgerung – dem zufolge wir anerkennen müssen, dass sich an der irregulären Einwanderung selbst kein schlechter Charakter offenbart – spielt jedoch in der Entwicklung dieser Politik ebenfalls eine eigenständige Rolle. Ich werde nun bloß zwei der Implikationen untersuchen, die aus der Anerkennung dieser Tatsachen folgen.

Die erste Implikation besagt, dass der Akt undokumentierter Einwanderung kein Ausweis eines moralischen Fehlers ist, selbst wenn dabei ein Gesetz gebrochen wird; entsprechend kann das reguläre System des Strafrechts schlecht auf den Fall undokumentierter Immigrantinnen angewendet werden. Ich werde mich hier auf Joel Feinbergs Analyse beziehen, der zufolge ein Großteil des Strafrechts zu dem Zweck entwickelt wurde, einen Bereich zu schaffen, in dem der Staat die Handlung des Kriminellen autoritativ verurteilen kann. So stellt insbesondere die Gefangenschaft nicht nur ein unangenehmes Resultat der kriminellen Handlung dar, sondern auch ein Schauspiel, in dem diese Handlung demonstrativ und allgemeinverbindlich abgelehnt wird. Im Gegensatz dazu tendieren andere Formen unliebsamer Behandlungen – wie Bußgelder, selbst wenn sie sehr hoch ausfallen sollten – nicht zu einer solch symbolischen Form öffentlicher Verurteilung. Wir behalten uns Gefangenschaft für diejenigen Fälle vor, in denen wir dem Kriminellen und der Welt mitteilen wollen, dass die begangene Tat nicht nur illegal, sondern auch moralisch verwerflich ist.[35] (Diese Tatsache kann, denke ich, auch die von vielen empfundene Wut darüber erklären, dass nach der Finanzkrise von 2008 kein Repräsentant einer Bank auch nur einen Tag in einem Gefängnis der Vereinigten Staaten verbracht hat.) Daraus folgt jedoch, dass die steigende Bereitschaft, undokumentierte Einwanderinnen mit Gefängnis zu bestrafen, in einem grundlegenden Sinne falsch ist. Migrantinnen werden zunehmend inhaftiert, entweder in gefängnisähnlichen Einrichtungen oder in Gefängnissen selbst. Das mag sicherlich allein schon aufgrund der oftmals unrechtmäßigen Zustände im US-amerikanischen Strafvollzug falsch sein.[36] Allerdings wäre es meiner Meinung nach auch schlicht aufgrund der unangemessenen Botschaft falsch, die durch das Ritual des Einsperrens kommuniziert wird. Wir stigmatisieren die undokumentierten Immigrantinnen anhand der gleichen Rituale als moralisch verkommen, wie wir es im Falle gewalttätiger Krimineller tun. Wenn wir anerkennen, dass die Migrantin durch ihre Entscheidung, das Gesetz zu brechen, kein Unrecht begeht, dass wir nicht auch begehen würden, wären wir dieser Art von Politik gegenüber stärker abgeneigt; wir haben kein Recht darauf, die Botschaft zu vermitteln, die aus einer solchen Politik folgt.

Die zweite Implikation besteht in der Idee, dass mit einer solchen Politik kriminelle Handlungen belohnt würden, eine Idee, die vor allem in konservativen Diskursen über Migration von ständiger Bedeutung ist. Dabei wird oft unter Verweis auf das moralische Unrecht argumentiert, das sich in der irregulären Migration selbst manifestieren würde. Wie wir in einem früheren Kapitel gesehen haben, nahm Steve King an, dass DACA Kriminelle belohnen würde. Er war mit dieser Ansicht nicht allein. So lehnte Tom Tancredo, ein konservativer Kongressabgeordneter aus Colorado, durchweg jedes Amnestie- und Einbürgerungsprogramm ab, da es „Menschen belohnt, die das Gesetz gebrochen haben."[37] Manchmal werden solche Ideen auch mit Blick auf die Zukunft entwickelt, wobei suggeriert wird, der Akt der irregulären Einwanderung würde auf einen bösartigen Charakter hinweisen, weshalb eine Reihe missliebiger Dinge von irregulären Migrantinnen zu erwarten seien – von mangelhafter Staatsbürgerschaft bis hin zu ausgewachsener Kriminalität. Jeff Sessions, Trumps erster Justizminister, führte wiederholt aus, dass Städte wie Chicago, die sich der Sanctuary Cities Bewegung angeschlossen haben, „kriminelle Ausländer" schützen würden – wobei er damit gleichzeitig zu meinen schien, dass diese Ausländer sowohl Kriminelle aufgrund ihrer irregulären Einreise seien, als auch zukünftig Gewaltverbrechen begehen würden. Auf ähnliche Art hat die Trump-Administration kürzlich eine Hotline eingeführt – genannt Victims of Immigration Crime Engagement (VOICE) –, bei der Informationen über „Verbrechen von Individuen mit einer Verbindung zu Einwanderung" mitgeteilt werden können.[38] Diese eher seltsam anmutende Formulierung blendet auf ähnliche Weise die Unterscheidung zwischen solchen Verbrechen aus, die einen deformierten moralischen Charakter offenbaren, und dem Verbrechen, das Einwanderungsgesetz zu brechen. Diese Ausblendung impliziert jedoch – zu Unrecht – dass der Status als undokumentierte Immigrantin auf die Bereitschaft zur Gewaltausübung hinweisen würde. Wie der Vater von Mollie Tibbets bemerkte, wurde seine Tochter nicht von einem undokumentierten Migranten getötet, sondern von einem Menschen, der zufälligerweise einen solchen Status innehatte.[39]

Am eindrücklichsten wurde die Trennung der Kinder irregulärer Einwanderer von ihren Eltern unter Verweis auf den Gedanken gerechtfertigt, dass die Eltern selbst moralisch außergewöhnlich verkommen wären. Katie Waldman, eine Sprecherin der Immigrations and Customs Enforcement (ICE), führte an, dass die Trennungen im Interesse der Kinder notwendig seien. Es wäre ihr zufolge für diese Kinder besser, in Gewahrsam genommen zu werden, als in der Obhut von Kriminellen zu verbleiben:

> „[ICE ist] gesetzlich verpflichtet die Interessen der Kinder zu schützen, sei es vor Menschenhandel, Drogenschmuggel oder bösartigen Personen, die willentlich unser Einwanderungsgesetz brechen."[40]

Diese Vorstellung beruht offensichtlich und explizit auf dem Gedanken, dass es einen bösartigen Akt darstellt, die Grenzen der USA ohne entsprechende Berechtigung zu überqueren. Demnach sind die das Einwanderungsgesetz brechenden Eltern schlechte Personen, weswegen wiederum das Kind vor seinen Eltern geschützt werden muss. Nichts von all dem ist jedoch wahr. Die Trennung von Familien an der Grenze beruht auf der Annahme, dass die Eltern dieser Familie schlechte Menschen sind, von denen die Kinder getrennt werden sollten. Aber ein solcher Gedanke setzt die Entscheidung, unter Bedingungen extremer Armut das Gesetz zu brechen, in eins mit der Demonstration eines außergewöhnlich verkommenen Charakters. Wenn schon nichts sonst, so kann uns das, was ich hier schreibe, zumindest einen Weg aufzeigen, wie dieser Gleichsetzung widersprochen werden kann. Selbst wenn das Gesetz, das Carla abweist, moralische Verpflichtungen schafft, handelte es sich dabei um ein Gesetz, das wir selbst – wären wir in der gleichen Situation – nicht befolgen könnten. Das sollte uns mit einem gewissen Grad an Demut in unseren politischen Debatten über Migration ausstatten – und mitunter mit dem Willen, Kompromisse einzugehen. Zumindest könnten wir in der Lage sein, eine recht weit verbreitete Version dieses Gedankens zu widerlegen – nämlich die, dass wir durch die Einbürgerung undokumentierter Einwohnerinnen eine außergewöhnlich unmoralische Gruppe von Personen belohnen würden, was angeblich die weitere Einwanderung solch unmoralischer Personen in der Zukunft erwarten ließe. Was die Einwanderung letztlich ausgelöst hat, könnte schlicht die Tatsache gewesen sein, dass eine solche Bewegung für die meisten, die sie unternommen haben, von großem Vorteil war.[41] Diejenigen, die sich für die Einwanderung entschieden haben – und diejenigen, die sich noch dafür entscheiden werden – reagieren allerdings nicht auf eine besonders unmoralische Weise, sondern so, dass die meisten vernünftigen Menschen ihre Entscheidung nachvollziehen können sollten. Die Migrationspolitik kann diese Fakten anerkennen und damit auch zugleich ihre ethischen Vorstellungen anpassen – oder sie kann sich dem verweigern und damit fortfahren, Migrantinnen und diejenigen, die für sie und mit ihnen arbeiten, zu dämonisieren.

Es ist selbstverständlich traurige Realität, dass der Großteil der Welt die zweite Option wählen wird. In einem Großteil der hoch entwickelten Länder sind die Rechte von Migrantinnen derzeit in Gefahr. Die Ansichten von Personen wie Trump sind klar im Vormarsch. Darauf antworte ich nicht damit, dass die Vorstellungen von Jeb Bush insgesamt korrekt sind. Derlei Vorstellungen könnten durchaus auf einer Konzeption des Rechts auf Ausschluss beruhen, die einer genaueren Untersuchung nicht standhält. Aber ich glaube, dass diejenigen, die ein solches Recht auf Ausschluss verteidigen – darunter auch ich – sehr viel vorsichtiger hinsichtlich dessen sein sollten, was dieses Recht implizieren könnte. Insbesondere denke ich, dass diejenigen, die gegen offene Grenzen argumentieren, anerkennen sollten, dass es vielerlei Einschränkungen dessen gibt, was undokumentierten Immigrantinnen berechtigterweise angetan werden darf. Unabhängig davon, ob die Entscheidung zur irregulären Einwanderung nun ein Akt der Liebe ist, so ist sie mit Sicherheit kein Akt moralischer Verkommenheit; und unsere Debatte über die Moral der Migration wird durch die Worte und Taten derer beschädigt, die darauf bestehen, dass dem so sei.

Eine Frage wurde in diesem Kapitel jedoch noch nicht beantwortet. Wenn es falsch ist, den undokumentierten Immigrantinnen einen moralisch mangelhaften Charakter vorzuwerfen – haben wir dann nicht auch Gründe dafür, nicht bloß Aussagen über ihre moralische Verkommenheit zu unterlassen, sondern sie auch nicht zurückzuweisen oder abzuschieben? Könnten wir, mit anderen Worten, nicht annehmen, dass wir Gründe haben, nicht nur das zu ändern, was wir über die Migrantinnen sagen, sondern auch das, was wir ihnen gegenüber tun? Ich denke, die Antwort lautet ja – aber zu diesem Schluss können wir nur gelangen, wenn wir uns die Tugend der Gnade und die aus ihr folgenden Konsequenzen genauer anschauen. Dieser Aufgabe werde ich mich nun zuwenden.

9
Über Gnade in der Politik

In den bisherigen Kapiteln habe ich oft erwähnt, dass die Gnade eine besondere Form der Tugend darstellt. Allerdings habe ich nicht viel darüber gesagt, was unter dieser Tugend zu verstehen ist. Es ist also nun an der Zeit, etwas genauer zu bestimmen, worum es sich bei der Gnade handelt und wie wir das Konzept im politischen Leben anwenden können. Dieses Kapitel versucht, diese Fragen zu beantworten und somit auch die notwendigen Vorarbeiten bereitzustellen, um die Beziehung zwischen Gnade und Migration im abschließenden Kapitel diskutieren zu können. Dabei gliedert sich dieses Kapitel in drei Teile. Zunächst möchte ich zeigen, weshalb ich meine, dass das Konzept der *Gnade* für unsere Diskussion von Nutzen sein kann. Dann möchte ich zeigen, warum der Rekurs auf die Gnade weder pervers noch spalterisch ist, obwohl es auf den ersten Blick so aussehen mag. Der letzte Teil dieses Kapitels beschäftigt sich dann mit dem Nachweis, dass Liberale – selbst politisch Liberale – Gründe dafür haben, eine Idee der Gnade als eigenständige politische Tugend anzusehen. Wie ich zeigen werde, sollte der Liberalismus davon ausgehen, dass ein gerechter Staat, in dem keine Gnade geübt wird, eine sehr ärmliche Form von Staat darstellt – und es tatsächlich unwahrscheinlich ist, dass er noch auf lange Zeit gerecht bleiben wird.

Bevor ich all dies zeigen kann, werde ich jedoch erklären müssen, weshalb es überhaupt plausibel ist, von der Gnade als einer politischen Tugend zu sprechen. Seit John Rawls' *Theorie der Gerechtigkeit* wurde ein Großteil der politischen Philosophie von Überlegungen hinsichtlich des Begriffs der Gerechtigkeit dominiert, also der Definition sowie der Verteidigung von Vorstellungen über Rechte und die gerechte politische Ordnung. Rawls selbst bestimmte Gerechtigkeit als die erste Tugend politischer Institutionen, ähnlich der Rolle, die dem Konzept der Wahrheit im Bereich der Mathematik zukommt. Demnach müssten auch „noch so gut funktionierende und wohlabgestimmte" Gesetze und Institutionen verändert oder gar abgeschafft werden, falls aus ihnen keine gerechten Verhältnisse entstünden.[1] Diese philosophischen Erwägungen spiegeln offensichtlich die politischen Kämpfe der 1960er Jahre, in denen viele soziale Konflikte – darunter die Kämpfe um Bürgerrechte in den Vereinigten Staaten – als Kämpfe um Gerechtigkeit verstanden werden konnten; also als Kämpfe derjenigen, denen Rechte verweigert wurden, für eine Welt, in denen diese Rechte geschützt werden. Dieses Erbe ist für das Verständnis der Philosophie und des Aktivismus der Gegenwart von essentieller Bedeutung. Allerdings kann es auch eine Art begriffliches Gefängnis darstellen. Der Staat kann auf vielerlei Weisen moralisch falsch handeln; allerdings besteht sein Versagen nicht in all diesen Fällen darin, Menschen vor den Institutionen, aus denen die jeweiligen gesellschaftlichen Konflikte erwachsen, als moralisch ungleich zu behandeln. Rawls selbst hat das verstanden. Während er Gerechtigkeit als erste Tugend politischer Institutionen bezeichnete, hat er niemals behauptet, dass es sich dabei um die *einzige* Tugend handeln würde. Tatsächlich stellte er später klar, dass viele andere Tugenden zu Recht Teil eines moralisch vertretbaren und stabilen Gefüges politischer Institutionen sein können. Ein Staat kann demnach ungerecht sein, wenn er individuellen Personen nicht diejenigen Rechte und Güter bereitstellt, die für eine effektive Ausübung menschlicher Handlungsfähigkeit vonnöten sind, oder wenn er auf andere Weise daran scheitert, Menschen mit dem zu versorgen, worauf sie einen moralischen Anspruch haben. Aber ein Staat der *nur* dies zu verhindern versucht – der also niemals Rechte verletzt, sich zugleich aber auch niemals bemüht, denjenigen zu helfen, deren Pläne und Projekte Unterstützung benötigen könnten – macht sich moralischer Defizite schuldig. Um Jeffrie Murphys Gedanken vom Beginn dieses Buchs zu wiederholen: Eine Person, die es stets vermeidet irgendwem Unrecht zu tun, kann nicht als ungerecht bezeichnet werden. Aber wenn das *alles* ist, was sie für ihre Mitmenschen tut, kann ihr sicherlich *etwas* vorgeworfen werden: Sie ist zumindest kaum ein geeignetes Beispiel dafür, wie ein guter Mensch handeln sollte.[2]

Ich verstehe Gnade also als die Tugend, eine Person nicht so hart zu behandeln, wie es uns durch die Gerechtigkeit gestattet wäre, und zwar aufgrund moralischer Bedenken hinsichtlich der Folgen einer solch harten Behandlung für die betroffene Person. Ganz ähnlich versteht auch Adam Perry den Begriff der Gnade.[3] In Fällen von Gnade, so Perry, gibt es zwei mögliche Optionen: So können wir uns entscheiden, eine bestimmte Person mit Härte oder Nachsicht zu behandeln – ohne uns dabei in einem der Fälle einer Ungerechtigkeit schuldig zu machen. Wir handeln der Tugend der Gnade gemäß, wenn wir uns für die nachsichtige Alternative entscheiden und von einer harten Behandlung Abstand nehmen. Allerdings ist dabei wichtig, dass die Tugend der Gnade bloß in den Fällen infrage kommt, in denen eine harte Behandlung nicht selbst moralisch falsch wäre. Wie Perry schreibt: eine Person *muss* nicht gnädig handeln. Nichtsdestotrotz, so Perry weiter, hat der Staat Grund dazu, in seinem Strafrecht die Tugend der Gnade walten zu lassen; Praktiken wie nachsichtige Bestrafungen – und vor allem Begnadigungen – können triftige moralische Bedenken widerspiegeln. Tatsächlich könnte die Demonstration von Gnade den Staat dazu befähigen, „die Tugend der Gnade für seine Bürgerinnen zu veranschaulichen."[4] Ich möchte diese Idee über das Strafrecht hinaus auf die Gesetze erweitern, die einzelne Personen daran hindern, Staatsgrenzen zu überqueren. Wir können demnach argumentieren, dass die Tugend der Gnade dem Staat vorschreibt, manche Menschen, die durch einen Ausschluss oder eine Abschiebung nicht ungerecht behandelt werden würden, nicht solchen Praktiken zu unterwerfen. Der Staat könnte also zugleich die Tugend der Gnade erfüllen und die Art von Tugend demonstrieren, der es für die Stabilität eines demokratischen Gemeinwesens bedarf.

Das ist denn auch der Kern der Tugend, die ich verteidigen möchte: Der Gedanke, dass sowohl Individuen als auch Staaten moralische Gründe dafür haben, den einzelnen Personen um sie herum dabei zu helfen, ihre Pläne zu entwickeln und zu verfolgen, selbst wenn sie nicht aufgrund von Gerechtigkeitserwägungen dazu verpflichtet sind, Verantwortung für deren Erfolg zu übernehmen. Wir haben also, mit anderen Worten, einen guten moralischen Grund, für den Erfolg des Lebens anderer Menschen zu arbeiten – selbst dann, wenn diese Personen uns nicht vorwerfen könnten, ungerecht zu handeln, würden wir ihnen nicht helfen. Ich selbst nutze die Sprache der Gnade, aber die gleiche Tugend könnte auch unter Verweis auf Ideen wie Freundlichkeit oder die Vermeidung von Grausamkeit beschrieben werden, oder vielleicht schlicht als eine Form moralischen Anstands. Wie auch immer wir diese Tugend beschreiben, sie beinhaltet prinzipiell, dass wir uns um die Bedürfnisse derjenigen kümmern sollen, die keinen besonderen Anspruch auf unsere Unterstützung geltend machen können. Die von diesen Personen angestrebten Güter sind so beschaffen, dass sie durch unsere Hilfe erlangt werden könnten. Diese Personen haben keinen bestimmten Anspruch auf unsere Unterstützung; sie können also weder auf eine drohende Vernichtung ihrer Handlungsfähigkeit noch auf eine besondere Beziehung, wie beispielsweise gleichberechtigte Mitbürgerschaft, verweisen, um uns zur Hilfe zu verpflichten. Wir könnten sie also harsch behandeln ohne dabei unfair zu sein. Sie sind demnach schlichtweg bedürftig und in einer Position, in der wir ihnen dabei helfen können, ihre Bedürfnisse zu erfüllen. Wir könnten diesen Menschen den Rücken zukehren, ohne dabei ungerecht zu handeln. Wenn wir uns jedoch kontinuierlich auf diese Weise gegenüber unseren Mitmenschen verhalten, offenbaren wir moralische Mängel; wir sind dann zwar nicht ungerecht, aber sicherlich ungnädig.

Letztlich möchte ich also die Idee verteidigen, dass eine Gesellschaft als ungnädig kritisiert werden kann, wenn sie sich weigert in ihrem Handeln über das hinauszugehen, was ihr die Gerechtigkeit vorschreibt. In meinen Augen werden viele den allgemeinen Gedanken akzeptieren, dass eine Tugend wie die Gnade innerhalb einer politischen Gesellschaft wirksam sein sollte – als Teil des diskursiven Instrumentariums, mit dem wir bestimmte Arten von Gesetzen bewerten, loben oder verurteilen. Allerdings werden wohl auch viele Menschen meinen, dass das Wort *Gnade* hier fehl am Platz ist, weshalb ich mich nun der Verteidigung dieses Begriffs zuwende.

1. Gnade und Perversität

Warum also sollten wir diese Tugend mit dem Wort *Gnade* überschreiben? Ich kann sicherlich nicht jede Person davon überzeugen, dass dies der richtige Begriff ist; aber ich kann zumindest erklären, warum dieser Begriff *in meinen Augen* richtig ist. Hierfür gibt es meiner Meinung nach drei Gründe. Der erste besteht darin, dass eine Person, die sich an einer staatlichen Grenze befindet, fast schon im wortwörtlichen Sinne *auf die Gnade* derjenigen angewiesen ist, die diese Grenze kontrollieren. Diese Vorstellung beinhaltet offensichtlich die Anerkennung eines radikalen Machtungleichgewichts. Ich bin auf Ihre Gnade angewiesen, wenn ich anhand einer durch Sie zu treffende Entscheidung vollkommen vernichtet werden könnte und Sie wiederum von mir nichts zu fürchten haben. Dieses gravierende Ungleichgewicht – wenn sich ein Individuum, als natürliche Person, einer politischen Gesellschaft gegenüberstellt, deren Grenzbeamte ihr mit Waffengewalt drohen, sollte sie versuchen, eine bestimmte Linie zu überschreiten – sollte nicht ignoriert oder durch ungenaue Formulierungen unkenntlich gemacht werden. Wir sollten uns daran erinnern, dass die Migrantin auf die Gnade des Landes angewiesen ist, in das sie Einlass begehrt und die Sprache der Gnade könnte ein Weg sein, uns diese Tatsache stets erneut bewusst zu machen.

Der zweite Grund besteht hingegen in der Anerkennung des Zusammenhangs zwischen Religion und der Moral der Migration. Viele Religionen verfügen über ein der Gnade ähnliches Konzept. Die Beschreibung der islamischen Idee des *rahmah* ähnelt auf frappierende Weise der christlichen Idee der Gnade. Gleiches gilt für den Buddhismus, in dem die Figur des Guanyin als der *Bodhisattva* des Mitgefühls und der Gnade verstanden wird. In der Geschichte des nordamerikanischen und europäischen Denkens über Migration hat das christliche Ideal der *Misericordia* – Barmherzigkeit – eine signifikante Rolle gespielt.[5] Ich behaupte selbstverständlich nicht, dass die Politik einer gerechten Gesellschaft spezielle christliche Normen widerspiegeln sollte. Vielmehr formuliere ich die eher kontextspezifische und moderate Behauptung, dass diejenigen, die für den rechtlichen Schutz von Migrantinnen gekämpft haben, dies in Europa und Nordamerika oft aus religiösen Gründen getan haben und politische Philosophinnen von der Arbeit dieser Menschen lernen können. So begann beispielsweise die Sanctuary-Bewegung mit der religiösen Verpflichtung, den vor Gewalt in Zentralamerika fliehenden Menschen eine Heimat innerhalb der Vereinigten Staaten zu geben – trotz der Tatsache, dass sich die USA der Gewährung eines solchen Schutzes verweigert hatten. John Fife, Pfarrer an der Southside Presbyterian Church in Tuscon, öffnete seine Kirche für diejenigen, die vor Unterdrückung flohen und zitierte zur Begründung die christlichen Ideale von Gerechtigkeit und Gnade:

> „Wir schreiben Ihnen, um Sie darüber zu informieren, dass die Southside United Presbyterian Church den Immigration and Nationality Act, Section 274(A), verletzen wird. […] Wir glauben, dass Gerechtigkeit und Gnade es von Menschen unseres Glaubens verlangen, das uns von Gott gegebene Recht, jeder Person zu helfen, die vor Verfolgung und Mord flieht, aktiv wahrzunehmen. […] Wir bitten Sie, in Gottes Namen, bei der Erfüllung ihres Amtes Gerechtigkeit und Gnade walten zu lassen.“[6]

Heute sind die von Fife und anderen gelegten Samen in einer neuen Sanctuary-Bewegung aufgegangen, die sich dem Widerstand gegen Abschiebungen durch die Trump-Administration verpflichtet hat. Auch hier ist der christliche Aktivismus durch den Gedanken inspiriert, dass Gerechtigkeit von Gnade begleitet werden muss – von dem Willen also, durch die Bedürfnisse von Personen außerhalb unserer Gemeinschaft bewegt zu werden, selbst wenn wir uns von ihnen abwenden könnten. Der Gedanke, dass christliche Gnade den Umgang eines Staates mit seinen Grenzen beeinflussen sollte, hat sich folglich bewährt und als wirkmächtig erwiesen. (Tatsächlich trägt eine historische Abhandlung über die katholische Lehre und den katholischen Aktivismus im Bereich der Migration schlicht den Titel *Mercy without Borders*.) Die United States Conference of Catholic Bishops hat erst kürzlich den Aufruf wiederholt, Gerechtigkeit mit Gnade zu verbinden:

> „Die Regulierung der Grenzen sowie die Einwanderungskontrollen eines Landes müssen von der Sorge um alle Menschen sowie von Gnade und Gerechtigkeit geleitet werden. Eine Nation kann nicht einfach entscheiden, dass ihre Ressourcen nur für ihre eigenen Landsleute und sonst niemanden zur Verfügung stehen. Es muss eine ehrliche Verpflichtung gegenüber den Bedürfnissen aller bestehen."[7]

Ich sollte hier erneut betonen, dass ich von politischen Philosophen nicht erwarte, dass sie diesen Begriff deshalb nutzen, *weil* er von der christlichen Soziallehre verwendet wurde. Vielmehr möchte ich, in einem moderateren Sinne, bewirken, dass politische Philosophinnen offen für die Möglichkeit sind, dass die christliche Lehre in diesem Falle einen wichtigen Punkt benennt – und dass wir die Sprache ihres Glaubens nutzen können, ohne uns dabei zugleich diesem Glauben verpflichten zu müssen. Demnach kann also selbst ein politisch Liberaler die rhetorische Kraft der christlichen Terminologie anerkennen, solange diese Terminologie genutzt werden kann, ohne dabei die umfassenderen Werte des christlichen Glaubens zu beschwören.

Im nächsten Teil dieses Kapitels werde ich zu zeigen versuchen, dass dies möglich ist. Aber der dritte und wichtigste Grund für die Nutzung des Konzepts der Gnade besteht darin, dass das Strafrecht selbst häufig auf dieses Konzept zurückgreift und dies auch in Kontexten tut, die den in diesem Buch betrachteten Fällen ähnlich sind. Rufen Sie sich noch einmal den Fall von Smilla ins Gedächtnis, oder den der Matrosen in *Dudley and Stephens*. Diesen Personen drohten bestimmte Strafen für bestimmte Verbrechen und es wäre nicht ungerecht gewesen, sie diesen Strafen auszusetzen. Allerdings erscheinen die zugewiesenen Strafen in beiden Fällen überzogen. Der Staat, der diese Personen einsperrt – oder im Falle der Matrosen hängt – würde ihnen zwar kein Unrecht tun, hat aber dennoch gute moralische Gründe dafür, diese Strafen zu vermeiden. (Ich möchte an dieser Stelle bemerken, dass die Matrosen durch Queen Victoria bedingt begnadigt wurden, und dass, obwohl die Queen Begnadigungen gegenüber im Allgemeinen abgeneigt war.) Wenn er Gnade gegenüber Kriminellen übt, verzichtet der Staat also auf eine Handlung, die ihm durch die Gerechtigkeit gestattet wäre und er tut dies im Namen des individuellen Wohls der Person, gegen die sich das staatliche Handeln richtet. Der Kriminelle erfährt also Gnade, wenn sein Verbrechen nicht mit einer Strafe geahndet wird, die durch die Gerechtigkeit gestattet wäre. Genauso verhält es sich auch im Falle der Migrantin. Die Tatsache, dass die Abschiebung oder der Ausschluss einer Person nicht ungerecht ist, gibt uns noch keine endgültige Antwort auf die moralische Frage. Kurzum: Es könnte sein, dass auch in diesem Fall die Tugend der Gnade den Staat von einer Handlung abhalten kann, die zwar gerecht, zugleich aber auch grausam wäre – ganz ähnlich wie im Strafrecht.

Diese Überlegungen bringen uns nun allerdings zum Vorwurf der Perversität. Demnach könnte meine Verteidigung des Begriffs der Gnade so verstanden werden, dass sie genau die Art von Beziehung voraussetzt, die eine Moraltheorie der Migration eigentlich verurteilen sollte. Es gibt drei Formen dieser Lesart. Die erste entspringt der Analogie mit dem Strafrecht. Ihr zufolge wird die gnädig behandelte Person als Verbrecherin identifiziert und erfährt deshalb Gnade, da sie gerechterweise verurteilt wurde. Damit scheint allerdings vorausgesetzt, dass eine Person, die irregulär die Grenzen eines Staates überquert – oder eine Person, die ohne einen Gerechtigkeitsanspruch auf Einreise an der Grenze ankommt –, auf irgendeine Art etwas getan hat, was der Vergebung bedarf. Das scheint, um es milde auszudrücken, pervers. Die zweite Form von Perversität beginnt mit der Idee der Dankbarkeit. So könnten wir meinen, dass die Person, der gegenüber wir gnädig sind, zu Dankbarkeit verpflichtet sei, denn schließlich wurde ihr etwas gewährt, was sie gewissermaßen nicht verdient hat – und wir könnten meinen, dass die moralisch angemessene Antwort darauf in Dankbarkeit bestehen sollte. Allerdings könnte ebenso gut eingewendet werden, dass es falsch ist, von Menschen wie Carla – um noch einmal Bezug auf das Beispiel aus dem vorherigen Kapitel zu nehmen – Dankbarkeit zu verlangen. Sie wird in relative Armut hineingeboren und überschreitet bei ihrer Einwanderung eine Linie hin zu relativem Wohlstand. Lassen Sie uns annehmen, sie könnte rechtmäßig abgeschoben werden, doch der Staat verzichtet auf diese Abschiebung. Erschiene es nicht zumindest ein bisschen selbstsüchtig, wenn dieser Staat auf Carlas Dankbarkeit bestehen würde? Erinnern wir uns: Sie wurde in Armut geboren, so wie viele derjenigen Bürgerinnen, die in einem solchen Falle Dankbarkeit von ihr erwarten würden, in relativen Wohlstand hineingeboren wurden. Ist die Erwartung nicht irgendwie pervers, dass die Armen den Wohlhabenden Dankbarkeit dafür schulden, wenn diese davon Abstand nehmen, das Leben der Armen auf profunde Weise zu verschlechtern? Abschließend können wir zum Thema des christlichen Glaubens zurückkehren und anführen, dass der Akt der Gnade Gott vorbehalten ist; diejenigen unter uns, die in dieser Welt Gnade üben, tun dies demnach in einer bewussten Nachahmung Gottes und seiner Liebe. Aber ist es nicht eine schreckliche Vorstellung, dass die Bürgerinnen wohlhabender Gesellschaften gleich Göttern Gnade gegenüber den relativ ärmeren Bürgerinnen der Entwicklungsländer üben? Besteht das Ziel der politischen Philosophie nicht darin, genau solche Formen von Ungleichheit zu überwinden, statt schlicht nach neuen Wegen zu suchen, solche Auswüchse einzudämmen?

All das sind starke Bedenken. Allerdings reichen sie in meinen Augen nicht aus, um auf den Begriff der Gnade zu verzichten. Wir können mit dem Problem beginnen, dass durch die Idee der Gnade die irreguläre Einwanderung als Unrecht betrachtet würde. Es ist wahr, dass die Norm der Gnade für gewöhnlich im Strafrecht beheimatet ist, was allerdings bloß die Tatsache widerspiegelt, dass das Strafrecht der Ort ist, an dem Staaten so direkt und offensichtlich wie sonst nirgends ihre Zwangsgewalt an Menschen ausüben. Die Idee der Gnade selbst scheint jedoch auf jeden Kontext anwendbar, in dem eine Person, die eine andere Person einer harten Bestrafung aussetzen dürfte, einen starken moralischen Grund hat, dies nicht zu tun – und zwar einen Grund, der darin liegt, wie wichtig das Wohl der betroffenen Person für die strafende Person ist und in welchem Maße wir das Wohlergehen der Person zerstören würden, würden wir tun, was die Gerechtigkeit uns gestattet. Wir sollten hierfür genauer betrachten, was der Migrantin durch den Staat widerfährt, würde dieser sie mit Gewalt ausweisen oder sie mit Gewalt davon abhalten, die Grenze zu überqueren. Der Staat nutzt Zwang, was bloß ein anderes Wort dafür ist, dass er gedenkt Gewalt anzuwenden oder mit Gewalt zu drohen. Der Zweck dieses Einsatzes staatlicher Zwangsgewalt weicht allerdings, wie der Supreme Court wiederholt bestätigt hat, vom Zweck der Bestrafung ab. Im Kontext der gegenwärtig geltenden Gesetze der Vereinigten Staaten mag eine Abschiebung unangenehm sein, aber sie ist „keine Strafe für ein Verbrechen."[8] Für diejenigen, die abgeschoben werden, mag diese juristische Nettigkeit allerdings eher irrelevant erscheinen. Durch die Abschiebung fügt der Staat einer einzelnen Person eine tiefgreifende Verletzung zu und diese individuelle Person mag daher so reagieren, als ob es sich dabei tatsächlich um eine Strafe handele. Und wenn das staatliche Handeln derart verstanden werden kann, scheint auch die Anwendung des Konzepts der Gnade in diesem Falle zulässig, und zwar um dem Staat aufzuzeigen, warum manche Abschiebungen oder Abschiebungsmuster moralisch verwerflich sind.

Das Problem der Dankbarkeit könnte vielleicht etwas komplizierter sein. Ich kann an dieser Stelle keine Theorie darüber präsentieren, wann wir Dankbarkeit empfinden sollten, oder wie eine solche Dankbarkeit auf die richtige Weise ausgedrückt werden sollte. Aber ich denke, dass zwei Argumente gegen die vorgebrachten Bedenken angeführt werden können. Zunächst könnte es sein, dass die angemessene Form der Dankbarkeit in einer politischen Gemeinschaft nichts verlangt, was über den eigenen Beitrag zur Erhaltung und Stabilität ebendieser Gemeinschaft hinausgeht. Gegenwärtig habe ich keine besondere Verpflichtung die Institutionen der französischen Demokratie zu bewahren – allerdings käme mir eine solche Pflicht zu, würde ich nach Frankreich ziehen, und ich könnte hier und jetzt stärkere Pflichten diesbezüglich haben, wenn diese Institutionen an Bedeutung verlieren oder in ihrer Existenz gefährdet wären. (Sicherlich habe ich eine Pflicht es zu *vermeiden* den politischen Institutionen Frankreichs zu schaden, allerdings hat diese Pflicht eher geringe Auswirkungen auf meinen Alltag.) Wir könnten also sagen, dass die von der Migrantin erwartete Dankbarkeit, gleich welcher Art sie auch sein mag, auf solche recht unstrittigen Fälle beschränkt ist. Die zweite Antwort besteht darin, dass die Tugend der Dankbarkeit wahrscheinlich ein Verfallsdatum hat. Als ich an meiner Universität angestellt wurde, empfand ich enorme Dankbarkeit all denjenigen gegenüber, die sich dazu entschieden hatten, mir diese seltene und wertvolle Möglichkeit zu geben und ich tat mein Bestes, um diese Dankbarkeit auch auszudrücken.

Ebenso fühlte ich eine Woge an Dankbarkeit, als ich Bürger der Vereinigten Staaten wurde, diesmal gegenüber dem Land, in dem ich ein Zuhause, eine Familie und eine Karriere gefunden habe. In beiden Fällen stand die von mir empfundene Dankbarkeit jedoch nicht in Konflikt mit dem – nach sehr kurzer Zeit – empfundenen Recht, die Praktiken und Entscheidungen sowohl meiner Universität als auch meines Landes zu kritisieren. Wenn eine Person sich dazu entscheidet, einer Gemeinschaft von Gleichen beizutreten, mag diese Person ihre Dankbarkeit zeigen. Allerdings muss diese Person sich auch nach kurzer Zeit der mühseligen Aufgabe widmen, als Gleiche zu *handeln*. Dankbarkeit könnte eine vorübergehende Tugend sein, zumindest wenn sie als etwas anderes verstanden wird als die andauernde Pflicht, die Institutionen der politischen Gemeinschaft zu bewahren, in der sich das eigene Leben abspielt.

Die letzte Sorge ist vielleicht auch die tiefgreifendste: Ist es nicht moralisch fragwürdig, wenn wir behaupten, dass beispielsweise die Vereinigten Staaten in ihren Angelegenheiten mit fremden Personen Gnade walten lassen sollten? Unser Ziel sollte doch in der Einrichtung einer gerechten Welt liegen, von der wir annehmen können – zumindest aus einer kosmopolitischen Perspektive –, dass sie keinerlei staatliche Grenzen aufweist. Auch hier gibt es wieder zwei Antworten. Die erste findet sich im früheren Teil dieses Buches und besagt, dass es keinen guten Grund für die Annahme gibt, der Liberalismus würde offene Grenzen verlangen. Sollte dies zutreffen, so wird es stets Fälle von Personen geben, die an Orte ziehen wollen ohne über ein entsprechendes Recht auf Einwanderung zu verfügen. Die zu beantwortende moralische Frage – hilft uns der Begriff der Gnade dabei, zu verstehen, was angesichts solcher Personen getan werden sollte? – scheint denn auch in einer absolut gerechten Welt aufzukommen. Sollten wir beispielsweise eine Welt vollkommener Verteilungsgerechtigkeit vorfinden – wie auch immer wir diesen Begriff definieren – scheint es immer noch einen Bedarf für den hier vorgestellten Ansatz geben. Die zweite Antwort zielt hingegen auf den ursprünglichen Zweck der politischen Philosophie. Selbst wenn die radikale Position offener Grenzen richtig sein sollte, sind wir doch mit der Tatsache konfrontiert, dass unsere Welt im Hier und Jetzt von Grenzen durchzogen ist, die allesamt recht sorgsam von bewaffneten Männern und Frauen bewacht werden. Wir bedürfen moralischer Prinzipien für genau diese Welt, mit diesen bewaffneten Männern und Frauen. Dass diese Prinzipien in einer Welt ohne Grenzen ihre

Bedeutung verlieren würden, negiert nicht ihre Kraft in dieser Welt. Bertolt Brechts „Die Nachtlager" drückt diese Überlegungen sehr gut aus. In diesem Lied ändert eine Person, die ein Nachtlager für Obdachlose bereitstellt, nichts an den materiellen Ursachen der Obdachlosigkeit. Dennoch tut sie etwas, hier und jetzt, das es wert ist, besungen zu werden:

> „Die Welt wird dadurch nicht anders
> Die Beziehungen zwischen den Menschen bessern sich nicht
> Das Zeitalter der Ausbeutung wird dadurch nicht verkürzt
> Aber einige Männer haben ein Nachtlager
> Der Wind wird von ihnen eine Nacht lang abgehalten
> Der ihnen zugedachte Schnee fällt auf die Straße."[9]

So verhält es sich auch mit der Idee der Gnade im Bereich der Migration. Wir können uns eine grundlegend reformierte Welt vorstellen, in der keine Migrantin von der Gnade des Staates abhängt, in den sie einwandern will – oder von der Gnade der globalen Kräfte, die Wohlstand und Armut produzieren. Aber während wir auf eine solche Welt warten, können wir damit fortfahren, herauszufinden, welchen Menschen geholfen werden muss und wie und warum wir diejenigen sind, die diesen Menschen zu helfen haben.

2. Tugenden im politischen Leben

Lassen Sie uns zu dem Gedanken zurückkehren, dass die Idee der Gnade eine unangemessene Grundlage für den öffentlichen Diskurs über die Ethik der Politik darstellen könnte – nicht, weil sie pervers, sondern weil sie *spalterisch* wirken könnte. Schließlich entwickelte John Rawls in seinem Konzept des öffentlichen Vernunftgebrauchs die Vorstellung einer Gesellschaft, die trotz der Präsenz einer Vielzahl verschiedener, umfassender ethischer Lehren stabil über die Zeit hinweg existieren kann. Laut Rawls sei eine solche Gesellschaft deshalb stabil, weil Fragen grundlegender Gerechtigkeit und wesentlicher Verfassungsinhalte unter Bezug auf eine politische Gerechtigkeitsvorstellung ausgehandelt würden – die, wie Rawls betont, nicht von der Wahrheit einer der einzelnen umfassenden Lehren abhänge.[10] Rawls fordert uns also auf, eine diskursive Praxis für das politische Leben zu entwickeln, die nicht auf den besonderen Tugenden oder Freveln, die von einer umfassenden Lehre bestimmt werden, beruht. In diesem und dem folgenden Kapitel möchte ich jedoch für eine diskursive Praxis argumentieren, in der ein bestimmtes Konzept der Gnade für die Bewertung der Migrationspolitik herangezogen werden kann. Müssen für den Erfolg dieses Arguments daher zunächst Rawls' Überlegungen widerlegt werden?

In meinen Augen ist das nicht der Fall. Es gibt für mein Vorhaben durchaus Raum in Rawls' politischem Liberalismus. Wenn wir recht nah an der von Rawls selbst verteidigten Vorstellung bleiben, können hierfür zwei Gründe angeführt werden. Der erste Grund besteht darin, dass Rawls seinen anspruchsvollsten Test der politischen Ethik auf Angelegenheiten grundlegender Gerechtigkeit und wesentliche Verfassungselemente beschränkt; sobald diese den Test bestanden haben, ist es für politische Akteure möglich, auch Argumente einzubringen, die in diesem grundlegenderen Kontext nicht angebracht wären. Um es einfach auszudrücken: Wir müssen nicht in allen politischen Kontexten im Einklang mit dem Konzept des öffentlichen Vernunftgebrauchs argumentieren – und auch wenn das eine recht komplexe interpretatorische Frage sein mag, ist es doch zumindest unklar, ob Rawls alle Aspekte der Migrationspolitik als Angelegenheiten grundlegender Gerechtigkeit betrachten würde. Der zweite Grund besteht hingegen darin, dass Rawls selbst die Bedeutung bestimmter politischer Tugenden anerkennt. Diese können innerhalb einer politischen Gerechtigkeitsvorstellung akzeptiert werden und damit zusätzliche Ressourcen bereitstellen, anhand derer die infrage stehende Gesellschaft sich hinsichtlich ihrer öffentlichen Institutionen ethisch orientiert:

> „Selbst wenn der politische Liberalismus jedoch eine gemeinsame Basis sucht und gegenüber Zielen neutral ist, ist es wichtig zu betonen, daß er dennoch die Überlegenheit bestimmter Formen des moralischen Charakters anerkennen und bestimmte moralische Tugenden fördern kann. So gehört zur Konzeption der Gerechtigkeit als Fairneß ein bestimmtes Verständnis politischer Tugenden, und zwar der Tugenden der fairen sozialen Kooperation: Höflichkeit und Toleranz, Vernünftigkeit und Sinn für Fairneß (4. Vorlesung, §§ 5–7). Der entscheidende Punkt dabei ist, daß die Einführung dieser Tugenden in eine politische Konzeption nicht zum perfektionistischen Staat einer umfassenden Lehre führt."[11]

Rawls betont an dieser Stelle einige bestimmte Tugenden, aber ich denke, dass wir diese Liste durchaus erweitern dürfen. In meinen Augen könnte die Idee der Gnade – oder eine politische Konzeption der Gnade, die von den Elementen einer umfassenden ethischen Lehre befreit wurde – eine solche Tugend darstellen.

Die Idee der Gnade könnte diese Rolle jedoch nur erfüllen, sofern gezeigt werden kann, dass sie potentiell mit einer Vielzahl umfassender Lehren vereinbar ist. Ich denke, dass es sich in der Tat so verhält. Im Rest dieses Abschnitts werde ich drei Formen umfassender Lehren besprechen, von denen eine jede Elemente besitzt, anhand derer eine politische Konzeption der Gnade begründet werden könnte. Das Ziel der folgenden Überlegungen besteht darin zu zeigen, dass wir, Rawls' eigener Einladung folgend, zeigen können, dass es selbst innerhalb seines Systems Platz für eine Tugend wie die der Gnade gibt. Sollte dem so sein, könnte das Konzept der Gnade die diskursive Bedeutung gewinnen, von der ich bereits gesprochen habe: Sie könnte dann eine Tugend darstellen, die wir im öffentlichen Diskurs für die Verteidigung oder Kritik bestimmter Formen der Migrationspolitik ins Feld führen können.

2.1 Christentum

Das Christentum habe ich zum Teil bereits weiter oben angesprochen und werde daher versuchen, mich im Folgenden kurz zu fassen. In der christlichen Tradition wird die Gnade explizit als Tugend benannt und auch auf das politische Leben angewandt. Wie diese Tugend gelebt und verstanden wird, variiert selbstverständlich stark; vielleicht kann für keine Angelegenheit von sozialer Bedeutung so etwas wie eine eindeutige christliche Position identifiziert werden. Das Konzept der Gnade ist jedoch von mehr oder weniger zentraler Bedeutung für jedes christliche Weltbild: Der Tod und die Wiederauferstehung Jesu Christi werden als eine Art Geschenk verstanden und nicht als Forderung der Gerechtigkeit. Der Gerechtigkeit nach steht dem gefallenen Sünder die Verdammnis zu; gerettet wird er durch die Gnade Gottes – nicht durch dessen Gerechtigkeit. Daher betont die christliche Theologie die Gnade als eine soziale Norm, die das Handeln der Kinder Gottes in der diesseitigen Welt leiten soll. Die päpstliche Enzyklika *Dives in Misericordia* – Über das göttliche Erbarmen – formuliert die Notwendigkeit, Gerechtigkeit durch Gnade zu mäßigen:

> „Die Erfahrung der Vergangenheit und auch unserer Zeit lehrt, daß die Gerechtigkeit allein nicht genügt, ja, zur Verneinung und Vernichtung ihrer selbst führen kann, wenn nicht einer tieferen Kraft – der Liebe – die Möglichkeit geboten wird, das menschliche Leben in seinen verschiedenen Bereichen zu prägen. Gerade die geschichtliche Erfahrung hat, unter anderem, zur Formulierung der Aussage geführt: *summum ius, summa iniuria* – höchstes Recht, höchstes Unrecht. Diese Behauptung entwertet die Gerechtigkeit nicht, noch verringert sie die Bedeutung der Ordnung, die sich auf sie aufbaut; sie weist nur unter einem anderen Aspekt auf die Notwendigkeit hin, aus jenen noch tieferen Quellen des Geistes zu schöpfen, denen sich die Ordnung der Gerechtigkeit selber verdankt."[12]

Die hier aufscheinende, fortdauernde Relevanz der Gnade hatte, wie bereits weiter oben gezeigt, einigen Einfluss auf die christliche Lehre hinsichtlich der Migration. Allerdings möchte ich betonen, dass sich diese Lehre nicht allein innerhalb des katholischen Glaubens entwickelte; vielmehr findet sie sich in einer Vielzahl christlicher Traditionen und an den vielen Orten des Widerstands gegen unbarmherzige Formen der Migrationspolitik. Roger Mielke, ein hochrangiger Vertreter der Evangelischen Kirche in Deutschland, führt aus, dass die Offenheit der Deutschen gegenüber Flüchtlingen (*Willkommenskultur*) die christliche Tugend der Gnade widerspiegelte:

> „[Wir] sollen die Verlorenen und Verwundeten aufnehmen und versorgen und auch diejenigen, die unseres Schutzes bedürfen. In jedem von Leid gezeichneten menschlichen Antlitz zeigt sich der Leib Jesu Christi selbst. Die Kirche hat jeder Person mit einer unbedingten Verpflichtung zu begegnen, die weit über den Kreis der christlichen Brüder und Schwestern hinausgeht, und hat die Hand auszustrecken, um jede Person als ein Geschöpf Gottes einzubeziehen, da dieser Mensch, obwohl verloren, ein Mensch ist, zu dessen Rettung Jesus erschien. Wie uns Matthäus 25,40 lehrt, ‚Wahrlich ich sage euch: Was ihr getan habt einem unter diesen meinen geringsten Brüdern, das habt ihr mir getan.' […] Im Einklang mit dieser Ethik gibt es nur eine mögliche Art für den Leib Christi, auf die Leiden der Verletzlichen und Marginalisierten zu antworten, darunter auch die Leiden der Flüchtlinge und Migrantinnen unserer Zeit: mit unbedingter Liebe. Kurz gesagt: Wir müssen uns in Barmherzigkeit üben."[13]

Papst Franziskus hat am 06. Juli 2018 in einer für die Migrantinnen dieser Welt gehaltenen Messe eine ähnliche Sprache benutzt. Am Jahrestag seines Besuches auf Lampedusa betonte der Papst die Notwendigkeit, Migrantinnen im Geiste der Gnade zu begegnen:

> „Der Herr verspricht allen Unterdrückten der Welt Erquickung und Befreiung, doch er braucht uns, um sein Versprechen wirksam werden zu lassen. Er braucht unsere Augen, um die Nöte der Brüder und Schwestern zu sehen. Er braucht unsere Hände, um zu helfen. Er braucht unsere Stimme, um die unter dem – zuweilen mitschuldigen – Stillschweigen vieler begangenen Ungerechtigkeiten anzuklagen. In der Tat müsste ich über viele Arten des Stillschweigens reden: das Stillschweigen des gesunden Menschenverstandes, das Stillschweigen des „Es war schon immer so", das Stillschweigen des „Wir" im steten Gegensatz zum „Ihr". Vor allem aber braucht der Herr unser Herz, um die barmherzige Liebe Gottes zu den Geringsten, zu den Ausgestoßenen, zu den Verlassenen, zu den Ausgegrenzten zum Ausdruck zu bringen. […] Angesichts der Herausforderungen durch die Migrationen heute besteht die einzige vernünftige Antwort in der Solidarität und Barmherzigkeit."[14]

An dieser Stelle könnte sicherlich noch Vieles mehr gesagt werden. Das christliche Konzept der Gnade ist komplex und ich kann nicht behaupten, über das theologische Wissen zu verfügen, um es vollumfänglich nachvollziehen zu können. Aber ich hoffe zumindest so viel gezeigt zu haben: Eine Christin wird, sofern sie von Ideen wie den hier besprochenen bewegt wird, fähig sein, in einen öffentlichen politischen Diskurs darüber einzutreten, wie und wann Gnade gegenüber Migrantinnen gezeigt werden sollte. Immerhin beginnt Rawls' *Politischer Liberalismus* damit, wie schwierig sich demokratische Deliberation über die Trennlinien religiösen Glaubens hinweg gestaltet, und er hoffte, gezeigt zu haben, dass eine solche Deliberation möglich ist. Die von mir verteidigte politische Konzeption der Gnade könnte also von einem vernünftigen Christen akzeptiert und im politischen Diskurs angeführt werden.

2.2 Die Care-Ethik

Ich glaube, dass etwas Ähnliches auch über die in der zweiten Hälfte des zwanzigsten Jahrhunderts entwickelte Care-Ethik gesagt werden kann. Allerdings möchte ich an dieser Stelle vorsichtig sein, denn ich schreibe in der Tradition von John Rawls, für den Erwägungen von Recht und Gerechtigkeit Vorrang hatten. Genau jene Haltung wurde jedoch durch Vertreterinnen der Care-Ethik oft zu modifizieren oder zu widerlegen versucht (was auch weiterhin geschieht). Viele Care-Ethikerinnen – obwohl keinesfalls alle – argumentieren, dass der Begriff der Gerechtigkeit zumindest die herausgehobene Position verlieren sollte, die er in den Werken von Philosophen wie Rawls einnimmt.[15] Es gibt zudem viele verschiedene Vorstellungen von Care-Ethik, die sich über wichtige Probleme uneins sind, darunter beispielsweise, ob Care in der Hauptsache einen emotionalen Zustand oder aber eine bestimmte Art von Arbeit oder Praxis bezeichnet.

Diese Unterschiede werden jedoch von gewissen Themen begleitet, die den verschiedenen Ansätzen gemeinsam sind und zeigen, dass eine öffentliche Norm der Gnade von vielen Care-Ethikerinnen als eine vernünftige Erweiterung unserer öffentlichen Diskussion über Migration akzeptiert werden könnte. Virginia Held beschreibt drei dieser Themen.[16] Das erste Thema ist die Forderung nach Anerkennung des Partikularen, statt seiner Subsumtion unter abstrakte Kategorien. Moralisch vorrangig sei die tatsächliche Begegnung mit der einzelnen Person, statt kaltblütiger Abstraktionen, die wir zur Ordnung und Kategorisierung von Menschen entwickeln könnten. Das zweite Thema zeigt sich in dem Gedanken, dass die Ethik selbst sich der Subsumtion unter allgemeinere Rechtfertigungsprinzipien widersetze. Ein Beispiel mag hier helfen: Held schreibt, dass die analytische politische Philosophie uns Freundschaften gestattet – solange das Prinzip, das diese Freundschaft rechtfertigt, nachgewiesenermaßen gegenüber allen Menschen gerechtfertigt werden kann, sowohl Freunden als auch Nicht-Freunden. Hierbei handele es sich allerdings, so Held, um eine unangemessene Beschreibung der Grundlagen des moralischen Lebens, ganz zu schweigen davon, dass sich auf einer solchen Basis bloß schwerlich eine gute Freundschaft aufbauen ließe. Die Care-Ethik steht insbesondere mit feministischen Ansätzen in Verbindung und entstammt der Überzeugung, dass die vor allem Frauen zugewiesenen sozialen Rollen – Hausfrau, Mutter,

Pflegende – mit der moralischen Forderung verbunden sind, für andere zu sorgen; die Macht dieser Forderungen war demnach für Männer oft unsichtbar. Das führt uns zum abschließenden Aspekt der Care-Ethik: In ihr nimmt die Anerkennung von Bedürftigkeit und Vulnerabilität sowie das allen Menschen gemeinsame Bedürfnis nach Sorge den Vorrang ein. Laut Held ist der in der Philosophie für gewöhnlich betrachtete Akteur der vereinzelte Mann, der als sich frei entscheidende Person beschrieben wird. Die Leben vulnerabler Personen – des Kindes oder älterer Menschen – würden dadurch beiseitegeschoben und entweder ignoriert oder als Problemfälle betrachtet, mit denen man sich zu einem späteren Zeitpunkt beschäftigen werde. Die Care-Ethik rückt demnach die Marginalisierten in den Fokus und argumentiert, dass es ethisch geboten sei, für diese verletzlichen Personen zu sorgen, statt zuerst abstrakte Prinzipien zu entwickeln, anhand derer ihre Fälle analysiert werden können.

Das alles stellt offensichtlich bloß eine grobe Skizze dieser komplexen Familie von Ansätzen dar, aber ich denke, dass sie für den Beginn ausreichen sollte. Die Care-Ethik scheint mit Überlegungen zu beginnen, die den Gedanken, die meinem Verständnis der Gnade zugrunde liegen, ganz ähnlich sind. Wie auch die Gnade versucht die Care-Ethik, über das Problem der bloßen Verwaltung formaler Gerechtigkeitsansprüche hinauszugehen. Stattdessen betrachtet sie die verschiedenen Arten, auf die bestimmte Menschen von den Handlungen anderer betroffen sind. Tatsächlich scheint sie sogar genau mit dem Gedanken zu beginnen, dass radikale Bedürftigkeit – ähnlich der Vulnerabilität mancher Personen, die an der Grenze unseres Staates erscheinen – eine eigenständige Basis für unsere moralischen Urteile bilden kann. Das Konzept der Gnade wiederum beginnt bei der Idee, dass wir in der Ethik nicht einfach auf abstrakte Prinzipien zurückgreifen, sondern vielmehr Untersuchungen darüber anstellen sollten, was wir für diejenigen tun können, die aufgrund ihrer Vulnerabilität durch unsere Entscheidungen besonders betroffen sind. Die besondere Vulnerabilität von Migrantinnen an staatlichen Grenzen ist auch tatsächlich schon von Feministinnen mithilfe des Instrumentariums der Care-Ethik untersucht worden; so hat beispielsweise Eva Kittay die Migration relativ armer Frauen aus Entwicklungsländern in entwickelte Länder untersucht und argumentiert, dass diese Art von Migration ein spezifisches moralisches Unrecht darstellt, da sie die Beziehung zwischen den Frauen und ihren Familien verletzt.[17]

Die Idee der Care-Ethik kann sicherlich auch als eine radikale Herausforderung für das gesamte Projekt liberaler Gerechtigkeit betrachtet werde. So können ihr beispielsweise genuin revolutionäre Implikationen hinsichtlich der globalen politischen Ordnung zugeschrieben werden. Indem sie ihre eigene Vorstellung der Care-Ehtik auf die internationale Sphäre anwendete, zeigte unter anderem Virginia Held, dass wir – in einer Welt, in der Individuen über internationale Grenzen hinweg wahrhaftig füreinander sorgen und in der diese Sorge Beziehungen und Praktiken gegenseitigen Respekts begründen würden – Konzepte wie die der Menschenrechte schlicht nicht mehr vonnöten wären. Wir sollten demnach eher die Idee des Rechts fallen lassen, als jeglichen ethischen Wert in Analogie zu irgendeiner Form von Gerechtigkeit zu verstehen.[18] Allerdings gehe ich davon aus, dass auch die Unterstützerinnen der Care-Ethik die politische Tugend der Gnade gutheißen könnten, wenn auch mitunter nur in dem vollen Bewusstsein, dass es sich dabei nicht um die von ihnen angestrebte radikalere Veränderung handelt. Die Tugend der Gnade spiegelt zumindest die Forderung wider, sich zuerst mit den konkreten Bedürfnissen des einzelnen Mitmenschen zu befassen, statt mit der abstrakten Frage zu beginnen, ob diese Person einen Gerechtigkeitsanspruch auf das hat, wonach sie ersucht. In diesem Sinne scheint es für diejenigen, die eine solche Form der Ethik unterstützen, zumindest potentiell möglich, die politische Konzeption der Gnade zu akzeptieren.

2.3. *Kantische Ethik*

Die Care-Ethik wird häufig in Abgrenzung zur Ethik in der Tradition Kants entwickelt und es gibt gute Gründe für die Ansicht, dass diese zwei Vorstellungen von unterschiedlichen Prämissen ausgehen. Die Care-Ethik beginnt mit dem Partikularen und einem Misstrauen gegenüber Prinzipien; die kantische Ethik hingegen greift in ihren moralischen Urteilen auf die Prinzipien des kategorischen Imperativs zurück und misstraut denjenigen Formen von Empathie und Partikularität, die von der Care-Ethik besonders hervorgehoben werden.[19]

Nichtsdestotrotz scheint es plausibel, dass die politische Tugend der Gnade auch von einem vernünftigen Kantianer akzeptiert werden könnte, und zwar als eine Möglichkeit, die Vorstellung von Wohltätigkeit, wie sie in der kantischen Ethik beschrieben wird, zu realisieren. Kants Theorie ist selbstredend nicht weniger komplex als irgendeine der hier besprochenen Theoretikerinnen und vieles von dem, was ich sagen werde, könnte von Kennerinnen seiner Theorie durchaus infrage gestellt werden. Aber vielleicht mag für unsere Zwecke dennoch ein grober Umriss der kantischen Ethik ausreichen.

Kant teilt seine ethischen Überlegungen in eine Rechts- sowie eine Tugendlehre.[20] Erstere beschreibe das Gebiet der Rechte und der Gerechtigkeit – Rechte, die prinzipiell mit ihrer zwangsweisen Durchsetzung kompatibel seien. Die Pflicht, nicht zu töten, sei demnach eine Pflicht der Gerechtigkeit, weshalb der Staat das Recht habe, diejenigen mittels seiner Zwangsgewalt zu bestrafen, die sie nicht befolgen. Die Tugendlehre hingegen beschreibt Kant zufolge eine Art moralischer Pflichten, die sich auf verschiedene Weisen von den Rechtspflichten unterscheidet. Tugendpflichten seien keine Pflichten der Gerechtigkeit, weshalb aus ihnen auch keine Rechtsansprüche abgeleitet werden könnten. So korrespondiere etwa mit der Pflicht zur Wohltätigkeit – der Pflicht also, aktiv zu den Plänen und Zwecke mancher Mitmenschen, nicht um unser-, sondern ihretwillen, beizutragen – daher kein Rechtsanspruch aufseiten des Empfängers unserer Wohltätigkeit. Diejenigen, denen gegenüber wir dieser Pflicht genüge täten, könnten also nicht behaupten, ihnen wäre ein Unrecht geschehen, hätten wir nicht entsprechend dieser Pflicht gehandelt. Stattdessen sei diese Pflicht *unvollkommen*. Zudem kann laut Kant diese Pflicht nicht auf legitime Weise mittels Zwang durchgesetzt werden; der Staat kann demnach also nicht vorschreiben, dass seine Bürgerinnen auf wohltätige Art miteinander umgehen sollen. Doch selbst wenn das praktisch möglich wäre – und es ist schwer vorstellbar, wie wir dazu gezwungen werden könnten, nach dieser Art von Tugend zu leben – handele es sich hierbei nicht um die richtige Art von Pflicht, um einen solchen Zwang begründen zu können. Stattdessen hätten wir ein eigenes Ermessen, innerhalb dessen wir für uns selbst entscheiden könnten, wie und wann wir einen Plan dafür entwickeln wollen, die Bedürftigen und Verletzlichen zu unterstützen. Wir hätten, wie Kant sagt, einen gewissen „Spielraum" (*latitudo*), in dem wir für uns selbst entscheiden dürften, auf welche Art wir wohltätig sind.[21] Dennoch bedeute die Verweigerung wohltätigen Handelns – die Verweigerung also, einen Plan zu entwickeln, in dem die Zwecke mancher Mitmenschen derart unterstützt werden, dass sie in unsere eigenen Zwecke integriert werden –

tatsächlich ein moralisches Versagen. Eine Person, die sich der Wohltätigkeit verweigere, folge nicht dem kategorischen Imperativ und die Maxime ihres Handelns könne entsprechend nicht gerechtfertigt werden:

> „Noch denkt ein vierter, dem es wohl geht, indessen er sieht, daß andere mit großen Mühseligkeiten zu kämpfen haben (denen er auch wohl helfen könnte): was geht's mich an? mag doch ein jeder so glücklich sein, als es der Himmel will, oder er sich selbst machen kann, ich werde ihm nichts entziehen, ja nicht einmal beneiden; nur zu seinem Wohlbefinden, oder seinem Beistande in der Not, habe ich nicht Lust, etwas beizutragen! Nun könnte allerdings, wenn eine solche Denkungsart ein allgemeines Naturgesetz würde, das menschliche Geschlecht gar wohl bestehen, und ohne Zweifel noch besser, als wenn jedermann von Teilnehmung und Wohlwollen schwatzt, auch sich beeifert, gelegentlich dergleichen auszuüben, dagegen aber auch, wo er nur kann, betrügt, das Recht der Menschen verkauft, oder ihm sonst Abbruch tut. Aber, obgleich es möglich ist, daß nach jener Maxime ein allgemeines Naturgesetz wohl bestehen könnte: so ist es doch unmöglich, zu wollen, daß ein solches Prinzip als Naturgesetz allenthalben gelte. Denn ein Wille, der dieses beschlösse, würde sich selbst widerstreiten, indem der Fälle sich doch manche eräugnen können, wo er anderer Liebe und Teilnehmung bedarf, und wo er, durch ein solches aus seinem eigenen Willen entsprungenes Naturgesetz, sich selbst alle Hoffnung des Beistandes, den er sich wünscht, rauben würde."[22]

Die Pflicht zur Wohltätigkeit ist folglich eine Pflicht, die uns allen zufällt. Sie ist nicht – wie in der Care-Ethik – im emotionalen Leben des Menschen begründet. Vielmehr misstraut Kant der Emotion und argumentiert, dass die einzige Form von Liebe, die durch eine Pflicht vorgeschrieben werden könne, aus der praktischen Handlungsfähigkeit hervorgehe:

> „Denn Liebe als Neigung kann nicht geboten werden, aber Wohltun aus Pflicht, selbst, wenn dazu gleich gar keine Neigung treibt, ja gar natürliche und unbezwingliche Abneigung widersteht, ist praktische und nicht pathologische Liebe, die im Willen liegt und nicht im Hange der Empfindung, in Grundsätzen der Handlung und nicht schmelzender Teilnehmung; jene aber allein kann geboten werden."[23]

Die beschriebenen Unterschiede verdecken jedoch eine tiefe Ähnlichkeit zwischen den beiden Ansätzen. Sowohl in der Care- als auch in der kantischen Ethik – wie übrigens auch im Christentum – wird geboten, sich um die vulnerablen Personen zu kümmern und einen Plan wohltätiger Handlungen zu entwickeln, durch den manchen Personen unsere prinzipielle Sorge zuteilwird. Wir sollen also manchen Personen dadurch helfen, dass wir uns ihre Projekte und Pläne zum Zweck nehmen. In keinem dieser Fälle haben diese Personen ein Anrecht auf unsere wohltätigen Handlungen. Wir haben folglich die Freiheit, zu entscheiden, wie wir helfen wollen und keiner potentiell durch uns begünstigen Person widerfährt ein Unrecht, wenn wir ihr schließlich doch nicht helfen sollten. Allerdings werden wir zu Recht kritisiert, sollten wir *niemals* über die Gebote bloßer Gerechtigkeit hinausgehen und stattdessen versuchen, unser Handeln allein durch den Verweis auf die Tatsache zu rechtfertigen, dass wir die Rechte anderer Personen achten. Für Kant – wie auch für die Care-Ethik und das Christentum – haben wir eine Pflicht, mehr zu tun als bloß Gerechtigkeit zu üben.[24]

Diese Gedanken werden selbstverständlich oft in einem Kontext entwickelt, in dem es darum geht, bestimmte Mitmenschen mit bestimmten Gütern zu versorgen und Kant konzentriert sich auf die Bereitstellung materieller Güter durch die Wohlhabenden für diejenigen, die weniger besitzen. Aber es gibt keinen Grund, davon auszugehen, dass diese Ideen auf den Kontext wohltätiger Spenden begrenzt sind. Stattdessen denke ich, dass sie auch für die Begründung des von mir beschriebenen politischen Konzepts der Gnade herangezogen werden können. All die soeben beschriebenen umfassenden Lehren könnten eine politische Gesellschaft akzeptieren, in der wir das Konzept der *Gnade* im öffentlichen Leben für die Kritik, die Verteidigung, Rechtfertigung oder Einführung bestimmter politischer Vorhaben nutzen. In diesem Buch konzentriere ich mich darauf, was aus der Idee der Gnade für unsere Migrationspolitik folgt; aber wir sollten uns stets daran erinnern, dass eine Gesellschaft die Tugend der Gnade mitunter in all ihren rechtlichen und politischen Institutionen ausüben muss – und nicht bloß an ihrer Grenze.

3. Zur Verteidigung der Gnade

Es ist jedoch weiterhin offen, *warum* der Staat gnädig handeln sollte. Perry hat uns sicherlich eine wichtige Antwort gegeben, die ich im Folgenden noch erweitern werde. Aber es lohnt sich, zunächst direkter zu fragen, aus welchen Eigenschaften des Staates folgen sollte, dass er die Tugend der Gnade zu demonstrieren hat. Immerhin gibt es plausible Gründe, die gegen eine solche Annahme sprechen. Nehmen Sie zum Beispiel das skeptische Argument von Xan, dem fiktiven Regenten über ein zukünftiges Vereinigtes Königreich, der sich mit einer Welt allgemeiner Sterilität konfrontiert sieht. Xans Bruder Theo argumentiert dafür, die Grenzen zu öffnen; Xan widerspricht diesem Gedanken:

> „Theo dachte: Jetzt reden sie sogar schon alle gleich. Aber *egal, wer spricht, man hört jedesmal* Xans Stimme. Laut sagte er: ‚Es geht hier doch nicht um Vergangenheit und Geschichte. Bei uns sind weder Ressourcen knapp noch Arbeitsstellen oder Wohnraum. In einer sterbenden und unterbevölkerten Welt die Einwanderungsquoten zu beschränken, das ist keine sonderlich noble Politik.' ‚Nobel war das zu keiner Zeit', entgegnete Xan. ‚Edelmut ist eine Tugend für Individualisten, aber nicht für Regierungen. Wenn eine Regierung großzügig ist, dann mit dem Geld, der Sicherheit und der Zukunft der ihr anvertrauten Menschen.'"[25]

Xans Ideen finden freilich starken Widerhall in konservativen Kreisen. So bemerkte Margaret Thatcher bekanntermaßen einmal, dass das Problem des Sozialismus darin bestanden habe, dass ihm schlussendlich das Geld fremder Leute ausgegangen sei. Warum sollten wir jedoch nicht auch von der Tugend der Gnade denken – oder der Wohltätigkeit oder irgendeiner ähnlichen Tugend –, dass sie einzelnen Personen, und nicht Staaten, zukommt?

Darauf gibt es zwei unterschiedliche Antworten. Ich bezeichne sie als die Argumente der *demokratischen Emotion* und des *demokratischen Handelns*, und werde sie in dieser Reihenfolge darlegen. Das erste Argument – das Argument der *demokratischen Emotion* – beginnt mit einer einfachen Tatsache: Demokratische Selbstbestimmung ist eine fragile Angelegenheit, insbesondere, da wir sehr gut wissen, was eine gedeihende Gesellschaft braucht, um fortzubestehen (wohingegen wir sehr wenig darüber wissen, wie wir eine solch gedeihende Gesellschaft von außen aufbauen). Eine Sache, die eine solche Gesellschaft zu benötigen scheint, ist das unter ihren Bürgerinnen verbreitete Gefühl, dass das Spiel einer auf Wahlen beruhenden Politik es wert ist, gespielt zu werden – und zwar auch dann, wenn die Ergebnisse dieser Politik nicht den eigenen Vorstellungen entsprechen sollten. In einer florierenden Demokratie wird Stabilität dadurch hergestellt, dass sich die Bürgerinnen zu einem gerechten Verhalten innerhalb des Staates verpflichtet fühlen, statt, bildlich gesprochen, schlicht die eigenen Spielsachen einzupacken und nach Hause zu gehen, wenn eine Wahl verloren wurde. John Rawls' Vorstellung von Stabilität veränderte sich im Verlauf seiner Karriere; allerdings änderte sich nichts an seiner Überzeugung, dass wir erst dann über eine wahrhaft demokratische Gesellschaft verfügen, wenn die einzelnen Bürgerinnen einen tatsächlichen und empfundenen moralischen Grund dafür haben, diese demokratische Gesellschaft auch weiterhin zu unterstützen und das selbst dann, wenn diese Gesellschaft nicht die Entscheidungen trifft, die sie selbst getroffen hätten. Aber wie ich bereits in der Besprechung des heuchlerischen Vetos betont habe, ist es innerhalb einer Gesellschaft – oder zumindest innerhalb großer Teile dieser Gesellschaft – durchaus möglich, dass diese Art von Verbundenheit nicht mehr empfunden wird. Sicherlich sind wir zu Recht besorgt über die Stabilität einer Gesellschaft, wenn die empathische Identifikation mit anderen Menschen schwieriger wird oder nicht mehr ersichtlich sein sollte.

Was also wird für eine solche Art von Stabilität benötigt? Auch in diesem Fall ist es einfacher zu bestimmen, was diese Stabilität bedroht, als was nötig ist, um eine solche Stabilität zu erreichen. Ein wichtiges Argument stammt von Susan Moller Okin, die in *Justice, Gender, and the Family* die Familie als zentralen Ort der Gerechtigkeit ausweist.[26] Sie nennt hierfür eine Vielzahl von Gründen; so würden etwa die bestehenden Normen des Scheidungsrechts sowie der geschlechtlichen Identität ungleiche Machtbeziehungen innerhalb der Familie schaffen, die wiederum die einzelnen Familienmitglieder auf unterschiedliche Weise zur gesellschaftlichen Teilhabe befähigten. Von größerer Bedeutung ist jedoch, dass laut Okin gerechtes Handeln eine *Fähigkeit* ist – dass Gerechtigkeit also moralischer Erziehung bedarf und die Familie selbst eine Art von Schule ist, in der diese Fähigkeit entweder gestärkt oder geschwächt wird. Vor allem ein Aspekt an Okins umfassender Kritik des patriarchalen Liberalismus ist für unsere Zwecke von Bedeutung: Uns sei es nicht möglich, in einer Sphäre unseres Lebens gerecht zu handeln, wenn diese Gerechtigkeit in anderen Sphären unserer Gesellschaft fortdauernd verweigert würde. Wenn es also der Familie erlaubt sei, in einer Sphäre der Ungerechtigkeit zu verbleiben, dann sei das sowohl aufgrund der konkreten Folgen für das Leben von Frauen falsch, aber auch aufgrund der Effekte, die es auf die moralischen Persönlichkeiten der Mädchen und Jungen habe, die in solch ungerechten Familien aufwachsen.

Durch Okins Argument können wir jedoch auch etwas über die Entwicklung der moralischen Persönlichkeit außerhalb der Familie lernen. Innerhalb einer Gesellschaft gerecht zu handeln ist laut Okin eine Fähigkeit; es handelt sich dabei jedoch um eine Fähigkeit, die letztendlich auf der fortdauernden Bereitschaft beruht, Menschen als einer moralischen Betrachtung würdig zu erachten.[27] Wir seien demnach Geschöpfe, die anfällig für moralische Korruption sind, also dafür, unsere moralischen Angelegenheiten so zu interpretieren, dass sie uns Gründe liefern, das zu tun, was wir ohnehin bereits vorhatten. Darüber hinaus seien wir auch anfällig für Vorurteile gegenüber den Erfahrungen anderer Menschen, seien diese Vorurteile implizit oder von anderer Art. Allerdings würden wir eben jene empathische Identifikation benötigen, um eine auf Wahlen beruhende Politik aufrecht erhalten zu können; wir müssten uns demnach beständig an das moralische Wesen unserer Gegnerinnen erinnern, wenn wir weiterhin mit ihnen eine solche Politik betreiben wollen – und wir müssten die moralische Realität derjenigen Bürgerinnen wahrnehmen, die wir niemals treffen werden, um zu vermeiden, dass wir aufgrund reinen Eigennutzes ihre Beschwerden ignorieren. Okins Argument verweist auf die Tatsache, dass die Lernerfahrungen aus einer gesellschaftlichen Sphäre sich auf unsere Kompetenzen in der politischen Sphäre auswirken. Es sei demnach schwierig, ein politischer Egalitarist zu sein, wenn die Familie patriarchal organisiert ist. Allerdings ist es in meinen Augen genauso schwierig, durch das Gut unserer Mitbürgerinnen motiviert zu werden, wenn unsere Gesellschaft das Gut Außenstehender als einer moralischen Erwägung nicht wert erachtet.

Was ich damit meine ist, dass es für eine florierende Gesellschaft sowohl die Tugend als auch das Gesetz braucht. Die Tugend muss jedoch ausgebildet werden. Sie verlangt zumindest eine konstante Anerkennung der moralischen Existenz anderer Menschen und eine Bereitschaft, sie als moralisch bedeutsame Geschöpfe zu betrachten, deren Anliegen es wert sind, von uns ernst genommen zu werden. In Abwesenheit einer solchen empathischen Identifikation geraten wir politisch mit einiger Wahrscheinlichkeit in eine Reihe pathologischer Zustände, von denen keiner zu einer stabilen Form demokratischer Selbstbestimmung zu führen scheint. Auf der Makroebene können wir recht simpel feststellen, dass eine Person dem Gedanken verpflichtet sein muss, dass Menschen mit anderer Meinung dennoch rationale Personen und keine Teufel sind, um weiterhin einen gemeinsamen Prozess politischer Entscheidungsfindung aufrecht erhalten zu können. Die politische Polarisierung des US-Wahlkampfs 2016 ist nicht per se problematisch; eine Demokratie kann derart polarisierte Meinungsverschiedenheiten aushalten, sofern die Bürgerinnen fähig sind, ihre Gegnerinnen zu respektieren und mit ihnen zu verhandeln. In meinen Augen ist vielmehr beunruhigend, wie in diesem Wahlkampf die jeweiligen Kontrahentinnen dämonisiert wurden, bis hin zu Haftandrohungen und der angedrohten Weigerung, die Wahl anzuerkennen, sollte sie verloren gehen.[28] Politische Deliberation verlangt eine kontinuierliche Einsicht darin, dass die eigenen Gegner keine Dämonen in Menschengestalt sind, sondern vernünftig urteilende Geschöpfe wie wir, woraus folgt, dass wir gute moralische Gründe haben, mit ihnen Politik zu betreiben. Auch Rawls bemerkt, dass die Tugend wie eine Pflanze ist, die der Pflege bedarf. Er vergleicht unseren Willen dazu, Politik im Geiste gegenseitigen Respekts zu betreiben, als eine Form von Kapital, das vermehrt oder bewahrt werden kann.[29] Ein Gemeinwesen kann demnach seinen Willen zur Gerechtigkeit verlieren, wenn es sich nicht fortdauernd daran erinnert, dass andere Menschen real existieren und wir ihnen die Formen moralischen Respekts schulden, die Geschöpfen wie ihnen zustehen.

Was hat all das nun mit Migration zu tun? In meinen Augen können uns die Handlungen des Staats gegenüber denjenigen, die sich außerhalb seines Territoriums befinden, einige Lektionen hinsichtlich unserer Pflichten gegenüber Menschen im Allgemeinen erteilen. Ein Gemeinwesen, dass das Wohl anderer Personen nicht ernsthaft in Betracht zieht – das es also bloß vermeidet, Rechte zu verletzen – wäre ein mangelhaftes Gemeinwesen, und zwar schlicht aufgrund der Tatsache, dass es seine Bevölkerung nicht daran erinnert, dass das Wohl anderer Personen es wert ist, von uns ernsthaft in Betracht gezogen zu werden. Wir müssen uns regelmäßig in der Erkenntnis üben, dass Menschen außerhalb unseres Gemeinwesens auch menschlich sind und somit genauso zählen wie unsere eigenen Nachbarn. Ich kann nicht behaupten, dass diese Art von Ritual die Kräfte eindämmen kann, die uns dazu gebracht haben, in Migrantinnen tendenziell Gefährder und Eindringlinge zu sehen. Aber die Abwesenheit des Rituals führt zum Verlust des Glaubens.[30] Wir sind manchmal skeptisch gegenüber einer solchen moralischen Erziehung; wir könnten uns fragen, ob irgendeine Person ihre moralischen Urteile an der Migrationspolitik ihres Gemeinwesens ausrichtet. Vielleicht nicht – aber es gibt bestimmte Formen der Migrationspolitik, von denen wir zuverlässig sagen können, dass sie unsere moralischen Kompetenzen untergraben können. So würde ein explizit rassistisches Migrationsregime rassistische Personen bestärken und Ideen unterstützen, die in den gegenwärtig existierenden demokratischen Gesellschaften bisher unter der Oberfläche verblieben. Genauso würde eine ungnädige Politik – die das Wohlergehen von Personen außerhalb des Gemeinwesens ignorieren und ihm entsprechend eine angemessene moralische Würdigung verweigern würde – unsere eigenen Tendenzen zu moralischer Korruption und Egoismus verstärken. Wir bedürfen der Fähigkeit zur Empathie, wenn wir auf bestimmte Weise gegenüber Neuankömmlingen zu handeln gedenken. Die gerichtlichen Klagen mehrerer republikanischer Gouverneure gegen die Aufnahme syrischer Flüchtlinge hat beispielsweise diejenigen moralisch unterstützt, die sich dem Aufbau von Formen gemeinschaftlicher Unterbringung dieser Flüchtlinge verweigerten.[31] Jedoch ist der Punkt in diesem Fall allgemeiner. Durch die Verweigerung, die Bitten derjenigen außerhalb des Staates anzuhören, erleichtert es der Staat Bürgerinnen, sich auch den Anliegen derjenigen zu verweigern, die eine andere politische Meinung haben.

An diesem Punkt möchte ich jedoch zu meinem zweiten Argument übergehen – dem Argument des *demokratischen Handelns*. Dieses Argument ist in seiner Formulierung schlichter, in seinen Details jedoch mitunter komplexer. Der Gedanke ist folgender: Wir sprechen im Allgemeinen so, als ob der Staat ein eigenständig handelnder Akteur sei. Selbst diejenigen unter uns, die kollektiven Entitäten oder Gruppen keinen metaphysischen Status zuschreiben wollen, halten es für richtig, wenn wir den Staat als eine Entität betrachten, die Steuern einzieht, regiert und so weiter. Demnach mache ich einen Fehler, wenn ich meinen Unmut über die von mir zu zahlenden Steuern an demjenigen auslasse, der die Steuern eintreibt; er ist bloß Angestellter des Staates und es ist der Staat, an den ich meinen Groll korrekterweise adressieren sollte.

Nun könnten wir diese Problematik dadurch auflösen, dass wir uns nur auf natürliche Personen beziehen; immerhin wurden die Steuergesetze von einzelnen Personen vorgeschlagen, einzelne Personen haben für diese Gesetze gestimmt und so weiter. Allerdings wollen wir oft davon ausgehen, dass der Staat gleich einem einzelnen Akteur über die Zeit hinweg und unabhängig von den speziellen Personen besteht, die zu einzelnen Zeitpunkten bestimmte Positionen innerhalb dieses Staates einnehmen. Hierzu müssen wir also imstande sein, den Staat als eigenständigen Akteur zu verstehen, woraus folgt, dass wir ihm auch die Bürden der Handlungsfähigkeit zuschreiben müssen, worunter dann auch ethische Verpflichtungen gegenüber denjenigen fallen, über die er seine Macht ausübt.

Das scheint nun ein langer Weg für eine schwache Schlussfolgerung gewesen zu sein – schließlich denkt niemand, dass der Staat nicht *ein paar* ethische Pflichten hat. Allerdings möchte ich das spezifischere Argument vorbringen, dass die meisten dieser dem Staat für gewöhnlich zugesprochenen Pflichten als politische Pflichten verstanden werden können. Der Staat täte mir beispielsweise Unrecht, würde er mein Recht auf Religionsfreiheit verletzen. Was ihn davon abhält, seine Zwangsgewalt mir gegenüber auf diese Weise einzusetzen, ist die Idee davon, was ein gerechter Staat ist. Ebenso tut mir der Staat in manchen Fällen Unrecht, wenn er mir nicht zukommen lässt, was ich für meine Handlungsfähigkeit benötige. So macht ein Staat, der keinen angemessenen Polizeidienst unterhält, mein Leben weniger sicher als ich berechtigterweise erwarten darf.

Beide Gedanken beziehen sich auf die Frage, ob der Staat seine Macht auf korrekte Weise ausübt. Sie beziehen sich somit auf die Legitimität oder Gerechtigkeit des Staates als Staat. Allerdings möchte ich auf einen allgemeineren Punkt hinaus. Ich denke, dass dem Staat einige ethische Pflichten nicht aufgrund seines Wesens als Staat zukommen, sondern allein aufgrund seiner *Existenz* als eine Art von Akteur, dem ethische Pflichten zukommen.

Dieser Punkt ist in meinen Augen deshalb von Bedeutung, da wir den Staat nicht als Akteur verstehen können, ohne gleichzeitig anzunehmen, dass er einige Pflichten gegenüber denjenigen hat, deren Wohl zu fördern oder zu verweigern in seiner Macht liegt. Dabei handelt es sich nicht um Rechtfertigungspflichten gegenüber denjenigen, über die er seine Macht ausübt; solche Pflichten kommen dem Staat zwar auch zu, allerdings möchte ich mich an dieser Stelle nicht mit ihnen beschäftigen. Stattdessen will ich bloß diese Tatsache feststellen: Wenn wir vom Staat als einem Akteur mit eigener Handlungsfähigkeit sprechen, dann müssen auch die mit dieser Handlungsfähigkeit einhergehenden Bürden akzeptiert werden, nämlich diejenigen Pflichten zu erfüllen, die allen Akteuren zukommen – nämlich, in einer Gemeinschaft mit anderen Akteuren zu existieren und Pläne zu entwickeln, in denen das Wohl einiger anderer Akteure befördert wird. Daraus folgt jedoch, dass der Staat kein vollständig moralischer Akteur sein kann ohne irgendeinen Plan zu entwickeln, in dem das Wohl anderer Menschen – derjenigen, gegenüber denen er keine Gerechtigkeitspflichten hat – Teil seiner eigenen Vorhaben wird.

Zu diesen Gedanken möchte ich noch eine letzte Idee hinzufügen: Unsere ethischen Pflichten treten oft dann am deutlichsten hervor, wenn sie von denjenigen vorgebracht werden, die über wenig Macht verfügen. In *Harry Potter und der Feuerkelch* fasst Sirius Black diesen Gedanken prägnant zusammen:

> „Wenn Du wissen willst, wie ein Mensch ist, dann sieh dir genau an, wie er seine Untergebenen behandelt, nicht die Gleichrangigen."[32]

Wir mögen blass werden bei dem Gedanken, dass ein potentieller Migrant in irgendeinem Sinne weniger Wert haben sollte als diejenigen, die bereits innerhalb einer Gesellschaft anwesend sind; aber die moralische Logik dieses Zitats hängt davon ab, dass wir *untergeben* hier schlicht im Sinne von Macht und nicht im Sinne von Tugend verstehen – und der Migrant verfügt, schlicht aufgrund seiner Situation, nicht über die Macht derjenigen, die bereits anwesend sind. Wie wir auf die Bedürfnisse solcher Menschen reagieren, offenbart unseren Charakter deshalb besonders deutlich, da der Erfüllung unserer politischen Pflichten im innerstaatlichen Bereich, trotz der Bedeutung von Empathie und Solidarität, oft ein gewisser Eigennutz beigemischt ist: Wir meinen vielleicht oft, dass wir unsere Kontrahentinnen deshalb gut behandeln sollten, weil sie eines Tages über uns regieren könnten. Daraus folgt jedoch, dass wir sowohl als Personen als auch als Mitglieder eines Staates eine bestimmte Verpflichtung haben – nämlich sehr sensibel hinsichtlich der Erfüllung unserer Pflichten gegenüber denjenigen zu sein, die uns gegenüber relativ machtlos sind. In meinen Augen definiert sich der Staat als moralischer Akteur am deutlichsten dadurch, wie er diejenigen behandelt, gegenüber denen er keine besonderen Verpflichtungen hat und die selbst kaum über die Macht verfügen, seinem Handeln Widerstand zu leisten. Daraus folgt allerdings, dass Migration tatsächlich einen bedeutenderen Gegenstand für die Idee der Gnade darstellen mag, als wir zunächst gedacht haben könnten. Wie ich zeigen werde, ist es möglich, Gnade auch über den innerstaatlichen Bereich hinaus zu üben; insbesondere im Bereich der Migration fällt jedoch auf, dass diejenigen, denen Migrationsrechte gewährt werden, im Allgemeinen nicht bereits Mitglieder dieses Staates und daher auch nicht besonders in der Lage sind, die Politik des Staates oder den Einsatz seiner Macht zum Nutzen oder Schaden Dritter zu beeinflussen. Sofern sich Akteure dadurch moralisch auszeichnen, wie sie diejenigen behandeln, die keinen Widerstand leisten können – die, wortwörtlich, *von der eigenen Gnade abhängen* – handelt es sich bei der Migration also um einen Bereich, in dem wir eine Menge über den Charakter wohlhabender Staaten und derjenigen, die diese Staaten anführen, erfahren können.

10
Migration und Gnade

Im vorangehenden Kapitel habe ich den Gedanken verteidigt, dass eine Gesellschaft mehr tun sollte, als bloß ungerechte Handlungen zu vermeiden. Stattdessen sollte sie auch im Einklang mit den Forderungen der Gnade handeln – worunter in meinen Augen auch fällt, dass es einen prinzipiellen Grund gibt, von bestimmten Formen harscher Behandlung abzusehen, selbst wenn diese Behandlung selbst nicht ungerecht sein sollte. Der gnadenvolle Akt wird in dem Bewusstsein getan, dass die von unserem Handeln abhängige Person ein Mensch ist; ihre Pläne bedeuten ihr genau so viel wie unsere Pläne uns selbst bedeuten. Durch die Praxis der Gnade vergegenwärtigen wir uns diese Tatsache, als Bürgerin und Person. Sie spiegelt die beständige Notwendigkeit, menschliche Wesen als wertvoll, als unserer Sorge wert zu betrachten – und nicht als bloße Belastungen oder Personen, die sich moralisch von uns unterscheiden. Das Konzept der Gnade kann darüber hinaus als eine öffentliche Norm verstanden werden, die in Auseinandersetzungen darüber, wie die Macht einer demokratischen Gesellschaft eingesetzt werden sollte, von Nutzen sein kann. Wie ich gezeigt habe, kann diese Norm von einer Vielzahl vernünftiger umfassender Lehren akzeptiert werden; die Norm selbst ist nicht spalterisch oder inkompatibel mit der politischen Form des Liberalismus, wie sie von John Rawls entwickelt wurde. Die Idee der Gnade kann uns somit helfen, die Ethik der Politik genauer zu verstehen, indem sie unser ethisches Instrumentarium über die Konzepte der Gerechtigkeit und der moralischen Rechte hinaus erweitert.

Nun bleibt uns jedoch noch die Frage zu beantworten, wie genau das Konzept der Gnade uns bei unseren Auseinandersetzungen im Bereich der Migrationspolitik helfen kann. Ich möchte an dieser Stelle vorsichtig sein, um nicht den Eindruck zu erwecken, dass nur eine einzige Form der Politik, dnämlich eine Politik, die eine Norm wie die der Gnade akzeptiert, für eine Gesellschaft zulässig ist. Was ich verteidigt habe, ist ein *Argument* und keine politisch verbindliche Schlussfolgerung. Darüber hinaus folgt aus der Norm der Gnade nicht, dass ihre Ansprüche bloß auf eine einzige Weise erfüllt werden können. Ihrem Wesen nach ist sie mit vielen verschiedenen Formen politischer Maßnahmen kompatibel. Folglich bin ich in dem, was ich hier sagen kann, eingeschränkt; es ist mir nicht möglich zu zeigen, dass eine bestimmte Politik der einzige Weg ist, auf dem eine politische Gemeinschaft der Norm der Gnade entsprechen kann und ich werde auch nicht versuchen, dies zu tun.

Es wird für uns dennoch möglich sein, ein gewisses Gespür dafür zu bekommen, wie die von mir verteidigten Ideen zur Rechtfertigung bestimmter Formen der Migrationspolitik genutzt werden könnten. Rufen Sie sich in Erinnerung, dass das Gewähren von Gnade mit der Zuteilung von Gütern oder bestimmten Vorteilen verbunden ist, auf die kein rechtlicher Anspruch besteht. Die Gnade besteht also in der Entscheidung, sich um das Wohl bestimmter Menschen zu kümmern; zwar besteht kein Anspruch auf eine solche Unterstützung, dennoch können wir diejenigen berechtigterweise kritisieren, die sich einer solchen Entscheidung verweigern. Wir können jedoch noch mehr sagen. Es scheint, als gebe es mindestens zwei Arten von Maßnahmen oder Formen von Politik, die der Norm der Gnade entsprechen. Gelegentlich sehen wir uns einer bestimmten Person gegenüber, die – ohne jegliche Ungerechtigkeit – durch unser Handeln völlig vernichtet werden könnte. Smilla könnte durch die Haft auf eine Art leiden, die angesichts ihrer geringen Vergehen überzogen wirkt. In diesem Falle finden wir eine einzelne Person vor und können das Konzept der Gnade auf ihren spezifischen Fall anwenden. Wir können dies als einen Fall *negativer Gnade* bezeichnen: Gnade bedeutet hier, etwas gegenüber einer bestimmten Person nicht zu tun, etwas, das uns zwar gemäß der Gerechtigkeit gestattet wäre, dieser Person jedoch fürchterlich schaden würde. Der Akt der Gnade liegt, im moralischen Sinne, immer noch in unserem eigenen Ermessen; wir könnten Smilla diesen Akt der Gnade verweigern, ohne dass wir uns einer Ungerechtigkeit schuldig machen würden. Aber in ihrem Fall würden wir zu Recht Gnade walten lassen – oder zumindest könnten wir schlicht durch einen Verweis auf ihre Behandlung erklären, warum es unsererseits ein Versagen wäre, nicht gnädig zu handeln.

Solche Fälle können wir von denjenigen unterscheiden, in denen potentiell einer beliebigen Anzahl von Menschen unsere Gnade zuteilwerden kann. Nehmen Sie zum Beispiel die Gruppe von Menschen, die in ein bestimmtes Land einwandern wollen, um ihren Karrieren Auftrieb zu verleihen. Wenn es stimmt, dass keine dieser Personen schlicht aufgrund dieses Wunsches ein Recht darauf hat, einzuwandern, wäre es nicht ungerecht, alle diese Menschen auszuschließen. Allerdings wären wir angesichts der Tatsache, wie sehr wir die Karrieren mancher dieser Menschen durch die Erlaubnis zur Einwanderung verbessern könnten, zutiefst ungnädig. Ich denke, wir können diesen Fall als *positive Gnade* bezeichnen: Die Gruppe von Menschen, denen gegenüber wir Gnade zeigen könnten, ist groß, und wir könnten uns sogar selbst dann als gnädig erweisen, wenn wir nicht allen Personen dieser Gruppe die von ihnen gewünschten Einwanderungsrechte zusprächen. Diese Art von Gnade ist in meinen Augen positiv, da sie versucht manchen – aber nicht allen – der infrage stehenden Personen einen positiven Vorteil zu verschaffen; es gibt in diesem Fall jedoch kein Individuum, dessen Pläne derart von unserer Entscheidung abhängen, dass wir sagen könnten, wir sollten insbesondere aufgrund dieser einen Person gnädig oder ungnädig handeln.

In meinen Augen gibt es keine scharfe Trennlinie zwischen diesen zwei Formen der Gnade. Es mag innerhalb einer Gruppe von Bewerberinnen auf positive Gnade immer Menschen geben, die sich als ein Fall zur Anwendung negativer Gnade herausstellen. Die zwei Kategorien sollten also eher als Idealtypen der Wirkungsweise von Gnade im Bereich der Migration verstanden werden. Es kommt vor, dass einzelne Personen um Rettung ersuchen; und sofern diese Rettung nicht durch die Gerechtigkeit gefordert wird, reicht es dennoch aus, diese Personen konkret anzutreffen, um die Norm der Gnade ins Spiel zu bringen. Es mag aber auch Zeiten geben, in denen wir Programme und Methoden einrichten wollen, anhand derer potentielle Migrantinnen unserer Gesellschaft beitreten könnten – selbst wenn wir nicht glauben, dass es ungerecht gegenüber irgendeiner bestimmten Person wäre, wenn wir uns dem verweigern würden. In beiden Fällen besteht der zentrale Punkt darin, dass wir denjenigen, denen wir die Aufnahme verweigern, kein Unrecht tun. Die Aufnahme sollte in bestimmten Fällen und durch bestimmte politische Instrumente gewährt werden – allerdings unterscheidet sich im Falle der Gnade das Versäumnis, bestimmten Geboten zu folgen, moralisch gesehen von den Arten der Ungerechtigkeit, die Staaten bisher traditionell zugesprochen wurden.

All das kann uns meiner Meinung nach bei der Diskussion der bereits zuvor besprochenen Formen der Migrationspolitik helfen. In Kapitel sieben habe ich drei Fälle betrachtet und untersucht, ob bestimmte Personen Rechtsansprüche auf Einwanderung geltend machen können: Personen mit einer besonderen Affinität zu einem Ort oder der dort lebenden Bevölkerung, die

Undokumentierten und diejenigen, die mit Bürgerinnen einer anderen Gesellschaft verheiratet sind. Ich habe argumentiert, dass eine Zurückweisung der Forderungen dieser Personen in keinem dieser Fälle ungerecht sei. Obwohl manche unter ihnen durchaus Gerechtigkeitsansprüche geltend machen könnten, entstammen diese Ansprüche doch nicht einfach den von mir jeweils beschriebenen Umständen. In diesem Kapitel will ich nun zeigen, wie das Konzept der Gnade uns helfen kann, zu erklären, warum bei diesen Personen – oder vielleicht manchen unter ihnen – das Konzept der Gnade für unsere moralischen Urteile besonders geeignet ist. Ich werde diese Beispiele der Reihe nach diskutieren und mit einigen Gedanken über die Gründe schließen, die für eine weitere Verfolgung meiner Untersuchungen sprechen.

1. Affinität

Wir fragen uns selten, warum es überhaupt Einwanderung geben sollte, abgesehen von solchen Situationen, in denen die Gerechtigkeit sie gebietet. Was also würde die Welt verlieren, wenn die Grenzen für alle Menschen, mit Ausnahme derjenigen, die vor Gewalt und Entbehrungen fliehen, geschlossen wären? Menschen überqueren Grenzen aus einer Vielzahl von Gründen, aber einer der bedeutendsten besteht in der simplen Tatsache, dass manche Dinge, ob nun überhaupt oder in einem größeren Ausmaß, allein innerhalb eines bestimmten Landes zu finden sind. So könnten wir beispielsweise in ein neues Land auswandern, da es für unsere Karrieren notwendig erscheint – oder, etwas allgemeiner formuliert, da diese Karrieren sich an diesem neuen Ort besser entwickeln. Wir könnten beispielsweise auch aufgrund der geografischen Besonderheiten eines Ortes einwandern wollen; Surfer etwa nach Australien oder Vulkanologinnen nach Hawaii. Was auch immer den Grund für einen Migrationswunsch darstellt, unsere Pläne würden sich durch eine Erlaubnis zur Einwanderung in einem solchen Fall erfolgreicher entwickeln. Eine vollkommene Abwesenheit von Einwanderungsrechten würde also die Abwesenheit solcher Güter darstellen, die in den beschriebenen Fällen durch eine Einwanderung überhaupt erst erreichbar wären.

Weiter oben habe ich argumentiert, dass die Zurückweisung solcher Migrantinnen keine Ungerechtigkeit darstellt. Personen, deren Handlungsfähigkeit durch eine Zurückweisung vernichtet würde, dürfen nicht abgeschoben werden; aber viele der von Zurückweisung oder Abschiebung bedrohten

Menschen können keinen solchen Grund vorbringen. Sie führen stattdessen an, dass ihre persönlichen Pläne durch die Zurückweisung beeinträchtigt würden. Dabei handelt es sich jedoch um einen schwächeren Grund. Führt eine Person eine besondere Affinität zu einem Ort als Grund für die Einwanderung an, so stellt sie sich als eine Person dar, deren Pläne von etwas abhängen, was nur innerhalb eines bestimmten Landes zugänglich ist. Dieses andere Land hat jedoch keine Pflicht im Sinne der Gerechtigkeit, ihr diesen Zugang zu gewähren. Allerdings würde sich dieses Land gemäß meiner Analyse als ungnädig erweisen, sollte es sich der Entwicklung eines Programms verweigern, durch das manche Personen ihr Leben durch die Nutzung von Ressourcen, die sich allein in diesem Land befinden, verbessern könnten.

In meinen Augen handelt es sich dabei um die bereits beschriebene positive Form der Gnade. Dem infrage stehenden Staat kommt demnach eine Pflicht zu, ein politisches Instrument zu entwickeln, durch das manchen Menschen ein Recht auf Einwanderung gewährt werden kann. Von einer solchen Politik könnten sowohl die Neuankömmlinge als auch das Aufnahmeland profitieren – und tatsächlich können wir genau das im Normalfall auch erwarten. Die einwandernde Person erleichtert den Austausch von Informationen, erwirbt neues Wissen, baut transnationale Beziehungen auf, stärkt transnationale Institutionen, und so weiter.[1] Allerdings hängt die Verpflichtung, einem Teil der Weltbevölkerung dadurch zu helfen, dass ihm Einwanderungsrechte zugesprochen wird, nicht von einem solchen Nutzen ab. Wenn sich eine bestimmte Gesellschaft dazu entscheiden sollte, außer Flüchtlingen keine Einwanderung zuzulassen, könnten wir meiner Meinung nach nicht sagen, dass dieses Land ungerecht handelt. Wohl aber würden viele von uns diese Gesellschaft als moralisch defizitär wahrnehmen. Zwar hat in diesem Fall keine einzelne Person ein Recht auf Einwanderung – genauso wenig wie ich ein solches Recht hatte, als ich in die Vereinigten Staaten auswandern wollte, um dort meine akademische Ausbildung abzuschließen und für eine Anstellung auch längerfristig zu verbleiben. Sollte sich ein Land allerdings weigern, eine solche Möglichkeiten überhaupt *irgendeiner* Person zu gewähren, kann das als eine Verweigerung von Gnade kritisiert werden. Um mit Kant zu sprechen: Wir könnten diesem Land die Verletzung seiner Wohltätigkeitspflicht vorwerfen. Sofern wir zu einer eher informellen Sprachebene tendieren, könnten wir eine solche Politik auch schlicht als *gemein* bezeichnen.

An dieser Stelle möchte ich allerdings auf einige wichtige Implikationen des soeben beschriebenen moralischen Arguments hinweisen. Es gibt eine Vielzahl unterschiedlicher Personen, die von der Einwanderung in ein bestimmtes Land profitieren würden. Konzentrieren wir uns zur Illustration kurz auf die Vereinigten Staaten: Angehende Schauspielerinnen könnten von einem Recht profitieren, nach Hollywood ziehen zu dürfen; angehende Paläontologen könnten von einem Recht profitieren, nach Utah ziehen zu dürfen.[2] Die Vereinigten Staaten müssen jedoch nicht alle Menschen mit Einwanderungsrechten ausstatten, die von einem dauerhaften Aufenthalt in den USA profitieren würden. Ich denke, dass die Vereinigten Staaten durchaus entscheiden dürfen, wen sie einwandern lassen wollen und wie sich die entsprechende Auswahlprozedur gestalten soll. Sie sind dabei nicht vollkommen frei: Wie bereits gezeigt, dürfen sie keine Entscheidungen treffen, die Menschen auf eine Weise dämonisieren, welche die Norm der Gleichbehandlung verletzt. Von unserer Entscheidungsfreiheit können wir schließlich immer noch guten oder schlechten Gebrauch machen. Aber da es nun mal kein Recht auf Zutritt gibt, kann mit einem gewissen Freiheitsgrad manchen Menschen die Einwanderung verwehrt, anderen wiederum gewährt werden. So könnten sich die Vereinigten Staaten beispielsweise dazu entscheiden, Schauspielerinnen gegenüber Paläontologinnen zu bevorzugen – oder umgekehrt. Ich nehme an, dass es sowohl mit der Idee der Gnade als auch mit der Idee der Gerechtigkeit in Einklang stehen würde, wenn die USA politische Instrumente entwickelten, die eine Gruppe der anderen vorzieht. Wie gesagt: Sie dürften dabei nicht eine ethnische Gruppe bevorzugen, da solche Kriterien auf impliziten und nicht zu rechtfertigenden Vorurteilen über die benachteiligte Gruppe beruhen würden. Paläontologen könnten allerdings keinen solchen Einwand erheben; wir könnten sie ausschließen, ohne ihre Rechte zu verletzen.

Das also ist die erste Implikation einer Begründung von Migrationsrechten anhand der Idee der Gnade: Es sind gesetzliche Rechte, wobei keine der mit ihnen ausgestatteten Personen ein moralisches Anrecht auf sie hat. Es wäre nicht ungerecht oder ungnädig, würden wir einer bestimmten Gruppe von Personen umfangreichere Rechte zusprechen, sofern damit nicht gegen das Prinzip der Gleichbehandlung verstoßen wird. Die zweite Implikation folgt aus der ersten: Prinzipiell ist es nicht ungerecht die Zahl von Personen zu begrenzen, die von der infrage stehenden Politik profitieren. So verteilen beispielsweise die USA pro Jahr nur 16.000 H-1B Visa, wobei noch einmal 20.000 Visa an diejenigen verteilt werden, die einen höheren Abschluss von US-amerikanischen Universitäten besitzen.[3] Das Kontingent dieser Aufenthaltstitel, die es dem Eigentümer erlauben, sich in den Vereinigten Staaten aufzuhalten und auf lange Sicht auch eine permanente Aufenthaltserlaubnis zu erlangen, ist oft bereits im April aufgebraucht. Dieser Umstand ist jedoch nicht ungerecht; diejenigen, die von diesem Programm profitieren, werden durch einen Akt der Gnade, oder etwas ihr sehr ähnlichem, begünstigt und diejenigen, die nicht von diesem Akt profitieren, können sich nicht darauf berufen, dass sie in ihren Rechten verletzt worden wären. Eine Begrenzung der Anzahl der Begünstigten scheint demnach mit dem Konzept der Gnade prinzipiell kompatibel zu sein. Die Trump-Administration hat jedoch erst kürzlich das Vorhaben verkündet, die Zahl akzeptierter Flüchtlinge auf 30.000 zu begrenzen.[4] Das scheint sich auf profunde Weise von dem soeben diskutierten Fall zu unterschieden. Wird eine schutzbedürftige Person aus dem Grund zurückgewiesen, dass bereits 30.000 andere vor ihr Zuflucht gefunden haben, kann sie meiner Meinung nach zu Recht behaupten, ungerecht behandelt zu werden. Sofern sich ihre Ansprüche auf die Genfer Flüchtlingskonvention gründen, erscheint es ungerecht, sie in die schrecklichen Umstände ihres Herkunftslandes zurückzuschicken. Es scheint, als sei die Gnade mit Kontingentregelungen kompatibel – die Gerechtigkeit jedoch nicht.

Ein letzter Unterschied zwischen Gnade und Gerechtigkeit besteht darin, dass ein ungerechter Staat im Prinzip Zwang von außen unterworfen werden kann. Sicherlich muss die Ungerechtigkeit für solcherlei Maßnahmen von der richtigen Art sein; Theoretikerinnen internationaler Gerechtigkeit waren stets sehr daran interessiert, zu vermeiden, dass es Staaten aufgrund trivialer oder kleinerer Verfehlungen anderer Staaten gestattet sein könnte, Zwangsgewalt gegen diese einzusetzen. Nichtsdestotrotz können ungerechte Handlungen prinzipiell den Einsatz von Zwangsgewalt legitimieren. Kant erhebt das zu einem zentralen Bestandteil seiner Rechtslehre: Die perfekten Pflichten dürften, zumindest prinzipiell, durch die Ausübung von Zwang verteidigt werden. Hingegen könnte eine Verweigerung der Gnade noch nicht einmal im Prinzip den Einsatz eines solchen Zwangs rechtfertigen. Kant wendet seine Überlegungen auf den innerstaatlichen Bereich an; aber sie können genauso gut auch auf die Migration angewendet werden. Ein Staat, der sich dem Schutz von Flüchtlingen verweigert, kann mitunter zu Recht Zwang unterworfen werden. (Ob ein solcher Zwang jemals das von ihm intendierte Ziel erreichen kann, steht auf einem anderen Blatt.) Ein Staat, der sich dem Prinzip der Gnade verweigert, verletzt jedoch keine Rechte – und keine externe Entität scheint daher dazu berechtigt, in diesem Falle einzuschreiten. Es gibt jedenfalls keine bestimmte Person, deren Rechte eine solche Intervention rechtfertigen könnten. Die anderen Staaten der Welt könnten diesen Staat als selbstbezogen, gemein und ungnädig bezeichnen – aber sie können ihn für diese Fehler genauso wenig zur Verantwortung ziehen, wie wir aufgrund der Verletzung unserer Wohltätigkeitspflichten innerhalb des Staates als Kriminelle behandelt werden können.

2. Undokumentierte Immigrantinnen

Ich möchte nun zu dem Fall der undokumentierten Immigrantinnen zurückkehren und an Morgan erinnern. Morgan hatte ohne gesetzliche Erlaubnis die Grenze zwischen Kanada und den Vereinigten Staaten überquert. Er wohnt nun in Portland und hat sich dort ein Leben aufgebaut; er hat eine Arbeit, Freunde, und all die Dinge, die ein an einem bestimmten Ort aufgebautes Leben ausmachen. Er scheint für mehrere Jahre in Portland gelebt zu haben – mehr Jahre jedenfalls, als nach Joseph Carens erforderlich wären, um ihm ein Recht darauf zuzusprechen, in den USA verbleiben zu dürfen. Morgans Leben und die dazugehörigen Pläne würden durch die Entscheidung, ihn abzuschieben, radikal zunichte gemacht. Dennoch kann, so mein Argument, diese Art des Schmerzes nicht herangezogen werden, um Morgan ein Recht auf Verbleib in den USA zuzusprechen. Morgan hatte die Entscheidung getroffen, jenes Gesetz zu brechen, das ihn davon abhalten sollte, ein Leben in Portland aufzubauen; der Schmerz, mit dem er im Falle einer Abschiebung konfrontiert wäre, kann daher nicht als ungerecht bezeichnet werden. Mit längerer Aufenthaltsdauer würde der Schmerz einer Abschiebung zweifellos größer werden. Dennoch ist es gerechtfertigt, ihm diesen Schmerz zuzufügen, würde er festgenommen und abgeschoben werden. Er hat darauf gewettet, nicht entdeckt zu werden; der Schmerz, der mit dem Verlieren dieser Wette verbunden ist, kann ihm zu Recht auferlegt werden.

Ich denke, dass diese Überlegungen zutreffen; allerdings bin ich auch davon überzeugt, dass die Vereinigten Staaten sich als ungnädig erwiesen, würden sie Morgan abschieben. (Ich möchte betonen, dass es sich hierbei nicht um einen hypothetischen Fall handelt: Die Vereinigten Staaten schieben Menschen wie Morgan, und auch Personen mit weitaus besser begründeten Ansprüchen, mit erschreckender Regelmäßigkeit ab.) Morgan scheint einen Fall negativer Gnade darzustellen. Personen, die sich ein Leben an einem Ort aufgebaut haben, können als Fälle betrachtet werden, bei denen das Gewähren der Gnade besonders angemessen wäre. Zumindest könnten wir einen Staat als ungnädig kritisieren, sollte er sich dazu entschließen, Menschen wie Morgan abzuschieben. Diese oder ähnliche Gedanken haben eine Vielzahl von Menschen inspiriert, darunter auch Personen, von denen es kaum erwartet werden würde: So verteidigte Ronald Reagan in seiner Debatte mit Walter Mondale im Jahr 1982 eine Amnestie für undokumentierte Personen, die schon seit langer Zeit in den USA leben:

> „Es ist wahr, unsere Grenzen sind außer Kontrolle. Es ist jedoch auch wahr, dass diese Situation bereits seit langer Zeit und unter vielen verschiedenen Regierungen bestand. [...] Ich glaube an die Idee der Amnestie für diejenigen, die hier gelebt und Wurzeln geschlagen haben, selbst wenn sie vor einiger Zeit illegal eingewandert sein sollten.“[5]

Diese Ideen führten dazu, dass Reagan die Bestimmungen des Immigration Reform and Control Act von 1986 unterstützte, der denjenigen legale Aufenthaltsrechte zusprach, die über vier Jahre hinweg in den Vereinigten Staaten gelebt hatten – unabhängig von den Gründen, die sie ursprünglich die Grenzen hatten überqueren lassen.[6]

Gegenwärtig sind konservative Politiker offensichtlich weniger willens als Reagan, die Abschiebung langjähriger Einwohnerinnen als grausam zu verurteilen. So argumentierte beispielsweise der Abgeordnete Steve King, dass undokumentierte Einwanderer die – seiner Meinung nach wohl zu respektierende – Entscheidung getroffen hätten, „im Schatten“ zu leben:

> „Sie kamen hierher, um im Schatten zu leben, und wir verweigern es ihnen nicht, im Schatten zu leben [...] Wenn sie aber durch einen Strafverfolgungsbeamten angetroffen werden, fordert es das Gesetz, dass die entsprechenden Abschiebungsprozeduren eingeleitet werden.“[7]

Kings Argument geht davon aus – statt argumentativ zu zeigen –, dass Amnestie unmoralisch ist; undokumentierte Einwanderer könnten demnach nicht berechtigterweise begnadigt werden, da sie gegen das Gesetz verstoßen hätten. Aber es gibt eine einfache Antwort auf Kings Aussage, welche derjenigen entspricht, die Reagan gab: Es gibt Umstände, unter denen eine Begnadigung oder eine Politik im Sinne der Gnade, die moralisch richtige Antwort darstellt.

King hat selbstredend auch vehement dafür argumentiert, dass jedes Einbürgerungs- oder Amnestieprogramm zu einer weiteren Welle irregulärer Einwanderung führen würde. Darauf gibt es in diesem Kontext zwei Antworten: eine empirische und eine philosophische. Die empirische Antwort lautet, dass, wie bereits in Kapitel acht gezeigt wurde, ein Teil undokumentierter Migration schlicht auf der Kluft zwischen Reichtum und Armut beruht; es ist unwahrscheinlich, dass die ausschlaggebende Überlegung für eine undokumentierte Einreise darin besteht, dass vielleicht, in einigen Jahrzehnten, ein den undokumentierten Aufenthalt legalisierendes Gesetz erlassen wird. Die philosophische Antwort entspringt jedoch der Annahme, dass King und die ihm Gleichgesinnten durchaus Recht haben könnten und die Folge der Einführung eines solchen Gesetzes tatsächlich in einem Anstieg der irregulären Migration bestehen würde. In meinen Augen lautet die Antwort in diesem Fall, dass wir kein Recht haben darauf haben, die Einhaltung des Gesetzes mit allen Mitteln zu erwirken, wenn die entsprechenden Maßnahmen unsere moralischen Pflichten verletzen würden. Wir können das gut an anderen Rechtskontexten erkennen: Könnten wir das Überqueren roter Ampeln einfach dadurch eliminieren, dass wir den ersten Verstoß gegen diese Vorschrift mit der Exekution bestraften, würden wir daraus nicht schließen, dass die Todesstrafe die richtige Antwort auf das Überqueren roter Ampeln darstelle. Selbstverständlich wäre eine solche Strafe ungerecht; was auch immer derjenige tut, der bei Rot die Straße überquert, es legitimiert nicht seinen staatlich verordneten Tod. Allerdings können ähnliche Überlegungen auch dann gelten, wenn es sich um eine legitime Durchsetzung der Gesetze handelt. Selbst wenn wir das Überqueren roter Ampeln durch öffentliche Demütigung derjenigen verhindern könnten, die rote Ampeln überqueren, würden wir nicht direkt daraus schließen, dass eine solche Demütigung moralisch akzeptabel sei. Es könnte sein, dass etwas sowohl effektiv als auch moralisch falsch ist – oder, etwas allgemeiner gefasst, dass etwas zwar effektiv, zugleich aber auch Zeichen eines schlechten moralischen Charakters sein kann, der diejenigen Tugenden geringschätzt, die ein menschliches Leben leiten sollten. Die Antwort auf King könnte also schlicht lauten, dass ein zukünftiger Anstieg irregulärer Migration der Preis wäre, den wir zu zahlen bereit sein sollten, um uns nicht eines Mangels an Gnade im Hier und Jetzt schuldig zu machen.

Ich möchte diesen Abschnitt mit der wiederholten Bemerkung schließen, dass all diese Überlegungen nur auf einen Teil der undokumentierten Immigrantinnen zutrifft; viele dieser Menschen entscheiden sich aber aufgrund ungerechter Umstände in ihren Herkunftsländern dafür, Grenzen zu überqueren, oder deshalb, weil sie mit solchen Umständen konfrontiert wären, würden sie unverzüglich zurückkehren. Diese Menschen haben einen Gerechtigkeitsanspruch und wir sollten uns an diesen Anspruch erinnern, wenn wir unsere Gesetze beschließen. Allerdings möchte ich abschließend noch einmal betonen, dass sich moralische Erwägungen auch darauf auswirken sollten, wie wir diejenigen behandeln, die keinen solchen Gerechtigkeitsanspruch anführen können.

3. Familienzusammenführung

In der Diskussion über die Familienzusammenführung werde ich mich auf die Zusammenführung von Ehegattinnen konzentrieren; minderjährige Kinder haben, wie ich bereits gezeigt habe, andere Ansprüche, die auf ihre Handlungsfähigkeit zurückzuführen sind und auf den Bedarf an bestimmten Personen, die jene Handlungsfähigkeit anleiten. Meinem Argument zufolge ist es einem Staat zumindest gestattet, dem ausländischen Ehegatten einer einheimischen Bürgerin die Einwanderung zu verweigern. In meinen Augen findet sich an einer solchen Handlung keine Ungerechtigkeit gegenüber dem Ehegatten. Sollte etwas daran ungerecht sein, so würde es darin zu finden sein, wie sich die Forderungen des einheimischen Ehegatten zu den Forderungen der jeweiligen Mitbürgerinnen verhalten. Aber selbst in diesem Fall bin ich nicht davon überzeugt, dass ein Ausschluss als ungerecht ausgewiesen werden könnte: Ein Staat, der schlicht verkündet, dass er zukünftig keine derart begründeten Einwanderungsrechte mehr zuspricht, würde seinen Bürgerinnen gegenüber bloß verkünden, dass sie ab sofort gute Gründe hätten, eheliche Verbindungen nur mit solchen Personen einzugehen, die bereits im Besitz von Aufenthaltsrechten sind. Das ist sicherlich grausam – dennoch denke ich nicht, dass es ungerecht ist. Auch in diesem Kontext können wir erwarten, dass Menschen Verantwortung für ihre Liebesbeziehungen übernehmen; uns wird etwa zu Recht gesagt, dass wir Liebesbeziehungen mit unseren Studentinnen, Angestellten oder – weiter gefasst – mit Personen, über die wir Macht haben, vermeiden sollten. Um es einfach auszudrücken, Liebe widerfährt Ihnen nicht bloß; und wenn wir diese Tatsache anerkennen, können wir auch gleichsam anerkennen, dass Ihre Liebe zu einer bestimmten Person uns noch nicht dazu verpflichtet, dieser Person eine Heimat zu geben.

Allerdings wirkt eine solche Verweigerung zutiefst ungnädig; und tatsächlich scheint die ausländische Ehegattin der offensichtlichere Fall negativer Gnade zu sein. Die Ehegattinnen formulieren eine bestimmte Art von Forderung gegenüber uns, die den einzigartigen und starken Charakter einer Liebesbeziehung widerspiegelt. Sollten wir jedoch versuchen, diese Forderungen als Forderungen der Gerechtigkeit auszuweisen, würden wir in die von Luara Ferracioli beschriebenen Schwierigkeiten geraten: Was zeichnet diese Beziehungen aus, dass sie als einzige diese Art von Wert ausdrücken können? Könnte es nicht andere Beziehungen geben – Co-Autoren, beste Freunde –, die gleichermaßen bedeutend und stark sind?

Eine mögliche Reaktion auf Ferracioli besteht in meinen Augen schlicht in Zustimmung; allerdings wird dadurch nicht angefochten, dass die Bevorzugung von Ehegattinnen, vor beispielsweise besten Freunden, moralisch richtig ist. Wir können dies damit rechtfertigen, dass wir uns bei der Zusammenführung von Ehegattinnen für einen Akt der Gnade entscheiden können – sowohl gegenüber der eigenen Mitbürgerin als auch der ausländischen Person, mit der sie zusammenleben möchte. Wenn Gnade die Bereitschaft zur Einführung bestimmter politischer Maßnahmen beinhaltet – ein Einwanderungsprogramm oder ein bestimmtes politisches Instrument –, anhand derer der Staat die Welt besser macht, indem er Menschen Rechte zuspricht, die auf derlei Rechte keinen Anspruch im Sinne der Gerechtigkeit haben, dann erscheint die Zusammenführung von Ehegattinnen als ein außergewöhnlich vielversprechender Gegenstand, um damit zu beginnen.

Diesen Überlegungen zufolge sind eheliche Beziehungen nicht deshalb besonders, weil sich in ihnen etwas finden ließe, das auf diese Weise in keiner anderen Beziehungsform vorhanden sei – das ist nicht der Fall. Eheliche Verbindungen sind vielmehr aufgrund der simplen Tatsache besonders, dass sie *für gewöhnlich* von größerer Bedeutung für unser Leben sind als beispielsweise beste Freunde. Wenn wir einen Gerechtigkeitsanspruch vorbringen, haben wir gute Gründe, gegenüber dem Ausdruck „*für gewöhnlich*" misstrauisch zu sein. Meiner Meinung nach können wir Menschen nicht gerecht behandeln, wenn wir ihre Ansprüche anhand von prognostischen Kategorien bewerten. Gerechtigkeit verlangt nach aussagekräftigeren Grundlagen. Wenn Personen zum Zwecke des Asyls aus den Vereinigten Staaten nach Kanada auswandern, haben sie für gewöhnlich kein Recht auf diese Form von Schutz. Allerdings würden wir in dem seltenen Fall, dass eine Person einen solchen Anspruch berechtigterweise vorbringt, diese Person ungerecht behandeln, würden wir sie schlicht unter die Kategorie der US-Amerikanerin subsumieren. (Das ist kein ausgedachter Fall: Der US-Amerikanerin Denise Harvey wurde erst kürzlich in Kanada der Flüchtlingsstatus zugesprochen und ihre Abschiebung nach Florida ausgesetzt.)[8] Wenn wir uns mit Fragen der Gerechtigkeit befassen, dürfen wir nicht auf Vorhersagen und Tendenzen zurückgreifen; wir müssen, ganz im Sinne Ferraciolis, genau bestimmen, wie jede einzelne Beziehung und jeder einzelne Fall tatsächlich von den beteiligten Personen erlebt wird.

Wenn es jedoch um Gnade geht, sind wir eher zur Nutzung von Vorhersagen berechtigt und zwar deshalb, weil keine einzelne Person einen Anspruch darauf geltend machen kann, dass wir ihr Leben verbessern. Stellen Sie sich vor, dass eine Gruppe von Ehegattinnen um Einwanderung in ein bestimmtes Land ersucht und keine von ihnen über ein bereits bestehendes Recht auf Einwanderung verfügt. Alle diese Personen würden von einer Einwanderung profitieren; manche würden sogar sehr stark davon profitieren und zugleich durch ihre Anwesenheit auch bestimmten Bürgerinnen des Ziellandes von großem Nutzen sein. Diese Menschen könnten wir als natürlichen Anwendungsfall der Gnade ansehen. Jedoch würde ein Staat, der sich weigerte, diesen Personen Einwanderungsrechte zuzusprechen, ihnen sowie den mit ihnen verheirateten Bürgerinnen kein Unrecht tun. Ein derart handelnder Staat mag sich dadurch zwar eines Mangels an Gnade schuldig machen, nicht aber eines Mangels an Gerechtigkeit. Tatsächlich wäre er mitunter noch nicht einmal eines Mangels an Gnade schuldig: Sofern er zeigen könnte, dass seine Migrationspolitik an anderer Stelle dem Wert der Gnade angemessen Ausdruck verleiht – vielleicht durch eine außergewöhnlich generöse Einwanderungspolitik für eine andere Gruppe von Ausländerinnen –, könnte es durchaus sein, dass er sich gar keines Vergehens schuldig macht.

Selbstredend würde in der Realität jedes Land, das Ehepartnerinnen die Zusammenführung verweigerte, mit hoher Wahrscheinlichkeit die Tugend der Gnade verletzen; die Zusammenführung von Ehegattinnen ist sozusagen die am tiefsten hängende Frucht der Gnade angesichts des quasi-universalen Respekts, den die monogame Liebesbeziehung in verschiedenen menschlichen Kulturen genießt. Allerdings ist das nicht gleichbedeutend mit der Aussage, dass die Zusammenführung von Ehegattinnen eine Forderung der Gerechtigkeit sei. Stattdessen können wir auf Ferraciolis Herausforderung dadurch reagieren, dass wir die ihr zugrunde liegenden Bedingungen zurückweisen. Wenn wir das gesetzliche Recht auf die Vereinigung von Ehegattinnen anhand des Konzepts der Gnade und nicht anhand des Konzepts der Gerechtigkeit begründen, macht es keinen bedeutsamen Unterschied, ob es andere Arten von Beziehungen gibt, die etwas ähnlich Wertvolles darstellen wie eine Ehe. Wird einer besten Freundin die Einwanderung verweigert, verletzt der sie zurückweisende Staat keine Rechte, da weder sie noch eine ausländische Ehegattin einen Anspruch auf Einwanderung hat. So lange jedoch keine zu verurteilende Unterscheidung zwischen Ehegattinnen und besten Freundinnen vorgenommen wird, besteht auch kein Unrecht darin, die Pflicht der Gnade bloß im Hinblick auf erstere zu erfüllen.

Sicherlich könnten wir immer noch zu zeigen versuchen, dass eine solche Unterscheidung nicht legitim ist. Wir könnten zum Beispiel anführen, dass der Vorzug von Ehegattinnen ein nicht zu rechtfertigendes Vorurteil zugunsten einer monogamen und dauerhaften Form von Beziehungen darstelle, wodurch andere Formen von Beziehungen oder Gemeinschaft benachteiligt würden.[9] Das erscheint ausreichend, um eine Diskussion darüber zu entfachen, ob die den Ehegattinnen zugesprochenen Rechte moralisch tragbar sind. Ich denke, dass dieser Einwand tatsächlich zutreffen würde, wenn beispielsweise diese Rechte ausschließlich und ausnahmslos heterosexuellen Ehegattinnen zugesprochen würden; homosexuelle Männer und Frauen würden sich berechtigterweise als Bürgerinnen zweiter Klasse behandelt fühlen, wenn ihre Beziehungen als weniger wertvoll als die ihrer heterosexuellen Mitbürgerinnen betrachtet würden. Eine ähnliche Forderung könnten wir nun auch mit Blick auf alternative Formen menschlicher

Beziehungen jenseits der Ehe anführen; aber ich gebe zu, dass ich einen Erfolg solcher Argumente für unwahrscheinlich halte. Die umfassenderen Rechte, die ehelichen Gemeinschaften im Vergleich zu anderen Beziehungsformen gewährt werden – seien sie geschlechtlicher oder sonstiger Natur – könnten schlicht die Vermutung ausdrücken, dass solche alternativen Beziehungsformen mit einer geringeren Wahrscheinlichkeit so wertvoll sind wie eine eheliche Beziehung. Das gesetzliche Recht auf Vereinigung von Ehegattinnen auf dieser Wahrscheinlichkeit zu begründen scheint mir weniger Feindseligkeit, als vielmehr eine empirische Tatsache auszudrücken. (In jedem Fall scheint sich eine solche Begründung von der Art von Feindseligkeit zu unterscheiden, die der Weigerung zugrunde liegt, homosexuellen Beziehungen den gleichen Wert wie heterosexuellen Beziehungen zuzusprechen.) All das schließt jedoch nicht aus, dass meine Überlegungen von mir bisher unerkannten Vorurteilen beeinflusst werden, die es in diesem Falle freizulegen gelte. Allerdings könnten wir dadurch, dass wir die Rechte auf Familienzusammenführung anhand des Konzepts der Gnade begründen, zumindest erkennen, wo genau diese Vorurteile am Werk sein mögen.

4. Gnade und Gerechtigkeit im öffentlichen Diskurs

All das sollte zeigen, wie das Konzept der Gnade unsere Diskussion über solche Maßnahmen der Migrationspolitik befördern kann, die selbst nicht durch die Gerechtigkeit gefordert werden. Allerdings kann die Idee der Gnade auch eine Rolle bei der Kritik ungerechter politischer Maßnahmen spielen: Neben ihrem philosophischen Wert kommt der Gnade auch eine rhetorische Anziehungskraft zu. Sie zeigt uns, wie wir nicht sein wollen; wie es uns zwar durch die Gerechtigkeit gestattet wäre zu sein, wir aber nicht sein sollten. Diese Überlegungen könnten uns bei der Argumentation gegen politische Maßnahmen helfen, die sowohl ungerecht als auch ungnädig sind.

Nehmen Sie zum Beispiel die Antwort der früheren First Lady Laura Bush auf die Entscheidung der Trump-Administration, die Kinder irregulär eingewanderter Personen an der Grenze von ihren Eltern zu trennen und in Einrichtungen unterzubringen, die Straflagern ähneln. Bush argumentierte, dass die Verurteilung dieser Politik am ehesten durch unsere Abscheu gegenüber Grausamkeit zu verstehen sei:

> „Ich erkenne die Notwendigkeit an, unsere staatlichen Grenzen zu kontrollieren und sie zu schützen, aber diese Null-Toleranz-Politik ist grausam. Sie ist unmoralisch. Und es bricht mir das Herz. [...] Kürzlich besuchte Colleen Kraft, Vorsitzende der American Academy of Pediatrics, eine der Unterbringungen, die durch das U.S. Office of Refugee Resettlement geführt werden. Wie sie berichtete, gab es zwar Betten, Spielzeug, Farbstifte, einen Spielplatz und die Windeln wurden gewechselt, allerdings waren die Mitarbeiterinnen der Unterkunft angewiesen, die Kinder nicht auf den Arm zu nehmen oder anzufassen, um sie zu trösten. Stellen Sie sich vor, dass Sie ein Kind nicht auf den Arm nehmen dürfen, das noch Windeln trägt. [...] Können wir im Jahre 2018 keine freundlichere, mitfühlendere und moralischere Antwort auf die gegenwärtige Krise finden? Ich glaube, dass wir das können."[10]

Was diesen Kindern angetan wurde – und was ihnen zum Zeitpunkt der Fertigstellung dieses Buches weiterhin angetan wird – ist zutiefst ungerecht. Aber wir können auch die Sprache der Gnade nutzen, die zugleich die Sprache der Tugend ist, um die Aspekte dieser Angelegenheit zu betonen, die sich auf unsere Identität beziehen: Wir wollen kein Land werden, das *so etwas* tut. Bush weist auch auf die Bedeutung der Vulnerabilität hin; das Kind, dessen Bedürfnisse nach menschlicher Anleitung und Liebe so umfassend sind, wird durch eine Politik, die ihm die Befriedigung dieser Bedürfnisse verweigert, in seinen grundlegenden Rechten verletzt. Die Sprache der Gnade mag uns in manchen Fällen helfen, das zu verstehen. Wir müssen für unser Argument keine Theorie der Gerechtigkeit aufstellen. Stattdessen können wir schlicht verkünden, wie wir nicht sein wollen. Als Senator Ted Cruz vorschlug, dass die Vereinigten Staaten christliche Flüchtlinge gegenüber muslimischen bevorzugen sollten, begann Präsident Barack Obama keine Diskussion über politische Gerechtigkeit oder die konstitutionelle Bedeutung der Gleichheit vor dem Gesetz. Er bemerkte bloß schlicht und einfach: So sind wir nicht.[11]

Es kann selbstverständlich nach wie vor als problematisch angesehen werden, eine durch die Gerechtigkeit gebotene Handlung anhand von Tugenden wie der Gnade zu verteidigen. Es hat einen Hauch von Government-House-Utilitarismus: Die tatsächliche Rechtfertigung einer bestimmten Politik unterscheidet sich demnach auf beunruhigende Weise von den Überlegungen, die der Öffentlichkeit präsentiert werden.[12] Aber ich denke, dass der Anschein der Unangemessenheit in diesem Falle nicht richtig ist.

Was den Kindern angetan wird, ist schließlich sowohl ungerecht als auch gnadenlos; es nimmt den Kindern, worauf sie ein Recht haben und die Entscheidung, Kinder derart zu behandeln, zeigt einen deutlichen Mangel an Sorge für diejenigen, die unseren Entscheidungen gegenüber besonders verletzlich sind. Unter diesen Umständen können wir manchmal, so scheint es, jegliche Gründe anführen, die dabei helfen, die Forderungen der Gerechtigkeit zu verteidigen. Zumindest hat John Rawls gezeigt, dass die Tradition des politischen Liberalismus es zulässt, verschiedene Arten von Gründen für bestimmte politische Vorhaben vorzubringen: So griff Martin Luther King Jr. griff sowohl auf abstrakte Erwägungen politischer Gerechtigkeit als auch auf theologische Rechtfertigungen aus der Bibel zurück – oftmals in ein und demselben Absatz.[13] Rawls bemerkt, dass diese Art von Argumentation seinem Ansatz nach erlaubt ist; was zählt, ist, dass die aus dem öffentlichen Vernunftgebrauch entnommenen Gründe mit diesen anderen Gründen kompatibel sind. Das Konzept der Gnade ist, wenn überhaupt, sogar noch weniger bedenklich. Die Erwägungen der Gnade sind, wie ich gezeigt habe, durchaus mit dem öffentlichen Vernunftgebrauch in Einklang zu bringen und es ist ebenso möglich, eine politische Konzeption der Gnade zu entwickeln – was ich hier vorschlage bedeutet also bloß, verschiedene Formen des öffentlichen Vernunftgebrauchs zu nutzen.

Das führt allerdings zu einer Frage, die an diesem Punkt gestellt werden muss: Für wen genau schreibe ich dieses Buch? Wem biete ich meinen Rat an? Ich denke, angesichts des bisher Gesagten soll dieses Buch zwei verschiedenen Gruppen von Leserinnen dienen. Die erste Gruppe sind philosophische Autorinnen – oder zumindest philosophische Autorinnen, die, wie ich, nicht davon ausgehen, dass aus der Theorie des Liberalismus zwangsläufig offene Grenzen folgen. Manche von uns argumentieren gegen offene Grenzen, weil sie denken, dass diese Grenzen etwas Wertvolles beschützen und daher verteidigt werden sollten. Andere von uns betrachten diese Grenzen weniger emotional; ihren Gedanken folgend schließe der Liberalismus staatliche Grenzen zwar nicht aus, eine Welt ohne sie sei jedoch besser. Wie auch immer wir über Grenzen denken, so gibt es doch viele Philosophinnen, die nicht bereit sind, die Öffnung der Grenzen im Namen einer liberalen Gerechtigkeitsvorstellung zu verteidigen; und dieses Buch wendet sich zunächst speziell an sie. In meinen Augen beginnt die eigentliche Debatte erst mit der Tatsache, dass eine liberale Gerechtigkeitsvorstellung es uns erlaubt, Migrantinnen auszuschließen. Wir können also ein Recht auf Ausschluss haben, ohne dabei zu meinen, dass es keine moralischen Prinzipien gibt, anhand derer bestimmt werden kann, ob von diesem Recht guter oder schlechter Gebrauch gemacht wird. Die philosophische Literatur würde von einer direkteren Auseinandersetzung damit, wie bestimmte Formen der Migrationspolitik die von uns geschätzten politischen Tugenden widerspiegeln, durchaus profitieren, und sollte sich nicht bloß mit der Gerechtigkeit jener politischen Maßnahmen beschäftigen.

Die zweite Gruppe von Leserinnen ist selbstverständlich die Gruppe derer, die fähig und willens sind, sich mit der Politik innerhalb der Grenzen eines liberalen Staats auseinanderzusetzen. Ich sehe meine hier geäußerten Überlegungen als eine Art Erinnerung für diese Menschen – zu denen ich mich selbstverständlich auch selbst zähle – dass wir Gründe haben, in unseren politischen Maßnahmen sowohl Grausamkeit als auch Ungerechtigkeit zu vermeiden. Wie die jüngsten Ereignisse zeigen, droht uns die Gefahr, in eine überaus schreckliche Situation hineinzugeraten: Wenn schon nicht direkt Illiberalismus, so droht uns doch eine größere Feindseligkeit gegenüber den Ansprüchen bedürftiger Personen.[14] Sofern wir uns nicht permanent an die Menschlichkeit der Marginalisierten erinnern, laufen wir Gefahr, eben jene Menschlichkeit zu vergessen und zwar schlicht deshalb, weil es uns das Leben leichter machen würde. Ich glaube, dass eine öffentliche Norm der Gnade eine Möglichkeit darstellt, ein solches Abgleiten zu verhindern. Auch wenn ich nicht glaube, dass eine solche Norm ausreicht, so kann sie doch den kleinen Nutzen haben, die Flamme des moralischen Bewusstseins am Leben zu halten.

5. Schlussfolgerungen

Diese Ideen, oder ihre Vorläufer, habe ich zum ersten Mal an der University of Durham präsentiert. Vor meiner Präsentation verließ ich das Gebäude und lief zur Kathedrale. Am Tor der Kathedrale befindet sich ein prächtiger bronzener Türknauf, der während des Mittelalters dazu genutzt wurde, anzukündigen, dass um Asyl in der Kathedrale ersucht wird. Ich erinnere mich, wie wunderbar ich das fand. Der Türknauf, durch Generationen von Händen poliert, war eine lebhafte Erinnerung an den Willen, die Verletzlichen zu schützen. Es war eine vielfältige Gruppe von Personen, die um Zuflucht in der Kathedrale bat: Manche versuchten einer berechtigten Bestrafung zu entkommen, während andere um Immunität vor zutiefst ungerechten Formen des Missbrauchs durch ihre feudalen Herren ersuchten. Der Türknauf der Kathedrale stellt für mich bis heute den Willen dar, aufmerksam gegenüber den besonderen Umständen derjenigen zu sein, die um Hilfe bitten, und das Gewähren von Asyl schließlich als die tugendhafte Antwort auf diese Bitten zu verstehen.

Ich begann erst später über die Tradition des Kirchenasyls (*sanctuary*) zu lesen und dabei zu lernen, welche Bedingungen mit der Bitte um diesen Schutz verbunden waren. Die um Asyl in der Kathedrale von Durham ersuchende Person musste ein schwarzes Gewand tragen, auf dem das gelbe Kreuz von St. Cuthbert abgebildet war, und England schließlich innerhalb von 37 Tagen verlassen.[15] Eine Wiedereinreise war nicht gestattet. Kehrte eine Person zurück, drohte ihr die Exekution. Die Kathedrale bot also Asyl – allerdings bloß vorübergehend und verbunden mit einer anschließenden Ausweisung, und das Vergehen, nach Gewährung des Asyls noch ohne entsprechendes Recht innerhalb Englands präsent zu sein, war in gewisser Hinsicht schlimmer als jeder Gesetzesverstoß, der dem Bedürfnis nach Schutz ursprünglich vorausgegangen war.

Diese zwei Seiten der Geschichte des Kirchenasyls erzählen von zwei verschiedenen Bestrebungen, die sowohl im Mittelalter als auch in unserer heutigen Zeit gefunden werden können. Auf der einen Seite besteht der Wunsch, zu helfen; der bedürftigen Person, die außerhalb unserer Gemeinschaft steht, zu helfen, sie in Sicherheit zu bringen und mit dem zu versorgen, was sie braucht, um die Vernichtung ihrer Person und ihrer Pläne zu verhindern. Auf der anderen Seite gibt es jedoch auch den Wunsch, nicht mit der Sorge um Menschen belastet zu werden; den Wunsch also, nicht die Person zu sein, die dazu verpflichtet ist, in eine besondere Beziehung mit dem Außenstehenden zu treten. Ich habe in den vorherigen Kapiteln den Gedanken verteidigt, dass dieser zweite Wunsch nicht an sich falsch ist; in meinen Augen ist es uns in gewissem Maße gestattet, zu verhindern, diejenigen Personen zu werden, die zur Gewährleistung der Rechte anderer Personen verpflichtet sind – sofern diese Rechte im Herkunftsland bereits auf angemessene Weise garantiert werden. Allerdings ist diese Erlaubnis zu Recht durch die Bedingungen der Gerechtigkeit eingeschränkt. Im weiteren Sinne scheint es zudem so, als ob in den letzten Jahren die Stärke und Vitalität des ersten Wunsches – zu tun, was wir können, um einer Person zu helfen, selbst wenn wir dazu nicht verpflichtet sind – langsam schwindet. Unsere politischen Führungskräfte sind zunehmend willens, Indifferenz – selbst Feindseligkeit – gegenüber den Forderungen der Bedürftigen jenseits unserer Grenzen zu demonstrieren. Noch schlimmer ist die wahrscheinlich eintretende Entwicklung, dass die Zahl der Schutzbedürftigen in naher Zukunft ansteigen wird. So wird der Klimawandel, den ich in diesem Kontext nicht diskutiert habe, wahrscheinlich sowohl die globale wirtschaftliche Ungleichheit verstärken, als auch die Zahl derjenigen vergrößern, die in die wohlhabenden Gesellschaften einwandern wollen.[16] Was das betrifft haben die Fortschritte in der Kommunikations- und Transporttechnologie mehr Menschen als je zuvor die Entscheidung zur Migration ermöglicht. Um das Zitat von Kishore Mahbubani zu wiederholen, mit dem dieses Buch begann: Das Mittelmeer hat sich in einen Teich verwandelt – und wer würde nicht einen Teich überqueren, wenn mit dieser Überquerung der Armut entkommen werden könnte?

Kurzum: wir haben eine Konstellation von Umständen, in denen der Wunsch zu helfen mit zunehmender Wahrscheinlichkeit durch Behauptungen erstickt wird, dass wir nicht helfen *müssen*; dass diese anderen uns nichts anzubieten haben; dass die Qualität ihres Lebens uns nichts angehen muss. Diese Tendenzen werden fortleben und ich denke nicht, dass irgendein philosophischer Text diese Entwicklung verhindern könnte. Aber dadurch wird es selbstverständlich nicht falsch, es zumindest zu versuchen.

Ich möchte dieses Buch mit einem Zitat von Kurt Vonnegut Jr. beschließen, dessen Buch *Gott segne Sie, Mr. Rosewater* eine Zukunft beschreibt, in der wir die Fähigkeit verloren haben, Menschen als Geschöpfe zu betrachten, die es wert sind, verteidigt zu werden. Anhand des Protagonisten Kilgore Trout zeigt Vonnegut, dass wir bald einem grundlegenden Problem gegenüberstehen könnten, vermutlich dem bedeutendsten Problem unserer Zukunft: Wie wir Menschen lieben können, die *uns keinerlei Nutzen bringen.* Vonnegut meint, dass wir uns nicht vorstellen können zu überleben, wenn wir nicht damit anfangen, Menschen zu lieben – sie nicht deshalb zu wertschätzen, weil sie etwas für uns tun können, *sondern weil sie menschliche Wesen sind.* Seine Charaktere – in diesem Buch, aber auch in anderen seiner Werke – gelangen regelmäßig an einen Punkt, der für Vonnegut die zentrale Einsicht in den Wert eines Menschen darstellt: Das wir aus dem Grund freundlich zu Menschen sein sollten, weil sie Menschen sind:

> „Hello Babys. Willkommen auf Erden. Im Sommer ist es heiß und im Winter kalt. Es ist rund und nass und voll hier. Grob geschätzt habt ihr etwa hundert Jahre hier. Es gibt nur eine Regel, Babys, die ich kenne –: Verdammt noch mal, ihr müßt anständig sein."[17]

Ich glaube, dass Vonnegut recht hat; und dass ein ernsthaftes Risiko besteht, die moralische Realität derjenigen aus den Augen zu verlieren, die unsere Grenzen überqueren wollen, sollte sich ihre Zahl in den kommenden Jahren vergrößern. Diese Menschen sind handelnde Personen, die sich durch ihre Handlungsfähigkeit ein Leben aufgebaut haben und das Gelingen dieses Lebens ist ihnen genauso wichtig wie uns das Gelingen unseres eigenen Lebens. Wir müssen uns in meinen Augen beständig an diese moralischen Tatsachen erinnern; und sei es auch nur, um die Gefahr zu vermeiden, diese Personen als Last statt als Menschen zu betrachten. Den Wohlhabenderen unter uns droht die Gefahr, grausam zu werden – also damit aufzuhören, sich um Gnade, Freundlichkeit oder irgendeine andere der Tugenden zu sorgen, die uns davon abhalten, an die Grenze dessen zu gehen, was das Recht auf Ausschluss uns erlaubt. Kurzum: Ich glaube, dass wir taub gegenüber denjenigen werden, die um Einlass bitten, selbst wenn sie noch lauter bitten und ihre Nöte noch dringender werden sollten. Ich würde gerne glauben, dass wir ihnen zuhören werden, und dass dieses Buch eine kleine Hilfe dabei sein kann, uns an die Gründe für ein solches Zuhören zu erinnern. Allerdings glaube ich nicht, dass dies der Fall sein wird. Genauso wenig glaube ich, dass wir den dunkleren Motiven unserer Natur widerstehen werden oder die von mir hier verteidigte Tugend einen Platz innerhalb unseres politischen Diskurses finden wird. Wenn aber schon nichts sonst, so können wir doch weiterhin die Bedeutung von Gerechtigkeit und Gnade geltend machen und in der Hoffnung auf eine Welt leben, in der diese Werte Wirklichkeit werden.

Anmerkungen

1. Über Moral und Migration

1 Die Literatur über den Kolonialismus ist umfangreich, wobei sich vergleichsweise wenige Philosophinnen systematisch mit dem Thema auseinandergesetzt haben. Zwei wichtige Ausnahmen sind Ypi 2003 und Lu 2017.

2 Mahbubani 2001, S. 61.

3 Viele philosophische Arbeiten jüngeren Datums gehen der Frage nach, inwiefern die Entscheidungen wohlhabender Konsumentinnen und politischer Akteure kausal mit Unterdrückung im Ausland verbunden sind. Siehe hierzu beispielsweise Wenar 2017 und Pogge 2008.

4 Oberman 2016b, Oberman 2015 und Cole 2002 haben damit begonnen, die aus dem Erbe des Kolonialismus folgenden Probleme in die politische Philosophie der Migration zu integrieren.

5 Es gibt mittlerweile umfassende Literatur zum Themenkomplex globaler Gerechtigkeit. Siehe beispielsweise Beitz 1999, Blake 2013, Blake 2002, Brock 2009, Caney 2006, Hassoun 2012, Miller 2007, Miller 1995, Nussbaum 2006, Pogge 2008, Rawls 1999, Risse 2012b, Tan 2014, Tan 2000 und Valentini 2012.

6 Diese Probleme werden in Ngai 2010 diskutiert.

7 Bacon 2009 bietet einen guten Überblick hinsichtlich dieser Effekte.

8 Diese Idee wird am stärksten mit der Arbeit von Joseph Carens verbunden. Siehe Carens 1987. Ayelet Shachar hat die Idee einer ungerechten Geburtsrechts-Lotterie in ihrer Arbeit weiterentwickelt. Siehe Shacher 2009.

9 Für einen Überblick siehe Collier 2013 und Borjas 2001.

10 United States Commission on Civil Rights 1980.

11 Selbst hier erweisen sich die Folgen jedoch als komplex. Siehe Borjas 2017 und Borjas 2001.

12 Diese Effekte werden in Card 1990 besprochen.

13 Die Verfassungsmäßigkeit dieses Gesetzes wurde in *Chan Chae Ping v. United States*, 130 U.S. 581 1889, angefochten, der oft auch einfach als Chinese Exclusion Case bezeichnet wird.

14 Siehe Pérez-Peña 2017 für eine Übersicht der Vergehen Arpaios.

15 Siehe Mendoza 2017, Mendoza 2011, Reed-Sandoval 2020, Reed-Sandoval 2015 und Reed-Sandoval 2013 für eine Diskussion dieses Zusammenhangs.

16 Weit verbreitete Metaphern beziehen sich auf Krankheiten, militärische Invasionen und Naturkatastrophen. Siehe Shariatmadari 2015.

17 Dieser Bericht wird in Alexijewitsch 2015, S. 453–456 wiedergegeben.

18 Alex Sager 2018 ist eine der wenigen philosophischen Arbeiten, die sich ernsthaft mit der Ich-Perspektive von Migrantinnen auseinandersetzt.

19 Zitiert in United States Commission on Civil Rights 1980.

20 Améry 1977, S. 81–82.

21 Eine detailliertere Besprechung der Hintergründe des Tehcir-Gesetzes findet sich in Bioxham 2005.

22 Das Wannsee-Protokoll verwendet durchgehend den Begriff des Transports. Unter dem folgenden Link finden sich das deutsche Originalprotokoll sowie eine Übersetzung ins Englische: https://www.ghwk.de/fileadmin/Redaktion/PDF/Konferenz/Wannsee_Protocol_German_English_200214es.pdf

23 Walzer 1994, S. 89.

24 Ich bespreche meine Bedenken in Blake 2014.

25 Murphy und Hampton 1988, S. 176.

26 Die sogenannten Jim-Crow-Gesetze bezeichnen die im späten 19. und frühen 20. Jahrhundert in den Südstaaten der USA eingeführten Gesetze der Rassentrennung, die im Nachgang der Abschaffung der Sklaverei erlassen wurden. Sie wurden bis 1965 angewandt. (Anm. d. Übers.)

27 Daher Thomas Hills Gedanke, dass eine Person einen zu beanstandenden Mangel an Selbstachtung zeigt, wenn sie ein Gut, auf das sie einen Gerechtigkeitsanspruch geltend machen kann, als Geschenk akzeptiert. Siehe Hill 1973.

28 Siehe beispielsweise Cabrera 2004.

29 Fisher 2013, S. xi.

30 Siehe Blake 2013 und Rawls 1999.

31 Risse bespricht den Weltstaat in Risse 2012a und bietet in meinen Augen schlüssig Gründe gegen das Ideal eines solchen globalen Staates.

32 Zu ersterem siehe Cohen 2009; zum zweiten Punkt siehe die Diskussion über Gastarbeiter in Carens 2013, Lenard und Straehle 2010 sowie Walzer 1986. Ich gehe auch über die ziemlich wichtige Frage hinsichtlich pluraler und transnationaler Formen von Staatsbürgerschaft hinweg. Siehe hierzu die Arbeiten von Rainer Bauböck, darunter die in Bauböck 2018a und Bauböck 2018b zusammengestellten Diskussionen.

33 Siehe hierzu Carens 2013 und Miller 2016.

34 Gillian Brock und ich besprechen dieses Problem in Brock und Blake 2014.

35 Meine Idee der Autonomie ist beeinflusst durch Raz 1988.

36 Siehe hierzu generell Rawls 1979 und Rawls 1993.

37 Bratman 1987 bietet eine starke Erläuterung davon, wie Pläne unsere Handlungsfähigkeit über die Zeit hinweg strukturieren.

38 Rawls 1979, S. 446.

39 Blake 2013.

40 Scarry 1985 zeigt eindrücklich, wie der Körper gegen die Person verwendet werden kann, die diesen Körper bewohnt.

41 Siehe Hightower 1997.

2. Gerechtigkeit und die Ausgeschlossenen, Teil 1: Offene Grenzen

1 Diese Rede wird in Ross 2016 beschrieben.

2 Donald Trumps Antrittsrede findet sich unter https://www.whitehouse.gov/briefings-statements/the-inaugural-address/.

3 Die Sprache der Illegalität wird in Gambino 2015 und Chomsky 2014 zurückgewiesen.

4 Der Bundesstaat Washington versuchte, das Inkrafttreten des Verbots zu verhindern und argumentierte, dass er (neben vielen anderen Mängeln) wahrscheinlich gegen den 14. Verfassungszusatz verstoße. Siehe das Dokument unter http://agportal-s3bucket.s3.amazonaws.com/uploadedfiles/Another/News/Press_Releases/Complaint%20as%20Filed.pdf.

5 Ähnlich wie das Kirchenasyl (engl. sanctuary) versucht diese Bewegung, Schutzformen jenseits staatlicher Stellen auf der Ebene von Kommunen und Individuen zu realisieren. (Anm. d. Übers.)

6 Gil Martinez wird in Adler 2007 zitiert.

7 McWorther 2012 beschreibt die Entstehung des Konzepts Juan Crow.

8 Die Zurückweisung von Flüchtlingen ist zu einem zentralen Teil der politischen Identität Matteo Salvinis geworden. Siehe Mead 2018.

9 Scoones et al. 2018 beschreibt einige dieser Bewegungen und bietet Überlegungen, wie ihnen begegnet werden könnte.

10 Miller 2019, S. 10.

11 Rawls 1979, S. 21–22.

12 Rawls 1979, S. 23.

13 Dieses sogenannte egalitaristische Plateau wird in Arbeiten wie Dworkin 2000 und Sen 1992 besprochen.

14 Die Bedeutung von Lincoln für Rawls kann kaum überbewertet werden. Für einen Überblick über Rawls' Leben siehe Pogge 2007.

15 Aus diesem Grund wurde die Begnadigung Arpaios von Personen lateinamerikanischer Herkunft als so entmutigend wahrgenommen. Siehe Romero 2017.

16 Siehe Reed-Sandoval 2005.

17 Die rechtlichen Pflichten der Polizei sind eine komplexe Angelegenheit; es ist nicht klar, ob die Polizei rechtlich belangt werden kann, wenn sie keinen angemessenen Schutz gewährleistet. Von einem moralischen Standpunkt aus scheint es jedoch klar, dass es moralisch falsch ist, einen solchen Schutz nicht zu gewährleisten.

18 *Plyler v. Doe*, 457 U.S. 202 1982, interne Zitate wurden nicht übernommen.

19 Ich möchte anmerken, dass es sich dabei nicht um die einzig mögliche, historische Erzählung über die Entstehung der internationalen Menschenrechte handelt. Siehe Moyn 2010.

20 Die *St. Louis* war ein Passagierschiff, das im Jahr 1939 mit hunderten aus Deutschland vor den Nationalsozialisten fliehenden Juden versuchte, in Kuba, Kanada sowie den USA anzulanden. Alle Länder verweigerten dem Schiff jedoch die Landeerlaubnis, woraufhin es nach monatelanger Irrfahrt nach Antwerpen zurückkehrte und die Flüchtenden auf europäische Länder verteilt wurden (Anm. d. Übers.).

21 Selbstverständlich handelt es sich nicht bei jeder Form von Migration um eine Bewegung aus den ehemaligen Kolonien hin zu den Metropolen dieser Welt. So erlebt Afrika derzeit tiefgreifende Veränderungen durch die Einwanderung (oder den temporären Aufenthalt) chinesischer Bürgerinnen in Ostafrika. Siehe French 2014 für eine Beschreibung dieser Entwicklungen.

22 Kukathas 2012, S. 656. Siehe auch Kukathas 2010 und Kukathas 2005.

23 Nett 1971.

24 Dieser Gedanke wird von Christine Korsgaard in Korsgaard 1996 sehr gut ausgedrückt.

25 Verschiedene Versionen dieses Arguments finden sich in Carens 2013, Carens 1987, Dummett 2001, Kukathas 2012, Kukathas 2005, Oberman 2016a, Schotel 2012 und Wilcox 2009. Bertram 2018 bietet eine besondere Version dieses Arguments. Leider erschien dieses Buch zu spät, sodass ich es hier nicht mehr mit aufnehmen konnte. (Das gleiche gilt für Owen 2019). Ich ignoriere Huemer 2010, da sein Argument sich auf das Recht auf Bewegungsfreiheit zur Vermeidung des Hungertods bezieht – wie zu sehen sein wird, müssen wir aber keine offenen Grenzen verteidigen, um ein Recht zu begründen, Grenzen zum Zwecke der Erhaltung der eigenen Handlungsfähigkeit zu überqueren.

26 Carens 2019, S. 9.

27 Cole 2002, S. xii.

28 Clemens (2011) bietet die wissenschaftlich anerkannte Größenordnung der ökonomischen Gewinne, die durch offene Grenzen erzielt werden könnten.

29 Die moralische Relevanz von Geldtransfers ist ein zentraler Punkt, in dem Gillian Brock und ich nicht übereinstimmen. Siehe Brock und Blake 2014.

30 In diesem Zusammenhang ist Obermans Arbeit über Armut und Migration eine Betrachtung wert. Siehe Oberman 2016b, Oberman 2015 und Oberman 2011. Thomas Pogge 1997 bietet eine mögliche Entgegnung zu Obermans Analyse.

31 Carens 2013, S. 239.

32 Oberman 2016a; siehe auch Freiman und Hidalgo 2013. Der Titel von Freimans und Hidalgos Aufsatz fasst ihre Perspektive gut zusammen; er lautet „Liberalism or Immigration Restrictions, but Not Both."

33 Gillian Brock widerspricht dem in Brock und Blake 2014 zu einem gewissen Grad; so auch Stilz 2016.

34 Ypi 2008, S. 393

35 Wellman und Cole 2011, S. 198–199. Cole zitiert hier ein früheres Argument von Ann Dummett.

36 Siehe Blake 2001.

37 Carens benutzt diese Ausdrucksweise zuerst in Carens 1987.

38 Ich sage „regelmäßig", um der Auseinandersetzung zwischen David Miller und Arash Abizadeh über das Wesen der Zwangsgewalt an den Grenzen Rechnung zu tragen. Für den Moment möchte ich einfach nur bemerken, dass staatlicher Ausschluss *oft* mit einer impliziten Androhung von Gewalt einhergeht, und dass diese Tatsache ausreichen sollte, um das Argument an dieser Stelle in Gang zu bekommen. Siehe Abizadeh 2008, Miller 2010 und Abizadeh 2010.

39 Abizadeh 2008.

40 Carens 2013, S. 258–259.

41 Blake 2013.

42 Rodríguez (2013), S. 2-3, Kursivdruck hinzugefügt.

43 Es gibt eine außergewöhnlich große Anzahl ökonomischer Untersuchungen der Migration. Siehe Borjas 2017, Borjas 2001, Chiswick 1988, Clemens 2011 und Collier 2013.

44 Higgins 2013.

45 Oberman 2016a, S. 38.

46 Oberman 2016a, S. 39.

47 Hosein 2013 entwickelt dieses Argument gut.

48 John Rawls unterscheidet entsprechend zwischen der Freiheit und dem Wert dieser Freiheit. Erstere repräsentiert das, woran wir nicht gehindert werden dürfen; letzterer bezeichnet die Mittel, die uns dafür zur Verfügung stehen. Siehe Rawls 1979.

49 *Masterpice Cakeshop v. Colorado Civil Rights Commissions*, 581 U.S. __ 2017.

50 Siehe Calhoun 2018 und Calhoun 2015 für Calhouns Verständnis des Zusammenhangs zwischen Zeit und Handlungsfähigkeit.

51 Ich sollte hier, um Missverständnissen vorzubeugen, anmerken, dass innerhalb der Vereinigten Staaten zwei Städte nicht weiter auseinander liegen könnten als Miami und meine Heimatstadt Seattle.

52 Carens 2013, S. 239. Siehe auch Bauböck 2011

53 Carens 2013, S. 242.

54 Carens 2013, S. 243.

55 Brock und Blake 2014.

56 Miller 2005; siehe auch Miller 2016.

57 Wellman 2008

58 Cole in Wellman und Cole 2011, S. 204.

59 Carens 2013, S. 250.

60 Carens 2013, S. 257

3. Gerechtigkeit und die Ausgeschlossenen, Teil 2: Geschlossene Grenzen

1 Es ist kein neuer Gedanke, dass Kinder nicht im vollen Sinne als moralische Akteure bezeichnet werden können. So verneint etwa die juristische Idee des *doli incapax* (Schuldunfähigkeit) die Möglichkeit, dass Kinder unter 14 Jahren über die Art von Unterscheidungsvermögen und Intelligenz verfügen, die es für kriminelle Schuldfähigkeit braucht.

2 Brian Kilmeade ist der Ko-Gastgeber von *Fox and Friends*. Eine Diskussion seiner Bemerkungen findet sich unter https://nymag.com/intelligencer/2018/06/fox-host-defends-trump-these-arent-our-kids.html

3 Siehe hierzu beispielsweise Bacon 2009 und Faux 2017.

4 Mendoza 2015 und Mendoza 2017 beinhalten Diskussionen dieses Problems.

5 Cafaro 2014 argumentiert, dass das globale Ökosystem durch einen Anstieg der Zahl an Menschen, die einen US-amerikanischen Lebensstil pflegen und entsprechende Mengen Kohlenstoffdioxid in die Atmosphäre emittieren, Schaden nehmen würde.

6 Siehe Macedo 2011 und Macedo 2012.

7 Die fortdauernde Bedeutung von Kant und Grotius wird in Risse 2012a hervorgehoben.

8 Sidgwick bespricht diese Gedanken in seinen *Elements of Politics*, Abschnitt XV. Siehe Sigdwick 2013 [1891].

9 Walzer 1994, S. 85–86.

10 Blake 2003.

11 Risse 2012a, S. 154.

12 Risse 2012a, S. 165.

13 Blake und Risse 2009.

14 Dieser Trend wird wahrscheinlich in absehbarer Zukunft anhalten. Siehe https://www.un.org/en/development/desa/news/population/world-urbanization-prospects-2014.html.

15 Die Geschichte von Elisha Otis wird in Bernard 2014 wiedergegeben.

16 Eine Besprechung der Bevölkerungsdichte Manhattans findet sich unter https://wagner.nyu.edu/files/rudincenter/dynamic_pop_manhattan.pdf. Informationen über die Bevölkerungsdichte der Mongolei finden sich unter.
https://data.worldbank.org/indicator/EN.POP.DNST?locations=MN

17 Risse 2016, S. 268–269.

18 Am bedeutendsten hierzu ist Miller 1995.

19 Miller 2019, S. 26.

20 Diese Ideen werden auf unterschiedliche Art in Kymlicka 1995 und Walzer 1986 verteidigt.

21 Song 2012, S. 41. Siehe auch Song 2018 und Song 2017.

22 Dieses Argument wird von verschiedenen Autoren entwickelt, dasjenige von Collier 1999 ist jedoch besonders wichtig.

23 Song 2009 zeigt, wie sich Solidarität aus Diversität entwickeln könnte.

24 Pevnick 2011, S. 38.

25 Okin geht davon aus, dass Locke mit seiner Position nicht alleine steht; sie schreibt, dass viele philosophische Positionen sich ins Absurde verkehren, sobald sie Reproduktionsarbeit in den Blick nehmen. Okin 1989a.

26 Walzer 1994, S. 418–419.

27 *People ex rel. Moloney v. Pullman's Palace-Car Co.*, 175 Ill. 125, 51 N.E. 664, 1898 (Ill., 01.01.1898).

28 Coakley 1985.

29 Pevnick 2011, S. 63.

30 Siehe Wellman 2008 sowie Wellman in Wellman und Cole 2011.

31 Siehe beispielsweise Fine 2010 und Blake 2012a.

32 *Roberts v. United States Jaycees*, 486 U.S. 609 1984. Anm. d. Übers.: Als Jaycees werden die Mitglieder der Organisation Junior Chamber International bezeichnet.

33 Lister 2010.

4. Gerechtigkeit, Gebietshoheit und Migration

1 Robert Nozick 1974 bietet detaillierte Ausführungen hinsichtlich dieser Methode.

2 Der Text der Montevideo-Konvention über die Rechte und Pflichten von Staaten findet sich unter https://www.jus.uio.no/english/services/library/treaties/01/1-02/rights-duties-states.xml.

3 Montevideo-Konvention, Artikel IX.

4 Staaten beanspruchen auch dann Hoheitsrechte über ihre eigenen Bürgerinnen, wenn diese ins Ausland reisen; etwa das Recht, die Rechte ihrer Bürgerinnen zu verteidigen, sollten diese Rechte bedroht sein. Diese Komplikation kann im vorliegenden Kontext jedoch außer Acht gelassen werden, auch wenn sie in der Praxis von Bedeutung ist. Dass es mehr als einen Weg gibt, hoheitliche Autorität herzustellen, ändert nichts an der Tatsache, dass diejenigen, die territoriale Grenzen überqueren, anderen durch eben diese Überquerung neue Pflichten auferlegen.

5 Einige Antworten auf diese Fragen finden sich in Beitz 2001, Ignatieff 2003 und Talbott 2005.

6 Für eine historische Einführung in die Entwicklung dieser Trichotomie siehe Kock 2012.

7 Es ist wahr, dass manche Verbrechen von solcher Schwere sind, dass sie überall verfolgt werden können; das Konzept universaler Hoheitsrechte existiert innerhalb des internationalen Rechts bereits seit langem und einige Autoren haben in jüngster Zeit auf eine Erweiterung des Verständnisses dieses Konzepts gedrängt. Es besteht jedoch als Ausnahme zur generellen Regel territorial verankerter Hoheitsrechte; meines Wissens hat bisher kein Autor darüber nachgedacht, dieses Konzept auf andere Verbrechen jenseits schwerster Menschenrechtsverletzungen anzuwenden. Siehe hierzu das Outcome Document des United Nations World Summit 2005, /RES/60/1, zu finden unter https://www.un.org/en/development/desa/population/migration/generalassembly/docs/globalcompact/A_RES_60_1.pdf.

8 Allerdings liegt alledem die Annahme zugrunde, dass die Besonderheit der Beziehung zwischen dem Staat und denjenigen, die sich innerhalb seines Hoheitsgebiets aufhalten, selbst wiederum moralisch legitim ist. Zwar gehe ich von dieser Annahme aus, allerdings habe ich sie nicht bewiesen. A. John Simmons' Arbeiten über die Besonderheiten dieses Problems stellen daher eine notwendige Vorarbeit für meine Diskussion dar: Simmons fragt, ob diese besonderen Beziehungen zwischen einzelnen Staaten und ihren Bürgerinnen auf angemessene Weise anhand eines allgemeinen Ansatzes moralischer Pflichten begründet werden könnten (so wie beispielsweise, ganz zentral, die Pflicht, gerechte Institutionen aufrechtzuerhalten). Ich gehe an dieser Stelle davon aus, dass eine Lösung für dieses Problem gefunden werden kann; sollte dem nicht so sein, könnte das, was ich in diesem Buch sage, ebenfalls moralisch problematisch sein. Siehe Simmons 1979.

9 Die stärkste Formulierung dieses Prinzips kann in der Internationale Konvention zum Schutz der Rechte aller Gastarbeiter und ihrer Familienangehörigen, GA Res. 45/158 vom 18. Dezember 1990, gefunden werden, in der die Diskriminierung derjenigen untersagt wird, die sich ohne legale Aufenthaltspapiere innerhalb eines Hoheitsgebiets befinden. Die meisten westlichen Staaten haben sich geweigert, diese Konvention zu unterschrieben. Dass die Rechte derjenigen, die sich innerhalb eines Hoheitsgebiets befinden, auch durch die Institutionen dieses Hoheitsgebiets zu schützen sind, ist weniger umstritten und sowohl Teil der Montevideo-Konvention als auch des Internationalen Pakts über bürgerliche und politische Rechte, Teil II, Artikel 2.

10 *Plyler v. Doe*, 457 U.S. 202 1982 auf Seite 215

11 Eine interessante Ausnahme hierzu ist Jeremy Waldron, der John Lockes Begründung von Eigentumsrechten aufgrund des Umstands kritisiert, dass es anhand dieses Ansatzes möglich wäre, das Auferlegen von Verpflichtungen ohne die Zustimmung der Verpflichteten zu rechtfertigen. Siehe Waldron 1990. Ich danke Wayne Summer für Diskussionen zu diesem Punkt.

12 Das Beispiel findet sich in Thomson 1971.

13 Terry Pratchett führt das Konzept des Anti-Verbrechens ein, um Fälle wie diesen abzubilden: Sie beinhalten Ideen wie Einbrechen-und-Dekorieren und das Erpressen durch gute Taten (wie beispielsweise der Androhung, die geheimen Spenden eines Mafiabosses an Wohltätigkeitsorganisationen zu veröffentlichen). Terry Pratchett besteht darauf, dass ein Anti-Verbrechen den Wunsch beinhalten muss, das Opfer zu stören oder bloßzustellen; meine eigene Anwendung des Konzepts beinhaltet keine solche Einschränkung. Siehe Pratchett 2002.

14 Kates und Pevnick 2014, S. 189. Ein ähnliches Argument findet sich in Arrildt 2016.

15 Ich bespreche diesen Fall in Blake 2003.

16 Eine mögliche Antwort auf dieses Argument könnte darin bestehen, dass mein Recht, meine Karriere zu verfolgen, einen ausreichenden Grund für eine Selbstverpflichtung der Vereinigten Staaten darstellt, mich einzulassen. Ich denke jedoch nicht, dass diese Antwort richtig ist: Das Recht darauf, sich seinen Aufenthaltsort auszusuchen, wird falsch interpretiert, wenn darunter ein Recht auf die größtmögliche Zahl der verfügbaren Optionen für meinen Aufenthaltsort verstanden wird. Es gibt schließlich auch in Kanada ausgezeichnete Graduiertenprogramme in der Philosophie; hätten die Vereinigten Staaten darauf bestanden, dass ich aus diesen Programmen wählen müsse, hätten sie nicht in Freiheiten eingegriffen, die meinen moralischen Rechten zugrunde liegen. Ich behandele diese Ideen auch in Kapitel sieben.

17 Eine gute Diskussion der aus diesen Ideen folgenden Einschränkungen bietet Mendoza 2011.

18 Ich bespreche manche dieser Ideen in Blake 2008.

19 Hier handelt es sich um eine Reinterpretation des Arguments von Arash Abizadeh in Abizadeh 2008.

20 *Edwards v. California*, 314 U.S. 160 1941, S. 173-174 zitiert in *Baldwin v. Seelig*, 294 U.S. 511 1935, S. 522.

21 Stephenson 1992.

22 *Saenz v. Roe*, 526 U.S. 489 1999.

23 Siehe hierzu allgemein Dinan 2010.

24 Das Versprechen – und die Gefahr – dieser Art von finanzieller Intervention wird in Wellman und Cole 2011 sowie Lister 2013 besprochen.

25 Rawls 1979.

26 Die Dublin-III-Verordnung findet sich unter https://eur-lex.europa.eu/eli/reg/2013/604/oj?locale=de.

27 So verteidigt beispielsweise Gillian Brock es als legitim, dass Staaten ihre eigenen hochtalentierten Bürgerinnen an der Auswanderung hindern. Siehe Brock 2009, S. 190-212.

28 Für eine entgegengesetzte Perspektive siehe Miller 2007.

29 Beispielsweise habe ich nicht die Idee der Familienzusammenführung betrachtet. Ich glaube jedoch, dass der hier vorgestellte Ansatz mit etwas der Familienzusammenführung Ähnlichem kompatibel ist; dieses Recht müsste aber daraus abgeleitet werden, dass es für einen Staat unzulässig ist, seine politischen Rechte zu nutzen, um die familiären Interessen seiner derzeitigen Mitglieder zu untergraben. Selbst wenn Staaten ein Recht auf Ausschluss hätten, müssten sie demnach ihren eigenen Mitgliedern gegenüber rechtfertigen, wie sie dieses Recht nutzen. Ich bedauere, dass ich in diesem Kontext nicht den Platz habe, um diese Angelegenheiten genauer zu betrachten.

30 Kates und Pevnick 2014, S. 192.

31 Oberman 2016a.

32 Brezger und Cassee 2016, S. 368; siehe auch Hidalgo 2014.

33 Volker Heins hat bei einer Diskussion dieses Arguments betont, dass mein Ansatz einer Theorie der Abwägung bedürfe, anhand derer ich zeigen könne, dass das Recht auf körperliche Integrität gewichtiger sei als das Recht darauf, keine neuen Pflichten auferlegt zu bekommen. Ich denke, dass er Recht haben könnte; aber ich bevorzuge es, diese zwei Rechte nicht nur im Hinblick auf ihr Gewicht als unterschiedlich zu betrachten, sondern auch hinsichtlich ihrer Art. Wir können die Diskussion der relativen Gewichtung verschiedener Rechte in meinen Augen dadurch vermeiden, dass wir manche Rechte aufgrund ihrer Art als vorrangig identifizieren können. Ich danke Heins für die Diskussion zu diesem Punkt.

34 Siehe bezüglich der Rechte gleichgeschlechtlicher Paare Macedo 2015.

35 Lamey 2016.

36 Lamey 2016.

37 Meine Lesart von Kant folgt Murphy 1994.

38 Im Jahre 1724 veröffentlichte ein Autor, von dem generell angenommen wurde, dass er selbst ein Pirat war, unter dem Pseudonym Charles Johnson eine Beschreibung der Gewohnheiten und Sitten der Piraten. Siehe Defoe 1982.

39 Defoe 1982, S. 208–209.

5. Zwang und Zuflucht

1 Viele Diskussionen der zeitgenössischen Regime des Flüchtlingsschutzes sind historisch sehr detailliert; siehe Maley 2016 und Gibney 2004. Price 2009 nennt außergewöhnlich viele historische Details zur prä- und frühmodernen Erfahrung mit Schutz und Asyl.

2 Die Argumente dieser Autoren sind übersichtlich zusammengefasst in Price 2009, S. 24–58.

3 Jeffersons Worte werden in *American State Papers, vol. 1, (1789–1852)* auf Seite 258 zitiert.

4 Es ist angebracht, anzumerken, dass der Zusammenhang zwischen der Genfer Flüchtlingskonvention und innerstaatlichem Recht ziemlich kompliziert ist. So waren die Vereinigten Staaten ursprünglich kein Unterzeichnerstaat, hatten die Definition der Konvention allerdings in ihr innerstaatliches Recht aufgenommen. Zudem ist die Interpretation der Genfer Flüchtlingskonvention weder statisch noch ohne Schwierigkeiten. Die Konvention sollte ursprünglich nur auf diejenigen angewendet werden, die vor 1951 vertrieben wurden, wurde dann jedoch auch in späteren Migrationskrisen als rechtlich wirksam anerkannt. Diese komplexen Entwicklungen werden sehr gut in Maley 2016 besprochen.

5 Der Text der Genfer Flüchtlingskonvention sowie des Protokolls findet sich unter https://www.unhcr.org/dach/wp-content/uploads/sites/27/2017/03/Genfer_Fluechtlingskonvention_und_New_Yorker_Protokoll.pdf.

6 Der Geltungsbereich der Norm des Non-Refoulement ist ebenfalls ungeklärt. Während sie ursprünglich nur auf diejenigen angewendet werden sollte, die einen Status gemäß der Genfer Flüchtlingskonvention für sich reklamieren konnten, wird die Norm nun so verstanden, dass sie auch auf eine größere Gruppe von Migrantinnen Anwendung findet. Siehe DeAngelo 2009 über den Geltungsbereich des Non-Refoulements.

7 Der Text der Cartagena Declaration on Refugees findet sich unter https://www.unhcr.org/about-us/background/45dc19084/cartagena-declaration-refugees-adopted-colloquium-international-protection.html.

8 Kritiker sind unter anderem Gibney 2004, Shacknove 1985 und Shue 1980.

9 Betts schlägt diesen Begriff vor um die Schwierigkeiten mit dem rechtlichen Konzept des Flüchtlings zu überwinden. Siehe Betts 2013; siehe ebenso Collier und Betts 2017, S. 68.

10 Siehe Lister 2013 und Price 2009.

11 Ich habe mich an dieser Unterscheidung bereits in Blake 2016 versucht.

12 Dieser geschichtliche Hintergrund wird auf lebendige Weise in Lewis 1964 beschrieben. Anm. d. Übers.: In Deutschland gibt es im Strafrecht das Rechtsinstitut der Pflichtverteidigung, wonach dem Beklagten in bestimmten Fällen ein Anwalt von Amts wegen gestellt werden muss. Allerdings spielt die finanzielle Situation des Beklagten dabei keine Rolle.

13 Pogge selbst unterstützt diese Unterscheidung nicht; er erkennt schlicht an, dass eine dieser Unterscheidung ähnliche Prämisse von den meisten seiner Gesprächspartnerinnen angenommen wird.

14 Carens 2013, S. 235.

15 Das ist eine der vielen Möglichkeiten, den Zuzug aufzuhalten. Italien hat kürzlich libyschen Schleppern mitunter finanzielle Anreize angeboten, damit sie aufhören, Migrantinnen dabei zu helfen, auf ihrem Weg nach Italien libysches Gebiet zu durchqueren. Es ist eine komplexe Frage, ob ein solches Vorgehen Zwang beinhaltet; vieles hängt dabei davon ab, ob diejenigen, die bezahlt

wurden, in Zukunft schlicht keine Anstrengungen mehr unternehmen werden, potentiellen Migrantinnen zu helfen – oder ob es sich dabei um Warlords handelt, die dafür bezahlt werden, die Weiterreise derjenigen zu verhindern, die auszuwandern versuchen. Siehe Walsh und Horowitz 2017.

16 Siehe Yenginsu und Hartocollis 2015. Die Preise für Personen mit einem Visum lassen sich unter https://www.ferrybodrum.com/ verifizieren.

17 Ähnliche Überlegungen bilden auch den Hintergrund für die Anstrengungen der Trump-Administration, mexikanische Bürgerinnen daran zu hindern, an der Grenze zu erscheinen und um Asyl zu bitten. Siehe Meyer 2018.

18 Bloom und Risse 2014.

19 Siehe Miller 2016, Miller 2010, Abizadeh 2010 und Abizadeh 2008.

20 Meine Zustimmung besteht nur teilweise, da ich argumentieren würde, dass Zwang selbst dann vorliegen kann, wenn eine Option entfernt wird – wenn etwa Wächter auf einer Linie im Sand positioniert und mit Schusswaffen ausgestattet werden, um auf diejenigen zu schießen, die sich über diese Linie bewegen wollen, ist es in meinen Augen fair, das, was sie tun, als Zwang anzusehen, selbst wenn mir eine Vielzahl von Optionen auf meiner Seite der Linie bleiben. Ich kann diesen Gedanken hier jedoch nicht angemessen verteidigen.

21 Verschiedene wichtige Rechtsdokumente haben dabei geholfen, diesen Geltungsbereich zu erweitern – darunter die Antifolterkonvention und der Internationale Pakt über bürgerliche und politische Rechte. Siehe generell Maley 2016 und Gibney 2004.

22 Unter anderem David Owen hat über die leidige Frage der Integration von Flüchtlingen in die sie aufnehmenden Gesellschaften geschrieben; an dieser Stelle werde ich diese Fragen nicht weiter untersuchen. Siehe Owen 2019.

23 McAdam und Chong 2014 besprechen das australische Vorgehen und sehen es als starken Widerspruch zu sowohl Text als auch Geist der internationalen Menschenrechtsgesetzgebung.

24 Benner und Dickerson 2018.

25 Bier 2016.

26 Eine Beschreibung dieses Vorschlags findet sich unter http://www.voanews.com/a/trump-border-wall-proposal/4194974.html.

27 Siehe Axelrod 2018.

28 Rawls 1999.

29 Blake 2013.

30 Ich sollte hier anmerken, dass sich die Umstände hier rasch verändern; so hat die Europäische Union damit begonnen, die Vereinigten Staaten dadurch unter Druck zu setzen, dass sie den Export tödlicher Medikamente in die Vereinigten Staaten strenger reguliert, um zu erreichen, dass die USA ihre Praxis der Todesstrafe ändern. Siehe Ford 2014. Es sollte hier auch angemerkt werden, dass mindestens ein Bundesstaat auf die Unverfügbarkeit von Arzneimitteln, die

zur Exekution eingesetzt werden können, mit der Wiedereinführung des Todes durch den elektrischen Stuhl reagiert hat. Siehe Raymond 2018.

31 Rawls bezieht sich hier auf Soper 1984.

32 Die Dokumente, die den Rahmen des R2P bilden, finden sich unter https://www.un.org/en/genocideprevention/about-responsibility-to-protect.shtml. Ich bin nicht der erste, der bemerkt, wie sich diese Konzepte auf das Thema der Migration beziehen lassen; siehe Welsh 2014 und Straehle 2012.

33 Lister 2013. Max Cherem erwähnt einen ähnlichen Gedanken: „Flüchtlinge fliehen die Verfolgungsgefahr, die aus der Auflösung einer politischen Mitgliedschaft folgt; ein Problem, das wiederum oftmals nur durch eine neue Mitgliedschaft dauerhaft gelöst werden kann." Cherem 2016, S. 192.

34 Améry 1977, S. 85–86.

35 Für eine Einführung dazu, wie schwierig es sein wird, sich an den Klimawandel anzupassen, siehe Gardiner 2011. Für die Beziehung zwischen Klimawandel und Migration siehe Lister 2014.

36 Operation Provision: http://www.forces.gc.ca/en/operations-abroad/op-provision-page§.

6. Auswahl und Zurückweisung: Über Migration, Ausschluss und das Veto des Heuchlers

1 Das australische Punktesystem ist außerordentlich deutlich; es gibt Punkte für Ausbildung, sprachliche Fertigkeiten, Alter und andere erwünschte Eigenschaften – und es schließt die Einwanderung von Menschen aus, die älter als 45 Jahre sind. Siehe die Webseite der australischen Regierung unter http://www.visabureau.com/australia/immigration-points-test.aspx.

2 Die Idee wird in Scanlon 2000 dargestellt.

3 Bump 2018 bietet einen Überblick über die Gefahren der Rhetorik Trumps.

4 Das Dokument des Weißen Hauses, in dem das Wort „Tiere" häufiger verwendet wird, als erwartet werden könnte, heißt „What you Need to Know about the Violent Animals of MS-13". Es ist verfügbar unter http://www.whitehouse.gov/articles/need-know-violent-animals-ms-13/.

5 https://www.whitehouse.gov/briefings-statements/remarks-president-trump-members-angel-families-immigration/.

6 https://www.presidency.ucsb.edu/documents/republican-party-platform-1888.

7 Die Regierung von Dominica ist hinsichtlich der Anforderungen also recht einfach zufriedenzustellen: Siehe https://cbiu.gov.dm. In den Vereinigten Staaten hingegen stellt es sich komplizierter dar – aber es ist generell möglich, ein permanentes Aufenthaltsrecht unter dem EB-5-Programm zu erhalten, sofern eine Geschäftsinvestition von einer Millionen Dollar getätigt und zehn Personen in den USA angestellt werden. Siehe https://www.uscis.gov/working-united-states/

permanent-workers/employment-based-immigration-fifth-preference-eb-5/about-eb-5-visa-classification.

8 http://www.lbjlibrary.org/lyndon-baines-johnson/timeline/lbj-on-immigration

9 Guerrero 2014 verteidigt das, was er eine „Lottokratie" nennt, in der die Herrschenden nicht durch Wahlen, sondern eine zufällige Auswahl bestimmt werden.

10 Siehe Lewis 1989 für David Lewis' Analyse der Beziehung zwischen Zufall und Strafrecht.

11 Siehe Coogan 2002, S. 261–269.

12 Philip Pettit hat mehr als jeder andere dafür getan, diesen Ansatz der politischen Theorie zu verteidigen und zu entwickeln. Siehe Pettit 2012.

13 Margalit nutzt die Idee der Demütigung, um eine bestimmte Form des sozialen Übels zu beschreiben, das nicht einfach auf die Sprache der Gerechtigkeit reduziert werden kann. Siehe Margalit 1996.

14 Carens 2013, S. 174.

15 Reed-Sandoval 2005.

16 Romero 2017.

17 *Yick Wo v. Hopkins*, 118 U.S: 356 1886, S. 375.

18 Holmes 1883, S. 3

19 Die Geschichte des Travel-Ban kann in Lowery und Dawsey 2018 nachgelesen werden.

20 *Trump v. Hawaii*, 585 U.S. __ 2018, S. 34.

21 *Trump v. Hawaii*, 585 U.S. __ 2018, S. 1.

22 Eine frühe Analyse der Ziele dieser Gruppe findet sich in Wudunn 1995.

23 Siehe Kennedy 1999; siehe auch Hosein 2018.

24 Der Jackson-Varner-Zusatz verweigerte der Sowjetunion den Status als meistbegünstigten Handelspartner, bis es denjenigen, denen die Ausreise zuerst verweigert worden war, erlaubt wurde, nach Israel auszuwandern.

25 Sorkin 2015 beschreibt Cruz' Anmerkungen.

26 Dabei handelt es sich um das EB-1-Programm, dessen verschiedene Kategorien hier zu finden sind: https://www.uscis.gov/working-united-states/permanent-workers/employment-based-immigration-first-preference-eb-1.

27 Brock und Blake 2014.

28 Bacon 2009.

29 Watson 2008.

30 Es gibt einen enormen Umfang an Forschung darüber, ob und wie Demokratien untergehen. Eine nützliche Übersicht dieser Literatur kann in Lührman und Wilson 2018 gefunden werden.

31 Harris und Lieberman 2017.

32 Dieses Programm hat selbstverständlich eine Menge Kritik auf sich gezogen. Siehe Berry und Sorenson 2018.

33 Schultheis 2018 bespricht die Aussichten für die schwedische Politik.

34 Das Pew Research Center hat einige seiner jüngsten Erkenntnisse hier zusammengefasst: http://www.people-press.org/2018/04/26/the-public-the-political-system-and-american-democracy/.

35 Carens 2013, S. 235.

36 Pratchett 1996.

37 Wintour 2018.

38 Es gibt einige begrenzte Erkenntnisse darüber, dass schon der Konsum von Fernsehshows, in denen gleichgeschlechtliche Paare auftreten, homophobe Empfindungen reduzieren könnte. Siehe Schiappa et al. 2006.

7. Menschen, Orte und Pläne: Über Liebe, Migration und Aufenthaltspapiere

1 Siehe Walzer 1986; Walzers Ansichten über den Vietnam-Krieg werden in Walzer 1977 dargelegt.

2 Für dieses Konzept siehe Stilz 2013.

3 Selbstverständlich war ich als Doktorand kein Migrant; das F-1-Visum, das ein Student erhält, ist, ziemlich explizit, rein temporärer Natur. Allerdings verlangt die Fertigstellung einer Doktorarbeit so viel Zeit, dass sie nur schwerlich in Analogie zum temporären Status eines Touristen verstanden werden kann – ich jedenfalls habe nach meiner Ankunft als Doktorand in den Vereinigten Staaten niemals außerhalb des Landes gelebt.

4 Makino 2004.

5 Pai 2014.

6 Carens 2010, S. 17.

7 Das Argument wird in Carens 2010 ausgeführt.

8 Hosein 2014. Hosein 2016 bietet eine sehr nützliche Klassifizierung anderer Argumente, die angeführt werden könnten.

9 Der Fall ist der von Eugen Karl Mullerschon, dessen Überziehug seines Visums sich angeblich auf 7.059 Tage beläuft. Siehe den Bericht auf https://www.bangkokpost.com/news/general/1509898/35-foreigners-nabbed-for-visa-offences.

10 Siehe „Council Orders Banksy Art Removal." BBCNews, 24. Oktober 2008. Abrufbar unter https://news.bbc.co.uk/2/hi/uk_news/england/london/7688251.stm.

11 Ich sollte für den Fall, dass einer meiner Arbeitgeber das hier liest, betonen, dass es sich hierbei um rein hypothetische Überlegungen handelt.

12 Der Fall wird in Slovic 2008 beschrieben.

13 Die Unterscheidung findet sich an verschiedenen Stellen in Dworkins Werk; siehe vor allem Dworkin 2000.

14 Die Worte des Abgeordneten King finden sich unter https://steveking.house.gov/issues/immigration.

15 Arendts Überlegungen waren einflussreich und ich kann an dieser Stelle nicht alle Komplexitäten ihrer Analyse untersuchen. Die ursprünglichen Konzepte finden sich in Arendt 1958.

16 Barbash 2015.

17 Bort 2018 bespricht diese Einrichtungen.

18 Siehe Maley 2016.

19 Carens 2013, S. 147–149.

20 Der Fall wird in Warikoo 2018 beschrieben.

21 Das Recht und die Moralität des Familienlebens sind ziemlich kompliziert. Ich habe sehr viel dazu von Brake und Miller 2018, Brake 2012, Macedo 2015 und Lister 2010 gelernt.

22 Carens 2013, S. 78–80.

23 Ferracioli 2016.

24 Die Entscheidung findet sich unter http://cdn.ca9.uscourts.gov/datastore/general/2017/09/07/17-16426%20Opinion%20Filed.pdf

25 Diese Analyse beruht auf Macedo 2015.

26 Siehe Lister 2010.

8. Reziprozität, die Undokumentierten und Jeb Bush

1 Diese verhängnisvollen Worte werden in O'Keefe 2014 diskutiert.

2 Diese Argumentationslinie wird häufig in informellen Diskussionsforen wie den sozialen Medien oder Leserinnenbriefen angeführt. Siehe erneut Chomsky 2014 und Gambino 2015.

3 Ich sollte anmerken, dass diese Antwort sich teilweise auf die linguistische Reduktion konzentriert, die den Begriff *illegal* so versteht, als ob er sich auf eine Person als Ganzes bezieht – wie ein Aktivist zu Recht bemerkte, tendieren wir auch nicht dazu, Menschen, die bei Rot über die Ampel gehen, als *illegale Fußgänger* zu bezeichnen.

4 Lister nutzt diese Ideen, um das Wesen der Gerechtigkeit zu besprechen; ich nutze sie in diesem Kontext, um die verwandte Idee zu diskutieren, wie (und warum) wir verpflichtet sind gerecht *zu handeln*. Siehe Lister 2011. Ich möchte anmerken, dass es sich dabei keinesfalls um die einzigen Perspektiven darauf handelt, wie die Pflicht zum Befolgen von Gesetzen verstanden werden kann. So werde ich beispielsweise nicht den Anarchismus von A. John Simmons und auch nicht Joseph Raz' Ansatz zur Rechtfertigung von Autorität besprechen. Siehe Simmons 1979 und Raz 1988.

5 Das Gesetz der Vereinigten Staaten betrachtet den Eintritt, oder den versuchten Eintritt, in die Vereinigten Staaten ohne vorherige Kontrolle als ein Verbrechen, dass mit bis zu zwei Jahren Haft bestraft werden kann. Siehe 8 U.S.C. §1325(a). Sich in den USA widerrechtlich aufzuhalten – wie es für gewöhnlich der Fall ist, wenn Personen legal eingereist sind und dann ihr Visum überziehen – wird hingegen bloß als zivilrechtliches Vergehen betrachtet. 8 U.S.C. §1324d(a)(2).

6 Diese Fälle unterscheiden sich. Joseph Carens verteidigt den ersten, David Bacon den zweiten Fall. Siehe Carens 2010 und Bacon 2009.

7 Abizadeh 2008.

8 John Rawls 1999, S. 136–137.

9 John Rawls 2001, S. 126–127.

10 A. John Simmons bespricht eine solche Perspektive und widerspricht ihr, da er sie als „transaktionale Reziprozität“ bezeichnet. Er bemerkt, dass eine solche Ansicht entweder auf einer Verpflichtung zur Fairness oder auf einer Verpflichtung zur Dankbarkeit für erhaltene Vorteile beruht. Da nichts von dem, was ich hier sage, auf diesen Unterscheidungen beruht, werde ich sie in diesem Kontext ignorieren. Wellman und Simmons 2005, S. 118–120.

11 Murphy 1973, S. 242.

12 Der Gedanke, dass diese Migrantinnen für einen in der Vergangenheit genossenen Vorteil dankbar sein sollten, ist nur eine mögliche Vorstellung von Reziprozität, der gemäß die Reziprozität eng mit wechselseitigen Vorteilen verbunden ist. Ich behaupte nicht, dass es sich dabei um die einzige oder gar beste Interpretation der Idee der Reziprozität handelt. Generell geht diese Idee jedoch davon aus, dass dieser Vorteil an einem moralischen Schwellenwert statt unter Bezug auf historische Verteilungen zu messen ist. Siehe Lister 2017.

13 Siehe beispielsweise Nussbaum 2006, S. 96–154.

14 Siehe beispielsweise Richardson 2006, S. 419–462.

15 Ramsay 2007. Zu finden unter https://www.gamespot.com/articles/blitz-banned-in-australia/1100-6164484/.

16 Dies würde aus Jonathan Quongs Ansatz kooperativer Aktivitäten folgen, wobei er nicht die Schlüsse hinsichtlich moralischer Verpflichtungen ableiten würde, denen ich hier folge. Quong 2007, S. 75–105.

17 In seinem frühen Werk beschreibt John Rawls eine natürliche Pflicht, „vorhandene und für uns geltende gerechte Institutionen zu unterstützen und ihre Regeln zu beachten.“ Rawls 1979, S. 137.

18 Siehe beispielsweise Valentini 2017, James 2012, Buchanan 2003.

19 Sharp und Gonzalez 2013.

20 Siehe Song 2017.

21 Rawls 1999.

22 McAdam und Chong 2014.

23 Oberman 2016b.

24 Cole 2002.

25 Peter Hoeg 1994, S. 117

26 Dieser Anstieg ist abhängig von Veränderungen in den politischen Maßnahmen zur Durchsetzung der Einwanderungsgesetze sowie der Nachfrage nach migrantischer Arbeitskraft. Siehe Borjas 2017.

27 Es gibt in Mexiko eine blühende Gemeinschaft wohlhabender Auswanderer, die aus Staaten wie Südkorea oder den Vereinigten Staaten kommen. Siehe Cave 2013.

28 *R. v. Dudley and Stephens*, 14 QBBD 273 (Queen's Bench Division, 1884), Hervorhebungen hinzugefügt.

29 Bekanntlich verurteilten die Richter die Mörder zum Tode, allerdings mit der Empfehlung, die Bestrafung aufgrund ihrer Gnadenlosigkeit umzuwandeln. Rawson 2000.

30 Ganz ähnlich demonstriert der Kobayashi-Maru-Test, vorgestellt in *Star Trek II: Der Zorn des Khan*, solch ein unvermeidliches Scheitern; James T. Kirk „gewinnt" nur durch Betrug und dadurch, dass er den Code des Tests umschreibt. Siehe Stemwedel 2015. Zu finden unter https://www.forbes.com/sites/janetstemwedel/2015/08/23/the-philosophy-of-star-trek-the-kobayashi-maru-no-win-scenarios-and-ethical-leadership/.

31 Littel 2008, S. 33–34. Es sollte an dieser Stelle erwähnt werden, dass das Argument des Erzählers im Verlauf der Geschichte an Überzeugungskraft einbüßt.

32 Rawls 1993, S. 17.

33 Warren 2015.

34 Rubin 2016.

35 Feinberg 1965.

36 Im Jahr 2012 bereitete Human Rights First einen Bericht für den UN-Sonderberichterstatter über die Menschenrechte von Migrantinnen vor, in dem detailliert festgehalten wurde, auf welche Arten viele industrialisierte Staaten ihre Gefängnissysteme mit den Systemen zur Einwanderungsregulation zusammenführen. Der Report findet sich unter https://www.humanrightsfirst.org/resource/new-detention-report-be-presented-special-rapporteur-human-rights-migrants. Siehe auch Parekh (2016).

37 Tom Tancredo verteidigte während seiner Zeit als Vorsitzender des House Immigration Caucus sowohl die zwangsweise Inhaftierung undokumentierter Immigrantinnen als auch die Maßnahme, dass neugeborenen Kindern dieser Personengruppe die Staatsbürgerschaft qua Geburt verweigert werden sollte. Das Interview findet sich unter http://www.washingtonpost.com/wp-dyn/content/discussion/2006/03/29/DI2006032901468.html.

38 Nixon und Robins 2017.

39 Tibbets 2018.

40 Sachetti 2018.

41 Daher Kishore Mahbubanis prophetische Worte aus dem Jahr 1992: „In den Augen der nordafrikanischen Bevölkerung ist das Mittelmeer, das einst Zivilisationen trennte, zu einem bloßen Teich geworden. Welcher Mensch würde nicht einen Teich überqueren, wenn es dadurch sein Leben verbessern könnte?“ Mahbubani 2001, S. 60.

9. Über Gnade in der Politik

1 Rawls 1979, S. 19.

2 Siehe Murphys Ansatz in Murphy und Hampton 1988.

3 Sein eigener Ansatz konzentriert sich jedoch auf den Sinneswandel der Person, die gnädig handelt; mein eigener Ansatz tut dies nicht.

4 Perry 2018. Es gibt mittlerweile eine relativ reichhaltige Literatur zum Konzept der Gnade, wobei ein Großteil diese Tugend vornehmlich auf das Strafrecht anwendet. Meine Überlegungen sind beeinflusst von Murphy und Hampton 1988, Card 1972, Rainbolt 1990 sowie Staihar und Macedo 2011.

5 Siehe beispielsweise die christliche Perspektive auf das Thema der Migration in Werken wie Kerwin und Gerschutz 2009 sowie Myers und Colwell 2012.

6 Siehe generell Rose 2012, in dem die Schwierigkeiten der frühen Sanctuary-Bewegung beschrieben werden.

7 Die Aussagen der Bischöfe finden sich unter http://www.usccb.org/issues-and-action/human-life-and-dignity/immigration/catholic-teaching-on-immigration-and-the-movement-of-peoples.cfm

8 *Fong Yue Ting v. U.S.*, 149 U.S. 698 1893.

9 Siehe Brecht 1981. Brecht schrieb diese Worte im Jahre 1931.

10 Rawls 1993.

11 Rawls 1998, S. 291.

12 Der Text der Enzyklika findet sich unter http://www.vatican.va/content/john-paul-ii/de/encyclicals/documents/hf_jp-ii_enc_30111980_dives-in-misericordia.html

13 Mielke 2016.

14 Heilige Messe für die Migrantinnen, Predigt von Papst Franziskus, 06.07.2018. http://www.vatican.va/content/francesco/de/homilies/2018/documents/papa-francesco_20180706_omelia-migranti.html

15 Sara Ruddick 1989, deren Werk manchmal als Beginn der Tradition der Care-Ethik angesehen wird, präsentiert diese Ethik nicht als möglichen Ersatz für die Ethik der Gerechtigkeit. Slote 2007 ist hingegen bereit in Betracht zu ziehen, dass die Idee der Sorge möglicherweise alles ist, was die Ethik benötigt.

16 Held 2006.

17 Kittay 2009.

18 „Wir müssen der traditionellen Versuchung widerstehen, die Reichweite der Gerechtigkeit derart zu erweitern, dass sie fälschlicherweise so verstanden wird, als ob sie uns eine umfassende Moral und die Antworten auf alle moralischen Fragen bereitstelle." Held 2006, S. 17.

19 Diese Dichotomie ist übertrieben; Kant war nicht so, wie seine Kritikerinnen ihn wahrgenommen haben, was genauso auch für viele derjenigen gilt, die in der Tradition der Care-Ethik schreiben. Aber der Kern dieser Dichotomie wird zu Recht hervorgehoben und es ist schwer zu leugnen, dass Kant weniger als etwa Virginia Held dazu bereit war, die ethische Relevanz emotionaler Verbindungen anzuerkennen.

20 Ich beziehe mich in meiner Analyse an dieser Stelle sehr stark auf Murphys Rekonstruktion von Kant. Siehe Murphy 1994.

21 Die Idee eines „Spielraums" wird in Gilabert 2010 besprochen.

22 GMS AA 423, S. 17–35.

23 GMS AA 399, S. 28–34.

24 Diese Formulierung an dieser Stelle ist aus Baier 2013 entliehen.

25 Siehe James 1993, S. 133–134. Es sollte betont werden, dass der Charakter Xan sich in der Verfilmung etwas von der Buchversion unterscheidet.

26 Okin 1989a.

27 Okin 1989b erweitert diese Überlegungen und diskutiert, inwiefern die Praxis der Empathie für soziale Gerechtigkeit erforderlich sein könnte.

28 Healy und Martin 2016.

29 Rawls 1993, S. 157.

30 Darin könnte das Gesetz etwas mit der christlichen Tradition gemein haben: Die Idee des *lex orandi, lex credenda* argumentiert, dass der Glaube manchmal eher aus einer Praxis heraus entsteht, als ihr voranzugehen.

31 Costa und Philips 2015 besprechen die koordinierten Anstrengungen der republikanischen Gouverneure, die Aufnahme syrischer Flüchtlinge in ihre Staaten zu unterbinden.

32 Rowling 2000, S. 548.

10. Migration und Gnade

1 Diese verschiedenen Formen möglichen Nutzens durch Einwanderung werden in Brock und Blake 2014 diskutiert.

2 Der kanadische Film Roadkill von 1989 beinhaltet einen Charakter, der sowohl ein Serienmörder ist, als auch die Tatsache nicht leiden kann, dass jeder ambitionierte Kanadier schließlich in die Vereinigten Staaten emigriert; er gibt bekannt, dass er ein Patriot sei und daher zu Hause bleiben und Kanadier töten wolle.

3 Das H1-B-Programm wird beschrieben unter https://www.uscis.gov/working-united-states/temporary-workers/h-1b-specialty-occupations-and-fashion-models/h-1b-fiscal-year-fy-2019-cap-season.

4 Davis 2018.

5 Ronald Reagan in *Public Papers of the Presidents of the United States: Ronald Reagan, June 30 to December 31, 1984*, S. 1600.

6 Der Text des IRCA findet sich unter https://www.uscis.gov/sites/default/files/ocomm/ilink/0-0-0-15.html.

7 Kings Worte finden sich unter https://www.gpo.gov/fdsys/pkg/CREC-2013-07-24/html/CREC-2013-07-24-pt1-PgH5044.htm.

8 Clarke 2014 erzählt die Geschichte von Denise Harvey.

9 Brake 2012 und Macedo 2015 bieten eine alternative Beantwortung der Frage an, ob eine demokratische Gesellschaft gerechterweise für die Monogamie eintreten kann.

10 Siehe Bush 2018.

11 Obamas Worte werden in Winston 2015 wiedergegeben.

12 Williams 1995 führt den Gedanken ein, dass zweistufige Formen des Utilitarismus die Gleichheit von Personen verletzen.

13 Rawls 1993.

14 Während ich dies schreibe, bewegt sich eine Karawane von Migrantinnen, die vor der Gewalt in Honduras fliehen, mit der Intention, die Vereinigten Staaten zu betreten, durch Mexiko. Der Präsident hat versprochen, Truppen an die Grenze zu schicken. Noch verstörender ist der Fakt, dass paramilitärische Gruppen zur Grenze ziehen könnten, um die Karawane am Grenzübertritt zu hindern. Die US Border Patrol zeigt sich besorgt, dass es zu Gewalt kommen könnte, sollten beide Seiten aufeinandertreffen. Siehe Merchant 2018.

15 Die Geschichte des Kirchenasyls ist selbstverständlich weitaus komplexer, als ich gedacht hatte – oder ich hier wiedergebe. Siehe Shoemaker 2011.

16 Siehe Lister 2014 für einige Überlegungen hinsichtlich des Zusammenhangs zwischen Klimawandel und Flucht; siehe Collier und Betts 2017 für einige Gedanken darüber, wie das derzeitige Schutzregime angesichts solcher Bedrohungen scheitern könnte.

17 Vonnegut 1990, S. 102.

Literaturverzeichnis

Abizadeh, Arash. 2010. „Democratic Legitimacy and State Coercion: A Reply to DavidMiller“, in: *Political Theory*, 38(1): S. 121–130. (Abizadeh, Arash. 2017. „Demokratietheoretische Argumente gegen die staatliche Grenzhoheit.“, in: *Ethik der Migration: Philosophische Schlüsseltexte*, hrsg. von Frank Dietrich, S. 98–120, Frankfurt a. M.: Suhrkamp.)

Abizadeh, Arash. 2008. „Democratic Theory and Border Coercion: No Right to Unilaterally Control Your Own Borders“, in: *Political Theory*, 36(1): S. 37–65.

Adler, Margot. 2007. „Churches May Help in Fight against Deportations“, *National Public Radio*. Verfügbar unter: https://www.npr.org/templates/story/story.php?storyId=10098237.

Alexijewitsch, Swetlana. 2015. *Secondhand-Zeit: Leben auf den Trümmern des Sozialismus*. Frankfurt a. M.: Suhrkamp.

American State Papers: Documents, Legislative and Executive, of the Congress of the United States, vol. 1 (1789–1815). 1832. Washington: Gales and Seaton.

Améry, Jean. 1977. *Jenseits von Schuld und Sühne: Bewältigungsversuche eines Überwältigten*. Stuttgart: Klett-Cotta.

Arendt, Hannah. 1958. *The Origins of Totalitarianism*. New York: World Publishing Company. (Arendt, Hannah. 1986. *Elemente und Ursprünge totaler Herrschaft. Antisemitismus, Imperialismus, totale Herrschaft*. München/Zürich: Piper.)

Arrildt, Julie. 2016. „State Borders as Defining Lines of Justice: Why the Right to Exclude Cannot Be Justified“, in: *Critical Review of International Social and Political Philosophy*, 21(4): S. 500–520.

Axelrod, Tal. 2018. „Trump: Border Troops Authorized to Use Lethal Force ‚If They Have To‘“, in: *The Hill*, 22.11.2018. Verfügbar unter: https://thehill.com/homenews/administration/417985-trump-border-troops-authorized-to-use-force-if-they-have-to. [letzter Abruf: 05.07.2020].

Bacon, David. 2009. *Illegal People: How Globalization Creates Migration and Criminalizes Immigrants*. Boston: Beacon Press.

Baier, Annette. 2013. „The Need for More Than Justice“, in: *Canadian Journal of Philosophy*, Zusatzband 13: S. 41–56.

Barbash, Fred. 2015. „Federal Judge in Texas Blocks Obama Immigration Orders“, in: *Washington Post*, 17.02.2015. Verfügbar unter: https://www.washingtonpost.com/news/morning-mix/wp/2015/02/17/federal-judge-in-texas-blocks-obama-immigration-orders/ [letzter Abruf 05.07.2020].

Bauböck, Rainer (Hrsg.). 2018a. *Debating European Citizenship*. New York: Springer.

Bauböck, Rainer (Hrsg.). 2018b. *Debating Transformations of National Citizenship*. New York: Springer.

Bauböck, Rainer. 2011. „Citizenship and Freedom of Movement", in: *Citizenship, Borders, and Human Needs*, hrsg. von Roger Smith, S. 343–376. Philadelphia: University of Pennsylvania Press.

Beitz, Charles. 2001. *The Idea of Human Rights*. Princeton: Princeton University Press.

Beitz, Charles. 1999. *Political Theory and International Relations*. Revised edition. Princeton: Princeton University Press.

Benner, Katie, and Caitlin Dickerson. 2018. „Sessions Says Domestic and Gang Violence Are Not Grounds for Asylum", in: *New York Times*, 11.06.2018. Verfügbar unter: https://www.nytimes.com/2018/06/11/us/politics/sessions-domestic-violence-asylum.html [letzter Abruf 05.07.2020].

Bernard, Andreas. 2014. *Lifted: A Cultural History of the Elevator*. New York: New York University Press. (Bernard, Andreas. 2006. *Die Geschichte des Fahrstuhls*. Frankfurt a. M.: Fischer-Taschenbuch-Verlag.)

Berry, Ellen und Martin Selsoe Sorensen. 2018. „In Denmark, Harsh New Laws for Immigrant ‚Ghettos'", in: *New York Times*, 01.07.2018. Verfügbar unter https://www.nytimes.com/2018/07/01/world/europe/denmark-immigrant-ghettos.html [letzter Abruf 05.07.2020].

Bertram, Chris. 2018. *Do States Have the Right to Exclude Immigrants?* Cambridge: Polity Press.

Betts, Alexander. 2013. *Survival Migration, Failed Governance, and the Crisis of Migration*. Ithaca: Cornell University Press.

Bier, David. 2016. „A Wall Is an Impractical, Expensive, and Ineffective Border Plan", in: *Cato at Liberty*, 26.11.2016. Verfügbar unter: https://www.cato.org/blog/border-wall-impractical-expensive-ineffective-plan [letzter Abruf 05.07.2020].

Bioxham, Donald. 2005. *The Great Game of Genocide: Imperialism, Nationalism, and the Destruction of the Ottoman Armenians*. New York: Oxford University Press.

Blake, Michael. 2016. „Positive and Negative Rights of Migration: A Reply to My Critics", in: *Ethics and Global Politics*, 9(1). Verfügbar unter: https://www.tandfonline.com/doi/pdf/10.3402/egp.v9.33553 [letzter Abruf 14.09.2020].

Blake, Michael. 2014. „The Costs of War: Justice, Liability, and the Pottery Barn Rule", in: *The Ethics of Armed Humanitarian Intervention*, hrsg. von Don Scheid, S. 133–147. Cambridge University Press.

Blake, Michael. 2013. *Justice and Foreign Policy*. Oxford: Oxford University Press.

Blake, Michael. 2012a. „Immigration, Association, and Anti-discrimination", in *Ethics*, 122(4): S. 1–16.

Blake, Michael. 2012b. „Equality without Documents: Political Justice and the Right to Amnesty", in: *Canadian Journal of Philosophy*, Zusatzband 36: S. 99–122.

Blake, Michael. 2008. „Immigration and Political Equality", in: *San Diego Law Review*, 45(4): S. 963–980.

Blake, Michael. 2003. „Immigration", in: *The Blackwell Companion to Applied Ethics*, hrsg. von Christopher Heath Wellman und R. G. Frey, S. 224–237. Oxford: Blackwell.

Blake, Michael. 2001. „Distributive Justice, State Coercion, and Autonomy", in: *Philosophy & Public Affairs*, 30(3): S: 257–296.

Blake, Michael und Mathias Risse. 2009. „Immigration and Original Ownership of the Earth", in: *Notre Dame Journal of Law, Ethics and Public Policy*, 23(1): S. 133–167.

Bloom, Tendayi und Verena Risse. 2014. „Examining Hidden Coercion at the Borders: Why Carrier Sanctions Cannot Be Justified", in *Ethics & Global Politics* 7(2): S. 65–82.

Borjas, George. 2017. „The Earnings of Undocumented Immigrants.", in: *NBER Working Paper,* Number 23236. Verfügbar unter: www.nber.org/papers/w23236 [letzter Abruf 14.09.2020].

Borjas, George. 2001. *Heaven's Door: Immigration Policy and the American Economy.* Princeton: Princeton University Press.

Bort, Ryan. 2018. „This Is the Prison-Like Border Facility Holding Migrant Children", in: *Rolling Stone*, 14.06.2018. Verfügbar unter: https://www.rollingstone.com/politics/politics-news/this-is-the-prison-like-border-facility-holding-migrant-children-628728/ [05.07.2020].

Brake, Elizabeth. 2012. *Minimizing Marriage: Marriage, Morality, and the Law.* New York: Oxford University Press.

Brake, Elizabeth und Lucinda Ferguson (Hrsg.). 2018. *Philosophical Foundations of Children's and Family Law*. New York: Oxford University Press.

Bratman, Michael. 1987. *Intentions, Plans, and Practical Reasons*. Stanford: Center for the Study of Language and Information.

Brecht, Bertolt. 1981. *Die Gedichte von Bertolt Brecht in einem Band*. Frankfurt a. M.: Suhrkamp.

Brezger, Jan und Andreas Cassee. „Debate: Immigrants and Newcomers by Birth—Do Statist Arguments Imply a Right to Exclude Both?", in: *Journal of Political Philosophy*, 24(1): S. 367–378.

Brock, Gillian. 2009. *Global Justice*. New York: Oxford University Press.

Brock, Gillian und Michael Blake. 2014. *Debating Brain Drain: May Governments Restrict Emigration?* New York: Oxford University Press.

Buchanan, Allen. 2003. *Justice, Legitimacy, and Self-Determination*. New York: Oxford University Press.

Bump, Philip. 2018. „The Slippery Slope of the Trump Administration's Political Embrace of Calling MS-13 ‚Animals'", in: *Washington Post*, 21.05.2018. Verfügbar unter: https://www.washingtonpost.com/news/politics/wp/2018/05/21/the-slippery-slope-of-the-trump-administrations-political-embrace-of-calling-ms-13-animals/ [letzter Abruf 05.07.2020].

Bush, Laura. 2018. „Laura Bush: Separating Children from Their Parents at the Border ‚Breaks My Heart'", in: *Washington Post*, 17.06.2018. Verfügbar unter: https://www.washingtonpost.com/opinions/laura-bush-separating-children-from-their-parents-at-the-border-breaks-my-heart/2018/06/17/f2df517a-7287-11e8-9780-b1dd6a09b549_story.html [letzter Abruf 05.07.2020].

Cabrera, Luis. 2004. *Political Theory of Global Justice: A Cosmopolitan Case for the World State*. Abingdon: Routledge.

Cafaro, Philip. 2014. *How Many Is Too Many? The Progressive Argument for Reducing Migration to the United States*. Chicago: University of Chicago Press.

Calhoun, Cheshire. 2018. *Doing Valuable Time*. New York: Oxford University Press.

Calhoun, Cheshire. 2015. *Getting It Right*. New York: Oxford University Press.

Caney, Simon. 2006. *Justice beyond Borders*. New York: Oxford University Press.

Card, Claudia. 1972. „On Mercy", in: *Philosophical Review*, 81(2): S. 182–207.

Card, David. 1990. „The Impact of the Mariel Boatlift on the Miami Labor Market", in: *Industrial and Labor Relations Review*, 43(2): S. 245–257.

Carens, Joseph H. 2019. *Fremde und Bürger*. Ditzingen: Reclam.

Carens, Joseph H. 2013. *The Ethics of Immigration*. New York: Oxford University Press.

Carens, Joseph H. 2010. *Immigrants and the Right to Stay*. Boston: MIT Press.

Cave, Damien. 2013. „For Migrants, New Land of Opportunity Is Mexico", in: *New York Times*, 23.09.2013. Verfügbar unter: https://www.nytimes.com/2013/09/22/world/americas/for-migrants-new-land-of-opportunity-is-mexico.html [letzter Abruf 05.07.2020].

Cherem, Max. 2016. „Refugee Rights: Against Expanding the Definition of a 'Refugee' and Unilateral Protection Elsewhere", in: *Journal of Political Philosophy*, 24(2): S. 183–205.

Chiswick, Barry R. 1988. „Illegal Immigration and Immigration Control", in: *Journal of Economic Perspectives*, 2(3): S. 101–115.

Chomsky, Aviva. 2014. *Undocumented: How Immigration Became Illegal*. Boston: Beacon Press.

Clarke, Katrina. 2014. „Florida Sex-Offender Who Had Relations with 16-Year-Old Granted Refugee Status in Canada“, in: *National Post*, 14.05.2014. Verfügbar unter: https://nationalpost.com/news/canada/florida-sex-offender-who-had-relations-with-16-year-old-granted-refugee-status-in-canada [letzter Abruf 05.07.2020].

Clemens, Michael. 2011. „Economics and Emigration: Trillion-Dollar Bills on the Sidewalk?“, in: *Journal of Economic Perspectives*, 25: S. 83–106.

Coakley, Michael. 1985. „Mayflower Descendants a Breed Apart“, in: *Chicago Tribune*, 05.09.1985. Verfügbar unter: https://www.chicagotribune.com/news/ct-xpm-1985-09-05-8502280039-story.html [letzter Abruf 05.07.2020].

Cohen, Elizabeth F. 2009. *Semi-citizenship in Democratic Politics*. Cambridge: Cambridge University Press.

Cole, Phillip. 2002. *Philosophies of Exclusion*. Edinburgh: Edinburgh University Press.

Collier, Paul. 2013. *Exodus: How Migration Is Changing Our World*. New York: Oxford University Press. (Collier, Paul. 2016. *Exodus: Warum wir Einwanderung neu regeln müssen*. München: Pantheon.)

Collier, Paul. 1999. „Why Has Africa Grown Slowly?“, in: *Journal of Economic Perspectives*, 13(3): S. 3–22.

Collier, Paul, und Alexander Betts. 2017. *Refuge: Rethinking Refugee Policy in a Changing World*. New York: Oxford University Press. (Collier, Paul und Alexander Betts. 2017. *Gestrandet: Warum unsere Flüchtlingspolitik allen schadet – und was jetzt zu tun ist*. München: Siedler.)

Coogan, Tim Pat. 2002. *Wherever Green Is Worn: The Story of the Irish Diaspora*. Basingstroke: Palgrave Macmillan.

Costa, Robert und Abby Philips. 2015. „Republican Governors and Candidates Move to Keep Muslim Immigrants Out“, in: *Washington Post*, 16.11.2015. Verfügbar unter: https://www.washingtonpost.com/politics/republican-governors-and-candidates-move-to-keep-muslim-migrants-out/2015/11/16/17adafaa-8c7c-11e5-baf4-bdf37355da0c_story.html [letzter Abruf 05.07.2020].

„Council Orders Banksy Art Removal“ 2008. *BBC News*, 24.10.2008. Verfügbar unter: http://news.bbc.co.uk/2/hi/uk_news/england/london/7688251.stm [letzter Abruf 13.09.2020].

Davis, Julie Hirschfeld. 2018. „Trump to Cap Refugees Allowed into U.S. at 30,000, a Record Low“, in: *New York Times*, 17.09.2018. Verfügbar unter: https://www.nytimes.com/2018/09/17/us/politics/trump-refugees-historic-cuts.html [letzter Abruf 05.07.2020].

DeAngelo, Ellen. 2009. „Non-Refoulement: The Search for a Consistent Interpretation of Article 33“, in: *Vanderbilt Journal of Transnational Law*, 42: S. 279–315.

Defoe, Daniel. 1982. *Umfassende Geschichte der Räubereien und Mordtaten der berüchtigten Piraten*. Frankfurt a. M.: Robinson Verlag.

Dinan, Desmond. 2010. *Ever Closer Union: An Introduction to European Integration*. Boulder: Lynne Rienner.

Dummett, Michael. 2001. *On Immigration and Refugees*. Abingdon: Routledge.

Dworkin, Ronald. 2000. *Sovereign Virtue: The Theory and Practice of Equality*. Cambridge: Harvard University Press. (Dworkin, Ronald. 2011. *Was ist Gleichheit?*, Frankfurt a. M.: Suhrkamp.)

„Europe Is Coddling Arab Strongmen to Keep Out Refugees." 2018. *The Economist*, 16.08.2018. Verfügbar unter: https://www.economist.com/middle-east-and-africa/2018/08/16/europe-is-coddling-arab-strongmen-to-keep-out-refugees [letzter Abruf 05.07.2020].

Faux, Jeff. 2017. „How U. S. Foreign Policy Helped Create the Immigration Crisis", in: *The Nation*, 18.10.2017. Verfügbar unter: https://www.thenation.com/article/archive/how-us-foreign-policy-helped-create-the-immigration-crisis/ [letzter Abruf 05.07.2020].

Feinberg, Joel. 1965. „The Expressive Function of Punishment", in: *The Monist*, 49(3): S. 397–423.

Ferracioli, Luara. 2016. „Family Migration Schemes and Liberal Neutrality: A Dilemma", in: *Journal of Moral Philosophy*, 13(5): S. 553–575.

Fine, Sarah. 2010. „Freedom of Association Is Not the Answer", in *Ethics* 120(2): S. 338–356. (Fine, Sarah. 2017. Assoziationsfreiheit ist nicht die Lösung, in: *Ethik der Migration: Philosophische Schlüsseltexte*, hrsg. von Frank Dietrich, S. 148–165. Frankfurt a. M.: Suhrkamp.)

Fisher, Michael H. 2013. *Migration: A New World History*. New York: Oxford University Press.

Ford, Matt. 2014. „Can Europe End the Death Penalty in America?", in: *The Atlantic*, 18.02.2014. Verfügbar unter: https://www.theatlantic.com/international/archive/2014/02/can-europe-end-the-death-penalty-in-america/283790/ [letzter Abruf 05.07.2020].

Freiman, Christopher und Javier Hidalgo. 2013. „Liberalism or Immigration Restrictions, but Not Both", in: *Journal of Ethics and Social Philosophy*, 10(2): S. 1–22.

French, Howard. 2014. *China's Second Continent: How a Million Migrants Are Building a New Empire in Africa*. New York: Random House.

Gambino, Lauren. 2015. „‚No Human Being Is Illegal': Linguists Argue against Mislabeling of Immigrants", in: *The Guardian*, 06.12.2015. Verfügbar unter: https://www.theguardian.com/us-news/2015/dec/06/illegal-immigrant-label-offensive-wrong-activists-say [letzter Abruf 05.07.2020].

Gardiner, Stephen M. 2011. *A Perfect Moral Storm: The Ethical Tragedy of Climate Change*. New York: Oxford University Press.

Gibney, Matthew. 2018. „The Ethics of Refugees.“, in: *Philosophy Compass* 13(10). Verfügbar unter: https://onlinelibrary.wiley.com/doi/abs/10.1111/phc3.12521 [letzter Abruf 05.07.2020].

Gibney, Matthew. 2004. *The Ethics and Politics of Asylum*. Cambridge: Cambridge University Press.

Gilabert, Pablo. 2010. „Kant and the Claims of the Poor “, in: *Philosophy and Phenomenological Research*, 81(2): S. 382–418.

Gilgan, Chloë M. 2017. „Exploring the Link between R2P and Refugee Protection: Arriving at Resettlement“, in: *Global Responsibility to Protect*, 9(4): S. 366–394.

Grotius, Hugo. 1979. *Rights of War and Peace*. New York: Hyperion Press. (Grotius, Hugo. 1869/1870. *Des Hugo Grotius drei Bücher über das Recht des Krieges und Friedens: In welchem das Natur- und Völkerrecht und das Wichtigste aus dem öffentlichen Recht erklärt werden*. Leipzig: Meiner.)

Guerrero, Alexander A. 2014. „Against Elections: The Lottocratic Alternative“, in: *Philosophy and Public Affairs*, 42(2): S. 125–178.

Harris, Frederick C. und Robert C. Lieberman. 2017. „The Return of Racism?“, in: *Foreign Affairs*, 21.08.2017. Verfügbar unter: https://www.foreignaffairs.com/articles/united-states/2017-08-21/return-racism [letzter Abruf 05.07.2020].

Hassoun, Nicole. 2012. *Globalization and Global Justice: Shrinking Distance, Expanding Obligations*. Cambridge: Cambridge University Press.

Healy, Patrick, and Jonathan Martin. 2016. „Donald Trump Won't Say If He'll Accept Result of Election“, in: *New York Times*, 19.10.2016. Verfügbar unter: https://www.nytimes.com/2016/10/20/us/politics/presidential-debate.html [letzter Abruf 05.07.2020].

Held, Virginia. 2006. *The Ethics of Care and Empathy*. New York: Oxford University Press.

Hidalgo, Javier. 2014. „Immigration Restrictions and the Right to Avoid Unwanted Obligations“, in: *Journal of Ethics and Social Philosophy*, 8(2): S. 1–9.

Higgins, Peter. 2013. *Immigration Justice*. New York: Oxford University Press.

Hightower, Jim. 1997. *There's Nothing in the Middle of the Road but Yellow Stripes and Dead Armadillos*. New York: HarperCollins.

Hill, Thomas. 1973. „Servility and Self-Respect“, in: *The Monist*, 57(1): S. 87–104.

Hoeg, Peter. 1994. *Fräulein Smillas Gespür für Schnee*. Reinbek bei Hamburg: Rowohlt Taschenbuch Verlag.

Holmes, Oliver Wendell. 1881. *The Common Law*. New York: Little, Brown. (Holmes, Oliver Wendell. 1912. *Das gemeine Recht Englands und Nordamerikas: In elf Abhandlungen dargestellt*. Berlin: Duncker und Humblot.)

Hosein, Adam. 2018. „Racial Profiling and a Reasonable Sense of Inferior Political Status“, in: *Journal of Political Philosophy*, 26(3): S. 1–20.

Hosein, Adam. 2016. „Arguments for Regularization“, in: *The Ethics and Politics of Immigration: Core Issues and Emerging Trends*, hrsg. von Alex Sager, S. 159–179. Lanham, MD: Rowman and Littlefield.

Hosein, Adam. 2014. „Immigration: The Case for Legalization“, in: *Social Theory and Practice*, 40(4): S. 609–630.

Hosein, Adam. 2013. „Immigration and Freedom of Movement“, in *Ethics and Global Politics*, 6(1): S. 25–37.

Huemer, Michael. 2010. „Is There a Right to Immigrate?“, in: *Social Theory and Practice*, 36(3): S. 429–461.

Ignatieff, Michael. 2001. *Human Rights as Politics and Idolatry*. Princeton: Princeton University Press. (Ignatieff, Michael. 2002. *Die Politik der Menschenrechte*. Hamburg: Europäische Verlagsanstalt.)

James, Aaron. 2012. *Fairness in Practice: A Social Contract for a Global Economy*. New York: Oxford University Press.

James, P. D. 1993. *Im Land der leeren Häuser*. München: Droemer Knaur.

Kant, Immanuel. 1785. „Grundlegung zur Metaphysik der Sitten“, in: *Gesammelte Schriften*, hrsg. von der Königlich Preußischen Akademie der Wissenschaften, Bd. IV: S. 385-463. [= GMS]

Kant, Immanuel. 1991. *Political Writings*. Übersetzt und hrsg. von Hans Reiss. Cambridge: Cambridge University Press.

Kates, Michael und Ryan Pevnick. 2014. „Immigration, Jurisdiction, and History“, in: *Philosophy and Public Affairs*, 42(2): S. 179–194.

Kennedy, Randall. 1999. „Suspect Policy“, in: *New Republic*, 13.09.1999. Verfügbar unter: https://newrepublic.com/article/63137/suspect-policy [letzter Abruf 05.07.2020].

Kerwin, Donald und Jill Marie Gerschutz (Hrsg.). 2009. *And You Welcomed Me: Migration and Catholic Social Teaching*. Lanham, MD: Lexington Press.

Kittay, Eva. 2009. „The Moral Harm of Migrant Carework: Realizing a Global Right to Care“, in: *Philosophical Topics*, 37(2): S. 53–73.

Kock, Ida Elizabeth. 2012. „Dichotomies, Trichotomies, or Waves of Duties?“, in: *Human Rights Law Review* 12(3): S. 81–103.

Korsgaard, Christine. 1996. *Creating the Kingdom of Ends*. Cambridge: Cambridge University Press.

Kukathas, Chandran. 2012. „Why Open Borders?“, in: *Ethical Perspectives*, 19(3): S. 649–675.

Kukathas, Chandran. 2010. „Expatriatism: The Theory and Practice of Open Borders“, in: *Citizenship, Borders and Human Needs*, hrsg. von Roger M. Smith, S. 324–342. Philadelphia: University of Pennsylvania Press.

Kukathas, Chandran. 2005. „The Case for Open Immigration", in: *Contemporary Debates in Applied Ethics*, hrsg. von Andrew I. Cohen und Christopher Heath Wellman, S. 207–220. Oxford: Blackwell.

Kullgren, Ian. 2016. „Trump Misrepresents Clinton's Position on ‚Open Borders'", in: *Politico*, 10.10.2016. Verfügbar unter: https://www.politico.com/blogs/2016-presidential-debate-fact-check/2016/10/trump-misrepresents-clintons-position-on-open-borders-230045 [letzter Abruf 05.07.2020].

Kymlicka, Will. 1995. *Multicultural Citizenship*. New York: Oxford University Press.

Lamey, Andy. 2016. „The Jurisdiction Argument for Migration Control: A Critique", in: *Social Theory and Practice*, 42(3): S. 581–604.

Lenard, Patti und Christine Straehle (Hrsg.). 2012. *Legislated Inequality: Temporary Labour Migration in Canada*. Montreal: McGill-Queens University Press.

Lenard, Patti und Christine Straehle. 2010. „Temporary Labour Migration: Exploitation, Tool for Development, or Both?", in: *Policy and Society*, 29(4): S. 283–294.

Lewis, Anthony. 1964. *Gideon's Trumpet*. New York: Random House.

Lewis, David. 1989. „The Punishment That Leaves Something to Chance", in: *Philosophy and Public Affairs*, 18(1): S. 53–67.

Lister, Andrew. 2017. „Public Reason and Reciprocity", in: *Journal of Political Philosophy*, 25(2): S. 155–172.

Lister, Andrew. 2011. „Justice as Fairness and Reciprocity", in: *Analyse & Kritik*, 33(1): S. 93–112.

Lister, Matthew. 2014. „Climate change refugees", in: *Critical Review of International Social and Political Philosophy*, 17(5): S. 618–634.

Lister, Matthew. 2013. „Who Are Refugees?", in: *Law and Philosophy*, 32(2): S. 645–651.

Lister, Matthew. 2010. „Immigration, Association and the Family", in: *Law and Philosophy*, 29: 717–745.

Littell, Jonathan. 2008. *Die Wohlgesinnten*. Berlin: Berliner Taschenbuch Verlag.

Lowery, Wesley und Josh Dawsey. 2018. „Early Chaos of Trump's Travel Ban Set Stage for a Year of Immigration Policy Debates", in: *Washington Post*, 06.02.2018. Verfügbar unter: https://www.washingtonpost.com/national/early-chaos-of-trumps-travel-ban-set-stage-for-a-year-of-immigration-policy-debates/2018/02/06/f5386128-01d0-11e8-8acf-ad2991367d9d_story.html [letzter Abruf 05.07.2020].

Lu, Catherine. 2017. *Justice and Reconciliation in World Politics*. Cambridge: Cambridge University Press.

Lührmann, Anna und Matthew Wilson. 2018. „One-Third of the World's Population Lives in a Declining Democracy. That Includes the United States", in: *Washington Post*, 04.07.2018. Verfügbar unter: https://www.washingtonpost.com/news/monkey-cage/wp/2018/07/03/one-third-of-the-worlds-population-lives-in-a-declining-democracy-that-includes-americans/ [letzter Abruf 05.07.2020].

Macedo, Stephen. 2015. *Just Married: Same-Sex Couples, Monogamy, and the Future of Marriage*. Princeton: Princeton University Press.

Macedo, Stephen. 2012. „The Moral Dilemma of U. S. Immigration Policy Revised", in: *Debating Immigration*, hrsg. von Carol Swaim, S- 286–310. Cambridge: Cambridge University Press.

Macedo, Stephen. 2011. „When and Why Should Liberal Democracies Restrict Immigration?", in: *Citizenship, Borders, and Human Needs*, hrsg. von Rogers M. Smith, S. 301–323. Philadelphia: University of Pennsylvania Press.

Mahbubani, Kishore. 2001. *Can Asians Think*? Hanover: Steerforth Publishing.

Makino, Catherine. 2004. „Japan Gets Tough on Visa Violators. 1-Day Overstay Can Bring Time in Cell, 5-Year Banishment", in: *San Francisco Chronicle*, 10.05.2004. Verfügbar unter: https://www.sfgate.com/news/article/Japan-gets-tough-on-visa-violators-1-day-2780955.php [letzter Abruf 05.07.2020].

Maley, William. 2016. *What Is a Refugee?* New York: Oxford University Press.

Mandle, Jon. 2009. *Rawls's „A Theory of Justice"*. Cambridge: Cambridge University Press.

Margalit, Avishai. 1996. *The Decent Society*. Cambridge: Harvard University Press. (Margalit, Avishai. 2012. *Politik der Würde: Über Achtung und Verachtung*. Berlin: Suhrkamp.)

Massey, Douglas und Kerstin Gentsch. 2014. „Undocumented Migration and the Wages of Mexican Immigrants", in: *International Migration Review*, 48(2): S. 482–499.

McAdam, Jane und Fiona Chong. 2014. *Refugees: Why Seeking Asylum Is Legal and Australia's Policies Are Not*. Randwick: University of New South Wales Press.

McWhorter, Diane. 2012. „The Strange Career of Juan Crow", in: *New York Times*, 16.06.2012. Verfügbar unter: https://www.nytimes.com/2012/06/17/opinion/sunday/no-sweet-home-alabama.html [letzter Abruf 05.07.2020].

Mead, Walter Russell. 2018. „Why Italy Dares to Turn Away Refugees", in: *Wall Street Journal*, 18.06.2018. Verfügbar unter: https://www.wsj.com/articles/why-italy-dares-to-turn-away-refugees-1529363157 [letzter Abruf 05.07.2020].

Mendoza, José Jorge. 2017. *The Moral and Political Philosophy of Immigration: Liberty, Security, and Equality*. Lanham, MD: Lexington Press.

Mendoza, José Jorge. 2015. „Enforcement Matters: Reframing the Philosophical Debate over Immigration", in: *Journal of Speculative Philosophy*, 29(1): S. 73–90.

Mendoza, José Jorge. 2011. „The Political Philosophy of Unauthorized Immigration", in: *APA Newsletter on Hispanic/Latino Issues in Philosophy* 10.2: S. 1–6.

Merchant, Nomaan. 2018. „Border Patrol Warns Landowners of 'Armed Civilians' after Militia Offers to Help Stop Caravan", in: *Chicago Tribune*, 26.10.2018. Verfügbar unter: https://www.chicagotribune.com/nation-world/ct-border-militia-caravan-20181026-story.html [letzter Abruf 05.07.2020].

Meyer, David. 2018. „Trump Moves to Limit Asylum Claims at U.S.-Mexico Border", in: *Fortune*, 09.11.2018. Verfügbar unter: https://fortune.com/2018/11/09/trump-administration-asylum-rules-southern-border/ [letzter Abruf 05.07.2020].

Mielke, Roger. 2016. „Stewarding Mercy: The Role of Churches in the Refugee Crisis", in: *Plough Quarterly*, 10. Verfügbar unter: https://www.plough.com/en/topics/justice/social-justice/immigration/stewarding-mercy [letzter Abruf 05.07.2020].

Miller, David. 2016. *Strangers in Our Midst: The Political Philosophy of Immigration*. Cambridge: Harvard University Press. (Miller, David. 2019. *Fremde in unserer Mitte: Politische Philosophie der Einwanderung*. Berlin: Suhrkamp.)

Miller, David. 2010. „Why Immigration Controls Are Not Coercive: A Reply to Arash Abizadeh", in: *Political Theory*, 38(1): S. 111–120.

Miller, David. 2007. *National Responsibility and Global Justice*. New York: Oxford University Press.

Miller, David. 2005. „Immigration: The Case for Limits", in: *Contemporary Debates in Applied Ethics*, hrsg. von Andrew I. Cohen und Christopher Heath Wellman, S. 193-206. Oxford: Blackwell.

Miller, David. 1995. *On Nationality*. New York: Oxford University Press.

„Morecambe Bay Cockling Disaster's Lasting Impact." 2014. *BBC News*, 03.02.2014. Verfügbar unter: https://www.bbc.co.uk/news/uk-england-lancashire-25986388 [letzter Abruf 05.07.2020].

Moyn, Samuel. 2010. *The Last Utopia: Human Rights in History*. Cambridge: Harvard University Press.

Murphy, Jeffrie. 1994. *Kant: The Philosophy of Right*. Macon, GA: Mercer University Press.

Murphy, Jeffrie G. 1973. „Marxism and Retribution", in: *Philosophy and Public Affairs*, 2(3): S. 217–243.

Murphy, Jeffrie G. und Jean Hampton. 1988. *Forgiveness and Mercy*. Cambridge: Cambridge University Press.

Myers, Ched und Matthew Colwell. 2012. *Our God Is Undocumented: Faith and Immigrant Justice*. Maryknoll, NY: Orbis Publishing.

Nett, Roger. 1971. „The Civil Right We Are Not Ready For: The Right of Free Movement of People on the Face of the Earth", in *Ethics*, 81(3): S. 212–227.

Ngai, Mae. 2010. „The Civil Rights Origin of Illegal Immigration", in: *International Labor and Working Class History*, 78: S. 93–99.

Nixon, Ron und Liz Robbins. 2017. „Office to Aid Crime Victims Is Latest Step in Crackdown on Immigrants", in: *New York Times*, 26.04.2017. Verfügbar unter: https://www.nytimes.com/2017/04/26/us/politics/trump-voice-immigrants-crime.html [letzter Abruf 05.07.2020].

Nozick, Robert. 1974. *Anarchy, State and Utopia*. New York: Basic Books. (Nozick, Robert. 2011. *Anarchie, Staat, Utopia*. München: Olzog.)

Nussbaum, Martha. 2006. *Frontiers of Justice: Disability, Nationality, Species Membership*. Cambridge: Harvard University Press. (Nussbaum, Martha. 2014. *Die Grenzen der Gerechtigkeit: Behinderung, Nationalität und Spezieszugehörigkeit*. Berlin: Suhrkamp.)

O'Keefe. 2014. „Jeb Bush: Many Illegal Immigrants Come out of an ‚Act of Love'", in: *Washington Post*, 06.04.2014. Verfügbar unter: https://www.washingtonpost.com/news/post-politics/wp/2014/04/06/jeb-bush-many-illegal-immigrants-come-out-of-an-act-of-love/ [letzter Abruf 05.07.2020].

Oberman, Kieran. 2016a. „Immigration as a Human Right", in: *Political Theory*, hrsg. von Sarah Fine und Lea Ypi, S. 32–56. New York: Oxford University Press.

Oberman, Kieran. 2016b. „Refugees and Economic Migrants: A Morally Spurious Distinction.", in: *The Critique*, 06.01.2016. Verfügbar unter: http://www.thecritique.com/articles/refugees-economic-migrants-a-morally-spurious-distinction-2/ [letzter Abruf 05.07.2020].

Oberman, Kieran. 2015. „Poverty and Immigration Policy", in: *American Political Science Review*, 109(2): S. 239–251.

Oberman, Kieran. 2011. „Immigration, Global Poverty and the Right to Stay", in: *Political Studies*, 59: S. 253–268.

Okin, Susan Moller. 1989a. *Justice, Gender and the Family*. New York: Basic Books.

Okin, Susan Moller. 1989b. „Reason and Feeling in Thinking about Justice", in *Ethics*, 99(2): S. 229–249.

Osterloh, Margit und Bruno S. Frey. 2018. „Cooperatives Instead of Migration Partnerships", in: *Analyse & Kritik*, 40(2): S. 201–225.

Owen, David. 2019. *Migration and Political Theory*. Abingdon: Routledge.

Owen, David. 2013. „Citizenship and the marginalities of migrants", in: *Critical Review of International Social and Political Philosophy*, 16(3): S. 326–343.

Pai, Hsiao-Hung. 2014. „The lessons of Morecambe Bay have not been learned", in: *The Guardian*, 03.02.2014. Verfügbar unter: https://www.theguardian.

com/commentisfree/2014/feb/03/morecambe-bay-cockle-pickers-tragedy [letzter Abruf 05.07.2020].

Parekh, Serena. 2016. *Refugees and the Ethics of Forced Displacement*. Abingdon: New York: Routledge.

Pérez-Peña, Richard. 2017. „Former Arizona Sheriff Joe Arpaio Is Convicted of Criminal Contempt", in: *New York Times*, 31.07.2017. Verfügbar unter: https://www.nytimes.com/2017/07/31/us/sheriff-joe-arpaio-convicted-arizona.html [letzter Abruf 05.07.2020].

Perry, Adam. 2018. „Mercy", in: *Philosophy and Public Affairs*, 46(1): S. 60–89.

Pettit, Philip. 2012. *On the People's Terms: A Republican Theory and Model of Democracy*. New York: Oxford University Press.

Pevnick, Ryan. 2011. *Immigration and the Constraints of Justice*. Cambridge: Cambridge University Press.

Pogge, Thomas. 2008. *World Poverty and Human Rights*. London: Polity Press. (Pogge, Thomas. 2011. *Weltarmut und Menschenrechte: Kosmopolitische Verantwortung und Reformen*. Berlin: De Gruyter.)

Pogge, Thomas. 2007. *John Rawls: His Life and Theory of Justice*. New York: Oxford University Press. (Pogge, Thomas. 1994. *John Rawls*. München: Beck.)

Pogge, Thomas W. 1997. „Migration and Poverty", in: *Citizenship and Exclusion*, hrsg. von Veit Bader, S. 12–27. London: Macmillan.

Pratchett, Terry. 2002. *Reaper Man*. New York: Harper Torch. (Pratchett, Terry. 2015. *Alles Sense: Ein Scheibenwelt-Roman*. München: Goldmann.)

Pratchett, Terry. 1996. *Feet of Clay*. London: Victor Gollancz. (Pratchett, Terry. 2018. *Hohle Köpfe: Ein Scheibenwelt-Roman*. München: Goldmann.)

Price, Matthew E. 2009. *Rethinking Asylum: History, Purpose, and Limits*. Cambridge: Cambridge University Press.

Public Papers of the Presidents of the United States: Ronald Reagan, 30.06.1984 bis 31.12.1984. Verfügbar unter: https://www.reaganlibrary.gov/sspeeches [letzter Abruf 05.07.2020].

Pufendorf, Samuel. 1934. *De Jure Naturae et Gentium*. Oxford: Clarendon Press. (Pufendorf, Samuel. 1998. *Acht Bücher vom Natur- und Völkerrecht*. Nachdruck der Ausgabe Frankfurt a. M.: Knochen von 1711. Hildesheim: Olms.)

Quong, Jonathan. 2007. „Contractualism, reciprocity, and egalitarian justice", in: *Philosophy, Politics, and Economics*, 6(1): S. 75–105.

Rainbolt, George. 1990. „Mercy: An Independent, Imperfect Virtue", in: *American Philosophical Quarterly*, 27(2): S. 169–173.

Ramsay, Rudolph. 2007. „Blitz Banned in Australia", in: *Gamespot*, 22.01.2007. Verfügbar unter: https://www.gamespot.com/articles/blitz-banned-in-australia/1100-6164484/ [letzter Abruf 05.07.2020].

Raymond, Adam K. 2018. „Tennessee Man Is First U.S. Inmate Killed by Electric Chair Since 2013", in: *New York Magazine*, 02.11.2018. Verfügbar unter: https://nymag.com/intelligencer/2018/11/tennessee-man-is-first-killed-in-electric-chair-since-2013.html [letzter Abruf 05.07.2020].

Rawls, John. 2001. *Justice as Fairness: A Restatement*. Hrsg. von Erin Kelly. Cambridge: Belknap Press of Harvard University Press. (Rawls, John. 2006. *Gerechtigkeit als Fairneß: Ein Neuentwurf*. Frankfurt a. M.: Suhrkamp.)

Rawls, John. 1999. *The Law of Peoples*. Cambridge: Harvard University Press. (Rawls, John. 2002. *Das Recht der Völker*. Berlin: De Gruyter.)

Rawls, John. 1993. *Political Liberalism*. New York: Columbia University Press (Rawls, John. 1998. *Politischer Liberalismus*. Frankfurt a. M.: Suhrkamp.)

Rawls, John. 1979. *Eine Theorie der Gerechtigkeit*. Frankfurt a. M.: Suhrkamp.

Rawson, Claude. 2000. „The Ultimate Taboo", in: *New York Times*, 13.04.2000. Verfügbar unter: https://www.nytimes.com/2000/04/16/books/the-ultimate-taboo.html [letzter Abruf 05.07.2020].

Raz, Joseph. 1988. *The Morality of Freedom*. New York: Oxford University Press.

Reed-Sandoval. 2020. *Socially Undocumented: Identity and Immigration Justice*. New York: Oxford University Press.

Reed-Sandoval, Amy. 2015. „Deportations as Theaters of Inequality", in: *Public Affairs Quarterly*, 29(2): S. 201–215.

Reed-Sandoval, Amy. 2013. „Locating the Injustice of Undocumented Migrant Oppression", in: *Journal of Social Philosophy*, 47(4): S. 372–398.

Richardson, Henry S. 2006. „Rawlsian Social-Contract Theory and the Severely Disabled", in: *Journal of Ethics*, 10(4): S. 419–462.

Risse, Mathias. 2016. „On Where We Differ: Sites Versus Grounds of Justice, and Some Other Reflections on Michael Blake's Justice and Foreign Policy", in: *Law and Philosophy*, 35(3): S. 251–270.

Risse, Mathias. 2012a. *On Global Justice*. Princeton: Princeton University Press.

Risse, Mathias. 2012b. *Global Political Philosophy*. Basingstoke: Palgrave Publishing.

Rodríguez, Cristina M. 2013. „Immigration, Civil Rights, and the Evolution of the People", in: *Daedalus*, 142(3): S. 228–241.

Romero, Simon. 2017. „Latinos Express Outrage as Trump Pardons Arpaio", in: *New York Times*, 25.08.2017. Verfügbar unter: https://www.nytimes.com/2017/08/25/us/joe-arpaio-pardon-latinos.html [letzter Abruf 05.07.2020].

Rose, Ananda. 2012. *Showdown in the Sonoran Desert: Religion, Law, and the Immigration Controversy*. New York: Oxford University Press.

Ross, Janell. 2016. „From Mexican rapists to bad hombres: the Trump campaign in two moments.", in: *Washington Post*, 20.10.2016. Verfügbar unter: https://

www.washingtonpost.com/news/the-fix/wp/2016/10/20/from-mexican-rapists-to-bad-hombres-the-trump-campaign-in-two-moments/ [letzter Abruf 05.07.2020].

Rowling, J. K. 2000. *Harry Potter und der Feuerkelch*. Hamburg: Carlsen.

Rubin, Jennifer. 2016. „Don't forget that, for a year, Trump called for deporting all illegal immigrants", in: *Washington Post*, 23.08.2016. Verfügbar unter: https://www.washingtonpost.com/blogs/right-turn/wp/2016/08/23/dont-forget-that-for-a-year-trump-called-for-deporting-all-illegal-immigrants/ [letzter Abruf 05.07.2020].

Ruddick, Sara. 1989. *Maternal Thinking: Towards a Politics of Peace*. New York: Ballentyne Books. (Ruddick, Sara. 1993. *Mütterliches Denken: Für eine Politik der Gewaltlosigkeit*. Frankfurt a. M.: Campus.)

Sacchetti, Maria. 2018. „Top Homeland Security officials urge criminal prosecution of parents crossing border with children", in: *Washington Post*, 26.04.2018. Verfügbar unter: https://www.washingtonpost.com/local/immigration/top-homeland-security-officials-urge-criminal-prosecution-of-parents-who-cross-border-with-children/2018/04/26/a0bdcee0-4964-11e8-8b5a-3b1697adcc2a_story.html [letzter Abruf 05.07.2020].

Sager, Alex. 2018. *Toward a Cosmopolitan Ethics of Mobility: The Migrant's-Eye View of the World*. New York: Springer.

Scanlon, T. M. 2000. *What We Owe to Each Other*. Cambridge: Harvard University Press.

Scarry, Elaine. 1985. *The Body in Pain*. New York: Oxford University Press. (Scarry, Elaine. 1992. *Der Körper im Schmerz: Die Chiffren der Verletzlichkeit und die Erfindung der Kultur*. Frankfurt a. M.: Fischer.)

Schiappa, E. et al. 2006. „Can one TV show make a difference? Will & Grace and the Parasocial Contact Hypothesis", in: *Journal of Homosexuality*, 52(4): S. 15–37.

Schotel, Bas. 2012. *On the Right of Exclusion: Law, Ethics and Immigration Policy*. Abingdon: Routledge Press.

Schultheis, Emily. 2018. „Is Sweden Ungovernable?", in: *Foreign Policy*, 10.10.2018. Verfügbar unter: https://foreignpolicy.com/2018/10/10/is-sweden-ungovernable/ [letzter Abruf 05.07.2020].

Scoones, Ian et al. 2018. „Emancipatory Rural Politics: Confronting Authoritarian Populism", in: *Journal of Peasant Studies*, 45(1): S. 1–20.

Sen, Amartya. 1992. *Inequality Re-examined*. New York: The Russell Sage Foundation.

Shachar, Ayelet. 2009. *The Birthright Lottery: Citizenship and Global Inequality*. Cambridge: Harvard University Press.

Shacknove, Andrew. 1985. „Who Is a Refugee?", in *Ethics*, 95(2): S. 274–284.

Shariatmadari, David. 2015. „Swarms, floods, and marauders: the toxic metaphors of the migration debate“, in: *The Guardian*, 10.08.2015. Verfügbar unter: https://www.theguardian.com/commentisfree/2015/aug/10/migration-debate-metaphors-swarms-floods-marauders-migrants [letzter Abruf 05.07.2020].

Sharp, Gene und Jaime Gonzalez. 2013. *How Nonviolent Struggle Works*. Boston: Albert Einstein Institution.

Shoemaker, Karl. 2011. *Sanctuary and Crime in the Middle Ages*. New York: Fordham University Press.

Shue, Henry. 1980. *Basic Right: Subsistence, Affluence, and U. S. Foreign Policy*. Princeton: Princeton University Press.

Sidgwick, Henry. 2013. *The Elements of Politics*. Cambridge: Cambridge University Press.

Simmons, A. John. 1979. *Moral Principles and Political Obligations*. Princeton: Princeton University Press.

Singer, Peter. 1971. „Famine, Affluence, and Morality“, in: *Philosophy and Public Affairs*, 1(3): S. 229–243.

Slote, Michael. 2007. *The Ethics of Care and Empathy*. Routledge.

Slovic, Beth. 2008. „He's an … Illegal Eh-lien.“, in: *Willamette Week*, 19.01.2008. Verfügbar unter: https://www.wweek.com/portland/article-8470-hes-an-illegal-eh-lien.html [letzter Abruf 05.07.2020].

Song, Sarah. 2018. *Immigration and Democracy*. New York: Oxford University Press.

Song, Sarah. 2017. „Why Does the State Have the Right to Control Immigration?“, in: *NOMOS LVII: Immigration, Emigration, and Migration*, hrsg. von Jack Knight: S. 3–50. New York University Press.

Song, Sarah. 2012. „The Boundary Problem in Democratic Theory: Why the Demos Should Be Bounded by the State“, in: *International Theory*, 4(1): S. 39–68.

Song, Sarah. 2009. „What Does It Mean to Be an American?“, in: *Daedalus*, 138(2): S. 31–40.

Soper, Philip 1984. *A Theory of Law*. Cambridge: Harvard University Press.

Sorkin, Amy Davidson. 2015. „Ted Cruz's Religious Test for Syrian Refugees“, in: *New Yorker*, 16.11.2015. Verfügbar unter: https://www.newyorker.com/news/amy-davidson/ted-cruzs-religious-test-for-syrian-refugees [letzter Abruf 05.07.2020].

Staihar, Jim und Stephen Macedo. 2011. „Defending a Role for Mercy in a Criminal Justice System“, in: *Merciful Judgments and Contemporary Society*, hrsg. von Austin Sarat, S. 138–194. Cambridge: Cambridge University Press.

Stemwedel, Janet T. 2015. „The Philosophy of Star Trek: The Kobayashi Maru, No-Win Scenarios, and Ethical Leadership“, in: *Forbes*, 23.08.2015. Verfügbar unter: https://www.forbes.com/sites/janetstemwedel/2015/08/23/the-philosophy-of-star-trek-the-kobayashi-maru-no-win-scenarios-and-ethical-leadership/ [letzter Abruf 05.07.2020].

Stephenson, Neal. 1992. *Snow Crash*. New York: Bantam Books. (Stephenson, Neal. 1994. *Snow Crash*. München: Goldmann.)

Stilz, Anna. 2016. „Is There an Unqualified Right to Leave?“, in: *Migration in Political Theory: The Ethics of Movement and Membership*, hrsg. von Lea Ypi und Sarah Fine, S. 57–79. New York: Oxford University Press.

Stilz, Anna. 2013. „Occupancy Rights and the Wrong of Removal“, in: *Philosophy and Public Affairs*, 41(4): S. 324–356.

Straehle, Christine. 2012. „Thinking about Protecting the Vulnerable When Thinking about Immigration: Is There a ‚Responsibility to Protect‘ in Immigration Regimes?“, in: *Journal of International Political Theory*, 8(1–2): S. 159–171.

Talbott, William. 2005. *Which Rights Should Be Universal?* New York: Oxford University Press.

Tan, Kok-Chor. 2014. *Justice, Institutions, and Luck*. New York: Oxford University Press.

Tan, Kok-Chor. 2000. *Toleration, Diversity, and Global Justice*. University Park, PA: Penn State University Press.

„Thirty-Five Foreigners Nabbed for Visa Offenses.“ 2018. Bankgok Post, 25.07.2018. Verfügbar unter: https://www.bangkokpost.com/thailand/general/1509898/35-foreigners-nabbed-for-visa-offences [letzter Abruf 05.07.2020].

Thomson, Judith Jarvis. 1971. „A Defense of Abortion“, in: *Philosophy and Public Affairs*, 1(1): S. 47–66.

Tibbetts, Ron. 2018. „From Mollie Tibbett's Father: Don't Distort Her Death to Advance Racist Views“, in: *Des Moines Register*, 01.09.2018. Verfügbar unter: https://www.desmoinesregister.com/story/opinion/columnists/2018/09/01/mollie-tibbetts-father-common-decency-immigration-heartless-despicable-donald-trump-jr-column/1163131002/ [letzter Abruf 05.07.2020].

United States Commission on Civil Rights. 1980. *The Tarnished Golden Door: Civil Rights Issues in Immigration*. US Government Printing Office.

Valentini, Laura. 2017. „The Natural Duty of Justice in Non-ideal Circumstances: On the Moral Demands of Institution-Building and Reform“, in: *European Journal of Political Theory*.

Valentini, Laura. 2012. *Justice in a Globalized World: A Normative Framework*. New York: Oxford University Press.

Vonnegut, Kurt. 1990. *Gott segne Sie, Mr. Rosewater*. München: Wilhelm Goldmann.

Waldron, Jeremy. 1990. *The Right to Private Property*. Oxford: Clarendon Press.

Walsh, Declan und Jason Horowitz. 2017. „Italy, Going It Alone, Stalls the Flow of Migrants. But at What Cost?", in: *New York Times*, 17.09.2017. Verfügbar unter: https://www.nytimes.com/2017/09/17/world/europe/italy-libya-migrant-crisis.html [letzter Abruf 05.07.2020].

Walzer, Michael. 1986. *Spheres of Justice*. New York: Basic Books. (Walzer, Michael. 1994. *Sphären der Gerechtigkeit: Ein Plädoyer für Pluralität und Gleichheit*. Frankfurt a.M.: Campus.)

Walzer, Michael. 1977. *Just and Unjust Wars*. New York: Basic Books.

Warikoo, Niraj. 2018. „Deaf, Disabled Detroit Immigrant in US for 34 Years Faces Deportation", in: *Detroit Free Press*, 08.09.2018. Verfügbar unter: https://eu.freep.com/story/news/local/michigan/detroit/2018/09/08/deaf-immigrant-deportation-francis-anwana/1226728002/ [letzter Abruf 05.07.2020].

Warren, Michael. 2015. „Trump Hits Jeb on ‚Act of Love'", in: *Washington Examiner*, 31.08.2015. Verfügbar unter: https://www.washingtonexaminer.com/weekly-standard/trump-hits-jeb-on-act-of-love [letzter Abruf 05.07.2020].

Watson, Lori. 2008. „Equal Justice: A Commentary on Michael Blake's 'Immigration and Political Equality'", in: *San Diego Law Review*, 45(4): S. 981–988.

Wellman, Christopher Heath. 2008. „Immigration and Freedom of Association", in *Ethics*, 119: S. 109–141. (Wellman, Christopher Heath. 2017. „Immigration und Assoziationsfreiheit", in: *Ethik der Migration: Philosophische Schlüsseltexte*, hrsg. von Frank Dietrich, S. 121–147. Frankfurt a. M.: Suhrkamp.)

Wellman, Christopher Heath und Phillip Cole. 2011. *Debating the Ethics of Immigration: Is There a Right to Exclude?* New York: Oxford University Press.

Wellman, Christopher Heath und A. John Simmons. 2005. *Is There a Duty to Obey the Law?* Cambridge: Cambridge University Press.

Welsh, Jennifer. 2014. „Fortress Europe and the Responsibility to Protect: Framing the Issue", in: *EUI Forum, European University Institute*, Florence. Verfügbar unter: https://www.eui.eu/Documents/RSCAS/PapersLampedusa/FORUM-Welshfinal.pdf [letzter Abruf 05.07.2020].

Wenar, Leif. 2017. *Blood Oil*. New York: Oxford University Press.

Wilcox, Shelly. 2009. „The Open Borders Debate on Immigration", in: *Philosophy Compass*, 4(5): S. 813–821.

Williams, Bernard. 1995. „The Point of View of the Universe: Sidgwick and the Ambitions of Ethics", in: *Cambridge Review*, 7. Mai.

Winston, Kimberly. 2015. „Obama Denounces Religious Test for Refugees: ‚That's Not Who We Are.'", in: *Washington Post*, 16.11.2015 Verfügbar unter: https://www.washingtonpost.com/national/religion/obama-denounces-religious-test-for-refugees-thats-not-who-we-are/2015/11/16/333e2acc-8cb0-11e5-934c-a369c80822c2_story.html [letzter Abruf 05.07.2020].

Wintour, Patrick. 2018. „Hillary Clinton: Europe Must Curb Immigration to Stop Rightwing Populists“, in: *The Guardian*, 22.11.2018. Verfügbar unter: https://www.theguardian.com/world/2018/nov/22/hillary-clinton-europe-must-curb-immigration-stop-populists-trump-brexit [letzter Abruf 05.07.2020].

WuDunn, Sheryl. 1995. „Secretive Japan Sect Evokes Both Loyalty and Hostility“, in: *New York Times*, 24.03.1995. Verfügbar unter: https://www.nytimes.com/1995/03/24/world/secretive-japan-sect-evokes-both-loyalty-and-hostility.html [letzter Abruf 05.07.2020].

Yenginsu, Ceylan und Anemona Hartocollis. 2015. „Amid Perilous Mediterranean Crossings, Migrants Find a Relatively Easy Path to Greece“, in: *New York Times*, 15.08.2015. Verfügbar unter: https://www.nytimes.com/2015/08/17/world/europe/turkey-greece-mediterranean-kos-bodrum-migrants-refugees.html [letzter Abruf 05.07.2020].

Yong, Caleb. 2018. „Caring Relationships and Family Migration Schemes“, in: *The Ethics and Politics of Immigration: Core Issues and Emerging Trends*, hrsg. von Alex Sager, S. 61–84. Lanham, MD: Rowman and Littlefield.

Ypi, Lea. 2013. „What's Wrong with Colonialism“, in: *Philosophy and Public Affairs*, 41(2): S. 158–191.

Ypi, Lea. 2008. „Justice in Migration: A Closed Borders Utopia?“, in: *Journal of Political Philosophy*, 16(4): S. 391–418.

Register